Handbook of Mathematical Models and Algorithms in Computer Vision and Imaging

Ke Chen · Carola-Bibiane Schönlieb ·
Xue-Cheng Tai · Laurent Younes
Editors

Handbook of Mathematical Models and Algorithms in Computer Vision and Imaging

Mathematical Imaging and Vision

Volume 2

With 553 Figures and 72 Tables

Editors
Ke Chen
Department of Mathematical Sciences
The University of Liverpool
Liverpool, UK

Xue-Cheng Tai
Hong Kong Center for
Cerebrocardiovascular Health
Engineering (COCHE)
Shatin, Hong Kong, China

Carola-Bibiane Schönlieb
Department of Applied Mathematics and
Theoretical Physics
University of Cambridge
Cambridge, UK

Laurent Younes
Department of Applied Mathematics and Statistics
Johns Hopkins University
Baltimore, MD, USA

ISBN 978-3-030-98660-5   ISBN 978-3-030-98661-2 (eBook)
https://doi.org/10.1007/978-3-030-98661-2

© Springer Nature Switzerland AG 2023
This work is subject to copyright. All rights are reserved by the Publisher, whether the whole or part of the material is concerned, specifically the rights of translation, reprinting, reuse of illustrations, recitation, broadcasting, reproduction on microfilms or in any other physical way, and transmission or information storage and retrieval, electronic adaptation, computer software, or by similar or dissimilar methodology now known or hereafter developed.
The use of general descriptive names, registered names, trademarks, service marks, etc. in this publication does not imply, even in the absence of a specific statement, that such names are exempt from the relevant protective laws and regulations and therefore free for general use.
The publisher, the authors, and the editors are safe to assume that the advice and information in this book are believed to be true and accurate at the date of publication. Neither the publisher nor the authors or the editors give a warranty, expressed or implied, with respect to the material contained herein or for any errors or omissions that may have been made. The publisher remains neutral with regard to jurisdictional claims in published maps and institutional affiliations.

This Springer imprint is published by the registered company Springer Nature Switzerland AG.
The registered company address is: Gewerbestrasse 11, 6330 Cham, Switzerland.

Preface

The rapid development of new imaging hardware, the advance in medical imaging, the advent of multi-sensor data fusion and multimodal imaging, as well as the advances in computer vision have sparked numerous research endeavours leading to highly sophisticated and rigorous mathematical models and theories. Motivated by the increasing use of variational models, shapes and flows, differential geometry, optimisation theory, numerical analysis, statistical/Bayesian graphical models, machine learning, and deep learning, we have invited contributions from leading researchers and publish this handbook to review and capture the state of the art of research in Computer Vision and Imaging.

This constantly improving technology that generates new demands not readily met by existing mathematical concepts and algorithms provides a compelling justification for such a book to meet the ever-growing challenges in applications and to drive future development. As a consequence, new mathematical models have to be found, analysed and realised in practice. Knowing the precise state-of-the-art developments is key, and hence this book will serve the large community of mathematics, imaging, computer vision, computer sciences, statistics, and, in general, imaging and vision research. Our primary audience are

- Graduate students
- Researchers
- Imaging and vision practitioners
- Applied mathematicians
- Medical imagers
- Engineers
- Computer scientists

Viewing discrete images as data sampled from functional surfaces enables the use of advanced tools from calculus, functions and calculus of variations, and optimisation and provides the basis of high-resolution imaging through variational models. No other framework can provide the comparable accuracy and precision to imaging and vision.

Although our initial emphasis is on the variational methods, which represent the optimal solutions to class of imaging and vision problems, and on effective algorithms, which are necessary for the methods to be translated to practical use in various applications, the editors recognise that the range of effective and efficient methods for solving problems from computer vision and imaging go beyond variational methods and have enlarged our coverage to include mathematical models and algorithms. So, the book title reflects this viewpoint and a big vision for the reference book.

All chapters will have introductions so that the book is readily accessible to graduate students. We have divided the 53 chapters of this book into 3 sections, namely

(a) Convex and Non-convex Large-Scale Optimisation in Imaging
(b) Model- and Data-Driven Variational Imaging Approaches
(c) Shape Spaces and Geometric Flows

to facilitate browsing the content list. However, such a division is artificial because, these days, research becomes increasingly intra-disciplinary as well as inter-disciplinary, and ideas from one topic often directly or indirectly inspire or transpire another topic. This is very exciting.

For newcomers to the field, the book provides a comprehensive and fast track introduction to the core research problems, to save time and get on with tackling new and emerging challenges, rather than running the risk of reproducing/comparing to some old works already done or reinventing same results. For researchers, exposure to the state of the art of research works leads to an overall view of the entire field so as to guide new research directions and avoid pitfalls in moving the field forward and looking into the next 25 years of imaging and information sciences.

The dreadful Covid-19 pandemic starting from 2020 has affected lives of everyone, of course including all researchers. We are still not out of the woods. The editors are very much grateful to the book authors who have endured much hardship during the last 3 years and overcome many difficulties to have completed their chapters on time. We are also indebted to many anonymous reviewers who provided valuable reviews and helpful criticism to improve presentations of our chapters.

The original gathering of all editors was in 2017 when the first three editors co-organised the prestigious Isaac Newton Institute programme titled "*Variational methods and effective algorithms for imaging and vision*" (https://www.newton.ac.uk/event/vmv/), partially supported by UK EPSRC GR/EP F005431 and Isaac Newton Institute for Mathematical Sciences. During the programme, Mr Jan Holland from Springer-Nature kindly suggested the idea of a book. We are grateful to his suggestion which sparked the editors' fruitful collaboration in the last few years. The large team of publishers who have offered immense help to us include Michael Hermann (Springer), Allan Cohen (Palgrave) and Salmanul Faris Nedum Palli (Springer). We thank them all.

Finally, we wish all readers a happy reading.

The editorial team:

Liverpool, UK	Ke Chen (Lead)
Cambridge, UK	Carola-Bibiane Schönlieb
Shatin, Hong Kong	Xue-Cheng Tai
Baltimore, USA	Laurent Younes
February 2023	

Contents

Volume 1

Part I Convex and Non-convex Large-Scale Optimization in Imaging 1

1 Convex Non-convex Variational Models 3
   Alessandro Lanza, Serena Morigi, Ivan W. Selesnick, and Fiorella Sgallari

2 Subsampled First-Order Optimization Methods with Applications in Imaging 61
   Stefania Bellavia, Tommaso Bianconcini, Nataša Krejić, and Benedetta Morini

3 Bregman Methods for Large-Scale Optimization with Applications in Imaging 97
   Martin Benning and Erlend Skaldehaug Riis

4 Fast Iterative Algorithms for Blind Phase Retrieval: A Survey 139
   Huibin Chang, Li Yang, and Stefano Marchesini

5 Modular ADMM-Based Strategies for Optimized Compression, Restoration, and Distributed Representations of Visual Data 175
   Yehuda Dar and Alfred M. Bruckstein

6 Connecting Hamilton-Jacobi Partial Differential Equations with Maximum a Posteriori and Posterior Mean Estimators for Some Non-convex Priors 209
   Jérôme Darbon, Gabriel P. Langlois, and Tingwei Meng

7 Multi-modality Imaging with Structure-Promoting Regularizers 235
   Matthias J. Ehrhardt

8	Diffraction Tomography, Fourier Reconstruction, and Full Waveform Inversion	273
	Florian Faucher, Clemens Kirisits, Michael Quellmalz, Otmar Scherzer, and Eric Setterqvist	
9	Models for Multiplicative Noise Removal	313
	Xiangchu Feng and Xiaolong Zhu	
10	Recent Approaches to Metal Artifact Reduction in X-Ray CT Imaging	347
	Soomin Jeon and Chang-Ock Lee	
11	Domain Decomposition for Non-smooth (in Particular TV) Minimization	379
	Andreas Langer	
12	Fast Numerical Methods for Image Segmentation Models	427
	Noor Badshah	
13	On Variable Splitting and Augmented Lagrangian Method for Total Variation-Related Image Restoration Models	503
	Zhifang Liu, Yuping Duan, Chunlin Wu, and Xue-Cheng Tai	
14	Sparse Regularized CT Reconstruction: An Optimization Perspective	551
	Elena Morotti and Elena Loli Piccolomini	
15	Recent Approaches for Image Colorization	585
	Fabien Pierre and Jean-François Aujol	
16	Numerical Solution for Sparse PDE Constrained Optimization	623
	Xiaoliang Song and Bo Yu	
17	Game Theory and Its Applications in Imaging and Vision	677
	Anis Theljani, Abderrahmane Habbal, Moez Kallel, and Ke Chen	
18	First-Order Primal–Dual Methods for Nonsmooth Non-convex Optimization	707
	Tuomo Valkonen	

Volume 2

Part II	Model- and Data-Driven Variational Imaging Approaches	749
19	Learned Iterative Reconstruction	751
	Jonas Adler	

20	**An Analysis of Generative Methods for Multiple Image Inpainting**	773

Coloma Ballester, Aurélie Bugeau, Samuel Hurault, Simone Parisotto, and Patricia Vitoria

21	**Analysis of Different Losses for Deep Learning Image Colorization**	821

Coloma Ballester, Hernan Carrillo, Michaël Clément, and Patricia Vitoria

22	**Influence of Color Spaces for Deep Learning Image Colorization**	847

Aurélie Bugeau, Rémi Giraud, and Lara Raad

23	**Variational Model-Based Deep Neural Networks for Image Reconstruction**	879

Yunmei Chen, Xiaojing Ye, and Qingchao Zhang

24	**Bilevel Optimization Methods in Imaging**	909

Juan Carlos De los Reyes and David Villacís

25	**Multi-parameter Approaches in Image Processing**	943

Markus Grasmair and Valeriya Naumova

26	**Generative Adversarial Networks for Robust Cryo-EM Image Denoising**	969

Hanlin Gu, Yin Xian, Ilona Christy Unarta, and Yuan Yao

27	**Variational Models and Their Combinations with Deep Learning in Medical Image Segmentation: A Survey**	1001

Luying Gui, Jun Ma and Xiaoping Yang

28	**Bidirectional Texture Function Modeling**	1023

Michal Haindl

29	**Regularization of Inverse Problems by Neural Networks**	1065

Markus Haltmeier and Linh Nguyen

30	**Shearlets: From Theory to Deep Learning**	1095

Gitta Kutyniok

31	**Learned Regularizers for Inverse Problems**	1133

Sebastian Lunz

32	**Filter Design for Image Decomposition and Applications to Forensics**	1155

Robin Richter, Duy H. Thai, Carsten Gottschlich, and Stephan F. Huckemann

33 Deep Learning Methods for Limited Data Problems in X-Ray Tomography 1183
Johannes Schwab

34 MRI Bias Field Estimation and Tissue Segmentation Using Multiplicative Intrinsic Component Optimization and Its Extensions 1203
Samad Wali, Chunming Li, and Lingyan Zhang

35 Data-Informed Regularization for Inverse and Imaging Problems 1235
Jonathan Wittmer and Tan Bui-Thanh

36 Randomized Kaczmarz Method for Single Particle X-Ray Image Phase Retrieval 1273
Yin Xian, Haiguang Liu, Xuecheng Tai, and Yang Wang

37 A Survey on Deep Learning-Based Diffeomorphic Mapping 1289
Huilin Yang, Junyan Lyu, Roger Tam, and Xiaoying Tang

Volume 3

Part III Shape Spaces and Geometric Flows 1323

38 Stochastic Shape Analysis 1325
Alexis Arnaudon, Darryl Holm, and Stefan Sommer

39 Intrinsic Riemannian Metrics on Spaces of Curves: Theory and Computation 1349
Martin Bauer, Nicolas Charon, Eric Klassen, and Alice Le Brigant

40 An Overview of SaT Segmentation Methodology and Its Applications in Image Processing 1385
Xiaohao Cai, Raymond Chan, and Tieyong Zeng

41 Recent Development of Medical Shape Analysis via Computational Quasi-conformal Geometry 1413
Hei-Long Chan and Lok-Ming Lui

42 A Survey of Topology and Geometry-Constrained Segmentation Methods in Weakly Supervised Settings 1437
Ke Chen, Noémie Debroux, and Carole Le Guyader

43	**Recent Developments of Surface Parameterization Methods Using Quasi-conformal Geometry**............................. Gary P. T. Choi and Lok Ming Lui	1483
44	**Recent Geometric Flows in Multi-orientation Image Processing via a Cartan Connection**....................................... R. Duits, B. M. N. Smets, A. J. Wemmenhove, J. W. Portegies, and E. J. Bekkers	1525
45	**PDE-Constrained Shape Optimization: Toward Product Shape Spaces and Stochastic Models** .... Caroline Geiersbach, Estefania Loayza-Romero, and Kathrin Welker	1585
46	**Iterative Methods for Computing Eigenvectors of Nonlinear Operators** .... Guy Gilboa	1631
47	**Optimal Transport for Generative Models** .... Xianfeng Gu, Na Lei, and Shing-Tung Yau	1659
48	**Image Reconstruction in Dynamic Inverse Problems with Temporal Models** .... Andreas Hauptmann, Ozan Öktem, and Carola Schönlieb	1707
49	**Computational Conformal Geometric Methods for Vision** .... Na Lei, Feng Luo, Shing-Tung Yau, and Xianfeng Gu	1739
50	**From Optimal Transport to Discrepancy**....................... Sebastian Neumayer and Gabriele Steidl	1791
51	**Compensated Convex-Based Transforms for Image Processing and Shape Interrogation** .... Antonio Orlando, Elaine Crooks, and Kewei Zhang	1827
52	**The Potts Model with Different Piecewise Constant Representations and Fast Algorithms: A Survey** .... Xuecheng Tai, Lingfeng Li, and Egil Bae	1887
53	**Shape Spaces: From Geometry to Biological Plausibility** .... Nicolas Charon and Laurent Younes	1929
Index	....	1959

About the Editors

Prof. Ke Chen, PhD received his BSc, MSc and PhD degrees in applied mathematics, respectively, from the Dalian University of Technology (China), the University of Manchester (UK) and the University of Plymouth (UK). He is a computational mathematician specialised in developing novel and fast numerical algorithms for various scientific computing (especially imaging) applications. He has been the director of multidisciplinary research at the Centre for Mathematical Imaging Techniques (CMIT) since 2007, and the director of the EPSRC Liverpool Centre of Mathematics in Healthcare (LCMH) since 2015. He heads a large group of computational imagers, tackling novel analysis of real-life images. His group's imaging work in variational modelling and algorithmic development is mostly interdisciplinary, strongly motivated by emerging real-life problems and their challenges: image restoration, image inpainting, tomography, image segmentation and registration.

Carola-Bibiane Schönlieb is Professor of Applied Mathematics at the University of Cambridge. There, she is head of the Cambridge Image Analysis group and co-director of the EPSRC Cambridge Mathematics of Information in Healthcare Hub. Since 2011, she is a fellow of Jesus College Cambridge and since 2016 a fellow of the Alan Turing Institute, London. She also holds the chair of the Committee for Applications and Interdisciplinary Relations (CAIR) of the EMS. Her current research interests focus on variational methods, partial differential equations and machine learning for image analysis, image processing and inverse imaging problems. She has active interdisciplinary collaborations with clinicians, biologists and physicists on biomedical imaging topics, chemical engineers and plant scientists on image sensing, as well as collaborations with artists and art conservators on digital art restoration.

Her research has been acknowledged by scientific prizes, among them the LMS Whitehead Prize 2016, the Philip Leverhulme Prize in 2017, the Calderon Prize 2019, a Royal Society Wolfson fellowship in 2020 and a doctorate honoris causa from the University of Klagenfurt in 2022, and by invitations to give plenary lectures at several renowned applied mathematics conferences, among them the SIAM conference on Imaging Science in 2014, the SIAM conference on Partial Differential Equations in 2015, the SIAM annual meeting in 2017, the Applied Inverse Problems Conference in 2019, the FOCM 2020 and the GAMM 2021.

Carola graduated from the Institute for Mathematics, University of Salzburg (Austria), in 2004. From 2004 to 2005, she held a teaching position in Salzburg. She received her PhD degree from the University of Cambridge (UK) in 2009. After 1 year of postdoctoral activity at the University of Göttingen (Germany), she became a lecturer at Cambridge in 2010, promoted to reader in 2015 and promoted to professor in 2018.

About the Editors

Prof. Xue-Cheng Tai is a chief research scientist and executive programme director at Hong Kong Center for Cerebro-cardiovascular Health Engineering (COCHE), Hong Kong Science Park. He is a professor and head of the Department of Mathematics at Hong Kong Baptist University (China) since 2017. Before 2017, he served as a professor in the Department of Mathematics at Bergen University (Norway). His research interests include numerical PDEs, optimisation techniques, inverse problems and image processing. He has done significant research work in his research areas and published more than 250 top-quality international conference and journal papers. He is the winner of the 8th Feng Kang Prize for scientific computing. Prof Tai has served as organising and programme committee member for a number of international conferences and has been often invited at international conferences. He has served as referee and reviewer for many premier conferences and journals.

Laurent Younes is a professor in the Department Applied Mathematics and Statistics, Johns Hopkins University (USA), which he joined in 2003, after 10 years as a researcher for the CNRS in France. He is a former student of Ecole Normale Supérieure (Paris) and of the University of Paris 11 from which he received his PhD in 1988. His work includes contributions to applied probability, statistics, graphical models, shape analysis and computational medicine. He is a fellow of the IMS and of the AMS.

Contributors

Jonas Adler Department of Mathematics, KTH – Royal Institute of Technology, Stockholm, Sweden

Alexis Arnaudon Department of Mathematics, Imperial College, London, UK

Blue Brain Project, École polytechnique fédéral de Lausanne (EPFL), Geneva, Switzerland

Jean-François Aujol Univ. Bordeaux, Bordeaux INP, CNRS, IMB, UMR 5251, Talence, France

Noor Badshah Department of Basic Sciences, University of Engineering and Technology, Peshawar, Pakistan

Egil Bae Norwegian Defence Research Establishment (FFI), Kjeller, Norway

Coloma Ballester IPCV, DTIC, University Pompeu Fabra, Barcelona, Spain

Martin Bauer Department of Mathematics, Florida State University, Tallahassee, FL, USA

E. J. Bekkers Amsterdam Machine Learning Lab, University of Amsterdam, Amsterdam, The Netherlands

Stefania Bellavia Dipartimento di Ingegneria Industriale, Università degli Studi di Firenze (INdAM-GNCS members), Firenze, Italia

Martin Benning The School of Mathematical Sciences, Queen Mary University of London, London, UK

Tommaso Bianconcini Verizon Connect, Firenze, Italia

Alfred M. Bruckstein Computer Science Department, Technion – Israel Institute of Technology, Haifa, Israel

Aurélie Bugeau LaBRI, CNRS, Université de Bordeaux, Talence, France

Tan Bui-Thanh Department of Aerospace Engineering and Engineering Mechanics, The Oden Institute for Computational Engineering and Sciences, UT Austin, Austin, TX, USA

Xiaohao Cai School of Electronics and Computer Science, University of Southampton, Southampton, UK

Hernan Carrillo LaBRI, CNRS, Bordeaux INP, Université de Bordeaux, Bordeaux, France

Hei-Long Chan Chinese University of Hong Kong, Hong Kong, China

Raymond Chan Department of Mathematics, College of Science, City University of Hong Kong, Kowloon Tong, Hong Kong, China

Huibin Chang School of Mathematical Sciences, Tianjin Normal University, Tianjin, China

Nicolas Charon Department of Applied Mathematics and Statistics, Johns Hopkins University, Baltimore, MD, USA

Center for Imaging Science, Johns Hopkins University, Baltimore, MD, USA

Ke Chen Department of Mathematical Sciences, Centre for Mathematical Imaging Techniques, University of Liverpool, Liverpool, UK

Yunmei Chen Department of Mathematics, University of Florida, Gainesville, FL, USA

Gary P. T. Choi Department of Mathematics, Massachusetts Institute of Technology, Cambridge, MA, USA

Michaël Clément LaBRI, CNRS, Bordeaux INP, Université de Bordeaux, Bordeaux, France

Elaine Crooks Department of Mathematics, Swansea University, Swansea, UK

Yehuda Dar Electrical and Computer Engineering Department, Rice University, Houston, TX, USA

Jérôme Darbon Division of Applied Mathematics, Brown University, Providence, RI, USA

Noémie Debroux Pascal Institute, University of Clermont Auvergne, Clermont-Ferrand, France

Juan Carlos De los Reyes Research Center for Mathematical Modelling (MODEMAT), Escuela Politécnica Nacional, Quito, Ecuador

Yuping Duan Center for Applied Mathematics, Tianjin University, Tianjin, China

R. Duits Applied Differential Geometry, Department of Mathematics and Computer Science, Eindhoven University of Technology, Eindhoven, The Netherlands

Matthias J. Ehrhardt Institute for Mathematical Innovation, University of Bath, Bath, UK

Florian Faucher Faculty of Mathematics, University of Vienna, Vienna, Austria

Xiangchu Feng School of Mathematics and Statistics, Xidian University, Xi'an, China

Caroline Geiersbach Weierstrass Institute, Berlin, Germany

Guy Gilboa Technion – IIT, Haifa, Israel

Rémi Giraud Univ. Bordeaux, CNRS, IMS UMR5251, Bordeaux INP, Talence, France

Carsten Gottschlich Institute for Mathematical Stochastics, University of Göttingen, Göttingen, Germany

Markus Grasmair NTNU, Trondheim, Norway

Hanlin Gu Hong Kong University of Science and Technology, Hong Kong, China

Xianfeng Gu Stony Brook University, Stony Brook, NY, USA

Luying Gui Department of Mathematics, Nanjing University of Science and Technology, Nanjing, China

Abderrahmane Habbal Modeling and Data Science, Mohammed VI Polytechnic University Benguerir, Morocco

Université Côte d'Azur, Inria, Sophia Antipolis, France

Michal Haindl Institute of Information Theory and Automation, Czech Academy of Sciences, Prague, Czechia

Markus Haltmeier Department of Mathematics, University of Innsbruck, Innsbruck, Austria

Andreas Hauptmann Research Unit of Mathematical Sciences, University of Oulu, Oulu, Finland

Darryl Holm Department of Mathematics, Imperial College, London, UK

Stephan F. Huckemann Felix-Bernstein-Institute for Mathematical Statistics in the Biosciences, University of Göttingen, Göttingen, Germany

Samuel Hurault Bordeaux INP, CNRS, IMB, Université de Bordeaux, Talence, France

Soomin Jeon Department of Radiology, Massachusetts General Hospital and Harvard Medical School, Boston, MA, USA

Moez Kallel Laboratory for Mathematical and Numerical Modeling in Engineering Science (LAMSIN), University of Tunis El Manar, National Engineering School of Tunis, Tunis-Belvédère, Tunisia

Clemens Kirisits Faculty of Mathematics, University of Vienna, Vienna, Austria

Eric Klassen Department of Mathematics, Florida State University, Tallahassee, FL, USA

Nataša Krejić Department of Mathematics and Informatics, Faculty of Sciences, University of Novi Sad, Novi Sad, Serbia

Gitta Kutyniok Ludwig-Maximilians-Universität München, Mathematisches Institut, München, Germany

Andreas Langer Centre for Mathematical Sciences, Lund University, Lund, Sweden

Gabriel P. Langlois Division of Applied Mathematics, Brown University, Providence, RI, USA

Alessandro Lanza Department of Mathematics, University of Bologna, Bologna, Italy

Alice Le Brigant Department of Applied Mathematics, University Paris, Paris, France

Carole Le Guyader INSA Rouen Normandie, Laboratory of Mathematics, Normandie University, Rouen, France

Chang-Ock Lee Department of Mathematical Sciences, KAIST, Daejeon, Republic of Korea

Na Lei Dalian University of Technology, Dalian, China

Chunming Li School of Information and Communication Engineering, University of Electronic Science and Technology of China, Chengdu, China

Lingfeng Li Department of Mathematics, Hong Kong Baptist University, Kowloon Tong, Hong Kong, China

Department of Mathematics, Southern University of Science and Technology, Shenzhen, China

Haiguang Liu Microsoft Research-Asian, Beijing, China

Zhifang Liu School of Mathematical Sciences, Tianjin Normal University, Tianjin, China

Estefania Loayza-Romero Institute for Analysis and Numerics, University of Münster, Münster, Germany

Lok Ming Lui Department of Mathematics, The Chinese University of Hong Kong, Hong Kong, China

Sebastian Lunz Department of Applied Mathematics and Theoretical Physics, University of Cambridge, Cambridge, UK

Feng Luo Rutgers University, Piscataway, NJ, USA

Junyan Lyu Department of Electronic and Electrical Engineering, Southern University of Science and Technology, Shenzhen, Guangdong, China

Jun Ma Department of Mathematics, Nanjing University of Science and Technology, Nanjing, China

Stefano Marchesini SLAC National Laboratory, Menlo Park, CA, USA

Tingwei Meng Division of Applied Mathematics, Brown University, Providence, RI, USA

Serena Morigi Department of Mathematics, University of Bologna, Bologna, Italy

Benedetta Morini Dipartimento di Ingegneria Industriale, Università degli Studi di Firenze (INdAM-GNCS members), Firenze, Italia

Elena Morotti Department of Political and Social Sciences, University of Bologna, Bologna, Italy

Valeriya Naumova Machine Intelligence Department, Simula Consulting and SimulaMet, Oslo, Norway

Sebastian Neumayer Institute of Mathematics, TU Berlin, Berlin, Germany

Linh Nguyen Department of Mathematics, University of Idaho, Moscow, ID, USA

Ozan Öktem Department of Information Technology, Division of Scientific Computing, Uppsala University, Uppsala, Sweden

Antonio Orlando CONICET, Departamento de Bioingeniería, Universidad Nacional de Tucumán, Tucumán, Argentina

Simone Parisotto DAMTP, University of Cambridge, Cambridge, UK

Elena Loli Piccolomini Department of Computer Science and Engineering, University of Bologna, Bologna, Italy

Fabien Pierre LORIA, UMR CNRS 7503, Université de Lorraine, INRIA projet Tangram, Nancy, France

J. W. Portegies Center for Analysis, Scientific Computing and Applications, Department of Mathematics and Computer Science, Eindhoven University of Technology, Eindhoven, The Netherlands

Michael Quellmalz Institute of Mathematics, Technical University Berlin, Berlin, Germany

Lara Raad LIGM, CNRS, Univ Gustave Eiffel, Marne-la-Vallée, France

Robin Richter Felix-Bernstein-Institute for Mathematical Statistics in the Biosciences, University of Göttingen, Göttingen, Germany

Erlend Skaldehaug Riis The Department of Applied Mathematics and Theoretical Physics, Cambridge, UK

Otmar Scherzer Faculty of Mathematics, University of Vienna, Vienna, Austria

Carola Schönlieb Department of Applied Mathematics and Theoretical Physics, University of Cambridge, Cambridge, UK

Johannes Schwab Department of Mathematics, University of Innsbruck, Innsbruck, Austria

Ivan W. Selesnick Department of Electrical and Computer Engineering, New York University, New York, NY, USA

Eric Setterqvist Johann Radon Institute for Computational and Applied Mathematics (RICAM), Linz, Austria

Fiorella Sgallari Department of Mathematics, University of Bologna, Bologna, Italy

B. M. N. Smets Applied Differential Geometry, Department of Mathematics and Computer Science, Eindhoven University of Technology, Eindhoven, The Netherlands

Stefan Sommer Department of Computer Science (DIKU), University of Copenhagen, Copenhagen E, Denmark

Xiaoliang Song School of Mathematical Sciences, Dalian University of Technology, Dalian, Liaoning, China

Gabriele Steidl Institute of Mathematics, TU Berlin, Berlin, Germany

Xuecheng Tai Hong Kong Center for Cerebro-cardiovascular Health Engineering (COCHE), Shatin, Hong Kong, China

Roger Tam School of Biomedical Engineering, The University of British Columbia, Vancouver, BC, Canada

Xiaoying Tang Department of Electronic and Electrical Engineering, Southern University of Science and Technology, Shenzhen, Guangdong, China

Duy H. Thai Department of Mathematics, Colorado State University, Fort Collins, CO, USA

Anis Theljani Department of Mathematical Sciences, University of Liverpool Mathematical Sciences Building, Liverpool, UK

Ilona Christy Unarta Hong Kong University of Science and Technology, Hong Kong, China

Tuomo Valkonen Center for Mathematical Modeling, Escuela Politécnica Nacional, Quito, Ecuador

Department of Mathematics and Statistics, University of Helsinki, Helsinki, Finland

David Villacís Research Center for Mathematical Modelling (MODEMAT), Escuela Politécnica Nacional, Quito, Ecuador

Patricia Vitoria IPCV, DTIC, University Pompeu Fabra, Barcelona, Spain

Samad Wali School of Information and Communication Engineering, University of Electronic Science and Technology of China, Chengdu, China

Yang Wang Hong Kong University of Science and Technology, Hong Kong, SAR, China

Kathrin Welker Faculty of Mechanical Engineering and Civil Engineering, Helmut-Schmidt-University/University of the Federal Armed Forces Hamburg, Hamburg, Germany

A. J. Wemmenhove Applied Differential Geometry, Department of Mathematics and Computer Science, Eindhoven University of Technology, Eindhoven, The Netherlands

Jonathan Wittmer Department of Aerospace Engineering and Engineering Mechanics, UT Austin, Austin, TX, USA

Chunlin Wu School of Mathematical Sciences, Nankai University, Tianjin, China

Yin Xian Hong Kong Applied Science and Technology Research Institute (ASTRI), Hong Kong, China

TCL Research Hong Kong, Hong Kong, SAR, China

Huilin Yang Department of Electronic and Electrical Engineering, Southern University of Science and Technology, Shenzhen, Guangdong, China

Li Yang School of Mathematical Sciences, Tianjin Normal University, Tianjin, China

Xiaoping Yang Department of Mathematics, Nanjing University, Nanjing, China

Yuan Yao Hong Kong University of Science and Technology, Hong Kong, China

Shing-Tung Yau Harvard University, Cambridge, MA, USA

Xiaojing Ye Department of Mathematics and Statistics, Georgia State University, Atlanta, GA, USA

Laurent Younes Center for Imaging Science, Johns Hopkins University, Baltimore, MD, USA

Bo Yu School of Mathematical Sciences, Dalian University of Technology, Dalian, Liaoning, China

Tieyong Zeng Department of Mathematics, The Chinese University of Hong Kong, Shatin, Hong Kong, China

Qingchao Zhang Department of Mathematics, University of Florida, Gainesville, FL, USA

Lingyan Zhang School of Information and Communication Engineering, University of Electronic Science and Technology of China, Chengdu, China

Kewei Zhang School of Mathematical Sciences, University of Nottingham, Nottingham, UK

Xiaolong Zhu School of Mathematics and Statistics, Xidian University, Xi'an, China

Part II
Model- and Data-Driven Variational Imaging Approaches

Learned Iterative Reconstruction

19

Jonas Adler

Contents

Introduction.. 752
   Deep Learning.. 754
Architectures... 754
   Gradient-Based Architectures.. 755
   Proximal-Based Architectures... 757
   Primal-Dual Networks... 758
   Other Schemes.. 759
Training Procedure.. 760
Engineering Aspects... 763
   Architectures for Learned Operator.. 763
   Initialization.. 763
   Parameter Sharing... 764
   Further Memory.. 764
   Preconditioning... 765
   Learned Step Length.. 765
   Scalable Training... 765
   Putting It All Together.. 766
Conclusions.. 766
References... 767

Abstract

Learned iterative reconstruction methods have recently emerged as a powerful tool to solve inverse problems. These deep learning techniques for image reconstruction achieve remarkable speed and accuracy by combining hard knowledge

---

J. Adler (✉)
Department of Mathematics, KTH – Royal Institute of Technology, Stockholm, Sweden
e-mail: jonasadl@kth.se

Now with DeepMind, London, UK

© Springer Nature Switzerland AG 2023
K. Chen et al. (eds.), *Handbook of Mathematical Models and Algorithms in Computer Vision and Imaging*, https://doi.org/10.1007/978-3-030-98661-2_67

about the physics of the image formation process, represented by the forward operator, with soft knowledge about how the reconstructions should look like, represented by deep neural networks. A diverse set of such methods have been proposed, and this chapter seeks to give an overview of their similarities and differences, as well as discussing some of the commonly used methods to improve their performance.

Keywords

Inverse Problems · Deep Learning · Iterative reconstruction · Architectures

Introduction

Inference problems are ubiquitous in the sciences, medicine, and engineering. In these problems, we are given some form of data $y \in Y$ and aim to infer a result $x \in X$ from it. Typical examples include image classification where y is an image and x is a label and image segmentation where y is an image and x is a pointwise label. Inverse problems are a specific class of inference problems where we have access to additional structure. In particular, we assume the existence of a known forward operator $\mathcal{T} : X \to Y$ such that

$$\mathcal{T}x = y + \delta$$

where $\delta \in Y$ is a noise term with known distribution. The inference problem is hence reduced to inverting this relationship, a process we call reconstruction.

Deep learning techniques (LeCun et al. 2015; Goodfellow et al. 2016) using convolutional neural networks (CNN) (LeCun et al. 1989) have recently achieved state-of-the art results in almost all fields of image processing (Krizhevsky et al. 2012), but until recently their application to image reconstruction has been limited. Several practical reasons for this can be claimed, notably lack of data, but perhaps one of the strongest reasons is that image reconstruction does not fit snugly into the standard problem formulation common to most image processing methods. In these problems, the input is an image, typically two-dimensional, and the output is also an image. The input and output images have a strong spatial relationship: A point in the input image corresponds to the same point in the output image, and if we translate the image, then the result should be translated as well (equivariance). These properties align perfectly with convolutions, whose use has been a major component in the deep learning revolution.

Neither of these properties hold in image reconstruction. Here, a point in the output image often depends globally on data from the input, and there is no trivial spatial relationship to use. In fact, in most interesting inverse problems, the input and output do not even belong to the same space. For example, in computed tomography, the input is a function on some set of lines through space, while the reconstruction should be a scalar field in space. Since the input and output lives in different spaces, we cannot even perform standard linear operations on them, such as addition, much less hope that a convolution would take us from one to the other.

One way to solve this would be to generalize the concept of convolutions, and a significant effort has actually been spent on how to connect these spaces in mathematically rigorous ways. Notably the field of Fourier integral operators (FIO) (Hörmander 1971) has been developed, and these operators can be seen as generalizations of convolutions. However, the simple point-correspondence of convolutions breaks down, and instead we get a point-to-set correspondence, the canonical relation. Fourier integral operators are also notoriously complicated to work with and often computationally expensive. For this reason, the generalization of convolutional neural networks, perhaps FIO-neural networks (Feliu-Faba et al. 2019; Alizadeh et al. 2019), has so far not been applied to inverse problems.

While some have gone for a fully learned approach, ignoring the inherent symmetries, this does not seem to scale to realistic problem sizes (Zhu et al. 2018). Instead researchers have taken a middle way of incorporating more knowledge about the forward operator in a separate non-learned way into their learned reconstruction techniques. A very successful early approach has been to somehow convert the reconstruction problem into an image processing problem, which is easier than one might expect. Simply start by applying *any* reconstruction operator to the data to obtain a suboptimal initial reconstruction and then train a convolutional neural network to map the initial reconstruction to a more high-quality reconstruction (Jin et al. 2017; Kang et al. 2017).

While such methods incorporate significant components of the physics of the problem, encapsulated in the initial reconstruction, this also gives the methods a strong bias toward the result of the initial reconstruction, and in particular if there is any information lost in the initial reconstruction, it cannot be recovered by the post-processing.

An alternative, *learned iterative reconstruction*, has been developed in recent years. In learned iterative reconstruction, the physics of the problem is not seen as a separate component to be done prior to applying learning, but rather it is seen as an integral component of the learned reconstruction operator of equal footing with other commonly used components in neural networks such as convolutions and pointwise nonlinearities, thus allowing us to learn a reconstruction method acting on measured raw data.

This chapter will survey the development of these learned iterative reconstruction schemes and try to give an overview of architectures, training procedures, and practical and theoretical results. We note that several other high-quality review papers have looked at deep learning for inverse problems (Wang et al. 2018; McCann and Unser 2019; Arridge et al. 2019; Hammernik and Knoll 2020) and invite the reader to look at them for a broader overview of other techniques to use deep learning for image reconstruction.

Deep Learning

A (deep) neural network is a highly overparametrized function composed of several relatively simple parametrized components, often called layers, where both the components and their composition are differentiable. The exact choice of these components and how they are composed are called the architecture of the neural network, and the archetypical example is the standard feed-forward network which is a composition of parametrized affine operators and pointwise nonlinearities.

Training of a neural network $N_\theta : Y \to X$ is another term for selecting the parameters θ, and for inference problems, it is typically performed using some set of supervised training data $(y_i, x_i) \in Y \times X$ where the parameters are selected in order to minimize the empirical risk function

$$L(\theta) = \sum_i \ell(N_\theta(y_i), x_i)$$

where $\ell : X \times X \to \mathbb{R}$ is a loss function describing how close the result is to the ground truth. The networks are trained using some form of stochastic gradient descent over the parameters θ, which is made possible by the back-propagation algorithm (LeCun et al. 1989) which exploits the compositional structure of the networks to compute the gradient of the loss using only knowledge about the derivatives of the components. There is however a wide range of variations in how to train neural networks, as we will explore in section "Training Procedure".

Architectures

Over the last years, a range of architectures for learned iterative reconstruction have been investigated, and although there have been steps toward it (Leuschner et al. 2019; Zbontar et al. 2018; Ramzi 2019), there is as of yet no consistent comparison of their performance in a benchmark, with each architecture sporting different upsides and downsides. We'll here give a broad overview of the most common architectures used in the literature.

The core idea of learned iterative reconstruction is to interlace application of knowledge-driven operators, e.g., the forward operator, with learned operators such as convolutional neural networks. There are multiple ways to motivate specific learned iterative reconstruction architectures, but the most popular is to see them as neural network architectures inspired by unrolling of optimization solvers (Hershey et al. 2014). Specifically one notes that an optimization solver stopped after a finite number of iterations almost satisfies our conditions for a neural network (Banert et al. 2018). It is an operator that takes the data as input, processes it with simple components such as computing linear combinations and gradients, and returns a reconstruction. The individual components are also often differentiable, so the only thing missing is parametrizing the scheme so that there is something to learn.

19 Learned Iterative Reconstruction

There are many optimization problems to be inspired by and even more solvers. Learned iterative reconstruction methods can be broadly classified according to what type of optimization solver they were inspired by, and by now most commonly used classes of optimization solvers have been converted into reconstruction schemes. The learning on the other hand is introduced by replacing certain components, such as gradients or proximals, with learned counterparts in the form of neural networks.

Here we must stop and stress that the architectures are merely *inspired* by optimization solvers. Learned iterative reconstruction schemes do not actually try to solve any optimization problem as part of computing the reconstruction, not even approximately.

We'll now introduce some of the most common such constructions in a structured manner. We'll then follow up with various engineering tricks that have been found to sometimes vastly improve performance before finally turning to the training.

Gradient-Based Architectures

A set of very well-studied optimization problems are those associated with the maximum a posteriori solution given some prior. These optimization problems have been extensively explored over the years, including in both Tikhonov and total variation (TV) regularization. It can be studied using Bayes' theorem, according to which the posterior distribution $P(x \mid y)$ can be decomposed into components

$$P(x \mid y) = \frac{P(y \mid x) P(x)}{P(y)}.$$

Assuming that the posterior is differentiable, a gradient-based method to find the maximum of the log posterior can thus be written as in Algorithm 1.

Algorithm 1 Gradient Descent

1: Select $x^0 \in X$
2: **for** $n = 1, \ldots$ **do**
3: $\quad x^n \leftarrow x^{n-1} + \alpha \left(\nabla_x \log P(y \mid x^{n-1}) + \nabla_x \log P(x^{n-1}) \right)$
4: **end for**

Here we note that the data likelihood $P(y \mid x)$ is exactly known in inverse problems, whereas the prior $P(x)$ has to be chosen by the practitioner. The idea of gradient-based learned iterative reconstruction schemes is to unroll a gradient descent scheme to a finite number of iterations and then replace the gradient of the log-prior with a learned component $\Lambda_\theta : X \to X$ parametrized by parameters θ as in Algorithm 2.

As long as the operators $x \to \Lambda_\theta(x)$ and $x \to \log P(y \mid x)$ are differentiable, this yields a deep neural network which is end-to-end trainable. Interestingly,

Algorithm 2 Learned Gradient Descent

1: Select $x^0 \in X$
2: **for** $n = 1, \ldots, N$ **do**
3: $\quad x^n \leftarrow x^{n-1} + \alpha \left(\nabla_x \log P(y \mid x^{n-1}) + \Lambda_\theta(x^{n-1}) \right)$
4: **end for**
5: $N_\theta(y) \leftarrow x^N$

this very basic form of gradient-based learned iterative reconstruction was never published on its own, but a wide range of closely related schemes have been considered (Hauptmann et al. 2019; Chen et al. 2018).

Variational Networks

Variational networks (Hammernik et al. 2018) are a widely used class of gradient-based learned iterative reconstruction methods that more closely follow the inspiration from optimization than other schemes. In particular, the learned operator Λ_θ is required to be the gradient of some function which is learned

$$\Lambda_\theta(x) = \nabla_x h_\theta(x).$$

In the original papers, the functional $h_\theta : X \to \mathbb{R}$ was chosen to be of the form

$$h_\theta(x) = \sum_{k=1}^{K} \phi_{\theta_k}(K_{\theta_k} x)$$

where $\phi : X \to \mathbb{R}$ is a learnable nonlinear function averaged over the domain and K is a convolution kernel. Similar schemes could be obtained by using more expressive forms, e.g., a multilayer perceptron.

The use of an actual gradient gives some further interpretability of the scheme as minimization of a specific functional and the additional inductive bias helps reducing overfitting. However, since the functional h_θ is typically highly non-convex and since we stop after a finite number of steps, it is hard to exploit this to analyze, e.g., stability of the solution.

Variational networks have been applied to MRI reconstruction, both in the simplified setting of Fourier inversion and for the real nonlinear setting of multi-coil data (Knoll et al. 2019; Schlemper et al. 2019). In addition to this, it has been applied to a range of other imaging modalities such as CT (Hammernik et al. 2017; Vishnevskiy et al. 2019; Kobler et al. 2018) and ultrasound imaging (Vishnevskiy et al. 2018).

Proximal-Based Architectures

The proximal gradient algorithm (Parikh et al. 2014) is a method for solving convex optimization problems given by the sum of two functionals where only one of the functional is required to be differentiable; the other needs only to have a proximal operator defined. The method is an excellent fit for inverse problems since the log data likelihood is typically smooth while the prior is not.

Given a specific (log-)prior, the proximal operator can be seen as a backward gradient step and is given by

$$\text{prox}_{x \to -\alpha \log P(x)}(\hat{x}) = \arg\min_{x \in X} \left(\frac{1}{2} \|x - \hat{x}\|^2 - \alpha \log P(x) \right).$$

Using this, the proximal gradient algorithm, given in the setting of Bayesian inversion, is given in Algorithm 3.

Algorithm 3 Proximal Gradient

1: Select $x^0 \in X$
2: **for** $n = 1, \ldots$ **do**
3: $\quad x^n \leftarrow \text{prox}_{-\alpha \log P}\left(x^{n-1} + \alpha \nabla_x \log P(y \mid x^{n-1})\right)$
4: **end for**

As an opportunity for learning, we note that this is very similar to the gradient ascent scheme except that instead of an additive gradient, the proximal of the log-prior acts on the updated point. The corresponding learned iterative reconstruction scheme can be obtained by replacing the proximal operator by a learned component.

Algorithm 4 Learned Proximal Gradient

1: Select $x^0 \in X$
2: **for** $n = 1, \ldots, N$ **do**
3: $\quad x^n \leftarrow \Lambda_\theta \left(x^{n-1} + \alpha \nabla_x \log P(y \mid x^{n-1})\right)$
4: **end for**
5: $N_\theta(y) \leftarrow x^N$

This type of scheme was first published under the name *recurrent inference machines* with applications to image processing problems (Putzky and Welling 2017). Several other papers extended the methods by adding further components but also by applying the method to CT (Adler and Öktem 2017; Gupta et al. 2018) MRI (Lønning et al. 2018) and photoacoustic tomography (Hauptmann et al. 2018; Yang et al. 2019).

We should however note that there is a different way to use the proximal gradient scheme. In particular, it is sometimes the case that the proximal of the data log-likelihood $x \to -\log P(y \mid x)$ is easy to compute. This is the case in most image processing problems but also in MRI. If the data likelihood is differentiable, one can use the proximal gradient scheme given in Algorithm 5.

Algorithm 5 Projected Gradient

1: Select $x^0 \in X$
2: **for** $n = 1, \ldots$ **do**
3: $\quad x^n \leftarrow \text{prox}_{x \to -\alpha \log P(y|x)} \left(x^{n-1} + \alpha \nabla_x \log P(x^{n-1}) \right)$
4: **end for**

This scheme often has very fast convergence since the proximal (depending on the data likelihood) can be seen as a projection onto the feasible set in a single iteration. For this reason, the algorithm is sometimes called projected gradient descent, and we'll adopt that name here for disambiguation purposes. Algorithms such as ADMM can also be seen as variations of the general idea.

One can then introduce learning as usual by replacing the knowledge-driven prior with a data-driven prior as in Algorithm 6.

Algorithm 6 Learned Projected Gradient

1: Select $x^0 \in X$
2: **for** $n = 1, \ldots, N$ **do**
3: $\quad x^n \leftarrow \text{prox}_{x \to -\alpha \log P(y|x)} \left(x^{n-1} + \Lambda_\theta (x^{n-1}) \right)$
4: **end for**
5: $N_\theta(y) \leftarrow x^N$

This class of algorithms has become very popular in MRI reconstruction due to their ease of implementation and speed improvements over gradient-based schemes. In this domain, the proximal step is often called a *data consistency* term, since the proximal enforces the result to be (approximately) consistent with the data (Schlemper et al. 2017; Aggarwal et al. 2018; Kofler et al. 2018). One of the first learned iterative reconstruction schemes, ADMM-Net (Sun et al. 2016) used a related approach for MRI reconstruction, and a range of works have followed with some interesting variations (Mardani et al. 2017a,b, 2018), and there has even been some analysis of their convergence (Schwab et al. 2018).

Primal-Dual Networks

Primal-dual proximal splitting optimization methods are another class of optimization schemes applicable to inverse problems. Specifically, if the data likelihood can be written

Algorithm 7 Primal-Dual

1: Select $x^0 \in X, z^0 \in Y$
2: **for** $n = 1, \ldots$ **do**
3:      $y^n \leftarrow \text{prox}_{-\alpha(\log \mathcal{L})^*} \left(y^{n-1} + \mathcal{T}x^{n-1} \right)$
4:      $x^n \leftarrow \text{prox}_{-\alpha \log P} \left(x^{n-1} - \mathcal{T}^* y^n \right)$
5: **end for**

Algorithm 8 Learned Primal-Dual

1: Select $x^0 \in X, z^0 \in Y$
2: **for** $n = 1, \ldots$ **do**
3:      $y^n \leftarrow \Gamma_\theta \left(y^{n-1} + \mathcal{T}x^{n-1} \right)$
4:      $x^n \leftarrow \Lambda_\theta \left(x^{n-1} - \mathcal{T}^* y^n \right)$
5: **end for**
6: $N_\theta(y) \leftarrow x^N$

$$P(y \mid x) = \mathcal{L}(y \mid \mathcal{T}x),$$

then the problem can be solved using a proximal-based scheme with only knowledge about the proximal of the functional $\mathcal{T}x \to -\log P(y \mid \mathcal{T}x)$. The most simple of such scheme, the Arrow-Hurwich algorithm (Arrow et al. 1958), is given in Algorithm 7. Accelerated versions of the scheme using momentum, including the primal-dual hybrid gradient algorithm (Chambolle and Pock 2011), are very popular for optimization in inverse problems due to their speed and versatility.

Following the recipe from before, we can convert the primal-dual algorithm into a learned scheme by replacing the proximals with learned operators. Here one could replace only the proximal related to the prior or both proximals, but most authors prefer to learn both and this gives rise to the learned primal dual scheme, as in Algorithm 8.

Given the versatility of this kind of algorithm, practically only requiring access to the forward operator, it can be applied to almost any inverse problem. So far, applications have been to CT (Adler and Öktem 2018b; Wu et al. 2018, 2019b), possibly with incomplete data (Zhang et al. 2019), and image processing (Vogel and Pock 2017).

Other Schemes

To round off our expose on classical methods that have been converted into learned iterative reconstruction schemes, we note that some authors have found their inspiration in iterative schemes outside of optimization. One such idea is Neumann networks (Gilton et al. 2019), which gain inspiration from the Neumann series for the inverse

$$\mathcal{T}^{-1} = \sum_{n=1}^{\infty} (I - \eta \mathcal{T}^* \mathcal{T})^n \eta \mathcal{T}^*$$

where $\eta < \|\mathcal{T}^*\mathcal{T}\|$ is a step length. The authors view the partial sums as an iteration and add a learning component as a small offset to the update, which leads to Algorithm 9.

Algorithm 9 Neumann Network

1: $x^0 \leftarrow \eta \mathcal{T}^* y$
2: $\hat{x}^0 \leftarrow x^0$
3: **for** $n = 1, \ldots$ **do**
4: $\quad x^n \leftarrow x^{n-1} - \eta \mathcal{T}^* \mathcal{T} x^{n-1} - \eta \Lambda_\theta(x^{n-1})$
5: $\quad \hat{x}^n \leftarrow \hat{x}^n + x^n$
6: **end for**
7: $N_\theta(y) \leftarrow \hat{x}^N$

We note that the algorithm is very similar to a gradient-based scheme, but that the result is given as the sum of all partial iterates, and that the data only enters in the beginning.

Others have taken inspiration from classical iterative reconstruction schemes, e.g., the Landweber algorithm (Aspri et al. 2018), and there is nothing to stop researchers from using other methods such as conjugate gradient in the future.

Training Procedure

Given an architecture, the next step is to select the optimal parameters. The definition of what's meant by "optimal" is however a hot area for both research and debate. By far the most popular definition of "optimal" for neural networks in general and learned iterative reconstruction schemes in particular is to view the problem as an inference problem where the data is seen as a sample from a random variable **y**, and we seek to infer the unknown signal which is a sample from another random variable, **x**. Our training data is seen as N samples (y_n, x_n) from the joint random variable (**y**, **x**). Further, as in the introduction, we introduce a loss function $\ell : X \to X$ which characterizes how good a single reconstruction is. Given all of this, the optimal parameter choice is defined as the parameters which minimize the risk function

$$L(\theta) = \mathbb{E}\,\ell(N_\theta(\mathbf{y}), \mathbf{x}).$$

Since the risk involves an expectation over the random variables **y** and **x**, which we don't have access to since they should represent all possible inputs/outputs, we need to approximate it using our training data. Thankfully, the sample mean is an unbiased estimator for the expectation, so we can instead chose to minimize the empirical risk function

$$\hat{L}(\theta) = \sum_{n=1}^{N} \ell(N_\theta(y_n), x_n).$$

This paradigm is known as empirical risk minimization. The next problem is to find a minimizer to $\hat{L}(\theta)$ and this is nontrivial, especially given the scale of both training data and networks in current practice. The machine learning community has converged on approximately solving the optimization problem using variations of stochastic gradient descent (LeCun et al. 1989; Kingma and Ba 2014).

A significant problem in implementing backpropagation for learned iterative reconstruction is that while there are well-maintained libraries for computing gradients of standard neural network components using automatic differentiation (Abadi et al. 2016; Paszke et al. 2017), these very rarely implement, e.g., the Radon transform. Many researchers solved this by wrapping other implementations of these operators such as ASTRA (van Aarle et al. 2015) using some glue library, e.g., ODL (Adler et al. 2017a). While this is very versatile and allows easy comparison to classical reconstruction algorithms, there are some performance downsides. Some have therefore implemented tomographic operators with native backpropagation (Syben et al. 2019), and there is considerable interest toward differentiable programming, a paradigm that would allow backpropagation through any operator (Innes et al. 2019; Bradbury et al. 2018).

Since we only have access to a finite amount of training data, empirical risk minimization will lead to *overfitting* to our available data, e.g., the optimal parameter choice for our training data will differ to the optimal choice for all possible inputs, and parameters that minimize the empirical risk will not be optimal for the expected risk. Classical statistical learning theory (and intuition) tells us that the more parameters we have, the more we will overfit to our training data, although this relation has been called into question for deep learning. Learned iterative reconstruction is often seen to have advantageous properties here since the number of parameters is typically much smaller than fully learned methods.

The choice of loss function can also have a significant impact on the learned operator, and authors have proposed a wide range of options. Thankfully, there is actually quite some theory related to the properties of various optimal reconstruction operators under a choice of loss which we can use to guide our choice of loss function. For example, with the squared norm $\ell(N_\theta(y), x) = \|N_\theta(y) - x\|^2$, the optimal will be the minimum mean squared error estimator, which is simply the conditional expectation $\mathbb{E}[x \mid y]$. This implies that a neural network trained with this loss should approximate the conditional expectation. Likewise, it is known that the optimal reconstruction given the 1-norm loss is the conditional median. Even more intricate losses have been investigated in the literature, e.g., when training with a Wasserstein loss, the optimal reconstruction is a *spatial* average of the posterior (Adler et al. 2017b).

Some authors have looked further than these relatively simple losses and have looked toward using neural networks to define a loss function. The earliest such attempts were to use perceptual losses (Johnson et al. 2016), which consider an image as good if it looks like the true image according to a neural network. The definition of "looks like the true image" is taken to have similar intermediate activations and the neural network typically taken to be a ImageNet classifier. This approach has been applied to CT and MRI denoising, where it gave more visually appealing results (Yang et al. 2017a,b, 2018).

A related type of loss is adversarial losses (Goodfellow et al. 2014), where one trains a neural network to judge how good a reconstruction is. In the most simple setting, a *discriminator* network is trained to determine if an image is a reconstruction or a true image, and the reconstruction operator is trained to generate true-looking images. In order to make sure that the network returns a reconstruction that is related to the input, one typically combines this with some form of classical loss and sometimes a cycle-consistency (data-fit) condition. The latter case is especially interesting, since it allows training without paired training data (Mardani 2017; Lei et al. 2019).

Another way of using a neural network to define the loss is to ask "how useful is the reconstruction?", where we define usefulness by how well another network can be trained on the reconstruction to solve some task (Adler et al. 2018). This general and straightforward idea can be applied to practically any downstream task, but initial work has focused on segmentation (Boink et al. 2019), object detection (Wu et al. 2018), and classification (Effland et al. 2018; Diamond et al. 2017).

All of the above methods (possibly excluding adversarial losses) require supervised training data. However, access to this kind of data, especially in large amounts, is often a luxury. Many hence see training using unsupervised data as something of a grand challenge in order to get truly scalable learned iterative reconstruction that is applicable in practice. Some algorithmic advances have been made in this direction, notably the Noise2Noise (Lehtinen et al. 2018) method which uses the fact that when trained with squared norm loss, the result should only depend on the conditional mean of the data. Hence, it is possible to train using noisy ground truth samples, and the learned reconstruction should approximate their mean. Other methods have been developed with the same goal, e.g., the SURE estimator (Raphan and Simoncelli 2007). These methods have just started being used for image reconstruction, but with promising results (Soltanayev and Chun 2018; Cha et al. 2019).

Finally there is great potential in combining learned reconstruction with advances in deep generative models in order to achieve true Bayesian reconstruction methods where one can sample from the posterior distribution instead of computing a single estimator (Adler and Öktem 2018a; Anonymous 2020). Such methods are especially relevant in the low signal/high noise setting, such as ultralow dose CT and dynamic imaging or for highly complicated imaging modalities such as seismic imaging (Herrmann et al. 2019).

To conclude, supervised training with simple losses is still by far the most popular way to train learned iterative reconstruction schemes, but their combination of expressive power, speed, and versatility allows a huge range of other options for training, and we can only expect this field to grow in the future.

Engineering Aspects

While the inspiration from optimization is important to the performance of learned iterative reconstruction, experience (Hessel et al. 2018) tells us that deep learning is highly sensitive to engineering and implementation choices and that including these can significantly improve performance. Learned iterative reconstruction methods have not turned out to be an exception, and considerable effort has been put into finding the best implementations. We'll here try to give a broad overview of these methods.

Architectures for Learned Operator

All learned iterative reconstruction schemes reduce learning the $Y \to X$ reconstruction operator into learning a $X \to X$ (and possibly also $Y \to Y$) operator such as a learned gradient or a learned proximal. This type of operator can be represented by a standard "off-the-shelf" convolutional neural network without any problems. Many authors have found a small, e.g., one to three layer, neural network to be sufficient for the task (Adler and Öktem 2017; Chen et al. 2018; Diamond et al. 2017; Mardani et al. 2017a), and some have decided to use more complicated architectures (Putzky and Welling 2017; Hauptmann et al. 2018), typically converging on some reduced version of the U-Net (Ronneberger et al. 2015). These networks are almost universally combined with the technique of residual learning (He et al. 2016; Jin et al. 2017) where the learned operators are of the form $\Lambda_\theta(x) = x + \hat{\Lambda}_\theta(x)$ with $\hat{\Lambda}_\theta$ a feed forward network.

As a general rule of thumb, for "simple" inverse problems such as fully sampled MRI or CT, small networks seem to work very well, while for more complicated inverse problems such as photoacoustic tomography or ultrasound, a larger network might be needed. However, larger networks typically require more training data to avoid overfitting.

Initialization

Just like optimization, all learned iterative reconstruction methods begin with an initial estimate x^0 which is then refined. Since only a finite number of steps are used, it's reasonable to expect this choice to have quite significant impact on the final result. Authors have converged on two different initialization schemes. These are zero-initialization (Adler and Öktem 2018b), $x^0 = 0$, and pseudo-inverse initialization $x^0 = \mathcal{T}^\dagger y$, where $\mathcal{T}^\dagger : Y \to X$ is some pseudo-inverse, e.g., zero filled Fourier inversion (Hammernik et al. 2018) or filtered back projection (Adler et al. 2017b). In some cases where the forward operator is approximately unitary, e.g., in photoacoustic tomography, the adjoint has been used in place of a pseudo-inverse (Hauptmann et al. 2018). Some have also tried learning some parameters of the initial reconstruction, e.g., learning the filters in filtered back projection (Hammernik et al. 2017). These more advanced initialization schemes have possible speed and accuracy advantages over zero-initialization since the learned operator only needs to learn a correction from the initial reconstruction, but they run a risk of overfitting to the initial reconstruction, giving worse generalization.

Parameter Sharing

The algorithms as presented here have been shown with a single learned gradient/proximal operator that is used in all iterations. However, it has been found by several authors that a significant improvement can be obtained by relaxing this requirement and instead learning a different operator Λ_{θ^n} for each iteration, where the full parameter vector is $\theta = [\theta^1, \theta^2, \ldots, \theta^N]$. For example, Adler and Öktem (2018b) reports a very noticeable 4.5 dB uplift when learning ten different proximals instead of one.

The reason for this uplift has not been thoroughly explained, but the most simple explanation is that it gives the network ten times more learned parameters. However, making a single proximal ten times larger has not been found to give the same uplift, so perhaps the explanation lies in the ability of different parts of the network to focus on different tasks, with early iterations focusing on large-scale structure while the last iterations finalize the finer structures.

Further Memory

Several optimization algorithms contain some concept of memory, e.g., momentum, which helps the algorithms by giving information from more points than the current point. Given the very high representative power of deep learning methods, one would expect that this type of additional information would be very useful to learned iterative reconstruction methods as well.

Several authors have explored this concept (Putzky and Welling 2017; Adler et al. 2017b; Adler and Öktem 2018b), typically having an extra "momentum" term in X^n for $n \approx 5$ which is updated alongside the reconstruction. These papers claim improvements, but it is also clear that many others opt not to use any further memory in their algorithms (Hammernik et al. 2018). It is hence not fully clear how large the benefit of using memory is.

Preconditioning

Since learned iterative reconstruction is inspired by optimization, it is perhaps not surprising that improvements to optimization schemes can be applied here as well. One particular such method is preconditioning, which is widely used to speed up optimization solvers. Several authors have investigated using such ideas in learned iterative reconstruction typically using preconditioners of the form of a regularized inverse (Gilton et al. 2019; Diamond et al. 2017; Aggarwal et al. 2018)

$$(\mathcal{T}^*\mathcal{T} + \lambda I)^{-1}.$$

However, this is only feasible when the above operator is easily computed, which is only really the case for image processing problems and Fourier inversion. Others have used approximations by, e.g., filtering (Hauptmann et al. 2019) or diagonal approximations to the Hessian (Ravishankar et al. 2019). Finally, some have investigated other optimization-based ways of speeding up convergence, e.g., Nesterov momentum (Li et al. 2018).

Learned Step Length

While learned iterative reconstruction exploits knowledge about the gradient or proximal of the data likelihood, the standard derivations typically give rise to algorithms with a step length that has to be selected by the user. Given that we're already learning large parts of the reconstruction, several authors have looked into learning this step length as well. There are two main ways of doing this. The most simple is to simply consider the step length as part of the learnable parameters and learn it along with the other parameters (Sun et al. 2016; Hammernik et al. 2018). A somewhat more intricate method is to learn to combine the gradient with the current iteration (Putzky and Welling 2017; Adler et al. 2017b). For example, in the learned proximal gradient scheme, one could use an update of the form

$$x^n \leftarrow \Lambda_\theta \left(x^{n-1}, \nabla_x \log P(y \mid x^{n-1}) \right)$$

where $\Lambda_\theta : X^2 \to X$ in this case. This should have some upsides in that the network could in theory learn, e.g., a preconditioner. Similar ideas can be applied to most proximal-based learned iterative schemes, e.g., learned primal-dual (Adler and Öktem 2018b).

Scalable Training

While learned iterative reconstruction schemes are at least an order of magnitude faster to evaluate than classical optimization-based reconstruction methods, training them using the backpropagation algorithm (LeCun et al. 1989) is extremely memory intense since every step of the algorithm has to be stored in memory. For this reason, researchers have had significant issues in scaling the algorithms beyond slice-by-slice cases of roughly 512^2 pixels.

A method to train on full 3d volumes of about 512^3 voxels hence either needs a very expensive supercomputer (Laanait et al. 2019) or to be trained without standard backpropagation. Several researchers have investigated the latter. One such method is to train the network one iteration at a time, which significantly reduces the amount of memory needed (Hauptmann et al. 2018; Wu et al. 2019a). Another method is to use gradient checkpointing (Chen et al. 2016) which reduces the amount of memory used by recomputing on the fly. An extreme case of this is invertible networks (Dinh et al. 2014; Jacobsen et al. 2018) which totally remove the need for storing intermediate results, enabling 3d reconstruction (Putzky et al. 2019).

Putting It All Together

It is common to combine several, if not all, of the above ideas in a single algorithm. To give a more practical example in CT, let us assume that $\mathcal{T}$ is the radon transform and that we have Gaussian noise, in which case $\log P(y \mid x) = \frac{1}{2}\|y - \mathcal{T}x\|^2$. A learned iterative reconstruction scheme for this inverse problem using the learned proximal gradient method can be obtained by combining pseudo-inverse initialization with initialization with avoiding parameter sharing, learned steps, extra memory, and preconditioning which should give a state-of-the-art reconstruction method. Most parts are straightforward, except for the choice of preconditioner. Here one could use that due to the Fourier slice theorem, the inverse Hessian $(\mathcal{T}^*\mathcal{T})^{-1}$ can be approximated by a convolution with a sharpening kernel K. Using this, we arrive at Algorithm 10 which is a state-of-the art learned iterative reconstruction algorithm.

Algorithm 10 Learned Proximal Gradient with engineering improvements

1: $x^0 \leftarrow [\mathcal{T}^\dagger y, 0, \ldots, 0] \in X^M$
2: **for** $n = 1, \ldots, N$ **do**
3: $\quad x^n \leftarrow \Lambda_{\theta^n}\left(x^{n-1}, K\mathcal{T}^*(\mathcal{T}x_1^{n-1} - y)\right)$
4: **end for**
5: $N_\theta(y) \leftarrow x_1^N$

Conclusions

Learned iterative reconstruction has attracted significant interest in just a few years, and research has quickly gone from a wild-west of architecture exploration to a more structured view. Given the enormous success of deep learning methods in general in solving supervised learning problems, research has started shifting toward new frontiers. The first is moving into more practicably applicable domains, where we need to learn from large amounts of data without a ground truth and with various artifacts. The second frontier is the ability to solve previously unsolvable problems such as reconstructing the posterior distribution or integrating reconstruction with image analysis tasks. A final frontier is to gain a theoretical understanding of why these algorithms work so well. Some steps toward this has been taken (Effland et al. 2019; Mardani et al. 2019), but there is still a huge gap between theory and practice. I suspect that we will see an explosive development in this field in the coming years and can only hope that this chapter can serve as an introduction to its many possibilities in the future.

References

Abadi, M., Barham, P., Chen, J., Chen, Z., Davis, A., Dean, J., Devin, M., Ghemawat, S., Irving, G., Isard, M., et al.: Tensorflow: a system for large-scale machine learning. In: 12th {USENIX} Symposium on Operating Systems Design and Implementation ({OSDI} 16), pp. 265–283 (2016)

Adler, J., Öktem, O.: Solving ill-posed inverse problems using iterative deep neural networks. Inverse Prob. 33(12), 124007 (2017)

Adler, J., Öktem, O.: Deep Bayesian Inversion. arXiv1811.05910 (2018a)

Adler, J., Öktem, O.: Learned primal-dual reconstruction. IEEE Trans. Med. Imaging 37(6), 1322–1332 (2018b)

Adler, J., Kohr, H., Öktem, O.: ODL-A Python Framework for Rapid Prototyping in Inverse Problems. Royal Institute of Technology (2017a)

Adler, J., Ringh, A., Öktem, O., Karlsson, J.: Learning to Solve Inverse Problems Using Wasserstein Loss. arXiv1710.10898 (2017b)

Adler, J., Lunz, S., Verdier, O., Schönlieb, C.B., Öktem, O.: Task Adapted Reconstruction for Inverse Problems. arXiv1809.00948 (2018)

Aggarwal, H.K., Mani, M.P., Jacob, M.: MoDL: model-based deep learning architecture for inverse problems. IEEE Trans. Med. Imaging 38(2), 394–405 (2018)

Alizadeh, K., Farhadi, A., Rastegari, M.: Butterfly Transform: An Efficient FFT Based Neural Architecture Design. arXiv1906.02256 (2019)

Anonymous: Closed loop deep Bayesian inversion: uncertainty driven acquisition for fast MRI. In: Submitted to International Conference on Learning Representations (2020). https://openreview.net/forum?id=BJlPOlBKDB. Under review

Arridge, S., Maass, P., Öktem, O., Schönlieb, C.B.: Solving inverse problems using data-driven models. Acta Numer. 28, 1–174 (2019)

Arrow, K.J., Hurwicz, L., Uzawa, H.: Studies in Linear and Non-linear Programming. Stanford University Press, Stanford (1958)

Aspri, A., Banert, S., Öktem, O., Scherzer, O.: A Data-Driven Iteratively Regularized Landweber Iteration. arXiv1812.00272 (2018)

Banert, S., Ringh, A., Adler, J., Karlsson, J., Öktem, O.: Data-Driven Nonsmooth Optimization. arXiv1808.00946 (2018)

Boink, Y.E., Manohar, S., Brune, C.: A Partially Learned Algorithm for Joint Photoacoustic Reconstruction and Segmentation. arXiv1906.07499 (2019)

Bradbury, J., Frostig, R., Hawkins, P., Johnson, M.J., Leary, C., Maclaurin, D., Wanderman-Milne, S.: JAX: composable transformations of Python+NumPy programs (2018). http://github.com/google/jax

Cha, E., Jang, J., Lee, J., Lee, E., Ye, J.C.: Boosting CNN Beyond Label in Inverse Problems. arXiv1906.07330 (2019)

Chambolle, A., Pock, T.: A first-order primal-dual algorithm for convex problems with applications to imaging. J. Math. Imaging Vis. **40**(1), 120–145 (2011)

Chen, T., Xu, B., Zhang, C., Guestrin, C.: Training Deep Nets with Sublinear Memory Cost. arXiv1604.06174 (2016)

Chen, H., Zhang, Y., Chen, Y., Zhang, J., Zhang, W., Sun, H., Lv, Y., Liao, P., Zhou, J., Wang, G.: LEARN: learned experts' assessment-based reconstruction network for sparse-data CT. IEEE Trans. Med. Imaging **37**(6), 1333–1347 (2018)

Diamond, S., Sitzmann, V., Boyd, S., Wetzstein, G., Heide, F.: Dirty Pixels: Optimizing Image Classification Architectures for Raw Sensor Data. arXiv1701.06487 (2017)

Dinh, L., Krueger, D., Bengio, Y.: Nice: Non-linear Independent Components Estimation. arXiv1410.8516 (2014)

Effland, A., Hölzel, M., Klatzer, T., Kobler, E., Landsberg, J., Neuhäuser, L., Pock, T., Rumpf, M.: Variational networks for joint image reconstruction and classification of tumor immune cell interactions in melanoma tissue sections. In: Bildverarbeitung für die Medizin 2018, pp. 334–340. Springer (2018)

Effland, A., Kobler, E., Kunisch, K., Pock, T.: An Optimal Control Approach to Early Stopping Variational Methods for Image Restoration. arXiv preprint arXiv:1907.08488 (2019)

Feliu-Faba, J., Fan, Y., Ying, L.: Meta-learning Pseudo-differential Operators with Deep Neural Networks. arXiv1906.06782 (2019)

Gilton, D., Ongie, G., Willett, R.: Neumann Networks for Inverse Problems in Imaging. arXiv1901.03707 (2019)

Goodfellow, I., Pouget-Abadie, J., Mirza, M., Xu, B., Warde-Farley, D., Ozair, S., Courville, A., Bengio, Y.: Generative adversarial nets. In: Advances in Neural Information Processing Systems, pp. 2672–2680 (2014)

Goodfellow, I., Bengio, Y., Courville, A.: Deep Learning. MIT Press, London (2016)

Gupta, H., Jin, K.H., Nguyen, H.Q., McCann, M.T., Unser, M.: CNN-based projected gradient descent for consistent CT image reconstruction. IEEE Trans. Med. Imaging **37**(6), 1440–1453 (2018)

Hammernik, K., Knoll, F.: Machine learning for image reconstruction. In: Handbook of Medical Image Computing and Computer Assisted Intervention, pp. 25–64. Elsevier, London (2020)

Hammernik, K., Würfl, T., Pock, T., Maier, A.: A deep learning architecture for limited-angle computed tomography reconstruction. In: Bildverarbeitung für die Medizin 2017, pp. 92–97. Springer (2017)

Hammernik, K., Klatzer, T., Kobler, E., Recht, M.P., Sodickson, D.K., Pock, T., Knoll, F.: Learning a variational network for reconstruction of accelerated mri data. Magn. Reson. Med. **79**(6), 3055–3071 (2018)

Hauptmann, A., Lucka, F., Betcke, M., Huynh, N., Adler, J., Cox, B., Beard, P., Ourselin, S., Arridge, S.: Model-based learning for accelerated, limited-view 3-d photoacoustic tomography. IEEE Trans. Med. Imaging **37**(6), 1382–1393 (2018)

Hauptmann, A., Adler, J., Arridge, S., Öktem, O.: Multi-Scale Learned Iterative Reconstruction. arXiv1908.00936 (2019)

He, K., Zhang, X., Ren, S., Sun, J.: Deep residual learning for image recognition. In: Proceedings of the IEEE Conference on Computer Vision and Pattern Recognition, pp. 770–778 (2016)

Herrmann, F.J., Siahkoohi, A., Rizzuti, G.: Learned Imaging with Constraints and Uncertainty Quantification. arXiv1909.06473 (2019)

Hershey, J.R., Roux, J.L., Weninger, F.: Deep Unfolding: Model-Based Inspiration of Novel Deep Architectures. arXiv1409.2574 (2014)

Hessel, M., Modayil, J., Van Hasselt, H., Schaul, T., Ostrovski, G., Dabney, W., Horgan, D., Piot, B., Azar, M., Silver, D.: Rainbow: combining improvements in deep reinforcement learning. In: Thirty-Second AAAI Conference on Artificial Intelligence (2018)

Hörmander, L.: Fourier integral operators. I. Acta Math. **127**(1), 79–183 (1971)

Innes, M., Edelman, A., Fischer, K., Rackauckus, C., Saba, E., Shah, V.B., Tebbutt, W.: Zygote: A Differentiable Programming System to Bridge Machine Learning and Scientific Computing. arXiv1907.07587 (2019)

Jacobsen, J.H., Smeulders, A., Oyallon, E.: i-Revnet: Deep Invertible Networks. arXiv1802.07088 (2018)

Jin, K.H., McCann, M.T., Froustey, E., Unser, M.: Deep convolutional neural network for inverse problems in imaging. IEEE Trans. Image Process. **26**(9), 4509–4522 (2017)

Johnson, J., Alahi, A., Fei-Fei, L.: Perceptual losses for real-time style transfer and super-resolution. In: European Conference on Computer Vision, pp. 694–711. Springer (2016)

Kang, E., Min, J., Ye, J.C.: A deep convolutional neural network using directional wavelets for low-dose x-ray CT reconstruction. Med. Phys. **44**(10), e360–e375 (2017)

Kingma, D.P., Ba, J.: Adam: A Method for Stochastic Optimization. arXiv1412.6980 (2014)

Knoll, F., Hammernik, K., Zhang, C., Moeller, S., Pock, T., Sodickson, D.K., Akcakaya, M.: Deep Learning Methods for Parallel Magnetic Resonance Image Reconstruction. arXiv1904.01112 (2019)

Kobler, E., Muckley, M., Chen, B., Knoll, F., Hammernik, K., Pock, T., Sodickson, D., Otazo, R.: Variational deep learning for low-dose computed tomography. In: 2018 IEEE International Conference on Acoustics, Speech and Signal Processing (ICASSP), pp. 6687–6691. IEEE (2018)

Kofler, A., Haltmeier, M., Kolbitsch, C., Kachelrieß, M., Dewey, M.: A u-nets cascade for sparse view computed tomography. In: International Workshop on Machine Learning for Medical Image Reconstruction, pp. 91–99. Springer (2018)

Krizhevsky, A., Sutskever, I., Hinton, G.E.: Imagenet classification with deep convolutional neural networks. In: Advances in Neural Information Processing Systems, pp. 1097–1105 (2012)

Laanait, N., Romero, J., Yin, J., Young, M.T., Treichler, S., Starchenko, V., Borisevich, A., Sergeev, A., Matheson, M.: Exascale Deep Learning for Scientific Inverse Problems. arXiv1909.11150 (2019)

LeCun, Y., Boser, B., Denker, J.S., Henderson, D., Howard, R.E., Hubbard, W., Jackel, L.D.: Backpropagation applied to handwritten zip code recognition. Neural Comput. **1**(4), 541–551 (1989)

LeCun, Y., Bengio, Y., Hinton, G.: Deep learning. Nature **521**(7553), 436–444 (2015)

Lehtinen, J., Munkberg, J., Hasselgren, J., Laine, S., Karras, T., Aittala, M., Aila, T.: Noise2noise: Learning Image Restoration Without Clean Data. arXiv1803.04189 (2018)

Lei, K., Mardani, M., Pauly, J.M., Vasawanala, S.S.: Wasserstein GANs for MR Imaging: From Paired to Unpaired Training. arXiv1910.07048 (2019)

Leuschner, J., Schmidt, M., Baguer, D.O., Maaß, P.: The LoDoPaB-CT Dataset: A Benchmark Dataset for Low-Dose CT Reconstruction Methods. arXiv1910.01113 (2019)

Li, H., Yang, Y., Chen, D., Lin, Z.: Optimization Algorithm Inspired Deep Neural Network Structure Design. arXiv1810.01638 (2018)

Lønning, K., Putzky, P., Caan, M.W., Welling, M.: Recurrent Inference Machines for Accelerated MRI Reconstruction. arXiv (2018)

Mardani, L.L.M.: Semi-supervised super-resolution GANs for MRI. In: 31st Conference on Neural Information Processing Systems (NIPS 2017), Long Beach (2017)

Mardani, M., Gong, E., Cheng, J.Y., Vasanawala, S., Zaharchuk, G., Alley, M., Thakur, N., Han, S., Dally, W., Pauly, J.M., et al.: Deep Generative Adversarial Networks for Compressed Sensing Automates MRI. arXiv1706.00051 (2017a)

Mardani, M., Monajemi, H., Papyan, V., Vasanawala, S., Donoho, D., Pauly, J.: Recurrent Generative Adversarial Networks for Proximal Learning and Automated Compressive Image Recovery. arXiv1711.10046 (2017b)

Mardani, M., Gong, E., Cheng, J.Y., Vasanawala, S.S., Zaharchuk, G., Xing, L., Pauly, J.M.: Deep generative adversarial neural networks for compressive sensing MRI. IEEE Trans. Med. Imaging **38**(1), 167–179 (2018)

Mardani, M., Sun, Q., Papyan, V., Vasanawala, S., Pauly, J., Donoho, D.: Degrees of Freedom Analysis of Unrolled Neural Networks. arXiv preprint arXiv:1906.03742 (2019)

McCann, M.T., Unser, M.: Algorithms for Biomedical Image Reconstruction. arXiv1901.03565 (2019)

Parikh, N., Boyd, S., et al.: Proximal algorithms. Found. Trends Optim. **1**(3), 127–239 (2014)

Paszke, A., Gross, S., Chintala, S., Chanan, G., Yang, E., DeVito, Z., Lin, Z., Desmaison, A., Antiga, L., Lerer, A.: Automatic differentiation in pytorch (2017)

Putzky, P., Welling, M.: Recurrent Inference Machines for Solving Inverse Problems. arXiv 1706.04008 (2017)

Putzky, P., Karkalousos, D., Teuwen, J., Miriakov, N., Bakker, B., Caan, M., Welling, M.: i-RIM Applied to the fastMRI Challenge. arXiv 1910.08952 (2019)

Ramzi, Z.: fastMRI reproducible benchmark. https://github.com/zaccharieramzi/fastmri-reproducible-benchmark (2019)

Raphan, M., Simoncelli, E.P.: Learning to be Bayesian without supervision. In: Advances in Neural Information Processing Systems, pp. 1145–1152 (2007)

Ravishankar, S., Ye, J.C., Fessler, J.A.: Image Reconstruction: From Sparsity to Data-Adaptive Methods and Machine Learning. arXiv 1904.02816 (2019)

Ronneberger, O., Fischer, P., Brox, T.: U-net: convolutional networks for biomedical image segmentation. In: International Conference on Medical Image Computing and Computer-Assisted Intervention, pp. 234–241. Springer (2015)

Schlemper, J., Caballero, J., Hajnal, J.V., Price, A., Rueckert, D.: A deep cascade of convolutional neural networks for mr image reconstruction. In: International Conference on Information Processing in Medical Imaging, pp. 647–658. Springer (2017)

Schlemper, J., Salehi, S.S.M., Kundu, P., Lazarus, C., Dyvorne, H., Rueckert, D., Sofka, M.: Nonuniform variational network: deep learning for accelerated nonuniform MR image reconstruction. In: International Conference on Medical Image Computing and Computer-Assisted Intervention, pp. 57–64. Springer (2019)

Schwab, J., Antholzer, S., Haltmeier, M.: Deep null space learning for inverse problems: convergence analysis and rates. Inverse Prob. https://iopscience.iop.org/article/10.1088/1361-6420/aaf14a (2018)

Soltanayev, S., Chun, S.Y.: Training deep learning based denoisers without ground truth data. In: Advances in Neural Information Processing Systems, pp. 3257–3267 (2018)

Sun, J., Li, H., Xu, Z., et al.: Deep ADMM-Net for compressive sensing MRI. In: Advances in Neural Information Processing Systems, pp. 10–18 (2016)

Syben, C., Michen, M., Stimpel, B., Seitz, S., Ploner, S., Maier, A.K.: PYRO-NN: Python Reconstruction Operators in Neural Networks. arXiv 1904.13342 (2019)

van Aarle, W., Palenstijn, W.J., De Beenhouwer, J., Altantzis, T., Bals, S., Batenburg, K.J., Sijbers, J.: The ASTRA Toolbox: a platform for advanced algorithm development in electron tomography. Ultramicroscopy **157**, 35–47 (2015)

Vishnevskiy, V., Sanabria, S.J., Goksel, O.: Image reconstruction via variational network for real-time hand-held sound-speed imaging. In: International Workshop on Machine Learning for Medical Image Reconstruction, pp. 120–128. Springer (2018)

Vishnevskiy, V., Rau, R., Goksel, O.: Deep Variational Networks with Exponential Weighting for Learning Computed Tomography. arXiv 1906.05528 (2019)

Vogel, C., Pock, T.: A primal dual network for low-level vision problems. In: German Conference on Pattern Recognition, pp. 189–202. Springer (2017)

Wang, G., Ye, J.C., Mueller, K., Fessler, J.A.: Image reconstruction is a new frontier of machine learning. IEEE Trans. Med. Imaging **37**(6), 1289–1296 (2018)

Wu, D., Kim, K., Dong, B., El Fakhri, G., Li, Q.: End-to-end lung nodule detection in computed tomography. In: International Workshop on Machine Learning in Medical Imaging, pp. 37–45. Springer (2018)

Wu, D., Kim, K., El Fakhri, G., Li, Q.: Computational-efficient cascaded neural network for CT image reconstruction. In: Medical Imaging 2019: Physics of Medical Imaging, vol. 10948, p. 109485Z. International Society for Optics and Photonics (2019a)

Wu, D., Kim, K., Kalra, M.K., De Man, B., Li, Q.: Learned primal-dual reconstruction for dual energy computed tomography with reduced dose. In: 15th International Meeting on Fully Three-Dimensional Image Reconstruction in Radiology and Nuclear Medicine, vol. 11072, p. 1107206. International Society for Optics and Photonics (2019b)

Yang, G., Yu, S., Dong, H., Slabaugh, G., Dragotti, P.L., Ye, X., Liu, F., Arridge, S., Keegan, J., Guo, Y., et al.: DAGAN: deep de-aliasing generative adversarial networks for fast compressed sensing MRI reconstruction. IEEE Trans. Med. Imaging **37**(6), 1310–1321 (2017a)

Yang, Q., Yan, P., Kalra, M.K., Wang, G.: CT Image Denoising with Perceptive Deep Neural Networks. arXiv1702.07019 (2017b)

Yang, Q., Yan, P., Zhang, Y., Yu, H., Shi, Y., Mou, X., Kalra, M.K., Zhang, Y., Sun, L., Wang, G.: Low-dose CT image denoising using a generative adversarial network with Wasserstein distance and perceptual loss. IEEE Trans. Med. Imaging **37**(6), 1348–1357 (2018)

Yang, C., Lan, H., Gao, F.: Accelerated photoacoustic tomography reconstruction via recurrent inference machines. In: 2019 41st Annual International Conference of the IEEE Engineering in Medicine and Biology Society (EMBC), pp. 6371–6374. IEEE (2019)

Zbontar, J., Knoll, F., Sriram, A., Muckley, M.J., Bruno, M., Defazio, A., Parente, M., Geras, K.J., Katsnelson, J., Chandarana, H., et al.: FastMRI: An Open Dataset and Benchmarks for Accelerated MRI. arXiv1811.08839 (2018)

Zhang, H., Dong, B., Liu, B.: JSR-Net: a deep network for joint spatial-radon domain CT reconstruction from incomplete data. In: ICASSP 2019–2019 IEEE International Conference on Acoustics, Speech and Signal Processing (ICASSP), pp. 3657–3661. IEEE (2019)

Zhu, B., Liu, J.Z., Cauley, S.F., Rosen, B.R., Rosen, M.S.: Image reconstruction by domain-transform manifold learning. Nature **555**(7697), 487 (2018)

An Analysis of Generative Methods for Multiple Image Inpainting

20

Coloma Ballester, Aurélie Bugeau, Samuel Hurault, Simone Parisotto, and Patricia Vitoria

Contents

Introduction	774
A Walk Through the Image Inpainting Literature	775
How to Achieve Multiple and Diverse Inpainting Results?	778
Generative Adversarial Networks	780
Variational Autoencoders and Conditional Variational Autoencoders	784
Autoregressive Models	788
Image Transformers	790
From Single-Image Evaluation Metrics to Diversity Evaluation	793
Experimental Results	795
Experimental Settings	796
Quantitative Performance	797
Qualitative Performance	803
Conclusions	809
Appendix	809
Additional Quantitative Results	809
Additional Qualitative Results	810
References	813

C. Ballester (✉) · P. Vitoria
IPCV, DTIC, University Pompeu Fabra, Barcelona, Spain
e-mail: coloma.ballester@upf.edu; patricia.vitoria@upf.edu

A. Bugeau
LaBRI, CNRS, Université de Bordeaux, Talence, France

Institut universitaire de France (IUF), Paris, France
e-mail: aurelie.bugeau@labri.fr

S. Hurault
Bordeaux INP, CNRS, IMB, Université de Bordeaux, Talence, France
e-mail: samuel.hurault@math.u-bordeaux.fr

S. Parisotto
DAMTP, University of Cambridge, Cambridge, UK
e-mail: sp751@cam.ac.uk

© Springer Nature Switzerland AG 2023
K. Chen et al. (eds.), *Handbook of Mathematical Models and Algorithms in Computer Vision and Imaging*, https://doi.org/10.1007/978-3-030-98661-2_119

Abstract

Image inpainting refers to the restoration of an image with missing regions in a way that is not detectable by the observer. The inpainting regions can be of any size and shape. This is an ill-posed inverse problem that does not have a unique solution. In this work, we focus on learning-based image completion methods for multiple and diverse inpainting which goal is to provide a set of distinct solutions for a given damaged image. These methods capitalize on the probabilistic nature of certain deep generative models to sample various solutions that coherently restore the missing content. Throughout the chapter, we will analyze the underlying theory and analyze the recent proposals for multiple inpainting. To investigate the pros and cons of each method, we present quantitative and qualitative comparisons, on common datasets, regarding both the quality and the diversity of the set of inpainted solutions. Our analysis allows us to identify the most successful generative strategies in both inpainting quality and inpainting diversity. This task is closely related to the learning of an accurate probability distribution of images. Depending on the dataset in use, the challenges that entail the training of such a model will be discussed through the analysis.

Keywords

Inverse problems · Inpainting · Multiple inpainting · Diverse inpainting · Deep learning · Generative methods

Introduction

Image inpainting, also called *amodal completion* or *disocclusion* in early days, is an active research area in many fields including applied mathematics and computer vision, with foundations in the Gestalt theory of shape perception. Inpainting relates to the virtual reconstruction of missing content in images in a way that is non-detectable by the observer (Bertalmío et al. 2000). It is an ill-posed inverse problem that can have multiple plausible solutions. Indeed, the fact that the inpainted image is not unique can be understood both mathematically and also because the reconstruction quality is judged by independent humans. On top of that, it has a strong impact on many real-life applications, e.g., in medical imaging (sinograms (Tovey et al. 2019), CT scans (Chen et al. 2012)), 3D surface data (Biasutti et al. 2019; Bevilacqua et al. 2017; Hervieu et al. 2010; Parisotto et al. 2020), art conservation (frescoes (Baatz et al. 2008), panel paintings (Ružić et al. 2011) and manuscripts (Calatroni et al. 2018)), image compression (Peter and Weickert 2015), camera artifact removal (Vitoria and Ballester 2019), and the restoration of old movies and videos (Grossauer 2006; Newson et al. 2014), just to name a few.

State-of-the-art image inpainting methods have achieved amazing results regarding the complex work of filling large missing areas in an image. However, most of the methods generally attempt to generate one single result from a given image, ignoring many other plausible solutions. In this chapter, we focus on analyzing recent advances in the inpainting literature, concentrating on the learning-based approaches for *multiple* and *diverse* inpainting. The goal of those methods is to estimate multiple plausible inpainted solutions while being as much diverse as possible. Those methods mainly focus on the idea of exploiting image coherency at several levels along with the power of neural networks trained on large datasets of images. Unlike previous one-to-one methods, multiple-image inpainting offers the advantage of exploring a large space of possible solutions. This procedure gives capacity to the user to eventually choose the preferred fit under his/her judgment instead of leaving the task of singling out one solution to the algorithm itself.

This chapter is structured as follows: Section "A Walk Through the Image Inpainting Literature" provides a brief overview of both model-based and learning-based inpainting methods in the literature. Section "How to Achieve Multiple and Diverse Inpainting Results?" presents the underlying theory of several approaches for multiple and diverse inpainting together with a review of the most representative (to the best of our knowledge) state-of-the-art proposals using those particular strategies. Section "From Single-Image Evaluation Metrics to Diversity Evaluation" presents the evaluation metrics for both inpainting quality and diverse inpainting. The multiple inpainting results of the methods of section "How to Achieve Multiple and Diverse Inpainting Results?" are presented and compared in section "Experimental Results" both quantitatively and qualitatively, on common datasets and masks, concerning three aspects: proximity to ground truth, perceptual quality, and inpainting diversity. Finally, section "Conclusions" concludes the presented analysis.

A Walk Through the Image Inpainting Literature

In the literature, inpainting methods can fall under different categories, e.g., *local vs. nonlocal* depending on the ability to capture and exploit non-nearby content, or *geometric vs. exemplar-based methods* depending on the action on points or patches. For our purposes, it is more convenient to distinguish between learning- and model-based approaches, according to the usage or not of machine learning techniques. For extensive reviews of existing inpainting methods, we refer the reader to the works in Guillemot and Le Meur (2014), Schonlieb (2015), Buyssens et al. (2015), and Parisotto et al. (2022).

Model-Based Inpainting
Model-based inpainting methods are designed to manipulate an image by exploiting its regularity and coherency features with an explicit model governing the inpainting workflow. One approach for restoring geometric image content is to locally propagate the intensity values and regularity of the image level lines inward the inpainting domain with curvature-driven (Nitzberg et al. 1993; Masnou and

Morel 1998; Ballester et al. 2001; Chan and Shen 2001; Esedoglu and Shen 2002; Shen et al. 2003) and diffusion-based (Caselles et al. 1998; Shen and Chan 2002; Tschumperle and Deriche 2005) evolutionary partial differential equations (PDEs), possibly of fluid-dynamic nature (Bertalmío et al. 2000, 2001; Tai et al. 2007) or with coherent transport mechanisms (Bornemann and März 2007), also by invoking variational principles (Grossauer and Scherzer 2003; Bertozzi et al. 2007) and regularization (possibly of higher order) priors (Papafitsoros and Schönlieb 2013). The filling-in of geometry, especially of small scratches and homogeneous content in small inpainting domains, is the most effective scenario of these methods, which perform poorly in the recovery of texture. Such issue is overcome by considering a patch (a group of neighboring points in the image domain) as the imaging atom containing the essential texture element. The variational formulation of dissimilarity metrics based on the estimation of a correspondence map between patches (Efros and Leung 1999; Bornard et al. 2002; Demanet et al. 2003; Criminisi et al. 2004; Aujol et al. 2010) has led to the design of optimal copying-pasting strategies for inpainting large domains. However, these methods still fail, e.g., in the presence of different scale-space features. Thus, some researchers have exploited, also using a variational approach, the efficiency of PatchMatch (Barnes et al. 2009) in computing a probabilistic approximation of correspondence maps between patches to average the contribution of multiple-source patches during the synthesis step. For example, Arias et al. (2011) and Newson et al. (2014) use it in a non-local mean fashion (Wexler et al. 2004), to inpaint rescaled versions of the original image with results propagated from the coarser to the finer scale; Cao et al. (2011) to guide the inpainting with geometric-sketches; Sun et al. (2005) to guide structures; or Mansfield et al. (2011), Eller and Fornasier (2016), and Fedorov et al. (2016) to account for geometric transformations of patches. However, these mathematical and numerical advances may result to be computationally expensive while suffering from having only one single-imaging source as input, and dependence on the initialization quality and the selection of associated parameters (e.g., the size of the patch). Thus, it seems natural to study if image coherency, smoothness, and self-similarity patterns can be further exploited by augmenting the dataset of source images and eventually synthesize multiple inpainting solutions: this is where diverse inpainting with deep learning-based generative approaches is a significant step forward.

One of the earliest model-based inpainting works dealing with multiple-source images is Kang et al. (2002), where salient landmarks are extracted in a scene under different perspectives and then synthesized by interpolation, guiding the imaging restoration. As said, model-based models are sensitive to initializations and chosen parameters: One way to diminish these drawbacks is to perform inpainting of the input image multiple times, by varying parameters like the patch size, the number of pyramid scales, initializations, and inpainting methodologies. Thus, a final assembling step will produce an inpainted image, which encodes locally the most coherent content (Hays and Efros 2007; Le Meur et al. 2013; Kumar et al. 2016). Still, the computational effort of estimating several solutions with different parameters and their fine-tuning is a keypoint, leading to the need for a one-encompassing strategy that can locally adapt the synthesis step from multiple-source images. This task can be solved with learning-based methods.

Learning-Based Methods

Learning-based methods address image inpainting by learning a mapping from a corrupted input to the estimated restoration by training on a large-scale dataset. Besides capturing local or non-local regularities and redundancy inside the image or the entire dataset, those methods also exploit high-level information inherent in the image itself, such as global regularities and patterns, or perceptual clues and semantics over the images.

Early learning-based methods tackled the problem as a blind inpainting problem (Ren et al. 2015; Cai et al. 2015) by minimizing the distance between the predicted image and the ground truth. This type of methods behaved as an image denoising algorithm and was limited to tiny inpainting domains. To deal with bigger and more realistic inpainting regions, later approaches incorporated in the model the information provided by the mask, e.g., Köhler et al. (2014), Ren et al. (2015), Pathak et al. (2016), and Lempitsky et al. (2018). Also, several modifications to vanilla convolutions have been proposed to explicitly use the information of the mask, like partial onvolutions (Liu et al. 2018) and gated convolutions (Yu et al. 2019), where the output of those layers only depends on non-corrupted points. Additionally, attempts to increase the receptive field without increasing the number of layers have been proposed with dilated convolutions (Iizuka et al. 2017; Wang et al. 2018) and contextual attention (Yu et al. 2018, 2019). Learning to inpaint in a single step has shown to be a complex endeavor. Progressive learning approaches have also been introduced to split the learning into several steps: for instance, Zhang et al. (2018a) progressively fills the holes from outside to inside; similarly, Guo et al. (2019), Zeng et al. (2020), and Li et al. (2020) also learn how to update the inpainting mask for next iteration, and Li et al. (2019) learns jointly structure and feature information.

To train the network, early approaches minimized some distance between the ground-truth and the predicted image. But this approach takes into account just one of the several possible plausible solutions to the inpainting problem. Several approaches have been proposed to overcome this drawback. Some works use perceptual metrics based on generative adversarial networks (GANs) aiming to generate more perceptually realistic results (Pathak et al. 2016; Yeh et al. 2017; Iizuka et al. 2017; Yu et al. 2018; Vitoria et al. 2019, 2020; Dapogny et al. 2020; Liu et al. 2019; Lahiri et al. 2020). Other works tackle the problem in the feature space by minimizing distances at feature space level (Fawzi et al. 2016; Yang et al. 2017; Vo et al. 2018) by using an additional pre-trained network, or by directly inpainting those features (Yan et al. 2018; Zeng et al. 2019). Also, two-step approaches have been proposed. They are based on a first coarse inpainting (Yang et al. 2017; Yu et al. 2018; Liu et al. 2019), edge learning (Liao et al. 2018; Nazeri et al. 2019; Li et al. 2019), or structure prediction (Xiong et al. 2019; Ren et al. 2019) and followed by a refinement step adding finer texture details. Furthermore, Liu et al. (2020) aimed to ensure consistency between structure and texture generation. Another big problem of early deep learning methods is that deep models treat input images with limited resolution. While first approaches were able to deal with images of maximum size 64 × 64, the latest methods can deal with 1024 × 1024 resolution images by using, for example, a multiscale approach (Yang et al. 2017; Zeng et al. 2019), or even to 8K resolution by generating first a low-resolution solution and second its high-frequency residuals (Yi et al. 2020b).

Recent works (e.g., Zheng et al. 2019; Zhao et al. 2020b; Cai and Wei 2020; Peng et al. 2021; Wan et al. 2021; Liu et al. 2021) deal with the ill-posed nature of the problem by allowing more than one possible plausible solution to a given image. They aim to generate multiple and diverse solutions by using deep probabilistic models based on variational autoencoders (VAEs), GANs, autoregressive models, transformers, or a combination of them. Note that those types of methods have been also used for real case applications such as diverse fashion image inpainting (Han et al. 2019) and Cosmic microwave background radiation (CMB) image inpainting (Yi et al. 2020a). Besides, it is worth mentioning that there are several single-image generation methods that estimate complete images with some variations. For instance, SinGAN (Rott Shaham et al. 2019) produces several random images which are deviations of an input image by learning the distribution of its patches. Park et al. (2019) synthesizes new images by controlling style and semantics. However, these strategies do not completely fit within the multiple inpainting problems where regions of the image are known and should not be changed. In this chapter, we will focus on the study of multiple-image inpainting methods. More precisely, we will review, analyze, and compare, theoretically as well as experimentally, the different approaches proposed on the literature to generate inpainting diversity.

How to Achieve Multiple and Diverse Inpainting Results?

In this section, we will describe the different tools and methods that successfully addressed multiple image inpainting. Later in section "Experimental Results", we will conduct a thorough experimental study comparing these methods visually and quantitatively.

As previously mentioned, image inpainting is an inverse problem with multiple plausible solutions. Generally, ill-posed problems are solved by incorporating some knowledge or priors into the solution. Mathematically, this is frequently done using a variational approach where a prior is added to a data-fidelity term to create an overall objective functional that is lastly optimized. The selected prior promotes the singling out of a particular solution. Traditionally, the incorporated priors were model-based, founded on properties of the expected solution.

More recently, data-driven proposals have emerged where the prior knowledge on the image distribution is implicitly or explicitly learned via neural networks optimization (we refer to the recent survey Arridge et al. 2019 and references therein). Among them, generative methods have been used to learn the underlying geometric and semantic priors of a set of non-corrupted images. Indeed, generative methods aim to estimate the probability distribution of a large set, $\mathcal{X}$, of data. In other words, any $x \in \mathcal{X}$ is assumed to come from an underlying and unknown probability distribution $\mathbb{P}_\mathcal{X}$, and the goal is to learn it from the data in $\mathcal{X}$. Due to its capacity to produce several outcomes given a single output, some authors have proposed to address the multiplicity of solutions by leveraging the probabilistic nature of generative models.

Through this chapter, we will assume that $\mathcal{X}$ is a set of images. Images will be assumed to be functions defined on a bounded domain $\Omega \subset \mathbb{R}^2$ with values in $\mathbb{R}^C$, with $C = 1$ for gray-level images and $C = 3$ for color images. With a slight abuse of notation, we will use the same notation to refer to the continuous setting, where $\Omega \subset \mathbb{R}^2$ is an infinite resolution image domain and $x : \Omega \to \mathbb{R}^C$ represents a continuous image, and to the discrete setting where Ω stands for a discrete domain given by a grid of $H \times W$ pixels, $H, W \in \mathbb{N}$, and x is a function defined on this discrete Ω and with values in $\mathbb{R}^C$. In the latter case, x is usually given in the form of a real-valued matrix of size $H \times W \times C$ representing the image values.

In the context of image inpainting, the inpainting domain, denoted here by O, represents the region of the image domain Ω where the image data is missing, and thus to be restored. Its complementary set, $O^c = \Omega \setminus O$, represents the region of Ω where the values of the image to be inpainted are known. The inpainting mask M will be defined as equal to 1 on the missing pixels of O and equal to 0 on $\Omega \setminus O$.

The space $\mathcal{X}$ of (complete) natural images is a high-dimensional space, and its distribution can be very complex. However, natural images contain local regularities, non-local self-similarities, global coherency, and even semantic structure. This is one of the reasons that inspired the use of latent-based models. These models use latent variables $z \in \mathcal{Z}$ in a lower-dimensional space $\dim(\mathcal{Z}) \leq \dim(\mathcal{X})$, associated with a probability distribution $\mathbb{P}_\mathcal{Z}$. Generative latent-based models aim to learn a generative model $G_\theta : \mathcal{Z} \to \mathcal{X}$, with parameters θ, mapping a latent variable z to an image x. With a slight abuse of notation and if it is understood from the context, we will forget about the θ subindex and simply write G. The main goal of this strategy is twofold: (1) to be able to generate samples $G(z)$ hoping that $G(z) \in \mathcal{X}$ for $z \sim \mathbb{P}_\mathcal{Z}$ (or that $\mathbb{P}_G$ is close to $\mathbb{P}_\mathcal{X}$) and (2) to use it for density estimation

$$p_\mathcal{X}(x) \approx p_G(x) = \int p(x|z) p_\mathcal{Z}(z) dz, \qquad (1)$$

where p_G stands for the parametric density of $\mathbb{P}_G = G_\# \mathbb{P}_\mathcal{Z}$, the pushforward measure of $\mathbb{P}_\mathcal{Z}$ through G (defined in brief as $G_\# \mathbb{P}_\mathcal{Z}(B) = \mathbb{P}_\mathcal{Z}(\{z \in \mathcal{Z} | G(z) \in B\})$, for any B in the Borel σ-algebra associated to $\mathcal{X}$). Let us notice that the likelihood $p(x|z)$ depends on G and can be interpreted as a measure of how close $G(z)$ is to x.

Numerous strategies have been developed to parametrize G (or G_θ) as a neural network model and to learn the appropriate parameters θ by making $\mathbb{P}_G$ as close as possible to $\mathbb{P}_\mathcal{X}$ for some probability distance $d(\mathbb{P}_G, \mathbb{P}_\mathcal{X})$. Among these strategies, we quote variational autoencoders, normalizing flows, generative adversarial networks, or autoregressive models.

The problem of image inpainting can also be naturally formulated in a probabilistic manner. Let y denote an observed incomplete image, which is unknown in O.

Table 1 Generative methods used in the analyzed state-of-the-art proposals for diverse inpainting

Method	VAE	Autoregressive	GAN	Transformers
PIC (Zheng et al. 2019)	✓		✓	
PiiGAN (Cai and Wei 2020)			✓	
UCTGAN (Zhao et al. 2020b)	✓		✓	
DSI-VQVAE (Peng et al. 2021)	✓	✓	✓	
ICT (Wan et al. 2021)			✓	✓
PD-GAN (Liu et al. 2021)			✓	
BAT (Yu et al. 2021)			✓	✓

We are interested in modeling the conditional distribution, $p(x|y)$, over the values of the variable x (corresponding to the complete image) conditioned on the value of the observed variable y. As possibly many plausible images are consistent with the same input image y, the distribution $p(x|y)$ will likely be multimodal. Then, each of the multiple solutions can be generated by sampling from that distribution using a given sampling strategy. Thus, the goal is not only to obtain a generative model that minimizes $d(\mathbb{P}_G, \mathbb{P}_{X_s})$, where $X_s \subset X$ is the set of possible solutions, but also to design a mechanism able to sample the conditional distribution $p(x|y)$, i.e., for a given damaged incomplete image y, output a set of plausible completions x of y.

In this section, we will analyze the different families of generative models proposed in the literature to realize diverse image inpainting. We will in particular describe generative adversarial networks (GAN), variational autoencoders (VAE), autoregressive models, and transformers. We will also detail the different objective losses proposed to train these networks. Finally, for each family of models, we will review several state-of-the-art diverse inpainting methods that relate to this model. Table 1 lists all the methods that will be reviewed in this section.

Generative Adversarial Networks

Generative adversarial networks (GANs) are a type of generative models that have received a lot of attention since the seminal work of Goodfellow et al. (2014). The GAN strategy is based on a game theory scenario between two networks, a generator network and a discriminator network, that are jointly trained competing against each other in the sense of a Nash equilibrium. The generator maps a vector from the latent space, $z \sim \mathbb{P}_Z$, to the image space trying to trick the discriminator, while the discriminator receives either a generated or a real image and must distinguish between both. The parameters of the generator and the discriminator are learned jointly by optimizing a GAN objective by a min-max procedure. This procedure leads the probability distribution of the generated data to be as close as possible, for some distance, to the one of the real data. Several GAN variants have appeared. They mainly differ on the choice of the distance $d(\mathbb{P}_1, \mathbb{P}_2)$ between two probability distributions $\mathbb{P}_1$ and $\mathbb{P}_2$. The first GAN by Goodfellow et al. (2014) (also referred to as vanilla GAN) makes use of the Jensen–Shannon divergence, which is defined from the Kullback–Leibler divergence (KL), by

$$d_{\text{JS}}(\mathbb{P}_1, \mathbb{P}_2) = \frac{1}{2}\left[\text{KL}\left(\mathbb{P}_1 \middle\| \frac{\mathbb{P}_1 + \mathbb{P}_2}{2}\right) + \text{KL}\left(\mathbb{P}_2 \middle\| \frac{\mathbb{P}_1 + \mathbb{P}_2}{2}\right)\right], \qquad (2)$$

where the KL is defined, for discrete probability densities, as

$$\text{KL}(\mathbb{P}_1, \mathbb{P}_2) = \sum_x \mathbb{P}_1(x) \log\left(\frac{\mathbb{P}_1(x)}{\mathbb{P}_2(x)}\right). \qquad (3)$$

and, for continuous densities, as

$$\text{KL}(\mathbb{P}_1, \mathbb{P}_2) = \int_{\mathcal{X}} \mathbb{P}_1(x) \log \frac{\mathbb{P}_1(x)}{\mathbb{P}_2(x)} dx. \qquad (4)$$

The Wasserstein GAN (Arjovsky et al. 2017) uses the Wasserstein-1 distance, given by

$$\mathbb{W}_1(\mathbb{P}_1, \mathbb{P}_2) = \inf_{\pi \in \Pi(\mathbb{P}_1, \mathbb{P}_2)} \mathbb{E}_{x,y \sim \pi}(\|x - y\|), \qquad (5)$$

where $\Pi(\mathbb{P}_1, \mathbb{P}_2)$ is the set of all joint distributions π whose marginals are, respectively, $\mathbb{P}_1$ and $\mathbb{P}_2$. By Kantorovitch–Rubenstein duality, the Wasserstein-1 distance can be computed as

$$\mathbb{W}_1(\mathbb{P}_1, \mathbb{P}_2) = \sup_{D \in \mathcal{D}} \left(\mathbb{E}_{x \sim \mathbb{P}_1}[D(x)] - \mathbb{E}_{y \sim \mathbb{P}_2}[D(y)]\right), \qquad (6)$$

where $\mathcal{D}$ denotes the set of 1-Lipschitz functions. In practice, the dual variable D is parametrized by a neural network and it represents the so-called discriminator.

Both the generator and discriminator are jointly trained to solve

$$\min_{G} \sup_{D \in \mathcal{D}} \left(\mathbb{E}_{x \sim \mathbb{P}_X}[D(x)] - \mathbb{E}_{y \sim \mathbb{P}_G}[D(y)]\right), \qquad (7)$$

in the case of the Wasserstein GAN, and

$$\min_{G} \max_{D} \mathbb{E}_{x \sim \mathbb{P}_X}\left[\log D(x)\right] + \mathbb{E}_{y \sim \mathbb{P}_G}\left[\log(1 - D(y))\right] \qquad (8)$$

for the vanilla GAN. In (8), the discriminator D is simply a classifier that tries to distinguish samples in the training set $\mathcal{X}$ (real samples) from the generated samples $G(z)$ (fake samples) by designing a probability $D(x) \in [0, 1]$ for its likelihood to be from the same distribution as the samples in $\mathcal{X}$.

GANs are sometimes referred to as implicit probabilistic models due to the fact that they are defined through a sampling procedure where the generator learns to generate new image samples. This is in contrast to variational autoencoders, autoregressive models, and methods that explicitly maximize the likelihood.

For the task of inpainting, several proposals set the problem as a conditioned one. The GAN approach is modified such that the input of the generator G is both an incomplete image y and a latent vector $z \sim \mathbb{P}_Z$, and G performs conditional image synthesis where the conditioning input is y. In the GAN-based works that we present in this section (Cai and Wei 2020; Liu et al. 2021), the authors focus on multimodal conditioned generation where the goal is to generate multiple plausible output images for the same given incomplete image.

Finally, let us mention that in these works, and in general in some works described in this chapter, the used generative methods are combined with consistency losses that encourage the inpainted images to be close to the ground truth. Examples of those consistency losses include value and feature reconstruction losses and perceptual losses. Nonetheless, multiple inpainting researchers acknowledge that it can be counterproductive to rely on consistency losses due to the fact that the ground truth is only one of the multiple solutions.

PiiGAN: Generative Adversarial Networks for Pluralistic Image Inpainting (Cai and Wei 2020)

One of the first methods that use GANs in order to generate pluralistic results is PiiGAN (Cai and Wei 2020). PiiGAN is a deep generative model that incorporates a style extractor that can extract the style features, in the form of a latent vector, from the ground-truth image.

To be more precise, the network is composed of one *generator* and *extractor* network. The training follows two different paths.

First, given a ground-truth image x_{gt}, the extractor is used to estimate the style feature z_{gt}. Once the style feature z_{gt} is obtained, the ground-truth image x_{gt} is masked and concatenated with the computed style feature z_{gt} and used as input of the *generator* network. The generator network will estimate the inpainted version x of the masked ground-truth image y. The estimated inpainted image $x_{out,1}$ is passed through the extractor network to estimate the corresponding style feature $z_{out,1}$. This path is supervised using the KL-divergence between the style features z_{gt} and $z_{out,1}$.

In parallel, another path estimates inpainted images from masked images without ground truth. That is, masked images without ground truth y_{raw} are fed to the generator with a random vector z_{raw}. An inpainted image $x_{out,2}$ is predicted followed by style feature prediction $z_{out,2}$. Additionally, they frame the inpainting of y_{raw} in an adversarial approach equipped with a local (that focuses just in the inpainted area) and a global discriminator applied to the inpainted image x_{raw}. This path is supervised using the L^1 norm of the difference between the style features x_{raw} and $x_{out,2}$ and adversarial loss applied to the inpainted image $x_{out,2}$ based on the Wasserstein loss (7) with gradient penalty as defined in Gulrajani et al. (2017).

The authors claim that their results are diverse and natural, especially for images with large missing areas. Figure 1 shows an overview of the algorithm pipeline.

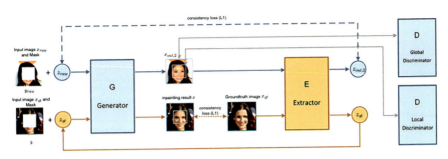

Fig. 1 Overview of the architecture of PiiGAN: Generative Adversarial Networks for Pluralistic Image Inpainting (Cai and Wei 2020). (Figure from Cai and Wei 2020)

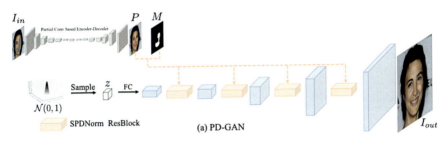

Fig. 2 Overview of the architecture of PD-GAN: Probabilistic Diverse GAN for Image Inpainting (Liu et al. 2021). (Figure from Liu et al. 2021)

PD-GAN: Probabilistic Diverse GAN for Image Inpainting (Liu et al. 2021)

The authors of Liu et al. (2021) propose a method to perform diverse image inpainting called PD-GAN. PD-GAN takes advantage of the benefits of GANs in generating diverse content from different random noise inputs. Figure 2 displays an overview of the algorithm pipeline. In contrast to the original vanilla GAN, in PD-GAN all the decoder deep features are modulated from coarse to fine by injecting prior information at each scale. This prior information is extracted from an initially restored image at a coarser resolution together with the inpainting mask. For that purpose, they introduce a probabilistic diversity normalization (SPDNorm) module based on the Spatially-adaptive denormalization (SPADE) module proposed in Park et al. (2019). SPDNorm works by modeling the probability of generating a pixel conditioned on the context information. It allows more diversity toward the center of the inpainted hole and more deterministic content around the inpainting boundary.

The objective loss is a combination of several losses, including a diversity loss, a reconstruction loss, an adversarial loss, and a feature matching loss (difference in the output feature layers computed with the learned discriminator). In general, in the context of multiple-image synthesis, *diversity losses* aim at ensuring that the different reconstructed images are diverse enough. In particular, the authors of PD-GAN (Liu et al. 2021) use the so-called *perceptual diversity loss*, defined as

$$\mathcal{L}_{pdiv}(x_{out_a}, x_{out_b}) = \frac{1}{\sum_l \|M \odot (\Phi_l(x_{out_a}) - \Phi_l(x_{out_b}))\|_1 + \epsilon}. \tag{9}$$

where x_{out_a} and x_{out_b} are two inpainted results, and M the inpainting mask (with 1 values on the missing pixels and 0 elsewhere). The minimization of (9) favors the maximization of the perceptual distance of inpainted regions in x_{out_a} and x_{out_b}. Notice that the non-masked pixels are not affected by this loss. A similar diversity loss was proposed in Mao et al. (2019).

Variational Autoencoders and Conditional Variational Autoencoders

Variational autoencoders (VAE) (Kingma and Welling 2013) are generative models for which the considered distance between probability distributions is the Kullback–Leibler divergence. Maximization of the log-likelihood criterion is equivalent to the minimization of a Kullback–Leibler divergence between the data and model distributions. In the VAE context, the generator G_θ is referred to as the *decoder*.

Let us first derive the vanilla VAE formulation in the general context of non-corrupted images $x \in \mathcal{X}$. Using Bayes rule, the likelihood $p_{G_\theta}(x)$, for $x \sim \mathbb{P}_X$ and $z \sim \mathbb{P}_Z$, is given by

$$p_\theta(x) = \frac{p_\theta(x,z)}{p_\theta(z|x)} = \frac{p_\theta(x|z)p_Z(z)}{p_\theta(z|x)} \tag{10}$$

where, to simplify notations, we have denoted p_{G_θ} simply by p_θ. In order to bypass the intractability of the posterior $p_\theta(z|x)$, variational autoencoders introduce a second neural network, $q_\psi(z|x)$, to parametrize an approximation of the true posterior. This neural network is referred to as the *encoder*. Let us now derive the VAE objective function. Following Kingma et al. (2019),

$$\log p_\theta(x) = \mathbb{E}_{q_\psi(z|x)}\left[\log \frac{p_\theta(x,z)}{q_\psi(z|x)}\right] + \mathbb{E}_{q_\psi(z|x)}\left[\log \frac{q_\psi(z|x)}{p_\theta(z|x)}\right] \tag{11}$$

$$= \mathbb{E}_{q_\psi(z|x)}\left[\log p_\theta(x,z) - \log q_\psi(z|x)\right] + \mathrm{KL}\left(q_\psi(z|x) \| p_\theta(z|x)\right) \tag{12}$$

$$= \mathcal{L}_{\theta,\psi}(x) + \mathrm{KL}\left(q_\psi(z|x) \| p_\theta(z|x)\right). \tag{13}$$

$\mathcal{L}_{\theta,\psi}$ is the so-called evidence lower bound (ELBO). By positivity of the KL, it verifies

$$\mathcal{L}_{\theta,\psi}(x) = \log p_\theta(x) - \mathrm{KL}\left(q_\psi(z|x) \| p_\theta(z|x)\right) \le \log p_\theta(x), \tag{14}$$

and $\mathcal{L}_{\theta,\psi}(x) = \log p_\theta(x)$ if and only if $q_\psi(z|x)$ is equal to $p_\theta(z|x)$.

VAE training consists in maximizing $\mathcal{L}_{\theta,\psi}$ in (14) with respect to the parameters $\{\theta, \psi\}$ of the encoder and of the decoder, simultaneously. The goal is to obtain a good approximation $q_\psi(z|x)$ of the true posterior $p_\theta(z|x)$ while maximizing the marginal likelihood $p_\theta(x)$.

The work Sohn et al. (2015) extends VAEs by proposing conditional variational autoencoders (CVAE). Their targeted distribution is the conditional distribution of x given an input "conditional" variable c and the maximization of the log-likelihood criterion becomes

$$\max_\theta \mathbb{E}_{x \sim \mathbb{P}_X} \log p_{G_\theta}(x|c). \tag{15}$$

CVAE loss is obtained with a similar argument as in (11), (12), (13), and (14) by maximizing the conditional log-likelihood, which gives the variational lower bound of the conditional log-likelihood

$$\mathbb{E}_{q_\psi(z|x)} \left[\log p_\theta(x|c,z)\right] - \mathrm{KL}\left(q_\psi(z|x,c) \| p_\theta(z|x)\right) \leq \log p_\theta(x|c). \tag{16}$$

Then, the idea of the deep conditional generative modeling is simple: given an observation (input) x, z is drawn from a prior distribution $p_\theta(z|x)$. Then, the output is generated from the distribution $p_\theta(x|z,c)$. Bao et al. (2017) combines a CVAE with a GAN (CVAE-GAN) for fine-grained category image generation. Even if inpainting results are shown, the network is not trained explicitly for inpainting but for image generation conditioned on image labels.

In the context of multiple-image inpainting, or more generally of multiple-image restoration, a straightforward idea is to condition the generative model on the input degraded image y and to generate multiple images x sampling from $p_\theta(x|z, c = y)$. BicycleGAN (Zhu et al. 2017) uses this idea for diverse image-to-image translation. Their goal is to learn a bijective mapping between two image domains with a multimodal conditional distribution. They combine CVAE-GAN with latent regressors and show that their method can produce both diverse and realistic results across various image-to-image translation problems. However, their method is not explicitly applied for image inpainting. Moreover, as observed by several authors (see, e.g., Zheng et al. 2019; Wan et al. 2021), using standard conditional VAEs or CVAE-GAN for the specific task for image inpainting still leads to minimal diversity and quality. Several extensions of these models have recently appeared for diverse image inpainting. They are presented below with more details.

Finally, let us notice that VAE model has been extended in van den Oord et al. (2017) and Razavi et al. (2019) to the so-called vector quantized–variational autoencoder (VQ-VAE) that uses vector quantization to model discrete latent variables. Such discretization is done to avoid posterior collapse. The quantization codebook is trained at the same time as the autoencoder with an objective loss made of a reconstruction term and a regularization term that ensures that the embedding fits the encoder and outputs, respectively. The work Razavi et al. (2019) is a hierarchical extension of van den Oord et al. (2017). In particular, the authors of Razavi et al.

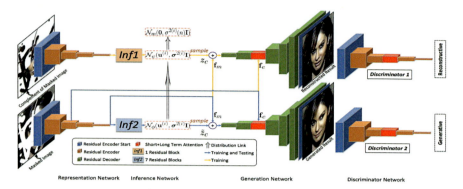

Fig. 3 Overview of the PIC architecture of pluralistic image completion (Zheng et al. 2019). (Figure from Zheng et al. 2019)

(2019) show that, by only considering two levels of a multiscale hierarchical organization of VQ-VAE (van den Oord et al. 2017), the information about image texture is disentangled from the information about the structure and geometry of the objects in an image. By combining the obtained hierarchical multiscale latent data with an autoregressive model as prior (see section "Autoregressive Models" below), they show an improved ability for generating high-resolution images.

Pluralistic Image Completion (Zheng et al. 2019)

The work Zheng et al. (2019) aims to estimate a probability distribution $p(x|y)$ from which to sample, where y represents an incomplete image and x one of its possible completions. They propose to use the conditional variational autoencoder (CVAE) (Sohn et al. 2015) approach described above which estimates a parametric distribution over a latent space, as in equation (16), from which sampling is possible. However, in Zheng et al. (2019), the authors observe that if they explicitly promote the inpainted output to be similar to the ground-truth image (either by any error-based loss such as, for instance, the L^1 distance, or as the authors show, by maximizing $\mathbb{E}_{q_\psi(z|x)}\left[\log p_\theta(x|y, z)\right]$ in (16) while KL $\left(q_\psi(z|x, y) \| p_\theta(z|x)\right)$ tends to zero), it results in a lack of diverse outputs. Alternatively, one could impose to fit the distribution of the training dataset by an adversarial approach including a discriminator as described in section "Generative Adversarial Networks". However, this approach is highly unstable. Instead, they propose a probabilistic framework with a dual pipeline composed of two paths. See a detailed pipeline in Fig. 3. One is the *reconstructive path* which is a VAE-based model that utilizes the ground truth to get a prior distribution of missing parts, $x|_O$, with the variance on the latent variables' prior depending on the hole area, and rebuild the exact same ground-truth image from this distribution. The other is a *generative path* for which the conditional prior, based only on the visible regions, is coupled to the distribution obtained in the reconstructive path to generate multiple and diverse samples. Both parts are framed in an adversarial approach to fit the distribution of the training dataset. Accordingly,

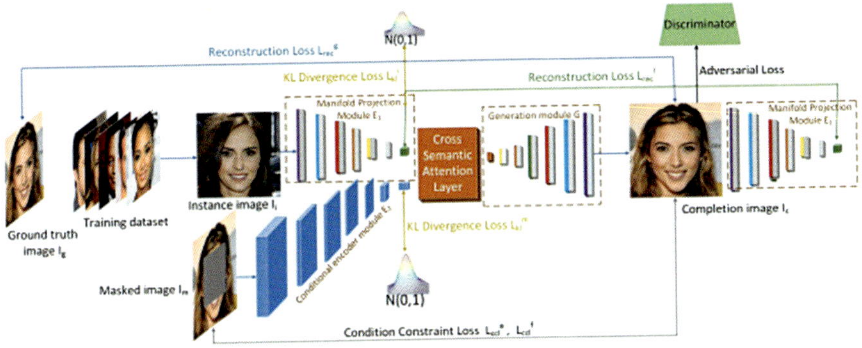

Fig. 4 Overview of the architecture of UCTGAN: Diverse inpainting based on unsupervised cross-space translation (Zhao et al. 2020b). (Figure from Zhao et al. 2020b)

the whole training loss is a combination of three types of terms. First, they use the KL divergences between the mentioned distributions. Second, the appearance terms based on the L^1 norm of the error, where in the generative path it only has into account the visible pixels. And lastly, the third term is an adversarial discriminator-based term. It is based on the L^1 difference among the discriminator features of the ground-truth and the reconstructed image for the reconstructive path and on the discriminator value on the generated image for the generative path. Additionally, to exploit the distant relation among the encoder and decoder, they use a modified self-attention layer that captures fine-grained features in the encoder and more semantic generative features in the decoder.

UCTGAN: Diverse Inpainting Based on Unsupervised Cross-Space Translation (Zhao et al. 2020b)

The authors of Zhao et al. (2020b) aim to produce multiple and diverse solutions conditioned by an instance image that guides the reconstruction, again aiming to maximize the conditional log-likelihood involving the variational lower bound (16) on the training dataset. They call their proposal UCTGAN. The pipeline is presented in Fig. 4. They use a two-encoder network that transforms the instance image and the corrupted image to a low-dimensional manifold space. A cross semantic attention layer combines the information in both low-dimensional spaces. Consecutively, a generator is used to compute the conditional reconstructed image.

The objective loss is composed of four terms. First, a constraint loss in the uncorrupted pixels is applied by minimizing the L^1 norm of the difference both at pixel and feature levels. Second, the KL divergence is used to project the low-dimensional manifold space of the instance image and masked image into a multivariate normal distribution space. Additionally, the L^1 norm of the difference in the low-dimensional manifold space of the instance image and the ground-truth image is added. Finally, all the training is framed in an adversarial approach using the vanilla GAN (8), where the discriminator works in the image space.

Generating Diverse Structure for Image Inpainting with Hierarchical VQ-VAE (Peng et al. 2021)

The multiple inpainting proposal in Peng et al. (2021) leverages three generative strategies, namely, variational autoencoders, generative adversarial methods, and autoregressive models. We first review below the main ideas of autoregressive models and then describe the proposal (Peng et al. 2021).

Autoregressive Models

In autoregressive models (Van Oord et al. 2016; Oord et al. 2016; Chen et al. 2018), the likelihood $p_\theta(x)$ is learned by choosing an order of the data variables $x = (x_1, x_2, \ldots, x_n) \in X$, frequently related to values on the n pixels of an image, and exploiting the fact that the joint distribution can be decomposed as

$$p(x) = p(x_1, x_2, \ldots, x_n) = p(x_1) \prod_{i=2}^{n} p(x_i | x_1, \ldots, x_{i-1}). \tag{17}$$

More generally, a similar decomposition to (17) can be obtained by splitting the set of variables in smaller disjoint subsets. In this case, and considering the variable order of $x_1, x_2, \ldots, x_n$ to be represented by a directed and noncyclic graph, one has

$$p(x) = p(x_1, x_2, \ldots, x_n) = p(x_1) \prod_{i=2}^{m} p(x_i | S(x_i)), \tag{18}$$

where $S(x_i)$ is the set of parent variables of variable i and $m \leq n$.

Autoregressive models have been used to learn a probability distribution or a conditional distribution, for instance, in the context of VAEs (cf. section "Variational Autoencoders and Conditional Variational Autoencoders") to model the prior or the decoder and also to tackle several problems in imaging such as image generation (e.g., Razavi et al. 2019), super resolution (e.g., Dahl et al. 2017), inpainting (e.g., Peng et al. 2021) or image colorization (e.g., Zhao et al. 2020a; Guadarrama et al. 2017; Royer et al. 2017), and also for other types of data such as audio and speech synthesis (e.g., Oord et al. 2018) or text (e.g., Bowman et al. 2015).

Generating Diverse Structure for Image Inpainting with Hierarchical VQ-VAE (Peng et al. 2021)

Inspired by the hierarchical vector quantized variational autoencoder (VQ-VAE) (Razavi et al. 2019) whose hierarchical architecture disentangles structural and textural information, the authors of Peng et al. (2021) propose a two-stage pipeline (cf. Fig. 5). As already pointed out by Razavi et al. (2019), by using a two-step approach instead of directly computing the final inpainted image, they aim to generate richer structure and texture images than previous VAE-based methods that often produce a distorted structure or blurry textures.

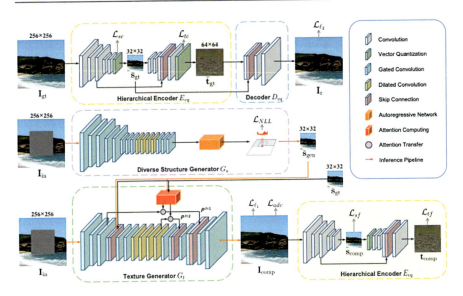

Fig. 5 Overview of the architecture of generating diverse structure for image inpainting with hierarchical VQ-VAE (DSI-VQVAE) (Peng et al. 2021). (Figure from Peng et al. 2021)

The first stage of Peng et al. (2021), known as *diverse structure generator*, generates multiple low-resolution results, each of which has a different structure by sampling from a conditional autoregressive distribution. The second stage, known as *texture generator*, uses an encoder–decoder architecture with a structural attention module that refines each low-resolution result separately by augmenting texture. The structural information module facilitates the capture of distant correlations. They further reuse the VQ-VAE to calculate two feature losses, which help improve structure coherence and texture realism, respectively.

The authors first train the hierarchical VQ-VAE and, afterward, the diverse structure generator (G_s depending on parameters θ) and the texture generator (G_t depending on parameters φ) are trained separately. These generators are later on used for inference. The structure generator G_s is constructed via a conditional autoregressive network for the distribution over structural features. In inference, it will generate different structural features via sampling. Its objective loss is defined as the negative log-likelihood

$$\mathcal{L}_\ell(\theta) = -\mathbb{E}_{x_{\text{gt}} \sim \mathbb{P}_X}[\log(p_\theta(s_{\text{gt}}|y, M)] \quad (19)$$

where y is the input image to be inpainted on the points of O where the hole mask M is equal to 1, $\mathbb{P}_X$ denotes the distribution of the training dataset, s_{gt} denote the vector quantized structural features of the ground truth at the coarser scale given by the hierarchical VQ-VAE, and θ the parameters of G_s.

Besides, the objective loss for the texture generator G_t is composed by: (i) the L^1 norm comparing the inpainted solution to the ground truth at pixel level, (ii) an adversarial loss using the discriminator trained with the SN-PatchGAN hinge version (Yu et al. 2019) applied to the resulting image and, moreover, (iii) a structural feature loss $\mathcal{L}_{sf}(\varphi)$, and (iv) a textural feature loss $\mathcal{L}_{tt}(\varphi)$. These last two losses are defined similarly using a multiclass cross-entropy loss. In particular, the structural feature loss is defined as

$$\mathcal{L}_{sf}(\varphi) = -\sum_{k,j} \alpha_{k,j} \log\left(\text{softmax}(\lambda_2\, \delta_{k,j})\right), \tag{20}$$

where $\delta_{k,j}$ denotes the truncated distance similarity score between the k-th feature vector of s_{comp} (computed from the inpainted image using the trained encoder) and the j-th prototype vector of the structural codebook of VQ-VAE, λ_2 is a parameter set to 10, and $\alpha_{k,j}$ is an indicator of the prototype vector class. That is, $\alpha_{k,j} = 1$ when the k-th feature vector of s_{gt} belongs to the j-th class of the structural codebook; otherwise, $\alpha_{k,j} = 0$. The authors define the textural feature loss $\mathcal{L}_{tt}(\varphi)$ in an analogous way.

As mentioned, in section "Experimental Results", we will experimentally analyze this method. It will be denoted there as DSI-VQVAE.

Image Transformers

Self-attention-based architectures, in particular transformers (Vaswani et al. 2017) are well-explored architectures in natural language processing (NLP). Transformers use a self-attention mechanism to model long-range relationships between the elements of an input sequence (for instance, in a text) that has shown to be more efficient than recurrent neural networks. They have achieved state-of-the-art results in several tasks not only in the field of NLP but also more recently for computer vision problems. The vanilla transformer (Vaswani et al. 2017) and its variants have been successfully applied in computer vision to, e.g., inpainting (Wan et al. 2021; Yu et al. 2021), object detection (Carion et al. 2020), image classification (Dosovitskiy et al. 2020), colorization (Kumar et al. 2021), and super resolution (Yang et al. 2020).

Instead of using inductive local biases like CNNs, transformers in imaging aim to have a global receptive field. For this, the image is first transformed by, as in the most basic approach, flattening the spatial dimensions of the input feature map into a sequence of features of size $M \times N \times F$, where $M \times N$ represents the flattened spatial dimensions and F the depth of a feature map. Then self-attention is applied over the extracted sequence. To ease the associated high computational cost, some authors substitute spatial pixels by patches. The attention mechanism looks at the input sequence and decides for each position which other parts of the sequence or image are important. More specifically, the transformers will transform the set of inputs, called *tokens*, using sequential blocks of multiheaded self-attention, which relate embedded inputs to each other. It is worth noticing that transformers will maintain the number of tokens throughout all computations. If tokens were related to pixels, each pixel would have a one-to-one correspondence with the output, thus, maintaining the spatial resolution of the original input image. Since transformers are set-to-set functions, they do not intrinsically retain the information of the spatial position for each individual token; thus, the embedding is concatenated to a learnable position embedding to add the positional information to the representation.

One advantage of using a transformer for image restoration is that it naturally supports pluralistic outputs by directly optimizing the underlying data distribution. One drawback is the computational complexity that increases quadratically with the input length, thus making it difficult to directly synthesize high-resolution images.

High-Fidelity Pluralistic Image Completion with Transformers (Wan et al. 2021)

The authors of Wan et al. (2021) exploit the benefits of both transformers and CNNs. The use of transformers will enforce a global structural understanding and pluralism support in the inpainted region, at a coarse resolution. On the other hand, the use of CNNs will allow working with high-resolution images without a high computational cost due to its capacity of estimating local textures efficiently.

Concretely, in this work, image completion is performed in two steps as shown in Fig. 6. In the first step, given a corrupted image, the authors use transformers to produce the probability distribution of structural appearance of complete images given the incomplete one. Low-resolution results can be obtained by sampling from this distribution with diversities that recover pluralistic coherent image structures. In the second step, guided by the computed image structures together with the available pixels of the input image, another upsampling CNN model is used to render high-fidelity textures for missing regions meanwhile ensuring coherence with neighboring pixels.

If X_Π denotes the set of masked tokens x_{π_k} (where $\Pi = \{\pi_1, \ldots, \pi_K\}$ denote their indexes), and $X_{-\Pi}$ denotes the set of unmasked tokens (corresponding to the

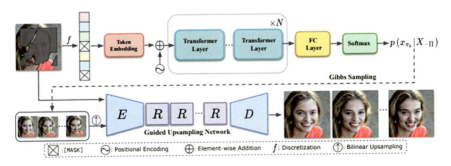

Fig. 6 Overview of the architecture of high-fidelity pluralistic image completion with transformers (Wan et al. 2021), referred to as ICT. (Figure from Wan et al. 2021)

visible regions), then the transformer is optimized by minimizing the negative log-likelihood of the masked tokens x_{π_k}, conditioned the visible regions $X_{-\Pi}$, that is,

$$\mathcal{L}_{\text{MLM}}(\theta) = \mathbb{E}_X \frac{1}{K} \sum_{k=1}^{K} -\log p(x_{\pi_k}|X_{-\Pi}; \theta) \qquad (21)$$

where θ contains the parameters of the transformer and the subindex MLM stands for the *masked language model* which is similar to the one in BERT (Devlin et al. 2018). One particularity of the ICT model is that each token attends simultaneously to all positions thanks to bidirectional attention. This enables the generated distribution to capture the full context, thus leading to a consistency between generated contents and unmasked region.

Once the transformer is trained, instead of directly sampling the entire set of masked positions which would lead to non-plausible results due to the independence property, they apply Gibbs sampling to iteratively sample tokens at different locations. To do so, in each iteration, a grid position is sampled from $p(x_{\pi_k}|X_{-\Pi}, X_{<\pi_k}, \theta)$ with the top-$\mathcal{K}$ predicted elements, where $X_{<\pi_k}$ denotes the previous generated tokens.

The second step is to perform texture refinement at the original resolution using a CNN, which is optimized by minimizing the L^1 loss between the predicted image and the ground truth, together with an adversarial loss based on the vanilla GAN (cf. (8) in section "Generative Adversarial Networks").

Diverse Image Inpainting with Bidirectional and Autoregressive Transformers (Yu et al. 2021)

This proposal exploits, as in Wan et al. (2021), a two-step strategy where transformers will encode global structure understanding and high-level semantics at the first stage, followed by a CNN-based generation of additional texture. While Wan et al. (2021) leverages bidirectional attention with the masked language model (MLM) as in BERT (Devlin et al. 2018), the authors of Yu et al. (2021) propose BAT-Fill that combines autoregressive models and bidirectional models (cf. Fig. 7). The first transformer-based step estimates the distribution of inpainted low-resolution structures from which to sample, from an input damaged image, a set $\{s_1, \ldots, s_J\}$ of plausible complete structures. Instead of only using a masked language model like BERT and Wan et al. (2021) (see above) that use bidirectional contextual information but predicts each masked token separately and independently (which can result in inconsistency in the generated result), BAT-Fill incorporates autoregressive modeling (factorizing the predicted tokens with the product rule). The input sequence of tokens is sorted by first having the visible tokens (permuted) and then the missing tokens (with the original order). In this way, the autoregressive model starts at the position of the first missing pixel. The BAT training objective is given by

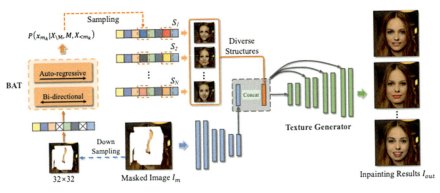

Fig. 7 Overview of the BAT architecture of diverse image inpainting with bidirectional and autoregressive transformers (Yu et al. 2021). (Figure from Yu et al. 2021)

$$\mathcal{L}_{\text{BAT}}(\theta) = \mathbb{E}_X \frac{1}{K} \sum_{k=1}^{K} -\log p(x_{\pi_k} | X_{-\Pi}, M, X_{<\pi_k}; \theta). \quad (22)$$

where we have used the same notations as in (21), namely, K is the length of masked tokens, and $X_{-\Pi}$ are all the unmasked tokens (corresponding to the visible regions). Finally, $X_{<\pi_k}$ denote the previous predicted tokens, and M the masked positions.

Finally, they construct a texture generator based on CNN-based synthesis, which is optimized by minimizing the L^1 loss between the predicted image and the ground truth together with an adversarial loss and a perceptual loss (Johnson et al. 2016).

In inference, each masked token is predicted bidirectionally and autoregressively. As in Wan et al. (2021), they iteratively use top-$\mathcal{K}$ sampling to randomly sample from the $\mathcal{K}$ most likely next tokens.

From Single-Image Evaluation Metrics to Diversity Evaluation

Currently, there is no consensus on automatic evaluation methods for single or diverse inpainting. As the problem is to recover a visually plausible image, performing quantitative evaluation is not trivial as the solution is not unique and the plausibility is a subjective term. Nevertheless, several evaluation metrics have been proposed through the years. We first detail those used for evaluating inpainting methods, one image at a time, before presenting the metric used as diversity scores.

Full-reference metrics compare the ground-truth image with the inpainted result. Famous measures in this category include L^1, L^2 distances, PSNR, or SSIM (Wang et al. 2004). These metrics analyze the ability of the model to reconstruct the original image content. Nevertheless, it is easy to demonstrate that they do not well characterize the realism of an image. Being close to the ground-truth image does not ensure being realistic. Other perceptual metrics have been proposed and are supposed to be more consistent with human judgment. In particular, Learned Perceptual Image Patch Similarity (LPIPS) (Zhang et al. 2018b) has been demonstrated to correlate well with the human perceptual similarity. It relies on the observation that hidden activations in CNNs trained for image classification are indeed a space where distance can strongly correlate with human judgment. Precisely, LPIPS computes a weighted L^2 norm between deep features of pair of images:

$$\text{LPIPS}(x, x_{gt}) = \sum_l \frac{1}{M_l N_l} \sum_{ij} \| w_l \odot (\Phi_r^l(i,j) - \Phi_{gt}^l(i,j)) \|_2^2 \qquad (23)$$

where x is the reconstructed image, x_{gt} is the ground truth, l is a layer number, (i,j) is a pixel, w_l are weights for each features, and Φ^l and $\Phi_{gt}^l \in \mathbb{R}^{M_l \times N_l \times C_l}$ are features unit-normalized in the channel dimension. LPIPS has been used in the context of inpainting when generating one image (e.g., Zheng et al. 2021). In Kettunen et al. (2019), it was shown that standard adversarial attack techniques can easily fool LPIPS. Therefore, a slightly different metric called E-LPIPS (Ensemble LPIPS) is proposed by applying random simple image transformations and dropout. Nonetheless, up to our knowledge, it has never been used in the context of inpainting.

When, apart from the set of images, there is available corresponding image categories, other metrics, that are also supposed to be following human judgment, can be used. The inception score (IS) (Salimans et al. 2016) was designed to measure how realistic the output from a GAN is. This score measures the variety of a set of generated images as well as the probability distribution of each image classification. This is done by comparing the class distribution of each image, which should have a low entropy, with the marginal distribution of the whole set, which should have high entropy:

$$\text{IS}(G) = \exp\left(\mathbb{E}_{x \sim p_g} \text{KL}\left(p(y|x) \| p(y) \right) \right) \qquad (24)$$

where p_g is the model distribution of the whole set given by the generative model G; x, an image sampled from p_g; $p(y|x)$, the conditional class distribution; KL, the Kullback–Leibler divergence; and $p(y)$, the marginal class distribution. As detailed in Barratt and Sharma (2018), inception score has its own limitations: sensitivity to small changes in network weights, misleading results when used beyond the ImageNet dataset (Rosca et al. 2017), and adversarial examples when used for model optimization. The IS score was adapted to diverse inpainting in Zhao et al. (2020b), leading to the Modified Inception Score (MIS). When performing inpainting, there is only one kind of image, and so $p(y)$ can be removed. The MIS is then defined as

$$\text{MIS(G)} = \exp\left(\mathbb{E}_{x \sim p_g} \sum_i \left(p(y_i|x) \log p(y_i|x)\right)\right), \quad (25)$$

where y_i of is the class label of the ith generated sample. Another improvement of the IS is the Fréchet Inception Distance (FID) (Heusel et al. 2017) that compares the statistics of generated images to the ones of original images. FID uses the inception pre-trained model to extract the feature vectors of real images and fake images and compare their feature-wise means (μ_r, μ_f) and covariances (Σ_r, Σ_f):

$$\text{FID} = \|\mu_r - \mu_f\|^2 + Tr(\Sigma_r + \Sigma_f + 2(\Sigma_r \Sigma_f)^{1/2}). \quad (26)$$

Fréchet Inception Distance has been widely used for validating single and diverse inpainting results in recent papers (e.g., Peng et al. 2021; Liu et al. 2021; Yu et al. 2021).

Measuring Diversity

In the context of pluralistic inpainting, following the idea proposed for image-to-image translation in Zhu et al. (2017), LPIPS has been used as a *diversity score* to measure how perceptually different the generated images are (Cai and Wei 2020; Zhao et al. 2020b; Liu et al. 2021). The higher the LPIPS, the more diversity is present in the results. For instance, in Cai and Wei (2020), they compute the average distance between the 10,000 pairs randomly generated from the 1000 center-masked image samples. LPIPS is computed on the full-inpainting results and mask-region inpainting results, respectively.

Experimental Results

In this section, we present a quantitative and qualitative comparison of several existing methods for multiple-image inpainting. We include an assessment of both the quality and the diversity of the inpainted solutions. All the results shown in this section are thanks to publicly available code together with pre-trained weights provided by the authors. In Table 2, we summarize, for all the methods previously reviewed, the conditions in which the experiments were conducted. Note that, among these methods, only PIC (Zheng et al. 2019), PiiGAN (Cai and Wei 2020), DSI-VQVAE (Peng et al. 2021), ICT (Wan et al. 2021), and BAT (Yu et al. 2021) provide source code and pre-trained models. In the rest of this section, we describe the experimental settings in section "Experimental Settings" including datasets and used masks; quantitative results in section "Quantitative Performance" including proximity to ground truth, perceptual quality, and inpainting diversity; and finally, a qualitative analysis is provided in section "Qualitative Performance".

Table 2 Generative methods for diverse inpainting: experimental conditions. Random regular and irregular masks are generated as in Zheng et al. (2019)

Method	Input size	Train datasets	Training masks	Code
PIC	256 × 256	Celeba-HQ ImageNet Paris Places2	Regular (center 128 × 128 + random) Irregular (random)	✓
PiiGAN	128 × 128	CelebA Mauflex Agricultural Disease	center 64 × 64	✓
UCTGAN	256 × 256	Celeba-HQ ImageNet Paris Places2	Regular (center 128 × 128 + random) Irregular (random)	✗
DSI-VQVAE	256 × 256	Celeba-HQ ImageNet Places2	Regular (center 128 × 128 + random) Irregular (random)	✓
ICT	256 × 256	FFHQ ImageNet Places2	Irregular Pconv (Liu et al. 2018)	✓
PD-GAN	256 × 256	Celeba-HQ Paris StreetView Places2	Irregular Pconv (Liu et al. 2018)	✗
BAT	256 × 256	CelebA-HQ Paris StreetView Places2	Irregular Pconv (Liu et al. 2018)	✓

Experimental Settings

Table 2 lists all the explained methods together with the training dataset and corresponding training masks. Aiming for a fair comparison, we compare and test the methods trained on the same training images, i.e., the VAE-based model PIC (Zheng et al. 2019), the VQVAE-based model DSI-VQVAE (Peng et al. 2021), and the two transformer-based models ICT (Wan et al. 2021) and BAT (Yu et al. 2021). Notice that we do not analyze the performance of PiiGAN (Cai and Wei 2020), as the training datasets and size images are different from the other methods.

Datasets

We evaluate the methods on the three datasets Celeba-HQ (Karras et al. 2018), Places2 (Zhou et al. 2017), and ImageNet (Russakovsky et al. 2015). All the evaluated models take as input images of resolution 256 × 256. Due to the long inference time of DSI-VQVAE and ICT methods (see Table 7), quantitative experiments are made on 100 randomly selected images from each training dataset. For each kind of mask (see below) and for each image, we sample 25 different results.

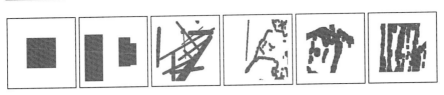

Fig. 8 Example for each kind of mask considered for evaluation. In gray are the hidden pixels. From left to right: center, random regular, random irregular, and irregular Pconv masks from Liao et al. (2018) with <20%, [40%, 60%] and [40%, 60%] hidden pixels

For Celeba-HQ, the 1024 × 1024 resolution images are resized to 256 × 256. For Places2 and ImageNet, the compared methods were trained on 256 × 256 patches either by resizing the input images (PIC), by cropping them, randomly (DSI) or to the center patch (BAT), or by both cropping and resizing (ICT). We will both consider center-cropped and resized versions of the input images to ensure a fair comparison among the trained models.

Note that ICT is not trained on Celeba-HQ but on the FFHQ face dataset (Karras et al. 2019). FFHQ contains higher variation than Celeba-HQ in terms of age, ethnicity, and image background. It also has a good coverage of accessories. Images from both datasets are, however, similarly aligned and cropped. Therefore, we still give the results of the ICT method tested on the Celeba-HQ dataset, but the reader should remember this difference when analyzing the results.

Inpainting Masks
We use the following type of masks: center, random regular, random irregular, and irregular masks from Liu et al. (2018) with different proportions of hidden pixels. Figure 8 shows an example of each kind of mask. The random masks are generated once for each test image so that all the methods are evaluated on the same degradation.

We would like to highlight that the methods PIC and DSI-VQVAE train a different model for regular and irregular holes. Testing on centered or random regular masks is realized with the former model, and testing on irregular masks with the latter. The transformer-based methods ICT and BAT only train on "irregular Pconv" holes given by Liu et al. (2018). Testing on each type of mask will be done with this unique model.

Quantitative Performance

We first analyze the numerical performance of each method. Table 3 shows quantitative results on Celeba-HQ dataset. Additionally, results on Places2 and ImageNet are, respectively, shown in Tables 4 and 5.

In Tables 4 and 5, we give our results obtained by center-cropping the images on Places2 and ImageNet, respectively. For fair comparison, we also give, in Appendix

Table 3 Quantitative comparison of four pluralistic image inpainting methods (PIC, DSI-VQVAE, ICT and BAT) on **Celeba-HQ** and for different kind of masks (central, random regular, random irregular and from Liu et al. 2018). Best and second best results by column are in bold and italics, respectively

Mask	Method	Similarity to GT			Realism		Diversity
		PSNR↑	SSIM↑	L^1 ↓	MIS↑	FID↓	LPIPS↑
Irregular <20%	PIC	34.63	0.964	**1.17**	0.0206	16.8	0.0009
	DSI-VQVAE	*35.49*	0.968	1.41	**0.0216**	11.0	*0.0081*
	ICT	34.72	0.968	2.09	0.0200	**9.84**	**0.0084**
	BAT	**36.25**	**0.974**	*1.20*	0.0208	*9.90*	0.0056
Irregular 20%–40%	PIC	26.69	0.879	4.19	*0.0216*	34.2	0.0091
	DSI-VQVAE	**27.36**	0.888	4.06	**0.0223**	28.8	*0.0357*
	ICT	26.83	*0.891*	4.71	0.0189	26.7	**0.0383**
	BAT	*27.28*	**0.900**	**3.85**	0.0214	**20.7**	0.0269
Irregular 40%–60%	PIC	21.47	0.745	10.36	0.0153	65.4	0.0527
	DSI-VQVAE	**22.53**	0.770	9.01	*0.0156*	51.9	*0.0916*
	ICT	21.92	*0.773*	9.82	0.0153	*50.7*	**0.0970**
	BAT	*22.35*	**0.787**	**8.91**	**0.0183**	**39.7**	0.0731
Central 128 × 128	PIC	24.46	0.868	5.26	*0.0212*	23.8	0.0288
	DSI-VQVAE	**25.25**	*0.880*	**5.08**	0.0210	*21.7*	0.0243
	ICT	24.45	0.872	6.06	0.0170	27.3	**0.0486**
	BAT	*25.10*	**0.882**	*5.21*	**0.0218**	**21.5**	*0.0365*
Random regular	PIC	24.16	0.840	7.23	0.0188	33.4	0.0402
	DSI-VQVAE	**24.98**	0.850	**6.46**	*0.0200*	*30.5*	0.0642
	ICT	24.51	*0.852*	7.24	0.0180	31.3	**0.0665**
	BAT	*24.85*	**0.860**	6.52	**0.0209**	**24.6**	0.0541
Random irregular	PIC	23.47	0.759	8.45	0.0161	73.5	0.0280
	DSI-VQVAE	*24.27*	*0.785*	7.56	*0.0167*	58.8	*0.0744*
	ICT	23.26	0.781	9.26	0.0148	*52.2*	**0.0855**
	BAT	**24.36**	**0.810**	**7.13**	**0.0186**	**40.8**	0.0495
Average	PIC	25.65	0.843	6.11	0.0189	41.2	0.0266
	DSI-VQVAE	*26.65*	*0.857*	*5.60*	0.0195	33.8	*0.0497*
	ICT	25.95	0.855	6.53	0.0173	*33.0*	**0.0575**
	BAT	**26.70**	**0.869**	**5.47**	**0.0203**	**26.2**	0.0410

Table 4 Quantitative comparison of four pluralistic image inpainting methods (PIC, DSI-VQVAE, ICT and BAT) on 256 × 256 **center-cropped** images from **Places2**, for different kind of masks (central, random regular, random irregular and from Liu et al. 2018)

Mask	Method	Similarity to GT			Realism		Diversity
		PSNR↑	SSIM↑	L^1 ↓	MIS↑	FID↓	LPIPS↑
Irregular <20%	PIC	30.48	0.937	2.02	**0.0507**	36.8	0.0050
	DSI-VQVAE	*31.58*	*0.952*	2.11	*0.0482*	*19.3*	*0.0187*
	ICT	29.86	0.943	3.64	0.0463	22.8	**0.0198**
	BAT	**32.20**	**0.957**	**1.83**	0.0463	**14.2**	0.0158
Irregular 20%–40%	PIC	23.88	0.820	6.46	0.0378	97.6	0.0344
	DSI-VQVAE	24.20	*0.844*	6.14	**0.0438**	*63.6*	*0.0707*
	ICT	23.08	0.831	8.05	*0.0428*	70.0	**0.0769**
	BAT	*24.10*	**0.853**	6.14	0.0423	**53.2**	0.0671
Irregular 40%–60%	PIC	19.92	0.667	13.75	0.0326	156.1	0.1309
	DSI-VQVAE	**20.34**	*0.703*	12.52	**0.0398**	*110.2*	0.1566
	ICT	19.49	0.686	14.66	*0.0371*	128.7	**0.1668**
	BAT	*19.98*	**0.705**	*13.10*	0.0364	**107.0**	*0.1610*
Central 128 × 128	PIC	20.98	0.812	9.00	0.0435	96.8	0.1080
	DSI-VQVAE	**21.41**	*0.819*	8.85	0.0416	**79.8**	**0.1234**
	ICT	20.93	0.812	10.22	**0.0476**	92.2	*0.1204*
	BAT	*21.20*	**0.822**	8.76	*0.0442*	*81.8*	0.1190
Random regular	PIC	21.70	0.783	10.14	*0.0425*	103.8	0.1124
	DSI-VQVAE	**22.36**	*0.805*	9.21	0.0412	**75.8**	0.1167
	ICT	21.75	0.796	10.77	0.0405	87.1	**0.1237**
	BAT	*22.34*	**0.808**	9.15	**0.0436**	*76.6*	*0.1200*
Random irregular	PIC	20.86	0.658	12.80	0.0255	165.4	0.0979
	DSI-VQVAE	**21.18**	*0.701*	11.78	0.0360	*114.4*	0.1450
	ICT	20.07	0.681	14.14	0.0334	131.9	**0.1548**
	BAT	*20.85*	**0.708**	*12.00*	**0.0374**	**103.2**	*0.1454*
Average	PIC	22.97	0.780	9.02	0.0388	109.2	0.0814
	DSI-VQVAE	**23.51**	*0.804*	8.44	**0.0418**	*76.9*	*0.1052*
	ICT	22.53	0.792	10.25	0.0413	88.9	**0.1107**
	BAT	*23.44*	**0.809**	8.97	*0.0417*	**72.7**	0.1047

(Tables 8 and 9), the results on these two datasets for resized images. The ICT method is run, as proposed in the original paper, with its top-$\mathcal{K}$ parameter (cf. section "Image Transformers") set to 50. We investigate the influence of the top-$\mathcal{K}$ parameter in Table 6. Note that for fair quantitative comparison, unlike Zheng

Table 5 Quantitative comparison of three pluralistic image inpainting methods (PIC, DSI-VQVAE and ICT) on 256 × 256 **center-cropped** images from **ImageNet**, for different kind of masks (central, random regular, random irregular and from Liu et al. 2018)

Mask	Method	Similarity to GT			Realism		Diversity
		PSNR↑	SSIM↑	L^1 ↓	MIS↑	FID↓	LPIPS↑
Irregular <20%	PIC	*30.33*	0.941	**2.02**	**0.2416**	20.2	0.0036
	DSI-VQVAE	**30.44**	**0.946**	2.38	*0.2361*	*12.1*	**0.0199**
	ICT	29.23	0.940	3.98	0.2323	**10.7**	*0.0185*
Irregular 20%–40%	PIC	**23.02**	0.797	7.37	0.1709	83.7	0.0289
	DSI-VQVAE	*22.98*	**0.809**	*7.56*	**0.2015**	53.4	**0.0855**
	ICT	22.24	*0.802*	9.23	*0.1970*	**24.9**	*0.0771*
Irregular 40%–60%	PIC	18.33	0.623	*16.34*	0.0792	183.9	0.1269
	DSI-VQVAE	**18.92**	**0.651**	**14.82**	*0.1192*	*126.3*	**0.1907**
	ICT	*18.52*	*0.646*	16.41	**0.1329**	**101.7**	*0.1700*
Central 128 × 128	PIC	19.87	0.794	**9.77**	0.1591	95.8	0.1067
	DSI-VQVAE	*20.06*	**0.795**	9.99	*0.1754*	*85.6*	**0.1291**
	ICT	**20.34**	0.795	10.76	*0.1753*	**73.8**	*0.1162*
Random regular	PIC	19.81	0.737	13.24	0.0934	129.2	0.1027
	DSI-VQVAE	**20.54**	**0.756**	**11.52**	*0.1305*	*89.3*	**0.1540**
	ICT	*20.32*	*0.752*	*12.76*	**0.1420**	**77.6**	*0.1360*
Random irregular	PIC	*19.51*	0.598	*14.78*	0.0645	193.0	0.0982
	DSI-VQVAE	**19.81**	**0.636**	**14.04**	*0.1147*	*136.8*	**0.1757**
	ICT	19.02	*0.628*	16.08	**0.1363**	**108.5**	*0.1574*
Average	PIC	*21.81*	0.748	*10.59*	0.1347	117.6	0.0735
	DSI-VQVAE	**22.13**	**0.765**	**10.07**	*0.1629*	*83.9*	**0.1258**
	ICT	21.61	*0.761*	11.54	**0.1693**	**66.2**	*0.1125*

et al. (2019), we do not use any discriminator score to select the best generated samples.

To measure inpainting quality, we take into account three factors: the similarity to the ground truth, the realism of inpainting outputs, and the diversity of those outputs. Definitions and details on the metrics for each factor can be found in section "From Single-Image Evaluation Metrics to Diversity Evaluation". Note that, contrary to Zheng et al. (2019), we do not use here any discriminator score to select the best samples before evaluation.

In each table, the best and second-best results by column are in bold and italics, respectively.

Table 6 Influence of the top-$\mathcal{K}$ parameter on the ICT results. Results obtained on Places2 dataset, with central mask

top-$\mathcal{K}$	Similarity to GT			Realism		Diversity
	PSNR↑	SSIM↑	L^1 ↓	MIS↑	FID↓	LPIPS↑
5	**21.76**	**0.820**	**6.52**	**0.0510**	**87.6**	0.0854
25	*21.16*	*0.813*	*10.03*	*0.0495*	*90.2*	*0.1146*
50	20.93	0.812	10.22	0.0476	92.2	**0.1204**

Proximity to Ground Truth

First, to measure the similarity between the inpainting results and the ground truth (GT), we use the following metrics: peak signal-to-noise ratio (PSNR), L^1 loss, and structural similarity (SSIM). For each input image to be inpainted, those metrics are averaged on the set of inpainted results.

Note that all the compared methods enforce somehow, in their training loss, similarity between the reconstructed image and the ground truth, either at pixel or feature levels. From the results in Tables 3, 4, 5, 8 and 9, we observe that, on all datasets, ICT and PIC obtained slightly lower scores than BAT and DSI-VQVAE in terms of GT similarity. A possible explanation for this performance gap is that, these two methods, contrary to the two others, consider a reconstruction loss only at the image level and not at the feature level. Being similar at feature levels encourages generating images having similar low-level (pixels, contours, etc.) and higher-level semantics to the ground truth.

Perceptual Quality

Second, to measure realism in the outputs, we measure perceptual quality by using Modified Inception Score (MIS) and Fréchet Inception Distance (FID) metrics (defined by (25) and (26), respectively). These two metrics are computed directly on the whole sets of generated or ground truth images.

BAT, ICT, and DSI-VQVAE are the methods that provide the best scores on average on all datasets. On the opposite, PIC gives the worst results quantitatively and, as we will see later, also qualitatively. We argue that a possible reason for the superior performance of BAT, ICT, and DSI-VQVAE is that, with different strategies, they separate the tasks of texture and structure recovery. Each task is handled with a specific subnetwork, first reconstructing structures that then guide the texture recovery. From a more practical point of view, BAT and ICT use transformers for global structure understanding and high-level semantics at a coarse resolution and CNNs for generating textures at the original resolution. DSI-VQVAE incorporates the multiscale hierarchical organization of VQ-VAE where the information corresponding to the texture is disentangled from the one about structure and geometry. Accordingly, DSI-VQVAE incorporates two different generators respectively devoted to both levels (cf. section "How to Achieve Multiple and Diverse Inpainting Results?"). Although DSI-VQVAE and PIC are VAE-based methods, DSI-VQVAE has the advantage that first, at low resolution, it proposes diverse completions of structure inside the hole. These different structures then guide the completion of texture at high resolution. PIC does not have this global structure completion (at least, not explicitly). All in all, splitting the estimation of coarse and fine details in two distinct steps seems like a successful approach for high-quality image inpainting.

Note also that BAT is the method that achieves the best scores in terms of realism. Indeed, as explained before, autoregressive transformers have the ability to model longer dependencies across the image than CNN-based methods, which can be crucial for image inpainting. Note that BAT outperforms the other transformer-based method ICT, especially on irregular masks and large holes. As explained in section "Autoregressive Models", one can explain this difference by the fact that BAT was trained, not only with bidirectional attention but also with autoregressive sampling. Therefore, it creates better consistency of the reconstructed structures, especially for large missing regions. The very good results of the DSI-VQVAE method also prove that autoregressive modeling performs well for realistic image inpainting.

Finally, one can observe the influence of the complexity of the training dataset on the performance. Notice that the underlying probability distribution of CelebA-HQ dataset is semantically less complex and diverse than the one of Places2 and ImageNet, and, thus, training is more difficult in the latter cases. We hypothesize that this affects both inpainting quality and inpainting diversity. Regarding quality the average FID score on all the studied methods trained on CelebA is equal to 33.55, while in the case of Places2 and Imagenet, it is equal to 86.92 and 128.43, respectively. This gives us an idea of the difference in complexity for each particular dataset.

Inpainting Diversity

To measure diversity, we rely on the LPIPS metric. The higher the LPIPS is, the more diverse are the outputs. For each generated sample, we compute the LPIPS distance with another sample randomly selected from the other 24 results from the same corrupted image. The reported LPIPS score corresponds to this distance averaged over the 2500 selected pairs.

First and foremost, from the range of LPIPS values on the different datasets, one can again observe the influence of the complexity of the training dataset. CelebA-HQ dataset is semantically more constraint and less complex than the one of Places2 and ImageNet, leading to lower diversity in the outputs. Indeed, the LPIPS is, in average, ∼2 times smaller on CelebA-HQ than on Places2 or ImageNet. Similarly, as expected, all the methods create more diverse samples on larger holes than on smaller holes.

These observations argue for the existence of a trade-off between inpainting quality and inpainting diversity. The harder the inpainting problem gets (on a more complex dataset or for a larger hole), the more diverse outputs will be created. This trade-off, already highlighted in Yu et al. (2021), also arises when parametrizing a method itself. We study in Table 6 the influence of the top-$\mathcal{K}$ parameter on the performance of the ICT algorithm. One can observe that using a smaller $\mathcal{K}$ creates outputs that are, on the one hand, closer to GT and more realistic but, on the other hand, less diverse.

PIC is the method giving the less diverse results on all datasets. One reason could be the aforementioned disentanglement of structure and texture of BAT, DSI-VQVAE, and ICT. In practice, these three methods first attempt to produce a multiplicity of coherent structures and then fill each of the sampled structure with a deterministic texture generator. This divide-and-conquer approach makes easier the creation of diversity as it is only performed on low-resolution structures and not on the whole reconstructed output.

ICT slightly outperforms DSI-VQVAE and BAT in terms of LPIPS on the Celeba-HQ testing images. Recall that for this experiment ICT was trained on the more diverse FFHQ dataset. This observation highlights again the influence of the training dataset on the capacity of the model to create diverse outputs.

Qualitative Performance

Similar to Zheng et al. (2019), Peng et al. (2021), Wan et al. (2021), and Yu et al. (2021), for *qualitative* comparison, we select for each method the 5 samples with the highest discriminator score out of the 25 generated samples. We use pretrained discriminators given by each of the models, i.e., for PIC, the discriminator of the generative pipeline; for DSI-VQVAE, the discriminator of the texture generation module; and for ICT and BAT, the discriminator of the upsampling module. We perform this comparison on a representative selection of testing images and masks. Figures 9, 10, 11, and 12 show some results on CelebA-HQ, Places2, and ImageNet datasets for the methods PIC, DSI-VQVAE, ICT, and BAT. BAT does not provide weights for ImageNet. Remember that ICT was not trained on Celeba-HQ but on FFHQ. Additional visual results are also given in the Appendix.

At first glance, we observe that DSI-VQVAE, ICT, and BAT provide more plausibly visual results than PIC. PIC struggles to recover information on less constrained datasets, like Places2 and Imagenet, and creates strong artifacts when applied to large missing regions (see second examples in Figs. 10 and 12). Among these methods, BAT and ICT propose the most realistic outputs. For instance, in Fig. 9, PIC generates results that do not maintain the proportions and harmony of a face (see the second example). DSI-VQVAE does not have a full understanding of the image either: for example, in the second example in Fig. 9 and the third example in Fig. 12, one eye is visible in the input image, but the other is not. On the opposite, transformer-based methods are able to reconstruct a left eye similar to the right. This can be explained by the capability of transformers to have a global structure understanding and high-level semantics. Other examples strengthening this observation are the first images of Fig. 10, where the inpainting of the snow is sometimes not realistic, and all the ImageNet results in Fig. 12.

When images contain strong structures, like Figs. 10 and 11, transformer-based methods again estimate more realistic reconstructions. This can be explained by

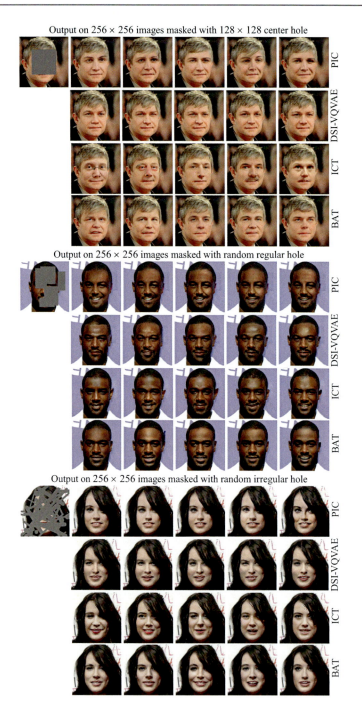

Fig. 9 Diverse inpainting output on 256× 256 images from Celeba dataset with center, random regular, and random irregular masks. For each method, out of 25 generated samples, the five samples with highest discriminator score are displayed

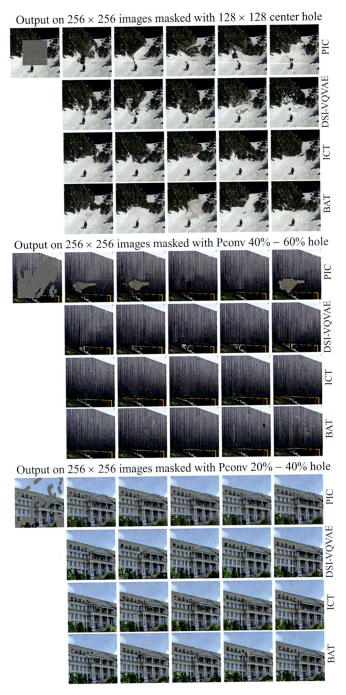

Fig. 10 Diverse inpainting output on 256 × 256 images from Places2 dataset with center and irregular masks with various proportion of hidden pixels. For each method, out of 25 generated samples, the 5 samples with highest discriminator score are displayed

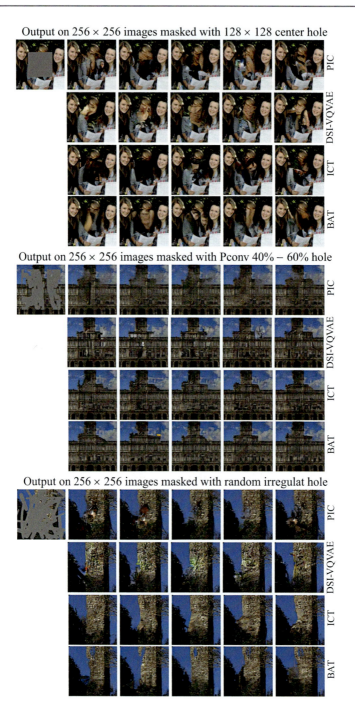

Fig. 11 Diverse inpainting output on 256 × 256 images from Places2 dataset with center and irregular masks with various proportion of hidden pixels. For each method, out of 25 generated samples, the five samples with highest discriminator score are displayed

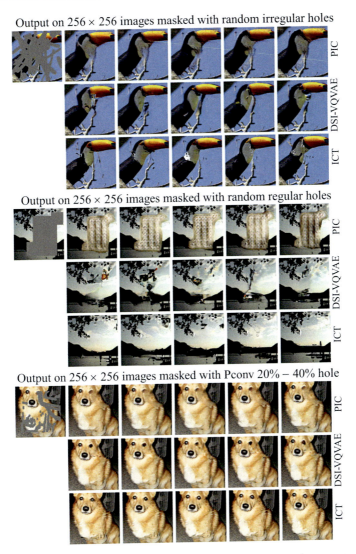

Fig. 12 Diverse inpainting output on 256 × 256 images from ImageNet dataset with center and irregular masks with various proportion of hidden pixels. For each method, out of 25 generated samples, the five samples with highest discriminator score are displayed

the fact that they include previously predicted tokens in the training objective, and thus, global consistency is imposed over the results. This consistency shall avoid problems in the center of big holes. In some situations, such as the middle example in Fig. 10, the structure and texture disentanglement of DSI-VQVAE also provides good reconstructions.

In terms of diversity, transformer-based methods are visually more diverse. For example, in Fig. 9, each transformer-based inpainted face corresponds to a different expression or different person, while in the case of DSI-VQVAE all generated faces are very similar. Also, in Fig. 11, even if one could imagine the result quite deterministic, ICT and BAT aim to propose multiple possibilities. Note the multiplicity of structures obtained by ICT compared to DSI–VQVAE and PIC in the chest of the dog or the skyline in Fig. 12.

Regarding the difference across datasets, while methods trained on CelebA-HQ all obtain satisfactory results (Fig. 9), results on Places2 (Fig. 11) and ImageNet (Fig. 12) are often not visually satisfactory. Also, as already noticed numerically, diversity is less visible on these two datasets. This is probably because the models have difficulties learning the underlying multimodal distribution of these complex and diverse datasets. This demonstrates the need for further research on the topic to be able to deal with real inpainting scenarios.

Influence of the Ccclusion Type. We summarize here our observations related to the influence of the shape of the missing region on the reconstruction quality and diversity. First, as expected, for all methods, when the hole is larger, the generated reconstructions are more diverse but farther away to the ground-truth image. Additionally, we visually observed that PIC may produce strong artifacts when tested on large missing regions, which is also quantitatively attested by its bad realism (MIS,FID) scores Table 5 on irregular and central holes. The superior performance of BAT on this kind of degradation seem to acknowledge the advantage of autoregressive sampling for filling large missing regions. Also notice that, although ICT and BAT were only trained on irregular masks, we do not observe a drop in performance while performing inpainting on regular masks, for instance, on the central one. This shows the capacity of those methods to generalize to unseen type of missing regions.

Computational Time. Despite image quality, an important aspect that should be considered when choosing an inpainting method is its inference time. In Table 7, for the four analyzed methods, we give the average runtime to sample one inpainting result from a central hole on a 256 × 256 input image. We run the experiments on a single P100 GPU. Despite showing lower inpainting quality or diversity (see before), PIC is tremendously faster to run than all the other methods (∼100 times faster than DSI-VQVAE and ICT and ∼50 times faster than BAT). While providing good results, inference time on autoregressive or transformer-based methods can be prohibitive for time-restricted applications.

Table 7 Average runtime to sample one inpainting result on a single P100 GPU for the four compared methods. Experiments conducted for **central** masks

Method	Time (s)
PIC	0.4
DSI-VQVAE	55
ICT	43
BAT	21

Conclusions

In this chapter, we have tackled the question of whether generative methods are a suitable strategy to obtain multiple solutions to problems that do not have a unique solution. By focusing on the inpainting problem, we have reviewed the main generative models and recent learning-based image completion methods for multiple and diverse inpainting. We have compared the methods with available code and model weights on three public datasets. We have shown that the transformer-based method BAT (or BAT-Fill) and the VQ-VAE-based method DSI-VQVAE provide the best results in both inpainting quality and multiple inpainting diversity. This is true both quantitatively and qualitatively. Our analysis highlights that their advantageous results are due to their strategy that consists in, first, sampling multiple structures inside the missing regions, and, second, generating textures at higher resolution in a deterministic way. The PIC method is, however, computationally way faster than the concurrence. Moreover, our analysis shows that the multiple inpainting problem is not solved yet, as the results lack of diversity or in general visually satisfactory results. The difficulty of learning the probability distribution depending on the training dataset is also evident from our study. Therefore, we argue that most efforts should be made on improving and exploring new generative strategies to enhance both the quality and diversity of the solutions of such ill-posed inverse problem with multiple solutions. For instance, following the spirit of structure/texture division, one could further separate the problem into different subtasks or tackle different regions of the scene separately. Another way to improve inpainting quality would be to have a control of the solution by bounding it through an input condition such as the semantic of the object you want to fill-in or by a reference image, among others. Finally, the computational burden of some of the transformer-based or autoregressive methods is prohibitive for sampling a high number of solutions in reasonable time. We think that this limitation has been overviewed for the purpose of image quality but should be now primarily addressed.

Appendix

Additional Quantitative Results

We provide in this section additional quantitative results on Places2 and ImageNet. Results from Tables 8 and 9 were conducted in the same conditions as Tables 4 and 5 but with 256×256 **resized** images instead of center-cropped images. Note that, in average, on both Places2 and ImageNet, the difference between methods is very similar when computed on resized or cropped images. The modification in aspect ratio due to the resize operation does not impede the results, even for models that were trained on "real" aspect ratios. The main reason for this is that the aspect ratio is not drastically changed when resizing Places2 and ImageNet images.

Table 8 Quantitative comparison of three pluralistic image inpainting methods (PIC, DSI-VQVAE, ICT) on 256 × 256 **resized** images from **Places2**

Mask	Method	Similarity to GT			Realism		Diversity
		PSNR↑	SSIM↑	L^1 ↓	MIS↑	FID↓	LPIPS↑
Irregular <20%	PIC	29.86	0.934	**2.14**	*0.0489*	32.3	0.0055
	DSI-VQVAE	**30.64**	**0.948**	*2.30*	**0.0533**	**20.0**	*0.0214*
	ICT	29.05	*0.939*	3.83	0.0450	*23.0*	**0.0224**
Irregular [20%, 40%]	PIC	22.98	0.808	*7.06*	0.0394	91.0	0.0375
	DSI-VQVAE	**23.04**	**0.832**	*6.92*	**0.0443**	**64.4**	*0.0789*
	ICT	22.11	*0.818*	8.81	*0.0423*	72.8	**0.0831**
Irregular [40%, 60%]	PIC	*19.01*	0.649	*14.71*	0.0273	144.2	0.1357
	DSI-VQVAE	**19.15**	**0.684**	**13.90**	0.0287	**115.0**	*0.1700*
	ICT	18.50	**0.669**	15.78	**0.0330**	*127.4*	**0.1755**
Central 128 × 128	PIC	**19.50**	**0.797**	**10.27**	0.0335	104.5	0.1129
	DSI-VQVAE	*19.46*	*0.797*	*10.60*	**0.0387**	**94.6**	**0.1364**
	ICT	19.42	0.796	11.72	*0.0352*	*101.0*	*0.1284*
Random regular	PIC	20.80	0.773	*10.95*	0.0359	93.8	0.1152
	DSI-VQVAE	**21.15**	**0.791**	**10.48**	**0.0426**	**79.0**	*0.1233*
	ICT	*21.03*	*0.787*	11.51	*0.0382*	*84.3*	**0.1239**
Random irregular	PIC	*19.91*	0.640	*13.85*	0.0246	157.7	0.1023
	DSI-VQVAE	**20.05**	**0.682**	**12.98**	**0.0329**	**116.5**	*0.1539*
	ICT	19.10	*0.662*	15.41	*0.0285*	*131.4*	**0.1607**
Average	PIC	*22.01*	0.767	**9.83**	0.0349	103.9	0.0848
	DSI-VQVAE	**22.25**	**0.789**	*9.53*	**0.0401**	**81.6**	*0.1140*
	ICT	21.54	*0.779*	11.18	*0.0370*	*90.0*	**0.1157**

Another explanation is that the training datasets are large enough and the models have enough capacity for being robust to such a transformation.

Additional Qualitative Results

In Figs. 13 and 14 we show additional inpainting visual results on Celeba-HQ and ImageNet datasets.

Table 9 Quantitative comparison of three pluralistic image inpainting methods (PIC, DSI-VQVAE, ICT) on 256 × 256 **resized** images from **ImageNet**

Mask	Method	Similarity to GT			Realism		Diversity
		PSNR↑	SSIM↑	L^1 ↓	MIS↑	FID↓	LPIPS↑
Irregular <20%	PIC	*31.37*	0.944	**1.82**	0.1885	21.5	0.0028
	DSI-VQVAE	**31.83**	**0.952**	*2.08*	*0.1913*	*12.8*	*0.0175*
	ICT	30.21	*0.946*	3.41	**0.2002**	**12.2**	**0.0203**
Irregular [20%, 40%]	PIC	*23.13*	0.807	*6.91*	0.1401	93.7	0.0323
	DSI-VQVAE	**23.45**	**0.825**	**6.72**	*0.1617*	*61.8*	*0.0790*
	ICT	22.36	*0.817*	8.34	**0.1739**	**52.1**	**0.0810**
Irregular [40%, 60%]	PIC	*18.39*	0.636	15.84	0.0497	198.0	0.1314
	DSI-VQVAE	**18.95**	**0.672**	**14.14**	*0.0737*	*147.9*	**0.1901**
	ICT	18.34	*0.663*	*15.85*	**0.0822**	**120.4**	*0.1764*
Central 128 × 128	PIC	19.31	0.795	*10.35*	0.0583	*153.9*	0.1091
	DSI-VQVAE	*19.47*	**0.800**	**10.25**	*0.0700*	172.1	**0.1293**
	ICT	**19.91**	*0.796*	11.27	**0.0790**	**120.3**	*0.1247*
Random regular	PIC	19.63	0.745	13.13	0.0690	150.5	0.1071
	DSI-VQVAE	**20.13**	**0.769**	**11.59**	**0.1048**	*113.8*	**0.1457**
	ICT	**20.13**	*0.766*	*12.56*	*0.1028*	**101.7**	*0.1376*
Random irregular	PIC	*19.70*	0.618	*14.04*	0.0457	194.6	0.1021
	DSI-VQVAE	**20.11**	**0.665**	**12.85**	*0.0642*	*155.9*	*0.1648*
	ICT	18.94	*0.649*	15.33	**0.0859**	**131.2**	**0.1652**
Average	PIC	*21.92*	0.758	*10.35*	0.0919	*135.4*	0.0949
	DSI-VQVAE	**22.32**	**0.781**	**9.61**	*0.1110*	107.7	**0.1211**
	ICT	21.64	*0.773*	11.13	**0.1207**	**89.7**	*0.1175*

Acknowledgments PV, CB, and AB acknowledge the EU Horizon 2020 research and innovation program NoMADS (Marie Skłodowska-Curie grant agreement No 777826). SP acknowledges the Leverhulme Trust Research Project Grant "Unveiling the invisible: Mathematics for Conservation in Arts and Humanities." CB and PV also acknowledge partial support by MICINN/FEDER UE project, ref. PGC2018-098625-B-I00, and RED2018-102511-T. AB also acknowledges the French Research Agency through the PostProdLEAP project (ANR-19-CE23-0027-01). SH acknowledges the French Ministry of Research through a CDSN grant of ENS Paris-Saclay.

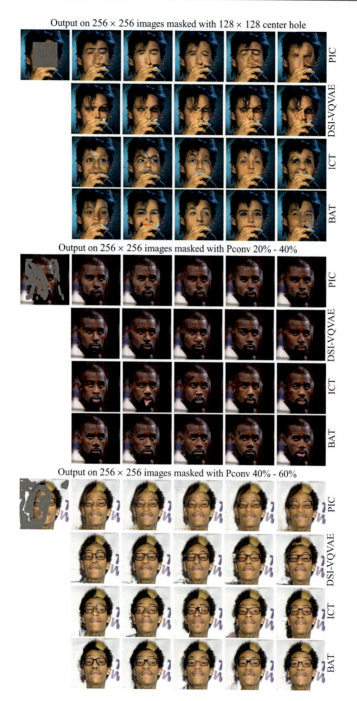

Fig. 13 Diverse inpainting output on 256 × 256 images from Celeba dataset with center, and irregular masks. For each method, out of 25 generated samples, the 5 samples with highest discriminator score are displayed

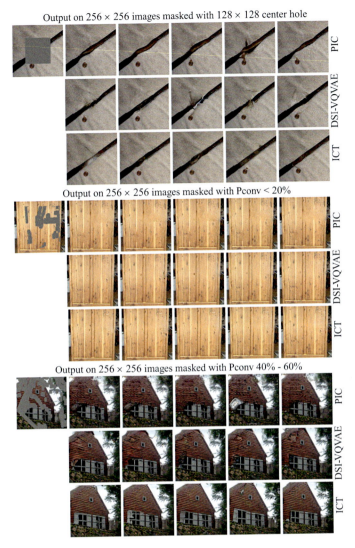

Fig. 14 Diverse inpainting output on 256 × 256 images from ImageNet dataset with centered and irregular masks with different hidded proportions. For each method, out of 25 generated samples, the 5 samples with highest discriminator score are displayed

References

Arias, P., Facciolo, G., Caselles, V., Sapiro, G.: A variational framework for exemplar-based image inpainting. Int. J. Comput. Vis. **93**(3), 319–347 (2011)

Arjovsky, M., Chintala, S., Bottou, L.: Wasserstein generative adversarial networks. In: International Conference on Machine Learning, pp. 214–223. PMLR (2017)

Arridge, S., Maass, P., Öktem, O., Schönlieb, C.-B.: Solving inverse problems using data-driven models. Acta Numer. **28**, 1–174 (2019)

Aujol, J.-F., Ladjal, S., Masnou, S.: Exemplar-based inpainting from a variational point of view. SIAM J. Math. Anal. **42**(3), 1246–1285 (2010)

Baatz, W., Fornasier, M., Markowich, P.A., bibiane Schönlieb, C.: Inpainting of ancient austrian frescoes. In: Conference Proceedings of Bridges, pp. 150–156 (2008)

Ballester, C., Bertalmío, M., Caselles, V., Sapiro, G., Verdera, J.: Filling-in by joint interpolation of vector fields and gray levels. IEEE Trans. Image Process. **10**(8), 1200–1211 (2001)

Bao, J., Chen, D., Wen, F., Li, H., Hua, G.: Cvae-gan: fine-grained image generation through asymmetric training. In: Proceedings of the IEEE International Conference on Computer Vision, pp. 2745–2754 (2017)

Barnes, C., Shechtman, E., Finkelstein, A., Goldman, D.B.: PatchMatch. In: ACM SIGGRAPH 2009 papers on – SIGGRAPH'09. ACM Press (2009)

Barratt, S., Sharma, R.: A note on the inception score (2018). arXiv preprint arXiv:1801.01973

Bertalmío, M., Bertozzi, A., Sapiro, G.: Navier-stokes, fluid dynamics, and image and video inpainting. In: Proceedings of the 2001 IEEE Computer Society Conference on Computer Vision and Pattern Recognition. CVPR 2001. IEEE Computer Society (2001)

Bertalmío, M., Sapiro, G., Caselles, V., Ballester, C.: Image inpainting. In: Proceedings of the 27th Annual Conference on Computer Graphics and Interactive Techniques, SIGGRAPH'00, pp. 417–424. ACM Press/Addison-Wesley Publishing Co (2000)

Bertozzi, A.L., Esedoglu, S., Gillette, A.: Inpainting of binary images using the cahn–hilliard equation. IEEE Trans. Image Process. **16**(1), 285–291 (2007)

Bevilacqua, M., Aujol, J.-F., Biasutti, P., Brédif, M., Bugeau, A.: Joint inpainting of depth and reflectance with visibility estimation. ISPRS J. Photogram. Rem. Sens. **125**, 16–32 (2017)

Biasutti, P., Aujol, J.-F., Brédif, M., Bugeau, A.: Diffusion and inpainting of reflectance and height LiDAR orthoimages. Comput. Vis. Image Underst. **179**, 31–40 (2019)

Bornard, R., Lecan, E., Laborelli, L., Chenot, J.-H.: Missing data correction in still images and image sequences. In: Proceedings of the Tenth ACM International Conference on Multimedia – MULTIMEDIA'02. ACM Press (2002)

Bornemann, F., März, T.: Fast image inpainting based on coherence transport. J. Math. Imag. Vis. **28**(3), 259–278 (2007)

Bowman, S.R., Vilnis, L., Vinyals, O., Dai, A.M., Jozefowicz, R., Bengio, S.: Generating sentences from a continuous space (2015). arXiv preprint arXiv:1511.06349

Buyssens, P., Daisy, M., Tschumperle, D., Lezoray, O.: Exemplar-based inpainting: Technical review and new heuristics for better geometric reconstructions. IEEE Trans. Image Process. **24**(6), 1809–1824 (2015)

Cai, N., Su, Z., Lin, Z., Wang, H., Yang, Z., Ling, B.W.-K.: Blind inpainting using the fully convolutional neural network. Vis. Comput. **33**(2), 249–261 (2015)

Cai, W., Wei, Z.: Piigan: generative adversarial networks for pluralistic image inpainting. IEEE Access **8**, 48451–48463 (2020)

Calatroni, L., d'Autume, M., Hocking, R., Panayotova, S., Parisotto, S., Ricciardi, P., Schönlieb, C.-B.: Unveiling the invisible: mathematical methods for restoring and interpreting illuminated manuscripts. Herit. Sci. **6**(1), 56 (2018)

Cao, F., Gousseau, Y., Masnou, S., Pérez, P.: Geometrically guided exemplar-based inpainting. SIAM J. Imag. Sci. **4**(4), 1143–1179 (2011)

Carion, N., Massa, F., Synnaeve, G., Usunier, N., Kirillov, A., Zagoruyko, S.: End-to-end object detection with transformers. In: European Conference on Computer Vision, pp. 213–229. Springer (2020)

Caselles, V., Morel, J.-M., Sbert, C.: An axiomatic approach to image interpolation. IEEE Trans. Image Process. **7**(3), 376–386 (1998)

Chan, T.F., Shen, J.: Nontexture inpainting by curvature-driven diffusions. J. Vis. Commun. Image Rep. **12**(4), 436–449 (2001)

Chen, X., Mishra, N., Rohaninejad, M., Abbeel, P.: Pixelsnail: An improved autoregressive generative model. In: International Conference on Machine Learning, pp. 864–872. PMLR (2018)

Chen, Y., Li, Y., Guo, H., Hu, Y., Luo, L., Yin, X., Gu, J., Toumoulin, C.: CT metal artifact reduction method based on improved image segmentation and sinogram in-painting. Math. Probl. Eng. **2012**, 1–18 (2012)

Criminisi, A., Perez, P., Toyama, K.: Region filling and object removal by exemplar-based image inpainting. IEEE Trans. Image Process. **13**(9), 1200–1212 (2004)

Dahl, R., Norouzi, M., Shlens, J.: Pixel recursive super resolution. In: Proceedings of the IEEE International Conference on Computer Vision, pp. 5439–5448 (2017)

Dapogny, A., Cord, M., Pérez, P.: The missing data encoder: cross-channel image completion with hide-and-seek adversarial network. Proc. AAAI Conf. Artif. Intell. **34**(07), 10688–10695 (2020)

Demanet, L., Song, B., Chan, T.: Image inpainting by correspondence maps: a deterministic approach. Appl. Comput. Math. **1100**, 217–50 (2003)

Devlin, J., Chang, M.-W., Lee, K., Toutanova, K.: Bert: pre-training of deep bidirectional transformers for language understanding (2018). arXiv preprint arXiv:1810.04805

Dosovitskiy, A., Beyer, L., Kolesnikov, A., Weissenborn, D., Zhai, X., Unterthiner, T., Dehghani, M., Minderer, M., Heigold, G., Gelly, S., et al.: An image is worth 16x16 words: Transformers for image recognition at scale (2020). arXiv preprint arXiv:2010.11929

Efros, A., Leung, T.: Texture synthesis by non-parametric sampling. In: Proceedings of the Seventh IEEE International Conference on Computer Vision. IEEE (1999)

Eller, M., Fornasier, M.: Rotation invariance in exemplar-based image inpainting. In: Variational Methods: In Maitine, B., Gabriel, P., Christoph, S., Jean-Baptiste, C., Thomas, H. (eds.), Imaging and Geometric Control, pp. 108–183. De Gruyter, Berlin, Boston (2017). https://doi.org/10.1515/9783110430394-004

Esedoglu, S., Shen, J.: Digital inpainting based on the mumford–shah–euler image model. Eur. J. Appl. Math. **13**(04), 353–370 (2002)

Fawzi, A., Samulowitz, H., Turaga, D., Frossard, P.: Image inpainting through neural networks hallucinations. In: IEEE 12th Image, Video, and Multidimensional Signal Processing Workshop, pp. 1–5. IEEE (2016)

Fedorov, V., Arias, P., Facciolo, G., Ballester, C.: Affine invariant self-similarity for exemplar-based inpainting. In: Proceedings of the 11th Joint Conference on Computer Vision, Imaging and Computer Graphics Theory and Applications. SCITEPRESS – Science and Technology Publications (2016)

Goodfellow, I.J., Pouget-Abadie, J., Mirza, M., Xu, B., Warde-Farley, D., Ozair, S., Courville, A.C., Bengio, Y.: Generative adversarial nets. Adv. Neural Inf. Process. Syst. **27**, 2672–2680 (2014)

Grossauer, H.: Inpainting of movies using optical flow. In: Mathematics in Industry, pp. 151–162. Springer, Berlin/Heidelberg (2006)

Grossauer, H., Scherzer, O.: Using the complex ginzburg-landau equation for digital inpainting in 2d and 3d. In: Scale Space Methods in Computer Vision, pp. 225–236. Springer, Berlin/Heidelberg (2003)

Guadarrama, S., Dahl, R., Bieber, D., Norouzi, M., Shlens, J., Murphy, K.: Pixcolor: Pixel recursive colorization (2017). arXiv preprint arXiv:1705.07208

Guillemot, C., Le Meur, O.: Image inpainting: overview and recent advances. IEEE Sig. Process. Mag. **31**(1), 127–144 (2014)

Gulrajani, I., Ahmed, F., Arjovsky, M., Dumoulin, V., Courville, A.: Improved training of wasserstein gans (2017). arXiv preprint arXiv:1704.00028

Guo, Z., Chen, Z., Yu, T., Chen, J., Liu, S.: Progressive image inpainting with full-resolution residual network. In: Proceedings of the 27th ACM International Conference on Multimedia, MM'19, New York, pp. 2496–2504. Association for Computing Machinery (ACM) (2019)

Han, X., Wu, Z., Huang, W., Scott, M.R., Davis, L.S.: Finet: compatible and diverse fashion image inpainting. In: Proceedings of the IEEE/CVF International Conference on Computer Vision, pp. 4481–4491 (2019)

Hays, J., Efros, A.A.: Scene completion using millions of photographs. ACM Trans. Graph. **26**(3), 87–94 (2007)

Hervieu, A., Papadakis, N., Bugeau, A., Gargallo, P., Caselles, V.: Stereoscopic image inpainting: distinct depth maps and images inpainting. In: 2010 20th International Conference on Pattern Recognition, pp. 4101–4104. IEEE (2010)

Heusel, M., Ramsauer, H., Unterthiner, T., Nessler, B., Hochreiter, S.: GANs trained by a two time-scale update rule converge to a local nash equilibrium. Adv. Neural Inf. Process. Syst. **30**, 6629–6640 (2017)

Iizuka, S., Simo-Serra, E., Ishikawa, H.: Globally and locally consistent image completion. In: ACM Transactions on Graphics (ToG), vol. **36**(4), pp. 1–14. ACM, New York, NY, USA (2017)

Johnson, J., Alahi, A., Fei-Fei, L.: Perceptual losses for real-time style transfer and super-resolution. In: European Conference on Computer Vision, pp. 694–711. Springer (2016)

Kang, S.H., Chan, T., Soatto, S.: Inpainting from multiple views. In: Proceedings. First International Symposium on 3D Data Processing Visualization and Transmission. IEEE Computer Society (2002)

Karras, T., Aila, T., Laine, S., Lehtinen, J.: Progressive growing of gans for improved quality, stability, and variation. In: International Conference on Learning Representations (2018)

Karras, T., Laine, S., and Aila, T.: A style-based generator architecture for generative adversarial networks. In: Proceedings of the IEEE/CVF Conference on Computer Vision and Pattern Recognition, pp. 4401–4410 (2019)

Kettunen, M., Härkönen, E., Lehtinen, J.: E-lpips: robust perceptual image similarity via random transformation ensembles (2019). arXiv preprint arXiv:1906.03973

Kingma, D.P., Welling, M.: Auto-encoding variational bayes. In: International Conference on Learning Representations (2013)

Kingma, D.P., Welling, M., et al.: An introduction to variational autoencoders. Found. Trends® Mach. Learn. **12**(4), 307–392 (2019)

Köhler, R., Schuler, C., Schölkopf, B., Harmeling, S.: Mask-specific inpainting with deep neural networks. In: Jiang, X., Hornegger, J., Koch, R. (eds.) Pattern Recognition, pp. 523–534, Springer International Publishing, Cham (2014)

Kumar, M., Weissenborn, D., Kalchbrenner, N.: Colorization transformer (2021). arXiv preprint arXiv:2102.04432

Kumar, V., Mukherjee, J., Mandal, S.K.D.: Image inpainting through metric labeling via guided patch mixing. IEEE Trans. Image Process. **25**(11), 5212–5226 (2016)

Lahiri, A., Jain, A.K., Agrawal, S., Mitra, P., Biswas, P.K.: Prior guided GAN based semantic inpainting. In: Proceedings of the IEEE/CVF Conference on Computer Vision and Pattern Recognition, pp. 13696–13705 (2020)

Le Meur, O., Ebdelli, M., Guillemot, C.: Hierarchical super-resolution-based inpainting. IEEE Trans. Image Process. **22**(10), 3779–3790 (2013)

Lempitsky, V., Vedaldi, A., Ulyanov, D.: Deep image prior. In: 2018 IEEE/CVF Conference on Computer Vision and Pattern Recognition, pp. 9446–9454. IEEE (2018)

Li, J., He, F., Zhang, L., Du, B., Tao, D.: Progressive reconstruction of visual structure for image inpainting. In: 2019 IEEE/CVF International Conference on Computer Vision. IEEE (2019)

Li, J., Wang, N., Zhang, L., Du, B., Tao, D.: Recurrent feature reasoning for image inpainting. In: Proceedings of the IEEE/CVF Conference on Computer Vision and Pattern Recognition, pp. 7760–7768 (2020)

Liao, L., Hu, R., Xiao, J., Wang, Z.: Edge-aware context encoder for image inpainting. In: 2018 IEEE International Conference on Acoustics, Speech and Signal Processing, pp. 3156–3160. IEEE (2018)

Liu, G., Reda, F.A., Shih, K.J., Wang, T.-C., Tao, A., Catanzaro, B.: Image inpainting for irregular holes using partial convolutions. In: European Conference on Computer Vision, pp. 89–105 (2018)

Liu, H., Jiang, B., Song, Y., Huang, W., Yang, C.: Rethinking image inpainting via a mutual encoder-decoder with feature equalizations. In: Computer Vision – ECCV 2020, pp. 725–741. Springer International Publishing (2020)

Liu, H., Jiang, B., Xiao, Y., Yang, C.: Coherent semantic attention for image inpainting. In: 2019 IEEE/CVF International Conference on Computer Vision. IEEE (2019)

Liu, H., Wan, Z., Huang, W., Song, Y., Han, X., Liao, J.: Pd-gan: probabilistic diverse gan for image inpainting. In: Proceedings of the IEEE/CVF Conference on Computer Vision and Pattern Recognition, pp. 9371–9381 (2021)

Mansfield, A., Prasad, M., Rother, C., Sharp, T., Kohli, P., Gool, L.V.: Transforming image completion. In: Procedings of the British Machine Vision Conference 2011. British Machine Vision Association (2011)

Mao, Q., Lee, H.-Y., Tseng, H.-Y., Ma, S., Yang, M.-H.: Mode seeking generative adversarial networks for diverse image synthesis. In: Conference on Computer Vision and Pattern Recognition, pp. 1429–1437 (2019)

Masnou, S., Morel, J.-M.: Level lines based disocclusion. In: Proceedings 1998 International Conference on Image Processing. ICIP98 (Cat. No.98CB36269). IEEE Computer Society (1998)

Nazeri, K., Ng, E., Joseph, T., Qureshi, F., Ebrahimi, M.: EdgeConnect: generative image inpainting with adversarial edge learning. In: The IEEE International Conference on Computer Vision Workshops (2019)

Newson, A., Almansa, A., Fradet, M., Gousseau, Y., Pérez, P.: Video inpainting of complex scenes. SIAM J. Imag. Sci. **7**(4), 1993–2019 (2014)

Nitzberg, M., Mumford, D., Shiota, T.: Filtering, Segmentation and Depth. Springer, Berlin/Heidelberg (1993)

Oord, A., Li, Y., Babuschkin, I., Simonyan, K., Vinyals, O., Kavukcuoglu, K., Driessche, G., Lockhart, E., Cobo, L., Stimberg, F., et al.: Parallel wavenet: Fast high-fidelity speech synthesis. In: International Conference on Machine Learning, pp. 3918–3926. PMLR (2018)

Oord, A.V.D., Kalchbrenner, N., Vinyals, O., Espeholt, L., Graves, A., Kavukcuoglu, K.: Conditional image generation with pixelcnn decoders. In: Proceedings of the 30th International Conference on Neural Information Processing Systems, pp. 4797–4805 (2016)

Papafitsoros, K., Schönlieb, C.B.: A combined first and second order variational approach for image reconstruction. J. Math. Imag. Vis. **48**(2), 308–338 (2013)

Parisotto, S., Lellmann, J., Masnou, S., Schönlieb, C.-B.: Higher-order total directional variation: imaging applications. SIAM J. Imag. Sci. **13**(4), 2063–2104 (2020)

Parisotto, S., Vitoria, P., Ballester, C., Bugeau, A., Reynolds, S., Schonlieb, C.-B.: The Art of Inpainting – A Monograph on Mathematical Methods for the Virtual Restoration of Illuminated Manuscripts (2022) (submitted)

Park, T., Liu, M.-Y., Wang, T.-C., Zhu, J.-Y.: Semantic image synthesis with spatially-adaptive normalization. In: Proceedings of the IEEE/CVF Conference on Computer Vision and Pattern Recognition, pp. 2337–2346 (2019)

Pathak, D., Krahenbuhl, P., Donahue, J., Darrell, T., Efros, A.A.: Context encoders: feature learning by inpainting. In: 2016 IEEE Conference on Computer Vision and Pattern Recognition, pp. 2536–2544. IEEE (2016)

Peng, J., Liu, D., Xu, S., Li, H.: Generating diverse structure for image inpainting with hierarchical vq-vae. In: Proceedings of the IEEE/CVF Conference on Computer Vision and Pattern Recognition, pp. 10775–10784 (2021)

Peter, P., Weickert, J.: Compressing images with diffusion- and exemplar-based inpainting. In: Lecture Notes in Computer Science, pp. 154–165. Springer International Publishing (2015)

Razavi, A., van den Oord, A., Vinyals, O.: Generating diverse high-fidelity images with vq-vae-2. In: Advances in Neural Information Processing Systems, pp. 14866–14876 (2019)

Ren, J.S., Xu, L., Yan, Q., Sun, W.: Shepard convolutional neural networks. In: Proceedings of the 28th International Conference on Neural Information Processing Systems. *NIPS'15*, Cambridge, MA, vol. 1, pp. 901–909. The MIT Press (2015)

Ren, Y., Yu, X., Zhang, R., Li, T.H., Liu, S., Li, G.: StructureFlow: image inpainting via structure-aware appearance flow. In: 2019 IEEE/CVF International Conference on Computer Vision, pp. 181–190. IEEE (2019)

Rosca, M., Lakshminarayanan, B., Warde-Farley, D., Mohamed, S.: Variational approaches for auto-encoding generative adversarial networks (2017). arXiv preprint arXiv:1706.04987

Rott Shaham, T., Dekel, T., Michaeli, T.: Singan: Learning a generative model from a single natural image. In: International Conference on Computer Vision (2019)

Royer, A., Kolesnikov, A., Lampert, C.H.: Probabilistic image colorization (2017). arXiv preprint arXiv:1705.04258

Russakovsky, O., Deng, J., Su, H., Krause, J., Satheesh, S., Ma, S., Huang, Z., Karpathy, A., Khosla, A., Bernstein, M., et al.: Imagenet large scale visual recognition challenge. Int. J. Comput. Vis. **115**(3), 211–252 (2015)

Ružić, T., Cornelis, B., Platiša, L., Pižurica, A., Dooms, A., Philips, W., Martens, M., Mey, M.D., Daubechies, I.: Virtual restoration of the ghent altarpiece using crack detection and inpainting. In: Advanced Concepts for Intelligent Vision Systems, pp. 417–428. Springer, Berlin/Heidelberg (2011)

Salimans, T., Goodfellow, I., Zaremba, W., Cheung, V., Radford, A., Chen, X.: Improved techniques for training gans. In: Advances in Neural Information Processing Systems (2016)

Schonlieb, C.-B.: Partial Differential Equation Methods for Image Inpainting. Cambridge University Press, New York (2015)

Shen, J., Chan, T.F.: Mathematical models for local nontexture inpaintings. SIAM J. Appl. Math. **62**(3), 1019–1043 (2002)

Shen, J., Kang, S.H., Chan, T.F.: Euler's elastica and curvature-based inpainting. SIAM J. Appl. Math. **63**(2), 564–592 (2003)

Sohn, K., Lee, H., Yan, X.: Learning structured output representation using deep conditional generative models. Adv. Neural Inf. Process. Syst. **28**, 3483–3491 (2015)

Sun, J., Yuan, L., Jia, J., Shum, H.-Y.: Image completion with structure propagation. ACM Trans. Graph. **24**(3), 861–868 (2005)

Tai, X.-C., Osher, S., Holm, R.: Image inpainting using a TV-stokes equation. In: Image Processing Based on Partial Differential Equations, pp. 3–22. Springer, Berlin/Heidelberg (2007)

Tovey, R., Benning, M., Brune, C., Lagerwerf, M.J., Collins, S.M., Leary, R.K., Midgley, P.A., Schönlieb, C.-B.: Directional sinogram inpainting for limited angle tomography. Inverse Probl. **35**(2), 024004 (2019)

Tschumperle, D., Deriche, R.: Vector-valued image regularization with PDEs: a common framework for different applications. IEEE Trans. Pattern Anal. Mach. Intell. **27**(4), 506–517 (2005)

van den Oord, A., Vinyals, O., Kavukcuoglu, K.: Neural discrete representation learning. In: Proceedings of the 31st International Conference on Neural Information Processing Systems, pp. 6309–6318 (2017)

Van Oord, A., Kalchbrenner, N., Kavukcuoglu, K.: Pixel recurrent neural networks. In: International Conference on Machine Learning, pp. 1747–1756. PMLR (2016)

Vaswani, A., Shazeer, N., Parmar, N., Uszkoreit, J., Jones, L., Gomez, A.N., Kaiser, Ł., Polosukhin, I.: Attention is all you need. In: Advances in Neural Information Processing Systems, pp. 5998–6008 (2017)

Vitoria, P., Ballester, C.: Automatic flare spot artifact detection and removal in photographs. J. Math. Imag. Vis. **61**(4), 515–533 (2019)

Vitoria, P., Sintes, J., Ballester, C.: Semantic image inpainting through improved Wasserstein generative adversarial networks. In: Proceedings of the 14th International Joint Conference on Computer Vision, Imaging and Computer Graphics Theory and Applications. VISAPP, vol. 4, pp. 249–260. INSTICC, SciTePress (2019)

Vitoria, P., Sintes, J., Ballester, C.: Semantic image completion through an adversarial strategy. In: Communications in Computer and Information Science, pp. 520–542. Springer International Publishing (2020)

Vo, H.V., Duong, N.Q.K., Pérez, P.: Structural inpainting. In: 2018 ACM Multimedia Conference on Multimedia Conference, MM'18, New York, pp. 1948–1956. Association for Computing Machinery (ACM) (2018)

Wan, Z., Zhang, J., Chen, D., Liao, J.: High-fidelity pluralistic image completion with transformers (2021). arXiv preprint arXiv:2103.14031

Wang, Z.B., Alan, C.S., Hamid, R.S.: Image quality assessment: From error visibility to structural similarity. IEEE Trans. Image Process 13(4), 600–612 (2004)

Wang, Y., Tao, X., Qi, X., Shen, X., Jia, J.: Image inpainting via generative multi-column convolutional neural networks. In: Proceedings of the 32nd International Conference on Neural Information Processing Systems, pp. 329–338. Curran Associates Inc., Montréal, Canada (2018)

Wexler, Y., Shechtman, E., Irani, M.: Space-time video completion. In: Proceedings of the 2004 IEEE Computer Society Conference on Computer Vision and Pattern Recognition, CVPR 2004. IEEE (2004)

Xiong, W., Yu, J., Lin, Z., Yang, J., Lu, X., Barnes, C., Luo, J.: Foreground-aware image inpainting. In: 2019 IEEE/CVF Conference on Computer Vision and Pattern Recognition, pp. 5840–5848. IEEE (2019)

Yan, Z., Li, X., Li, M., Zuo, W., Shan, S.: Shift-net: image inpainting via deep feature rearrangement. In: Computer Vision – ECCV 2018, pp. 3–19. Springer International Publishing (2018)

Yang, C., Lu, X., Lin, Z., Shechtman, E., Wang, O., Li, H.: High-resolution image inpainting using multi-scale neural patch synthesis. In: 2017 IEEE Conference on Computer Vision and Pattern Recognition, pp. 6721–6729. IEEE (2017)

Yang, F., Yang, H., Fu, J., Lu, H., Guo, B.: Learning texture transformer network for image super-resolution. In: Proceedings of the IEEE/CVF Conference on Computer Vision and Pattern Recognition, pp. 5791–5800 (2020)

Yeh, R.A., Chen, C., Lim, T.Y., Schwing, A.G., Hasegawa-Johnson, M., Do, M.N.: Semantic image inpainting with deep generative models. In: 2017 IEEE Conference on Computer Vision and Pattern Recognition, pp. 5485–5493. IEEE (2017)

Yi, K., Guo, Y., Fan, Y., Hamann, J., Wang, Y.G.: Cosmovae: variational autoencoder for CMB image inpainting (2020a). arXiv preprint arXiv:2001.11651

Yi, Z., Tang, Q., Azizi, S., Jang, D., Xu, Z.: Contextual residual aggregation for ultra high-resolution image inpainting. In: 2020 IEEE/CVF Conference on Computer Vision and Pattern Recognition, pp. 7505–7514. IEEE (2020b)

Yu, J., Lin, Z., Yang, J., Shen, X., Lu, X., Huang, T.: Free-form image inpainting with gated convolution. In: International Conference on Computer Vision, pp. 4470–4479 (2019)

Yu, J., Lin, Z., Yang, J., Shen, X., Lu, X., Huang, T.S.: Generative image inpainting with contextual attention. In: 2018 IEEE/CVF Conference on Computer Vision and Pattern Recognition, pp. 5505–5514. IEEE (2018)

Yu, Y., Zhan, F., Wu, R., Pan, J., Cui, K., Lu, S., Ma, F., Xie, X., Miao, C.: Diverse image inpainting with bidirectional and autoregressive transformers (2021). arXiv preprint arXiv:2104.12335

Zeng, Y., Fu, J., Chao, H., Guo, B.: Learning pyramid-context encoder network for high-quality image inpainting. In: 2019 IEEE/CVF Conference on Computer Vision and Pattern Recognition, pp. 1486–1494. IEEE (2019)

Zeng, Y., Lin, Z., Yang, J., Zhang, J., Shechtman, E., Lu, H.: High-resolution image inpainting with iterative confidence feedback and guided upsampling. In: European Conference on Computer Vision, pp. 1–17. Springer (2020)

Zhang, H., Hu, Z., Luo, C., Zuo, W., Wang, M.: Semantic image inpainting with progressive generative networks. In: 2018 ACM Multimedia Conference on Multimedia Conference, MM'18, pp. 1939–1947. ACM Press (2018a)

Zhang, R., Isola, P., Efros, A.A., Shechtman, E., Wang, O.: The unreasonable effectiveness of deep features as a perceptual metric. In: Conference on Computer Vision and Pattern Recognition (2018b)

Zhao, J., Han, J., Shao, L., Snoek, C.G.: Pixelated semantic colorization. Int. J. Comput. Vis. **128**(4), 818–834 (2020a)

Zhao, L., Mo, Q., Lin, S., Wang, Z., Zuo, Z., Chen, H., Xing, W., Lu, D.: UCTGAN: diverse image inpainting based on unsupervised cross-space translation. In: Conference on Computer Vision and Pattern Recognition, pp. 5741–5750 (2020b)

Zheng, C., Cham, T.-J., Cai, J.: Pluralistic image completion. In: Conference on Computer Vision and Pattern Recognition, pp. 1438–1447 (2019)

Zheng, C., Cham, T.-J., Cai, J.: Tfill: image completion via a transformer-based architecture (2021). arXiv preprint arXiv:2104.00845

Zhou, B., Lapedriza, A., Khosla, A., Oliva, A., Torralba, A.: Places: a 10 million image database for scene recognition. IEEE Trans. Pattern Anal. Mach. Intell. **40**(6), 1452–1464 (2017)

Zhu, J.-Y., Zhang, R., Pathak, D., Darrell, T., Efros, A.A., Wang, O., Shechtman, E.: Multimodal image-to-image translation by enforcing bi-cycle consistency. In: Advances in Neural Information Processing Systems, pp. 465–476 (2017)

Analysis of Different Losses for Deep Learning Image Colorization

21

Coloma Ballester, Hernan Carrillo, Michaël Clément, and Patricia Vitoria

Contents

Introduction	822
Losses in the Colorization Literature	823
Error-Based Losses	824
Generative Adversarial Network-Based Losses	826
Distribution-Based Losses	828
Proposed Colorization Framework	831
Detailed Architecture	831
Quantitative Evaluation Metrics Used in Colorization Methods	833
Experimental Analysis	836
Quantitative Evaluation	836
Qualitative Evaluation	838
Generalization to Archive Images	844
Conclusion	844
References	844

Abstract

Image colorization aims to add color information to a grayscale image in a realistic way. Recent methods mostly rely on deep learning strategies. While learning to automatically colorize an image, one can define well-suited objective functions related to the desired color output. Some of them are based on a specific type of error between the predicted image and ground truth one, while other

---

C. Ballester · P. Vitoria
University Pompeu Fabra, Barcelona, Spain
e-mail: coloma.ballester@upf.edu; patricia.vitoria@upf.edu

H. Carrillo · M. Clément (✉)
LaBRI, CNRS, Bordeaux INP, Université de Bordeaux, Bordeaux, France
e-mail: hernan.carrillo-lindado@u-bordeaux.fr; michael.clement@labri.fr

© Springer Nature Switzerland AG 2023
K. Chen et al. (eds.), *Handbook of Mathematical Models and Algorithms in Computer Vision and Imaging*, https://doi.org/10.1007/978-3-030-98661-2_127

losses rely on the comparison of perceptual properties. But, is the choice of the objective function that crucial, i.e., does it play an important role in the results? In this chapter, we aim to answer this question by analyzing the impact of the loss function on the estimated colorization results. To that goal, we review the different losses and evaluation metrics that are used in the literature. We then train a baseline network with several of the reviewed objective functions, classic L1 and L2 losses, as well as more complex combinations such as Wasserstein GAN and VGG-based LPIPS loss. Quantitative results show that the models trained with VGG-based LPIPS provide overall slightly better results for most evaluation metrics. Qualitative results exhibit more vivid colors when trained with Wasserstein GAN plus the L2 loss or again with the VGG-based LPIPS. Finally, the convenience of quantitative user studies is also discussed to overcome the difficulty of properly assessing on colorized images, notably for the case of old archive photographs where no ground truth is available.

Keywords

Image colorization · Deep learning · Loss functions · Color spaces

Introduction

Color is acknowledged to be captured by the human visual system at the first milliseconds. Color perception allows to highly increase the perceived diversity of real scenes since more than 2 million colors are identified by humans. Besides, although humans are interested in color and have used it since the dawn of humanity, full comprehension of the chromatic aspect of color is still an open problem. Color images capturing a real scene indeed include both structure information (edges, textures) which is mostly contained in the so-called black-and-white component of the image and chromatic information which, when added to the achromatic black-and-white component, provides the rich color vision of the scene image. This achromatic and chromatic dichotomy is also palpable in works of art: artists often slide between drawing strength from the massive richness of the variations on black and white and exploiting the infinite power of color, even using it as an actor on its own.

Image colorization aims to hallucinate the missing color information of a given grayscale image by, as in the case of learning-based methods, directly learning a mapping from the grayscale to the color information by minimizing a chosen objective function. The objective function favors the desired properties the estimated colorization should satisfy. Due to the ill-posed nature of the problem, in most cases, one does not aim to recover the actual ground truth color – that is, the real color of the actual scene captured in the grayscale image – but rather to produce a plausible colorization for a human observer. Accordingly, choosing the right way to train such networks is not trivial. The network could end up penalizing a good solution far away from the ground truth data or estimating an average of all possible correct solutions. Alternatively, instead of directly learning the per-pixel chrominance information, some methods learn a per-pixel color distribution to, afterward, sample from it the color at each pixel. In principle, this could encourage the mapping to be one to many, which can be desirable. However, how to properly capitalize and train such networks to account for the different possible solutions having, both, geometric and semantic meaning remains an open problem.

This chapter aims to analyze the influence of the optimized objective function on the results of automatic deep learning methods for image colorization. Some of the chosen objective functions favor colorization results perceptually as plausible as the associated color ground truth image, no matter the pixel-wise color differences between them, while others aim to recover the ground truth values. To the best of our knowledge, there is currently no study about their influence over the results.

Additionally, besides the selected objective function used to train the model, another important choice is the color space we will work on. Almost all colorization methods work either on a Luminance–Chrominance or on the RGB color space. Only a few of them, such as Larsson et al. (2016), work on Hue-Saturation-based color spaces. Thus, together with this chapter, another chapter of the current handbook, called ▶ Chap. 22, "Influence of Color Spaces for Deep Learning Image Colorization" has been added for completeness. It focuses on the influence of color spaces. It also contains a more detailed review of the literature on image colorization and of the used datasets. We refer the reader to the mentioned chapter for these reviews.

The rest of this chapter is organized as follows. In Section "Losses in the Colorization Literature," we first make a review of the loss functions that have been used in the field of image colorization while connecting them with the colorization-related works. Section "Proposed Colorization Framework" details the framework used to analyze the influence of the different losses, including both the chosen architecture and evaluation metrics. Finally, in Section "Experimental Analysis," we present quantitative and qualitative colorization results on a classical image dataset, and Section "Generalization to Archive Images" shows extended results on archive images. Conclusions can be found in Section "Conclusion."

Losses in the Colorization Literature

The objective loss function summarizes the desired properties that we want the estimated outcome to satisfy. In this section, we review the losses and evaluation methods used in the literature.

Along this chapter, a color image is assumed to be defined on a bounded domain Ω, a subset of $\mathbb{R}^2$. With a slight abuse of notation, we will both use the same notation to refer to the continuous setting, where $\Omega \subset \mathbb{R}^2$ is an infinite resolution image domain and $u : \Omega \to \mathbb{R}^C$, and to the discrete setting, where Ω represents a discrete domain given by a grid of $M \times N$ pixels, $M, N \in \mathbb{N}$, and u is a function defined on this discrete Ω and with values in $\mathbb{R}^C$. In the latter case, u is usually given by a real-valued matrix of size $M \times N \times C$ representing the image values. Finally, C

Table 1 Losses used to train deep learning methods for image colorization. CE stands for cross-entropy and KL for Kullback–Leibler divergence

	Using GANs						Histogram prediction			User guided	Diverse			Object aware			Survey
	Cheng et al. (2015)	Iizuka et al. (2016)	Vitoria et al. (2020)	Nazeri et al. (2018)	Cao et al. (2017)	Yoo et al. (2019)	Antic (2019)	Larsson et al. (2016)	Zhang et al. (2016)	Mouzon et al. (2019)	Zhang et al. (2017)	He et al. (2018)	Deshpande et al. (2017)	Guadarrama et al. (2017)	Royer et al. (2017)	Kumar et al. (2021)	Su et al. (2020)	Pucci et al. (2021)	Kong et al. (2021)	winner of Gu et al. (2019)
MAE				•										•						•
smooth-L1						•					•	•				•				
MSE	•	•	•	•				•	•	•			•				•	•		•
GANs			•	•	•	•	•											•		
KL on distributions						•														
CE on distributions									•	•	•							•		
KL for classification			•																	
CE for classification				•															•	
neg log-likelihood															•	•	•	•		
Perceptual							•					•								

can be either equal to 3 if u is a color image or equal to 2 if the goal is to reconstruct the two chrominance channels and, thus, the input grayscale image is not modified during colorization.

Error-Based Losses

In the following, the different losses used in the literature of image colorization are described and related to some representative works that capitalize on them. Table 1 summarizes it.

MSE or squared L2 loss. Given two functions u and v defined on Ω and with values in $\mathbb{R}^C$, $C \in \mathbb{N}$, the so-called Mean Square Error (MSE) between u and v is defined as the squared L2 loss of their difference. That is

$$\mathrm{MSE}(u, v) = \|u - v\|^2_{L^2(\Omega;\mathbb{R}^C)} = \int_\Omega \|u(x) - v(x)\|^2_2 dx, \qquad (1)$$

where $\|\cdot\|_2$ denotes the Euclidean norm in $\mathbb{R}^C$. In the discrete setting, it is equal to the sum of the square differences between the image values, that is

$$\text{MSE}(u, v) = \sum_{i=1}^{M} \sum_{j=1}^{N} \sum_{k=1}^{C} (u_{i,j,k} - v_{i,j,k})^2. \quad (2)$$

It has been extensively used for image colorization methods (Cheng et al. 2015; Larsson et al. 2016; Zhang et al. 2016; Iizuka et al. 2016; Isola et al. 2017; Nazeri et al. 2018; Vitoria et al. 2020) (see also Table 1), where $C = 3$ if u and v are color images (usually the predicted and the ground truth data) or $C = 2$ in the case that u and v are chrominance images. Although while the training with this loss can lead to a more stable solution, it is not robust to outliers in the data and penalizes large errors while being more tolerant to small errors.

MAE or L1 loss with l^1-coupling. The Mean Absolute Error is defined as the L1 loss with l^1-coupling, that is

$$\text{MAE}(u, v) = \int_{\Omega} \|u(x) - v(x)\|_{l^1} dx = \int_{\Omega} \sum_{k=1}^{C} |u_k(x) - v_k(x)| dx. \quad (3)$$

In the discrete setting, it coincides with the sum of the absolute differences $|u_{i,j,k} - v_{i,j,k}|$. Some authors use a l^2-coupled version of it:

$$\text{MAE}^c(u, v) = \sum_{i=1}^{M} \sum_{j=1}^{N} \sqrt{\sum_{k=1}^{C} (u_{i,j,k} - v_{i,j,k})^2}. \quad (4)$$

Both MAE and MAE^c losses are robust to outliers.

To ease the non-differentiability issue in the minimization of the MAE and MAE^c, some authors use the **Smooth L1** or **Huber loss**. It is simply defined by substituting the absolute value $|\cdot|$ in (3) by

$$l_H(g) = \begin{cases} \frac{1}{2}g^2 & \text{if } |g| \leq \delta \\ \delta(|g| - \frac{1}{2}\delta) & \text{otherwise} \end{cases} \quad (5)$$

for $g \in \mathbb{R}$. Several works Su et al. (2020), Cao et al. (2017), Yoo et al. (2019), Zhang et al. (2017), He et al. (2018), and Guadarrama et al. (2017) use MAE, MAE^c, or Smooth L1 losses either alone or combined with other losses (cf. Table 1).

Previous error-based losses aim to find a solution close to the ground truth. This is counterproductive to the idea that image colorization has multiple possible solutions. Additionally, both metrics are poorly related to perceptual quality. Nonetheless, both metrics are the most used ones to train deep learning approaches. In Section "Experimental Analysis," we present some numerical results together with a comparison with other kinds of losses.

Aiming at favoring a solution keeping from the ground truth not the exact values but more perceptual or style features, the following error losses have been proposed and used for colorization purposes.

Feature Loss. The feature reconstruction loss (Gatys et al. 2016; Johnson et al. 2016) is a perceptual loss that encourages images to have similar feature representations as the ones computed by a pretrained network, denoted here by Φ. Let $\Phi_l(u)$ be the activation of the l-th layer of the network Φ when processing the image u; if l is a convolutional layer, then $\Phi_l(u)$ will be a feature map of size $C_l \times W_l \times H_l$. The *feature reconstruction* loss is the normalized squared Euclidean distance between feature representations, that is

$$\mathcal{L}^l_{\text{feat}}(u, v) = \frac{1}{C_l W_l H_l} \left\| \Phi_l(u) - \Phi_l(v) \right\|_2^2. \tag{6}$$

It penalizes the output reconstructed image when it deviates in feature content from the target.

In our experimental analysis in Section "Experimental Analysis," we analyze the influence of the perceptual loss given by the VGG-based LPIPS (21), which was introduced in Ding et al. (2021) as a generalization of the perceptual loss above (Johnson et al. 2016).

Generative Adversarial Network-Based Losses

Aiming to favor more diverse and perceptually plausible colorization results, losses based on *Generative Adversarial Networks* (GANs) (Goodfellow et al. 2014) have been introduced in the colorization literature (Isola et al. 2017; Cao et al. 2017; Nazeri et al. 2018; Yoo et al. 2019; Vitoria et al. 2020). GANs are a kind of generative methods where the goal is to learn the probability distribution of the considered dataset by learning to generate new samples as if they where coming from that dataset. In the case of GANs, the learning is done by an adversarial learning strategy.

Vanilla GAN. The first GAN proposal by Goodfellow et al. (2014) is based on a game theory scenario between two networks competing one against another. The first network called generator, denoted by G, aims to generate samples of data as similar as possible to the ones of real data $\mathcal{P}_r$. The second network, called discriminator, aims to classify between real and generated data. To do so, the discriminator, denoted here by D, is trained to maximize the probability of correctly distinguishing between real examples and samples created by the generator. On the other hand, G is trained to fool the discriminator by generating realistic examples. The adversarial loss of the vanilla GAN is defined as

$$\mathcal{L}_{\text{adv}}(G_\theta, D_\phi) = \mathbb{E}_{u \sim \mathcal{P}_r}[\log D_\phi(u)] + \mathbb{E}_{v \sim \mathcal{P}_{G_\theta}}[\log(1 - D_\phi(v))], \tag{7}$$

and the min-max adversarial optimization problem is

$$\min_{G_\theta} \max_{D_\phi} \mathcal{L}_{\text{adv}}(G_\theta, D_\phi). \tag{8}$$

Wasserstein GAN. Although vanilla GANs have achieved good results in many domains, they have some drawbacks like convergence, vanishing gradients, and mode collapse problems. Therefore, some modifications from the original GAN have been proposed. For example, the *Wasserstein GAN* (WGAN), proposed by Arjovsky et al. (2017), replaces the underlying Jensen–Shannon divergence from the original proposal with the Wasserstein−1 distance (or Earth Mover distance) between two probability distributions. Then, the WGAN loss, $\mathcal{L}_{\text{adv,wgan}}$, and WGAN optimization problem can be defined as

$$\min_{G_\theta} \max_{D_\phi \in \mathcal{D}} \mathcal{L}_{\text{adv,wgan}}(G_\theta, D_\phi) = \min_{G_\theta} \max_{D_\phi \in \mathcal{D}} \left(\mathbb{E}_{u \sim \mathcal{P}_r}[D_\phi(u)] - \mathbb{E}_{v \sim \mathcal{P}_{G_\theta}}[D_\phi(v)] \right) \tag{9}$$

where $\mathcal{D}$ denotes the set of 1-Lipschitz functions. To enforce the 1-Lipschitz condition, in Gulrajani et al. (2017), the authors propose a *Gradient Penalty* (GP) term constraining the L2 norm of the gradient while optimizing the original WGAN during training. The resulting loss for the WGAN-GP can be defined as

$$\min_{G_\theta} \max_{D_\phi} \left(\mathbb{E}_{u \sim \mathcal{P}_r}[D_\phi(u)] - \mathbb{E}_{v \sim \mathcal{P}_{G_\theta}}[D_\phi(v)] - \lambda \mathbb{E}_{\widehat{u} \sim \widehat{\mathcal{P}}}[(\|\nabla_{\widehat{u}} D(\widehat{u})\|_2 - 1)^2] \right) \tag{10}$$

where $\widehat{u}$ is a sample defined as

$$\widehat{u} = tu + (1-t)v,$$

with t uniformly sampled in $[0, 1]$ and $u \sim \mathcal{P}_r, v \sim \mathcal{P}_{G_\theta}$. The last term in (10) provides a tractable approximation to enforce the norm of the gradient of D to be less than 1. The authors of Gulrajani et al. (2017) motivated it by a theoretical result showing that the optimal discriminator D contains straight lines connecting samples in the ground truth space and samples in the space of generated data. Moreover, they experimentally observed that this technique exhibits good performance in practice. Finally, let us observe that the minus before the gradient penalty term in (10) corresponds to the fact that the WGAN min-max objective (10) implies maximization with respect to the discriminator parameters.

In our experimental results in Section "Experimental Analysis," we will present a comparison of several losses, and we will include a combination of WGAN loss and a VGG-based LPIPS loss. To the best of our knowledge, it has not been proposed yet.

Distribution-Based Losses

As mentioned in Section "Introduction," some authors colorize an image after learning a certain probability distribution such as a color probability distribution (Larsson et al. 2016; Zhang et al. 2016, 2017; Royer et al. 2017), or a distribution of semantic classes (Vitoria et al. 2020), or directly using it for classification purposes (Iizuka et al. 2016). The remaining of this section describes the corresponding measures of the difference between two probability distributions that have been used in the mentioned related work (see also Table 1).

Kullback–Leibler loss. The *Kullback–Leibler* (KL) loss is the directed divergence between two probability densities ρ and $\widehat{\rho}$ defined in the same space $\mathcal{Y}$. It is defined as the relative entropy from $\widehat{\rho}$ to ρ which, for discrete probability densities, is given by

$$KL(\rho||\widehat{\rho}) = \sum_{y \in \mathcal{Y}} \rho(y) \log \frac{\rho(y)}{\widehat{\rho}(y)}. \tag{11}$$

Here, ρ is usually taken as the ground truth density (sometimes as a Dirac delta or a one-hot vector on the ground truth value, or a regularized one) and $\widehat{\rho}$ the predicted one.

Some works predict a color distribution density per pixel where the color bins are associated to a fixed 2D grid in a chrominance space (e.g., CIE Lab in Zhang et al. 2016). In Zhang et al. (2016), the final color of each pixel in the inferred color image is given by the expectation (sum over the color bin centroids weighted by the histogram). Others, such as Larsson et al. (2016), learn Hue-Saturation-based color distributions. More precisely, Larsson et al. (2016) learn the marginal distributions $\widehat{\rho}^{Hue}$ and $\widehat{\rho}^{Chroma}$ of Hue and Chroma, per pixel, where chroma is related to saturation by the formula Saturation=$\frac{Chroma}{Value}$ and Value=Luminance+$\frac{Chroma}{2}$. They use the KL divergence to measure the deviation between the estimated distributions and the ground truth ones. The marginal ground truth distributions, ρ^{Chroma}, ρ^{Hue}, are again defined as either a one-hot vector on the bin associated to the ground truth color or regularized version of it. Then, their loss is

$$\mathcal{L}(\rho||\widehat{\rho}) = KL(\rho^{Chroma}||\widehat{\rho}^{Chroma}) + \lambda c KL(\rho^{Hue}||\widehat{\rho}^{Hue}) \tag{12}$$

where $c \in [0, 1]$ is the ground truth Chroma of the considered pixel and $\lambda = 5$ in Larsson et al. (2016). The authors introduce this weight depending on the Chroma multiplying the KL term on ρ^{Hue} to avoid Hue instability issues when Chroma approaches zero. For inference and to sample a color value per pixel from the estimated marginal distributions, they experimentally tested that a median-based selection (a periodically modified version in the case of Hue) gives the best results.

Besides, Vitoria et al. (2020) uses the KL loss (11) to learn, for each image, the distribution density of semantic classes, for a fixed number of classes. It provides information about the semantic content and objects present in the image.

In particular, they define the ground truth probability density ρ of semantic classes to be the output distribution of a pre-trained VGG-16 model applied to the grayscale image and $\widehat{\rho}$ the estimated class distribution density.

Cross-Entropy Loss. Cross-entropy loss is used for classification problems, and it is sometimes referred to as logistic loss. For discrete densities, it is defined as

$$CE(\rho, \widehat{\rho}) = -\sum_{y \in \mathcal{Y}} \rho(y) \log \widehat{\rho}(y), \tag{13}$$

where, again, ρ is usually taken as the ground truth density and $\widehat{\rho}$ the predicted one. In the classification context, ρ is often a one-hot vector equal to 1 on the ground truth class, or a regularized version of it. Let us also note, from (11) and (13), that there is a relationship between the Kullback–Leibler and the cross-entropy losses given by

$$CE(\rho, \widehat{\rho}) = E(\rho) + KL(\rho || \widehat{\rho}), \tag{14}$$

where $E(\rho)$ denotes the entropy of ρ.

Cross-entropy is used as a classification loss in Iizuka et al. (2016) where the network is trained on a large-scale dataset. The architecture is made of two encoding networks that learn local and global features and a decoder that learns the color image from these features. The classification loss is used to guide the training of the global feature network from image label estimation. It is combined with a MSE loss that compares estimated color image with the ground truth.

In Zhang et al. (2016, 2017), CE is applied on color distributions. Zhang et al. (2016) treat the colorization problem as multinomial classification by learning a mapping from the input grayscale image to a probability distribution over possible discrete chrominance values. CE compares the estimated distribution with the one of the ground truth. Zhang et al. (2017) build upon this framework and incorporate user interaction. Finally, Mouzon et al. (2019) and Pierre and Aujol (2020) stem from the resulting distributions from Zhang et al. (2016) that, in a subsequent step, are incorporated in a variational approach (Pierre et al. 2015).

Log-likelihood Maximization for Diversity. Some works propose to generate several possible colorizations, for the same input gray-level image, by sampling over possible color distributions that are often learned by maximizing the log likelihood conditioned to the grayscale image (Guadarrama et al. 2017; Royer et al. 2017; Kumar et al. 2021).

The work *Pixcolor: Pixel recursive colorization* (Guadarrama et al. 2017) colorizes an image by first learning the color distribution of images conditioned to a grayscale input. It stems from autoregressive models (Van Oord et al. 2016; Oord et al. 2016; Chen et al. 2018) that exploit the fact that a color probability distribution $p(u)$ can be in principle learned by choosing an order of the data variables $u = (u_1, u_2, \ldots, u_n) \in X$, associated with the color values of a discrete color image u at its n pixels (where X denotes the space of discrete color images), and exploiting the fact that the joint distribution can be decomposed as

$$p(u) = p(u_1, u_2, \ldots, u_n) = p(u_1) \prod_{i=2}^{n} p(u_i | u_1, \ldots, u_{i-1}). \tag{15}$$

As claimed by Guadarrama et al. (2017), this ordering tends to capture dependencies between pixels to ensure that, at inference, colors will be consistently selected. By working in the $YCbCr$ color space and by discretizing the Cb and Cr channels separately into 32 bins, they propose to model the conditional distribution of u given the grayscale image Y by

$$p(u^{b,r}|Y) = \prod_i p(u_i^r | u_1^{b,r}, \ldots, u_{i-1}^{b,r}, Y) p(u_i^b | u_i^r, u_1^{b,r}, \ldots, u_{i-1}^{b,r}, Y), \tag{16}$$

where u_i^b denotes the Cb value for pixel i, u_i^r its Cr value, and $u_i^{b,r}$ its (Cb, Cr) chrominance. They train the model using maximum likelihood, with a cross-entropy loss per pixel. Afterward, they perform high-resolution refinement to upscale the chrominance image at the dimensions of the original grayscale image.

In Royer et al. (2017), a feed-forward network followed by an autoregressive network is used to predict for each pixel a probability distribution over all possible chrominances conditioned to the luminance. They work in the Lab color space. $p(u^{a,b}|L)$ is factorized again as in (15) and (16) as the product of terms of the form $p(u_i^{a,b} | u_1^{a,b}, \ldots, u_{i-1}^{a,b}, L)$, which are learned on a set of training images D by minimizing negative log-likelihood of the chrominance channels in the training data:

$$\arg\min -\sum_{u \in D} \log p(u^{a,b}|L). \tag{17}$$

L and $u^{a,b}$ denote the luminance and chrominance channels, respectively. In order to speed up the learning, Royer et al. (2017) approximate each distribution $p(u_i^{a,b} | u_1^{a,b}, \ldots, u_{i-1}^{a,b}, L)$ with a mixture of ten logistic distributions.

Kumar et al. (2021) also address the generation of multiple outputs for a given grayscale image, in this case using transformers. They use a conditional autoregressive transfomer (a conditional variant of Axial Transformer particular self-attention with Ho et al. 2019) to first produce a low-resolution colorization of the grayscale image (both spatial and color low resolution) that is then upsampled with two parallel networks for upsampling the spatial and color resolutions. The model is trained to minimize the negative log-likelihood of the distributions that are estimated by each network.

Several works combine distribution-based losses with error-based ones. For instance, aiming to learn the distribution of color images conditioned to a grayscale version $p(u|L)$, Deshpande et al. (2017) uses a VAE approach and log-likelihood maximization to learn a low-dimensional (latent variables) embedding of color images, combined with error losses on the output of the decoder that favor to keep color specificity (with a L2 loss that compares the projection of the generated color and ground truth images along a top-k principal components), colorfulness (with a loss that encourages rare colors to appear), and similar gradients to the ground truth color image (with a loss that compares the gradients of the generated images with the ones of the ground truth). Moreover, the conditional distribution $p(z|L)$ of the latent variables given the grayscale image is assumed to be a Gaussian mixture and learned minimizing the conditional negative log likelihood.

The authors of Pucci et al. (2021) capitalize on capsule networks (Sabour et al. 2017) to learn a color distribution over a set of quantized colors. To that goal, they use a weighted cross-entropy loss where the weights are used to weight more rare colors, with a MSE loss on the (a, b) channels.

Kong et al. (2021) propose a multitask network in an adversarial manner that uses a MSE loss on hue, saturation, and lightness channels to perform colorization and a cross-entropy loss to learn a semantic segmentation.

Finally, it is worth mentioning that Ding et al. (2021) compare different cost functions to train a deep neural network on four low-level vision tasks, denoising, blind image deblurring, single image super resolution, and lossy image compression, although it is not done for image colorization.

In the following sections, we will present a comparison of the different loss functions for the colorization task. To do so, we propose a baseline colorization network architecture (presented in the next section) and show experimental results for the different loss functions on the same dataset.

Proposed Colorization Framework

In this section, we present the framework used to study the influence of the chosen objective loss on the estimated images colorization results. First we detail the architecture and second the dataset used for both training and testing. Note that the same architecture and training procedure is used in ▶ Chap. 22, "Influence of Color Spaces for Deep Learning Image Colorization" of this handbook.

Detailed Architecture

The architecture used in our experiments is an encoder–decoder U-Net composed of five stages. Figure 1 displays a summary of the whole architecture. All convolutional blocks are composed of two 2D convolutional layers with kernels of kernel size equal to 3×3, each one followed by 2D batch normalization and a ReLU activation. For the encoder, downsampling is done by using a max pooling operator after each convolutional block. After downsampling, the number of filters is doubled in the following block. For the decoder, upsampling is done by using 2D transpose convolutions (with 4×4 kernels with stride 2). At a given stage, the corresponding

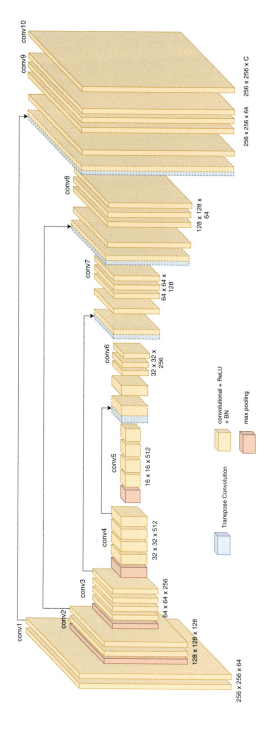

Fig. 1 Summary of the baseline U-Net architecture used in our experiments. It outputs a $256 \times 256 \times C$ image, where C stands for the number of channels, being equal to 2 when estimating the missing chrominance channels and to 3 when estimating the RGB components

21 Analysis of Different Losses for Deep Learning Image Colorization

Table 2 Detailed architecture and output resolution for each block

Layer type	Output resolution
Input	3 × H × W
Conv1 + Max-pooling	64 × H/2 × W/2
Conv2 + Max-pooling	128 × H/4 × W/4
Conv3 + Max-pooling	256 × H/8 × W/8
Conv4 + Max-pooling	512 × H/16 × W/16
Conv5 + Conv. Transpose (I)	512 × H/8 × W/8
Conv6 + Conv. Transpose (II)	256 × H/4 × W/4
Conv7 + Conv. Transpose (III)	128 × H/2 × W/2
Conv8 + Conv. Transpose (IV)	64 × H × W
Conv9	64 × H × W
Conv10	C × H × W

encoder and decoder blocks are linked with skip connections: feature maps from the encoder are concatenated with the ones from the corresponding upsampling path and fused using 1 × 1 convolutions. More details can be found in Table 2.

The encoder architecture is identical to the CNN part of a VGG network (Simonyan and Zisserman 2015). It allows us to start from pretrained weights initially used for ImageNet classification.

The training settings are described as follows:

- Optimizer: Adam
- Learning rate: 2e-5.
- Batch size: 16 images (10–11 GB RAM on Nvidia Titan V).
- All images are resized to 256 × 256 for training which enables using batches. In practice, to keep the aspect ratio, the image is resized such that the smallest dimension matches 256. If the other dimension remains larger than 256, we then apply a random crop to obtain a square image. Note that the random crop is performed using the same seed for all trainings.

More details regarding this framework are given in ▶ Chap. 22, "Influence of Color Spaces for Deep Learning Image Colorization".

Quantitative Evaluation Metrics Used in Colorization Methods

For the last 20 years, colorization methods have mostly been evaluated with MAE, MSE, Peak Signal-to-Noise Ratio (PSNR), and Structural Similarity Index (SSIM) metrics (Wang et al. 2004).

In the context of colorization, the PSNR measures the ratio between the maximum value of a color target image $u : \Omega \to \mathbb{R}^C$ and the Mean Square Error (MSE) between u and a colorized image $v : \Omega \to \mathbb{R}^C$ with $\Omega \in \mathbb{Z}^2$ a discrete grid of size $M \times N$. That is

$$\text{PSNR}(u, v) = 20 \log_{10}(\max u)$$
$$- 10 \log_{10}\left(\frac{1}{CMN} \sum_{k=1}^{C} \sum_{i=1}^{M} \sum_{j=1}^{N} (u(i, j, k) - v(i, j, k))^2\right), \tag{18}$$

where $C = 3$ when working in the RGB color space and $C = 2$ in any luminance–chrominance color space as YUV, Lab, and YCbCr. The PSNR score is considered as a reconstruction measure tending to favor methods that will output results as close as possible to the ground truth image in terms of the MSE.

SSIM intends to measure the perceived change in structural information between two images. It combines three measures to compare images, color (l), contrast (c), and structure (s):

$$\text{SSIM}(u, v) = l(u, v)c(u, v)s(u, v) = \frac{(2\mu_u \mu_v) + c_1}{\mu_u^2 + \mu_v^2 + c_1} \frac{(2\sigma_u \sigma_v + c_2)}{\sigma_u^2 + \sigma_v^2 + c_2} \frac{(\sigma_{uv} + c_3)}{\sigma_u \sigma_v + c_3} \tag{19}$$

where μ_u (resp. σ_u) is the mean value (resp. the variance) of image u values and σ_{uv} the covariance of u and v. c_1, c_2, and c_3 are regularization constants that are used to stabilize the division for images with mean or standard deviation close to zero.

More recently, other perceptual metrics based on deep learning have been proposed: the Fréchet Inception Distance (FID) (Heusel et al. 2017) and a Learned Perceptual Image Patch Similarity (LPIPS) (Zhang et al. 2018). They have been widely used in image editing for their ability to correlate well with human perceptual similarity. FID (Heusel et al. 2017) is a quantitative measure used to evaluate the quality of the outputs' generative model and which aims at approximating human perceptual evaluation. It is based on the Fréchet distance (Dowson and Landau 1982) which measures the distance between two multivariate Gaussian distributions. FID is computed between the feature-wise mean and covariance matrices of the features extracted from an Inception v3 neural network applied to the input images (μ_r, Σ_r) and those of the generated images (μ_g, Σ_g):

$$\text{FID}((\mu_r, \Sigma_r), (\mu_g, \Sigma_g)) = \|\mu_r - \mu_g\|_2^2 + Tr(\Sigma_r + \Sigma_g - 2\Sigma_r \Sigma_g)^{1/2}. \tag{20}$$

LPIPS (Zhang et al. 2018) computes a weighted L2 distance between deep features of a pair of images u and v:

$$\text{LPIPS}(u, v) = \sum_l \frac{1}{H_l W_l} \sum_{i=1}^{H_l} \sum_{j=1}^{W_l} \|\omega_l \odot (\Phi_l(u)_{i,j} - \Phi_l(v)_{i,j})\|_2^2, \tag{21}$$

where H_l (resp. W_l) is the height (resp. the width) of feature map Φ_l at layer l and ω_l are weights for each features. Note that features are unit-normalized in the channel dimension.

Other quantitative metrics can be found in the literature for image colorization. Accuracy (Nazeri et al. 2018) measures the ratio between the number of pixels that have the same color information as the source and the total number of pixels. Raw accuracy (AuC) (Zhang et al. 2016) computes the percentage of predicted pixel colors within a threshold of the L2 distance from the ground truth in ab color space. The result is then swept across thresholds from 0 to 150 to produce a cumulative mass function. Deshpande et al. (2017) evaluate colorfulness as the MSE on histograms. Royer et al. (2017) verify if the framework produces vivid colors by computing the average perceptual saturation (Lübbe 2010). Other works evaluate the capability of a classification network to infer the right class to the generated image (Zhang et al. 2016; He et al. 2018). Zhang et al. (2016) feed the generated image to a classification network and observe if the classifier performs well.

Table 3 Evaluation metrics used by deep learning methods for image colorization

	Quantitative							User study		
	L1/MAE	L2/MSE	PSNR	SSIM	LPIPS	FID	Other	AMT Fooling Rate	Naturalness	Other
Cheng et al. (2015)				•						
Iizuka et al. (2016)							•		•	
Using GANS										
Vitoria et al. (2020)			•						•	
Nazeri et al. (2018)	•						•			
Cao et al. (2017)		•	•							•
Yoo et al. (2019)					•					•
Histograms Prediction										
Larsson et al. (2016)		•	•				•			
Zhang et al. (2016)							•	•		
User Guided										
Zhang et al. (2017)			•					•		
He et al. (2018)			•				•	•		
Diverse										
Deshpande et al. (2017)		•								•
Guadarrama et al. (2017)				•				•		
Royer et al. (2017)								•		•
Kumar et al. (2021)						•		•		
Object Aware										
Su et al. (2020)			•	•	•					
Pucci et al. (2021)			•		•					
Kong et al. (2021)			•	•			•			
Survey										
Gu et al. (2019)			•	•						•

Note that all models that are trained with a L2 loss will more likely get better PSNR or MSE as the L2 loss is correlated with the evaluation.

Table 3 summarizes the quantitative evaluation metrics more generally used in the literature of image colorization. In our experiments, we choose to rely on the more generally used and more recent ones, namely, L1 (MAE), L2 (MSE), PSNR, SSIM, LPIPS, and FID.

Experimental Analysis

To compare the influence of the objective loss in the resulting colorization results, we train the network described in Section "Proposed Colorization Framework" by changing the objective loss. In particular, we train the network with the L1 loss, the L2 loss, the VGG-based LPIPS, the combination of WGAN plus L2 losses, and the combination of WGAN and VGG-based LPIPS. To the best of our knowledge, the combination of the VGG-based LPIPS loss with a WGAN training procedure is novel and has not been proposed in the recent literature.

For each of these losses, depending on the chosen color space, we estimate:

- either the two (a, b) chrominance channels given the luminance channel L as input;
- or the three (R, G, B) color channels given a grayscale image as input.

In this section, we present a quantitative and qualitative comparison for all of these combinations. Note that to compute the VGG-based LPIPS loss, the output colorization always has to be converted to RGB (in a differentiable way), even for Lab color space, because this loss is computed with a pre-trained VGG expecting RGB images as input. To this end, we have used the Kornia implementation of differentiable color space conversions (Riba et al. 2020).

Throughout our experiments, we use the COCO dataset (Lin et al. 2014), containing various natural images of different sizes. COCO is divided into three sets that approximately contain 118k, 5k, and 40k images that, respectively, correspond to the training, validation, and test sets. Note that we carefully remove all grayscale images, which represent around 3% of the overall amount of each set. Although larger datasets such as ImageNet have been regularly used in the literature, COCO offers a sufficient number and a good variety of images so we can efficiently train and compare numerous models. While the training is done on batches of square 256×256 images, for testing, we apply the network to images at their original resolution.

Quantitative Evaluation

Table 4 shows the quantitative results comparing five losses, namely, the L1 loss, the L2 loss, the VGG-based LPIPS, the combination of WGAN plus L2 losses, and

Table 4 Quantitative evaluation of colorization results for different loss functions. Metrics are used to compare ground truth to every images in the 40k test set. Best and second best results by column are in bold and italics, respectively

Color space	Loss function	MAE ↓	MSE ↓	PSNR ↑	SSIM ↑	LPIPS ↓	FID ↓
Lab	L1	0.04407	0.00589	22.3020	**0.9268**	0.1587	8.8109
Lab	L2	0.04488	0.00585	22.3283	*0.9250*	0.1613	8.1517
Lab	LPIPS	**0.04374**	**0.00566**	**22.4699**	0.9228	**0.1403**	*3.2221*
Lab	WGAN+L2	0.04459	0.00582	22.3512	0.9243	0.1609	7.6127
Lab	WGAN+LPIPS	*0.04383*	*0.00568*	*22.4541*	0.9223	*0.1406*	**3.1045**
RGB	L1	**0.04385**	0.00587	22.3119	**0.9268**	0.1583	8.0125
RGB	L2	*0.04458*	*0.00587*	22.3136	*0.9255*	0.1606	7.4223
RGB	LPIPS	0.04573	**0.00577**	**22.3892**	0.9196	**0.1429**	*3.0576*
RGB	WGAN+L2	0.05256	0.00651	21.8667	0.8559	0.2469	15.4780
RGB	WGAN+LPIPS	0.04901	0.00679	21.6806	0.9137	*0.1495*	**2.6719**

the combination of WGAN and VGG-based LPIPS (denoted in Table 4 as L1, L2, LPIPS, WGAN+L2, and WGAN+LPIPS, respectively). The first five rows display this assessment when the used color space is Lab (i.e., the model estimates the two ab chrominance channels), while for the last five rows, the used color space is RGB (i.e., the model estimates the three RGB color channels). In particular, let us remark that the quantitative evaluations are always performed in the final RGB color space. Thus, even when the model is trained to estimate the ab chrominance channels, the resulting Lab color image is converted to the RGB color space to compute the evaluation metrics.

From the results in Table 4, we observe that for the analyzed dataset, the models trained with the VGG-based LPIPS loss function provide overall better quantitative results, for both Lab and RGB color spaces. This is especially true for the perceptual metrics LPIPS and FID, as they are strongly correlated to this loss function. The fact that the VGG-based LPIPS training loss is computed on RGB color space (as this loss is computed with a pre-trained VGG expecting RGB images as input) and also is a quantitative result might be related to the performance (see also ▶ Chap. 22 "Influence of Color Spaces for Deep Learning Image Colorization"). In the same spirit, we can observe a slight correlation between the used training loss and the quantitative metric. For instance, when training with L1, MAE results are better. However, we can see that L2 loss is not at the top in any of the metrics, while we could have expected in the case of MSE or PSNR, but this is not the case.

Nevertheless, no strong tendency clearly emerges from this table: for many metrics, the different losses do not differ so much from one another and could be in the margin of error. From our analysis, we hypothesize that, apart from the chosen objective function, the network architecture design, and the training process, may play a very important role as a prior on the colorization operator. Further analysis will be done on that matter.

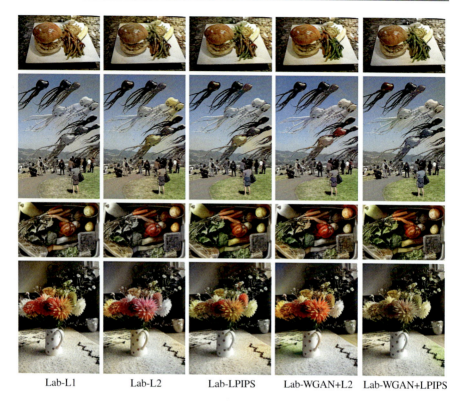

Fig. 2 Examples where multiple objects are in the same image. Five losses are compared, namely, L1, L2, LPIPS, WGAN+L2, and WGAN+LPIPS. The used color space is Lab for all the cases (i.e., the model estimates two ab chrominance channels)

Finally, let us mention the importance of user-based quantitative studies to properly assess colorizations results. It is not just important in cases where no ground truth is available, such as the ones of old archive photographs, but also due to the fact that multiple colorizations are always possible. Several works propose different user-based metrics, e.g., *naturalness* or *fooling rate*. Nevertheless, efforts should be made on a widely accepted protocol and a widespread user study metric.

Qualitative Evaluation

Figures 2, 3, and 4 show a qualitative experimental comparison of the five losses, namely, L1, L2, VGG-based LPIPS, WGAN+L2, and WGAN+VGG-based LPIPS. In all cases, the models were trained on the Lab color space. Still, we recall that any model based on VGG-based LPIPS loss requires to convert the predicted image to the RGB color space in a differentiable way (i.e., with Kornia (Riba et al. 2020)).

21 Analysis of Different Losses for Deep Learning Image Colorization

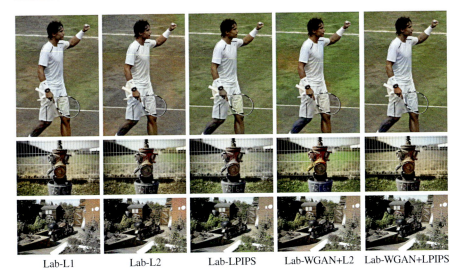

Fig. 3 Examples to evaluate shininess of the results. Five losses are compared, namely, L1, L2, LPIPS, WGAN+L2, and WGAN+LPIPS. The used color space is Lab for all the cases

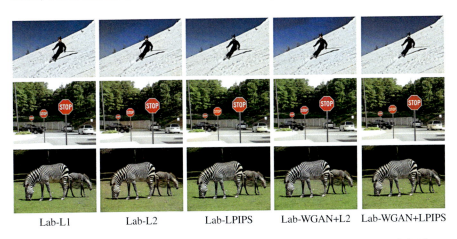

Fig. 4 Colorization results on images that contain objects have strong structures and that have been seen many times in the training set. Five losses are compared, namely, L1, L2, LPIPS, WGAN+L2, and WGAN+LPIPS. The used color space is Lab for all the cases

In Fig. 2, we can see some results obtained for each of the studied losses in images with multiple objects. We can observe that each loss brings slightly different colors to objects. Overall, VGG-based LPIPS and WGAN losses generate shinier and more colorful images (it can be seen, for instance, in the sky, grass, and vegetables), although we can observe colorful examples in the case of the L2 loss in the example of the flowers or vegetables. However, WGAN hallucinates more unrealistic colors as can be seen on the table or the wall on the image with a flower

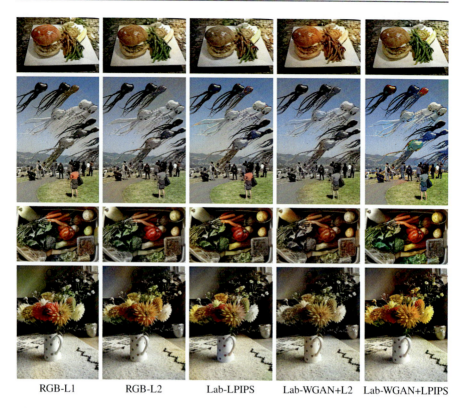

Fig. 5 Examples where multiple objects are in the same image. Five losses are compared, namely, L1, L2, LPIPS, WGAN+L2, and WGAN+LPIPS perceptual. The used color space is RGB for all the cases (i.e., the model estimates three RGB color channels)

of the last row of Fig. 2. This effect can be reduced by improving architecture and semantic features (e.g., Vitoria et al. 2020) or by introducing spatial localization (e.g., Su et al. 2020). Besides, by comparing the two last columns obtained with the models trained with the adversarial strategy WGAN combined with, respectively, the L2 or the VGG-based LPIPS, one can observe that WGAN+VGG-based LPIPS tends to homogenize colors (e.g., some of the balloons take similar color to the sky on the second row; the flowers on the fifth have grayish colors, more similar to the wall). WGAN+VGG-based LPIPS also tends to have less bleeding than WGAN+L2.

The generation of more vivid colors with VGG-based LPIPS and WGAN losses in also visible on Fig. 3. The grass and bushes are more green and look more natural. However, none of the losses give consistency to all the limbs of the tennis player on the first row (e.g., the right leg).

Figure 4 shows results on objects, here zebra and stop sign, with strong contours that were highly present in the training set. The colorization of this object is impressive for any loss. None of the losses manage to properly colorize the person

21 Analysis of Different Losses for Deep Learning Image Colorization 841

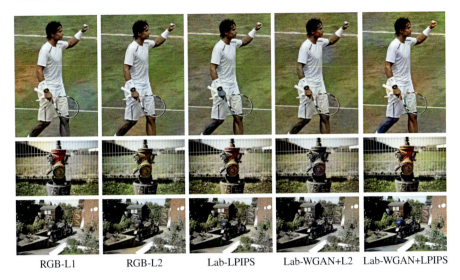

Fig. 6 Examples to evaluate shininess of the results. Five losses are compared, namely, L1, L2, LPIPS, WGAN+L2, and WGAN+LPIPS perceptual. The used color space is RGB for all the cases

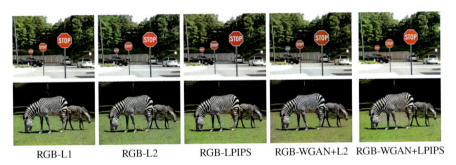

Fig. 7 Colorization results on images that contain objects which have strong structures and that have been seen many times in the training set. Four losses are compared, namely, L1, L2, LPIPS, and WGAN+L2. The used color space is RGB for all the cases

near the center car on the first row. This type of examples could be improved by learning high-level semantics on the image content.

Figures 5, 6, and 7 show an additional experimental comparison of five losses, namely, L1, L2, VGG-based LPIPS, WGAN+L2, and WGAN+VGG-based LPIPS, but when the network is trained to learn the three RGB color channels for all the cases. For these test images, more realistic and consistent results are obtained in general for this configuration. Let us notice from the results in these three figures that more colorful images are obtained compared to the ones of Figs. 2, 3, and 4, although less textured. Further analysis on the influence of the chosen color space can be found in ▶ Chap. 22, "Influence of Color Spaces for Deep Learning Image Colorization".

    Lab-L1   Lab-L2   Lab-LPIPS   Lab-WGAN+L2   Lab-WGAN+LPIPS

Fig. 8 Examples in original black and white Images. These colorization results have been obtained using the five networks trained, respectively, with L1, L2, LPIPS, WGAN+L2, and WGAN+LPIPS losses, and learning the two ab chrominance channels

21 Analysis of Different Losses for Deep Learning Image Colorization

Fig. 9 Examples in original black and white Images. These colorization results have been obtained using the five networks trained, respectively, with L1, L2, LPIPS, WGAN+L2, and WGAN+LPIPS losses, and learning the three RGB color channels

Generalization to Archive Images

Finally, in Figs. 8 and 9, we can see additional colorization results on real black and white images from the Pascal VOC dataset. Those results have been obtained using the network trained with the five different losses, respectively, with L1, L2, VGG-based LPIPS, WGAN+L2, and WGAN+VGG-based LPIPS. For Fig. 8, only the two ab chrominance channels are learned, while in Fig. 9, the three RGB color channels are learned. Again, none of the losses manage to consistently colorize the skin of all the people of the image at the first, second, and fourth rows of Fig. 8, although possibly it is slightly better when using perceptual and GAN losses. Notice that also in these cases, the colors are slightly more vivid, specially visible in the first two rows of Fig. 8. However, color inconsistency and failures in spatial localization appear, more visible in the first four rows. As mentioned, this effect can be reduced by introducing semantic information (e.g., Vitoria et al. 2020) or spatial localization (e.g., Su et al. 2020).

Conclusion

In this chapter, we have studied the role of loss functions on automatic colorization with deep learning methods. Using a fixed standard network, we have shown that the choice of the right loss does not seem to play a crucial role in the colorization results. We therefore argue that most efforts should be made on the influence of the architecture design, as it is related to the type of colorization operator one can expect to obtain. Indeed, in our analysis, we used a U-Net-based architecture which has shown to have a strong impact on the experimental results. For the employed architecture, the models including the VGG-based LPIPS loss function provide overall slightly better results, especially for the perceptual metrics LPIPS and FID. Likewise, the role of both architectures and losses for obtaining a real diversity of colorization results could be explored in future works.

Acknowledgments This study has been carried out with financial support from the French Research Agency through the PostProdLEAP project (ANR-19-CE23-0027-01) and from the EU Horizon 2020 research and innovation programme NoMADS (Marie Skłodowska-Curie grant agreement No 777826). The first and fourth authors acknowledge partial support by MICINN/FEDER UE project, ref. PGC2018-098625-B-I00, and RED2018-102511-T. This chapter was written together with another chapter of the current handbook, called ▶ Chap. 22, "Influence of Color Spaces for Deep Learning Image Colorization". All authors have contributed to both chapters.

References

Antic, J.: Deoldify. https://github.com/jantic/DeOldify (2019)
Arjovsky, M., Chintala, S., Bottou, L.: Wasserstein Generative Adversarial Networks. In: International Conference on Machine Learning, vol 70, pp. 214–223 (2017)

Cao, Y., Zhou, Z., Zhang, W., Yu, Y.: Unsupervised diverse colorization via Generative Adversarial Networks. In: Joint European Conference on Machine Learning and Knowledge Discovery in Databases, pp. 151–166 (2017)

Chen, X., Mishra, N., Rohaninejad, M., Abbeel, P.: Pixelsnail: an improved autoregressive generative model. In: International Conference on Machine Learning, pp. 864–872 (2018)

Cheng, Z., Yang, Q., Sheng, B.: Deep colorization. In: IEEE International Conference on Computer Vision, pp. 415–423 (2015)

Deshpande, A., Lu, J., Yeh, M.-C., Jin Chong, M., Forsyth, D.: Learning diverse image colorization. In: IEEE Conference on Computer Vision and Pattern Recognition, pp. 6837–6845 (2017)

Ding, K., Ma, K., Wang, S., Simoncelli, E.P.: Comparison of full-reference image quality models for optimization of image processing systems. Int. J. Comput. Vis. **129**(4), 1258–1281 (2021)

Dowson, D., Landau, B.: The Fréchet distance between multivariate normal distributions. J. Multivar. Anal. **12**(3), 450–455 (1982)

Gatys, L.A., Ecker, A.S., Bethge, M.: A neural algorithm of artistic style. J. Vis. **16**(12), 326 (2016)

Goodfellow, I., Pouget-Abadie, J., Mirza, M., Xu, B., Warde-Farley, D., Ozair, S., Courville, A., Bengio, Y.: Generative adversarial nets. In: Advances in Neural Information Processing Systems (2014)

Guadarrama, S., Dahl, R., Bieber, D., Norouzi, M., Shlens, J., Murphy, K.: Pixcolor: pixel recursive colorization. In: British Machine Vision Conference (2017)

Gu, S., Timofte, R., Zhang, R.: Ntire 2019 challenge on image colorization: report. In: Conference on Computer Vision and Pattern Recognition Workshops (2019)

Gulrajani, I., Ahmed, F., Arjovsky, M., Dumoulin, V., Courville, A.: Improved training of Wasserstein GANs. In: Advances in Neural Information Processing Systems, pp. 5769–5779 (2017)

He, M., Chen, D., Liao, J., Sander, P.V., Yuan, L.: Deep exemplar-based colorization. ACM Trans. Graph. **37**(4), 1–16 (2018)

Heusel, M., Ramsauer, H., Unterthiner, T., Nessler, B., Hochreiter, S.: GANs trained by a two time-scale update rule converge to a local Nash equilibrium. In: Advances in Neural Information Processing Systems, vol. 30 (2017)

Ho, J., Kalchbrenner, N., Weissenborn, D., Salimans, T.: Axial attention in multidimensional transformers (2019). arXiv preprint arXiv:1912.12180

Iizuka, S., Simo-Serra, E., Ishikawa, H.: Let there be color!: joint end-to-end learning of global and local image priors for automatic image colorization with simultaneous classification. ACM Trans. Graph. **35**(4), 1–11 (2016)

Isola, P., Zhu, J.-Y., Zhou, T., Efros, A.A.: Image-to-image translation with conditional adversarial networks. In: IEEE Conference on Computer Vision and Pattern Recognition, pp. 1125–1134 (2017)

Johnson, J., Alahi, A., Fei-Fei, L.: Perceptual losses for real-time style transfer and super-resolution. In: European Conference on Computer Vision, pp. 694–711 (2016)

Kong, G., Tian, H., Duan, X., Long, H.: Adversarial edge-aware image colorization with semantic segmentation. IEEE Access **9**, 28194–28203 (2021)

Kumar, M., Weissenborn, D., Kalchbrenner, N.: Colorization transformer (2021). arXiv preprint arXiv:2102.04432

Larsson, G., Maire, M., Shakhnarovich, G.: Learning representations for automatic colorization. In: European Conference on Computer Vision, pp. 577–593 (2016)

Lin, T.-Y., Maire, M., Belongie, S., Hays, J., Perona, P., Ramanan, D., Dollár, P., Zitnick, C.L.: Microsoft COCO: common objects in context. In: European Conference on Computer Vision, pp. 740–755 (2014)

Lübbe, E.: Colours in the Mind-Colour Systems in Reality: A Formula for Colour Saturation. BoD–Books on Demand, Norderstedt (2010)

Mouzon, T., Pierre, F., Berger, M.-O.: Joint CNN and variational model for fully-automatic image colorization. In: Scale Space and Variational Methods in Computer Vision, pp. 535–546 (2019)

Nazeri, K., Ng, E., Ebrahimi, M.: Image colorization using Generative Adversarial Networks. In: International Conference on Articulated Motion and Deformable Objects, pp. 85–94 (2018)

Oord, A.V.D., Kalchbrenner, N., Vinyals, O., Espeholt, L., Graves, A., Kavukcuoglu, K.: Conditional image generation with PixelCNN decoders. In: Advances in Neural Information Processing Systems (2016)

Pierre, F., Aujol, J.-F.: Recent approaches for image colorization. In: Handbook of Mathematical Models and Algorithms in Computer Vision and Imaging (2020)

Pierre, F., Aujol, J.-F., Bugeau, A., Papadakis, N., Ta, V.-T.: Luminance-chrominance model for image colorization. SIAM J. Imag. Sci. **8**(1), 536–563 (2015)

Pucci, R., Micheloni, C., Martinel, N.: Collaborative image and object level features for image colourisation. In: IEEE Conference on Computer Vision and Pattern Recognition, pp. 2160–2169 (2021)

Riba, E., Mishkin, D., Ponsa, D., Rublee, E., Bradski, G.: Kornia: an open source differentiable computer vision library for PyTorch. In: Winter Conference on Applications of Computer Vision, pp. 3674–3683 (2020)

Royer, A., Kolesnikov, A., Lampert, C.H.: Probabilistic image colorization. In: British Machine Vision Conference (2017)

Sabour, S., Frosst, N., Hinton, G.E.: Dynamic routing between capsules. In: Advances in Neural Information Processing Systems, vol. 30 (2017)

Simonyan, K., Zisserman, A.: Very deep convolutional networks for large-scale image recognition. In: International Conference on Learning Representations (2015)

Su, J.-W., Chu, H.-K., Huang, J.-B.: Instance-aware image colorization. In: IEEE Conference on Computer Vision and Pattern Recognition, pp. 7968–7977 (2020)

Van Oord, A., Kalchbrenner, N., Kavukcuoglu, K.: Pixel recurrent neural networks. In: International Conference on Machine Learning, pp. 1747–1756 (2016)

Vitoria, P., Raad, L., Ballester, C.: ChromaGAN: adversarial picture colorization with semantic class distribution. In: Winter Conference on Applications of Computer Vision, pp. 2445–2454 (2020)

Wang, Z., Bovik, A.C., Sheikh, H.R., Simoncelli, E.P.: Image quality assessment: from error visibility to structural similarity. IEEE Trans. Image Process. **13**(4), 600–612 (2004)

Yoo, S., Bahng, H., Chung, S., Lee, J., Chang, J., Choo, J.: Coloring with limited data: few-shot colorization via memory augmented networks. In: IEEE Conference on Computer Vision and Pattern Recognition (2019)

Zhang, R., Isola, P., Efros, A.A.: Colorful image colorization. In: European Conference on Computer Vision, pp. 649–666 (2016)

Zhang, R., Zhu, J.-Y., Isola, P., Geng, X., Lin, A.S., Yu, T., Efros, A.A.: Real-time user-guided image colorization with learned deep priors. ACM Trans. Graph. **36**, 1–11 (2017)

Zhang, R., Isola, P., Efros, A.A., Shechtman, E., Wang, O.: The unreasonable effectiveness of deep features as a perceptual metric. In: IEEE Conference on Computer Vision and Pattern Recognition, pp. 586–595 (2018)

Influence of Color Spaces for Deep Learning Image Colorization

22

Aurélie Bugeau, Rémi Giraud, and Lara Raad

Contents

Introduction	848
Related Work	849
On Color Spaces	849
Review of Colorization Methods	851
Datasets Used in Literature	859
Proposed Colorization Framework	861
Detailed Architecture	861
Training and Testing Images	862
Learning Strategy for Different Color Spaces	863
Analysis of the Influence of Color Spaces	863
Quantitative Evaluation	866
Qualitative Evaluation	868
Generalization to Archive Images	872
Conclusion	874
References	875

A. Bugeau (✉)
LaBRI, CNRS, UMR5800, Univ. Bordeaux, F-33400 Talence, France
Institut universitaire de France (IUF), Paris, France
e-mail: aurelie.bugeau@labri.fr

R. Giraud
Univ. Bordeaux, CNRS, IMS UMR5251, Bordeaux INP, F-33400 Talence, France
e-mail: remi.giraud@ims-bordeaux.fr

L. Raad
LIGM, CNRS, Univ Gustave Eiffel, F-77454 Marne-la-Vallée, France
e-mail: lara.raadcisa@esiee.fr

© Springer Nature Switzerland AG 2023
K. Chen et al. (eds.), *Handbook of Mathematical Models and Algorithms in Computer Vision and Imaging*, https://doi.org/10.1007/978-3-030-98661-2_125

Abstract

Colorization is a process that converts a grayscale image into a colored one that looks as natural as possible. Over the years this task has received a lot of attention. Existing colorization methods rely on different color spaces: RGB, YUV, Lab, etc. In this chapter, we aim to study their influence on the results obtained by training a deep neural network, to answer the following question: "Is it crucial to correctly choose the right color space in deep learning-based colorization?" First, we briefly summarize the literature and, in particular, deep learning-based methods. We then compare the results obtained with the same deep neural network architecture with RGB, YUV, and Lab color spaces. Qualitative and quantitative analysis do not conclude similarly on which color space is better. We then show the importance of carefully designing the architecture and evaluation protocols depending on the types of images that are being processed and their specificities: strong/small contours, few/many objects, recent/archive images.

Keywords

Image colorization · Deep learning · Color spaces

Introduction

Image colorization consists in recovering a colored image from a grayscale one. This process attracts a lot of attention in the image-editing community in order to restore or colorize old grayscale movies or pictures. While turning a colored image into a grayscale one is only a matter of standard, the reverse operation is a strongly ill-posed problem as no information on which color has to be added is known. Therefore priors must be considered. In the literature, there exist three kinds of priors leading to different types of colorization methods. In the first category, initiated by Levin et al. (2004), the user manually adds initial colors through scribbles to the grayscale image. The colorization process is then performed by propagating the input color data to the whole image. The second category, called automatic or patch-based colorization, initiated by Welsh et al. (2002), consists in transferring color from one (or many) initial colored image considered as an example. The last category, which attracts most research nowadays, concerns deep learning approaches. The necessary color prior here is learned from large datasets.

Generally, in colorization methods, the initial grayscale image is considered as the luminance channel which is not modified during the colorization. The objective is then to reconstruct the two chrominance channels, before turning back to the RGB color space. Different luminance-chrominance spaces exist and have been used for image colorization. One common problem with all image colorization methods that aim at reconstructing the chrominances of the target image is that the recovered chrominances combined with the input luminance may not fall into the RGB cube when converting back to the RGB color space. Therefore, some works have decided to work directly on RGB to cope with this limitation by constraining the luminance channel (Pierre et al. 2014).

The objective of this chapter is to analyze the influence of color spaces on the results of automatic deep learning methods for image colorization. This chapter comes together with another chapter of this handbook. This other chapter, ▶ Chap. 21, "Analysis of Different Losses for Deep Learning Image Colorization", focuses on the influence of losses. We refer the reader to it for a review of the traditionally used different losses and evaluation metrics. Here, after reviewing existing works in image colorization and, in particular, works based on deep learning, we will focus on the influence of color spaces. Based on our analysis of the literature, a baseline architecture is defined and later used in all comparisons. Additionally, again based on the literature review, we set a uniform training procedure to ensure fair comparisons. Experiments encompass qualitative and quantitative analysis.

The chapter is organized as follows. Section "Related Work" first recalls some basics on color spaces and then provides a detailed survey of the literature on colorization methods and finally lists the datasets traditionally used. Next, in section "Proposed Colorization Framework", we present the chosen architecture and in section "Learning Strategy for Different Color Spaces" the learning strategy. Section "Analysis of the Influence of Color Spaces" presents the results of the different experiments. A discussion on the generalization of this work to archive images is later provided in section "Generalization to Archive Images" before a conclusion is drawn.

Related Work

On Color Spaces

This section presents the different color spaces that have been used for colorization in the literature. For more information about color theory and color constancy (i.e., the underlying ability of human vision to perceive colors very robustly with respect to changes of illumination), see, for instance, Ebner (2007) and Fairchild (2013).

Colored images are traditionally saved in the RGB color space. A grayscale image contains only one channel that encodes the luminosity (perceived brightness of that object by a human observer) or the luminance (absolute amount of light emitted by an object per unit area). A way to model this luminance Y which is close to the human perception of luminance is:

$$Y = 0.299R + 0.587G + 0.114B, \qquad (1)$$

where R, G, and B are, respectively, the amount of light emitted by an object per unit area in the low, medium, and high frequency bands that are visible by a human eye. Colorization aims to retrieve color information from a grayscale image. To do so, and to easily constrain the luminance channel, most methods propose to work in a luminance-chrominance space. The problem becomes the retrieval of two chrominance channels given the luminance Y. There exist several luminance-chrominance spaces. Two of them are mostly used for colorization. The first one, YUV, historically used for a specific analog encoding of color information in television systems, is the result of the linear transformation:

$$\begin{pmatrix} Y \\ U \\ V \end{pmatrix} = \begin{pmatrix} 0.299 & 0.587 & 0.114 \\ -0.14713 & -0.28886 & 0.436 \\ 0.615 & -0.51498 & -0.10001 \end{pmatrix} \begin{pmatrix} R \\ G \\ B \end{pmatrix}.$$

The reverse conversion from YUV and RGB is simply obtained by inverting the matrix. The other linear space that has been used for colorization is YCbCr.

The CIELAB color space, also referred to as Lab or La*b*, defined by the International Commission on Illumination (CIE) in 1976, is also frequently used for colorization. It has been designed such that the distances between colors in this space correspond to the perceptual distances of colors for a human observer. The three channels become uncorrelated. The transformation from RGB to Lab (and the reverse) is nonlinear. First, it is necessary to convert the RGB values to the CIEXYZ color space:

$$\begin{pmatrix} X \\ Y \\ Z \end{pmatrix} = \begin{pmatrix} 2.769 & 1.7518 & 0.13 \\ 1 & 4.5907 & 0.0601 \\ 0 & 0.0565 & 5.5943 \end{pmatrix} \begin{pmatrix} R \\ G \\ B \end{pmatrix}.$$

Then, the transformation to Lab is given by:

$$L = 116 f(Y/Y_n) - 16,$$
$$a = 500 \left[f(X/X_n) - f(Y/Y_n) \right],$$
$$b = 200 \left[f(Y/Y_n) - f(Z/Z_n) \right],$$

with

$$f(t) = \begin{cases} t^{1/3} & \text{if } t > (\frac{6}{29})^3, \\ \frac{1}{3} \left(\frac{29}{6} \right)^2 t + \frac{4}{29} & \text{otherwise,} \end{cases}$$

where X_n, Y_n, and Z_n describe a specified white achromatic reference illuminant. Obviously, the reverse operation from Lab to RGB is also nonlinear.

Despite RGB or luminance-chrominance color spaces, few methods relying on hue-based spaces have been proposed for colorization. For instance, Larsson et al. (2016) rely on a hue-chroma-luminance space.

Table 1 lists the color spaces used in deep learning colorization methods described in the next subsection. It distinctly appears that the Lab color space is the most widely used. We will further discuss this choice in section "Analysis of the Influence of Color Spaces".

Table 1 Color spaces used in deep learning methods for image colorization

	Cheng et al. (2015)	Iizuka et al. (2016)	Vitoria et al. (2020)	Nazeri et al. (2018)	Cao et al. (2017)	Yoo et al. (2019)	Antic (2019)	Larsson et al. (2016)	Zhang et al. (2016)	Mouzon et al. (2019)	Zhang et al. (2017)	He et al. (2018)	Deshpande et al. (2017)	Guadarrama et al. (2017)	Royer et al. (2017)	Kumar et al. (2021)	Su et al. (2020)	Pucci et al. (2021)	Kong et al. (2021)	Gu et al. (2019) (winner)
		Using GANs						Histogram prediction			User guided		Diverse				Object aware			Survey
RGB					•	•	•										•			•
YUV	•				•		•													
YCbCr															•					
Lab		•	•	•				•	•	•	•	•		•		•		•	•	•
hue/chroma								•												
Comparison		•			•			•												

In general terms, as can be seen in Table 1, most methods work in a luminance-chrominance space, and the cost functions to optimize are in general defined in the same space. Hence, converting from and to RGB to one of these luminance/chrominance spaces is not involved in the backpropagation step. Once the training is performed, at inference time the chrominance values given by the network together with the luminance component are converted back to the RGB color space. As mentioned earlier, this operation tends to perform an abrupt value clipping to fit in the RGB cube hence modifying both the original luminance values and the predicted chrominance values. Two libraries are most commonly used for the conversion step: the color module of scikit-image (Zhang et al. 2016, 2017; Larsson et al. 2016; Royer et al. 2017) and the color space conversion functions of OpenCV (Iizuka et al. 2016; Vitoria et al. 2020).

Review of Colorization Methods

This section presents an overview of the colorization methods in the three categories: scribble-based, exemplar-based, and deep learning. For a more detailed review with the same classification, we refer the reader to the recent review Li et al. (2020). Another survey focused on deep learning approaches proposes a taxonomy to separate these methods into seven categories (Anwar et al. 2020). The authors of this review have redrawn all networks architectures, thus allowing to easily compare architecture specificity. Comparisons of methods are made on a new Natural-Color Dataset made of objects with white background.

The NTIRE challenge is a competition for different computer vision tasks related to image enhancement and restoration. One of the tasks in 2019 was image colorization (Gu et al. 2019), with two tracks: colorization without or with guidance given by a second input that provides several color guiding points.

Scribble-Based Image Colorization

The first category of colorization methods relies on color priors coming from scribbles drawn by the user (see Fig. 1). These colors are propagated to all pixels by diffusion schemes.

The first manual colorization method based on scribbles was proposed by Levin et al. (2004). It solves an optimization problem to diffuse the chrominances of scribbles with the assumption that chrominances should have small variations where the luminance has small variations. To reduce the number of needed scribbles, Luan et al. (2007) first use scribbles to segment the image before diffusing the colors. Yatziv and Sapiro (2006) propose a simple yet fast method by using geodesic distances to blend the chrominances given by the scribbles. In Huang et al. (2005), edge information is extracted to reduce color bleeding. Heu et al. (2009) use pixel priorities to ensure that important areas end up with the right colors. Other propagation schemes include probabilistic distance transform (Lagodzinski and Smolka 2008), discriminative textural features (Kawulok et al. 2012), structure tensors (Drew and Finlayson 2011), nonlocal graph regularization (Lézoray et al. 2008), matrix completion (Wang and Zhang 2012; Yao and James 2015) or rank minimization (Ling et al. 2015). As often described in the literature, with these manual approaches, the contours are not well preserved. To cope with this issue, in Ding et al. (2012), scribbles are automatically generated after segmenting the image and the user only needs to provide one color per scribble. However, all manual methods suffer from the following drawback: if the target represents a complex scene, the user interaction becomes very important. On the other hand,

Fig. 1 Example of scribble-based image colorization taken from Levin et al. (2004). The user draws color that are successively diffused to neighbor pixels according under some constraints that depend on the different methods

these approaches propose a global optimization over the image, thus leading to spatial consistency in the result.

Exemplar-Based Image Colorization

The second category of colorization methods concerns exemplar-based methods which rely on a color reference image as prior. The first exemplar-based colorization method was proposed by Welsh et al. (2002). It makes the assumption that pixels with similar intensities or similar neighborhood should have similar colors. It extends the texture synthesis approach by Efros and Leung (1999): the final color of one pixel is copied from the most similar pixel in a reference input colored image. The similarity between pixels relies on patch-based metrics (see Fig. 2). This approach has given rise to many extension in the literature (Di Blasi and Reforgiato 2003; Liu and Zhang 2012). In particular, many works have focused on choosing or designing appropriate features for matching pixels (Chia et al. 2011; Gupta et al. 2012; Bugeau and Ta 2012; Cheng et al. 2015; Arbelot et al. 2016, 2017).

To overcome the spatial consistency and coupling problems in automatic methods, several works rely on image segmentation. For instance, Irony et al. (2005) propose to determine the best matches between the target pixels and regions in a pre-segmented source image. With these correspondences, micro-scribbles from the source are initialized on the target image and colors are propagated as in Levin et al. (2004). Tai et al. (2005) build a probabilistic segmentation of both images where one pixel can belong to many regions. They use it to transfer color between any two regions having similar statistics with an expectation-maximization scheme.

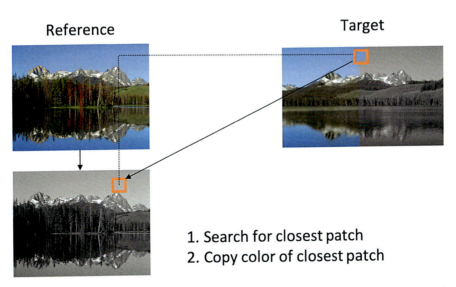

Fig. 2 Principle of exemplar-based image colorization. Methods in this category have proposed different similar patch search strategies and techniques to add spatial consistency when copying patch colors

Gupta et al. (2012) extract different features from the superpixels (Ren and Malik 2003) of the target image and match them with the source ones. The final colors are computed by imposing spatial consistency as in Levin et al. (2004). Li et al. (2017b) extract low- and high-level features on superpixels of the reference to form a dictionary then used as a dictionary-based sparse reconstruction problem. Sparse representation was previously used for colorization in Pang et al. (2013) where images are segmented from scribbles. These approaches incorporate local consistency into automatic methods via segmentation. In Charpiat et al. (2008), spatial consistency is solved with graph cuts after estimating for each pixel the conditional probability of colors. In Bugeau et al. (2014) and Pierre et al. (2014) each pixel can only take its chrominance (or RGB color) among a reduced set of possible candidates chosen from the reference image. The final color is chosen using a variational formulation. In the same trend, Fang et al. (2019) propose a superpixel-based variational model. In Li et al. (2017a), the distribution of intensity deviation for uniform and nonuniform regions is learned and used in a Markov random field (MRF) model for improved consistency. Finally, Li et al. (2019) propose cross-scale local texture matching, which are then fused using global graph-cut optimization.

A major problem of this family of methods is the high dependency on the reference image. Chia et al. (2011) therefore propose to rely on several reference images obtained from an Internet search based on semantic information.

Deep Learning Methods for Image Colorization

Since 2012, deep learning approaches, in particular convolutional neural networks (CNNs), have become very popular in the community of computer vision and computer graphics.

The first deep learning-based colorization methods were proposed in Cheng et al. (2015) and Deshpande et al. (2015). In Cheng et al. (2015), a fully automated system extracts handcrafted low and high features and feeds them as input to a three-layer fully connected neural network trained with a L2 loss. The network predicts the U and V channels of the YUV luminance-chrominance space. The authors also add an optional clustering stage where the images are divided in different types of scenes, according to the previously extracted semantic features. Then, a different neural network is trained for each of the clusters.

End-to-end approaches: Later on, papers focused more on *end-to-end approaches* (see Fig. 3).

For instance, the paper that won both tracks of the Gu et al. (2019) NTIRE 2019 Challenge on Image Colorization was the end-to-end method proposed by IPCV_IIMT. It implements an encoder-decoder structure that resembles to a U-Net with the encoder built using deep dense-residual blocks. Wan et al. (2020a) proposed to combine neural networks with color propagation. It first trains a neural network in order to colorize interest points of extracted superpixels. Then those colors are propagated by optimizing an objective function. In an older work, Iizuka et al. (2016) presented an end-to-end colorization framework based on CNNs to

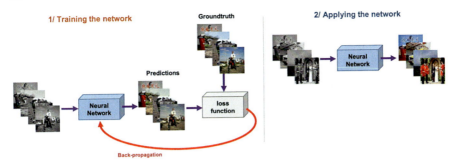

Fig. 3 Principle of basic end-to-end colorization networks

infer the *ab* channels of the CIE Lab color space. This work is built on the basis that a classification of the images can help to provide global priors that will improve the colorization performance. The network extracts global and local features and is jointly trained for classification and colorization in a labeled dataset.

Using GANs: Still being end-to-end, other methods use *generative adversarial networks* (GANs) (Goodfellow et al. 2014). Isola et al. (2017) propose the so-called image-to-image method pix2pix. It maps an input image to an output image using a U-Net generator and a patch GANs discriminator. The method is used in many applications including colorization. This method was extended in Nazeri et al. (2018) using deep convolutional GAN (DCGAN) (Radford et al. 2016). In Cao et al. (2017), a fully convolutional generator with a conditional GANs is considered. This architecture does not use downsampling to avoid extracting global features which are not suitable to recover accurate boundaries. To avoid noise attenuation and make the colorization results more diversified, they concatenate a noise channel onto the first half of the generator layers. GANs have also been used in chromaGAN (Vitoria et al. 2020) which extends Iizuka et al. (2016) by proposing to learn the semantic image distribution without any need of a labeled dataset. This method combines three losses: a color error loss by computing MSE on *ab* channels, a class distribution loss by computing the Kullback-Leibler divergence on VGG-16 class distribution vectors, and an adversarial Wasserstein GAN (WGAN) loss (Arjovsky et al. 2017). To prevent the need for training on a huge amount of data, Yoo et al. (2019) introduce MemoPainter, a few-shot colorization framework. MemoPainter is able to colorize an image with limited data by using an external memory network in addition to a colorization network. The memory network learns to retrieve a color feature that best matches the ground-truth color feature of the query image, while the generator-discriminator colorization network learns to effectively inject the color feature to the target grayscale image.

DeOldify (Antic 2019) is another end-to-end image and video colorization method mapping the missing chrominance values to the grayscale input image. A ResNet (ResNet101 or ResNet34) is used as the backbone of the generator of a U-Net architecture trained as follows: the generator is first trained with the perceptual loss (Johnson et al. 2016), followed by training the critic as a binary classifier distinguishing between real images and those generated by the generator, and finally the generator and critic are trained together in an adversarial manner on 1–3% of the ImageNet (Deng et al. 2009) data. The latter is the so-called NoGAN strategy which is enough to add color realism to the results and which also allows to avoid flickering across video frames while the colorization is applied individually frame per frame.

Predicting distributions instead of images: Regression does not handle multi-modal color distributions well (Larsson et al. 2016). Larsson et al. (2016) and Zhang et al. (2016) address this issue by *predicting distributions* over a set of bins, as it was initially done in the exemplar-based method (Charpiat et al. 2008). They therefore rely on a discretization of color spaces. In Larsson et al. (2016), the color space is binned with evenly spaced Gaussian quantiles. Experiments are run for hue/chroma and Lab color spaces with either separated or joint distributions. Inference of the colored image from the distribution uses expectation (sum over the color bin centroids weighted by the histogram). In Zhang et al. (2016), the *ab* output space is quantized into bins with grid size 10 and the 313 values which are in gamut are kept. The inference is the annealed mean of the distribution. In Mouzon et al. (2019) and Pierre and Aujol (2020), the resulting distributions from Zhang et al. (2016) are later used in a variational approach (Pierre et al. 2015a).

Considering user priors: Few methods give the possibility to add user inputs as additional priors. The architecture in Zhang et al. (2017) learns to propagate color hints by fusing low-level cues and high-level semantic information. He et al. (2018) uses a reference colored image to guide the output of their deep exemplar-based colorization method.

Generating diverse image colorizations: Some methods have been designed to generate diverse colorizations as there is not one unique solution to the colorization problem. Deshpande et al. (2017) relied on a variational auto-encoder (VAE) to learn a low-dimensional embedding of color spaces. The mapping from a grayscale input image to color distribution of the latent space is done by learning a mixture density network (MDN). At test time, it is possible to sample the conditional model and use the VAE decoder to generate diverse colored images. In their PixColor model, Guadarrama et al. (2017) first train a conditional PixelCNN (Oord et al. 2016) to generate multiple latent low-resolution colored images, and then train a second CNN to generate the final high-resolution images. Another method, called PIC, that uses PixelCNN++ (Salimans et al. 2017) (an extension to the original PixelCNN), was proposed in Royer et al. (2017). A feed-forward CNN first maps grayscale image to an embedding that encodes color information. This embedding is then fed to the autoregressive PixelCNN++ model which predicts a distribution of image chromacity. The colTran model proposed by Kumar et al. (2021) is based on an axial transformer (Ho et al. 2019) autoregressive model. ColTran includes three networks, all relying on column/row self-attention blocks: the autoregressive model that estimates low-resolution coarse colorization, a color upsampler, and a spatial upsampler.

Restoring and colorizing: Luo et al. (2020) propose to specifically restore and colorize old black and white portrait photos in a unified framework. It uses an additional high-quality color reference image (the sibling) automatically generated by first training a network that projects images into the StyleGAN2 (Karras et al. 2020) latent space and then uses the pretrained StyleGAN2 generator to create the sibling. Fine details and colors are extracted from the sibling. A latent code is then optimized through a three-term cost function and decoded by a StyleGAN2 generator yielding a high-quality color version of the antique input. The cost function is composed of a color term inspired by the style loss in Gatys et al. (2016a) between the features of the sibling and those of the generated high-quality colored image, a perceptual term (Johnson et al. 2016) between a degraded version of the generative model's output and the antique input, and a contextual term between the VGG features of the sibling and those of the generated high-quality colored image.

Decomposing the scene into objects: Recently, some methods try to explicitly deal with the decomposition of the scene into objects in order to tackle one of the main drawbacks of most deep learning-based colorization methods which is color bleeding across different objects. Su et al. (2020) proposed to colorize a grayscale image in an instance-aware fashion. They train three separate networks: a first one that performs global colorization, a second one that achieves instant colorization, and a third one that fuses both colorization networks. These networks are trained by minimizing the Huber loss (also called Smooth L1 loss). In general, after fusing both results the global colorization will be enhanced. The instances per image are obtained by using a standard pretrained object detection network, Mask R-CNN (He et al. 2017). Pucci et al. (2021) propose to improve Zhang et al. (2016) by using a network which is more aware of image instances, in the spirit of Su et al. (2020), by combining convolutional and capsule networks. They train from end to end a single network which first generates a per-pixel color distribution followed by a final convolutional layer that recovers the missing chrominance channels as opposed to Zhang et al. (2016) that computes the annealed mean on the per-pixel color distribution network's output. They train the network by minimizing the cross-entropy between per pixel color distributions and L2 loss on the chrominance channels. Kong et al. (2021) propose to colorize a grayscale image by training a multitask network for colorization and semantic segmentation in an adversarial manner. They train a U-Net-type network with a three-term cost function: a color regression loss in terms of hue, saturation, and lightness; the cross-entropy on the ground-truth and generated semantic labels; and a GANs term. The main objective of the proposal is to reduce color bleeding across edges.

Table 2 summarizes all these deep learning methods providing details on their particular inputs (other than the obvious grayscale image), their outputs, their

Table 2 Short description of deep networks for image colorization, their input, other than grayscale image, output. Here FCONV stands for fully convolutional, FC for fully connected, and U-Net for a U-Net-like network and not the vanilla U-Net

	Additional inputs	Network	Network's output	Post-processing
Cheng et al. (2015)	Handcrafted features	3 layers FC	UV	Joint bilateral filtering
Iizuka et al. (2016)	–	CNNs (local/global)	*ab*	Upsampling
Wan et al. (2020a)	Superpixels' handcrafted features	FC net	Interest points' color	Propagation and refinement
Using GANs				
Vitoria et al. (2020)	–	CNNs (local/global) + PatchGAN	*ab*	Upsampling
Nazeri et al. (2018)	–	U-Net (Isola et al. 2017) + DCGAN	Lab	–
Cao et al. (2017)	–	FCONV generator with multi-layer noise + PatchGAN	UV/RGB (diverse)	–
Yoo et al. (2019)	Color thief features	Colorization U-Net + memory nets noise	–	–
Antic (2019)	–	U-Net + self-attention + GAN	RGB	YUV conversion + cat(original Y/UV) + RGB conversion
Histogram prediction				
Larsson et al. (2016)	–	VGG-16 + FC layers	Distributions	Expectation
Zhang et al. (2016)	–	VGG-styled net	Distributions	Annealed mean
Mouzon et al. (2019)	–	Zhang et al. (2016)	Distributions	Variational model
User guided				
Zhang et al. (2017)	User point, global histograms, and average saturation	U-Net	Distributions + *ab*	–
He et al. (2018)	Color reference	Similarity sub-net + U-Net (gray VGG-19)	Bidirectional similarity maps + *ab*	–

(continued)

Table 2 (continued)

	Additional inputs	Network	Network's output	Post-processing
Diverse colorization and autoregressive models				
Deshpande et al. (2017)	–	cVAE + MDN	Diverse colorization	–
Guadarrama et al. (2017)	–	PixelCNN + CNN	Diverse colorization	–
Royer et al. (2017)	–	CNN + PixelCNN++	Diverse colorization	–
Kumar et al. (2021)	–	Axial transformer + color/spatial upsamplers (self-attention blocks)	Diverse colorization	–
Object aware				
Su et al. (2020)	Object bounding boxes	U-Net (global/instance) + CNN (fusion)	ab	–
Pucci et al. (2021)	–	CNN + capsule net	ab	–
Kong et al. (2021)	–	U-Net + PatchGAN	ab + semantic segmentation	–
Survey				
Gu et al. (2019)	–	U-Net	RGB	–

architectures, and pre- and post-processing steps. This summary table is only provided for deep learning-based methods since we focus on deep learning-based strategies in the remaining of the chapter.

Datasets Used in Literature

To train and test the deep learning methods presented in previous subsection, different datasets have been used. Table 3 summarizes the use of these datasets in colorization methods. They contain from one thousand (DIV2K (Agustsson and Timofte 2017)) to million of images (ImageNet (Deng et al. 2009)). Image dimensions also vary a lot, from 32×32 in CIFAR-10 (Krizhevsky et al. 2009) to 2K resolution in DIV2K.

Other differences concern the content of the images itself. Some datasets are very specific to a type of image: faces (LFW (Huang et al. 2007)) and bedrooms (LSUN (Yu et al. 2015)). Other present various scenes as Places (Zhou et al. 2017) with 205 scene categories, COCO (Lin et al. 2014) with 80 object categories and 91 stuff categories, and SUN (Xiao et al. 2010) with 899 scene categories.

Table 3 Datasets used in the literature for training or testing

	SUN	ImageNet/ILSVRC-2015	COCO	CIFAR-10	DIV2K	Pascal VOC	Places	LSUN bedroom or church	testing on historic BW photo	Remark/Other
Cheng et al. (2015)	•									
Iizuka et al. (2016)							•		•	
Using GANs										
Vitoria et al. (2020)		•							•	
Nazeri et al. (2018)				•		•				
Cao et al. (2017)							•			
Yoo et al. (2019)										Yumi, Monster, etc.
Antic (2019)		•								training on 1–3% of ImageNet images
Histograms prediction										
Larsson et al. (2016)	•	•								
Zhang et al. (2016)		•				•				training on 1.3M ImageNet images
User guided										
Zhang et al. (2017)		•								
He et al. (2018)		•								training on 700k ImageNet image/7 categories
Diverse										
Deshpande et al. (2017)		•						•		LFW
Guadarrama et al. (2017)		•								
Royer et al. (2017)		•		•						
Kumar et al. (2021)		•								
Object aware										
Su et al. (2020)		•	•				•			
Pucci et al. (2021)		•	•				•			
Kong et al. (2021)					•					
Survey										
Gu et al. (2019)					•					
Anwar et al. (2020)										Own Natural-Color Dataset

Proposed Colorization Framework

In this section, we present the framework that we will use for evaluating the influence of color spaces on image colorization results. First, we detail the architecture and, second, the dataset used for training and testing.

Note that the same architecture and training procedure are used in the ▶ Chap. 21, "Analysis of Different Losses for Deep Learning Image Colorization" of this handbook.

Detailed Architecture

The architecture used in our experiments is an encoder-decoder U-Net deep network composed of five stages (see Fig. 4). All convolutional blocks are composed of two 2D convolutional layers with 3 × 3 kernels, each one followed by 2D batch normalization (BN) and a ReLU activation. For the encoder, downsampling is done with max pooling layers after each convolutional block. After each downsampling, the number of filters is doubled in the following block. For the decoder, upsampling is done with 2D transpose convolutions (4 × 4 kernels with stride 2). At a given stage, the corresponding encoder and decoder blocks are linked with skip connections: feature maps from the encoder are concatenated with the ones from the corresponding upsampling path and fused using 1 × 1 convolutions. More details can be found in Table 4. The encoder architecture is identical to the CNN part of a VGG-19 network (Simonyan and Zisserman 2015). It allows us to start from pretrained weights initially used for ImageNet classification. Moreover, the encoder architecture choice was motivated by the fact that most deep learning-based approaches use a VGG-type architecture to generate the missing chrominances.

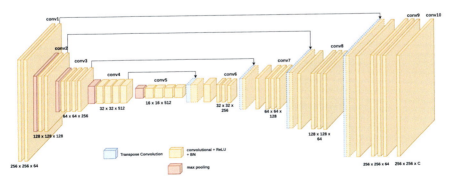

Fig. 4 Summary of the baseline U-Net architecture used in our experiments. It outputs a 256 × 256 × C image, where C stands for the number of channels, being equal to 2 when estimating the missing chrominance channels and to 3 when estimating the RGB components

Table 4 Detailed architecture and output resolution for each block

Layer type	Output resolution
Input	3 × H × W
Conv1 + Max-pooling	64 × H/2 × W/2
Conv2 + Max-pooling	128 × H/4 × W/4
Conv3 + Max-pooling	256 × H/8 × W/8
Conv4 + Max-pooling	512 × H/16 × W/16
Conv5 + Conv. Transpose (I)	512 × H/8 × W/8
Conv6 + Conv. Transpose (II)	256 × H/4 × W/4
Conv7 + Conv. Transpose (III)	128 × H/2 × W/2
Conv8 + Conv. Transpose (IV)	64 × H × W
Conv9	64 × H × W
Conv10	C × H × W

The training settings are described as follows:

- Optimizer: Adam
- Learning rate: 2e-5 as in ChromaGAN (Vitoria et al. 2020).
- Batch size: 16 images (approx. 11 GB RAM usage on Nvidia Titan V).
- All images are resized to 256 × 256 for training which enable using batches. In practice, to keep the aspect ratio, the image is resized such that the smallest dimension matches 256. If the other dimension remains larger than 256, we then apply a random crop to obtain a square image. Note that the random crop is performed using the same seed for all trainings.

When generating images, it is crucial to remain in the range of acceptable values of color spaces. In particular, we must ensure that the final image takes values between 0 and 255. In our implementation, we use simple clipping on final RGB values. Other strategies are sometimes considered as in Iizuka et al. (2016) where the a*b* components are globally normalized so they lie in the [0,1] range of the Sigmoid transfer function.

Training and Testing Images

Throughout our experiments we use the COCO dataset (Lin et al. 2014), containing various natural images of different sizes. COCO is divided into three sets that approximately contain 118k, 5k, and 40k images that, respectively, correspond to the training, validation, and test sets. Note that we carefully remove all grayscale images, which represent around 3% of the overall amount of each set. Although larger datasets such as ImageNet have been regularly used in the literature, COCO offers a sufficient number and a good variety of images so we can efficiently train and compare numerous models.

Learning Strategy for Different Color Spaces

The goal of the whole colorization process is to generate RGB images that look visually natural. When training on different color spaces, one must decide which color space is used to compute losses and when is the conversion back to RGB performed. In this chapter, we propose to experiment with three learning strategies to compare RGB, YUV, and Lab color spaces (see Fig. 5):

- *RGB*: in this case, the network takes as input a grayscale image L and directly estimates a three-channel RGB image of size $256 \times 256 \times 3$. The loss is done directly in the RGB color space. This strategy is illustrated in Fig. 5a.
- *YUV and Lab Luminance/chrominance*: in this case, the network takes as input a grayscale image considered as the luminance (L for Lab, Y for YUV) and outputs two chrominance channels (a, b or U, V). The loss compares the output with the corresponding chrominance channels of the ground-truth image converted to the luminance/chrominance space. After concatenating the initial luminance channel to the inferred chrominances, the image is converted back to RGB for visualization purposes. This strategy is illustrated in Fig. 5b.
- *LabRGB*: as in the previous case, the network takes as input the luminance and estimates the corresponding two chrominance channels. After concatenating with the corresponding luminance channel, they are converted to the RGB color space and the loss is computed directly there. Notice that in this last case, as the loss is computed on RGB color space, the conversion must be done in a way that is differentiable to be able to compute the gradient and allow the backpropagation step. We perform the color conversion using the color module in the `Kornia` library. Kornia (Riba et al. 2020) is a differentiable library that consists of a set of routines and differentiable modules to solve generic computer vision problems. It allows classical computer vision tasks to be integrated into deep learning models. Computing the loss on RGB images instead of chrominance ones enables to ensure images are similar to ground truth after the clipping operation needed to fit into the RGB cube. This strategy is illustrated in Fig. 5c.

Remark. During training, all images are resized to 256×256. One advantage of using luminance/chrominance spaces is that only chrominance channels are resized. It is therefore possible to keep the original content of the luminance channels without manipulating it with the resizing steps.

Analysis of the Influence of Color Spaces

This section presents quantitative and qualitative results obtained with the three strategies discussed above. For this analysis, we have considered, as loss function, the L2 loss and the VGG-based LPIPS which was introduced in Ding et al. (2021) as a generalization of the feature loss (Johnson et al. 2016). These loss functions are defined hereafter.

Fig. 5 Illustration of the different learning strategies for our proposed framework. (**a**) Learning strategy directly predicting the RGB colors. (**b**) Learning strategy predicting the two chrominance channels. (**c**) Learning strategy predicting the two chrominance channels and then converting to RGB

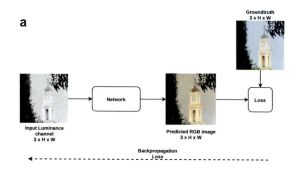

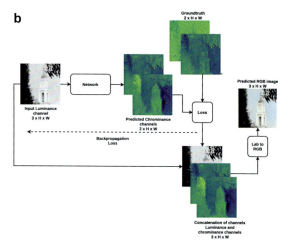

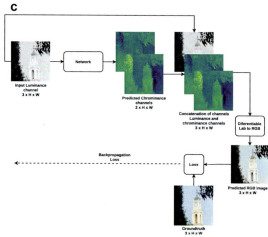

MSE or squared L2 loss. The L2 loss, between two functions u and v defined on Ω and with values in $\mathbb{R}^C$, $C \in \mathbb{N}$, is defined as the squared L2 loss of their difference. That is,

$$\text{MSE}(u, v) = \|u - v\|_{L^2(\Omega; \mathbb{R}^C)}^2 = \int_\Omega \|u(x) - v(x)\|_2^2 dx, \qquad (2)$$

where $\|\cdot\|_2$ denotes the Euclidean norm in $\mathbb{R}^C$. In the discrete setting, it is equal to the sum of the square differences between the image values, that is,

$$\text{MSE}(u, v) = \sum_{i=1}^{M} \sum_{j=1}^{N} \sum_{k=1}^{C} (u_{i,j,k} - v_{i,j,k})^2. \qquad (3)$$

Feature Loss. The feature reconstruction loss (Gatys et al. 2016b; Johnson et al. 2016) is a perceptual loss that encourages images to have similar feature representations as the ones computed by a pretrained network, denoted here by Φ. Let $\Phi_l(u)$ be the activation of the l-th layer of the network Φ when processing the image u; if l is a convolutional layer, then $\Phi_l(u)$ will be a feature map of size $C_l \times W_l \times H_l$. The *feature reconstruction* loss is the normalized squared Euclidean distance between feature representations, that is,

$$\mathcal{L}_{\text{feat}}^l(u, v) = \frac{1}{C_l W_l H_l} \|\Phi_l(u) - \Phi_l(v)\|_2^2. \qquad (4)$$

It penalizes the output reconstructed image when it deviates in feature content from the target.

LPIPS. LPIPS (Zhang et al. 2018) computes a weighted L2 distance between deep features of a pair of images u and v:

$$\text{LPIPS}(u, v) = \sum_l \frac{1}{H_l W_l} \sum_{i=1}^{H_l} \sum_{j=1}^{W_l} \|\omega_l \odot (\Phi_l(u)_{i,j} - \Phi_l(v)_{i,j})\|_2^2, \qquad (5)$$

where H_l (resp. W_l) is the height (resp. the width) of feature map Φ_l at layer l and ω_l is the weight for each feature. Note that features are unit-normalized in the channel dimension. We will denote VGG-based LPIPS when feature maps Φ_l are taken from a VGG network.

Note that to compute the VGG-based LPIPS loss, the output colorization always has to be converted to RGB, even for YUV and Lab color spaces (as in Fig. 5c), because this loss is computed with a pretrained VGG expecting RGB images as input. Since VGG-based LPIPS is computed on RGB images, the two strategies *Lab* and *LabRGB* are the same. For more details on the various losses usually used in colorization, we refer the reader to the ▶ Chap. 21, "Analysis of Different

Losses for Deep Learning Image Colorization." Our experiments have shown that same conclusions can be drawn with other losses.

For testing, we apply the network to images at their original resolution, while training is done on batches of square 256 × 256 images.

Quantitative Evaluation

There is no standard protocol for quantitative evaluation of automatic colorization methods. We refer the reader to the ▶ Chap. 21, "Analysis of Different Losses for Deep Learning Image Colorization" for a detailed survey of quantitative evaluation methods used in image colorization literature and analysis of correlation between losses and type of evaluation metrics used. We choose to rely on the more generally used and more recent ones: L1 (MAE), L2 (MSE), PSNR, SSIM (Wang et al. 2004), LPIPS (Zhang et al. 2018), and FID (Fréchet Inception Distance) (Dowson and Landau 1982), which are defined hereafter.

MAE or L1 loss with l^1-coupling. The mean absolute error is defined as the L1 loss with l^1-coupling, that is,

$$\text{MAE}(u,v) = \int_\Omega \|u(x) - v(x)\|_{l^1} dx = \int_\Omega \sum_{k=1}^{C} |u_k(x) - v_k(x)| dx. \quad (6)$$

In the discrete setting, it coincides with the sum of the absolute differences $|u_{i,j,k} - v_{i,j,k}|$. Some authors use a l^2-coupled version of it:

$$\text{MAE}^c(u,v) = \sum_{i=1}^{M} \sum_{j=1}^{N} \sqrt{\sum_{k=1}^{C} (u_{i,j,k} - v_{i,j,k})^2}. \quad (7)$$

Both MAE and MAE^c losses are robust to outliers.

PSNR. The PSNR measures the ratio between the maximum value of a color target image $u : \Omega \to \mathbb{R}^C$ and the mean square error (MSE) between u and a colorized image $v : \Omega \to \mathbb{R}^C$ with $\Omega \in \mathbb{Z}^2$ a discrete grid of size $M \times N$. That is,

$$\text{PSNR}(u,v) = 20 \log_{10}(\max u)$$

$$- 10 \log_{10} \left(\frac{1}{CMN} \sum_{k=1}^{C} \sum_{i=1}^{M} \sum_{j=1}^{N} (u(i,j,k) - v(i,j,k))^2 \right), \quad (8)$$

where $C = 3$ when working in the RGB color space and $C = 2$ in any luminance-chrominance color space as YUV, Lab, and YCbCr. The PSNR score is considered as a reconstruction measure tending to favor methods that will output results as close as possible to the ground-truth image in terms of the MSE.

SSIM. SSIM intends to measure the perceived change in structural information between two images. It combines three measures to compare image color (l), contrast (c), and structure (s):

$$\text{SSIM}(u, v) = l(u, v)c(u, v)s(u, v) = \frac{(2\mu_u\mu_v) + c_1}{\mu_u^2 + \mu_v^2 + c_1} \frac{(2\sigma_u\sigma_v + c_2)}{\sigma_u^2 + \sigma_v^2 + c_2} \frac{(\sigma_{uv} + c_3)}{\sigma_u\sigma_v + c_3}, \tag{9}$$

where μ_u (resp. σ_u) is the mean value (resp. the variance) of image u values and σ_{uv} the covariance of u and v. c_1, c_2, c_3 are regularization constants that are used to stabilize the division for images with mean or standard deviation close to zero.

FID. FID (Heusel et al. 2017) is a quantitative measure used to evaluate the quality of the outputs' generative model and which aims at approximating human perceptual evaluation. It is based on the Fréchet distance (Dowson and Landau 1982) which measures the distance between two multivariate Gaussian distributions. FID is computed between the feature-wise mean and covariance matrices of the features extracted from an Inception v3 neural network applied to the input images (μ_r, Σ_r) and those of the generated images (μ_g, Σ_g):

$$\text{FID}\left((\mu_r, \Sigma_r), (\mu_g, \Sigma_g)\right) = \|\mu_r - \mu_g\|_2^2 + Tr(\Sigma_r + \Sigma_g - 2\Sigma_r\Sigma_g)^{1/2}. \tag{10}$$

The results are presented in Table 5. In terms of these metrics, the best results are obtained with YUV color space except for L1 and Fréchet Inception Distance, even if not by much. The results in Table 5 also indicate that Lab does not outperform other color spaces when using a classic reconstruction loss (L2), while better results are obtained when using the VGG-based LPIPS. Thus, using a feature-based reconstruction loss is better suited as was already the case in exemplar-based image colorization methods where different features for patch-based metrics were proposed for matching pixels. LabRGB strategy gets the worst quantitative results based on Table 5. One would expect to get the "best of both" color spaces while recovering from the loss of information in the conversion process. However, this is not reflected with these particular evaluation metrics. The LabRGB line for VGG-based LPIPS is not included, as it would be identical to the Lab one. Also, note that the quantitative evaluation is performed on RGB images as opposed to training which is done for specific color spaces (RGB, YUV, Lab, and LabRGB).

Table 5 Quantitative evaluation of colorization results for different color spaces. Metrics are used to compare ground truth to every image in the 40k test set. Best and second best results by column are in bold and italicized respectively

Color space	Loss function	L1 ↓	L2 ↓	PSNR ↑	SSIM ↑	LPIPS ↓	FID ↓
RGB	L2	**0.04458**	0.00587	22.3136	*0.9255*	0.1606	**7.4223**
YUV	L2	*0.04469*	**0.00562**	22.5052	**0.9278**	**0.1593**	*7.6642*
Lab	L2	0.04488	*0.00585*	*22.3283*	0.9250	0.1613	8.1517
LabRGB	L2	0.04608	0.00589	22.2989	0.9209	0.1698	8.3413
RGB	LPIPS	0.04573	0.00577	22.3892	*0.9197*	0.1429	**3.0576**
YUV	LPIPS	*0.04460*	**0.00557**	22.5438	0.9097	**0.1400**	3.3260
Lab	LPIPS	**0.04374**	*0.00566*	22.4699	**0.9228**	*0.1403*	*3.2221*

Qualitative Evaluation

In this section, we qualitatively analyze the results obtained by training the network with different color spaces as explained in section "Learning Strategy for Different Color Spaces".

Figure 6 shows results on images and objects (here person skiing, stop sign and zebra) with strong contours that were highly present in the training set. The colorization of these images is really impressive for any color space. Nevertheless, YUV has the tendency to sometimes create artifacts that are not predictable. This is visible with the blue stain in the YUV-L2 zebra and the yellow spot in the YUV-LPIPS zebra. One can also notice that the overall colorization tends to be more homogeneous with LabRGB-L2 than with Lab-L2 as can be seen, for instance, on the wall behind the stop signs, the grass, and tree leaves in the zebra image which suggest that it might be better to compute losses over RGB images. A similar remark is valid for the VGG-based LPIPS results as can be seen, for instance, in the homogeneous colorization of the sky in the person skiing image where the loss is again computed over the RGB image. This indicates that there could be an additional influence on the results when using VGG-based LPIPS given that the predicted colored image is converted back to RGB before backpropagation.

Figure 7 presents results on images where the final colorization is not consistent over the whole image. On the first row, the color of the water is stopped by the chair legs. On the second row, the colors of the grass and the sky are not always similar on both sides of the hydrant. LabRGB seems to reduce this effect. This happens when strong contours seem to stop the colorization and are independent on the color space. Global coherency can only be obtained if the receptive field is large enough and that self-similarities present in natural images are preserved. These results highlight that efforts must be put on the design of architectures that would impose these constraints.

One major problem in automatic colorization results comes from color bleedings that occur as soon as contours are not strong enough. Figure 8 illustrates this problem in different contexts. On the first row, the color from the flowers bleeds

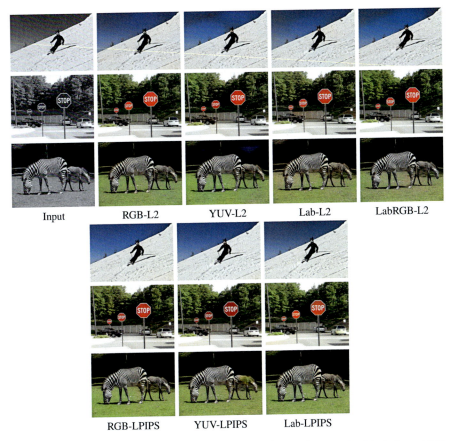

Fig. 6 Colorization results with different color spaces on images that contain objects, have strong structures, and have been seen many times in the training set. The three first rows are with L2 loss and the three last ones with VGG-based LPIPS

to the wall. On the second row, the green of the grass bleeds to the shorts. Finally, on the last row, the green of the grass bleeds to the neck of the background cow. These effects are independent from the color space or the loss. Some methods reduce this effect by introducing semantic information (e.g., Vitoria et al. 2020) or spatial localization (e.g., Su et al. 2020), while others achieve to reduce it by considering segmentation as an additional task (e.g., Kong et al. 2021). Note that with the VGG-based LPIPS, Lab color space provides more realistic result on the tennis man image.

Finally, Fig. 9 presents colorization of images containing many different objects. We see that final colors might be dependent on the color spaces and are more diverse and colorful with Lab color space. LabRGB strategy with L2 loss is probably the more realistic statement that holds with the VGG-based LPIPS.

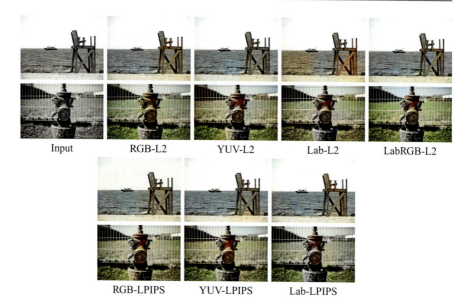

Fig. 7 Colorization results with different color spaces on images that exhibit strong structures that may lead to inconsistent spatial colors. The two first rows are with L2 loss and the two last ones with VGG-based LPIPS

The qualitative evaluation does not point to the same conclusion as the quantitative one. According to Table 5, the best colorization is obtained for YUV color space. However, the qualitative analysis shows that even if in some cases colors are brighter and more saturated in other ones, it creates unpredictable color stains (yellowish and blueish). This raises the question on the necessity to design specific metrics for the colorization task, which should be combined with user studies. Also, in the qualitative evaluation, one can observe that when working with LabRGB instead of Lab, the overall colorization result looks more stable and homogeneous as opposed to what is concluded in the quantitative evaluation.

Summary of qualitative analysis: Our analysis leads us to the following conclusions:

- There is no major difference in the results regarding the color space that is used.
- YUV color space sometimes generates color artifacts that are hardly predictable. This is probably due to clipping that is necessary to remain in the color space range of values.
- More realistic and consistent results are obtained when losses are computed in the RGB color space.
- There is no evidence justifying why most colorization methods in the literature choose to work with Lab. One can assume that this is mainly done to ease the colorization problem by working in a perceptual luminance-chrominance color space. In addition, differentiable color conversion libraries were not available up

Fig. 8 Colorization results with different color spaces on images that contain small contours which lead to color bleeding. The two first rows are with L2 loss and the two last ones with VGG-based LPIPS

to 2020 to apply a strategy as in Fig. 5c. In fact, the qualitative results show that when training on RGB, the luminance reconstruction is satisfying in all examples. Hence, there is no obvious reason why not to work directly in RGB color space.
- Same conclusions hold with different losses.

Fig. 9 Colorization results with different color spaces on images that contain several small objects which end up with different colors depending on the color spaces used. The three first rows are with L2 loss and the three last ones with VGG-based LPIPS

Generalization to Archive Images

Archive images present many artifacts due to acquisition methods (analog or numeric with different material qualities and manufacturing processes) and preservation conditions. They lead to images with different resolutions, film grains, scratches and holes, flickering, etc. Tools available for professional colorization

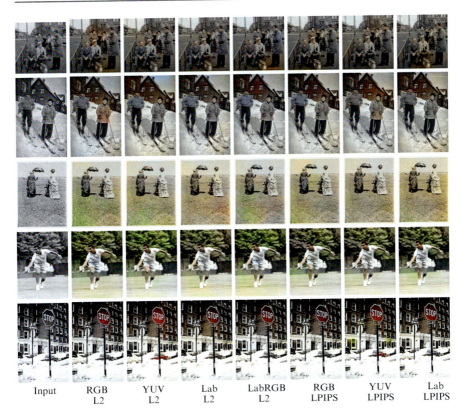

Fig. 10 Colorization results with different color spaces and L2 or VGG-based LPIPS on archive black and white images

enable artists to reach high-level quality images but require long human intervention. The current pipeline for professional colorization usually starts with restoration: denoising, deblurring, completion, super-resolution with off-the-shelf tools (e.g., Diamant) and manual correction. Next, images are segmented into objects and manually colorized by specialists with color spectrum that must be historically and artistically correct.

Automatic colorization methods could at least help professionals in the last step. Very few papers in the literature tackle old black and white images' colorization. In deep learning-based approaches, Vitoria et al. (2020) and Antic (2019) present some results on Legacy Black and White Photographs, while Luo et al. (2020) restore and colorize old black and white portraits. Wan et al. (2020b) focus on the restoration of old photos by training two variational autoencoders (VAE) to project clean and old photos to two latent spaces and to learn the translation between these latent spaces on synthetic paired data. Old photos are synthesized using Pascal VOC dataset's images.

Figure 10 presents some results obtained by applying the networks trained in this chapter on archive images. As we can observe on the second, third, and fourth rows, while on clean images sky and grass are often well colorized, it is not the case on archive images. This is probably due to the grain and noise in these images. Similarly the skin of persons is not as well colorized as in clean images. Color bleeding is here again a real issue. On the other hand, for objects with strong contours that were present in the database (e.g., stop sign), the colorization works very well. This indicates the importance on training or fine tuning on images that are related to the purpose of the network (many of the objects present in old black and white photos are not well represented with the most often used datasets).

Conclusion

This chapter has presented the role of the color spaces on automatic colorization with deep learning. Using a fixed standard network, we have shown, qualitatively and quantitatively, that the choice of the right color space is not straightforward and might depend on several factors such as the architecture or the type of images. With our architecture, the best quantitative results are obtained in YUV, while qualitative results rather teach us to compute losses in the RGB color space. We therefore argue that most efforts should be made on the architecture design. Furthermore, for all methods the final step consists in clipping final values to fit in the RGB color cube. This abrupt operation sometimes leads to artifacts with saturated pixels. An interesting topic for future research would be to learn a model that learns a projection into the color cube while preserving good image quality, similar to the geometric model from Pierre et al. (2015b). Future works should also include the development of methods that would give the possibility to produce several outputs in the same trend as HistoGAN (Afifi et al. 2021). Finally, if the purpose of colorization is often to enhance old black and white images, research papers rarely focus on this application. Strategies for better training or transfer learning must be developed in the future along with complete architectures that perform colorization together with other quality improvement methods such as super resolution, denoising, or deblurring.

Acknowledgments This study has been carried out with financial support from the French Research Agency through the PostProdLEAP project (ANR-19-CE23-0027-01) and from the EU Horizon 2020 research and innovation program NoMADS (Marie Skłodowska-Curie grant agreement No 777826). This chapter was written together with another chapter of the current handbook, ▶ Chap. 21, "Analysis of Different Losses for Deep Learning Image Colorization." All authors have contributed to both chapters.

References

Afifi, M., Brubaker, M.A., Brown, M.S.: HistoGAN: controlling colors of gan-generated and real images via color histograms. In: IEEE Conference on Computer Vision and Pattern Recognition, pp. 7941–7950 (2021)

Agustsson, E., Timofte, R.: Ntire 2017 challenge on single image super-resolution: dataset and study. In: Conference on Computer Vision and Pattern Recognition Workshops, pp. 126–135 (2017)

Antic, J.: Deoldify (2019). https://github.com/jantic/DeOldify

Anwar, S., Tahir, M., Li, C., Mian, A., Khan, F.S., Muzaffar, A.W.: Image colorization: a survey and dataset (2020). arXiv preprint arXiv:2008.10774

Arbelot, B., Vergne, R., Hurtut, T., Thollot, J.: Automatic texture guided color transfer and colorization. In: Expressive, Elsevier, pp. 21–32 (2016)

Arbelot, B., Vergne, R., Hurtut, T., Thollot, J.: Local texture-based color transfer and colorization. Comput. Graph. **62**, 15–27 (2017)

Arjovsky, M., Chintala, S., Bottou, L.: Wasserstein generative adversarial networks. In: International Conference on Machine Learning, vol. 70, pp. 214–223 (2017)

Bugeau, A., Ta, V.-T.: Patch-based image colorization. In: International Conference on Pattern Recognition, pp. 3058–3061 (2012)

Bugeau, A., Ta, V.-T., Papadakis, N.: Variational exemplar-based image colorization. IEEE Trans. Image Process. **23**(1), 298–307 (2014)

Cao, Y., Zhou, Z., Zhang, W., Yu, Y.: Unsupervised diverse colorization via generative adversarial networks. In: Joint European Conference on Machine Learning and Knowledge Discovery in Databases, pp. 151–166 (2017)

Charpiat, G., Hofmann, M., Schölkopf, B.: Automatic image colorization via multimodal predictions. In: European Conference on Computer Vision, pp. 126–139 (2008)

Cheng, Z., Yang, Q., Sheng, B.: Deep colorization. In: IEEE International Conference on Computer Vision, pp. 415–423 (2015)

Chia, A.Y.-S., Zhuo, S., Gupta, R.K., Tai, Y.-W., Cho, S.-Y., Tan, P., Lin, S.: Semantic colorization with internet images. In: ACM SIGGRAPH ASIA (2011)

Deng, J., Dong, W., Socher, R., Li, L.-J., Li, K., Fei-Fei, L.: Imagenet: a large-scale hierarchical image database. In: IEEE Conference on Computer Vision and Pattern Recognition, pp. 248–255 (2009)

Deshpande, A., Rock, J., Forsyth, D.: Learning large-scale automatic image colorization. In: IEEE International Conference on Computer Vision (2015)

Deshpande, A., Lu, J., Yeh, M.-C., Jin Chong, M., Forsyth, D.: Learning diverse image colorization. In: IEEE Conference on Computer Vision and Pattern Recognition, pp. 6837–6845 (2017)

Di Blasi, G., Reforgiato, D.: Fast colorization of gray images. In: Eurographics Italian, Eurographics Association (2003)

Ding, X., Xu, Y., Deng, L., Yang, X.: Colorization using quaternion algebra with automatic scribble generation. In: Advances in Multimedia Modeling, pp. 103–114 (2012)

Ding, K., Ma, K., Wang, S., Simoncelli, E.P.: Comparison of full-reference image quality models for optimization of image processing systems. Int. J. Comput. Vis. **129**(4), 1258–1281 (2021)

Dowson, D., Landau, B.: The Fréchet distance between multivariate normal distributions. J. Multivar. Anal. **12**(3), 450–455 (1982)

Drew, M.S., Finlayson, G.D.: Improvement of colorization realism via the structure tensor. Int. J. Image Graph. **11**(4), 589–609 (2011)

Ebner, M.: Color Constancy, vol. 7. Wiley, Hoboken (2007)

Efros, A., Leung, T.: Texture synthesis by non-parametric sampling. In: IEEE International Conference on Computer Vision, pp. 1033–1038 (1999)

Fairchild, M.D. : Color Appearance Models. Wiley, Hoboken (2013)

Fang, F., Wang, T., Zeng, T., Zhang, G.: A superpixel-based variational model for image colorization. IEEE Trans. Vis. Comput. Graph. 26(10), 2931–2943 (2019)

Gatys, L.A., Ecker, A.S., Bethge, M.: Image style transfer using convolutional neural networks. In: IEEE Conference on Computer Vision and Pattern Recognition, pp. 2414–2423 (2016a)

Gatys, L.A., Ecker, A.S., Bethge, M.: A neural algorithm of artistic style. J. Vis. 16(12), 326 (2016b)

Goodfellow, I., Pouget-Abadie, J., Mirza, M., Xu, B., Warde-Farley, D., Ozair, S., Courville, A., Bengio, Y.: Generative adversarial nets. In: Advances in Neural Information Processing Systems (2014)

Guadarrama, S., Dahl, R., Bieber, D., Norouzi, M., Shlens, J., Murphy, K.: Pixcolor: pixel recursive colorization. In: British Machine Vision Conference (2017)

Gu, S., Timofte, R., Zhang, R.: Ntire 2019 challenge on image colorization: report. In: Conference on Computer Vision and Pattern Recognition Workshops (2019)

Gupta, R.K., Chia, A.Y.-S., Rajan, D., Ng, E.S., Zhiyong, H.: Image colorization using similar images. In: ACM International Conference on Multimedia, pp. 369–378 (2012)

He, K., Gkioxari, G., Dollár, P., Girshick, R.: Mask R-CNN. In: Proceedings of the IEEE International Conference on Computer Vision, pp. 2961–2969 (2017)

He, M., Chen, D., Liao, J., Sander, P.V., Yuan, L.: Deep exemplar-based colorization. ACM Trans. Graph. 37(4), 1–16 (2018)

Heu, J., Hyun, D.-Y., Kim, C.-S., Lee, S.-U.: Image and video colorization based on prioritized source propagation. In: IEEE International Conference on Image Processing, pp. 465–468 (2009)

Heusel, M., Ramsauer, H., Unterthiner, T., Nessler, B., Hochreiter, S.: GANs trained by a two timescale update rule converge to a local Nash equilibrium. In: Advances in Neural Information Processing Systems, vol. 30 (2017)

Ho, J., Kalchbrenner, N., Weissenborn, D., Salimans, T.: Axial attention in multidimensional transformers. arXiv preprint arXiv:1912.12180 (2019)

Huang, Y.-C., Tung, Y.-S., Chen, J.-C., Wang, S.-W., Wu, J.-L.: An adaptive edge detection based colorization algorithm and its applications. In: ACM International Conference on Multimedia, pp. 351–354 (2005)

Huang, G.B., Ramesh, M., Berg, T., Learned-Miller, E.: Labeled faces in the wild: a database for studying face recognition in unconstrained environments. Technical Report 07-49, University of Massachusetts, Amherst (2007)

Iizuka, S., Simo-Serra, E., Ishikawa, H.: Let there be color!: joint end-to-end learning of global and local image priors for automatic image colorization with simultaneous classification. ACM Trans. Graph. 35(4), 1–11 (2016)

Irony, R., Cohen-Or, D., Lischinski, D.: Colorization by example. In: Eurographics Conference on Rendering Techniques (2005)

Isola, P., Zhu, J.-Y., Zhou, T., Efros, A.A.: Image-to-image translation with conditional adversarial networks. In: IEEE Conference on Computer Vision and Pattern Recognition, pp. 1125–1134 (2017)

Johnson, J., Alahi, A., Fei-Fei, L.: Perceptual losses for real-time style transfer and super-resolution. In: European Conference on Computer Vision, pp. 694–711 (2016)

Karras, T., Laine, S., Aittala, M., Hellsten, J., Lehtinen, J., Aila, T.: Analyzing and improving the image quality of stylegan. In: IEEE Conference on Computer Vision and Pattern Recognition, pp. 8110–8119 (2020)

Kawulok, M., Kawulok, J., Smolka, B.: Discriminative textural features for image and video colorization. IEICE Trans. Inf. Syst. 95-D(7), 1722–1730 (2012)

Kong, G., Tian, H., Duan, X., Long, H.: Adversarial edge-aware image colorization with semantic segmentation. IEEE Access 9, 28194–28203 (2021)

Krizhevsky, A., Hinton, G., et al.: Learning multiple layers of features from tiny images. Technical report, University of Toronto (2009)

Kumar, M., Weissenborn, D., Kalchbrenner, N.: Colorization transformer. arXiv preprint arXiv:2102.04432 (2021)

Lagodzinski, P., Smolka, B.: Digital image colorization based on probabilistic distance transformation. In: 50th International Symposium ELMAR, vol. 2, pp. 495–498 (2008)

Larsson, G., Maire, M., Shakhnarovich, G.: Learning representations for automatic colorization. In: European Conference on Computer Vision, pp. 577–593 (2016)

Levin, A., Lischinski, D., Weiss, Y.: Colorization using optimization. ACM Trans. Graph. **23**(3), 689–694 (2004)

Lézoray, O., Ta, V.-T., Elmoataz, A.: Nonlocal graph regularization for image colorization. In: International Conference on Pattern Recognition, pp. 1–4 (2008)

Li, B., Lai, Y.-K., Rosin, P.L.: Example-based image colorization via automatic feature selection and fusion. Neurocomputing **266**, 687–698 (2017a)

Li, B., Zhao, F., Su, Z., Liang, X., Lai, Y.-K., Rosin, P.L.: Example-based image colorization using locality consistent sparse representation. IEEE Trans. Image Process. **26**(11), 5188–5202 (2017b)

Li, B., Lai, Y.-K., John, M., Rosin, P.L.: Automatic example-based image colorization using location-aware cross-scale matching. IEEE Trans. Image Process. **28**(9), 4606–4619 (2019)

Li, B., Lai, Y.-K., Rosin, P.L.: A review of image colourisation. In: Handbook of Pattern Recognition and Computer Vision, p. 139. World Scientific, Singapore (2020)

Lin, T.-Y., Maire, M., Belongie, S., Hays, J., Perona, P., Ramanan, D., Dollár, P., Zitnick, C.L.: Microsoft COCO: common objects in context. In: European Conference on Computer Vision, pp. 740–755 (2014)

Ling, Y., Au, O.C., Pang, J., Zeng, J., Yuan, Y., Zheng, A.: Image colorization via color propagation and rank minimization. In: IEEE International Conference on Image Processing, pp. 4228–4232 (2015)

Liu, S., Zhang, X.: Automatic grayscale image colorization using histogram regression. Pattern Recogn. Lett. **33**(13), 1673–1681 (2012)

Luan, Q., Wen, F., Cohen-Or, D., Liang, L., Xu, Y.-Q., Shum, H.-Y.: Natural image colorization. In: Eurographics Conference on Rendering Techniques, pp. 309–320 (2007)

Luo, X., Zhang, X., Yoo, P., Martin-Brualla, R., Lawrence, J., Seitz, S.M.: Time-travel rephotography. arXiv preprint arXiv:2012.12261 (2020)

Mouzon, T., Pierre, F., Berger, M.-O.: Joint CNN and variational model for fully-automatic image colorization. In: Scale Space and Variational Methods in Computer Vision, pp. 535–546 (2019)

Nazeri, K., Ng, E., Ebrahimi, M.: Image colorization using generative adversarial networks. In: International Conference on Articulated Motion and Deformable Objects, pp. 85–94 (2018)

Oord, A.V.D., Kalchbrenner, N., Vinyals, O., Espeholt, L., Graves, A., Kavukcuoglu, K.: Conditional image generation with PixelCNN decoders. In: Advances in Neural Information Processing Systems (2016)

Pang, J., Au, O.C., Tang, K., Guo, Y.: Image colorization using sparse representation. In: IEEE International Conference on Acoustics, Speech, and Signal Processing, pp. 1578–1582 (2013)

Pierre, F., Aujol, J.-F.: Recent approaches for image colorization. In: Handbook of Mathematical Models and Algorithms in Computer Vision and Imaging, springer (2020)

Pierre, F., Aujol, J.-F., Bugeau, A., Ta, V.-T.: A unified model for image colorization. In: European Conference on Computer Vision Workshops, pp. 297–308 (2014)

Pierre, F., Aujol, J.-F., Bugeau, A., Papadakis, N., Ta, V.-T.: Luminance-chrominance model for image colorization. SIAM J. Imaging Sci. **8**(1), 536–563 (2015a)

Pierre, F., Aujol, J.-F., Bugeau, A., Ta, V.-T.: Luminance-Hue Specification in the RGB Space. In: Scale Space and Variational Methods in Computer Vision, pp. 413–424. Springer, Cham (2015b)

Pucci, R., Micheloni, C., Martinel, N.: Collaborative image and object level features for image colourisation. In: IEEE Conference on Computer Vision and Pattern Recognition, pp. 2160–2169 (2021)

Radford, A., Metz, L., Chintala, S.: Unsupervised representation learning with deep convolutional generative adversarial networks. In: International Conference on Learning Representations (2016)

Ren, X., Malik, J.: Learning a classification model for segmentation. In: IEEE International Conference on Computer Vision, pp. 10–17 (2003)

Riba, E., Mishkin, D., Ponsa, D., Rublee, E., Bradski, G.: Kornia: an open source differentiable computer vision library for PyTorch. In: Winter Conference on Applications of Computer Vision, pp. 3674–3683 (2020)

Royer, A., Kolesnikov, A., Lampert, C.H.: Probabilistic image colorization. In: British Machine Vision Conference (2017)

Salimans, T., Karpathy, A., Chen, X., Kingma, D.P.: PixelCNN++: improving the PixelCNN with discretized logistic mixture likelihood and other modifications. arXiv preprint arXiv:1701.05517 (2017)

Simonyan, K., Zisserman, A.: Very deep convolutional networks for large-scale image recognition. In: International Conference on Learning Representations (2015)

Su, J.-W., Chu, H.-K., Huang, J.-B.: Instance-aware image colorization. In: IEEE Conference on Computer Vision and Pattern Recognition, pp. 7968–7977 (2020)

Tai, Y.-W., Jia, J., Tang, C.-K.: Local color transfer via probabilistic segmentation by expectation-maximization. In: IEEE Conference on Computer Vision and Pattern Recognition, pp. 747–754 (2005)

Vitoria, P., Raad, L., Ballester, C.: ChromaGAN: adversarial picture colorization with semantic class distribution. In: Winter Conference on Applications of Computer Vision, pp. 2445–2454 (2020)

Wan, S., Xia, Y., Qi, L., Yang, Y.-H., Atiquzzaman, M.: Automated colorization of a grayscale image with seed points propagation. IEEE Trans. Multimedia **22**(7), 1756–1768 (2020a)

Wan, Z., Zhang, B., Chen, D., Zhang, P., Chen, D., Liao, J., Wen, F.: Bringing old photos back to life. In: IEEE Conference on Computer Vision and Pattern Recognition, pp. 2747–2757 (2020b)

Wang, S., Zhang, Z.: Colorization by matrix completion. In: AAAI Conference on Artificial Intelligence (2012)

Wang, Z., Bovik, A.C., Sheikh, H.R., Simoncelli, E.P.: Image quality assessment: from error visibility to structural similarity. IEEE Trans. Image Process. **13**(4), 600–612 (2004)

Welsh, T., Ashikhmin, M., Mueller, K. Transferring color to greyscale images. ACM Trans. Graph. **21**(3), 277–280 (2002)

Xiao, J., Hays, J., Ehinger, K.A., Oliva, A., Torralba, A.: Sun database: large-scale scene recognition from abbey to zoo. In: IEEE Conference on Computer Vision and Pattern Recognition, pp. 3485–3492 (2010)

Yao, Q., James, T.K.: Colorization by patch-based local low-rank matrix completion. In: AAAI Conference on Artificial Intelligence (2015)

Yatziv, L., Sapiro, G.: Fast image and video colorization using chrominance blending. IEEE Trans. Image Process. **15**(5), 1120–1129 (2006)

Yoo, S., Bahng, H., Chung, S., Lee, J., Chang, J., Choo, J.: Coloring with limited data: few-shot colorization via memory augmented networks. In: IEEE Conference on Computer Vision and Pattern Recognition (2019)

Yu, F., Seff, A., Zhang, Y., Song, S., Funkhouser, T., Xiao, J.: LSUN: construction of a large-scale image dataset using deep learning with humans in the loop. arXiv preprint arXiv:1506.03365 (2015)

Zhang, R., Isola, P., Efros, A.A.: Colorful image colorization. In: European Conference on Computer Vision, pp. 649–666 (2016)

Zhang, R., Zhu, J.-Y., Isola, P., Geng, X., Lin, A.S., Yu, T., Efros, A.A.: Real-time user-guided image colorization with learned deep priors. ACM Trans. Graph. **36**, 1–11 (2017)

Zhang, R., Isola, P., Efros, A.A., Shechtman, E., Wang, O.: The unreasonable effectiveness of deep features as a perceptual metric. In: IEEE Conference on Computer Vision and Pattern Recognition, pp. 586–595 (2018)

Zhou, B., Lapedriza, A., Khosla, A., Oliva, A., Torralba, A.: Places: a 10 million image database for scene recognition. IEEE Trans. Pattern Anal. Mach. Intell. **40**(6), 1452–1464 (2017)

Variational Model-Based Deep Neural Networks for Image Reconstruction

23

Yunmei Chen, Xiaojing Ye, and Qingchao Zhang

Contents

Introduction	880
Learned Algorithm for Specified Optimization Problem	882
Structured Image Reconstruction Networks	885
Proximal Point Network	886
ISTA-Net	888
ADMM-Net	891
Variational Network	894
Primal-Dual Network	896
Learnable Descent Algorithm	898
Concluding Remarks	901
References	905

Abstract

In recent years, we have witnessed unprecedented growth of research interests in deep learning approaches to image reconstruction. A majority of these approaches are inspired by the well-developed variational method and associated optimization algorithms for the inverse problem of image reconstruction. These approaches mimic the iterative schemes of the standard optimization algorithms but integrate learnable components to form structured deep neural networks and employ large amount of observation data to train the networks for the specific reconstruction tasks. They have demonstrated significantly improved

---

Y. Chen (✉) · Q. Zhang
Department of Mathematics, University of Florida, Gainesville, FL, USA
e-mail: yun@math.ufl.edu; qingchaozhang@ufl.edu

X. Ye
Department of Mathematics and Statistics, Georgia State University, Atlanta, GA, USA
e-mail: xye@gsu.edu

© Springer Nature Switzerland AG 2023
K. Chen et al. (eds.), *Handbook of Mathematical Models and Algorithms in Computer Vision and Imaging*, https://doi.org/10.1007/978-3-030-98661-2_57

empirical performance and require much lower computational cost compared to the classical methods in a variety of applications. We provide the details of the derivations, the network architectures, and the training procedures for several typical networks in this field.

Keywords

Image reconstruction · Variational method · Deep neural network · Optimization

Introduction

Variational method has been one of the most mature and effective approaches for solving inverse problems in imaging Aubert and Vese (1997), Dal Maso et al. (1992), Koepfler et al. (1994), and Scherzer et al. (2009). In the context of image reconstruction, the inverse problem can be formulated as an optimization in a general form as follows:

$$\min_{u} g(u) + h(u), \tag{1}$$

where u is the image to be reconstructed, $h(u)$ is the data fidelity that measures the discrepancy between u and the acquired data (often in the transformed domain), and $g(u)$ is a regularization term which imposes the prior knowledge or our preference on the solution u.

To instantiate the variational method (1), we may consider the image reconstruction problem with total-variation (TV) regularization for compressive sensing magnetic resonance imaging (CS-MRI) in the discretized form: Suppose that the gray-scale image u to be reconstructed is defined on the two-dimensional $\sqrt{n} \times \sqrt{n}$ mesh grid (thus a total of n pixels) representing its square domain $[0, 1]^2$. Then u can be interpreted as a vector in $\mathbb{R}^n$ where its ith component $u_i \in \mathbb{R}$ is the integral (or average) of the image intensity value over the ith pixel for $i = 1, \ldots, n$. MRI scanners can acquire the Fourier coefficients of u, from which one can recover u simply by applying inverse Fourier transform. For fast imaging in CS-MRI, we only acquire a fraction of Fourier coefficients $b \in \mathbb{C}^m$ with $m < n$, which relates to u by $b = P\mathcal{F}u + e$ where $\mathcal{F} \in \mathbb{C}^{n \times n}$ is the discrete Fourier transform matrix, $P \in \mathbb{R}^{m \times n}$ is a binary selection matrix (one entry as 1 and the rest as 0 in each row) indicating the indices of the sampled Fourier coefficients, and $e \in \mathbb{C}^m$ represents the unknown noise in data acquisition. Then the data fidelity term $h(u)$ in (1) can be set to $(1/2) \cdot \|P\mathcal{F}u - b\|_2^2$. For fast imaging, m is often much smaller than n and hence we need additional regularization $g(u)$ in (1) to ensure robust and stable recovery of u. TV is one of the most commonly used regularization in image reconstruction–the simplified version of TV in the discrete setting is $TV(u) = \sum_{i=1}^{n} \|D_i u\|_2$ where $D_i \in \mathbb{R}^{2 \times n}$ is binary and has only two nonzero entries (1 and -1) corresponding to the forward finite difference approximations to partial derivatives along the coordinate axes at pixel i. Hence the regularization can be set to $g(u) = \mu TV(u)$

for some user-chosen weight parameter $\mu > 0$ in (1). The motivation of using TV as regularization is that images with small TV tend to have distinct constant intensity values in different regions and sharp intensity change on the boundary between two regions, hence displaying the included objects with clear intensity contrasts. The minimization in (1) thus reflects the principle of the variational method for image recovery—we want to find the minimizer u such that it is consistent to the observed data (small value of $h(u)$) and meanwhile has desired regularity (small value of $g(u)$). To this point, (1) becomes an optimization problem of $u \in \mathbb{R}^n$, for which we can apply a proper numerical optimization algorithm and solve for u from (1).

The variational method yields a concise and elegant formulation of image reconstruction as in (1). It has achieved great success in image reconstruction thanks to the fast developments of numerical optimization techniques in the past decades. However, there are several main issues associated with this approach.

The first issue with (1) is the choice of regularization $g(u)$. There are numerous regularization terms proposed in the literature. Although many of them have proven robust in practice, they are often overly simplified and cannot capture the fine details in medical images which are critical in diagnosis and treatment. For example, TV regularization is known for its "staircase" effect due to its promotion of sparse gradients, such that the reconstructed images tend to be piecewise constant which are not ideal approximations to the real-world images. For example, important fine structures and minor contrast changes can be smeared in the reconstructed image using TV regularization, which is unacceptable for applications that require high image quality.

The second issue is the parameter tuning. To achieve desired balance between noise reduction and faithful structural reconstruction, the parameters of a reconstruction model (e.g., $\mu > 0$ mentioned above) and its associated optimization algorithm (such as step sizes) need to be carefully tuned. Unfortunately, the image quality is often very sensitive to these parameters; and the optimal parameters are also shown to be highly dependent on the specific acquisition settings and imaging datasets.

Last but not least, the reconstruction time of iterative optimization algorithms is also a major concern on their applications in real-world problems. Despite that the efficiency of optimization algorithms is continuously being improved, these algorithms, even for convex problems, often require hundreds of iterations or more to converge, which result in long computational time.

The issues with the classical variational methods and optimization algorithms mentioned above inspired a new class of deep learning-based approaches. Deep learning Goodfellow et al. (2016) with deep neural networks (DNNs) as the core component has achieved great success in a variety of real-world applications, including computer vision (He et al. 2016; Krizhevsky et al. 2012; Zeiler and Fergus 2014), natural language processing (Devlin et al. 1810; Hinton et al. 2012; Sarikaya et al. 2014; Socher et al. 2012; Vaswani et al. 2017), medical imaging (Hammernik et al. 2018; Schlemper et al. 2018; Sun et al. 2016), etc. DNNs have provable representation power and can be trained with little or no knowledge about the underlying functions. However, there are several major issues of such standard deep learning approaches: (i) Generic DNNs may fail to approximate the desired functions if the training data is scarce; (ii) the training of these DNNs is prone to overfitting, noises, and outliers; and (iii) the trained DNNs are mostly "blackboxes" without rigorous mathematical justification and can be very difficult to interpret.

To mitigate the aforementioned issues of DNNs, a class of *learnable optimization algorithms* (LOAs) has been proposed recently. In brief, the architectures of the neural networks in LOAs mimic the iterative scheme of the optimization algorithms, also known of "unrolling" the optimization algorithms. More specifically, these reconstruction networks are composed of a small number of phases, where each phase mimics one iteration of a classical, optimization-based reconstruction algorithm. In most cases, the terms corresponding to the manually designed regularization in the classical methods are parameterized by multilayer perceptrons whose parameters are to be learned adaptively in the offline training process with lots of imaging data. After training, these networks work as fast feedforward mappings with extremely low computational cost, so that the reconstruction of new images can be performed on the fly. These methods combine the best parts of variational methods and deep learning for fast and adaptive image reconstruction. In the next section, we first consider the algorithms that are designed to solve a prescribed model in the form of (1). Section "Structured Image Reconstruction Networks" is dedicated to the class of deep reconstruction networks that can learn the variational model or algorithm such that the outputs are high-quality reconstructions of the images.

Learned Algorithm for Specified Optimization Problem

Learned optimization algorithms are modifications of traditional optimization algorithms by including trainable components, such as deep neural networks or the layers, for fast and adaptive numerical solution. This approach is motivated by the viewing the iterative scheme in traditional optimization algorithm (e.g., gradient descent) as a feedforward neural network with repeated, predesigned layers. The main structures of these algorithms largely adopt those of the original optimization algorithms. To make these algorithms more adaptive to the given problem, learnable components are introduced so they can improve over the original algorithms using the available data.

In this section, we showcase several learned optimization algorithms for the well-known l_1 minimization problem as follows:

$$\min_u \mu \|u\|_1 + \frac{1}{2} \|Au - b\|^2, \tag{2}$$

where $A \in \mathbb{R}^{m \times n}$, $b \in \mathbb{R}^m$, and the parameter $\mu > 0$ are given. The solution of (2) is also known as the least absolute shrinkage and selection operator (lasso) or sparse recovery since the solution u fits the observed data b in the data fidelity term $h(u) := (1/2) \cdot \|Au - b\|^2$ and meanwhile tends to have only a small amount of nonzero components (hence sparse) due to the l_1 regularization $g(u) := \mu \|u\|_1$. A basic method for solving (2) is called the iterative shrinkage-threshold algorithm (ISTA). To solve (2), ISTA first approximates $h(u)$ by its first-order Taylor expansion at the previous iterate $u^{(k)}$ plus a quadratic penalty term with weight $1/(2\alpha)$ in each iteration k as follows:

$$h(u) \approx h(u^{(k)}) + \langle \nabla h(u^{(k)}), u - u^{(k)} \rangle + \frac{1}{2\alpha} \|u - u^{(k)}\|^2$$

$$= \frac{1}{2\alpha} \|u - (u^{(k)} - \alpha \nabla h(u^{(k)}))\|^2 + \text{const}, \tag{3}$$

where we completed the square to obtain the equality above, and the term "const" represents a constant independent of u. As a result, ISTA generates the next iterate $u^{(k+1)}$ by

$$u^{(k+1)} = \arg\min_u \left\{ g(u) + \frac{1}{2\alpha} \|u - (u^{(k)} - \alpha \nabla h(u^{(k)}))\|^2 \right\}, \tag{4}$$

where the constant term is omitted since it does not affect the result $u^{(k+1)}$ in (4). To obtain $u^{(k+1)}$ in (4), it is essential to find the solution of the proximity operator prox_g defined below for any given $z \in \mathbb{R}^n$:

$$\text{prox}_g(z) := \arg\min_x \left\{ g(x) + \frac{1}{2} \|x - z\|^2 \right\}. \tag{5}$$

With $g(x) := \mu \|x\|_1$, the proximity operator prox_g has a closed form solution, called the shrinkage operator S_μ. That is, the ith component of $S_\mu(z) = \text{prox}_g(z) \in \mathbb{R}^n$ is

$$[S_\mu(z)]_i = [\text{prox}_g(z)]_i = \text{sign}(z_i) \cdot \max\{|z_i| - \mu, 0\}. \tag{6}$$

Therefore, $S_\mu(z)$ "shrinks" the magnitude of each component of its argument z by μ; if the magnitude is smaller than μ, then it becomes 0 after the shrinkage. Combining (4), (5), and (6) yields the scheme of ISTA:

$$u^{(k+1)} = S_{\mu/L} \left(u^{(k)} - \frac{1}{L} A^\top (Au^{(k)} - b) \right), \tag{7}$$

where α is set to the optimal value $1/L$ in (7) and L is the largest eigenvalue of $A^\top A$ (i.e., the Lipschitz constant of $\nabla h(u) = A^\top (Au - b)$). It can be shown that, starting from any initial guess $u^{(0)}$, ISTA (7) generates a sequence $\{u^{(k)}\}$ that converges to a solution of (2) at a sublinear rate of $O(1/k)$ in function value.

However, the practical performance of ISTA is not satisfactory as it often requires hundreds to thousands of iterations to obtain an acceptable approximation to the solution. Although there are a variety of optimization techniques to improve the convergence of ISTA, the traditional variational formulation and optimization still fall short in real-world applications due to the relatively slow convergence and the issues mentioned in section "Introduction". Inspired by the great success of deep learning, for a fixed A, we may ask whether it is possible to learn the terms, such as μ, L, and even $A^\top$, in (7) adaptively if we have many instances of b and their corresponding solutions to (2). In Gregor and LeCun (2010), this approach is examined and results in the learned ISTA (LISTA) formed as a K-layer feedforward neural network:

$$u^{(k+1)} = \sigma_k(W_1^{(k)} b + W_2^{(k)} u^{(k)}) \tag{8}$$

for $k = 0, \ldots, K - 1$. In LISTA (8), the linear mappings $W_1^{(k)}$, $W_2^{(k)}$ and the nonlinear mapping (can also be a preselected nonlinear activation function) σ_k can be learned, such that the final output $u^{(K)}$, as a function of these parameters $\Theta := (\ldots, W_1^{(k)}, W_2^{(k)}, \sigma_k, \ldots)$, is close to a solution u^* of (2) for a given b. More specifically, given N pairs of training data $\{(b_j, u_j^*) : 1 \leq j \leq N\}$, where $b_j \in \mathbb{R}^m$ is the input data of the optimization problem (2) and $u_j^* \in \mathbb{R}^n$ is the corresponding ground truth (e.g., solution obtained by solving the minimization problem (2) with b_j using some classical optimization algorithm to high accuracy), then one can learn the optimal network parameter Θ^* by solving the minimization problem

$$\min_\Theta \frac{1}{N} \sum_{j=1}^N \|u^{(K)}(b_j; \Theta) - u_j^*\|^2$$

where $u^{(K)}(b; \Theta)$ denotes the output of the K-phase network with parameter Θ and input data b. By training the parameter Θ with various of b and the corresponding u^*, LISTA can find an effective path from $u^{(0)}$ to $u^{(K)}$ using the learned Θ^*. If training result is satisfactory with a small K (e.g., $K = 10$), then LISTA, as a feedforward neural network, is expected to compute good approximation of u^* given new input b on the fly. Note that LISTA (8) reduces to ISTA (7) if the parameters are not learned but pre-defined as $W_1^{(k)} = A^\top/L$, $W_2^{(k)} = I - A^\top A/L$, and $\sigma_k(\cdot) = S_{\mu/L}(\cdot)$ for all k. It is shown that LISTA can achieve similar solution accuracy with iteration number K 18 to 35 times fewer than that required in ISTA or FISTA for problems with dimension 100 to 400 (Gregor and LeCun 2010).

In recent years, there have been a number of follow-up research works that exploit the properties and variations of LISTA. In Chen et al. (2018), a simplified version of LISTA is proposed:

$$u^{(k+1)} = S_{\mu/L}\left(u^{(k)} - \frac{1}{L} W^\top (A u^{(k)} - b)\right), \tag{9}$$

with learnable W, and the convergence of (9) for solving (2) is also established in Chen et al. (2018) and Liu et al. (2019). In Sprechmann et al. (2015), LISTA is extended to learnable pursuit process architectures for structured sparse and robust low rank models derived from proximal gradient algorithm. It is shown that such network architecture can approximate the exact sparse or low rank representation at a fraction of the complexity of the standard optimization methods. In Xin et al. (2016), a learned iterative hard thresholding (IHT) algorithm where σ_k is replaced by a hard thresholding operator H_k is developed, and its potential to recover minimal l_0 norm solution is shown both theoretically and empirically. The work Borgerding et al. (2017) developed a learned approximate message passing (LAMP) algorithm for the lasso problem (2):

$$v^{(k+1)} = \beta_k v^{(k)} - Au^{(k)} + b, \tag{10a}$$

$$u^{(k+1)} = S_{\mu_k}(u^{(k)} + A^\top v^{(k+1)}). \tag{10b}$$

In contrast to LISTA, LAMP (10) includes a residual $v^{(k)}$ in each layer k, which performs shrinkage dependent on k. By the inclusion of the "Onsager correction" term $\beta_k v^{(k)}$ to decouple errors across layers, LAMP appears to outperform LISTA in accuracy empirically. For example, on synthetic data with Gaussian matrix A, LAMP takes 7 iteration numbers to obtain the normalized mean square error (NMSE) -34dB, whereas LISTA takes 15 iterations (Borgerding et al. 2017).

The aforementioned learned optimization algorithms are for unconstrained minimizations. Recently, the work in Xie et al. (1905) developed an algorithm, called the differentiable linearized alternating direction method of multipliers (D-LADMM), can be used to solve problems with linear equality constraints. D-LADMM is a K-layer linearized ADMM-inspired deep neural network, which is obtained by using learnable weights in the classical linearized ADMM and generalizing the proximal operator to learnable activation functions. It is proved that there exist a set of learnable parameters for D-LADMM to generate globally converged solutions.

To this point, we have seen several instances of modifying the ISTA (7) to obtain deep neural networks with trainable components to solve (2). Each iteration of ISTA is transformed into one layer of a neural network, the parameters of which are then trained using available imaging data. Once properly trained, these networks can often achieve more accurate approximations of the solution in much less time than the traditional approaches. Global convergence results, sometimes even better than the original optimization algorithms, have been established for several of these methods. However, most of these methods are restricted to the variational model (1) with l_1 or l_0 regularization, so that the proximity operators can yield closed-form shrinkage as the nonlinear activation function. It remains as an open problem on extending this type of methods to handle more general or learnable regularization.

Structured Image Reconstruction Networks

In this section, we introduce several deep neural networks inspired by classical optimization algorithms for image reconstruction. Unlike the learned algorithms discussed in section "Learned Algorithm for Specified Optimization Problem", these networks aim at solving the given reconstruction problem demonstrated by training dataset (often includes ground truth images), rather than any prescribed optimization problem such as the lasso (2). As a result, they do not require manually designed regularization and specified objective function but can implicitly learn an adaptive regularization using the training data. This class of methods has become the mainstream for deep learning-based image reconstruction research in recent years.

The optimization-inspired reconstruction networks in this section also share the same main feature: each phase of these networks corresponds to one iteration of the classical optimization. More specifically, the data fidelity term h in (1) that describes the relation between image and acquired data is largely preserved as in optimization algorithms. However, unlike the methods in section "Learned Algorithm for Specified Optimization Problem", the regularization term g is unknown but can be replaced by neural networks whose parameters are learned adaptively from data.

In the remainder of this section, we introduce several reconstruction neural networks developed along this line. Most of these networks can be applied to a wide range of image reconstruction problems as they are customized to learn from the training data directly rather than for any specific imaging application or modality. The training process can be time-consuming but is performed offline. Once trained properly, however, they serve as fast feedforward mappings that reconstruct high-quality images of the same type as those in the training dataset.

Proximal Point Network

A group of deep neural networks inspired by variational methods and optimization algorithms directly leverage the popular deep neural network structures into the optimization schemes. Considering the variational model (1) with general g and h, we can rewrite its proximal point algorithm (4) as an equivalent two-step scheme by introducing an auxiliary variable $r^{(k)} = u^{(k-1)} - \alpha \nabla h(u^{(k-1)})$ and using the definition of the proximity operator in (5):

$$r^{(k)} = u^{(k-1)} - \alpha \nabla h(u^{(k-1)}), \tag{11a}$$

$$u^{(k)} = \text{prox}_{\alpha g}(r^{(k)}). \tag{11b}$$

As the data fidelity h is formulated based on the definitive relation between image and acquired data, such as $h(u) = (1/2) \cdot \|P\mathcal{F}u - b\|^2$ in CS-MRI as shown in section "Introduction", it is often kept unmodified in (11a). Moroeover, the step size α can be set to α_k which is not manually chosen but learned during the training process. On the other hand, the proximal term in (11b) is due to the regularization g and performs as an image "denoiser" that modifies inputs $r^{(k)}$ to obtain an improved image $u^{(k)}$. Instead of choosing regularization g manually and solving (11b) in each iteration, we can directly parametrize its proximity operator $\text{prox}_{\alpha g}$ as a learnable denoiser parametrized as convolutional neural network (CNN) (Goodfellow et al. 2016). Moreover, we can use the residual network (ResNet) structure proposed in

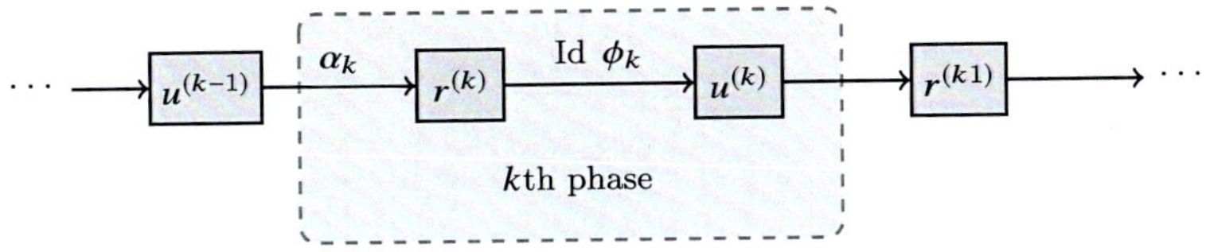

Fig. 1 Architecture of the proximal point network (11a) and (12). The kth phase updates $r^{(k)}$ and $u^{(k)}$. The dependencies of each variable on other variables are shown as incoming arrows, and the network parameters used for update are labeled next to the corresponding arrows

He et al. (2016) for the CNN which proves to be more effective for reducing training error in imaging applications. Namely, we replace the proximity operator $\text{prox}_{\alpha g}$ in (11b) by a denoising network (Zhang et al. 2017):

$$u^{(k)} = r^{(k)} + \phi_k(r^{(k)}) \tag{12}$$

where ϕ_k is a standard multiplayer CNN that maps $r^{(k)}$ to the residual between $u^{(k)}$ and $r^{(k)}$. The architecture of the proximal point network given by (11a) and (12) is illustrated in Fig. 1, where each arrow indicates a mapping from its input to the output with the required network parameters labeled next to it.

Let Θ denote the collection of learnable parameters in ϕ_k (e.g., the convolutional kernels and the biases) and algorithm parameters (e.g., $\alpha_k > 0$) for all $k = 1, \ldots, K$, and then the output after K cycles (phases) of (11a) and (12) is a function of Θ for any given imaging data b. Denote this output by $u^{(K)}(b; \Theta)$, which is the output of any given image data b passing through this network with parameter Θ; we can form the loss function of Θ by regression as:

$$L(\Theta; b, u^*) = \frac{1}{2} \|u^{(K)}(b; \Theta) - u^*\|^2, \tag{13}$$

where u^* is the ground truth image corresponding to the (possibly noisy and incomplete) imaging data b, both given in the training data. By feeding in a large amount of instances of form (b, u^*), we can solve for the minimizer Θ^* of the sum of L as in (13) over all of these instances. Then the deep reconstruction network with K phases, each consisting of (11a) and (12), is a feedforward neural network with parameters Θ^* for fast image reconstruction given any new coming data b.

The proximal point network can be applied to a variety of imaging applications, including image denoising, image deblurring, and image super-resolution by replacing the proximal operator by a denoiser network in regularization subproblem of half-quadratic splitting algorithm (Zhang et al. 2017). In Zhang et al. (2017), ϕ_k is designed to contain 7 dilated convolutions with 64 feature maps in each middle layer, where ReLU activation function is used after the first convolution, and both batch normalization (BN) and ReLU are used in every convolution thereafter. The training data is composed of 256×4000 image patches of size 35×35 cropped from the BSD400 (Martin et al. 2001), 400 images from ImageNet validation set

(Deng et al. 2009), and 4,744 Waterloo Exploration images (Ma et al. 2016). They evaluate their results on BSD68 (Roth and Black 2009), Set5, and Set14 (Timofte et al. 2014), respectively. In Zhang and Ghanem (2018), IRCNN is compared with several other methods on Set11 (Kulkarni et al. 2016) with various sampling ratios, and the results will be presented later in this section.

The work developed in Cheng et al. (2019), Chun et al. (2019), Meinhardt et al. (2017), Rick Chang et al. (2017), Wang et al. (2016), and Zhang et al. (2017) can all be considered as variations of the method described above. For instance, CNN denoiser has been placed in the proximal gradient descent algorithm in Meinhardt et al. (2017), subproblem in half-quadratic splitting in Zhang et al. (2017), subproblem in ADMM in Meinhardt et al. (2017) and Rick Chang et al. (2017), and subproblems in primal-dual algorithm in Cheng et al. (2019), Meinhardt et al. (2017), and Wang et al. (2016).

ISTA-Net

ISTA-Net Zhang and Ghanem (2018) is a deep neural network architecture for image reconstruction inspired by ISTA as given in (7). Recall that ISTA is originally derived to solve the l_1 minimization problem (2), i.e., (1) with $g(u) = \mu \|u\|_1$ and $h(u) = (1/2) \cdot \|Au - b\|^2$, as we showed in section "Learned Algorithm for Specified Optimization Problem". For image reconstruction, the sole l_1 norm is not a suitable regularization since almost all natural images are not sparse themselves. Instead, they are often sparse in certain transform domains. Let $\Psi \in \mathbb{R}^{n \times n}$ be a *sparsifying* operator (e.g., wavelet transform) that transforms u into a sparse vector Ψu. Then, we can modify lasso (2) and obtain a similar form as:

$$\min_{u} g(\Psi u) + h(u) . \qquad (14)$$

Although (14) does not exactly match the ISTA (2) due to the presence of Ψ, this can be easily resolved by using an orthogonal sparsifying operator Ψ and setting $x = \Psi u$ as the unknown for (2). For example, if we set Ψ to an orthogonal 2D wavelet transform. In this case, we just need to solve x from the exact form of (2) with $g(x) = \mu \|x\|_1$ and $\tilde{h}(x) := h(\Psi^\top x)$ as the data fidelity, and recover $u = \Psi^\top x$ using the output x of ISTA. Integrating this change of variables into the scheme (11), we obtain a slightly modified version of ISTA as follows:

$$r^{(k)} = u^{(k-1)} - \alpha \nabla h(u^{(k-1)}), \qquad (15a)$$

$$u^{(k)} = \Psi^\top \mathrm{prox}_{\alpha g}(\Psi r^{(k)}) = \Psi^\top S_\theta(\Psi r^{(k)}), \qquad (15b)$$

where $\theta = \alpha \mu$ combines the two parameters, and (15b) involves shrinkage due to the choice of $g(x) = \mu \|x\|_1$. The gradient ∇h in (15a) is due to the data fidelity h in (14). Therefore, we do not need to "learn" this part in the reconstruction. On the other hand, the use of the sparsifying transform Ψ and ℓ_1 regularization is rather heuristic. If there is sufficient amount of training data, it is likely that we can learn a better representation of this regularization using a deep learning technique.

Bearing this idea, ISTA-Net is proposed to replace the transform Ψ and $\Psi^\top$ in (15) by multilayer convolutional neural networks (CNN), while keeping the prox$_{\alpha g}$, i.e., the shrinkage due to the ℓ_1 norm, as it seems robust in suppressing noises. To this end, ISTA-Net follows the scheme of ISTA (15) and constructs a deep neural network of a prescribed K phases as in section "Proximal Point Network".

Unlike LISTA and its variations in section "Learned Algorithm for Specified Optimization Problem", the kth phase of ISTA-Net is to mimic the two steps in the kth iteration of ISTA in (15). Given the output $u^{(k-1)}$ of the previous phase, the update of $r^{(k)}$ follows (15a) directly since h is known to accurately describe the data formation. Therefore, only the parameter α in (15a), which behaves as the step size in ISTA, is set to α_k and is to be learned during the training process in ISTA-Net. After $r^{(k)}$ is updated, it is passed to (15b) with Ψ and $\Psi^\top$ replaced by two multilayer CNNs $H^{(k)}$ and $\tilde{H}^{(k)}$, respectively, and the shrinkage parameter θ is replaced by θ_k, which is to be learned as well. Namely, $u^{(k)}$ is updated by

$$u^{(k)} = \tilde{H}^{(k)}\left(S_{\theta_k}\left(H^{(k)}(r^{(k)})\right)\right). \tag{16}$$

In ISTA-Net Zhang and Ghanem (2018), $H^{(k)}$ and $\tilde{H}^{(k)}$ are set to simple two-layer CNNs as follows:

$$H^{(k)}(r) = w_2^{(k)} * \sigma(w_1^{(k)} * r^{(k)}) \quad \text{and} \quad \tilde{H}^{(k)}(\tilde{r}) = \tilde{w}_2^{(k)} * \sigma(\tilde{w}_1^{(k)} * \tilde{r}^{(k)}) \tag{17}$$

where $w_1^{(k)}$, $w_2^{(k)}$, $\tilde{w}_1^{(k)}$, and $\tilde{w}_2^{(k)}$ are convolutional kernels in the kth phase to be learned, and σ is a component-wise activation function such as ReLU, i.e., $\sigma(x) = \max(x, 0)$ component wisely. In the numerical implementation of ISTA-Net Zhang and Ghanem (2018), w_1 and $\tilde{w}_2$ are convolutions with d kernels of size 3×3; w_2 and $\tilde{w}_1$ are convolutions with d kernels of size $3 \times 3 \times d$ with d set to 32.

To this point, we can see that ISTA-Net is a deep neural network with a prescribed number of K phases. Each phase of ISTA-Net mimics one iteration (15) of ISTA and is formed as: $r^{(k)}$ and $u^{(k)}$ by

$$r^{(k)} = u^{(k-1)} - \alpha_k \nabla h(u^{(k-1)}), \tag{18a}$$

$$u^{(k)} = \tilde{H}^{(k)} S_{\theta_k}(H^{(k)} r^{(k)}), \tag{18b}$$

where we have omitted excessive parentheses for notation simplicity, i.e., $H^{(k)} r^{(k)}$ stands for $H^{(k)}(r^{(k)})$, etc. The K phases are concatenated in order, where the kth phase accepts the output $u^{(k-1)}$ of the previous phase, updates $r^{(k)}$ using (18a) with α_k, and finally outputs $u^{(k)}$ using (18b). Hence, the parameters to be learned are α_k, θ_k, and $w_1^{(k)}$, $w_2^{(k)}$ in $H^{(k)}$ and $\tilde{w}_1^{(k)}$ and $\tilde{w}_2^{(k)}$ in $\tilde{H}^{(k)}$ for $k = 1, 2, \ldots, K$. In the first phase, the input is the initial guess $u^{(0)}$, which can be set to $A^\top b$. The output of the last phase, $u^{(K)}$, is used in the loss function that measures its squared discrepancy

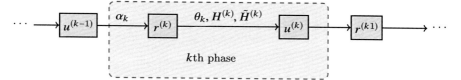

Fig. 2 Architecture of ISTA-Net (18). The kth phase updates $r^{(k)}$ and $u^{(k)}$. The dependencies of each variable on other variables are shown as incoming arrows, and the network parameters used for update are labeled next to the corresponding arrows

to the corresponding ground truth, high-quality image u^*:

$$L_{\text{dis}}(\Theta; b, u^*) = \frac{1}{2}\|u^{(K)}(b; \Theta) - u^*\|^2 \tag{19}$$

where (b, u^*) is a training pair as in the proximal point network in section "Proximal Point Network", and $\Theta := \{\alpha_k, \theta_k, w_1^{(k)}, w_2^{(k)}, \tilde{w}_1^{(k)}, \tilde{w}_2^{(k)} \mid k = 1, \ldots, K\}$. The structure of the ISTA-Net can be visualized in Fig. 2. For more details of the network structure and its relation to the back-propagation procedure, we refer to Wang et al. (2019).

In addition, since $H^{(k)}$ and $\tilde{H}^{(k)}$ in (17) are replacing Ψ and $\Psi^\top$, respectively, they are expected to satisfy $\tilde{H}^{(k)} H^{(k)} = I$, the identity mapping. To make this constraint approximately satisfied, the mismatch between $\tilde{H}^{(k)}(H^{(k)}(u^*))$ and u^* can be integrated into the following loss function, despite that it is much weaker than $\tilde{H}^{(k)} H^{(k)} = I$:

$$L_{\text{id}}(\Theta; u^*) = \frac{1}{2} \sum_{k=1}^{K} \|\tilde{H}^{(k)}(H^{(k)}(u^*)) - u^*\|^2. \tag{20}$$

The loss function for a particular training pair (b, u^*) is thus the sum of the losses in (19) and (20) with a balancing parameter $\gamma > 0$:

$$L(\Theta; b, u^*) = L_{\text{dis}}(\Theta; b, u^*) + \gamma\, L_{\text{id}}(\Theta; u^*), \tag{21}$$

and the total loss function during training is the sum of $L(\Theta; b, u^*)$ in (21) over all training pairs of form (b, u^*) in the training dataset.

The optimal parameter Θ^* can be obtained by minimizing the loss function (21), which can be accomplished using the stochastic gradient descent (SGD) method. The key in the implementation of SGD is the computation of the gradient of (21) with respect to each network parameter, i.e., $\alpha_k, \theta_k, w_1^{(k)}, w_2^{(k)}, \tilde{w}_1^{(k)}, \tilde{w}_2^{(k)}$ for $k = 1, \ldots, K$. More specifically, we first need to compute the gradient of L defined in (21) with respect to the main variables $u^{(k)}$ and $r^{(k)}$. Then we compute the gradients of $u^{(k)}$ with respect to its parameters, i.e., $\theta_k, w_1^{(k)}, w_2^{(k)}, \tilde{w}_1^{(k)}, \tilde{w}_2^{(k)}$, and the gradient of $r^{(k)}$ with respect to $\alpha^{(k)}$. Finally, the gradients of L with respect to these network

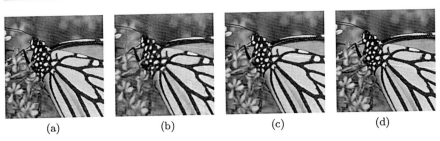

Fig. 3 Qualitative reconstruction results of ISTA-Net$^+$ (Zhang and Ghanem 2018) applied to the Butterfly image in Set11 (Kulkarni et al. 2016) with various sampling ratios. The numbers in the captions of (**b**)-(**d**) are the corresponding sampling ratios, and PSNR are shown in the parentheses. Results are generated by the code available at https://github.com/jianzhangcs/ISTA-Net. (**a**) True (**b**) 10% (25.91) (**c**) 25% (33.52) (**d**) 50% (40.18)

parameters can be built by multiplying the involved partial derivatives according to the chain rule. The derivations are fairly straightforward. For completeness, we provided the details of this back-propagation in the Appendix.

ISTA-Net (Zhang and Ghanem 2018) evaluated the reconstruction results on datasets BSD68 (Martin et al. 2001) and Set11 (Kulkarni et al. 2016), respectively. The training set contains $N = 88,912$ pairs (b, u^*), where u^* is 33×33 image patch randomly cropped from the images in 91Images dataset (Kulkarni et al. 2016) and b is the corresponding CS measurement. In Table 1, the reconstructed results are shown and compared with a traditional variational method TVAL3 (Li et al. 2013) and a non-iterative network IRCNN (Zhang et al. 2017), where the ISTA-Net$^+$ is the residual shortcut enhanced version ISTA-Net; for the detailed implementation of ISTA-Net$^+$, please refer to Zhang and Ghanem (2018). Some reconstructed images of Butterfly in Set11 (Kulkarni et al. 2016) by ISTA-Net$^+$ with various sampling ratios are displayed in Fig. 3.

ADMM-Net

ADMM-Net (Sun et al. 2016) is one of the earliest attempts to unroll a known optimization algorithm into a deep neural network. ADMM-Net is originated from the alternating minimization method of multipliers, or ADMM for short, which is a numerical algorithm particularly effective for convex optimization problems with linear equality constraints. Combined with the variable splitting technique, ADMM has been very popular and successful in solving variety of nonsmooth and/or constrained problems.

In its standard form, ADMM can solve constrained convex problems where the primal variable (i.e., the variable to be solved in the optimization problem) consists of two blocks related by a linear equality constraint. In addition, there is a dual variable, i.e., the Lagrangian multiplier, associated with the equality constraint. In each iteration, ADMM updates the two blocks of the primal variables in order, one at each time with the other one fixed and then the dual variable using the updated primal variable. ADMM yields more complex iterations due to the multiple-variable structure than ISTA.

We first recall the variable splitting and the original ADMM for image reconstruction problem, which is formulated as the one in ISTA as (14):

$$\min_u g(\Psi u) + \frac{1}{2}\|Au - b\|^2, \qquad (22)$$

but with more specific data fidelity $h(u) = (1/2) \cdot \|Au - b\|^2$. Here, we write the regularization in (22) as a composite function where g is simple (i.e., the proximity operator prox_g has closed form or is easy to compute) and Ψ as a linear operator. A typical example is the total variation regularization we mentioned in section "Introduction": $g(\Psi u) := \mu \sum_{i=1}^n \|D_i u\|_2$ with weight parameter $\mu > 0$. That is, Ψ is the discrete gradient operator (finite forward differences) D, and g is a slight variation of l_1 norm which takes sum of the l_2 norms of the gradients at all pixels. For ADMM to work efficiently, there is also requirement on the matrices Ψ and A, which we will specify later. To apply ADMM, we first use variable splitting by introducing an auxiliary variable w such that $w = Du$ and rewrite (22) as the following equivalent problem:

$$\min_{w,u} \left\{ g(w) + \frac{1}{2}\|Au - b\|^2 \right\}, \text{ subject to } w = Du. \qquad (23)$$

Then, we formulate its associated augmented Lagrangian:

$$L(u, w; \lambda) = g(w) + \frac{1}{2}\|Au - b\|^2 + \langle \lambda, w - Du \rangle + \frac{\rho}{2}\|w - Du\|^2, \qquad (24)$$

with Lagrangian multiplier λ. ADMM is then applied to solve (23) with the augmented Lagrangian (24). In each iteration of ADMM, the primal variables w and u are updated in order, and then the dual variable λ is updated. In the case of CS-MRI with $A = P\mathcal{F}$ mentioned in section "Introduction", the subproblems are given as follows:

$$w^{(k)} = S_\theta(Du^{(k-1)} - \lambda^{(k-1)}), \qquad (25a)$$

$$u^{(k)} = (\rho D^\top D + A^\top A)^{-1}(A^\top b + \rho D^\top w^{(k)} - D^\top \lambda^{(k-1)}), \qquad (25b)$$

$$\lambda^{(k)} = \lambda^{(k-1)} + \rho(w^{(k)} - Du^{(k)}), \qquad (25c)$$

where $\theta = \mu/\rho$. Given an initial guess $(w^{(0)}, u^{(0)}, \lambda^{(0)})$, ADMM repeats the cycle of the three steps (25) for iteration $k = 1, 2, \ldots$, until a stopping criterion is satisfied. As we can see, for ADMM to work efficiently, the inverse of $D^\top D + \rho A^\top A$ in (25b) must be easy to compute. In certain imaging applications, this is possible since both $D^\top D$ and $A^\top A$ can be diagonalized by fast transforms (such as Fourier), with which the update $u^{(k)}$ (25b) requires very low computational cost.

ADMM-Net (Sun et al. 2016) is a deep reconstruction network architecture that mimics the ADMM scheme (25). Similar to the case of ISTA-Net, each phase of ADMM-Net mimics one iteration of ADMM (25). More specifically, ADMM-Net sets a fixed iteration number K. The kth phase of ADMM-Net mimics the kth iteration of ADMM (25), but ADMM-Net replaces the gradient operator D by a parameterized filter (convolution) $H^{(k)}$ and the fixed parameters θ and ρ by θ_k and ρ_k to be learned through training. The original ADMM-Net (Sun et al. 2016) is designed to solve the single-coil CS-MRI problem with $A = P\mathcal{F}$, for which the kth phase of ADMM-Net reduces to:

$$w^{(k)} = S_{\theta_k}(H^{(k)} u^{(k-1)} - \lambda^{(k-1)}), \tag{26a}$$

$$u^{(k)} = \mathcal{F}^\top (P^\top P + \rho_k \mathcal{F} H^{(k)\top} H^{(k)} \mathcal{F}^\top)^{-1} (P^\top b + \rho_k \mathcal{F} H^{(k)\top} (w^{(k)} + \lambda^{(k-1)})), \tag{26b}$$

$$\lambda^{(k)} = \lambda^{(k-1)} + (w^{(k)} - H^{(k)} u^{(k)}), \tag{26c}$$

where S_θ is the shrinkage by $\theta > 0$ as in (18b).

In ADMM-Net (Sun et al. 2016), $H^{(k)}$ is set to a linear combination of a set of given filters $\{B_l\}$ with coefficients $\gamma^{(k)} = (\cdots, \gamma_l^{(k)}, \cdots) \in \mathbb{R}^{|\{B_l\}|}$, i.e., $H^{(k)} = \sum_l \gamma_l^{(k)} B_l$. Therefore, $H^{(k)}$ is completely determined by the coefficients $\gamma^{(k)}$ in the kth phase. Moreover, the shrinkage in (25a) is replaced by a piecewise linear function (PLF) determined by a set of control points and the associated function values. More specifically, let $\{p_0, \ldots, p_{N_c}\}$ be a set of $N_c + 1$ control points on $\mathbb{R}$. In Sun et al. (2016), these control points are simply chosen as uniform mesh grid points on the interval $[-1, 1]$, i.e., $p_0 = -1$ and $p_{N_c} = 1$, and $p_l - p_{l-1} = 2/N_c$ for $l = 1, \ldots, N_c$. Then, the PLF $S(h; \{p_l, q_l^{(k)}\})$ in $[-1, 1]$ is completely determined by the values $\{q_l^{(k)}\}$ at the corresponding control points $\{p_l\}$. Outside the interval $[-1, 1]$, the PLF $S(h; \{p_l, q_l^{(k)}\})$ is set to have slope 1 and concatenates with its part in $[-1, 1]$ at p_0 and p_{N_c} to form a continuous function. Then, instead of learning θ_k in the shrinkage operation S_{θ_k} in (25a), the original ADMM-Net learns the values $\{q_l^{(k)}\}$ as a part of the network parameters. The output $u^{(K)}$ is a function of the input b and network parameters $\Theta = \{\theta_k, \rho_k, \gamma^{(k)} \mid k = 1, \ldots, K\}$. The architecture of ADMM-Net is shown in Fig. 4. As usual, the loss function can be set to the squared error of $u^{(K)}$ from the ground truth, reference image u^* corresponding to data b:

$$L(\Theta; b, u^*) = \frac{1}{2} \|u^{(K)}(b; \Theta) - u^*\|^2. \tag{27}$$

The total loss function is the sum of the loss in (27) above over all training pairs (b, u^*) in the given training dataset. Then, the total loss function is minimized using the (stochastic) gradient descent method, and the minimizer Θ^* is the learned

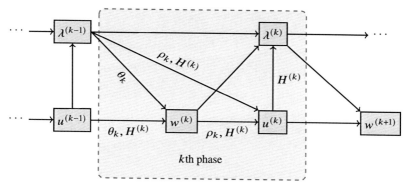

Fig. 4 Architecture of ADMM-Net (26). The kth phase updates $w^{(k)}$, $u^{(k)}$, and $\lambda^{(k)}$. The dependencies of each variable on other variables are shown as incoming arrows, and the network parameters used for update are labeled next to the corresponding arrows

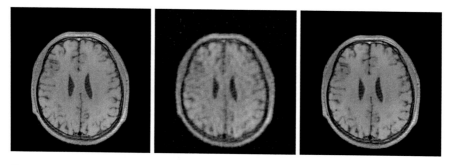

Fig. 5 Brain MR image reconstruction by ADMM-Net (Sun et al. 2016) with sampling ratio 20%. Left: ground truth. Middle: image reconstructed by zero filling. Right: reconstructed image by ADMM-Net. Results are generated by the code available at https://github.com/yangyan92/Deep-ADMM-Net

network parameters. More details about the derivation of the back-propagation and its relation to the network structure in Fig. 4 are provided in Wang et al. (2019). In Sun et al. (2016), ADMM-Net is applied to brain and chest MR image reconstruction, where the training and testing datasets are 100 and 50 images, respectively, randomly picked from MRI dataset (Bennett 2013). The qualitative results of a selected brain MR images reconstructed by ADMM-Net with CS ratio 20% are presented in Fig. 5.

Variational Network

As we have seen above, the proximal point network, ISTA-Net, and ADMM-Net all aim to solve the variational model of form:

$$\min_u f(u), \quad \text{where } f(u) := g(Du) + \lambda h(u), \tag{28}$$

where g, D, and even h can be learned from the training data adaptively. If we apply the well-known gradient descent method in numerical optimization to (28), we obtain:

$$u^{(k)} = u^{(k-1)} - \alpha_k (D^\top \nabla g(Du^{(k-1)}) + \lambda \nabla h(u^{(k-1)})) \tag{29}$$

where α_k is the step size in iteration k. Note that above we adopted a slight abuse of notation ∇g, since in image reconstruction g often represents the ℓ_1 norm or alike which is not differentiable. Hence, it is more rigorous to interpret ∇g as a subgradient of g, and the updating rule (29) is the subgradient descent. Nevertheless, this term will be replaced by a parameterized function to be learned in training, and thus its differentiability is not an important issue in the following derivation of the variational reconstruction network.

The variational network (Hammernik et al. 2018) was inspired by this concise updating rule (29). In Hammernik et al. (2018), the variational network is a fixed number of K phases, and each phase mimics one iteration of (29). The kth phase of variational network is built as

$$u^{(k)} = u^{(k-1)} - H^{(k)\top} \phi_k(H^{(k)} u^{(k-1)}) - \lambda_k \nabla h(u^{(k-1)}), \tag{30}$$

Here λ_k, $H^{(k)}$, and ϕ_k are all to be learned from data. The step size α_k is omitted since it is absorbed by the learnable terms. In particular, $H^{(k)}$ is a convolution to replace the manually chosen linear operator D (e.g., gradient in traditional image reconstruction) in (29), and ϕ_k is a parameterized function to replace ∇g.

In Hammernik et al. (2018), ϕ_k in (30) is represented as a linear combination of Gaussian functions. First of all, ϕ_k is to be applied to $H^{(k)} u^{(k-1)} \in \mathbb{R}^n$ component wisely, and hence it is sufficient to describe the component-wise operation of ϕ_k using a univariate function. To this end, we first determine a set of $N_c + 1$ control points $\{p_l : l = 0, \ldots, N_c\}$ uniformly spaced on a prescribed interval $[-I, I]$ such that $-I = p_0 < p_1 < \cdots < p_{N_c} = I$ and $p_l - p_{l-1} = 2I/N_c$ for $l = 1, \ldots, N_c$. For each point p_l, the Gaussian function with a prescribed standard deviation σ is given by

$$B_l(x) = e^{-(x-p_l)^2/(2\sigma^2)}. \tag{31}$$

Then, ϕ_k is set to a linear combination of $B_l(x)$ with coefficients $\gamma_l^{(k)}$ to be determined:

$$\phi_k(x) = \sum_{l=0}^{N_c} \gamma_l^{(k)} B_l(x). \tag{32}$$

One can also design other basis functions, instead of (31) or even parametrize ϕ_k as a generic neural network. For $H^{(k)}$, it is a convolution operation applied to $u^{(k-1)}$,

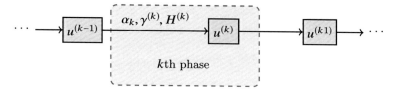

Fig. 6 Architecture of the variational network (30). The kth phase updates $u^{(k)}$. The dependencies of each variable on other variables are shown as incoming arrows, and the network parameters used for update are labeled next to the corresponding arrows

and hence it suffices to determine the convolution kernel. This is a very simplified case of convolution layers of CNNs, and we omit the details here.

Now we can see that the variational network consists of K phases, where each phase operates as (30). In particular, the first phase accepts $u^{(0)}$ as the input such as $A^\top b$. The last Kth phase outputs $u^{(K)}$, which is used in the loss function to compare with the reference image u^*:

$$L(\Theta; b, u^*) = \frac{1}{2} \|u^{(K)}(b; \Theta) - u^*\|^2. \tag{33}$$

where the network parameter $\Theta := \{\alpha_k, \gamma^{(k)}, H^{(k)} \mid k = 1, \ldots, K\}$. The total loss function is then the sum of (33) over all training pairs of form (b, u^*). The architecture of variational network is presented in Fig. 6. More details about the derivation of the back-propagation and its relation to the network structure in Fig. 6 are provided in Wang et al. (2019). Similar to the proximal point network and ISTA-Net introduced above, the variational network can be applied to problems where the data fidelity term h is differentiable with Lipschitz continuous gradient.

In Hammernik et al. (2018), the variational network considered above is applied to parallel imaging MR image reconstruction. In their experiment, $H^{(k)}$ is implemented as 48 real/imaginary filter pairs and N_c is prescribed to be 31. The network is trained on the dataset which contains 20 image slices from 10 patients and tested on reconstructing the whole image volume for 10 clinical patients that is non-overlapping with training set. The qualitative illustration of a reconstructed scan of variational network is visualized in Fig. 7.

Primal-Dual Network

Primal-dual network (PD-Net) is a deep neural network architecture for image reconstruction inspired by the primal-dual hybrid gradient algorithm (Chambolle and Pock 2011). There have been a number of work that developed PD-Nets and applied to image reconstruction (Adler and Öktem 2018; Cheng et al. 2019; Heide et al. 2014; Meinhardt et al. 2017).

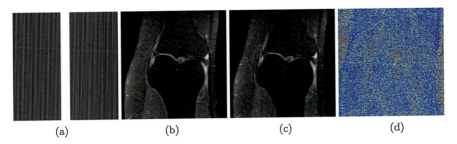

Fig. 7 The reconstruction result of an exemplified MR image by variational network (Hammernik et al. 2018) with sampling ratio 31.60. Results are generated by the code available at https://github.com/VLOGroup/mri-variationalnetwork. (**a**) Mask (**b**) Reference (**c**) VN (**d**) Error

As we discussed above, in the image reconstruction context, the variational models (1) are often represented with $g(u)$ as a regularization function and $\tilde{h}(u) = h(Au) := (1/2) \cdot \|Au - b\|^2$. In this case, we can rewrite (1) as an equivalent min-max problem by Fenchel transformation:

$$\min_{u} \max_{z,y} \langle Au, z \rangle - h^*(z) + \langle u, y \rangle - g^*(y) \tag{34}$$

where $h^*(z)$ and $g^*(y)$ are the conjugates (Fenchel dual) of $h(Au)$ and $g(u)$, respectively. Due to the Moreau's decomposition theorem:

$$\operatorname{prox}_{\tau f^*}(b) = b - \tau \operatorname{prox}_{\tau^{-1} f}(b/\tau) \tag{35}$$

for any $b \in \mathbb{R}^n$, $\tau > 0$, and convex function f, one can obtain the following iterative scheme by applying the primal-dual gradient algorithm to (34):

$$z^{(k+1)} = \arg\min_{z} \left\{ -\langle A\bar{u}^k, z \rangle + h^*(z) + \frac{1}{2\gamma}\|z - z^k\|^2 \right\}$$

$$= \operatorname{prox}_{\gamma h^*}(z^k + \gamma A\bar{u}^k) = z^k + \gamma A\bar{u}^k - \gamma \operatorname{prox}_{\gamma^{-1} h}(\frac{1}{\gamma} z^k + A\bar{u}^k) \tag{36a}$$

$$y^{(k+1)} = \arg\min_{y} \left\{ -\langle \bar{u}^k, y \rangle + g^*(y) + \frac{1}{2\gamma}\|y - y^k\|^2 \right\}$$

$$= \operatorname{prox}_{\gamma g^*}(y^k + \gamma \bar{u}^k) = y^k + \gamma \bar{u}^k - \gamma \operatorname{prox}_{\gamma^{-1} g}(\frac{1}{\gamma} y^k + \bar{u}^k) \tag{36b}$$

$$u^{(k+1)} = \arg\min_{u} \left\{ \langle Au, z^{(k+1)} \rangle + \langle u, y^{(k+1)} \rangle + \frac{1}{2\tau}\|u - u^{(k)}\|^2 \right\}$$

$$= u^k - \tau A^\top z^{(k+1)} - \tau y^{(k+1)} \tag{36c}$$

$$\bar{u}^{(k+1)} = u^{(k+1)} + \theta(u^{(k+1)} - u^{(k)}) \tag{36d}$$

Similar to the deep reconstruction networks introduced above, PD-Net also mimics the primal-dual algorithm above to construct K phases such that the kth phase in PD-Net corresponds to the kth iteration in (36). Then the proximity operator $\text{prox}_{\gamma^{-1}h}$ and $\text{prox}_{\gamma^{-1}g}$ in the updates (36a) and (36b) are replaced by CNN denoisers as in section "Proximal Point Network". The PD-Nets have been applied to natural image reconstruction in Meinhardt et al. (2017) and MRI compressive sensing in Adler and Öktem (2018); Cheng et al. (2019), which demonstrate promising performance in these applications.

Depending on which terms are designed to be learnable, three variants of the PD-Net architecture are provided in Cheng et al. (2019), which are PDHG-CSNet, CP-Net and PD-Net as follows. (i) The primal-dual hybrid gradient CS network (PDHG-CSNet) substitutes $\text{prox}_{\tau g}$ with a learned CNN denoiser in Chambolle-Pock algorithm (Chambolle and Pock 2011) which solves the (1) with $\tilde{h}(u) = h(Au) := (1/2) \cdot \|Au - b\|^2$ by iterating

$$z^{(k+1)} = \frac{z^{(k)} + \sigma(A\bar{u}^{(k)} - b)}{1 + \sigma}, \tag{37a}$$

$$u^{(k+1)} = \text{prox}_{\tau g}(u^{(k)} - \tau A^* z^{(k+1)}), \tag{37b}$$

$$\bar{u}^{(k+1)} = u^{(k+1)} + \theta(u^{(k+1)} - u^{(k)}), \tag{37c}$$

where σ, τ, and θ are algorithm parameters. (ii) The Chambolle-Pock network (CP-Net) learns a generalized Chambolle-Pock algorithm with the data fidelity term $(1/2) \cdot \|Au - b\|^2$ relaxed to $h(Au)$. Then the updating scheme of $z^{(k+1)}$ becomes $z^{(k+1)} = \text{prox}_{\sigma h^*}(z^{(k)} + \sigma A\bar{u}^{(k)})$ and CP-Net learns both $\text{prox}_{\tau g}$ and $\text{prox}_{\sigma h^*}$ with CNN denoisers. (iii) By breaking the linear combination parts in above iterates for $z^{(k+1)}$, $u^{(k+1)}$, and $\bar{u}^{(k+1)}$ in CP-Net, primal-dual net (PD-Net) further increases the network flexibility by freely learning those combinations in addition to the learnable proximal operators. In Cheng et al. (2019), the primal or dual proximal operators are substituted by learned CNN denoisers with 3 convolutional layers and 32 channels in each hidden layer. All these networks are trained and tested on 1400 and 200 images of size 256 × 256 and the corresponding k-space data undersampled by Poisson disk sampling mask. The qualitative reconstruction results of these three variations of the network on MR images are shown in Fig. 8, which are obtained from (Cheng et al. 2019).

Learnable Descent Algorithm

The LOAs conducted in the supervised learning framework are motivated by a disciplined bilevel optimization problem as follows:

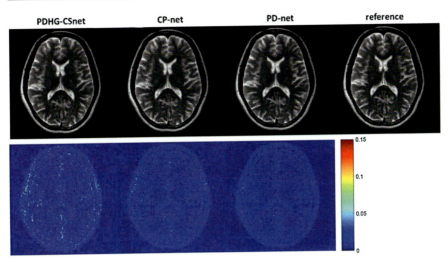

Fig. 8 Images reconstructed by primal-dual hybrid gradient CS network (PDHG-CSNet), Chambolle-Pock algorithm-inspired network (CP-Net), and primal dual net (PD-Net). The data was undersampled with a 6X Poisson disk mask

$$\min_{\Theta} \quad \frac{1}{N} \sum_{j=1}^{N} \mathcal{L}(u(b_j; \Theta), u_j^*) + R(\Theta), \tag{38a}$$

$$\text{s.t.} \quad u(b_j; \Theta) = \arg\min_{u \in \mathcal{U}} \{ f(u; b_j, \Theta) := g(u; \Theta) + h(u; b_j, \Theta) \} \tag{38b}$$

where h is the data fidelity term to ensure that the reconstructed image u is faithful to the given data b, and g is the regularization that may incorporate proper prior information of u. The regularization $g(\cdot; \Theta)$ (and possibly h also) is realized as a DNN with parameter Θ to be learned. The loss function $\mathcal{L}(u, u^*)$ is to measure the difference between a reconstruction u and the corresponding ground truth image u^* from the training data. The optimal parameter Θ of g (and h) is then obtained by solving the upper-level optimization (38a).

If the actual minimizer $u(b; \Theta)$ is replaced by the direct output of an LOA-based DNN (such as ISTA-Net etc. in the previous subsection) which mimics an iterative optimization scheme for solving the lower-level minimization in the constraint of (38) and then (38) reduces to the unrolling methods introduced in the previous subsections. However, the unrolled networks do not have any convergence guarantee, and the learned components do not represent g in (38) and can be difficult to interpret.

To obtain convergence guarantee with interpretable network structures, (Chen et al. 2020) proposed a novel learnable descent algorithm (LDA). Consider the case where the data fidelity term $h(u) := (1/2) \cdot \|Au - b\|^2$ (or any smooth but possibly nonconvex function) and $g(u)$ is a nonsmooth nonconvex regularization function which is design to be $g(u) = \|r(u)\|_{2,1} = \sum_{i=1}^{m} \|r_i(u)\|$. Here $r = (r_1, \ldots, r_m)$ is a smooth but nonconvex mapping realized by a deep neural network whose parameters are learned from training data, and $r_i(u) \in \mathbb{R}^d$ stands for a d-dimensional feature vector for $i = 1, \ldots, m$. To overcome the nondifferentiability issue of $g(u)$, a smooth approximation of g by applying Nesterov's smoothing technique (Nesterov 2005) is employed: $g_\varepsilon(u) = \sum_{i \in I_0} \frac{1}{2\varepsilon} \|r_i(u)\|^2 + \sum_{i \in I_1} \left(\|r_i(u)\| - \frac{\varepsilon}{2} \right)$, where the index set I_0 and its complement I_1 at u for the given r and ε are defined by $I_0 = \{i \in [m] \mid \|r_i(u)\| \le \varepsilon\}$, $I_1 = [m] \setminus I_0$. Denote $f_\varepsilon(u) = h(u) + g_\varepsilon(u)$ (we omit Θ for notation simplicity). Then LDA iterates

$$z_{k+1} = u_k - \alpha_k \nabla h(u_k), \tag{39a}$$

$$w_{k+1} = z_{k+1} - \tau_k \nabla g_{\varepsilon_k}(z_{k+1}), \tag{39b}$$

$$v_{k+1} = z_{k+1} - \alpha_k \nabla g_{\varepsilon_k}(u_k), \tag{39c}$$

where in each iteration $u_{k+1} = w_{k+1}$ if $f_{\varepsilon_k}(w_{k+1}) \le f_{\varepsilon_k}(v_{k+1})$ and v_{k+1} otherwise; and $\varepsilon_{k+1} = \lambda \varepsilon_k$ if $\|\nabla f_{\varepsilon_k}(u_{k+1})\| < \sigma \varepsilon_k$ and $\varepsilon_{k+1} = \varepsilon_k$ otherwise, where $\lambda \in (0, 1)$ is a prescribed hyperparameter. It is shown that ε_k will monotonically decrease to 0 such that f_{ε_k} approximates the original nonsmooth nonconvex function f, and any accumulation points of a particular subsequence of $\{u_k\}$ is a Clarke stationary point (analouge to the critical points of differentiable functions) of the nonsmooth nonconvex function f (Chen et al. 2020).

Since LDA follows the algorithm exactly, the convergence of the LDA network can be guaranteed. Moreover, the practical performance of LDA is very promising in a wide range of image reconstruction applications. For example, Table 1 shows the PSNR of the reconstructions obtained by LDA (with r parameterized by a simple generic 4-layer CNN and $K = 15$ total phases) on the dataset Set11 (Kulkarni et al. 2016) with a prefixed sampling matrix. Compared to the classical TV-based reconstruction method and several unrolling methods, LDA achieves the best reconstruction quality with highest PSNR. In addition, LDA uses much fewer parameters than the other networks as Θ is shared by all its phases. In Fig. 9, the qualitative reconstruction result of LDA is shown and compared with several state-

Table 1 Average PSNR (dB) of reconstructions obtained by the some methods on *Set11* dataset with various CS ratios and the number of learnable network parameters (#Param), where the PSNR data is quoted from Zhang and Ghanem (2018) and Chen et al. (2020)

Method	10%	25%	50%	#Param
TVAL3 Li et al. (2013)	22.99	27.92	33.55	NA
IRCNN Zhang et al. (2017)	24.02	30.07	36.23	185,472
ISTA-Net Zhang and Ghanem (2018)	25.80	31.53	37.43	171,090
ISTA-Net+ Zhang and Ghanem (2018)	26.64	32.57	38.07	336,978
LDA Chen et al. (2020)	27.42	32.92	38.50	27,967

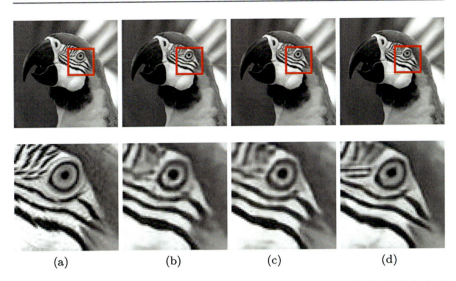

Fig. 9 Reconstruction of parrot image in Set11 (Kulkarni et al. 2016) with CS ratio 10% obtained by CS-Net (Shi et al. 2017), SCS-Net (Shi et al. 2019) and LDA (Chen et al. 2020). Images in the bottom row zoom in the corresponding ones in the top row. PSNR are shown in the parentheses. (**a**) Reference (**b**) CS-Net (28.00) (**c**) SCS-Net (28.10) (**d**) LDA (29.54)

of-the-art reconstruction networks. A more intriguing property of LDA is that the feature map r is explicitly learned and can be interpreted. In Fig. 10, the 2-norm of the learned feature map r at all pixels is shown and compared to the norm of gradient (forward differences at each pixel) used by the classical TV-based method. It can be seen that important details, such as the antennae of the butterfly, the lip of Lena, and the bill of the parrot, are faithfully recovered by LDA.

Concluding Remarks

We reviewed several typical deep neural networks inspired by the variational method and associated numerical optimization algorithms for the inverse problem of image reconstruction. These neural networks have architectures that mimic the well-known efficient optimization algorithms, such that each phase of a network corresponds to one iteration in the original numerical scheme. The algorithm parameters and other manually selected terms, such as the regularization, in the variational model and optimization algorithm are replaced by learnable components in the deep reconstruction network. The network output is thus a function of these parameters and learnable components. Given the ground truth or high-quality image data, we can form the loss function which measures the discrepancy between the network output and the ground truth and apply back-propagation and stochastic gradient descent method to optimize the parameters such that the loss function is minimized during the training procedure. After training, these networks with optimal parameters serve as fast feedforward networks that can reconstruct high-quality images on the fly.

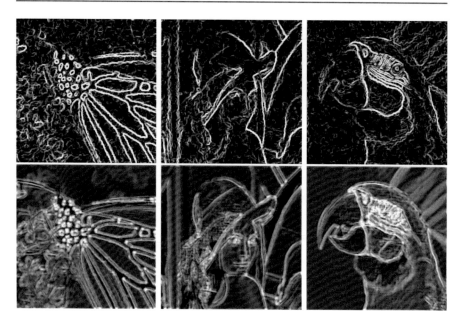

Fig. 10 The norm of the gradient at every pixel in TV based image reconstruction (top row) and the norm of the feature map r at every pixel learned in LDA (bottom row), where important details, such as the antennae of the butterfly, the lip of Lena, and the bill of the parrot, are faithfully recovered by LDA. (Images are obtained from Chen et al. 2020)

These methods have demonstrated significantly improved empirical performance and require much lower computational cost compared to the classical methods in a variety of applications.

Appendix: Back-Propagation in ISTA-Net

For completeness, we provide the details of derivations to obtain gradients of the loss function L in (21) with respect to the network parameters Θ for ISTA-Net. For more details of the network structure and its relation to the back-propagation procedure for ISTA-Net and ADMM-Net introduced in section "Structured Image Reconstruction Networks", we refer to Wang et al. (2019).

The process of back-propagation is essentially applying chain rule repeatedly, also called the "back-propagation" in deep learning. To obtain the gradient of the loss function L with respect to the parameters, it is helpful to consult the network structure for the dependency between the parameters and the inputs and outputs of nodes.

We first check the gradients of L defined in (21) with respect to $u^{(k)}$ and $r^{(k)}$. Note that L takes $u^{(k)}$ and $r^{(k)}$, which are vectors in $\mathbb{R}^n$, and output scalars, we know the gradients of L with respect to $u^{(k)}$ and $r^{(k)}$ are both vectors in $\mathbb{R}^n$ as well. We use partial derivatives to indicate spatial dependencies and compute the gradients here. First of all, we have

$$\frac{\partial L}{\partial r^{(k)}} = \frac{\partial L}{\partial u^{(k)}} \frac{\partial u^{(k)}}{\partial r^{(k)}}, \tag{40}$$

due to that $u^{(k)}$ is a function of $r^{(k)}$ as shown in Fig. 2. The gradient $\partial u^{(k)}/\partial r^{(k)}$ in (40) is straightforward to compute due to the relation between $r^{(k)}$ and $u^{(k)}$ in (18b) and the chain rule:

$$\frac{\partial u^{(k)}}{\partial r^{(k)}} = \nabla \tilde{H}^{(k)}(s_k) \cdot S'_{\theta_k}(h_k) \cdot \nabla H^{(k)}(r^{(k)}), \tag{41}$$

where the notations are simplified using the following definitions,

$$h_k := H^{(k)} r^{(k)} \quad \text{and} \quad s_k := S_{\theta_k}(h_k). \tag{42}$$

Substituting (41) into (40), we see that $\partial L/\partial r^{(k)}$ can be obtained once we have $\partial L/\partial u^{(k)}$. The gradient $\partial L/\partial u^{(k)}$ can also be computed by the chain rule:

$$\frac{\partial L}{\partial u^{(k)}} = \frac{\partial L}{\partial r^{(k+1)}} \frac{\partial r^{(k+1)}}{\partial u^{(k)}}, \tag{43}$$

where $\partial r^{(k+1)}/\partial u^{(k)}$ is obtained by (18a) for $k \leftarrow k+1$ as

$$\frac{\partial r^{(k+1)}}{\partial u^{(k)}} = I - \alpha_{k+1} \nabla^2 h(u^{(k)}). \tag{44}$$

Hence, we can get $\partial L/\partial u^{(k)}$ once $\partial L/\partial r^{(k+1)}$ is computed. Therefore, we can compute the gradients of L with respect to $u^{(k)}$ and $r^{(k)}$ for all k in the order from left to right using (40), (41), (43), and (44), starting from $\partial L/\partial u^{(K)} = u^{(K)} - u^*$, as follows:

$$\frac{\partial L}{\partial u^{(K)}} \to \frac{\partial L}{\partial r^{(K)}} \to \cdots \to \frac{\partial L}{\partial r^{(k+1)}} \to \frac{\partial L}{\partial u^{(k)}} \to \frac{\partial L}{\partial r^{(k)}} \to \cdots \to \frac{\partial L}{\partial u^{(0)}} \tag{45}$$

That is, we first compute $\partial L/\partial u^{(K)} = u^{(K)} - u^*$ according to the definition of L in (21), use it to compute $\partial L/\partial r^{(K)}$ according to (40) and (41), and then $\partial L/\partial u^{(K-1)}$ according to (43) and (44), and so on. This is the effect of back-propagation.

Now we compute the gradients of $r^{(k)}$ and $u^{(k)}$ with respect to the network parameters used in (18a) and (18b), respectively. The derivative of $r^{(k)}$ with respect to α_k is straightforward due to (18a):

$$\frac{\partial r^{(k)}}{\partial \alpha_k} = -\nabla h(u^{(k)}). \tag{46}$$

The gradient of $u^{(k)}$ with respect to $w_j^{(k)}$ in the jth layer of the CNN $H^{(k)}$ defined in (17) can be obtained by applying the chain rule to (18b):

$$\frac{\partial u^{(k)}}{\partial w_j^{(k)}} = \nabla \tilde{H}^{(k)}(s_k) \cdot S'_{\theta_k}(h_k) \cdot \frac{\partial h_k}{\partial w_j^{(k)}} \qquad (47)$$

for $j = 1, 2$, where h_k is the output of $H^{(k)}$ given the input $r^{(k)}$ and s_k is the output of S_{θ_k} given the input h_k defined in (42). The partial derivative $\partial h_k / \partial w_j^{(k)}$ is standard as in the back-propagation of CNN, which we omit the details here. Similarly, the gradient of $u^{(k)}$ with respect to $\tilde{w}_j^{(k)}$ in the jth layer of the CNN $\tilde{H}^{(k)}$ defined in (17) can be obtained since $u^{(k)}$ and s_k are the output and input of $\tilde{H}^{(k)}$, respectively. The gradient of $u^{(k)}$ with respect to θ_k is slightly different:

$$\frac{\partial u^{(k)}}{\partial \theta_k} = \nabla \tilde{H}^{(k)}(s_k) \cdot \frac{\partial S_{\theta_k}(h_k)}{\partial \theta_k}. \qquad (48)$$

In this case, we will need to treat $S_{\theta_k}(h_k) \in \mathbb{R}^n$ as a function of θ_k for given h_k, i.e., $S.(h_k) : \theta_k \mapsto S_{\theta_k}(h_k)$ defined by

$$[S_{\theta_k}(h_k)]_i = \begin{cases} -\theta_k + [h_k]_i & \text{if } 0 < \theta_k < h_k, \\ \theta_k - [h_k]_i & \text{if } 0 < \theta_k < -h_k, \\ 0 & \text{otherwise.} \end{cases} \qquad (49)$$

Hence, the derivative of $S_{\theta_k}(h_k)$ with respect to θ_k is

$$\left[\frac{\partial S_{\theta_k}(h_k)}{\partial \theta_k}\right]_i = \begin{cases} -1 & \text{if } 0 < \theta_k < h_k, \\ 1 & \text{if } 0 < \theta_k < -h_k, \\ 0 & \text{otherwise.} \end{cases} \qquad (50)$$

With all the partial derivatives obtained above, we can apply the chain rule to compute the gradient of L with respect to each of the network parameters. For example,

$$\frac{\partial L}{\partial \alpha_k} = \frac{\partial L}{\partial r^{(k)}} \frac{\partial r^{(k)}}{\partial \alpha_k}, \qquad (51)$$

where $\partial L / \partial r^{(k)}$ is obtained by (40) and (41) following the back-propagation process and $\partial r^{(k)} / \partial \alpha_k$ is obtained by (46). The partial derivatives with respect to the other parameters can be similarly computed as follows:

$$\frac{\partial L}{\partial \theta_k} = \frac{\partial L}{\partial u^{(k)}} \frac{\partial u^{(k)}}{\partial \theta_k}, \quad \frac{\partial L}{\partial w_j^{(k)}} = \frac{\partial L}{\partial u^{(k)}} \frac{\partial u^{(k)}}{\partial w_j^{(k)}}, \quad \frac{\partial L}{\partial \tilde{w}_j^{(k)}} = \frac{\partial L}{\partial u^{(k)}} \frac{\partial u^{(k)}}{\partial \tilde{w}_j^{(k)}} \quad (52)$$

where $\partial L/\partial u^{(k)}$ is obtained by (43) and (44) and the partial derivatives of $u^{(k)}$ with respect to θ_k, $w_j^{(k)}$, and $\tilde{w}_j^{(k)}$ are obtained similarly as explained above.

With these gradients of L with respect to the network parameters, we can employ a stochastic gradient descent (SGD) method and find the optimal parameters Θ^* that minimizes (21) over the entire training dataset. With the optimal Θ^*, ISTA-Net works as a feedforward mapping, which takes imaging data b and outputs a reconstructed image $u^{(K)}$. This feedforward mapping can be computed very fast since all operations in (18) are explicit given Θ^*.

References

Adler, J., Öktem, O.: Learned primal-dual reconstruction. IEEE Trans. Med. Imaging **37**(6), 1322–1332 (2018)

Aubert, G., Vese, L.: A variational method in image recovery. SIAM J. Numer. Anal. **34**(5), 1948–1979 (1997)

Bennett Landman, S.W.E.: 2013 diencephalon free challenge (2013). https://doi.org/10.7303/syn3270353

Borgerding, M., Schniter, P., Rangan, S.: Amp-inspired deep networks for sparse linear inverse problems. IEEE Trans. Signal Process. **65**(16), 4293–4308 (2017)

Chambolle, A., Pock, T.: A first-order primal-dual algorithm for convex problems with applications to imaging. J. Math. Imaging Vision **40**(1), 120–145 (2011)

Chen, X., Liu, J., Wang, Z., Yin, W.: Theoretical linear convergence of unfolded ista and its practical weights and thresholds. In: Advances in Neural Information Processing Systems, pp. 9061–9071 (2018)

Chen, Y., Liu, H., Ye, X., Zhang, Q.: Learnable descent algorithm for nonsmooth nonconvex image reconstruction. arXiv preprint arXiv:2007.11245 (2020)

Cheng, J., Wang, H., Ying, L., Liang, D.: Model learning: Primal dual networks for fast mr imaging. ArXiv **abs/1908.02426** (2019)

Chun, I.Y., Huang, Z., Lim, H., Fessler, J.A.: Momentum-net: Fast and convergent iterative neural network for inverse problems. arXiv preprint arXiv:1907.11818 (2019)

Dal Maso, G., Morel, J.M., Solimini, S.: A variational method in image segmentation: Existence and approximation results. Acta Math. **168**(1), 89–151 (1992)

Deng, J., Dong, W., Socher, R., Li, L.J., Li, K., Fei-Fei, L.: Imagenet: A large-scale hierarchical image database. In: 2009 IEEE Conference on Computer Vision and Pattern Recognition, pp. 248–255. IEEE (2009)

Devlin, J., Chang, M.W., Lee, K., Toutanova, K.: Bert: Pre-training of deep bidirectional transformers for language understanding. arXiv preprint arXiv:1810.04805 (2018)

Goodfellow, I., Bengio, Y., Courville, A., Bengio, Y.: Deep Learning, vol. 1. MIT Press, Cambridge (2016)

Gregor, K., LeCun, Y.: Learning fast approximations of sparse coding. In: J. Fürnkranz, T. Joachims (eds.) Proceedings of the 27th Internation Conference on Machine Learning (ICML 2010), pp. 399–406, Haifa (2010)

Hammernik, K., Klatzer, T., Kobler, E., Recht, M.P., Sodickson, D.K., Pock, T., Knoll, F.: Learning a variational network for reconstruction of accelerated MRI data. Magn. Reson. Med. **79**(6), 3055–3071 (2018)

He, K., Zhang, X., Ren, S., Sun, J.: Deep residual learning for image recognition. In: Proceedings of the IEEE Conference on Computer Vision and Pattern Recognition, pp. 770–778 (2016)

Heide, F., Steinberger, M., Tsai, Y.T., Rouf, M., Pająk, D., Reddy, D., Gallo, O., Liu, J., Heidrich, W., Egiazarian, K., et al.: Flexisp: A flexible camera image processing framework. ACM Trans. Graph. **33**(6), 231 (2014)

Hinton, G., Deng, L., Yu, D., Dahl, G., Mohamed, A.R., Jaitly, N., Senior, A., Vanhoucke, V., Nguyen, P., Kingsbury, B., et al.: Deep neural networks for acoustic modeling in speech recognition. IEEE Signal Process. Mag. **29**, 82–97 (2012)

Koepfler, G., Lopez, C., Morel, J.M.: A multiscale algorithm for image segmentation by variational method. SIAM J. Numer. Anal. **31**(1), 282–299 (1994)

Krizhevsky, A., Sutskever, I., Hinton, G.E.: Imagenet classification with deep convolutional neural networks. In: Advances in Neural Information Processing Systems, pp. 1097–1105 (2012)

Kulkarni, K., Lohit, S., Turaga, P., Kerviche, R., Ashok, A.: Reconnet: Non-iterative reconstruction of images from compressively sensed measurements. In: Proceedings of the IEEE Conference on Computer Vision and Pattern Recognition, pp. 449–458 (2016)

Li, C., Yin, W., Jiang, H., Zhang, Y.: An efficient augmented lagrangian method with applications to total variation minimization. Comput. Optim. Appl. **56**(3), 507–530 (2013)

Liu, J., Chen, X., Wang, Z., Yin, W.: Alista: Analytic weights are as good as learned weights in lista. In: ICLR (2019)

Ma, K., Duanmu, Z., Wu, Q., Wang, Z., Yong, H., Li, H., Zhang, L.: Waterloo exploration database: New challenges for image quality assessment models. IEEE Trans. Image Process. **26**(2), 1004–1016 (2016)

Martin, D., Fowlkes, C., Tal, D., Malik, J.: A database of human segmented natural images and its application to evaluating segmentation algorithms and measuring ecological statistics. In: Proceedings of 8th International Conference Computer Vision, vol. 2, pp. 416–423 (2001)

Meinhardt, T., Moller, M., Hazirbas, C., Cremers, D.: Learning proximal operators: Using denoising networks for regularizing inverse imaging problems. In: Proceedings of the IEEE International Conference on Computer Vision, pp. 1781–1790 (2017)

Nesterov, Y.: Smooth minimization of non-smooth functions. Math. Program. **103**(1), 127–152 (2005)

Rick Chang, J., Li, C.L., Poczos, B., Vijaya Kumar, B., Sankaranarayanan, A.C.: One network to solve them all–solving linear inverse problems using deep projection models. In: Proceedings of the IEEE International Conference on Computer Vision, pp. 5888–5897 (2017)

Roth, S., Black, M.J.: Fields of experts. Int. J. Comput. Vis. **82**(2), 205 (2009)

Sarikaya, R., Hinton, G.E., Deoras, A.: Application of deep belief networks for natural language understanding. IEEE/ACM Trans. Audio Speech Lang. Process. **22**(4), 778–784 (2014)

Scherzer, O., Grasmair, M., Grossauer, H., Haltmeier, M., Lenzen, F.: Variational Methods in Imaging. Springer, New York (2009)

Schlemper, J., Caballero, J., Hajnal, J.V., Price, A.N., Rueckert, D.: A deep cascade of convolutional neural networks for dynamic MR image reconstruction. IEEE Trans. Med. Imaging **37**(2), 491–503 (2018)

Shi, W., Jiang, F., Liu, S., Zhao, D.: Scalable convolutional neural network for image compressed sensing. In: The IEEE Conference on Computer Vision and Pattern Recognition (CVPR) (2019)

Shi, W., Jiang, F., Zhang, S., Zhao, D.: Deep networks for compressed image sensing. In: 2017 IEEE International Conference on Multimedia and Expo (ICME), pp. 877–882. IEEE (2017)

Socher, R., Bengio, Y., Manning, C.D.: Deep learning for nlp (without magic). In: Tutorial Abstracts of ACL 2012, pp. 5–5. Association for Computational Linguistics (2012)

Sprechmann, P., Bronstein, A.M., Sapiro, G.: Learning efficient sparse and low rank models. IEEE Trans. Pattern Anal. Mach. Intell. **37**(9), 1821–1833 (2015)

Sun, J., Li, H., Xu, Z., et al.: Deep admm-net for compressive sensing mri. In: Advances in Neural Information Processing Systems, pp. 10–18 (2016)

Timofte, R., De Smet, V., Van Gool, L.: A+: Adjusted anchored neighborhood regression for fast super-resolution. In: Asian Conference on Computer Vision, pp. 111–126. Springer (2014)

Vaswani, A., Shazeer, N., Parmar, N., Uszkoreit, J., Jones, L., Gomez, A.N., Kaiser, Ł., Polosukhin, I.: Attention is all you need. In: Advances in Neural Information Processing Systems, pp. 5998–6008 (2017)

Wang, G., Zhang, Y., Ye, X., Mou, X.: Machine Learning for Tomographic Imaging (2019). IOP Publishing. https://doi.org/10.1088/2053-2563/ab3cc4

Wang, S., Fidler, S., Urtasun, R.: Proximal deep structured models. In: Advances in Neural Information Processing Systems, pp. 865–873 (2016)

Xie, X., Wu, J., Zhong, Z., Liu, G., Lin, Z.: Differentiable linearized admm. arXiv preprint arXiv:1905.06179 (2019)

Xin, B., Wang, Y., Gao, W., Wipf, D., Wang, B.: Maximal sparsity with deep networks? In: Advances in Neural Information Processing Systems, pp. 4340–4348 (2016)

Zeiler, M.D., Fergus, R.: Visualizing and understanding convolutional networks. In: European Conference on Computer Vision, pp. 818–833. Springer (2014)

Zhang, J., Ghanem, B.: Ista-net: Interpretable optimization-inspired deep network for image compressive sensing. In: Proceedings of the IEEE Conference on Computer Vision and Pattern Recognition, pp. 1828–1837 (2018)

Zhang, K., Zuo, W., Gu, S., Zhang, L.: Learning deep cnn denoiser prior for image restoration. In: Proceedings of the IEEE Conference on Computer Vision and Pattern Recognition, pp. 3929–3938 (2017)

Bilevel Optimization Methods in Imaging

24

Juan Carlos De los Reyes and David Villacís

Contents

Introduction	910
Variational Inverse Problems Setting	912
Image Reconstruction as an Inverse Problem	912
Regularizers	912
Restoration Models	915
Optimality and Duality	915
Solution Methods	916
Bilevel Optimization in Imaging	917
Total Variation Gaussian Denoising	919
Solution Algorithms	924
Infinite-Dimensional Case	924
Existence and Other Properties	926
Stationarity Conditions	927
Dualization	929
Nonlocal Problems	930
Neural Network Optimization	932
Deep Neural Networks as a Further Regularizer	933
Deep Unrolling Within Optimization	933
Numerical Experiments	934
Conclusions	938
References	939

Abstract

Optimization techniques have been widely used for image restoration tasks, as many imaging problems may be formulated as minimization ones with the

---

J. C. De los Reyes (✉) · D. Villacís
Research Center for Mathematical Modelling (MODEMAT), Escuela Politécnica Nacional, Quito, Ecuador
e-mail: juan.delosreyes@epn.edu.ec; david.villacis01@epn.edu.ec

© Springer Nature Switzerland AG 2023
K. Chen et al. (eds.), *Handbook of Mathematical Models and Algorithms in Computer Vision and Imaging*, https://doi.org/10.1007/978-3-030-98661-2_66

recovered image as the target minimizer. Recently, novel optimization ideas also entered the scene in combination with machine learning approaches, to improve the reconstruction of images by optimally choosing different parameters/functions of interest in the models. This chapter provides a review of the latest developments concerning the latter, with special emphasis on bilevel optimization techniques and their use for learning local and nonlocal image restoration models in a supervised manner. Moreover, the use of related optimization ideas within the development of neural networks in imaging will be briefly discussed.

Keywords

Bilevel optimization · Machine learning · Variational models

Introduction

Several classical image processing tasks such as denoising, inpainting, and deblurring, among others, may be treated as minimization problems in suitable function spaces and using properly chosen energy functionals, typically nonsmooth ones. As a consequence, the historical connection between optimization and imaging has been very fruitful, and several analytical and algorithmic developments have originated from this close relationship. We refer to Chambolle and Pock (2016) and the references therein for a thorough review on these links and current developments.

More recently, new optimization ideas entered the scene hand in hand with modern data-driven approaches. Although machine learning techniques have years of tradition on solving inverse and imaging problems, its use in combination with structural properties of the mathematical models has proven to be of relevance, leading to state-of-the-art developments and applications (see, e.g., Calatroni et al. 2017; Arridge et al. 2019; Holler et al. 2018; Hintermüller and Papafitsoros 2019; Sherry et al. 2020).

A learning approach that combines practical and theoretical advantages is *bilevel optimization*. Within this setting, the imaging problems are considered as lower-level constraints, while on the upper-level a loss function, based on a training set, is used for estimating the different parameters involved in the models. The resulting mathematical problems pose different challenges that need to be addressed using sophisticated tools from variational and nonsmooth analysis (Outrata 2000; Mordukhovich 2018; Schirotzek 2007).

A prototypical problem in this direction is the parameter learning associated with image restoration models. An initial contribution in this respect was the paper by Tappen and coauthors Tappen (2007), where the parameters of a *Markov random fields* model were learned by means of variational optimization. Thereafter, Haber and coauthors Haber et al. (2008) considered a general learning approach for inverse problems and, although no mathematical theory was developed, made a case for the successful application of such methodology. A renewed interest took place around the year 2013, where on basis of developments on optimal control of variational inequalities, the learning of parameters for variational denoising models was carried out in function space (De los Reyes 2011) and in finite-dimensions (Kunisch and Pock 2013). Since then, the field has expanded, and several papers have been devoted to different theoretical and computational aspects: noise model learning (Calatroni et al. 2013; Calatroni and Papafitsoros 2019), higher-order regularizers (De los Reyes et al. 2017; Davoli and Liu 2018; Davoli et al. 2019; Hintermüller and Rautenberg 2017), blind deconvolution (Hintermüller and Wu 2015), inexact gradients (Ochs et al. 2016; Ehrhardt and Roberts 2020), and nonlocal models (d'Elia et al. 2019; Bartels and Weber 2020).

When confronted with variational imaging models, the bilevel optimization problem structure becomes quite involved to be analyzed, as classical nonlinear or bilevel programming results (see, e.g., Dempe 2002) cannot be directly utilized. As a remedy, tools from nonsmooth variational analysis have to be employed to cope with the difficulties related with the lack of differentiability of the solution mapping or the failure of standard constraint qualification conditions. In finite dimensions, for instance, generalized Mordukhovich tangential and normal cones (Mordukhovich 2018) have to be computed in order to obtain relatively sharp stationarity conditions. These aspects will be illustrated in section "Bilevel Optimization in Imaging" of this manuscript, targeting the parameter learning of image denoising problems.

The analysis of the infinite-dimensional counterpart becomes even harder, as topological properties of finite-dimensional spaces are in general missing and, therefore, variational analysis results on generalized normal cones are mostly inapplicable. The study of the function space setting, however, has proven to be of importance for deriving structural properties of the reconstructed images and optimal parameters (De los Reyes et al. 2016), as well as for devising mesh-independent solution algorithms. Moreover, the study of spatially dependent parameters in variational imaging problems has attracted increasing attention in recent years. Apart of the learning approach carried out in Van Chung et al. (2017), Hintermüller and coauthors have considered an alternative loss functional based on image statistics in combination with dualization of the lower-level problem (Hintermüller and Rautenberg 2017). Recently, also bilevel problems with infinite-dimensional nonlocal variational lower-level models have been investigated (d'Elia et al. 2019). A summary of these contributions will be presented in section "Infinite-Dimensional Case" of this chapter.

Although supervised bilevel learning has been usually presented as a competing approach to modern neural networks, theoretical results obtained for the variational optimization problems may be considered in the design of novel types of neural networks as well. This effort has been carried out in Lunz et al. (2018) and Kobler et al. (2020), where generative adversarial neural networks and multiscale convolutional neural networks are considered, respectively. Moreover, the use of neural networks for improving the efficiency of intermediate steps within an optimization method has also been studied (Adler and Öktem 2018; Sun et al.

2016; Kobler et al. 2017). A short discussion on these connections is provided in section "Neural Network Optimization".

Variational Inverse Problems Setting

Image Reconstruction as an Inverse Problem

Image reconstruction aims to restore or enhance a degraded image obtained by a given acquisition process. In general, images can be degraded due to poor imaging conditions and problems in the storage device or the communication channel, to name a few. A frequentist model used to analyze this phenomenon can be stated as

$$f = A(u) + n, \qquad (1)$$

where u is the original image, f is the observed degraded image, n is the noise contained in the observed image, and A is a possibly nonlinear forward operator that models the acquisition process. In most imaging problems, the operator A is rank deficient, leading to an ill-posed inverse problem. Therefore, nonuniqueness of solutions or instability of the direct inversion of such operator motivates the use of different solution techniques.

A classical way to solve such inverse problems is to make use of a variational "energy" formulation. Using this methodology, we can state the solution of (1) as the solution of the following optimization problem:

$$\hat{u} := \arg\min_{u} \mathcal{E}(u, \lambda, \alpha) := \mathcal{F}(u, \lambda) + \mathcal{R}(Hu, \alpha), \qquad (2)$$

where $\hat{u}$ is the reconstructed image, H a bounded linear operator, $\mathcal{F}$ the *data fidelity*, and $\mathcal{R}$ a *regularization* term. The parameters λ and α affect the contribution of the fidelity and regularization terms to the final solution, respectively. The choice of these two terms has a crucial impact on the solution. Indeed, the data fidelity term models the type of noise present in the image, while the regularization term promotes certain features which are known a priori about the image.

Regularizers

A seminal idea proposed by Tikhonov and Arsenin (1977) for the solution of inverse problems is to use the following type of regularization term:

$$\mathcal{R}(\nabla u, \alpha) = \alpha \int_{\Omega} \|\nabla u\|_2^2 dx, \qquad (3)$$

aiming at recovering certain smooth properties of the solution. In the context of image restoration, however, the solution obtained correspondingly is not desirable, precisely since the regularizer involved has very strong isotropic smoothing properties which leads to a loss of edge information in the reconstructed image.

In order to preserve the edge information as much as possible, Rudin et al. (1992) proposed the use of the *isotropic total variation* of the image:

$$TV_\alpha(u) := \alpha \int_\Omega \|\nabla u\|_2 dx. \tag{4}$$

This regularizer promotes solutions close to a piecewise constant image that is composed by homogeneous regions separated by sharp edges. Because one of the main characteristics of images are edges, as they define divisions between objects in a scene, their preservation seems like a good idea and a favorable feature of TV regularization. The drawback of such a regularization procedure becomes apparent as soon as it is applied to images that are not only consist of constant intensity regions and jumps but also possess more complicated structures, like smooth intensity variations or textures. A well-known artifact introduced by TV regularization in this case is called staircasing (Ring 2000).

One possibility to counteract such artifacts is the introduction of higher-order derivatives in the image regularization. Two main second-order total variation models have been introduced in the past: the infimal-convolution total variation (ICTV) model of Chambolle and Lions Chambolle and Lions (1997) and the total generalized variation (TGV) proposed by Bredies and coauthors (2010). Although higher-order models were also formally introduced, we focus on second-order ones, since these regularizers have received much more attention in recent relevant imaging applications (Knoll et al. 2011; Bredies et al. 2010). For an open and bounded image domain $\Omega \subset \mathbb{R}^2$, the ICTV regularizer reads:

$$\text{ICTV}_{\alpha,\beta}(u) := \min_{v \in W^{1,1}(\Omega),\, \nabla v \in BV(\Omega)} \alpha \|Du - \nabla v\|_{\mathcal{M}(\Omega;\mathbb{R}^2)} + \beta \|D\nabla v\|_{\mathcal{M}(\Omega;\mathbb{R}^{2\times 2})}. \tag{5}$$

On the other hand, second-order TGV (Bredies et al. 2010) reads:

$$\text{TGV}^2_{\alpha,\beta}(u) := \min_{w \in BD(\Omega)} \alpha \|Du - w\|_{\mathcal{M}(\Omega;\mathbb{R}^2)} + \beta \|Ew\|_{\mathcal{M}(\Omega;\text{Sym}^2(\mathbb{R}^2))}. \tag{6}$$

Here $BD(\Omega) := \{w \in L^1(\Omega;\mathbb{R}^n) \mid \|Ew\|_{\mathcal{M}(\Omega;\mathbb{R}^{n\times n})} < \infty\}$ is the space of vector fields of bounded deformation on Ω, and E denotes the *symmetrized gradient* and $\text{Sym}^2(\mathbb{R}^2)$ the space of symmetric tensors of order 2 with arguments in $\mathbb{R}^2$. The parameters α, β are fixed positive parameters. The main difference between (5) and (6) is that we do not generally have that $w = \nabla v$ for any function v. That results in some qualitative differences of ICTV and TGV regularization; compare, for instance De los Reyes et al. (2017).

Although TV-based regularizers perform well in many instances, for images with texture structures, neighborhood information is not enough to get good reconstructions. A remedy to this are nonlocal models, which consider similar intensity patterns between pixels or patches in a given spatial neighborhood or all over the whole image domain. Originally, the main concern within this framework was the design of direct nonlocal filters (Yaroslavsky 1986; Tomasi and Manduchi 1998; Buades et al. 2005), being the *nonlocal means filter* arguably the more popular regularizer in this context. The techniques diversified afterward with the consideration of different energy functionals to accomplish the denoising task (Gilboa and Osher 2007, 2008; Lou et al. 2010). In particular, the variational framework developed in Gilboa and Osher (2007) enabled the employment of additional modeling features that have been used already for image reconstruction tasks in local models. A modified variational nonlocal means regularizer, for instance, is given by

$$NL(u) := \int_{\Omega \cup \Omega_I} \int_{\Omega \cup \Omega_I} \big(u(\mathbf{x}) - u(\mathbf{y})\big)^2 \gamma(\mathbf{x}, \mathbf{y}) \, d\mathbf{y} d\mathbf{x}, \qquad (7)$$

with the localized integrable kernel

$$\gamma(\mathbf{x}, \mathbf{y}) = \exp\left\{-\int_{B_\rho(0)} w(\tau)\big(f(\mathbf{x}+\tau) - f(\mathbf{y}+\tau)\big)^2 d\tau\right\} \chi\big(\mathbf{y} \in B_\epsilon(\mathbf{x})\big),$$

Here, Ω_I stands for the interaction domain of a bounded region Ω consisting of all points outside of the domain that interact with points inside of it. The function $w(t)$ controls the intensity threshold at which the nonlocal filter acts and is the target of a learning scheme. For a comparison between total variation and nonlocal means, see Fig. 1.

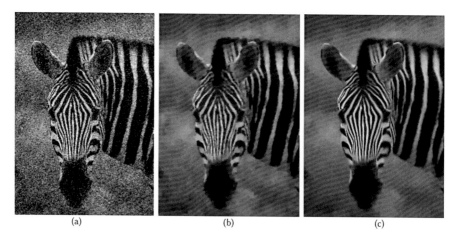

Fig. 1 Comparison of regularizers in variational image denoising. (**a**) Noisy (**b**) Total Variation (**c**) Nonlocal Means

Restoration Models

Three well-known image restoration tasks are denoising, deblurring, and inpainting. The goal of denoising is to recover a noise-free image u from a particular noise contaminated one f. This perturbation is usually modeled based on the statistical estimates or approximated by a proper noise model coming from the physics behind the acquisition of f. For a normally distributed f, the data term corresponds to a squared Euclidean norm (Rudin et al. 1992):

$$\mathcal{F}(u, \lambda) := \lambda \int_\Omega \|u - f\|^2 dx. \tag{8}$$

In the case of a Poisson noise distribution present in the damaged image, the data fidelity term was studied in Sawatzky et al. (2009) and Le et al. (2007) and has the form $\mathcal{F}(u, \lambda) := \lambda \int_\Omega (u - f) \log u \, dx$. In Nikolova (2004), the author studied impulse noise contaminated images and proposed the nonsmooth data fidelity term $\mathcal{F}(u, \lambda) := \lambda \int_\Omega \|u - f\|_1 dx$. Other convex and non-convex data fidelity models, as well as several combinations, have been investigated as well.

In the case of deblurring, the task consists in recovering a sharp image from its blurry observation. This blur usually comes as an *optical blur* from de deviation of the object from the focused imaging plane, *mechanical blur* from the rapid motion of either the target object or the imaging device, and of *medium-induced blur* due to the optical turbulence of the photonics media. Given a blur operator A, the image deblurring problem reads

$$\mathcal{F}(u, \lambda) := \lambda \int_\Omega \|A(u) - f\|^2 dx. \tag{9}$$

The remaining task, *image inpainting*, consists in recovering lost parts of a damaged image. If Ω corresponds to the original image domain, due to different problems in image acquisition, transmission, and numerous external factors, there usually exists a subdomain $\Omega_0 \subset \Omega$ where the information is missing. Moreover, the observable portion of the image $\Omega \setminus \Omega_0$ is often degraded with noise and blur. The final goal of this task, which also encompasses denoising and deblurring, is to reconstruct the image in the entire domain Ω from this degraded observation. The fidelity term takes typically the Gaussian form:

$$\mathcal{F}(u, \lambda) := \lambda \int_{\Omega \setminus \Omega_0} \|A(u) - f\|^2 dx. \tag{10}$$

Optimality and Duality

As described in the previous section, variational regularizers are typically nonsmooth, while fidelity terms are in many circumstances convex and differentiable. In both cases, however, convexity appears to be an important feature, which enables the use of convex analysis tools for characterizing the solution of the restoration models at hand.

By restating problem (2) for fixed parameters $\lambda \in \mathcal{P}_\lambda^+$ and $\alpha \in \mathcal{P}_\alpha^+$, we obtain

$$\min_{u \in X} \mathcal{F}_\lambda(u) + \mathcal{R}_\alpha(Hu), \tag{11}$$

where X, Y are two Banach spaces and $\mathcal{P}_\lambda^+, \mathcal{P}_\alpha^+$ are suitable positive sets in the parameters spaces. Assuming that $\mathcal{R}_\alpha : Y \to \mathbb{R}$ is a proper convex, lower semicontinuous, and possibly nonsmooth function; $\mathcal{F}_\lambda : X \to \mathbb{R}$ a smooth, proper convex, and lower semicontinuous function; and $H : X \to Y$ a bounded linear operator, the optimality condition for this primal problem reads

$$0 \in \partial(\mathcal{F}_\lambda(u) + \mathcal{R}_\alpha(Hu)) = \partial(\mathcal{F}_\lambda(u)) + H^*(\partial\mathcal{R}_\alpha(Hu)), \tag{12}$$

where $\partial(\cdot)$ denotes the standard convex analysis subdifferential. Introducing the dual multiplier $q \in Y$, the dual problem of (11) is given by

$$\max_{q \in Y} -\mathcal{F}_\lambda^\star(-H^*q) - \mathcal{R}_\alpha^\star(q), \tag{13}$$

where $\mathcal{F}_\lambda^\star$ and $\mathcal{R}_\alpha^\star$ stand for the convex conjugate of $\mathcal{F}_\lambda$ and $\mathcal{R}_\alpha$, respectively.

By satisfying some suitable hypotheses on $\mathcal{F}_\lambda$ and $\mathcal{R}_\alpha$, existence of optimal solutions for both the primal and dual problems can be guaranteed. Furthermore, it can be proven that the cost functional values coincide and that both solutions are linked through extremality conditions, i.e., the primal $\hat{u}$ and dual $\hat{q}$ optimal solutions satisfy

$$H^*\hat{q} \in \partial\mathcal{F}_\lambda(\hat{u}), \tag{14a}$$
$$-\hat{q} \in \partial\mathcal{R}_\alpha(H\hat{u}). \tag{14b}$$

In addition, we can formulate an equivalent primal-dual saddle point problem (Ekeland and Temam 1999) with the following structure:

$$\min_{u \in X} \max_{q \in Y} \langle H(u), q \rangle + \mathcal{F}_\lambda(u) - \mathcal{R}_\alpha^\star(q). \tag{15}$$

Solution Methods

Since the nonsmoothness of the function $\mathcal{R}_\alpha$ prevents the direct use of standard differentiable techniques, there are several numerical strategies for finding solutions to (2). A first idea consists in solving this type of problems by making use of subgradient-based methods for dealing with the primal problem directly. Although this appears to be the most natural approach, this option has the drawback of the classical slow convergence rate of subgradient methods (Beck 2017 Chapter 8).

By exploiting the differentiability of $\mathcal{F}_\lambda$ and the fact that in general the regularizer $\mathcal{R}_\alpha$ is a simple convex lower semicontinuous function, *forward-backward* splitting schemes were developed, where in each iteration a gradient descent step on $\mathcal{F}$ and a proximal step on $\mathcal{R}_\alpha$ are performed. The resulting algorithm behaves robustly and gets faster as the smoothness properties of $\mathcal{F}_\lambda$ improve. Moreover, accelerated versions of this scheme (like the FISTA algorithm) became quite popular in the last years.

Alternatively, the saddle point formulation (15) may be numerically exploited. A popular strategy considers an alternate update, where first a descent step for the primal variable u is performed and thereafter an ascent step in the dual variable p is carried out. This procedure, called *ADMM*, can further be speed up by considering a relaxation step (see, e.g., Chambolle and Pock 2011). These primal-dual update steps are well-suited for parallel computation, making these methods practical for high-resolution image denoising (Villacís 2017). Related popular primal-dual methods are the well-known Douglas-Rachford and the Chambolle-Pock algorithms. An extension to nonlinear operators H can be found in Valkonen (2014).

Another frequent numerical alternative consists in regularizing the non-differentiable term by means of a sufficiently smooth function. As a consequence, fast second-order methods, i.e., methods where both gradient and hessian information is used to define a descent direction, may me devised for the solution of the regularized problems. Indeed, Newton and semismooth Newton methods, along with globalization strategies, have been used to solve image restoration models (see, e.g., Hintermüller and Stadler 2006; De los Reyes and Schönlieb 2013).

Bilevel Optimization in Imaging

The parameters λ and α, considered as invariant in the previous section, actually play a crucial role in the quality of the reconstructed image. Instead of trying to tune them by trial-and-error, the natural question on wether is it possible to select them in an optimal way arises. Combining existing training sets with a supervised bilevel optimization framework, a rigorous learning approach has been developed for variational image restoration in recent years (De los Reyes and Schönlieb 2013; De los Reyes et al. 2017; Kunisch and Pock 2013; Hintermüller and Wu 2015).

Let us consider a training dataset of P pairs $(u_k^{\text{train}}, f_k)$, for $k = 1, \ldots, P$, where each u_k^{train} corresponds to ground-truth data and f_k to the corresponding corrupted one. To obtain the optimal parameters (λ, α), we consider the following class of *bilevel optimization* problems:

$$\min_{(\lambda,\alpha)} \sum_{k=1}^{P} J(u_k, u_k^{\text{train}}) \tag{16}$$

$$\text{s.t.} \quad u_k \in \arg\min_{u \in \mathbb{R}^n} \mathcal{E}(u, \lambda, \alpha, f_k), \tag{17}$$

where the upper-level problem handles the optimal parameter loss function J, while the lower-level problem corresponds to the restoration model of interest.

A general family of lower-level problems that allow us to learn the noise model, as described in De los Reyes and Schönlieb (2013) and Calatroni et al. (2013), as well as the weights for a family of regularizers (De los Reyes et al. 2017; Kunisch and Pock 2013) is given by the energy

$$\arg\min_{u\in\mathbb{R}^n} \mathcal{E}(u,\lambda,\alpha,f) := \sum_{j=1}^{M}\sum_{i=1}^{r_j} \lambda_{j,i}\phi_j(u;f)_i + \sum_{l=1}^{N}\sum_{i=1}^{s_l} \alpha_{l,i}\|(\mathbb{B}_l u)_i\|, \qquad (18)$$

where $\phi_j, j = 1, \ldots, M$, are different convex restoration (fidelity) models and $\mathbb{B}_l, l = 1, \ldots, N$, are bounded linear operators (matrices or tensors) related to different regularizers. The norm $\|\cdot\|$ corresponds to the Euclidean or the Frobenius norm, depending on the corresponding operators. The vector $u \in \mathbb{R}^n$ can be just the reconstructed image or an extended version that includes additional information (e.g., higher-order information). The abstract model (18) has indeed two sets of model parameters: λ for the different data terms available and α for the regularization terms considered. Moreover, these parameters may be considered *scale-dependent*, meaning that each parameter $\lambda_j \in \mathbb{R}_+^{r_j}, \alpha_l \in \mathbb{R}_+^{s_l}$, takes one scalar value for each component (pixel, patch, etc.) of the image model and regularizer, respectively.

In contrast, by assuming scalar parameters $\alpha_l, \lambda_j \in \mathbb{R}_+$, we will affect all components with the same intensity, yielding

$$\arg\min_{u\in\mathbb{R}^n} \mathcal{E}(u,\lambda,\alpha,f) := \sum_{j=1}^{M}\lambda_j \sum_{i=1}^{r_j}\phi_j(u;f)_i + \sum_{l=1}^{N}\alpha_l \sum_{i=1}^{s_l}\|(\mathbb{B}_l u)_i\|. \qquad (19)$$

Moreover, a further generalization for patch-dependent parameters can be made. Let us consider $\lambda_j \in \mathbb{R}^{m_j}, \alpha_l \in \mathbb{R}^{m_l}$, with $m_j, m_l \ll n$, and patch operators $P_j : \mathbb{R}^{m_j} \mapsto \mathbb{R}_+^{r_j}$ and $Q_l : \mathbb{R}^{m_l} \mapsto \mathbb{R}_+^{s_l}$. The lower-level problem energy may then be written as

$$\arg\min_{u\in\mathbb{R}^n} \mathcal{E}(u,\lambda,\alpha,f) := \sum_{j=1}^{M}\sum_{i=1}^{r_j} P_j(\lambda_j)_i \phi_j(u;f)_i + \sum_{l=1}^{N}\sum_{i=1}^{s_l} Q_l(\alpha_l)_i \|(\mathbb{B}_l u)_i\|. \qquad (20)$$

Most classical image denoising variational models (TV-l_2, TV-l_1, TGV-l_2, ICTV-l_2, etc.) as well as TV deblurring and inpainting are instances of the latter.

Also an essential component of The bilevel problem are equations (16) and (17) is the loss function J, which models the quality of the reconstruction when compared to the original image provided in the dataset. One classic approach is to compute the difference between a ground truth image u^{train} and its reconstruction u using a mean squared error (MSE) criteria $J(u, u^{\text{train}}) = MSE(u, u^{\text{train}}) := \frac{1}{2}\|u - u^{\text{train}}\|_2^2$, which is closely related to the peak signal-to-noise ratio quality measure $PSNR(u, u^{\text{train}}) := 10\log_{10}(255^2/MSE(u, u^{\text{train}}))$. Even though this measure is widely used in the imaging community due to its low computational complexity, it depends strongly on the image intensity scaling. Furthermore, PSNR does not necessarily coincide with a human visual response to the image quality.

A more reliable quality measure proposed is the structural similarity index (SSIM) (Wang et al. 2004), which can be casted as

$$J(u, u^{\text{train}}) = SSIM(u, u^{\text{train}}) = l(u, u^{\text{train}})c(u, u^{\text{train}})s(u, u^{\text{train}}),$$

where

$$l(u, u^{\text{train}}) = \frac{2\mu_u \mu_{u^{\text{train}}} + C_1}{\mu_u^2 + \mu_{u^{\text{train}}}^2 + C_1},$$

$$c(u, u^{\text{train}}) = \frac{2\sigma_u \sigma_{u^{\text{train}}} + C_2}{\sigma_u^2 + \sigma_{u^{\text{train}}}^2 + C_2},$$

$$s(u, u^{\text{train}}) = \frac{2\sigma_{uu^{\text{train}}} + C_3}{\sigma_u + \sigma_{u^{\text{train}}} + C_3},$$

and μ_u and σ_u correspond to the mean luminance and the standard deviation of the image u, respectively. The use of this quality measure in the bilevel optimization context is, however, restrictive due to its nonsmoothness and non-convexity.

An alternative loss function aimed at prioritizing jump preservation was proposed in De los Reyes et al. (2017), where the authors make use of a Huber regularization of a total variation cost:

$$J(u, u^{\text{train}}) := \sum_{j=1}^{m} |\mathbb{K}(u - u^{\text{train}})_j|_\epsilon.$$

This loss function is differentiable, convex, and it was proven advantageous for evaluating the quality of the reconstructed image.

Total Variation Gaussian Denoising

To simplify the exposition of the methodology, let us restrict the analysis to the bilevel problem (16) in the specific case of total variation denoising and a single image dataset (u^{train}, f). By considering a scale-dependent parameter $\lambda \in \mathbb{R}_+^n$, our bilevel problem then reads

$$\min_{\lambda \in \mathbb{R}_+^n} J(u(\lambda), u^{\text{train}}) \tag{21}$$

$$\text{s.t} \quad u(\lambda) = \arg\min_{u \in \mathbb{R}^n} \frac{1}{2} \sum_{i=1}^n \lambda_i \|u_i - f_i\|^2 + \sum_{i=1}^s \|(\mathbb{K}u)_i\| \tag{22}$$

where $\mathbb{K} : \mathbb{R}^n \to \mathbb{R}^{s \times 2}$ is the discrete gradient operator with respect to directions in x and y, i.e., $\mathbb{K}u = (K_x u, K_y u)$, where K_x and K_y correspond to the discrete partial derivative with respect to the horizontal and vertical direction, respectively. Thanks to the convexity of the energy function in the lower-level problem, we can replace the constraint by its necessary and sufficient optimality condition, yielding

$$\min_{\lambda \in \mathbb{R}_+^n} J(u(\lambda), u^{\text{train}}) \tag{23}$$

$$\text{s.t} \quad \langle \lambda \circ (u - f), v - u \rangle + \sum_{i=1}^s \|(\mathbb{K}v)_i\| - \sum_{i=1}^s \|(\mathbb{K}u)_i\| \geq 0, \quad \forall v \in \mathbb{R}^n, \tag{24}$$

where $\circ$ stands for the Hadamard product between vectors. This is an optimization problem constrained by a variational inequality of the second kind, along with non-negativity constraints for the parameter λ.

Moreover, using duality techniques, the variational inequality of the second kind in problem (23) can be equivalently written in primal-dual form, yielding the following reformulation of problem (21):

$$\begin{aligned}
& \underset{(\lambda, u, q) \in \mathbb{R}^n \times \mathbb{R}^n \times \mathbb{R}^{s \times 2}}{\text{minimize}} && J(u, u^{\text{train}}) \\
& \text{subject to} && \lambda \circ (u - f) + \mathbb{K}^T q = 0 \\
& && \langle q_j, (\mathbb{K}u)_j \rangle = \|(\mathbb{K}u)_j\|, && \forall i = 1, \ldots, s \\
& && \|q_j\| \leq 1, && \forall j = 1, \ldots, s \\
& && \lambda_j \geq 0, && \forall j = 1, \ldots, n.
\end{aligned} \tag{25}$$

Failure of Standard Constraint Qualification Conditions

A key goal in the study of an optimization problem is the derivation of optimality conditions that allow a proper characterization of stationary points. To do so, Lagrange multiplier's existence theorems are usually proven on basis of the so-called constraint qualification conditions (Nocedal and Wright 2006). Next, we show that in the case of problem (23), the situation is not standard at all and classical optimization theory typically fails.

Even though the primal-dual reformulation transforms problem (23) into a constrained nonlinear optimization one, the difficulties related to the nonsmoothness remain in the constraints. One may try to circumvent this by considering a smooth reformulation of the restrictions in order to use standard nonlinear programming techniques. One possibility consists in rewriting (25) in the equivalent differentiable form:

$$\min J(u, u^{\text{train}})$$
$$\text{s.t } \lambda \circ (u - f) + \mathbb{K}^\top q = 0$$
$$\langle q_i, (\mathbb{K}u)_i \rangle^2 - \|(\mathbb{K}u)_i\|^2 = 0, \quad \forall i = 1, \ldots, s$$
$$-\langle q_i, (\mathbb{K}u)_i \rangle \le 0, \quad \forall i = 1, \ldots, s$$
$$\|q_i\|^2 - 1 \le 0, \quad \forall i = 1, \ldots, s$$
$$-\lambda_i \le 0, \quad \forall i = 1, \ldots, n,$$

and trying to apply nonlinear programming results.

Considering a toy example where $u \in \mathbb{R}^2$, $\lambda \in \mathbb{R}$, $q \in \mathbb{R}^2$ and $\mathbb{K} : \mathbb{R}^2 \to \mathbb{R}^2$ is defined by $\mathbb{K} = \begin{pmatrix} 1 & -1 \\ 0 & 1 \end{pmatrix}$, we may indeed analyze case-by-case and verify whether a standard constraint qualification has a chance to hold. To verify either the *Linear Independence Constraint Qualification Condition (LICQ)* or the *Mangasarian-Fromowitz Constraint Qualification Condition (MFCQ)* Nocedal and Wright (2006), we have to analyze the rank of the matrix formed by the gradients of the equality constraints, which is given by

$$\nabla h(u, q, \lambda) := \begin{pmatrix} \lambda & 0 & 2(u_1 - u_2)(q_1^2 - 1) & 0 \\ 0 & \lambda & -2(u_1 - u_2)(q_1^2 - 1) & 2u_2(q_2^2 - 1) \\ 1 & -1 & 2q_1(u_1 - u_2)^2 & 0 \\ 0 & 1 & 0 & 2q_2 u_2^2 \\ u_1 - f_1 & u_2 - f_2 & 0 & 0 \end{pmatrix} \quad (26)$$

We then obtain the following cases:

$(\mathbb{K}u)_1 = 0, (\mathbb{K}u)_2 \neq 0$: In this case we know that $u_1 - u_2 = 0$ and the dual variable verifies $|q_2| = 1$. Consequently, $\nabla h_3(u, q, \lambda) = (0, 0, 0, 0, 0)^\top$ and, therefore, the columns of $\nabla h(u, q, \lambda)$ are not linearly independent, and neither LICQ nor MFCQ holds.

$(\mathbb{K}u)_1 \neq 0, (\mathbb{K}u)_2 = 0$: Similar than the previous case, we reach to the same violation of linear independence, with $\nabla h_4(u, q, \lambda)$ equal to zero.

$(\mathbb{K}u)_1 \neq 0, (\mathbb{K}u)_2 \neq 0$: In this case $|q_i| = 1$, $i = 1, 2$ and we obtain $\nabla h_3(u, q, \lambda) = (0, 0, 2q_1(u_1 - u_2)^2, 0, 0)^\top$ and $\nabla h_4(u, q, \lambda) = (0, 0, 0, 2q_2 u_2^2, 0)^\top$. The linear independence may be satisfied in this case, and existence of Lagrange multipliers may have a chance to be justified. This is, however, a case with scarce practical relevance. In the imaging setting, it would be related to completely smooth images (with no edges).

Alternative Optimality Conditions

From the discussion above, it becomes clear that standard constraint qualifications cannot be expected to hold for the type of bilevel problems at hand and, therefore, classical nonlinear programming results cannot be used for guaranteeing existence of Lagrange multipliers. As an alternative, nonsmooth analysis techniques may be used to derive stationarity conditions, at the price of being possibly weaker than the ones originally expected.

In that sense, a first idea consists in carrying out a nonsmooth analysis of the solution operator associated to the lower-level problem. Indeed, it can be shown (De los Reyes and Meyer 2016; Hintermüller and Wu 2015) that the solution mapping $S : \mathbb{R}^n_+ \to \mathbb{R}^n, \lambda \mapsto u$, for the lower-level problem is Bouligand differentiable, i.e., directionally differentiable and locally Lipschitz continuous. Using the chain rule for B-differentiable functions, the composite loss function is Bouligand differentiable as well (Dontchev and Rockafellar 2009). This implies that the problem (28) can be written in reduced form as

$$\min_{\lambda \in \mathbb{R}^n_+} \mathcal{G}(\lambda) = J(S(\lambda), \lambda),$$

and a stationarity condition for a local optimal solution λ^* is given by

$$\langle J_u(u^*, \lambda^*), \eta \rangle + \langle J_\lambda(u^*, \lambda^*), \lambda - \lambda^* \rangle \geq 0, \ \forall \lambda \in \mathbb{R}^n_+, \qquad (27)$$

where $u^* = S(\lambda^*)$ and $\eta := S'(\lambda^*; \lambda - \lambda^*)$ is the directional derivative of the solution mapping in direction $\lambda - \lambda^*$. Condition (27) is also known as *Bouligand (B-) stationarity*. Even though this stationarity condition is sharp, it is hardly usable due to the nonlinearity of the directional derivative.

A different approach is pursued in Outrata (2000), where the author reformulates problems such as (23) using a generalized equation:

$$\min_{\lambda \in \mathbb{R}^n_+} J(u(\lambda), u^{\text{train}}) \qquad (28)$$

$$\text{s.t.} \quad 0 \in \lambda \circ (u - f) + Q(u), \qquad (29)$$

with $Q : \mathbb{R}^n \rightrightarrows \mathbb{R}^n$ a multifunction with a closed graph defined by

$$Q(u) := \left\{ \mathbb{K}^\top q : q \in \mathbb{R}^{s \times 2} : \begin{cases} q_j = \frac{(\mathbb{K}u)_j}{\|(\mathbb{K}u)_j\|}, & \text{if } (\mathbb{K}u)_j \neq 0, \\ \|q_j\| \leq 1, & \text{if } (\mathbb{K}u)_j = 0. \end{cases} \right\}$$

This problem may be interpreted as a *Generalized Mathematical Program with Equilibrium Constraints*, and Mordukhovich variational analysis may be used to derive first-order necessary optimality conditions (Outrata 2000, Theorem 3.1). To this aim, let us introduce the computed Mordukhovich normal cone (see, e.g., Hintermüller and Wu 2015):

$N_{GphQ}^M(u, \mathbb{K}^\top q)$

$$= \left\{(\mathbb{K}^\top w, v) : \begin{cases} \|(\mathbb{K}u)_j\| w_j = (\mathbb{K}v)_j - \langle(\mathbb{K}v)_j, q_j\rangle q_j, & \text{if } (\mathbb{K}u)_j \neq 0, \\ (\mathbb{K}v)_j = 0, & \text{if } |q_j|_2 < 1, \\ \begin{array}{l}(\mathbb{K}v)_j = 0, \vee \\ (\mathbb{K}v)_j = cq_j (c \in \mathbb{R}), \langle w_j, q_j\rangle = 0 \vee \\ (\mathbb{K}v)_j = cq_j (c \geq 0), \langle w_j, q_j\rangle \geq 0.\end{array} & \text{if } (\mathbb{K}u)_j = 0, |q_j|_2 = 1 \end{cases}\right\}$$

Let (λ^*, u^*, q^*) be a local solution of problem (28), and let $(\mathbb{K}^\top w, v) \in N_{GphQ}^{(M)}(u^*, \mathbb{K}^\top q^*)$ be a solution of the system

$$\begin{pmatrix} 0 & -\text{diag}(u^* - f) \\ I & -\text{diag}(\lambda^*) \end{pmatrix} \begin{pmatrix} \mathbb{K}^\top w \\ v \end{pmatrix} \in \{0\} \times N_{\mathbb{R}_+^n}^M \quad (30)$$

The vector (λ^*, u^*, q^*) is said to satisfy the *constraint qualification* if $\mathbb{K}^\top w = 0$ and $v = 0$ is the unique solution to the problem above.

Under this constraint qualification, there exist Lagrange multipliers $(\mathbb{K}^\top \varphi, p, \vartheta)$ such that the following *Mordukhovich (M-) stationary* system holds true:

$$\lambda \circ (u^* - f) + \mathbb{K}^\top q^* = 0, \tag{31a}$$

$$\langle q_j^*, (\mathbb{K}u^*)_j \rangle = \|(\mathbb{K}u^*)_j\|, \quad \forall i = 1, \ldots, s, \tag{31b}$$

$$\|q_j^*\| \leq 1, \quad \forall j = 1, \ldots, s, \tag{31c}$$

$$\lambda \circ p + \mathbb{K}^\top \varphi = \nabla_u J(u^*), \tag{31d}$$

$$(u^* - f) \circ p + \vartheta = 0, \tag{31e}$$

$$\|(\mathbb{K}u^*)_j\| \varphi_j = (\mathbb{K}p)_j - \langle(\mathbb{K}p)_j, q_j^*\rangle q_j^*, \quad \text{if } (\mathbb{K}u^*)_j \neq 0, \tag{31f}$$

$$(\mathbb{K}p)_j = 0, \quad \text{if } (\mathbb{K}u^*)_j = 0, \|q_j^*\| < 1, \tag{31g}$$

$$\left.\begin{array}{l}(\mathbb{K}p)_j = 0 \vee \\ (\mathbb{K}p)_j = cq_j^* (c \in \mathbb{R}), \langle \varphi_j, q_j^*\rangle = 0 \vee \\ (\mathbb{K}p)_j = cq_j^* (c \geq 0), \langle \varphi_j, q_j^*\rangle \geq 0,\end{array}\right\} \quad \text{if } (\mathbb{K}u^*)_j = 0, \|q_j^*\| = 1, \tag{31h}$$

$$0 \leq \lambda \perp \vartheta \geq 0, \tag{31i}$$

The difference betweeen M-stationarity and strong stationarity systems concerns the information about the multipliers on the so-called biactive set $\mathcal{B} = j \in \{1, \ldots, s\} : (\mathbb{K}u)_j = 0, \|q_j\| = 1\}$. The biactive characterization of those multipliers in (31h) is actually weaker than in a strong stationarity system.

An even weaker stationarity system may be obtained by regularizing the Euclidean norm in (23) and then deriving optimality conditions for each regularized problem and afterward passing to the limit in the regularized optimality systems (De los Reyes 2011). In that case, a *Clarke (C-) stationary* system is obtained, where (31h) is replaced by

$$(\mathbb{K}p^*)_j = cq_j^*(c \in \mathbb{R}), \quad \langle \varphi_j^*, q_j^* \rangle \geq 0, \qquad \text{if } (\mathbb{K}u^*)_j = 0, \|q_j^*\| = 1. \qquad (32)$$

Finally, it can be proven that if *strict complementarity holds*, i.e., if the biactive set is empty, all strong, B-, M-, and C-stationarity conditions are equivalent (see, e.g., De los Reyes 2015; De los Reyes and Meyer 2016).

Solution Algorithms

When dealing with the numerical optimization of the bilevel problem, the solution of a regularized version of (28) appears to be the more frequent approach. In this line, the nonsmoothness is regularized by means of a differentiable function, and nonlinear optimization methods are then applied. In De los Reyes and Schönlieb (2013), for instance, the authors implement a BFGS algorithm with Armijo backtracking to solve a regularized bilevel problem for image denoising. Alternatively, the authors in Hintermüller and Wu (2015) propose a projected gradient method to find stationary points in the case of blind deconvolution.

For dealing with the nonsmooth bilevel problem, we point out the works (Outrata and Zowe 1995) and (Christof et al. 2020). In the first one, subgradients of the reduced cost function are computed by means of a generalized adjoint equation, while, in the second one, a trust-region method exploiting the nonsmooth Bouligand subdifferential properties of the solution operator is proposed. Both algorithms are precisely devised for optimization problems with variational inequality constraints, and convergence toward a C-stationary point is verified in the second one.

Infinite-Dimensional Case

The infinite-dimensional counterpart of the bilevel learning approach (16) poses additional difficulties in the analysis of the resulting nonsmooth problems, since properties like directional differentiability of the solution mapping cannot be derived in function spaces, unless very restrictive assumptions are made (De los Reyes and Meyer 2016).

The study of the infinite-dimensional problems becomes important, however, to derive properties which are resolution independent, as well as to shed light on the development of algorithms whose efficiency does not depend on the number of pixels of the image. Moreover, in the recent past, the use of parameter functions, instead of vectors, has proven to be superior for different imaging tasks, and, in order to consider spatially dependent parameters, the function space framework appears indeed to be the natural choice in this context.

Considering as image domain the open bounded convex set $\Omega \subset \mathbb{R}^2$ and assuming that the noisy image f lies in the Hilbert $Y = L^2(\Omega)$, the bilevel problem, for a single training pair, consists in searching for parameters $\lambda = (\lambda_1, \ldots, \lambda_M)$ and $\alpha = (\alpha_1, \ldots, \alpha_N)$ in abstract nonnegative parameter sets $\mathcal{P}_\lambda^+$ and $\mathcal{P}_\alpha^+$ that solve

$$\min_{\alpha \in \mathcal{P}_\alpha^+, \lambda \in \mathcal{P}_\lambda^+} J(u_{\alpha,\lambda}) \quad \text{s.t.} \quad u_{\alpha,\lambda} \in \arg\min_{u \in X} \mathcal{E}(u; \lambda, \alpha), \tag{P}$$

with

$$\mathcal{E}(u; \lambda, \alpha) := \sum_{i=1}^{M} \int_\Omega \lambda_i(x) \phi_i(x, [Au](x)) \, dx + \sum_{j=1}^{N} \int_\Omega \alpha_j(x) \, d|B_j u|(x).$$

where the loss functional $J : X \to \mathbb{R}$ is assumed to be convex, proper, and weak* lower semicontinuous. Our solution u lies in an abstract space X, mapped by the linear operator A to Y. Depending on B, A, and the ϕ_i, different problems as well as assumptions have to be made (De los Reyes et al. 2016). In general, convexity of $\mathcal{E}(\cdot; \lambda, \alpha)$ and compactness properties in the space of functions of bounded variation turn out to be crucial for proving existence of optimal solutions.

To overcome the difficulties related to the nonsmoothness of (P) and the lack of regularity of the solutions, smoothing terms are usually added within the bilevel framework in order to carry out the analysis. For one, we require Huber regularization of the Radon norms. This is needed for the single-valued differentiability of the solution map $(\lambda, \alpha) \mapsto u_{\alpha,\lambda}$. Secondly, we take a convex, proper, and weak* lower-semicontinuous smoothing functional $H : X \to [0, \infty]$. The typical choice is the elliptic energy $H(u) = \frac{1}{2} \|\nabla u\|^2$.

For parameters $\mu \geq 0$ and $\gamma \in (0, \infty]$, we consider as in De los Reyes et al. (2016) the problem

$$\min_{\alpha \in \mathcal{P}_\alpha^+, \lambda \in \mathcal{P}_\lambda^+} J(u_{\alpha,\lambda,\gamma,\mu}) \quad \text{s.t.} \quad u_{\alpha,\lambda,\gamma,\mu} \in \arg\min_{u \in X \cap \text{dom}\, \mu H} \mathcal{E}^{\gamma,\mu}(u; \lambda, \alpha) \tag{P$^{\gamma,\mu}$}$$

with the regularized energy

$$\mathcal{E}^{\gamma,\mu}(u; \lambda, \alpha) := \mu H(u) + \sum_{i=1}^{M} \int_\Omega \lambda_i(x) \phi_i(x, [Au](x)) \, dx$$

$$+ \sum_{j=1}^{N} \int_\Omega \alpha_j(x) \, d|B_j u|_\gamma(x).$$

We denote by $|B_j u|_\gamma$ the Huberised total variation measure, where

$$|g|_\gamma = \begin{cases} \|g\|_2 - \frac{1}{2\gamma}, & \|g\|_2 \geq 1/\gamma, \\ \frac{\gamma}{2}\|g\|_2^2, & \|g\|_2 < 1/\gamma, \end{cases}$$

for $\gamma \in (0, \infty]$. Considering the Lebesgue decomposition of $v \in \mathcal{M}(\Omega; \mathbb{R}^n)$ into the absolutely continuous part $f\mathcal{L}^n$ and the singular part v^s, we set

$$|v|_\gamma(V) := \int_V |f(x)|_\gamma \, dx + |v^s|(V), \quad (V \in \mathcal{B}(\Omega)).$$

The measure $|v|_\gamma$ corresponds to the Huber regularization of the total variation measure $|v|$.

Existence and Other Properties

The first questions to be answered concerning the bilevel problem (P) are related to the existence of optimal parameters as well as the structure of the optimizers. At least partially, some answers to these inquires have been given in De los Reyes et al. (2016) (see also the review paper Calatroni et al. 2017). We briefly summarize next the main results obtained in those references.

Considering the particular, but frequent, setup with quadratic loss functional and fidelity term

$$J(u) = \frac{1}{2}\|Au - u^{\text{train}}\|^2_{L^2(\Omega)}, \quad \text{and} \quad \phi_1(x, v) = \frac{1}{2}|f(x) - v|^2, \tag{33}$$

and with $M = 1$ and $\mathcal{P}_\lambda^+ = \{1\}$, we may obtain conditions for positivity of the parameters $\alpha = (\alpha_1, \ldots, \alpha_N) \in \mathcal{P}_\alpha^+ = [0, \infty]^N$. In fact, suppose that $f, f_0 \in BV(\Omega) \cap L^2(\Omega)$ satisfy

$$TV(f) > TV(u^{\text{train}}), \tag{34}$$

then there exist $\bar{\mu}, \bar{\gamma} > 0$ such that any optimal solution $\alpha_{\gamma,\mu} \in [0, \infty]$ to the problem

$$\min_{\alpha \in [0, \infty]} \frac{1}{2}\|u^{\text{train}} - u_\alpha\|^2_{L^2(\Omega)}$$

with

$$u_\alpha \in \arg\min_{u \in BV(\Omega)} \left(\frac{1}{2}\|f - u\|^2_{L^2(\Omega)} + \alpha|Du|_\gamma(\Omega) + \frac{\mu}{2}\|\nabla u\|^2_{L^2(\Omega;\mathbb{R}^n)} \right)$$

satisfies $\alpha_{\gamma,\mu} > 0$, whenever $\mu \in [0, \bar{\mu}]$ and $\gamma \in [\bar{\gamma}, \infty]$. The choice $\gamma = \infty$ should be understood as the standard unregularized total variation measure or norm.

For fixed values $\gamma < \infty$ and $\mu > 0$, existence of an optimal parameter can be proven by the direct method of the calculus of variations. What condition 34 guarantees is existence of an optimal interior solution $\alpha > 0$ to (P) without any additional box constraints. Moreover, condition (34) also guarantees convergence of optimal parameters of the numerically regularized H^1 problems ($P^{\gamma,\mu}$) to a solution of the original $BV(\Omega)$ problem (P).

A similar structural result may be obtained for second-order total generalized variation Gaussian denoising, again assuming that the noisy data has to oscillate more in terms of TGV^2 than the ground truth does. Specifically, if the data $f, u^{\text{train}} \in L^2(\Omega) \cap BV(\Omega)$ satisfies for some $\alpha_2 > 0$ the condition

$$\text{TGV}^2_{(\alpha_2,1)}(f) > \text{TGV}^2_{(\alpha_2,1)}(u^{\text{train}}), \tag{35}$$

then there exists $\bar{\mu}, \bar{\gamma} > 0$ such that any optimal solution $\alpha_{\gamma,\mu} = ((\alpha_{\gamma,\mu})_1, (\alpha_{\gamma,\mu})_2)$ to the problem

$$\min_{\alpha \in [0,\infty]^2} \frac{1}{2} \|f_0 - v_\alpha\|^2_{L^2(\Omega)}$$

with

$$(v_\alpha, w_\alpha) \in \arg\min_{\substack{v \in BV(\Omega) \\ w \in BD(\Omega)}} \left(\frac{1}{2} \|f - v\|^2_{L^2(\Omega)} + \alpha_1 |Dv - w|_\gamma(\Omega) + \alpha_2 |Ew|_\gamma(\Omega) \right.$$

$$\left. + \frac{\mu}{2} \|(\nabla v, \nabla w)\|^2_{L^2(\Omega; \mathbb{R}^n \times \mathbb{R}^{n \times n})} \right)$$

satisfies $(\alpha_{\gamma,\mu})_1, (\alpha_{\gamma,\mu})_2 > 0$, whenever $\mu \in [0, \bar{\mu}]$, $\gamma \in [\bar{\gamma}, \infty]$. Observe that we allow for infinite parameters α.

Additionally, a result on the approximation properties as $\gamma \nearrow \infty$ and $\mu \searrow 0$ is also obtained. In fact, for both previous settings, there exist $\bar{\gamma} \in (0, \infty)$ and $\bar{\mu} \in (0, \infty)$ such that the solution map $(\gamma, \mu) \mapsto \alpha_{\gamma,\mu}$ is outer semicontinuous within $[\bar{\gamma}, \infty] \times [0, \bar{\mu}]$. Roughly, the outer semicontinuity (Rockafellar and Wets 1998) of the solution map means that as the regularization vanishes, any optimal parameters for the regularized models ($P^{\gamma,\mu}$) tend to some optimal parameters of the original model (P).

Stationarity Conditions

The family of problems ($P^{\gamma,\mu}$) constitute PDE-constrained optimization instances, and, therefore, suitable techniques from this field may be utilized to derive optimality conditions. For the limiting cases $\gamma \to \infty$ or $\mu \to 0$, an additional asymptotic analysis needs to be performed in order to get stationarity conditions for the optimal solutions.

Several instances of the abstract problem ($P^{\gamma,\mu}$) have been individually considered in previous contributions. The case with total variation regularization was considered in De los Reyes and Schönlieb (2013) in presence of several noise models, and, after proving the Gâteaux differentiability of the solution operator, an optimality system was derived. Thereafter, an asymptotic analysis with respect to $\gamma \to \infty$ was carried out (with $\mu > 0$), obtaining an optimality system for the corresponding problem. In that case the optimization problem corresponds to one with variational inequality constraints, and the characterization concerns C-stationary points. Differentiability properties of higher-order regularization solution operators were also investigated in De los Reyes et al. (2017), with the corresponding first-order optimality conditions.

For the general problem ($P^{\gamma,\mu}$), using the Lagrangian formalism, the following optimality system is obtained:

$$\mu \int_\Omega \langle \nabla u, \nabla v \rangle \, dx + \sum_{i=1}^{M} \int_\Omega \lambda_i \phi_i'(Au) Av \, dx$$

$$+ \sum_{j=1}^{N} \int_\Omega \alpha_j \langle h_\gamma(B_j u), B_j v \rangle \, dx = 0, \quad \forall v \in V, \quad (36)$$

$$\mu \int_\Omega \langle \nabla p, \nabla v \rangle \, dx + \sum_{i=1}^{M} \int_\Omega \langle \lambda_i \phi_i''(Au) Ap, Av \rangle \, dx$$

$$+ \sum_{j=1}^{N} \int_\Omega \alpha_j \langle h_\gamma'^*(B_j u) B_j p, B_j v \rangle \, dx = -\nabla_u J(u) v, \quad \forall v \in V, \quad (37)$$

$$\int_\Omega \phi_i(Au) Ap(\zeta - \lambda_i) \, dx \geq 0, \quad \forall \zeta \geq 0, \ i = 1, \ldots, M, \quad (38)$$

$$\int_\Omega h_\gamma(B_j u) B_j p(\eta - \alpha_j) \, dx \geq 0, \quad \forall \eta \geq 0, \ j = 1, \ldots, N, \quad (39)$$

where V stands for the Sobolev space where the regularized image lives (typically a subspace of $H^1(\Omega; \mathbb{R}^m)$), $p \in V$ stands for the adjoint state, and h_γ is a regularized version of the TV subdifferential, e.g.,

$$h_\gamma(z) := \begin{cases} \frac{z}{|z|} & \text{if } \gamma|z| - 1 \geq \frac{1}{2\gamma} \\ \frac{z}{|z|}(1 - \frac{\gamma}{2}(1 - \gamma|z| + \frac{1}{2\gamma})^2) & \text{if } \gamma|z| - 1 \in (-\frac{1}{2\gamma}, \frac{1}{2\gamma}) \\ \gamma z & \text{if } \gamma|z| - 1 \leq -\frac{1}{2\gamma}. \end{cases} \quad (40)$$

The rigorous derivation of the optimality system has to be justified for each specific combination of spaces, regularizers, noise models, and cost functionals.

With help of the adjoint equation (37), also gradient formulas for the reduced cost functional $\mathcal{G}(\lambda, \alpha) := J(u_{\alpha,\lambda}, \lambda, \alpha)$ are derived:

$$(\nabla_\lambda \mathcal{G})_i = \int_\Omega \phi_i(Au) Ap \, dx, \qquad (\nabla_\alpha \mathcal{G})_j = \int_\Omega h_\gamma(B_j u) B_j p \, dx, \qquad (41)$$

for $i = 1, \ldots, M$ and $j = 1, \ldots, N$. The gradient information is of numerical importance in the design of solution algorithms. In the case of finite dimensional parameters, thanks to the structure of the minimizers reviewed previously, the corresponding variational inequalities (38)-(39) turn into equalities. This has important numerical consequences, since in such cases the gradient formulas (41) may be used without additional projection steps.

Dualization

An alternative technique for studying the bilevel problem, via duality, was proposed by Hintermüller and coauthors (2017), where the lower-level problem is replaced by its pre-dual version. In the case of total variation and with a weight solely on the regularizer, the bilevel problem becomes

$$\min_{\underline{\alpha} \leq \alpha(x) \leq \overline{\alpha}} J(R(\mathrm{div}\, p)) + \frac{\beta}{2} \|\alpha\|^2_{H^1(\Omega)} \qquad (D)$$

$$\text{s.t. } p \in \underset{p \in \mathbb{H}_0^1(\Omega):\, |p(x)|_\infty \leq \alpha(x)}{\arg\min} \left(\frac{\mu}{2} \|\nabla p\|^2_{L^2(\Omega)} + \frac{\gamma}{2} \|p\|^2_{L^2(\Omega)} + \frac{1}{2} \|\mathrm{div}\, p + f\|^2_{L^2(\Omega)} \right),$$

where $\mu, \gamma > 0$ are regularization parameters and R stands for the localized residual function. As a consequence, the necessary and sufficient optimality condition for the lower-level problem becomes a variational inequality of the first kind, which may be reformulated as a complementarity system as well. The abstract problem then constitutes a *mathematical program with equilibrium constraints* in function space.

The treatment of this problem is, however, by no means any easier than the primal bilevel one. In fact, in order to carry out the analysis, the authors have to penalize the pointwise box constraint by means of a Moreau-Yosida function $\mathcal{P}_\delta(p, \alpha)$, yielding the problem

$$\min_{\underline{\alpha} \leq \alpha(x) \leq \overline{\alpha}} J(R(\mathrm{div}\, p)) + \frac{\beta}{2} \|\alpha\|^2_{H^1(\Omega)}$$

$$\text{s.t. } p \in \underset{p \in \mathbb{H}_0^1(\Omega)}{\arg\min} \left(\frac{\mu}{2} \|\nabla p\|^2_{L^2(\Omega)} + \frac{\gamma}{2} \|p\|^2_{L^2(\Omega)} + \frac{1}{2} \|\mathrm{div}\, p + f\|^2_{L^2(\Omega)} + \frac{1}{\epsilon} \mathcal{P}_\delta(p, \alpha) \right),$$

For each penalized problem, existence of Lagrange multipliers is then proven using standard Karush-Kuhn-Tucker theory in function spaces. Although no limit analysis is carried out in order to get an optimality system for problem (D), the authors provide some useful density and stability results.

Nonlocal Problems

As mentioned in section "Variational Inverse Problems Setting", nonlocal models perform particularly well in problems where different textures are present in the image, as similar intensity patterns between pixels or patches in a given spatial subdomain are taken into account for the restoration (Yaroslavsky 1986; Tomasi and Manduchi 1998; Buades et al. 2005). In Gilboa and Osher (2007) and Gilboa and Osher (2008), an energy-based variational framework was introduced for nonlocal imaging models, allowing the analytical study of different underlying properties in function spaces. Moreover, nonlocal vector calculus has been developed in the last years, providing a very useful analytical toolbox for dealing with nonlocal models arising in different application areas (Gunzburger and Lehoucq 2010; Du et al. 2012).

Within this framework, a bilevel learning formulation for estimating the weights in nonlocal imaging problems was recently studied in d'Elia et al. (2019), considering both the case of a weight within the kernel and the one with the weight in front of the fidelity term. Assuming there is a single training pair of a clean and a noisy images (u^{train}, f), the general problem reads as follows:

$$\min_{0 \leq \lambda, w \leq U} J(u) \tag{42}$$

$$\text{s.t.} \quad u(\lambda, w) = \arg\min_u \frac{1}{2} \int_\Omega \int_\Omega (u(\mathbf{x}) - u(\mathbf{y}))^2 \gamma_w(\mathbf{x}, \mathbf{y}) \, d\mathbf{y} \, d\mathbf{x} + \int_\Omega \lambda (u - f)^2 \tag{43}$$

where

$$\gamma_w(\mathbf{x}, \mathbf{y}) := \exp\left\{-\int_{B_\rho(0)} w(\tau) \left(f(\mathbf{x} + \tau) - f(\mathbf{y} + \tau)\right)^2 d\tau\right\} \chi\left(\mathbf{y} \in B_\epsilon(\mathbf{x})\right)$$

corresponds to the modified nonlocal means kernel. Alternative nonlocal kernels, pixelwise or patchwise, may be considered as well.

In this case, the unique solution to the lower-level problem belongs to the space $V_c^w := \{v \in L^2(\Omega \cup \Omega_I) : \|v\|_{V^w} < \infty, v|_{\Omega_I} = 0\}$, where $\Omega_I := \{\mathbf{y} \in \mathbb{R}^d \setminus \Omega : \|\mathbf{x} - \mathbf{y}\| \leq \varepsilon, \ \forall \mathbf{x} \in \Omega\}$ is the so-called interaction domain where volume constraints are imposed, and

$$\|u\|_{V^w}^2 := \int_{\Omega \cup \Omega_I} \int_{\Omega \cup \Omega_I} (u(\mathbf{x}) - u(\mathbf{y}))^2 \gamma_w(\mathbf{x}, \mathbf{y}) \, d\mathbf{y} \, d\mathbf{x}$$

is the nonlocal energy norm. If w is a constant weight, the space is simply denoted as V.

Existence of an optimal solution for the bilevel problem in each of the settings has been proven in d'Elia et al. (2019), under the inclusion of box constraints for the parameters. For the case of a spatially dependent coefficient in front of the fidelity, an extra Tikhonov regularization term has to be added to the loss functional to get existence of an optimal solution.

In constrast to the variational regularizers reviewed before, for the nonlocal problem (43), Gâteaux differentiability of the solution operator can be demonstrated. As a consequence, necessary optimality systems that characterize strong stationary points can be established in each of the cases (see d'Elia et al. 2019 for further details).

For the case when a spatially dependent weight $\lambda \in H^1(\Omega)$ is optimized, while keeping the kernel fixed, a necessary optimality condition is given by the following complementarity problem:

$$\int_{\Omega \cup \Omega_I} \int_{\Omega \cup \Omega_I} (u(\mathbf{x}) - u(\mathbf{y}))(\psi(\mathbf{x}) - \psi(\mathbf{y}))\gamma_w(\mathbf{x}, \mathbf{y}) \, d\mathbf{y} \, d\mathbf{x}$$
$$+ \int_\Omega \lambda (u - f) \psi \, d\mathbf{x} = 0, \quad \forall \psi \in V_c, \quad (44a)$$

$$\int_{\Omega \cup \Omega_I} \int_{\Omega \cup \Omega_I} (p(\mathbf{x}) - p(\mathbf{y}))(\phi(\mathbf{x}) - \phi(\mathbf{y}))\gamma_w(\mathbf{x}, \mathbf{y}) \, d\mathbf{y} \, d\mathbf{x}$$
$$+ \int_\Omega \lambda p \phi \, d\mathbf{x} = -\nabla_u J(u)\phi, \quad \forall \phi \in V_c, \quad (44b)$$

$$-\beta \Delta \lambda + \beta \lambda = \sigma_\Omega^+ - \sigma_\Omega^- \quad \text{in } \Omega,$$
$$\beta \frac{\partial \lambda}{\partial \mathbf{n}} = \sigma_\Gamma^+ - \sigma_\Gamma^- \quad \text{on } \Gamma, \quad (44c)$$

$$0 \leq \sigma_\Omega^+(\mathbf{x}) \perp \lambda(\mathbf{x}) \geq 0, \quad 0 \leq \sigma_\Omega^-(\mathbf{x}) \perp (U - \lambda(\mathbf{x})) \geq 0, \quad \forall \mathbf{x} \in \Omega,$$
$$0 \leq \sigma_\Gamma^+(\mathbf{x}) \perp \lambda(\mathbf{x}) \geq 0, \quad 0 \leq \sigma_\Gamma^-(\mathbf{x}) \perp (U - \lambda(\mathbf{x})) \geq 0, \quad \forall \mathbf{x} \in \Gamma, \quad (44d)$$

where $\sigma_\Omega^+, -\sigma_\Omega^-, \sigma_\Gamma^+, \sigma_\Gamma^-$ are Karush-Kuhn-Tucker multipliers associated to the box constraints. As can be observed, in this case, the optimality system couples local and nonlocal systems of equations with additional pointwise complementarity conditions.

On the other hand, for the case when the weight within the kernel $w \in L^2(B_\rho(0))$ is optimized, the following optimality system is satisfied:

$$\int_{\Omega \cup \Omega_I} \int_{\Omega \cup \Omega_I} (u(\mathbf{x}) - u(\mathbf{y}))(\psi(\mathbf{x}) - \psi(\mathbf{y})) \gamma_w(\mathbf{x}, \mathbf{y}) \, d\mathbf{y} \, d\mathbf{x}$$
$$+ \int_{\Omega} \lambda (u - f) \psi \, d\mathbf{x} = 0, \quad \forall \psi \in V_c, \quad (45a)$$

$$\int_{\Omega \cup \Omega_I} \int_{\Omega \cup \Omega_I} (p(\mathbf{x}) - p(\mathbf{y}))(\phi(\mathbf{x}) - \phi(\mathbf{y})) \gamma_w(\mathbf{x}, \mathbf{y}) \, d\mathbf{y} \, d\mathbf{x}$$
$$+ \int_{\Omega} \lambda p \phi \, d\mathbf{x} = -J'(u)\phi, \quad \forall \phi \in V_c, \quad (45b)$$

$$\int_{\Omega \cup \Omega_I} \int_{\Omega \cup \Omega_I} \left[(u(w) - u(w)')(p - p') \widetilde{\gamma}_{(h-w)}(\mathbf{x}, \mathbf{y}) \right] d\mathbf{y} \, d\mathbf{x} \geq 0, \quad \forall h \in \mathcal{U}_{ad}. \quad (45c)$$

with $\mathcal{U}_{ad} := \{v \in L^2(B_\rho(\mathbf{0})) : 0 \leq w(\mathbf{t}) \leq U, \forall \mathbf{t} \in B_\rho(\mathbf{0})\}$ and the linearized kernel

$$\widetilde{\gamma}_h(\mathbf{x}, \mathbf{y}) = \gamma_w(\mathbf{x}, \mathbf{y}) \int_{B_\rho(\mathbf{0})} -h(\tau)\big(f(\mathbf{x}+\tau) - f(\mathbf{y}+\tau)\big)^2 \, d\tau. \quad (45d)$$

In this case, even if "only" a scalar is determined, the computational cost becomes high since in principle the kernel changes in each iteration of any solution algorithm and, with it, the assembly of the nonlocal interaction matrix, which is in principle a dense one.

Neural Network Optimization

In previous sections, we presented variational image restoration as a special class of ill-posed inverse problems and considered a bilevel learning framework for determining the different parameters involved. This approach allows to incorporate a priori information about the solution, enabling a deeper understanding of the models at hand. Even though this setting allows us to get insight into the structural properties of the variational models, finding the optimal solution is often highly computationally demanding and in some cases not suitable for real-world applications.

On the other hand, neural networks and, in particular, convolutional neural networks (CNN) have been widely used for image restoration tasks, such as denoising (Burger et al. 2012), blind and non-blind deblurring (Xu et al. 2014), demosaicking (Wang 2014), and super-resolution (Dong et al. 2014), among others. Despite such success in practical cases, these learning structures still lack explainability and reliability (Szegedy et al. 2013). Indeed, the incorporation of a priori knowledge in these models is very complicated, and in most applications, it is treated as a black box.

Recently, the gap between both frameworks has started to be bridged. By using bilevel optimization and optimal control ideas, some approaches that combine the best properties of variational models and neural networks have been proposed. We provide next a brief review on a few of them, with the sole purpose of highlighting the importance of these connections.

Deep Neural Networks as a Further Regularizer

Even though we have previously detailed bilevel learning strategies for variational problems, recently also bilevel optimization approaches that make use of neural networks have been proposed. In particular, *Deep Bilevel Optimization Neural Networks (BOONet)*, introduced by Antil and coworkers (2020), develop a strategy for finding optimal regularization parameters based on a bilevel optimization problem. Here, an upper-level optimization problem is used to measure the reconstruction error on a training dataset, while the lower-level problem measures the misfit of the data reconstruction. This reconstruction is based on a generalized regularizer that has a network-like structure, leading to insightful comparisons over regularizers and activation functions used in neural networks.

Now, regarding the regularization term in (2), it has been further improved recently by making use of a pretrained CNN. Indeed, in Lunz et al. (2018) a data-driven regularizer is built using modern generative adversarial network principles, leading to the *neural network Tikhonov* (NETT) approach, where a pretrained network is composed with a regularization functional (Li et al. 2020).

In Kobler et al. (2020), a different technique for learning regularizers is proposed, called *total deep variation*. In this case, the regularizer is built using a multi-scale convolutional neural network, which training is based on a sampled optimal control problem interpretation. This formulation allows the authors to provide a sensitivity analysis of the learned coefficients with respect to the training dataset. It is worth mentioning that this regularizer can be trained using a different dataset than the application at hand, resembling the properties of transfer learning strategies.

Deep Unrolling Within Optimization

Assuming we use an iterative scheme for solving (2) that is based on a proximal operator

$$\text{prox}_{\tau \mathcal{R}}(\hat{u}) = \arg\min_{u \in \mathbb{R}^n} \frac{1}{2\tau} \|u - \hat{u}\|^2 + \mathcal{R}(u), \tag{46}$$

it was observed in Venkatakrishnan et al. (2013) that this denoising subproblem may be replaced by a more sophisticated neural network such as *BM3D* (Dabov et al. 2007). However, this general purpose approach does not exploit the variational structure of the original problem, and, thus, the explainability provided by classical variational approaches is missing.

An alternative strategy consists in using deep neural network architectures to replace inner operators (such as prox, gradients, etc.) of an optimization scheme to solve the variational imaging task of interest. Even though this training process is computationally expensive, this procedure is often performed as an off-line batch operation. Once trained, the network evaluation is less expensive when used in the reconstruction process. Gregor and LeCun (2010) were able to incorporate these ideas into an ISTA iterative scheme for solving a sparse coding problem (LISTA). This procedure is based on "unrolling" the iterative scheme and replacing its explicit updates with learned ones. Hersey et al. (2014) propose to unfold the iterations into a layer-based structure similar to a neural network with application to learning optimal parameters of Markov random fields and nonnegative matrix factorization.

In the imaging context, these ideas have been considered in the unrolling of iterative schemes for problems with the structure presented in (11), such as proximal gradient (Adler and Öktem 2017), primal-dual hybrid gradient (Adler and Öktem 2018), or ADMM (Sun et al. 2016). This technique generates new tailor-designed deep neural network architectures that make use of the structure within the problem at hand. In the case of a Field of Experts (FoE) regularizer, this approach led to the development of *variational networks* (Kobler et al. 2017), where the authors rely on unrolling a proximal gradient descent step for a finite number of iterations and the connections to residual neural networks (He et al. 2016) are highlighted.

Numerical Experiments

In this section, we compare different bilevel parameter optimization techniques for image denoising, both from a reconstruction quality perspective and from a *learning* point of view. In particular, we will learn optimal scalar and scale-dependent parameters for image denoising models using total variation, total generalized variation, and nonlocal means regularizers. For the scalar experiment, we will make use of a semismooth Newton solver for the corresponding lower-level problem and the proposed BFGS method described in De los Reyes et al. (2017) for the bilevel problem. Moreover, the scale- and patch-dependent experiments will be solved using the primal-dual hybrid gradient (PDHG) method for the lower-level problem, and a trust-region strategy will be implemented for finding stationary points in the bilevel problem along the lines of Christof et al. (2020).

For obtaining such parameters, we will use a dataset of faces based on the popular CelebA Faces. This dataset will be split into a *training* dataset that will be used to learn the optimal parameters and a *validation* dataset that will be used to estimate the generalization error, i.e., to get an idea of the performance of the learned parameter in a set not previously trained on. Both datasets are generated by converting the original images in black and white, balancing its contrast and adding Gaussian noise of different intensities. A subset of the used training images is depicted in Fig. 2.

As a first experiment, we use a bilevel strategy for learning a scalar parameter for the variational formulation presented in (19) and taking the particular cases of total variation (TV) and total generalized variation (TGV). In Fig. 3, a subset of the validation dataset is shown with the optimal parameter and the corresponding SSIM measure. As extensively reported in De los Reyes et al. (2017), TGV is superior

Fig. 2 Sample of the training CelebA dataset

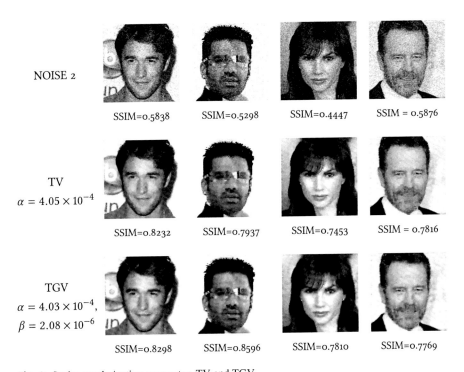

Fig. 3 Scalar regularization parameter: TV and TGV

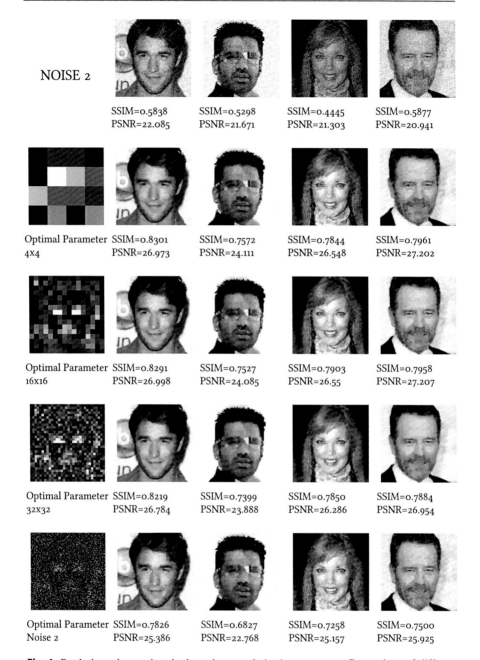

Fig. 4 Patch-dependent and scale-dependent regularization parameter Comparison of different learned patch-dependent and scale-dependent parameters used for denoising the validation dataset with noise 2

24 Bilevel Optimization Methods in Imaging

with respect to TV except in isolated instances. In the reconstruction of faces, as several gray scales are present within each structural component, TGV turns out to be a robust regularizer.

Moving further, we consider a scale-dependent parameter (18) and patch-dependent parameters (20), for the case of total variation denoising. The optimal learned parameters for the highest noise level in the training set are shown in Fig. 4. These learned parameters retrieve structural properties of the training dataset. In particular, as the number of degrees of freedom increases, a face structure can be identified in the weights.

It is of particular interest the behavior of these parameters in the validation dataset. Table 1 show the values of the optimal SSIM reconstruction (SD-TV) in both the training and validation datasets. Even though the more degrees of freedom for the regularization parameter allow for a better fitting in the training dataset, it performs poorly in the validation dataset according to the SSIM metric (see Fig. 5). This behavior is widely known in the machine learning community as *over-fitting*.

Table 1 SSIM quality measures Quality measures obtained in the training and validation dataset using the optimal parameter learned for different image denoising models

num	noisy	TRAINING TV	TGV	NL	SD-TV	PD-TV4	PD-TV16	PD-TV32
1	0.5838	0.8583	0.8715	0.7889	0.8441	0.8405	0.8433	0.8341
2	0.5298	0.8397	0.8463	0.7729	0.8226	0.8107	0.8121	0.8194
3	0.4447	0.8412	0.8612	0.8433	0.8713	0.8651	0.8655	0.8639
4	0.5877	0.8159	0.8270	0.8026	0.8531	0.8505	0.8544	0.8625
5	0.4865	0.7896	0.8234	0.8110	0.8398	0.8498	0.8457	0.8607
6	0.4699	0.8285	0.8469	0.7909	0.8343	0.8275	0.8283	0.8281
7	0.4827	0.8413	0.8564	0.7909	0.8218	0.7727	0.7785	0.8017
8	0.4884	0.8095	0.8325	0.7751	0.8389	0.8370	0.8381	0.8391
9	0.6144	0.8353	0.8654	0.7934	0.8505	0.8484	0.8484	0.8495
10	0.5029	0.8087	0.8366	0.7945	0.8298	0.7992	0.8087	0.8313
mean		**0.8268**	**0.8467**	**0.7963**	**0.8407**	**0.8298**	**0.8323**	**0.8391**
num	noisy	VALIDATION TV	TGV	NL	SD-TV	PD-TV4	PD-TV16	PD-TV32
1	0.6020	0.8232	0.8298	0.7847	0.7826	0.8301	0.8292	0.8219
2	0.5915	0.8557	0.8596	0.7094	0.6827	0.7572	0.7527	0.7399
3	0.5280	0.7480	0.7342	0.7707	0.7258	0.7844	0.7903	0.7850
4	0.5076	0.7816	0.7769	0.7221	0.7500	0.7961	0.7958	0.7884
5	0.4569	0.7944	0.7841	0.7728	0.7856	0.8254	0.8306	0.8284
6	0.5342	0.8215	0.8344	0.7258	0.7434	0.7847	0.7856	0.7783
7	0.4937	0.7865	0.7789	0.6591	0.7064	0.7628	0.7577	0.7375
8	0.5457	0.7453	0.7569	0.6903	0.7328	0.7780	0.7797	0.7708
9	0.4907	0.7567	0.7809	0.8092	0.7277	0.8036	0.7995	0.7855
10	0.5475	0.7937	0.8146	0.8359	0.8086	0.8586	0.8561	0.8452
mean		**0.7907**	**0.7950**	**0.7480**	**0.7445**	**0.7981**	**0.7977**	**0.7881**

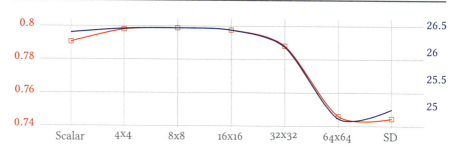

Fig. 5 Validation dataset reconstructions Average values of SSIM (red) and PSNR (blue) for the reconstruction of the validation dataset using different parameter models

To prevent the effect of *over-fitting* from happening and obtain better generalization properties, the patch-dependent regularization parameters (with few degrees of freedom) may be considered. To test this statement and realize how many degrees of freedom should serve that goal, we carry out an extra experiment. Specifically, the denoised results in the validation dataset for different dimensions are presented in Fig. 5. Indeed, the restriction on the degrees of freedom for the regularization parameter allows better generalization according to both the SSIM and the PSNR quality measures.

Conclusions

Bilevel optimization methods, in combination with energy models as lower-level problems, represent a state-of-the-art alternative for finding optimal quantities of interest in image processing tasks. Those quantities may be coefficients in the data fidelities, weights in the operators, or general functions involved in the different model terms. This methodology is particularly useful in combination with supervised learning techniques, where one can take advantage of the existence of training and validation sets.

These bilevel techniques have also the advantage that they can be mathematically analyzed and different results can be demonstrated, which allow an understanding of the structural characteristics of the problems under study. Issues such as the existence of optimal solutions and their regularity, and their characterization through first and second order optimality conditions, can be rigorously addressed. To achieve these objectives, however, the use of modern techniques of non-smooth optimization and variational analysis is important, in order to carry out a successful treatment of the non-convexities and non-differentiability of the problems.

On the basis of the studied properties, it is also possible to design efficient algorithms for solving the bilevel problems, as well as to design neural networks better adjusted to specific image processing tasks. These issues have already been addressed in the community, and represent a promising research direction for the future.

References

Adler, J., Öktem, O.: Solving ill-posed inverse problems using iterative deep neural networks. Inverse Probl. **33**(12), 124007 (2017)

Adler, J., Öktem, O.: Learned primal-dual reconstruction. IEEE Trans. Med. Imaging **37**(6), 1322–1332 (2018)

Antil, H., Di, Z.W., Khatri, R.: Bilevel optimization, deep learning and fractional Laplacian regularizatin with applications in tomography. Inverse Probl. **36**(6), 064001 (2020)

Arridge, S., Maass, P., Öktem, O., Schönlieb, C.-B.: Solving inverse problems using data-driven models. Acta Numer. **28**, 1–174 (2019)

Bartels, S., Weber, N.: Parameter learning and fractional differential operators: application in image regularization and decomposition. arXiv preprint arXiv:2001.03394 (2020)

Beck, A.: First-Order Methods in Optimization. Society for Industrial and Applied Mathematics, Philadelphia (2017)

Bredies, K., Kunisch, K., Pock, T.: Total generalized variation. SIAM J. Imaging Sci. **3**(3), 492–526 (2010)

Buades, A., Coll, B., Morel, J.-M.: A non-local algorithm for image denoising. In: IEEE CVPR, vol. 2, pp. 60–65 (2005)

Burger, H.C., Schuler, C.J., Harmeling, S.: Image denoising: can plain neural networks compete with BM3D? In: 2012 IEEE Conference on Computer Vision and Pattern Recognition, pp. 2392–2399 (2012)

Calatroni, L., De los Reyes, J.C., Schönlieb, C.-B.: Dynamic sampling schemes for optimal noise learning under multiple nonsmooth constraints. In: IFIP Conference on System Modeling and Optimization, pp. 85–95 (2013)

Calatroni, L., Papafitsoros, K.: Analysis and automatic parameter selection of a variational model for mixed Gaussian and salt-and-pepper noise removal. Inverse Probl. **35**(11), 114001 (2019)

Calatroni, L., Cao, C., De los Reyes, J.C., Schönlieb, C.-B., Valkonen, T.: Bilevel Approaches for Learning of Variational Imaging Models. Walter de Gruyter GmbH, pp. 252–290 (2017)

Chambolle, A., Lions, P.-L.: Image recovery via total variation minimization and related problems. Numer. Math. **76**(2), 167–188 (1997)

Chambolle, A., Pock, T.: A first-order primal-dual algorithm for convex problems with applications to imaging. JMIV **40**, 120–145 (2011)

Chambolle, A., Pock, T.: An introduction to continuous optimization for imaging. Acta Numer. **25**, 161–319 (2016)

Christof, C., De los Reyes, J.C., Meyer, C.: A nonsmooth trust-region method for locally Lipschitz functions with application to optimization problems constrained by variational inequalities. SIAM J. Optim. **30**(3), 2163–2196 (2020)

D'Elia, M., De los Reyes, J.C., Miniguano, A.: Bilevel parameter optimization for nonlocal image denoising models. arXiv preprint arXiv:1912.02347 (2019)

Dabov, K., et al.: Image denoising by sparse 3-D transform-domain collaborative filtering. IEEE Trans. Image Process. **16**(8), 2080–2095 (2007)

Davoli, E., Fonseca, I., Liu, P.: Adaptive image processing: first order PDE constraint regularizers and a bilevel training scheme. arXiv preprint arXiv:1902.01122 (2019)

Davoli, E., Liu, P.: One dimensional fractional order TGV: gamma-convergence and bilevel training scheme. Commun. Math. Sci. **16**(1), 213–237 (2018)

De los Reyes, J.C.: Optimal control of a class of variational inequalities of the second kind. SIAM J. Control Optim. **49**(4), 1629–1658 (2011)

De los Reyes, J.C.: Numerical PDE-Constrained Optimization. Springer, Cham (2015)

De los Reyes, J.C., Meyer, C.: Strong stationarity conditions for a class of optimization problems governed by variational inequalities of the second kind. J. Optim. Theory Appl. **168**(2), 375–409 (2016)

De los Reyes, J.C., Schönlieb, C.-B., Valkonen, T.: The structure of optimal parameters for image restoration problems. J. Math. Anal. Appl. **434**(1), 464–500 (2016)

De los Reyes, J.C., Schönlieb, C.-B., Valkonen, T.: Bilevel parameter learning for higher-order total variation regularisation models. J. Math. Imaging Vision **57**, 1–25 (2017)

De los Reyes, J.C., Schönlieb, C.-B.: Image denoising: learning the noise model via nonsmooth PDE-constrained optimization. Inverse Probl. Imaging **7**(4), 1183–1214 (2013)

Dempe, S.: Foundations of Bilevel Programming. Springer Science & Business Media, Boston (2002)

Dong, C., Loy, C.C., He, K., Tang, X.: Learning a deep convolutional network for image super-resolution. In: European Conference on Computer Vision, pp. 184–199 (2014)

Dontchev, A.L., Rockafellar, R.T.: Implicit Functions and Solution Mappings, vol. 543. Springer, New York (2009)

Du, Q., et al.: Analysis and approximation of nonlocal diffusion problems with volume constraints. SIAM Rev. **54**(4), 667–696 (2012)

Ehrhardt, M.J., Roberts, L.: Inexact Derivative-Free Optimization for Bilevel Learning. arXiv preprint arXiv:2006.12674 (2020)

Ekeland, I., Temam, R.: Convex Analysis and Variational Problems. Society for Industrial and Applied Mathematics, Philadelphia (1999)

Gilboa, G., Osher, S.: Nonlocal operators with applications to image processing. Multiscale Model. Simul. **7**(3), 1005–1028 (2008)

Gilboa, G., Osher, S.: Nonlocal linear image regularization and supervised segmentation. Multiscale Model. Simul. **6**(2), 595–630 (2007)

Gregor, K., LeCun, Y.: Learning fast approximations of sparse coding. In: Proceedings of the 27th International Conference on International Conference on Machine Learning, pp. 399–406 (2010)

Gunzburger, M., Lehoucq, R.B.: A nonlocal vector calculus with application to nonlocal boundary value problems. Multiscale Model. Simul. **8**, 1581–1598 (2010)

Haber, E., Horesh, L., Tenorio, L.: Numerical methods for experimental design of large-scale linear ill-posed inverse problems. Inverse Probl. **24**(5), 055012 (2008)

He, K., Zhang, X., Ren, S., Sun, J.: Deep residual learning for image recognition. In: Proceedings of the IEEE Conference on Computer Vision and Pattern Recognition, pp. 770–778 (2016)

Hershey, J.R., Le Roux, J., Weninger, F.: Deep unfolding: model-based inspiration of novel deep architectures. arXiv preprint arXiv:1409.2574 (2014)

Hintermüller, M., Papafitsoros, K.: Generating Structured Nonsmooth Priors and Associated Primal-Dual Methods, vol. 20, pp. 437–502. Elsevier (2019)

Hintermüller, M., Rautenberg, C.N.: Optimal selection of the regularization function in a weighted total variation model. Part I: Modelling and theory. J. Math. Imaging Vis. **59**(3), 498–514 (2017)

Hintermüller, M., Stadler, G.: An infeasible primal-dual algorithm for total bounded variation-based inf-convolution-type image restoration". SIAM J. Sci. Comput. **28**(1), 1–23 (2006)

Hintermüller, M., Wu, T.: Bilevel optimization for calibrating point spread functions in blind deconvolution. Inverse Probl. Imaging **9**(4), 1139–1170 (2015)

Holler, G., Kunisch, K., Barnard, R.C.: A bilevel approach for parameter learning in inverse problems. Inverse Probl. **34**(11), 115012 (2018)

Knoll, F., Bredies, K., Pock, T., Stollberger, R.: Second order total generalized variation (TGV) for MRI. Magn. Reson. Med. **65**(2), 480–491 (2011)

Kobler, E., Effland, A., Kunisch, K., Pock, T.: Total Deep Variation for Linear Inverse Problems. arXiv preprint arXiv:2001.05005 (2020)

Kobler, E., Klatzer, T., Hammernik, K., Pock, T.: Variational networks: connecting variational methods and deep learning. In: German Conference on Pattern Recognition, pp. 281–293 (2017)

Kunisch, K., Pock, T.: A bilevel optimization approach for parameter learning in variational models. SIAM J. Imaging Sci. **6**(2), 938–983 (2013)

Le, T., Chartrand, R., Asaki, T.J.: A variational approach to reconstructing images corrupted by Poisson noise. J. Math. Imaging Vis. **27**(3), 257–263 (2007)

Li, H., et al.: NETT: solving inverse problems with deep neural networks. Inverse Probl. **36**(6), 065005 (2020)

Lou, Y., et al.: Image recovery via nonlocal operators. J. Sci. Comput. **42**(2), 185–197 (2010)

Lunz, S., Öktem, O., Schönlieb, C.-B.: Adversarial regularizers in inverse problems. arXiv preprint arXiv:1805.11572 (2018)

Mordukhovich, B.S.: Variational Analysis and Applications. Springer, Cham (2018)

Nikolova, M.: A variational approach to remove outliers and impulse noise. J. Math. Imaging Vis. **20**(1), 99–120 (2004)

Nocedal, J., Wright, S.: Numerical Optimization. Springer Science & Business Media, New York (2006)

Ochs, P., Ranftl, R., Brox, T., Pock, T.: Techniques for gradient-based bilevel optimization with non-smooth lower level problems. J. Math. Imaging Vis. **56**(2), 175–194 (2016)

Outrata, J.V.: A generalized mathematical program with equilibrium constraints. SIAM J. Control Optim. **38**(5), 1623–1638 (2000)

Outrata, J., Zowe, J.: A numerical approach to optimization problems with variational inequality constraints. Math. Program. **68**(1), 105–130 (1995)

Ring, W.: Structural properties of solutions to total variation regularization problems. ESAIM: Math. Model. Numer. Anal. **34**(4), 799–810 (2000)

Rockafellar, R.T., Wets, R.J.-B.: Variational Analysis, vol. 317. Springer Science & Business Media, Berlin (1998)

Rudin, L.I., Osher, S., Fatemi, E.: Nonlinear total variation based noise removal algorithms. Phys. D: Nonlinear Phenom. **60**(1–4), 259–268 (1992)

Sawatzky, A., Brune, C., Müller, J., Burger, M.: Total Variation Processing of Images with Poisson Statistics, vol. 5702, pp. 533–540. Springer, Berlin/Heidelberg (2009)

Schirotzek, W.: Nonsmooth Analysis. Springer Science & Business Media, Berlin/Heidelberg (2007)

Sherry, F., et al.: Learning the sampling pattern for MRI. IEEE Trans. Med. Imaging **39**(12), 4310–4321 (2020)

Sun, J., Li, H., Xu, Z.: Deep ADMM-Net for compressive sensing MRI. Adv. Neural Inf. Process. Syst. **29** (2016)

Szegedy, C., et al.: Intriguing properties of neural networks. arXiv preprint arXiv:1312.6199 (2013)

Tappen, M.F.: Utilizing variational optimization to learn Markov random fields. In: 2007 IEEE Conference on Computer Vision and Pattern Recognition, pp. 1–8 (2007)

Tikhonov, A.N., Arsenin, V.: Solutions of Ill-Posed Problems, vol. 14. Winston, Washington, DC (1977)

Tomasi, C., Manduchi, R.: Bilateral filtering for gray and color images. In: ICCV, p. 2 (1998)

Valkonen, T.: A primal–dual hybrid gradient method for nonlinear operators with applications to MRI. Inverse Probl. **30**(5), 055012 (2014)

Van Chung, C., De los Reyes, J.C., Schönlieb, C.-B.: Learning optimal spatially-dependent regularization parameters in total variation image denoising. Inverse Probl. **33**(7), 074005 (2017)

Venkatakrishnan, S.V., C.A. Bouman, Wohlberg, B.: Plug-and-play priors for model based reconstruction. In: 2013 IEEE Global Conference on Signal and Information Processing, pp. 945–948 (2013)

Villacís, D.: First order methods for high resolution image denoising. Latin American Journal of Computing Faculty of Systems Engineering Escuela Politécnica Nacional Quito-Ecuador **4**(3), 37–42 (2017)

Wang, Y.-Q.: A multilayer neural network for image demosaicking. In: 2014 IEEE International Conference on Image Processing (ICIP), pp. 1852–1856 (2014)

Wang, Z., Bovik, A.C., Sheikh, H.R., Simoncelli, E.P., others: Image quality assessment: from error visibility to structural similarity. IEEE Trans. Image Process. **13**(4), 600–612 (2004)

Xu, L., Ren, J.S.J., Liu, C., Jia, J.: Deep convolutional neural network for image deconvolution. In: Advances in Neural Information Processing Systems, pp. 1790–1798 (2014)

Yaroslavsky, L.P.: Digital picture processing: an introduction. Appl. Opt. **25**, 3127 (1986)

Multi-parameter Approaches in Image Processing

25

Markus Grasmair and Valeriya Naumova

Contents

Introduction	944
PDE-Based Approaches	946
Dictionary-Based Approaches	950
Parameter Selection	953
Multiparameter Discrepancy Principle	953
Balancing Principle and Balanced Discrepancy Principle	954
L-Hypersurface	955
Generalized Lasso Path	955
Parameter Learning	956
Numerical Solution	957
Numerical Examples	958
Conclusion	963
References	965

Abstract

Natural images often exhibit a highly complex structure that is difficult to describe using a single regularization term. Instead, many variational models for image restoration rely on different regularization terms in order to capture the different components of the image in question. While the resulting multipenalty approaches have in principle a greater potential for accurate image reconstructions than single-penalty models, their practical performance relies heavily on a

---

M. Grasmair (✉)
NTNU, Trondheim, Norway
e-mail: markus.grasmair@ntnu.no

V. Naumova (✉)
Machine Intelligence Department, Simula Consulting and SimulaMet, Oslo, Norway
e-mail: valeriya@simula.no

© Springer Nature Switzerland AG 2023
K. Chen et al. (eds.), *Handbook of Mathematical Models and Algorithms in Computer Vision and Imaging*, https://doi.org/10.1007/978-3-030-98661-2_69

good choice of the regularization parameters. In this chapter, we provide a brief overview of existing multipenalty models for image restoration tasks. Moreover, we discuss different approaches to the problem of multiparameter selection. For the numerical examples, we will focus on the balanced discrepancy principle and the L-hypersurface method applied to PDE-based image denoising problems.

Keywords

Multiparameter regularization · Image restoration · Variational methods · Parameter selection · Discrepancy principle · L-hypersurface

Introduction

Image restoration aiming, for instance, at the recovery of lost information from noisy, blurred, and/or partially observed images plays an important role in many practical applications such as anomaly detection in medical images and galaxy analysis in astronomical images. With the massive production of digital images and videos, the need for efficient image restoration methods emerges even more. No matter how good cameras are, an improvement of the images is always desirable. Moreover, many image restoration tasks such as image denoising are necessary in many more applications than the ones mentioned above. Image denoising, being the simplest possible inverse problem, provides a useful and by now well-accepted framework in which different image processing ideas and techniques can be tested, compared, and perfected. Therefore, the field of image processing in general has received numerous contributions in the last decades from diverse scientific communities. Various statistical estimators, deep learning methods, adaptive filters, partial differential equations, transform-domain methods, splines, differential geometry-based methods, and regularization are only some of many areas and tools explored in studying image processing tasks.

This chapter does not intend to provide an overview of the vast amount of methods in image processing, but rather concentrate on variational multiparameter approaches for image restoration. These approaches have provided notable advances on different image restoration tasks in the last decades and continue to play an important role in this and other fields.

Mathematically speaking, we model an image restoration problem as follows:

$$y = Au + \delta, \qquad (1)$$

where u is a ground truth image affected by the action of the imaging operator A and is measured in the presence of a random noise δ. For simplicity, we assume here that the noise is additive, although most of the argumentation and methods below still remain valid for more complicated scenarios. In the simplest case of denoising, the operator A is the identity; other typical examples are convolution operators in the case of deblurring and masking operators in inpainting tasks.

Classical variational approaches for the solution of (1) aim at solving an optimization problem of the form:

$$\hat{u} = \arg\min_{u}\{\ell(Au, y) + \lambda \Psi(u)\}, \quad (2)$$

where ℓ is a loss function that penalizes mismatch to the measurements, $\Psi(u)$ is a regularization term that penalizes mismatch to the image class of interest, and $\lambda > 0$ is a regularization parameter that balances the two terms. Such simple variational models cannot easily account for the highly complex and heterogeneous structure of natural images. As a potential remedy for this, an alternative approach based on the idea of imposing different penalization on the image u or its components u_k has been proposed. This leads to the model:

$$\hat{u} = \arg\min_{u}\{\ell(Au, y) + \sum_{i=1}^{K} \lambda_i \Psi_i(u)\}.$$

In the specific case when we are interested in separating different components of the image, such as cartoon and texture, we impose different penalization terms on the different components. This results in the model:

$$\hat{u} = \arg\min_{u=u_1+\ldots+u_K} \{\ell(Au, y) + \lambda_1 \Psi_1(u_1) + \ldots + \lambda_K \Psi_K(u_K)\}. \quad (3)$$

Again, ℓ is a loss function penalizing the mismatch to the measurements. Moreover, each regularization term Ψ_k with corresponding regularization parameter $\lambda_k > 0$ penalizes a different aspect of the combined image $u = u_1 + \ldots + u_K$.

Based on the general formulations (2) or (3), one can differentiate at least two large classes of mathematical image restoration methods. On the one hand, there are PDE-based or, more general, variational methods for image restoration where the penalty terms use local (or in some recent approaches also nonlocal) potentially higher-order gradient information of the image. Typical approaches in that direction are variants of total variation regularization (cf. Rudin et al. 1992; Chambolle and Lions 1997) or, within a multi-penalty context, Mumford-Shah regularization (cf. Mumford and Shah 1989) or total generalized variation (cf. Bredies et al. 2010). A large overview of such methods can be, for instance, found in Scherzer et al. (2009) or Aubert and Kornprobst (2006).

On the other hand, there are approaches based on (generalized) wavelet decompositions or similar approaches based on computational harmonic analysis, which typically assume some type of sparsity of the images with respect to a suitable basis or dictionary. A classical example in that direction is the sub-quadratic waveletbased penalization promoted in Daubechies et al. (2004). Multi-penalty approaches based on a collection of different dictionaries have been studied in Bobin et al. (2007); see also the references therein. These approaches allow to separate several morphologies in the image; typical examples of which are again texture and cartoon.

PDE-Based Approaches

The simplest PDE-based approaches use quadratic regularization terms, leading to linear, elliptic PDEs. The most basic example is the single-penalty model:

$$\hat{u} = \arg\min_u \left(\frac{1}{2} \|Au - y\|^2 + \frac{\lambda}{2} \int_\Omega |\nabla u|^2 \right),$$

where $\Omega \subset \mathbb{R}^2$ is the imaging domain. This leads to the Euler-Lagrange equation (or optimality condition):

$$A^* A \hat{u} - \lambda \Delta \hat{u} = A^* y$$

with homogeneous Neumann boundary conditions. Multi-penalty approaches replace the H^1-norm in the regularization term by a composite of several terms of different orders. One of the simplest examples here uses in addition a squared norm of the Laplacian, leading to the model:

$$\hat{u} = \arg\min_u \left(\frac{1}{2} \|Au - y\|^2 + \frac{\lambda_1}{2} \int_\Omega |\nabla u|^2 + \frac{\lambda_2}{2} \int_\Omega (\Delta u)^2 \right), \tag{4}$$

or the corresponding Euler-Lagrange equation:

$$A^* A \hat{u} - \lambda_1 \Delta \hat{u} + \lambda_2 \Delta^2 \hat{u} = A^* y.$$

Such models have been attractive for a long time mainly because of their computational simplicity: they only require the solution of a linear system, which moreover has in many cases a very simple structure. However, the usage of the squared H^1-norm leads to very smooth, blurred results, a problem that may be made even worse by the addition of higher-order terms.

In Rudin et al. (1992), it has been argued that the "correct" way for treating image restoration problems is the usage of the total variation as the regularization term. There, one uses the L^1-norm of the image gradient as penalization term, that is, $\Psi(u) = TV(u) = \int |\nabla u| dx$. In contrast to a quadratic penalization of the gradient, this has the advantage of a much weaker penalization of large gradients, allowing edges to remain in the restored image. While the total variation is well suited for capturing large uniform regions in images, and also edges, it destroys the other important feature of natural images: textured patterns. In order to be able to reconstruct realistic images, it is therefore necessary to find a way for incorporating textures into the regularization functionals.

One possibility, suggested by Meyer (2001) (see also Vese and Osher (2003), which contains the first numerical implementation of the method), is to decompose the image into a geometry part u_1, which can be treated by the total variation, and a texture part u_2, for the treatment of which he introduced a norm that is dual to total variation. The resulting model has the form:

$$\hat{u} = \arg\min_{u}\left(\frac{1}{2}\|A(u_1 + u_2) - y\|^2 + \lambda_1 TV(u_1) + \lambda_2 \|u_2\|_G\right), \quad (5)$$

where the *G-norm* $\|\cdot\|_G$ is defined as follows:

$$\|v\|_G = \inf\{\|\mathbf{v}\|_\infty : v = \nabla \cdot \mathbf{v}\}.$$

Equivalently, this can be formulated as follows:

$$\hat{u} = \arg\min_{u,\mathbf{v}}\left(\frac{1}{2}\|A(u + \nabla \cdot \mathbf{v}) - y\|^2 + \lambda_1 TV(u) + \lambda_2 \|\mathbf{v}\|_\infty\right).$$

For a more precise definition of the involved spaces, see Meyer (2001). The intuition behind the introduction of the *G*-norm is the idea that textures mainly consist of rapidly oscillating, relatively uniform patterns. For such repeating structures, however, their *G*-norm is inversely proportional to the frequency of the oscillations: in the one-dimensional case, for instance, the *G*-norm of the function $u_2(x) = \sin(kx)$ would be $1/k$. More complex-related decomposition models, where an image is decomposed into more than two parts, have been suggested, for instance, in Bertalmio et al. (2003) and Aujol et al. (2005). An example of the resulting decomposition of an image into its cartoon part and texture part is given in Fig. 1. Note that only the positive part of the texture is shown and that it has been rescaled to fill the whole color range.

An alternative image decomposition approach can be derived from a model of image formation, which originates from the fact that natural images are projections of three-dimensional objects onto the two-dimensional image plane. Assuming that the depicted objects are up to a certain degree "homogeneous," this gives rise to the model of images consisting of several distinct, smooth regions, bordered by the different objects' silhouettes, which coincide with discontinuities in the image *u*. Based on this assumption, Mumford and Shah formulated their famous model

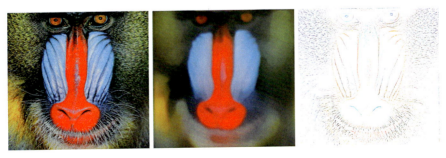

Fig. 1 Decomposition of an image into a cartoon and texture part according to Meyers model (5). *Left:* Original image. *Middle:* Resulting cartoon part. *Right:* Rescaled, positive texture part

(Mumford and Shah 1989):

$$\hat{u} = \arg\min_{u}\left(\frac{1}{2}\|Au - y\|^2 + \lambda_1 \int_{\Omega\setminus K} |\nabla u|^2 + \lambda_2 \operatorname{len}(K)\right), \quad (6)$$

where K denotes the (one-dimensional) edge set in the image u and $\operatorname{len}(K)$ its length. Originally, this model has been only formulated in the context of denoising, whereas its application to deblurring problems requires some additional constraints to be included (e.g., see Fornasier et al. 2013). Moreover, in contrast to the other models discussed in this chapter, it has the disadvantage of being highly non-convex. In addition, its numerical minimization requires in general some form of either approximation or parametrization of the edge set K. Different approaches to that end have been suggested using, e.g., phase-field approaches (see Ambrosio and Tortorelli 1990), nonlocal approximations (see Braides and Dal Maso 1997), singular perturbations (see Braides 1998), topological gradients (see Grasmair et al. 2013; Beretta et al. 2014), finite difference approximations (see Chambolle 1995; Gobbino 1998), or convex relaxations (see Pock et al. 2010). Note that this list is by no means exhaustive. In the numerical experiments in Section "Numerical Examples," we have used the phase-field approach due to Ambrosio and Tortorelli (1990); see also Aubert and Kornprobst (2006, Chap. 4.2.4). Here, the edge set K is approximated by a *phase-field* v, which is a function on Ω that is approximately 0 in a thin strip surrounding K and approximately 1 outside this strip. This yields the model:

$$\min_{u,v}\left(\frac{1}{2}\|Au - y\|^2 + \lambda_1 \int_{\Omega} v^2|\nabla u|^2 + \lambda_2 \int_{\Omega}\left(\epsilon|\nabla v|^2 + \frac{1}{4\epsilon}(v-1)^2\right)\right) \quad (7)$$

for some small parameter $\epsilon > 0$, which roughly corresponds to the width of the strip that approximates K. As ϵ tends to zero, the solutions (u_ϵ, v_ϵ) of (7) converge to solutions of (6) in the sense that the functions u_ϵ converge to a solution $\hat{u}$ of (6), whereas the phase fields v_ϵ converge to an indicator function of $\Omega \setminus K$. An example for this approximation of the Mumford-Shah model is presented in Fig. 2.

One of the drawbacks of total variation regularization is the so-called *stair-casing effect* that is often observed in the obtained results: in the reconstructed images,

Fig. 2 Application of the Mumford-Shah model to the parrots image. *Left:* Original image. *Middle:* Resulting cartoon part. *Right:* Resulting edge indicator according the Ambrosio-Tortorelli approximation (Ambrosio and Tortorelli 1990)

small edges are often inserted, and smooth changes of the intensities are broken up and replaced by gradual transitions. Convex approaches for improving this behavior are often formulated in terms of infimal convolutions of several convex functionals penalizing different smoothness properties of the component functions. The most basic approach in that direction is the infimal convolution of a total variation term and a quadratic penalization of the gradient, that is, the model:

$$\hat{u} = \arg\min_{u_1, u_2} \left(\frac{1}{2} \|A(u_1 + u_2) - y\|^2 + \lambda_1 \int_\Omega |\nabla u_1| + \lambda_2 \frac{1}{2} \int_\Omega |\nabla u_2|^2 \right).$$

Equivalently, this can be seen as a regularization method using the Huber-type functional:

$$\Psi_\varepsilon(u) = \int_{|\nabla u| \geq \varepsilon} |\nabla u| + \frac{1}{2\varepsilon} \int_{|\nabla u| < \varepsilon} |\nabla u|^2$$

with parameter $\varepsilon = \lambda_1/\lambda_2$. The quadratic term that becomes active at small gradients limits the stair-casing and allows for smooth, slow-intensity transitions, whereas the linear term penalizing large gradients allows for edges to remain in the reconstructed image.

Other common methods combine derivatives of several orders. The first idea in this direction can be found in Chambolle and Lions (1997), where the authors propose the model:

$$\hat{u} = \arg\min_{u_1, u_2} \left(\frac{1}{2} \|A(u_1 + u_2) - y\|^2 + \lambda_1 \int_\Omega |\nabla u_1| + \lambda_2 \int_\Omega |\nabla^2 u_2| \right). \tag{8}$$

This can be rewritten as follows:

$$\hat{u} = \arg\min_{u, v} \left(\frac{1}{2} \|Au - y\|^2 + \lambda_1 \int_\Omega |\nabla u - \nabla v| + \lambda_2 \int_\Omega |\nabla^2 v| \right). \tag{9}$$

Assuming that $\lambda_2 \gg \lambda_1$, we can interpret this model as a two-stage process, where we construct first a preliminary approximation ∇v of the gradient of the image which itself has a very low total variation and then construct the final approximation u in such a way that the total variation of the difference $u - v$ is small. As a consequence, the final result u will not be piecewise constant any more.

The paper Bredies et al. (2010) (see also Bredies and Holler 2014) introduced the concept of *total generalized variation* as a further generalization of total variation regularization with enhanced smoothing capabilities. In its second-order variant, it reduces to the model:

$$\hat{u} = \arg\min_{u, \mathbf{v}} \left(\frac{1}{2} \|Au - y\|^2 + \lambda_1 \int_\Omega |\nabla u - \mathbf{v}| + \lambda_2 \int_\Omega |\mathcal{E}\mathbf{v}| \right), \tag{10}$$

Fig. 3 Application of TGV regularization. *Left:* Resulting image $\hat{u}$. *Middle:* Norm of $D\hat{u}$. *Right:* Norm of **v**

where:

$$\mathcal{E}\mathbf{v} = \frac{1}{2}\bigl(\nabla \mathbf{v} + (\nabla \mathbf{v})^T\bigr)$$

is the symmetrized gradient of the vector function **v**. Here it is the gradient of u that is decomposed into two parts, the first being sparse, the second having a sparse symmetrized gradient. Compared to (9), the difference is that the second total variation has been replaced by the total deformation and the vector field **v** that forms the first approximation of the gradient of u is no longer required to be irrotational. In the one-dimensional case, the two models (9) and (10) are equivalent. Figure 3 shows an example of the application of TGV regularization and the resulting decomposition of the gradient.

Dictionary-Based Approaches

PDE-based approaches were among the first ones that have considered the separation of an image into several distinct components. However, the actual solution of the resulting models can be very demanding numerically, in particular for non-quadratic models. In order to overcome this bottleneck, a complementary direction of work, inspired by ideas and advances from signal processing, is to consider image reconstruction and separation from the point of view of sparsity and compressed sensing.

Sparsity has become important prior for many image processing applications. Since natural images typically are not sparse in their pixel domain, different transforms such as wavelet transforms and different generalizations have been proposed in the last decades with the goal of finding better and more efficient image representations. In the sparse model, each datum (signal) can be approximated by the linear combination of a small (sparse) number of elementary signals, called atoms, from a prespecified basis or frame, called dictionary. The natural next question is how to select or learn a dictionary Φ providing sparse representations for a given data class.

There are mainly two approaches: one is to employ some analytically defined dictionaries such as wavelets or an overcomplete discrete cosine transform, which are fast and easy to implement, but are suited only for a specific data type. The other approach is to learn a dictionary from the given training dataset for a specific task; see, for instance, Field and Olshausen (1996), Aharon et al. (2006), Gribonval and Schnass (2010), and Mairal et al. (2012). Even though the latter approach provides state-of-the-art results for many image processing tasks, it is very computationally and data demanding.

When one works with dictionary transforms, one can largely distinguish between synthesis-based approaches, which are purely formulated in terms of the dictionary coefficients, and analysis-based approaches, which essentially start from the resulting image. A single-penalty synthesis-based model has the form:

$$\min_{\alpha} \left(\frac{1}{2} \|A(\Phi \alpha) - y\| + \lambda \Psi(\alpha) \right),$$

and the resulting image is given as follows:

$$u = \Phi \alpha.$$

Here α are the (sparse) coefficients and Φ is the dictionary.

A corresponding analysis-based approach would take the form:

$$\min_{u} \left(\frac{1}{2} \|Au - y\| + \lambda \Psi(\Phi^{\dagger} u) \right),$$

with $\Phi^{\dagger}$ being the pseudo-inverse of the synthesis operator Φ. If one works with bases instead of frames, the matrix Φ is square and invertible, $\Phi^{\dagger} = \Phi^{-1}$, and the two approaches are equivalent. Moreover, in the case of orthonormal bases like the Fourier basis, we have $\Phi^{-1} = \Phi^{T}$.

We will first discuss two approaches that use a single analytic dictionary together with a multiparameter approach: the first approach, which is also the probably best known multiparameter approach, is the elastic net (Zou and Hastie 2005), which takes the form:

$$\min_{\alpha} \left(\frac{1}{2} \|A(\Phi \alpha) - y\|^2 + \lambda_1 \|\alpha\|_1 + \lambda_2 \frac{1}{2} \|\alpha\|_2^2 \right).$$

It has been widely used in statistics for robust regression and within imaging for various tasks like feature selection.

The second approach in this category uses the multi-scale nature of the wavelets and imposes different regularization parameters for different frequency bands of the wavelet regularization operator. This idea was pursued in Lu et al. (2007) for the recovery of the high-resolution images. The regularization operator for the ill-posed problem is decomposed in a multiscale manner by using bi-orthogonal wavelets or tight frames. Specifically, the authors propose a multi-resolution framework which introduces different regularization parameters for different frequency bands of the regularization operator resulting in:

$$\hat{u} = \arg\min_u \left(\frac{1}{2} \|Au - y\|^2 + \frac{1}{2} \sum_{s=0}^{p} \lambda_s \|R_s^T u\|_2^2 \right),$$

where R_s and $\bar{R}_s$ are obtained from a wavelet or frame system with:

$$\sum_{s=0}^{p} \bar{R}_s^T R_s = \mathbb{I}.$$

Here $\mathbb{I}$ is the identity matrix. This model has the explicit solution:

$$(A^T A + \sum_{s=0}^{p} \lambda_s \bar{R}_s^T R_s)\hat{u} = A^T y.$$

The authors demonstrate in extensive numerical examples the superiority of the method for high-resolution image recovery from a set of shifted low-resolution images compared to single-penalty methods such as H^1 and total variation. Moreover, the proposed multiparameter approach requires less computational resources than the single-penalty total variation method.

Another approach more common in harmonic analysis and signal processing literature is based on imposing different penalization for different image components for cartoon-texture separation task. These ideas were pursued by Daubechies and Teschke (2005), who proposed a multi-penalty formulation with an ℓ_1- and a weighted ℓ_2-norm as a numerically efficient substitute for the variational problem in Vese and Osher (2003). In particular, penalization in the wavelet shrinkage substitutes the total variation constraint considered in Vese and Osher (2003) and allows to recover cartoon, whereas the weighted ℓ_2-norm is used for texture recovery in the Fourier domain.

In signal processing, a wider class of potential transforms for recovery of different morphological components has been considered. In particular, morphological component analysis (MCA) (Starck et al. 2004, 2005; Bobin et al. 2007) has been devised to solve the problem of recovering the different components from their combination. MCA assumes that a dictionary of bases $\Phi_1 = [\phi_{11}, \ldots, \phi_{1K}]$ and $\Phi_2 = [\phi_{21}, \ldots, \phi_{2K}]$ exists such that u_1 is sparse in Φ_1 and not in Φ_2 and vice versa. In Starck et al. (2004, 2005) it was proposed to estimate the components u_1 and u_2 by solving the constrained optimization problem:

$$\min_{u_1, u_2} \|\Phi_1^\dagger u_1\|_1 + \|\Phi_2^\dagger u_2\|_1 \text{ s.t. } \|y - A(u_1 + u_2)\|_2^2 \leq \sigma. \tag{11}$$

Fig. 4 Decomposition of an image into piecewise smooth and texture parts according to MCA. The addition of the texture part and the piecewise smooth part reproduces the original image. *Left:* Original image. *Middle:* Piecewise smooth content part. *Right:* Texture part

The parameter σ should take into account both the noise level and the model inaccuracies in representing sparsely u_1 and u_2. Figure 4 illustrates the performance of the MCA method for image separation with two analytic dictionaries: curvelets for the cartoon part and the discrete cosine transform for the texture part.

The benefit of such a separation is obvious, as there is an agreement that images are in fact a mixture of cartoon and texture parts. By treating each of the parts separately using a proper dictionary, the image is processed much better and still efficiently by using analytic-based dictionaries. Moreover, MCA can be run either on the complete image or on small and overlapping patches. The immediate benefits of the latter mode are the locality of the processing, allowing for efficient parallel implementation, and the ability to incorporate learned dictionaries into the MCA.

Parameter Selection

Despite the promising performance of multiparameter models on various tasks, they lead to additional challenges related to the need for multiple parameter selection. This topic has been largely underexplored with some scarce efforts from different communities. Below we provide an overview of the existing approaches on the parameter selection, most of which were extended from a single-parameter to a multiparameter setting, whereas others are more data-driven and focus on learning the parameters from a given training dataset.

Multiparameter Discrepancy Principle

The authors of Lu and Pereverzev (2011) consider the multiparameter functional:

$$\hat{u} = \arg\min_{u} \|Au - y\|^2 + \sum_{i=1}^{K} \lambda_i \Psi_i(u) + \beta\|u\|^2, \tag{12}$$

where $\{\lambda_i\}_{i=1}^K > 0$ and $\beta > 0$ are the regularization parameters, $\Psi_i(u) = \|R_i u\|^2$, and R_i is a penalizing operator. They proposed an a posteriori strategy based on the extension of the classical discrepancy principle for choosing the parameter set $(\{\lambda_i\}_{i=1}^K, \beta)$. Specifically, the parameters are chosen as to satisfy the equation:

$$\|Au(\{\lambda_i\}_{i=1}^K, \beta) - y\| = c\delta,$$

where δ is the (assumed) noise level and $c \geq 1$ is some a priori specified parameter. Typically, c is chosen slightly larger than 1, e.g., $c = 1.2$, in order to obtain a stable solution in case of a slight underestimation of the noise level.

The authors also propose a numerical realization of this principle based on the model function approximation, which approximates the discrepancy term locally by means of some simple model function $m(\{\lambda_i\}_{i=1}^K, \beta)$ and allows to find subsequent parameters based on some simple equations. The scheme results in a nonunique parameter selection rule, which limits its applicability in practice. To overcome this issue, the follow-up work Lu et al. (2010) introduced a quasi-optimality criterion to facilitate a unique choice.

Balancing Principle and Balanced Discrepancy Principle

The nonuniqueness of the discrepancy principle was also addressed in Ito et al. (2014), where the authors consider augmented Tikhonov regularization and revisit the balancing principle for two parameter regularization. As a result, the balanced discrepancy principle was suggested, which incorporates the constraints into the augmented approach and allows to partially resolve the nonuniqueness issue.

As a first step, we consider the following balancing principle, where we choose the parameters λ_i in such a way that the system

$$\begin{cases} \hat{u} = \arg\min_u \|Au - y\|^2 + \lambda_1 \Psi_1(u) + \lambda_2 \Psi_2(u) \\ \lambda_i = \frac{1}{\gamma} \frac{\|y - A\hat{u}\|}{\Psi_i(\hat{u})}, \quad i = 1, 2, \end{cases}$$

is satisfied. That is, we are interested in finding parameters λ_i which balance the data fidelity with the respective penalty term. Here $\gamma > 0$ is a weighting parameter.

The balanced discrepancy principle combines this idea with the discrepancy principle. That is, we choose the weighting parameter γ in such a way that the residual satisfies $\|Au - y\| = c\delta$. For two-parameter regularization, this leads to the system:

$$\begin{cases} \|Au - y\| = c\delta, \ c \geq 1, \\ \lambda_1 \Psi_1(u) = \lambda_2 \Psi_2(u). \end{cases} \quad (13)$$

Again, $c \geq 1$ is a safety parameter tasked to ensure stability of the method and is usually chosen slightly larger than 1. Efficient algorithms based on Broyden's method and fixed point methods have been suggested in Ito et al. (2014) for the numerical solution of (13).

L-Hypersurface

In Belge et al. (1998, 2002), a parameter selection rule for functional (12) with $\beta = 0$ has been proposed, which is based on the generalization of the L-curve method (Hansen 1992) to the multiparameter setting. Similar to the one-dimensional case, one plots on the appropriate scale the residual norm

$$z(\lambda) = \|y - A\hat{u}(\lambda)\|^2$$

against the constraint norms

$$u_j(\lambda) = \Psi_j(\hat{u}), \qquad j = 1, \ldots, K.$$

Here $\lambda = [\lambda_1, \ldots, \lambda_K]^T$. Given an appropriate scaling function ϕ, e.g., $\phi(u) = \log(u)$, the L-hypersurface is defined as the graph of the map $\beta(\lambda) : \mathbb{R}_+^K \to \mathbb{R}^{K+1}$:

$$\beta(\lambda) = [\phi(u_1(\lambda)), \ldots, \phi(u_K(\lambda)), \phi(z(\lambda))].$$

A point on the L-hypersurface around which the surface is maximally warped corresponds to a point where the regularization and data-fitting errors are approximately balanced. The surface warpedness can be measured by calculating the Gaussian curvature. However, since evaluation of the Gaussian curvature for a large number of regularization parameters can be a computationally expensive task, which also might yield multiple extrema, the authors propose a surrogate minimum distance function (MDF) to approximate the curvature. However, the accuracy of the L-hypersurface approximation with MDF sometimes depends on the MDF origin. The authors provide some heuristic rule for the origin selection, which seems to work in specific cases. However, a robust means for selecting the origin is needed to promote practical usability of the method.

Generalized Lasso Path

In Grasmair et al. (2018) a fully adaptive approach for parameter selection was proposed for a multi-penalty functional of the form:

$$\min_{u,v} \frac{1}{2} \|A(u+v) - y\|^2 + \lambda_1 \|u\|_1 + \lambda_2 \|v\|_2^2,$$

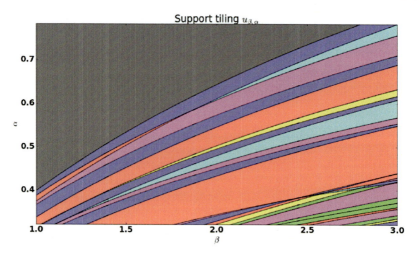

Fig. 5 Part of the parameter space detailing the different solutions. Each of the different tiles corresponds to a different support or sign pattern of the solution of interest

where u is a sparse signal of interest and v is some pre-measurement noise. The authors first extended the single-penalty lasso path algorithm (Efron et al. 2004) to a multiparameter lasso path algorithm, which partitions the parameter space $(\lambda_1, \lambda_2) \in \mathbb{R}^2_+$ into disjoint connected tiles such that the solution $\hat{u}(\lambda_1, \lambda_2)$ has a constant support and sign pattern on each tile; see Fig. 5. Such partitioning also allows to get additional insights into the problem at hand and understand the sensitivity of the solution with respect to the parameters. Once the tiles are constructed, one can employ a data-driven rule that adaptively selects a tile and corresponding parameters (λ_1, λ_2), based on maximizing the signal-to-noise ratio of the solution. The authors provide an efficient algorithm for tile construction. Moreover, the superiority of the algorithm with respect to the state-of-the-art methods such as orthogonal matching pursuit, iterative hard thresholding, and the lasso is demonstrated in extensive numerical results. The approach can be extended to other non-homogenous norms as soon as the sparsity of the solution is relatively low.

Parameter Learning

An alternative approach to the methods above is based on a learning framework, where one uses a training set $\{u_i, y_i\}_{i=1}^N$ of independent pairs of clean and noisy images to select the optimal parameter by minimizing a certain objective functional. A prominent example in regularization theory are bi-level optimization methods (Kunisch and Pock 2013; De los Reyes and Schönlieb 2013), where the lower-level problem is defined through a parametrized variational model such as (9), (10), or (12), and the upper-level problem measures the error of the lower-level solution with respect to the ground truth, i.e., $\|\hat{u} - u\|_2^2$. Parameters are then learnt by minimizing the mean error of the lower-level problem on the given training set. In many cases, the solution of the lower-level problem presupposes the PDE-based optimization and can be very computationally demanding.

In many real-life applications, the access to the ground truth image cannot be granted or might indeed be impossible, such as in X-ray tomography. Therefore, the recent efforts were dedicated to the development of an unsupervised parameter learning rule for different regularization methods (de Vito et al. 2018). The attractiveness of the suggested approach lies in the fact that one requires only a training set of noisy samples $\{y_i\}_{i=1}^{N}$ for learning the optimal parameter for given class of images or data in general. The idea behind the method is that the ground truth images follow intrinsically a lower dimensional geometry (i.e., they belong to a lower dimensional manifold), which can be approximated by using a training set of noisy samples. Once the proxy $\tilde{u}$ is calculated, one can use it to guide the selection of the parameter by minimizing, for instance, the discrepancy $\|\hat{u}-\tilde{u}\|^2$. The first step of finding a suitable proxy $\tilde{u}$ is completely independent of the regularization method and explores only the structure of the solution, whereas the second step of selecting the optimal parameter is dependent on the regularization method. The authors also showed that a learned parameter can be used on new images with similar structure without any retraining.

Numerical Solution

With the exception of the Mumford-Shah model, all approaches discussed above require the minimization of a convex functional composed of three or more subparts. However, many of the models include some non-smooth, sparsity-promoting terms, which make the application of non-smooth optimization algorithms necessary. In the recent years, convex analysis-based methods have been the method of choice in many imaging applications, particularly methods based on the augmented Lagrangian or alternating direction method of multipliers (ADMM) and various splitting methods. A large overview of different algorithms can be found in Bauschke and Combettes (2011), Combettes and Pesquet (2011), and Komodakis and Pesquet (2015). A specific mention here is deserved by the Chambolle-Pock algorithm (Chambolle and Pock 2011), which has been demonstrated to be very efficient in several total variation-based applications. We refer the reader not familiar with convex analysis to Komodakis and Pesquet (2015), which contains a succinct introduction into the main concepts and results necessary for the implementation of the different algorithms.

There are at least two notable differences between single-penalty and multi-penalty methods when it comes to their practical implementation: first, by their very nature, they require the minimization of functionals consisting of three or more separate terms. However, many of the more efficient primal-dual methods are primarily formulated only for a sum of two functionals, that is, a loss term and a single regularization term. In the situation of pure decomposition-based models like (8) or (11), which take the form:

$$\min_{u_1,\ldots,u_K} \left(\frac{1}{2} \|A(u_1 + \ldots + u_K) - y\|^2 + \lambda_1 \Psi(u_1) + \ldots \lambda_K \Psi(u_K) \right),$$

it is nevertheless possible to apply the standard approaches by splitting the whole model into the two subparts:

$$\begin{aligned} F_1(u_1,\ldots,u_K) &= \frac{1}{2}\|A(u_1 + \ldots + u_K) - y\|^2, \\ F_2(u_1,\ldots,u_K) &= \lambda_1 \Psi_1(u_1) + \ldots \lambda_K \Psi_K(u_K). \end{aligned} \qquad (14)$$

In this case, the *prox-operator* (see Komodakis and Pesquet 2015) for F_2^*, which is the central ingredient in all the aforementioned algorithms, decouples into the sum of the prox-operators for the regularization functionals $\Psi_1^*,\ldots,\Psi_K^*$. As long as the latter ones can be efficiently evaluated, an efficient implementation of these algorithms is possible.

In situations where this split is not possible, the direct application of many well-known splitting methods can be numerically more challenging. However, there exist generalizations to the sum of three of more convex functionals. For instance, examples of how ADMM and Douglas-Rachford splitting can be adapted to this more general setting can be found in Combettes and Pesquet (2011, Chap 10.7). In addition, there exists a growing number of algorithms specifically aimed at the minimization of a sum of three convex functionals. One notable example here is due to Condat (2013) and Vũ (2013). We refer again to Komodakis and Pesquet (2015), where a large number of similar algorithms are collected.

The second difference to single-parameter settings is the numerical realization of the parameter choice: heuristic rules for single-parameter regularization like balancing principle or the *L*-curve require the minimization of the regularization functional for a number of different regularization parameters in order to find the optimal choice. However, the situation becomes notably more complicated for multi-penalty methods, as one has to find optimal parameters within an at least two-dimensional set. Methods like *L*-hypersurfaces therefore require many more solutions to yield reasonable results than in the single-penalty case. In order to speed up computations, it is therefore necessary to implement good stopping criteria for the optimization algorithms that not only take into account the convergence of the algorithm but also the question whether the current parameter setting may be feasible or not; in the latter case, an early termination of the optimization algorithm can lead to a significant gain in efficiency.

Numerical Examples

In the following, we will demonstrate the behavior of several parameter choice rules for different approaches to image denoising. We restrict ourselves in this section to PDE-based models. The main reason is that dictionary-based approaches that provide state-of-the-art results are based on learned dictionaries rather than predefined ones (Starck et al. 2015). Learning a dictionary is a problem by itself, requiring a proper tuning of many parameters that influence the performance of the algorithm. The bi-level approaches presuppose the existence of training set for finding a parameter that minimizes the error between the ground truth and the reconstructed image. This type of setting falls within machine learning framework and is outside the scope of the current chapter.

We compare the performance of the balanced discrepancy principle, the L-hypersurface method, and the discrepancy principle without additional balancing. We chose to omit a comparison with the generalized lasso and with machine learning approaches, as the former is a method that is applicable only in very specialized settings and the latter require a large amount of training data of sufficiently good quality.

As a test example, we have used the "baboon" image, as it contains sharp edges and a high contrast between different image regions as well as parts characterized by a marked texture. For the denoising, we have added to each of the three color channels pixel-wise i.i.d. Gaussian noise with a standard deviation $\sigma = 50$, the true image taking values in the range [0, 255]. See Fig. 6 for the true and the noisy image.

We consider first the H^1-Laplacian model (4) applied to denoising, that is, the model:

$$\hat{u} = \arg\min_u \left(\frac{1}{2} \|u - y\|^2 + \frac{\lambda_1}{2} \int_\Omega |\nabla u|^2 + \frac{\lambda_2}{2} \int_\Omega (\Delta u)^2 \right), \tag{15}$$

for some given noisy image y. Here all terms are applied separately, but with the same regularization parameters λ_1 and λ_2, to the three color channels of the image. As discussed above, this is a quadratic optimization problem with the Euler-Lagrange equation (optimality condition):

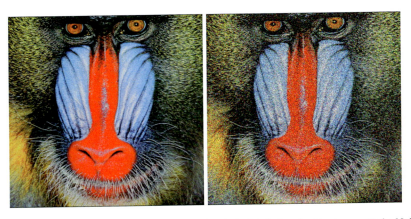

Fig. 6 Test image used for the numerical examples. *Left:* Original, noise-free image. *Right:* Noisy image

$$u - \lambda_1 \Delta u + \lambda_2 \Delta^2 u = y,$$

again applied separately to the different color channels.

Figures 7 and 8 show the results obtained by this approach together with an analysis of the parameter settings. Figure 7 shows the resulting L-hypersurface, the parameters for which the residual is smaller than the noise level, and results for the balanced discrepancy principle (13); here we have chosen $c = 1$, as the precise

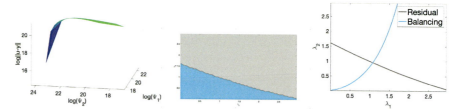

Fig. 7 Analysis of the parameter settings for the H^1-Laplace denoising model (15) applied to the noisy baboon image. *Left:* Resulting L-hypersurface. *Middle:* Admissible (blue) versus inadmissible (gray) parameter settings according to the discrepancy principle. *Right:* The gray curve depicts the parameters that satisfy the discrepancy principle with equality, the blue curve the parameters that satisfy the balancing principle. The parameter setting chosen according to the balanced discrepancy principle is the intersection of the two curves

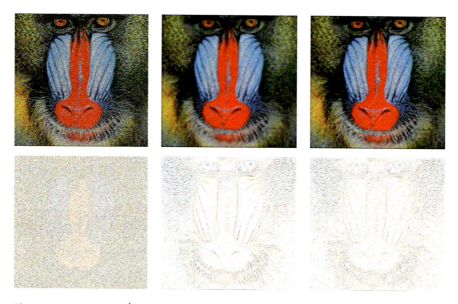

Fig. 8 Results of the H^1-Laplace denoising model for different parameter choices. *First row:* Resulting denoised image. *Second row:* Error, that is, difference between reconstruction and true, noise-free image. *Left:* Optimal reconstruction according to MSE, obtained by full grid search; PSNR = 15.18. *Middle:* Optimal reconstruction subject to discrepancy principle; PSNR = 20.59. *Right:* Result with balanced discrepancy principle; PSNR = 20.68

noise level was available. The latter yields a unique parameter pair, which has been used to obtain the right hand images in Fig. 8. In addition, we have performed a full grid search in order to find the parameter pair that minimizes the mean square error (MSE) as well as the pair minimizing the MSE subject to the constraint that the discrepancy principle is satisfied with equality. The resulting images as well as the PSNR for the different results are shown in Fig. 8. Note that the latter uses the knowledge of the actual noise-free image, which of course is not available in practice. Moreover, it is necessary to mention that both the MSE and the PSNR are somehow dubious quality measures for images, as they ignore all structural information that is present in the images and only consider point-wise discrepancies.

Next, we perform a similar numerical study for the Ambrosio-Tortorelli approximation (7) of the Mumford-Shah model (6), that is, the model:

$$\min_{u,v} \left(\frac{1}{2} \|u - y\|^2 + \lambda_1 \int_\Omega v^2 |\nabla u|^2 + \lambda_2 \int_\Omega \left(\epsilon |\nabla v|^2 + \frac{1}{4\epsilon}(v-1)^2 \right) \right).$$

This functional is non-convex because of the interaction between v and u in the second term, and thus the convex optimization methods discussed in the previous section are not readily applicable. Instead, we apply an alternating minimization procedure, where we alternate between minimizing with respect to u for fixed v and with respect to v for fixed u. This results in the iteration:

$$u_{k+1} \leftarrow \text{solution of } u - 2\lambda_1 \nabla \cdot (v_k^2 \nabla u) = y,$$

$$v_{k+1} \leftarrow \text{solution of } \left(1 + 4\lambda_1 \epsilon |\nabla u_{k+1}|^2\right) v - 4\lambda_2 \epsilon^2 \Delta v = 1,$$

where both PDEs are solved with homogeneous Neumann boundary conditions. The parameter ϵ was chosen to be 1 pixel-width; this results in an edge-indicator function that is highly localized around the detected edges.

The results for this approximation of the Mumford-Shah model are shown in Figs. 9 and 10. Again, we have compared the result for the balanced discrepancy principle with the optimal results according to MSE obtained by a full grid search. As can be expected from the Mumford-Shah model, which completely disregards texture, the results are more cartoon-like than with even the H^1-Laplace model, leading to a slightly lower PSNR. At the same time, the result includes distinct edges, which have been blurred in the other model.

We can also observe for the Mumford-Shah model that the balancing of the two regularization terms is crucial even in the presence of the discrepancy principle. This can be seen clearly in Fig. 11, where we have compared the results according to the balanced discrepancy principle with a result that satisfies the discrepancy principle, but where the second regularization parameter has been chosen to small. One can clearly see that this results in a general under-smoothing of the image.

As final example, we consider the Chambolle-Lions model (8) applied to the noisy parrots image, that is, the model:

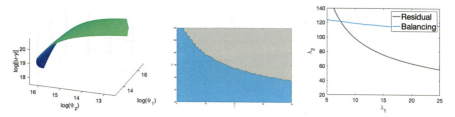

Fig. 9 Analysis of the parameter settings for the Mumford-Shah denoising model applied to the noisy baboon image. *Left:* Resulting L-hypersurface. *Middle:* Admissible (blue) versus inadmissible (gray) parameter settings according to the discrepancy principle. *Right:* The gray curve depicts the parameters that satisfy the discrepancy principle with equality, the blue curve the parameters that satisfy the balancing principle. The parameter setting chosen according to the balanced discrepancy principle is the intersection of the two curves

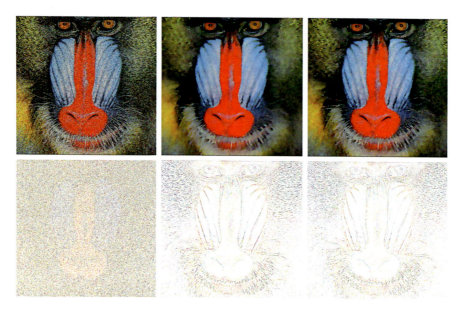

Fig. 10 Results of the Mumford-Shah denoising model for different parameter choices. *First row:* Resulting denoised image. *Second row:* Error, that is, difference between reconstruction and true, noise-free image. *Left:* Optimal reconstruction according to MSE, obtained by full grid search; PSNR = 15.28. *Middle:* Optimal reconstruction subject to discrepancy principle; PSNR = 19.93. *Right:* Result with balanced discrepancy principle; PSNR = 20.29

$$(\hat{u}_1, \hat{u}_2) = \arg\min_{u_1, u_2} \left(\frac{1}{2} \|u_1 + u_2 - y\|^2 + \lambda_1 \int_\Omega |\nabla u_1| + \lambda_2 \int_\Omega |\nabla^2 u_2| \right). \quad (16)$$

In this case, the result is decomposition of the restored image $\hat{u}$ into a part $\hat{u}_1$ mostly containing the cartoon-like components of $\hat{u}$ and a part $\hat{u}_2$ mostly containing the texture-like components. Moreover, we have a convex but non-smooth optimization

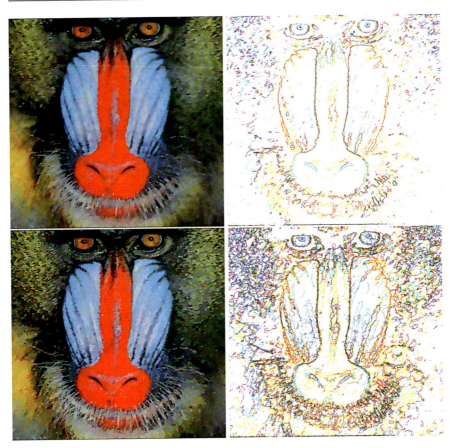

Fig. 11 Application of Mumford-Shah denoising to the noisy baboon image Fig. 6. *Upper row:* Denoised image and edge indicator using the balanced discrepancy principle. *Lower row:* Denoised image and edge indicator satisfying the discrepancy principle, but not the additional balancing principle

problem, which can be solved by any of the methods described in Section "Numerical Solution". Specifically, we have used the Chambolle-Pock algorithm (Chambolle and Pock 2011) using the splitting (14). We note here that the solution of (16) is not unique, as neither regularization term penalizes constant functions. In order to obtain a unique solution, we have therefore added the restriction $\int_\Omega \hat{u}_1 \, dx = 0$. The results for this model are shown in Figs. 12 and 13.

Conclusion

Multiparameter regularization is a theoretically sound and efficient framework for various image processing applications ranging from the basic task of image denoising to inpainting and deblurring. Both PDE-based and data-driven approaches

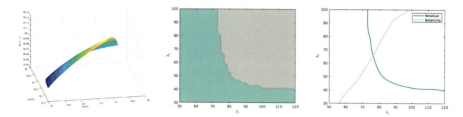

Fig. 12 Analysis of the parameter settings for the Chambolle-Lions model applied to the noisy parrots image. *Left:* Resulting L-hypersurface. *Middle:* Admissible (blue) versus inadmissible (gray) parameter settings according to the discrepancy principle. *Right:* The blue curve depicts the parameters that satisfy the discrepancy principle with equality, the gray curve the parameters that satisfy the balancing principle. The parameter setting chosen according to the balanced discrepancy principle is the intersection of the two curves

Fig. 13 Application of the Chambolle-Lions model to the denoising of the parrots image. *Upper row, left:* Noisy data. *Upper row, right:* Total result $\hat{u}_1 + \hat{u}_2$ using the balanced discrepancy principle. *Lower row, left:* Cartoon part $\hat{u}_1$ of the solution. *Lower row, right:* Texture part $\hat{u}_2$ of the solution

have been utilizing multiparameter regularization to obtain a good reconstruction quality and reduce the number of degrees of freedom. In this chapter, we provided an overview of the state of the art for multiparameter methods for image processing applications, also discussing aspects related to parameter selection and numerical realization. For clarification, we have also illustrated the performance of certain methods on simple denoising and decomposition examples.

There are several interesting open questions related to both numerical and theoretical aspects of multiparameter regularization. Specifically, further systematic studies of parameter learning from noisy data (unsupervised learning) not only could be beneficial for the specific methods but also could provide new insights into efficient construction of unsupervised deep learning algorithms.

References

Aharon, M., Elad, M., Bruckstein, A.M.: On the uniqueness of overcomplete dictionaries, and a practical way to retrieve them. J. Linear Algebra Appl. **416**, 48–67 (2006)

Ambrosio, L., Tortorelli, V.M.: Approximation of functionals depending on jumps by elliptic functionals via Γ-convergence. Commun. Pure Appl. Math. **43**(8), 999–1036 (1990)

Aubert, G., Kornprobst, P.: Mathematical Problems in Image Processing. Applied Mathematical Sciences, vol. 147, 2nd edn. Springer, New York. Partial differential equations and the calculus of variations, With a foreword by Olivier Faugeras (2006)

Aujol, J.-F., Aubert, G., Blanc-Féraud, L., Chambolle, A.: Image decomposition into a bounded variation component and an oscillating component. J. Math. Imaging Vis. **22**(1), 71–88 (2005)

Bauschke, H.H., Combettes, P.L.: Convex Analysis and Monotone Operator Theory in Hilbert Spaces. CMS Books in Mathematics/Ouvrages de Mathématiques de la SMC. Springer, New York (2011)

Belge, M., Kilmer, M.E., Miller, E.L.: Simultaneous multiple regularization parameter selection by means of the l-hypersurface with applications to linear inverse problems posed in the wavelet transform domain. In: Bayesian Inference for Inverse Problems, vol. 3459, pp. 328–336. International Society for Optics and Photonics (1998)

Belge, M., Kilmer, M.E., Miller, E.L.: Efficient determination of multiple regularization parameters in a generalized l-curve framework. Inverse Prob. **18**(4), 1161 (2002)

Beretta, E., Grasmair, M., Muszkieta, M., Scherzer, O.: A variational algorithm for the detection of line segments. Inverse Prob. Imaging **8**(2), 389–408 (2014)

Bertalmio, M., Vese, L., Sapiro, G., Osher, S.: Simultaneous structure and texture image inpainting. IEEE Trans. Image Process. **12**(8), 882–889 (2003)

Bobin, J., Starck, J.-L., Fadili, J.M., Moudden, Y., Donoho, D.L.: Morphological component analysis: an adaptive thresholding strategy. IEEE Trans. Image Process. **16**(11), 2675–2681 (2007)

Braides, A.: Approximation of Free-Discontinuity Problems. Lecture Notes in Mathematics, vol. 1694. Springer, Berlin (1998)

Braides, A., Dal Maso, G.: Non-local approximation of the Mumford-Shah functional. Calc. Var. Partial Differ. Equ. **5**, 293–322 (1997)

Bredies, K., Holler, M.: Regularization of linear inverse problems with total generalized variation. J. Inverse Ill-Posed Probl. **22**(6), 871–913 (2014)

Bredies, K., Kunisch, K., Pock, T.: Total generalized variation. SIAM J. Imaging Sci. **3**(3), 492–526 (2010)

Chambolle, A.: Image segmentation by variational methods: Mumford and Shah functional and the discrete approximations. SIAM J. Appl. Math. **55**(3), 827–863 (1995)

Chambolle, A., Lions, P.-L.: Image recovery via total variation minimization and related problems. Numer. Math. **76**(2), 167–188 (1997)

Chambolle, A., Pock, T.: A first-order primal-dual algorithm for convex problems with applications to imaging. J. Math. Imaging Vis. **40**(1), 120–145 (2011)

Combettes, P.L., Pesquet, J.-C.: Proximal splitting methods in signal processing. In: Fixed-Point Algorithms for Inverse Problems in Science and Engineering. Springer Optimization and Its Applications, vol. 49, pp. 185–212. Springer, New York (2011)

Condat, L.: A primal-dual splitting method for convex optimization involving lipschitzian, proximable and linear composite terms. J. Optim. Theory Appl. **158**(2), 460–479 (2013)

Daubechies, I., Teschke, G.: Variational image restoration by means of wavelets: Simultaneous decomposition, deblurring, and denoising. Appl. Comput. Harmon. Anal. **19**(1), 1–16 (2005)

Daubechies, I., Defrise, M., De Mol, C.: An iterative thresholding algorithm for linear inverse problems with a sparsity constraint. Commun. Pure Appl. Math. **57**(11), 1413–1457 (2004)

De los Reyes, J.C., Schönlieb, C.-B.: Image denoising: learning the noise model via nonsmooth PDE-constrained optimization. Inverse Probl. Imaging **7**(4), 1183–1214 (2013)

de Vito, E., Kereta, Z., Naumova, V.: Unsupervised parameter selection for denoising with the elastic net (2018)

Efron, B., Hastie, T., Johnstone, I., Tibshirani, R.: Least angle regression. Ann. Stat. **32**(2), 407–499 (2004)

Field, D.J., Olshausen, B.A.: Emergence of simple-cell receptive field properties by learning a sparse code for natural images. Nature **381**, 607–609 (1996)

Fornasier, M., March, R., Solombrino, F.: Existence of minimizers of the Mumford-Shah functional with singular operators and unbounded data. Ann. Mat. Pura Appl. **192**(3), 361–391 (2013)

Gobbino, M.: Finite difference approximation of the Mumford-Shah functional. Commun. Pure Appl. Math. **51**(2), 197–228 (1998)

Grasmair, M., Muszkieta, M., Scherzer, O.: An approach to the minimization of the Mumford-Shah functional using Γ-convergence and topological asymptotic expansion. Interfaces Free Bound. **15**(2), 141–166 (2013)

Grasmair, M., Klock, T., Naumova, V.: Adaptive multi-penalty regularization based on a generalized lasso path. Appl. Comput. Harmon. Anal. **49**(1), 30–55 (2018)

Gribonval, R., Schnass, K.: Dictionary identifiability – sparse matrix-factorisation via l_1-minimisation. IEEE Trans. Inf. Theory **56**(7), 3523–3539 (2010)

Hansen, P.C.: Analysis of discrete ill-posed problems by means of the L-curve. SIAM Rev. **34**(4), 561–580 (1992)

Ito, K., Jin, B., Takeuchi, T.: Multi-parameter tikhonov regularization – an augmented approach. Chin. Ann. Math. Ser. B **35**(3), 383–398 (2014)

Komodakis, N., Pesquet, J.: Playing with duality: an overview of recent primal-dual approaches for solving large-scale optimization problems. IEEE Sig. Process. Mag. **32**(6), 31–54 (2015)

Kunisch, K., Pock, T.: A bilevel optimization approach for parameter learning in variational models. SIAM J. Imaging Sci. **6**(2), 938–983 (2013)

Lu, S., Pereverzev, S.V.: Multi-parameter regularization and its numerical realization. Numer. Math. **118**(1), 1–31 (2011)

Lu, Y., Shen, L., Xu, Y.: Multi-parameter regularization methods for high-resolution image reconstruction with displacement errors. IEEE Trans. Circuits Syst. I: Regul. Pap. **54**(8), 1788–1799 (2007)

Lu, S., Pereverzev, S.V., Shao, Y., Tautenhahn, U.: Discrepancy curves for multi-parameter regularization. J. Inverse Ill-Posed Prob. **18**(6), 655–676 (2010)

Mairal, J., Bach, F., Ponce, J.: Task-driven dictionary learning. IEEE Trans. Pattern Anal. Mach. Intell. **34**(4), 791–804 (2012)

Meyer, Y.: Oscillating Patterns in Image Processing and Nonlinear Evolution Equations. University Lecture Series, vol. 22. American Mathematical Society, Providence (2001). The fifteenth Dean Jacqueline B. Lewis memorial lectures

Mumford, D., Shah, J.: Optimal approximations by piecewise smooth functions and associated variational problems. Commun. Pure Appl. Math. **42**(5), 577–685 (1989)

Pock, T., Cremers, D., Bischof, H., Chambolle, A.: Global solutions of variational models with convex regularization. SIAM J. Imaging Sci. **3**(4), 1122–1145 (2010)

Rudin, L.I., Osher, S., Fatemi, E.: Nonlinear total variation based noise removal algorithms. Physica D **60**(1–4), 259–268 (1992)

Scherzer, O., Grasmair, M., Grossauer, H., Haltmeier, M., Lenzen, F.: Variational Methods in Imaging. Applied Mathematical Sciences, vol. 167. Springer, New York (2009)

Starck, J.-L., Elad, M., Donoho, D.: Redundant multiscale transforms and their application for morphological component separation. In: Advances in Imaging and Electron Physics, pp. 287–348. Elsevier, London (2004)

Starck, J.-L., Elad, M., Donoho, D.L.: Image decomposition via the combination of sparse representations and a variational approach. IEEE Trans. Image Process. **14**(10), 1570–1582 (2005)

Starck, J.-L., Murtagh, F., Fadili, J.: Sparse Image and Signal Processing: Wavelets and Related Geometric Multiscale Analysis, 2nd edn. Cambridge University Press, New York (2015)

Vese, L., Osher, S.: Modeling textures with total variation minimization and oscillating patterns in image processing. J. Sci. Comput. **19**(1–3), 553–572 (2003). Special issue in honor of the sixtieth birthday of Stanley Osher

Vũ, B.C.: A splitting algorithm for dual monotone inclusions involving cocoercive operators. Adv. Comput. Math. **38**(3), 667–681 (2013)

Zou, H., Hastie, T.: Regularization and variable selection via the elastic net. J. R. Stat. Soc. Ser. B Stat. Methodol. **67**(2), 301–320 (2005)

Generative Adversarial Networks for Robust Cryo-EM Image Denoising

26

Hanlin Gu, Yin Xian, Ilona Christy Unarta, and Yuan Yao

Contents

Introduction	971
Robust Denoising in Deep Learning	971
Challenges of Cryo-EM Image Denoising	971
Outline	973
Background: Data Representation and Mapping	974
Autoencoder	974
GAN	975
Robust Denoising Method	976
Huber Contamination Noise Model	976
Robust Denoising Method	977
Robust Recovery via β-GAN	978
Stabilized Robust Denoising by Joint Autoencoder and β-GAN	981
Application: Robust Denoising of Cryo-EM Images	981
Datasets	981
Evaluation Method	984

This research made use of the computing resources of the X-GPU cluster supported by the Hong Kong Research Grant Council Collaborative Research Fund: C6021-19EF. The research of Hanlin Gu and Yuan Yao is supported in part by HKRGC 16308321, ITF UIM/390, the Hong Kong Research Grant Council NSFC/RGC Joint Research Scheme N_HKUST635/20, as well as awards from Tencent AI Lab and Si Family Foundation. We would like to thank Dr. Xuhui Huang for helpful discussions.

H. Gu · I. C. Unarta · Y. Yao (✉)
Hong Kong University of Science and Technology, Hong Kong, China
e-mail: hguaf@connect.ust.hk; icunarta@connect.ust.hk; yuany@ust.hk

Y. Xian
Hong Kong Applied Science and Technology Research Institute (ASTRI), Hong Kong, China

© Springer Nature Switzerland AG 2023
K. Chen et al. (eds.), *Handbook of Mathematical Models and Algorithms in Computer Vision and Imaging*, https://doi.org/10.1007/978-3-030-98661-2_126

Network Architecture and Hyperparameter.. 986
Results for RNAP... 986
Results for EMPIAR-10028.. 990
Conclusion... 991
Appendix.. 992
Influence of Parameter(α, β) Brings in β-GAN...................................... 992
Clustering to Solve the Conformational Heterogeneity............................ 992
Convolution Network.. 994
Test RNAP Dataset with PGGAN Strategy...................................... 995
Influence of the Regularization Parameter: λ................................... 997
References... 997

Abstract

The cryo-electron microscopy (cryo-EM) becomes popular for macromolecular structure determination. However, the 2D images captured by cryo-EM are of high noise and often mixed with multiple heterogeneous conformations and contamination, imposing a challenge for denoising. Traditional image denoising methods and simple denoising autoencoder cannot work well when the signal-to-noise ratio (SNR) of images is meager and contamination distribution is complex. Thus it is desired to develop new effective denoising techniques to facilitate further research such as 3D reconstruction, 2D conformation classification, and so on. In this chapter, we approach the robust denoising problem for cryo-EM images by introducing a family of generative adversarial networks (GANs), called β-GAN, which is able to achieve robust estimation of certain distributional parameters under Huber contamination model with statistical optimality. To address the denoising challenges, for example, the traditional image generative model might be contaminated by a small portion of unknown outliers, β-GANs are exploited to enhance the robustness of denoising autoencoder. Our proposed method is evaluated by both a simulated dataset on the *Thermus aquaticus* RNA polymerase (RNAP) and a real-world dataset on the *Plasmodium falciparum* 80S ribosome dataset (EMPIAR-10028), in terms of mean square error (MSE), peak signal-to-noise ratio (PSNR), structural similarity index measure (SSIM), and 3D reconstruction as well. Quantitative comparisons show that equipped with some designs of β-GANs and the robust ℓ_1-autoencoder, one can stabilize the training of GANs and achieve the state-of-the-art performance of robust denoising with low SNR data and against possible information contamination. Our proposed methodology thus provides an effective tool for robust denoising of cryo-EM 2D images and helpful for 3D structure reconstruction.

Keywords

Generative adversarial networks · Autoencoder · Robust statistics · Denoising · Cryo-electron microscopy

Introduction

Robust Denoising in Deep Learning

Deep learning technique has rapidly entered into the field of image processing. One of the most popular methods was the denoising autoencoder (DA) motivated by Vincent et al. (2008). It used the reference data to learn a compressed representation (encoder) for the dataset. One extension of DA was presented in Xie et al. (2012), which exploited the sparsity regularization and the reconstruction loss to avoid over-fitting. Other developments, such as Zhang et al. (2017), made use of the residual network architecture to improve the quality of denoised images. In addition, Agostinelli et al. (2013) combined several sparse denoising autoencoders to enhance the robustness under different noise.

The generative adversarial network (GAN) recently gained its popularity and provides a promising new approach for image denoising. GAN was proposed by Goodfellow et al. (2014) and was mainly composed of two parts: the generator (G: generate the new samples) and the discriminator (D: determine whether the samples are real or generated (fake)). Original GAN (Goodfellow et al. 2014) aimed to minimize the Jensen-Shannon (JS) divergence between distributions of the generated samples and the true samples, hence called JS-GAN. Various GANs were then studied, and in particular, Arjovsky et al. (2017) proposed the Wasserstein GAN (WGAN), which replaced the JS divergence with Wasserstein distance. Gulrajani et al. (2017) further improved the WGAN with the gradient penalty that stabilized the model training. For image denoising problem, GAN could better describe the distribution of original data by exploiting the common information of samples. Consequently, GANs were widely applied in the image denoising problem (Tran et al. 2020; Tripathi et al. 2018; Yang et al. 2018; Chen et al. 2018; Dong et al. 2020).

Recently, Gao et al. (2019, 2020) showed that a general family of GANs (β-GANs, including JS-GAN and TV-GAN) enjoyed robust reconstruction when the datasets contain outliers under Huber contamination models (Huber 1992). In this case, observed samples are drawn from a complex distribution, which is a mixture of contamination distribution and real data distribution. A particular example is provided by cryo-electron microscopy (cryo-EM) imaging, where the original noisy images are likely contaminated with outliers as broken or non-particles. The main challenges of cryo-EM image denoising are summarized in the subsequent section.

Challenges of Cryo-EM Image Denoising

The cryo-electron microscopy (cryo-EM) has become one of the most popular techniques to resolve the atomic structure. In the past, cryo-EM was limited to large complexes or low-resolution models. Recently the development of new detector hardware has dramatically improved the resolution in cryo-EM (Kühlbrandt 2014), which made cryo-EM widely used in a variety of research fields. Different from X-ray crystallography (Warren 1990), cryo-EM had the advantage of preventing the recrystallization of inherent water and recontamination. Also, cryo-EM was superior to nuclear magnetic resonance spectroscopy (Wüthrich 1986) in solving macromolecules in the native state. In addition, both X-ray crystallography and nuclear magnetic resonance spectroscopy required large amounts of relatively pure samples, whereas cryo-EM required much fewer samples (Bai et al. 2015). For this celebrated development of cryo-EM for the high-resolution structure determination of biomolecules in solution, the Nobel Prize in Chemistry in 2017 was awarded to three pioneers in this field (Shen 2018).

However, it is a computational challenge in processing raw cryo-EM images, due to heterogeneity in molecular conformations and high noise. Macromolecules in natural conditions are usually heterogeneous, i.e., multiple metastable structures might coexist in the experimental samples (Frank 2006; Scheres 2016). Such conformational heterogeneity adds extra difficulty to the structural reconstruction as we need to assign each 2D image to not only the correct projection angle but also its corresponding conformation. This imposes a computational challenge that one needs to denoise the cryo-EM images without losing the key features of their corresponding conformations. Moreover, in the process of generating cryo-EM images, one needs to provide a view using the electron microscope for samples that are in frozen condition. Thus there are two types of noise: one is from ice, and the other is from the electron microscope. Both of them are significant in contributing high noise in cryo-EM images and leave a difficulty to the detection of particle structures (Fig. 1 shows a typical noisy cryo-EM image with its reference image, which is totally non-identifiable to human eyes). In extreme cases, some experimental images even do not contain any particles, rendering it difficult for particle picking either manually or automatically (Wang et al. 2016). How to

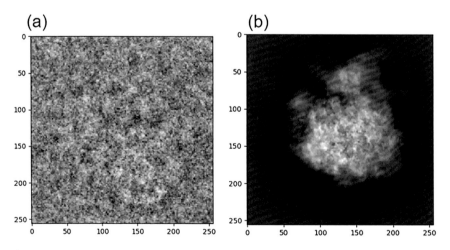

Fig. 1 (a) A noisy cryo-EM image; (b) a reference image

achieve robust denoising against such kind of contamination thus becomes a critical problem. Therefore, it is a great challenge to develop robust denoising methods for cryo-EM images to reconstruct heterogeneous biomolecular structures.

There are a plethora of denoising methods developed in applied mathematics and machine learning that could be applied to cryo-EM image denoising. Most of them in cryo-EM are based on unsupervised learning, which don't need any reference image data to learn. Wang and Yin (2013) proposed a filtering method based on nonlocal means, which made use of the rotational symmetry of some biological molecules. Also, Wei and Yin (2010) designed the adaptive nonlocal filter, which made use of a wide range of pixels to estimate the denoised pixel values. Besides, Xian et al. (2018) compared transform domain filtering method, BM3D (Dabov et al. 2007), and dictionary learning method, KSVD (Aharon et al. 2006), in denoising problem in cryo-EM. However, all of these didn't work well in low signal-to-noise ratio (SNR) situations. In addition, Covariance Wiener Filtering (CWF) (Bhamre et al. 2016) was proposed for image denoising. However, CWF needed large sample size of data in order to estimate the covariance matrix correctly, although it had an attractive denoising effect. Therefore, a robust denoising method in cryo-EM images was needed.

Outline

In this chapter, we propose a robust denoising scheme of cryo-EM images by exploiting joint training of both autoencoders and a new type of GANs β-GANs. Our main results are summarized as follows:

- Both autoencoder and GANs help each other for cryo-EM denoising in low signal-to-noise ratio scenarios. On the one hand, autoencoder helps stabilize GANs during training, without which the training processes of GANs are often collapsed due to high noise; on the other hand, GANs help autoencoder in denoising by sharing information in similar samples via distribution learning. For example, WGAN combined with autoencoder often achieve state-of-the-art performance due to its ability of exploiting information in similar samples for denoising.
- To achieve robustness against partial contamination of samples, one needs to choose both robust reconstruction loss for autoencoder (e.g., ℓ_1 loss) and robust β-GANs (e.g., (.5, .5)-GAN or (1, 1)-GAN,[1] which is proved to be robust against Huber contamination) that achieve competitive performance with WGANs in contamination-free scenarios, but do not deteriorate that much with data contamination.
- Numerical experiments are conducted with both a heterogeneous conformational dataset on the *Thermus aquaticus* RNA polymerase (RNAP) and a homogenous

---

[1] β-GAN has two parameters: α and β, written as (α, β)-GAN in this chapter.

dataset on the *Plasmodium falciparum* 80S ribosome dataset (EMPIAR-10028). The experiments on those datasets show the validity of the proposed methodology and suggest that while WGAN, (.5, .5)-GAN, and (1, 1)-GAN combined with ℓ_1-autoencoder are among the best choices in contamination-free cases, the latter two are overall the most recommended for robust denoising.

To achieve the goals above, this chapter is to provide an overview on various developments of GANs with their robustness properties. After that we focus on the application to the challenge of cryo-EM image robust denoising problem.

The chapter is structured as follows. In section "Background: Data Representation and Mapping," we provide a general overview of autoencoder and GAN. In section "Robust Denoising Method," we model the tradition denoising problem based on Huber contamination firstly and discuss β-GAN and its statistics. Finally, we give our algorithm based on combination of β-GAN and autoencoder, which is training stable. The evaluation of the algorithm in cryo-EM date is shown in the section "Application: Robust Denoising of Cryo-EM Images." The section "Conclusion" concludes the chapter. In addition, we implement the supplementary experiment in the section "Appendix."

Background: Data Representation and Mapping

Efficient representation learning of data distribution is crucial for many machine learning-based models. For a set of the real data samples X, the classical way to learn the probability distribution of this data (P_r) is to find P_θ by minimizing the distance between P_r and P_θ, such as Kullback-Leibler divergence $D_{KL}(P_r||P_\theta)$. This means we can pass a random variable through a parametric function to generate samples following a certain distribution P_θ instead of directly estimating the unknown distribution P_r. By varying θ, we can change this distribution and make it close to the real data distribution P_r. Autoencoder and GANs are two well-known methods in data representation. Autoencoder is good at learning the representation of data with low dimensions with an explicit characterization of P_θ, while GAN offers flexibility in defining the objective function (such as the Jensen-Shannon divergence) by directly generating samples without explicitly formulating P_θ.

Autoencoder

Autoencoder (Baldi 2012) is a type of neural network used to learn efficient codings of unlabeled data. It learns a representation (encoding) for a set of data, typically for dimensional reduction by training the network. An autoencoder has two main parts: encoder and decoder. The encoder maps the input data x ($\in X$) into a latent representation z, while the decoder maps the latent representation back to the data space:

$$z \sim \text{Enc}(x) \qquad (1)$$
$$\hat{x} \sim \text{Dec}(z). \qquad (2)$$

Autoencoders are trained to minimize reconstruction errors, such as $\mathcal{L}(x, \hat{x}) = ||x - \hat{x}||^2$.

Various techniques have been developed to improve data representation ability for autoencoder, such as imposing regularization on the encoding layer:

$$\mathcal{L}(x, \hat{x}) + \Omega(h), \qquad (3)$$

where h is the mapping function of the encoding layer and $\Omega(h)$ is the regularization term. The autoencoder is good at data denoising and dimension reduction.

GAN

The generative adversarial network (GAN), firstly proposed by Goodfellow (Goodfellow et al. 2014, called JS-GAN), is a class of machine learning framework. The goal of GAN is to learn to generate new data with the same statistics as the training set. Though original GAN is proposed as a form of generative model for unsupervised learning, GAN has proven useful for semi-supervised learning, fully supervised learning, and reinforcement learning (Hua et al. 2019; Sarmad et al. 2019; Dai et al. 2017).

Although GAN has shown great success in machine learning, the training of GAN is not easy and is known to be slow and unstable. The problems of GAN (Bau et al. 2019; Arjovsky et al. 2017) include:

- *Hard to achieve Nash equilibrium.* The updating process of the generator and the discriminator models are hard to guarantee a convergence.
- *Vanishing gradient.* The gradient update is slow when the discriminator is well trained.
- *Mode collapse.* The generator fails to generate samples with enough representative.

JS-GAN

JS-GAN proposed in Goodfellow et al. (2014) took Jensen-Shannon (JS) distance to measure the difference between different data distributions. The mathematics expression is follows:

$$\min_G \max_D \mathbb{E}_{x \sim P(X), z \sim P(Z)}[\log D(x) + \log(1 - D(G(z)))], \qquad (4)$$

where G is a generator which maps disentangled noise $z \sim P(Z)$ (usually Gaussian $N(0, I)$) to fake image data in a purpose to confuse the discriminator D from real data. The discriminator D is simply a classifier, which makes an effort to distinguish real data from the fake data generated by G. $P(X)$ is the input data distribution. z is noise. $P(Z)$ is the noise distribution, and it is used for data generation. Training of GANs is a minimax game by alternatively updating generators and discriminators, where the purpose of generators is to fool the discriminator as an adversarial process.

WGAN and WGANgp

Wasserstein GAN (Arjovsky et al. 2017) replaced the JS distance with the Wasserstein distance:

$$\min_G \max_D \mathbb{E}_{x \sim P(X), z \sim P(z)}\{D(x) - D(G(z))\}. \tag{5}$$

In reality, WGAN applied weight clipping of neural network to satisfy Lipschitz condition for discriminator. Moreover, Gulrajani et al. (2017) proposed WGANgp based on WGAN, which introduced a penalty in gradient to stabilize the training:

$$\min_G \max_D \mathbb{E}_{(x,z) \sim P(X,z)}\{D(x) - D(G(z)) + \mu \mathbb{E}_{\tilde{x}}(\|\nabla_{\tilde{x}} D(\tilde{x})\|_2 - 1)^2\}, \tag{6}$$

where $\tilde{x}$ is uniformly sampled along straight lines connecting pairs of the generated and real samples and μ is a weighting parameter. In WGANgp, the last layer of the sigmoid function in the discriminator network is removed. Thus D's output range is the whole real $\mathbb{R}$, but its gradient is close to 1 to achieve Lipschitz-1 condition.

Robust Denoising Method

Huber Contamination Noise Model

Let $x \in \mathbb{R}^{d_1 \times d_2}$ be a clean image, often called reference image in the sequel. The generative model of noisy image $y \in \mathbb{R}^{d_1 \times d_2}$ under the linear, weak phase approximation (Bhamre et al. 2016) could be described by

$$y = a * x + \zeta, \tag{7}$$

where $*$ denotes the convolution operation, a is the point spread function of the microscope convolving with the clean image, and ζ is an additive noise, usually assumed to be Gaussian noise that corrupts the image. In order to remove the noise the microscope brings, traditional denoising autoencoder could be exploited to learn from examples $(y_i, x_i)_{i=1,\ldots,n}$ the inverse mapping a^{-1} from the noisy image y to the clean image x.

However, this model is not sufficient in the real case. In the experimental data, the contamination will significantly affect the denoising efficiency if the denoising methods continuously depend on the sample outliers. Therefore we introduce the following Huber contamination model to extend the image formation model (see Eq. (7)).

Consider that the pair of reference image and experimental image (x, y) is subject to the following mixture distribution P_ϵ:

$$P_\epsilon = (1 - \epsilon)P_0 + \epsilon Q, \quad \epsilon \in [0, 1], \tag{8}$$

a mixture of true distribution P_0 of probability $(1 - \epsilon)$ and arbitrary contamination distribution Q of probability ϵ. P_0 is characterized by Eq. (7), and Q accounts for the unknown contamination distribution possibly due to ice, broken of data, and so on such that the image sample does not contain any particle information. This is called the Huber contamination model in statistics (Huber 1992). Our purpose is that given n samples $(x_i, y_i) \sim P_\epsilon$ $(i = 1, \ldots, n)$, possibly contaminated with unknown Q, learn a robust inverse map $a^{-1}(y)$.

Robust Denoising Method

Exploit a neural network to approximate the robust inverse mapping $G_\theta : \mathbb{R}^{d_1 \times d_2} \to \mathbb{R}^{d_1 \times d_2}$. The neural network is parameterized by $\theta \in \Theta$. The goal is to ensure that discrepancy between reference image x and reconstructed image $\widehat{x} = G_\theta(y)$ is small. Such a discrepancy is usually measured by some nonnegative loss function: $\ell(x, \widehat{x})$. Therefore, the denoising problem minimizes the following expected loss:

$$\arg\min_{\theta \in \Theta} \mathcal{L}(\theta) := \mathbb{E}_{x,y}[\ell(x, G_\theta(y))]. \tag{9}$$

In practice, given a set of training samples $S = \{(x_i, y_i) : i = 1, \ldots, n\}$, we aim to solve the following empirical loss minimization problem:

$$\arg\min_{\theta \in \Theta} \widehat{\mathcal{L}}_S(\theta) := \frac{1}{n} \sum_{i=1}^{n} \ell(x_i, G_\theta(y_i)). \tag{10}$$

The following choices of loss functions are considered:

- (ℓ_2-**Autoencoder**) $\ell(x, \widehat{x}) = \frac{1}{2}\|x - \widehat{x}\|_2^2 := \frac{1}{2}\sum_{i,j}(x_{ij} - \widehat{x}_{ij})^2$, or $\mathbb{E}\ell(x, \widehat{x}) = D_{KL}(p(x)\|q(\widehat{x}_\theta))$ equivalently, where $\widehat{x}_\theta \sim \mathcal{N}(x, \sigma^2 I_D)$;
- (ℓ_1-**Autoencoder**) $\ell(x, \widehat{x}) = \|x - \widehat{x}\|_1 := \sum_{i,j}|x_{ij} - \widehat{x}_{ij}|$, or $\mathbb{E}\ell(x, \widehat{x}) = D_{KL}(p(x)\|q(\widehat{x}_\theta))$ equivalently, where $\widehat{x}_\theta \sim \text{Laplace}(x, b)$;
- (**Wasserstein-GAN**) $\ell(x, \widehat{x}) = W_1(p(x), q_\theta(\widehat{x}))$, where W_1 is the 1-Wasserstein distance between distributions of x and $\widehat{x}$;
- (β-**GAN**) $\ell(x, \widehat{x}) = D(p(x)\|q_\theta(\widehat{x}))$, where D is some divergence function to be discussed below between distributions of x and $\widehat{x}$.

Both the ℓ_2 and ℓ_1 losses consider the reconstruction error of G_θ. The ℓ_2 loss above is equivalent to assume that $G_\theta(y|x)$ follows a Gaussian distribution $\mathcal{N}(x, \sigma^2 I_D)$, and the ℓ_1 loss instead assumes a Laplacian distribution centered at x. As a result, the ℓ_2 loss pushes the reconstructed image $\widehat{x}$ toward mean by averaging out the details and thus blurs the image. On the other hand, the ℓ_1 loss pushes $\widehat{x}$ toward the coordinate-wise median, keeping the majority of details while ignoring some large deviations. It improves the reconstructed image and is more robust than the ℓ_2 loss against large outliers. Although ℓ_1-autoencoder has a more robust loss than ℓ_2, both of them are not sufficient to handle the contamination. In the framework of the Huber contamination model (Eq. (8)), β-GAN is introduced below.

Robust Recovery via β-GAN

Recently Gao et al. (2019, 2020) came up with a more general form of β-GAN. It aims to solve the following minimax optimization problem to find the G_θ:

$$\min_{G_\theta} \max_{D} \mathbb{E}[S(D(x), 1) + S(D(G_\theta(y)), 0)], \tag{11}$$

where $S(t, 1) = -\int_t^1 c^{\alpha-1}(1-c)^\beta dc$, $S(t, 0) = -\int_0^t c^\alpha (1-c)^{\beta-1} dc$, $\alpha, \beta \in [-1, 1]$. For simplicity, we denote this family with parameters α, β by (α, β)-GAN in this chapter.

The family of (α, β)-GAN includes many popular members. For example, when $\alpha = 0$, $\beta = 0$, it becomes the JS-GAN (Goodfellow et al. 2014), which aims to solve the minmax problem (Eq. (4)) whose loss is the Jensen-Shannon divergence. When $\alpha = 1, \beta = 1$, the loss is a simple mean square loss; when $\alpha = -0.5, \beta = -0.5$, the loss is boost score.

However, the Wasserstein GAN (WGAN) is not a member of this family. By formally taking $S(t, 1) = t$ and $S(t, 0) = -t$, we could derive the type of WGAN as Eq. (5).

Robust Recovery Theory

Extend the traditional image generative model to a Huber contamination model, and exploit the β-GAN toward robust denoising under unknown contamination. Below includes a brief introduction to robust β-GAN, which achieves provable robust estimate or recovery under Huber contamination model. Recently, Gao establishes the statistical optimality of β-GANs for robust estimate of mean (location) and covariance (scatter) of the general elliptical distributions (Gao et al. 2019, 2020). Here we introduce the main results.

Definition 1 (Elliptical Distribution). A random vector $X \in \mathbb{R}^p$ follows an elliptical distribution if and only if it has the representation $X = \theta + \xi AU$, where

$\theta \in \mathbb{R}^p$ and $A \in \mathbb{R}^{p \times r}$ are model parameters. The random variable U is distributed uniformly on the unit sphere $\{u \in \mathbb{R}^r : \|u\| = 1\}$, and $\xi \geq 0$ is a random variable in $\mathbb{R}$ independent of U. The vector θ and the matrix $\Sigma = AA^T$ are called the location and the scatter of the elliptical distribution.

Normal distribution is just a member in this family characterized by mean θ and covariance matrix Σ. Cauchy distribution is another member in this family whose moments do not exist.

Definition 2 (Huber Contamination Model). $X_1, \ldots, X_n \stackrel{iid}{\sim} (1-\epsilon)P_{\text{ell}} + \epsilon Q$, where we consider the P_{ell} an elliptical distribution in its canonical form.

A more general data-generating process than Huber contamination model is called the strong contamination model below, as the TV neighborhood of a given elliptical distribution P_{ell}:

Definition 3 (Strong Contamination Model). $X_1, \ldots, X_n \stackrel{iid}{\sim} P$, for some P satisfying

$$TV(P, P_{\text{ell}}) < \epsilon.$$

Definition 4 (Discriminator Class). Let $\text{sigmoid}(x) = \frac{1}{1+e^{-x}}$, $\text{ramp}(x) = \max(\min(x+1/2, 1), 0)$, and $\text{ReLU}(x) = \max(x, 0)$. Define a general discriminator class of deep neural nets: firstly define the a ramp bottom layer

$$\mathcal{G}_{\text{ramp}} = g(x) = \text{ramp}(u^t x + b), u \in \mathbb{R}^p, b \in \mathbb{R}. \tag{12}$$

Then, with $\mathcal{G}_1(B) = \mathcal{G}_{\text{ramp}}$, inductively define

$$\mathcal{G}_{l+1}(B) = \left\{ g(x) = \text{ReLU}\left(\sum_{h \geq 1} v_h g_h(x) \right) : \sum_{h \geq 1} |v_h| \leq B, g_h \in \mathcal{G}_l(B) \right\}. \tag{13}$$

Note that the neighboring two layers are connected via ReLU activation functions. Finally, the network structure is defined by

$$\mathcal{D}^L(\kappa, B) = \left\{ D(x) = \text{sigmoid}\left(\sum_{j \geq 1} w_j g_j(x) \right) : \sum_{j \geq 1} |w_j| \leq \kappa, g_j \in \mathcal{G}_L(B) \right\}. \tag{14}$$

This is a network architecture consisting of L hidden layers.

Now consider the following β-GAN induced by a proper scoring rule $S : [0, 1] \times \{0, 1\} \to \mathbb{R}$ with the discriminator class above:

$$(\hat{\theta}, \hat{\Sigma}) = \arg\min_{(\theta, \Sigma)} \max_{D \in \mathcal{D}^L(\kappa, B)} \frac{1}{n} \sum_{i=1}^{n} S(D(x_i), 1) + \mathbb{E}_{x \sim P_{\text{ell}}(\Theta, \Sigma)} S(D((x)), 0). \quad (15)$$

The following theorem shows that such a β-GAN may give a statistically optimal estimate of location and scatter of the general family of elliptical distributions under strong contamination models.

Theorem 1 (Gao et al. 2020). *Consider the (α, β)-GANs with $|\alpha - \beta| < 1$. The discriminator class $D = \mathcal{D}^L(k, B)$ is specified by Eq. (14). Assume $\frac{p}{n} + \epsilon^2 \leq c$ for some sufficiently small constant $c > 0$. Set $1 \leq L = O(1), 1 \leq B = O(1)$, and $\kappa = O(\sqrt{\frac{p}{n}} + \epsilon)$. Then for any $X_1, \ldots X_n \overset{iid}{\sim} P$, for some P satisfying $TV(P, P_{\text{ell}}) < \epsilon$ with small enough ϵ, we have*

$$\|\hat{\theta} - \theta\|^2 < C(\frac{p}{n} \vee \epsilon^2),$$

$$\|\hat{\Sigma} - \Sigma\|_{op}^2 < C(\frac{p}{n} \vee \epsilon^2), \quad (16)$$

with probability at least $1 - e^{C'(p+n\epsilon^2)}$ (universal constants C and C') uniformly over all $\theta \in \mathbb{R}^p$ and all $\|\Sigma\|_{op} \leq M$.

The theorem established that for all $|\alpha - \beta| < 1$, (α, β)-GAN family is robust in the sense that one can learn a distribution P_{ell} from contaminated distributions P_ϵ such that $TV(P_\epsilon, P_{\text{ell}}) < \epsilon$, which includes Huber contamination model as a special case. Therefore a (α, β)-GAN with suitable choice of network architecture can robustly learn the generative model from arbitrary contamination Q when ϵ is small (e.g., no more than $1/3$).

In the current case, the denoising autoencoder network is modified to $G_\theta(y)$, providing us a universal approximation of the location (mean) of the inverse generative model as Eq. (7), where the noise can be any member of the elliptical distribution. Moreover, the discriminator is adapted to the image classification problem in the current case. Equipped with this design, the proposed (α, β)-GAN may help enhance the denoising autoencoder robustness against unknown contamination, e.g., the Huber contamination model for real contamination in the image data. The experimental results in fact confirm the efficacy of such a design.

In addition, Wasserstein GAN (WGAN) is not a member of this β-GAN family. Compared to JS-GAN, WGAN aims to minimize the Wasserstein distance between the sample distribution and the generator distribution. Therefore, WGAN is not robust in the sense of contamination models above as arbitrary ϵ portion of outliers can be far away from the main distribution P_0 such that the Wasserstein distance is arbitrarily large.

Stabilized Robust Denoising by Joint Autoencoder and β-GAN

Although β-GAN can robustly recover model parameters with contaminated samples, as a zero-sum game involving a non-convex-concave minimax optimization problem, training GANs is notoriously unstable with typical cyclic dynamics and possible mode collapse entrapped by local optima (Arjovsky et al. 2017). However, in this section we show that the introduction of autoencoder loss is able to stabilize the training and avoid the mode collapse. In particular, autoencoder can help stabilize GAN during training, without which the training processes of GAN are often oscillating and sometimes collapsed due to the presence of high noise.

Compared with the autoencoder, β-GAN can further help denoising by exploiting common information in similar samples during distribution training. In GAN, the divergence or Wasserstein distance between the reference image set and the denoised image set is minimized. The similar images can therefore help boost signals for each other.

For these considerations, a combined loss is proposed with both β-GAN and autoencoder reconstruction loss:

$$\widehat{\mathcal{L}}_{GAN}(x, \widehat{x}) + \lambda \|x - \widehat{x}\|_p^p, \tag{17}$$

where $p \in \{1, 2\}$ and $\lambda \geq 0$ is a trade-off parameter for ℓ_p reconstruction loss. Algorithm 1 summarizes the procedure of joint training of autoencoder and GAN, which will be denoted as "GAN+ℓ_p" in the experimental section depending on the proper choice of GAN and p. The main algorithm is shown in Algorithm 1.

Stability of Combining Autoencoder into GAN

We illustrate that autoencoder is indispensable to GANs in stabilizing the training in the joint training of autoencoder and GAN scheme.

As an illustration, Fig. 2 shows the comparison of training a JS-GAN and a joint JS-GAN + ℓ_1-autoencoder. Training and test mean square error curves are plotted against iteration numbers in the RNAP data under $SNR = 0.1$ as Fig. 2. It shows that JS-GAN training suffers from drastic oscillations, while joint training of JS-GAN + ℓ_1-autoencoder exhibits a stable process. In fact, with the aid of autoencoder here, one does not need the popular "log D trick" in JS-GAN.

Application: Robust Denoising of Cryo-EM Images

Datasets

RNAP: Simulation Dataset

We design a conformational heterogeneous dataset obtained by simulations. We use *Thermus aquaticus* RNA polymerase (RNAP) in complex with σ^A factor (*Taq* holoenzyme) for our dataset. RNAP is the enzyme that transcribes RNA

Algorithm 1 Joint training of (α, β)-GAN and ℓ_p-autoencoder

Input:
1. (α, β) for $S(t, 1) = -\int_t^1 c^{\alpha-1}(1-c)^\beta dc$, $S(t, 0) = -\int_0^t c^\alpha(1-c)^{\beta-1} dc$
   or $S(t, 1) = t$, $S(t, 0) = -t$ for WGAN
2. λ regularization parameter of the ℓ_p-Autoencoder
3. k_d number of iterations for discriminator, k_g number of iterations for generator
4. η_d learning rate of discriminator, η_g learning rate of generator
5. ω weights of discriminator, θ weights of generator

1: **for** number of training iterations **do**
2:   • Sample minibatch of m examples $\{(x^{(1)}, y^{(1)}), \ldots, (x^{(m)}, y^{(m)})\}$ from reference-noisy image pairs.
3:   **for** $k = 1, 2 \ldots, k_d$ **do**
4:     • Update the discriminator by gradient ascent:
5:     $g_\omega \leftarrow \frac{1}{m} \sum_{i=1}^m \nabla_\omega [S(D_\omega(x_i), 1) + S(D_\omega(G_\theta(y_i)), 0) + \mu(\|\nabla_{\tilde{x}} D_\omega(\tilde{x}_i)\|_2 - 1)^2]$
    where $\mu > 0$ for WGANgp only;
6:     $\omega \leftarrow \omega + \eta_d g_\omega$
7:   **end for**
8:   **for** $k = 1, 2 \ldots, k_g$ **do**
9:     • Update the generator by gradient descent:
10:     $g_\theta \leftarrow \frac{1}{m} \sum_{i=1}^m \nabla_\theta [S(D_\omega(G_\theta(y_i)), 0) + \lambda |G_\theta(y_i) - x_i|^p]$, $p \in \{1, 2\}$;
11:     $\theta \leftarrow \theta - \eta_g g_\omega$
12:   **end for**
13: **end for**

Return: Denoised image: $\hat{x}_i = G_\theta(y_i)$

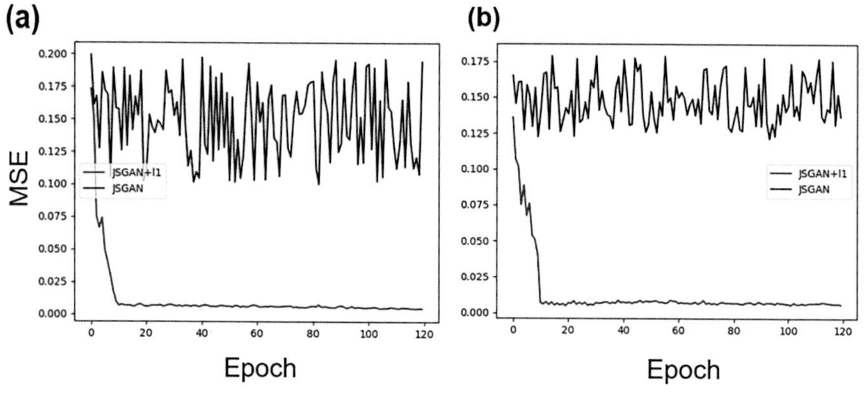

Fig. 2 Comparison between JS-GAN (black) and joint JS-GAN-ℓ_1-autoencoder (blue). (**a**) and (**b**) are the change of MSE in training and testing data. Joint training of JS-GAN-ℓ_1-autoencoder is much more stable than pure JS-GAN training that oscillates a lot

from DNA (transcription) in the cell. During the initiation of transcription, the holoenzyme must bind to the DNA and then separate the double-stranded DNA into single-stranded (Browning and Busby 2004). *Taq* holoenzyme has a crab-claw-like structure, with two flexible domains, the clamp and β pincers. The clamp, especially,

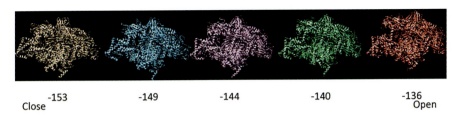

Fig. 3 Five conformations in RNAP heterogeneous dataset; from left to right are close conformation to open conformation of different angles

has been suggested to play an important role in the initiation, as it has been captured in various conformations by cryo-EM during initiation (Chen et al. 2020). Thus, we focus on the movement of the clamp in this study. To generate the heterogeneous dataset, we start with two crystal structures of *Taq* holoenzyme, which vary in their clamp conformation, open (PDB ID: 1L9U (Murakami et al. 2002)) and closed (PDB ID: 4XLN (Bae et al. 2015)) clamp. For the closed-clamp structure, we remove the DNA and RNA in the crystal structure, leaving only the RNAP and σ^A for our dataset. The *Taq* holoenzyme has about 370 kDa molecular weight. We then generate the clamp intermediate structures between the open and closed clamp using multiple-basin coarse-grained (CG) molecular dynamic (MD) simulations (Okazaki et al. 2006; Kenzaki et al. 2011). CG-MD simulations simplify the system such that the atoms in each amino acid are represented by one particle. The structures from CG-MD simulations are refined back to all-atom or atomic structures using PD2 ca2main (Moore et al. 2013) and SCRWL4 (Krivov et al. 2009). Five structures with equally spaced clamp opening angle are chosen for our heterogeneous dataset (shown in Fig. 3). Then, we convert the atomic structures to $128 \times 128 \times 128$ volumes using Xmipp package (Marabini et al. 1996) and generate the 2D projections with an image size of 128×128 pixels. We further contaminate those clean images with additive Gaussian noise at different signal-to-noise ratio (SNR): $SNR = 0.05$. The SNR is defined as the ratio of signal power and the noise power in the real space. For simplicity, we did not apply the contrast transfer function (CTF) to the datasets, and all the images are centered. Figure 3 shows the five conformation pictures.

Training data size is 25,000 paired images (noisy and reference images). Test data to calculate the MSE, PSNR, and SSIM is another 1500 paired images.

EMPIAR-10028: Real Dataset

This is a real-world experimental dataset that was firstly studied in the *Plasmodium falciparum* 80S ribosome dataset (EMPIAR-10028) (Wong et al. 2014). They recover the cryo-EM structure of the cytoplasmic ribosome from the human malaria parasite, *Plasmodium falciparum*, in complex with emetine, an anti-protozoan drug, at 3.2Å resolution. Ribosome is the essential enzyme that translates RNA to protein molecules, the second step of central dogma. The inhibition of ribosome activity of *Plasmodium falciparum* would effectively kill the parasite (Wong et al. 2014). We can regard this dataset to have homogeneous property. This dataset contains

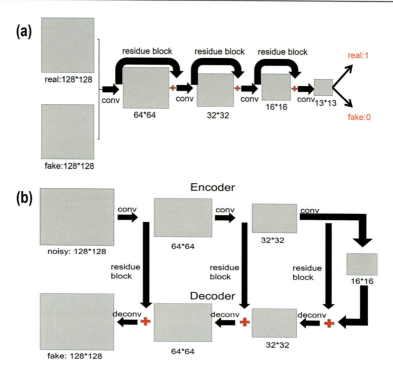

Fig. 4 The architectures of (**a**) discriminator D and (**b**) generator G, which borrow the residue structure. The input image size (128×128) here is adapted to RNAP dataset, while input image size of EMPIAR-10028 dataset is 256×256

105,247 noisy particles with an image size of 360×360 pixels. In order to decrease the complexity of the computing, we pick up the center square of each image with a size of 256×256, since the surrounding area of the image is entirely useless that does not lose information in such a preprocessing. Then the 256×256 images are fed as the input of the G_θ-network (Fig. 4). Since the GAN-based method needs clean images as reference, we prepare their clean counterparts in the following way: we first use cryoSPARC1.0 (Punjani et al. 2017) to build a $3.2 Å$ resolution volume and then rotate the 3D volume by the Euler angles obtained by cryoSPARC to get projected 2D images. The training data size we pick is 19,500, and the test data size is 500.

Evaluation Method

We exploit the following three metrics to determine whether the denoising result is good or not. They are the mean square error (MSE), the peak signal-to-noise ratio (PSNR), and the structural similarity index measure (SSIM).

- (**MSE**) For images of size $d_1 \times d_2$, the mean square error (MSE) between the reference image x and the denoised image $\widehat{x}$ is defined as

$$\text{MSE} := \frac{1}{d_1 d_2} \sum_{i=1}^{d_1} \sum_{j=1}^{d_2} (x(i,j) - \widehat{x}(i,j))^2.$$

  The smaller is the MSE, the better the denoising result is.
- (**PSNR**) Similarly, the peak signal-to-noise ratio (PSNR) between the reference image x and the denoised image $\widehat{x}$ whose pixel value range is $[0, t]$ (1 by default) is defined by

$$\text{PSNR} := 10 \log_{10} \frac{t^2}{\frac{1}{d_1 d_2} \sum_{i=1}^{d_1} \sum_{j=1}^{d_2} (x(i,j) - \widehat{x}(i,j))^2}.$$

  The larger is the PSNR, the better the denoising result is.
- (**SSIM**) The third criterion which is the structural similarity index measure (SSIM) between reference image x and denoised image $\widehat{x}$ is defined in (Wang et al. 2004):

$$\text{SSIM} = \frac{(2\mu_x \mu_{\widehat{x}} + c_1)(2\sigma_x \sigma_{\widehat{x}} + c_2)(\sigma_{x\widehat{x}} + c_3)}{(\mu_x^2 + \mu_{\widehat{x}}^2 + c_1)(\sigma_x^2 + \sigma_{\widehat{x}}^2 + c_2)(\sigma_x \sigma_{\widehat{x}} + c_3)}.$$

where μ_x ($\mu_{\widehat{x}}$) and σ_x ($\sigma_{\widehat{x}}$) are the mean and variance of x ($\widehat{x}$), respectively; $\sigma_{x\widehat{x}}$ is covariance of x and $\widehat{x}$; $c_1 = K_1 L^2$, $c_2 = K_2 L^2$, and $c_3 = \frac{c_2}{2}$ are three variables to stabilize the division with weak denominator ($K_1 = 0.01$, $K_2 = 0.03$ by default); and L is the dynamic range of the pixel value (1 by default). The value of SSIM lies in $[0, 1]$, where the closer it is to 1, the better the result is.

Although these metrics are widely used in image denoising, they might not be the best metrics for cryo-EM images. In Appendix "Influence of the Regularization Parameter: λ," it shows an example that the best-reconstructed images perhaps do not meet the best MSE/PSNR/SSIM metrics.

In addition to these metrics, we consider the 3D reconstruction based on denoised images. Particularly, we take the 3D reconstruction by RELION to validate the denoised result. The procedure of our RELION reconstruction is as follows: firstly creating the 3D initial model, then doing 3D classification, followed by operating 3D auto-refine. Moreover, for heterogeneous conformations in simulation data, we further turn the denoising results into a clustering problem to measure the efficacy of denoising methods, whose details will be discussed in Appendix "Clustering to Solve the Conformational Heterogeneity."

Network Architecture and Hyperparameter

In the experiments of this chapter, the best results come from the ResNet architecture (Su et al. 2018) shown in Fig. 4, which has been successfully applied to study biological problems such as predicting protein-RNA binding. The generator in such GANs exploits the autoencoder network architecture, while the discriminator is a binary classification ResNet. In Appendix "Convolution Network" and "Test RNAP Dataset with PGGAN Strategy," we also discuss a convolutional network without residual blocks and the PGGAN (Karras et al. 2018) strategy with their experimental results, respectively.

We chose Adam (Kingma and Ba 2015) for the optimization. The learning rate of the discriminator is $\eta_d = 0.001$, and the learning rate of the generator is $\eta_g = 0.01$. We choose $m = 20$ as our batch size, $k_d = 1$, and $k_g = 2$ in Algorithm 1.

For (α, β)-GAN, we report two types of choices, (1) $\alpha = 1, \beta = 1$ and (2) $\alpha = 0.5, \beta = 0.5$ since they show the best results in our experiments, while the others are collected in Appendix "Influence of Parameter(α, β) Brings in β-GAN." For WGAN, the gradient penalty with parameter $\mu = 10$ is used to accelerate the speed of convergence, and hence the algorithm is denoted as WGANgp below. The trade-off (regularization) parameter of ℓ_1 or ℓ_2 reconstruction loss is set to be $\lambda = 10$ throughout this section, while an ablation study on varying λ is discussed in Appendix "Influence of the Regularization Parameter: λ."

Results for RNAP

Denoising Without Contamination

In this part, we attempt to denoise the noisy image without the contamination (i.e., $\epsilon = 0$ in Eq. (8)). In order to present the advantage of GAN, we compare the denoising result in different methods. Table 1 shows the MSE and PSNR of different methods in SNR 0.05 and 0.1. We recognize the traditional methods such as KSVD, BM3D, nonlocal mean, and CWF can remove the noise partially and extract the general outline, but they still leave the unclear piece. However, deep learning methods can perform much better. Specifically, we observe that GAN-based methods, especially WGANgp $+\ell_1$ loss and $(.5, .5)$-GAN $+\ell_1$ loss, perform better than denoising autoencoder methods, which only optimizes ℓ_1 or ℓ_2 loss. The adversarial process inspires the generation process, and the additional ℓ_1 loss optimization speeds up the process of generation toward reference images. Notably, WGANgp and $(5, .5)$- or $(1, 1)$-GANs are among the best methods, where the best mean performances up to one standard deviation are all marked in bold font. Specifically, compared with $(.5, .5)$-GAN, the WGANgp get better PSNR and SSIM in SNR 0.1; the $(.5, .5)$-GAN shows the advantage in PSNR and SSIM in SNR 0.05, while $(1, 1)$-GAN is competitive within one standard deviation. Also, Fig. 5a presents the denoised images of denoising methods in SNR 0.05. For the convenience of comparison, we choose a clear open conformation (the rightmost

Table 1 Denoising result without contamination in simulated RNAP dataset: MSE, PSNR, and SSIM of different models, such as BM3D (Dabov et al. 2007), KSVD (Aharon et al. 2006), nonlocal means (Wei and Yin 2010), CWF (Bhamre et al. 2016), DA, and GAN-based methods

Method/SNR	MSE 0.1	MSE 0.05	PSNR 0.1	PSNR 0.05	SSIM 0.1	SSIM 0.05
BM3D	3.52e-2 (7.81e-3)	5.87e-2(9.91e-3)	14.54(0.15)	12.13(0.14)	0.20(0.01)	0.08(0.01)
KSVD	1.84e-2(6.58e-3)	3.49e-2(7.62e-3)	17.57(0.16)	14.61(0.14)	0.33(0.01)	0.19(0.01)
Nonlocal means	5.02e-2(5.51e-3)	5.81e-2(8.94e-3)	13.04(0.50)	12.40(0.65)	0.18(0.01)	0.09(0.01)
CWF	2.53e-2(2.03e-3)	9.28e-3(8.81e-4)	16.06(0.33)	20.31(0.41)	0.25(0.01)	0.08(0.01)
ℓ_2-Autoencoder[a]	3.13e-3(7.97e-5)	4.02e-3(1.48e-4)	25.10(0.11)	23.67(0.77)	0.79(0.02)	**0.79(0.01)**
ℓ_1-Autoencoder[b]	3.16e-3(7.05e-5)	4.23e-3(1.32e-4)	25.05(0.09)	23.80(0.13)	0.77(0.02)	0.76(0.01)
(0, 0)-GAN + ℓ_1 [c]	3.06e-3(5.76e-5)	**4.02e-3(5.67e-4)**	25.25(0.04)	24.00(0.06)	0.78(0.03)	**0.78(0.03)**
WGANgp + ℓ_1	**2.95e-3(1.41e-5)**	**4.00e-3(8.12e-5)**	**25.42(0.04)**	**24.06(0.05)**	**0.83(0.02)**	**0.80(0.03)**
(1, 1)-GAN + ℓ_1	2.99e-3(3.51e-5)	**4.01e-3(1.54e-4)**	25.30(0.05)	**24.07(0.16)**	**0.82(0.03)**	**0.79(0.03)**
(.5, .5)-GAN+ ℓ_1	3.01e-3(2.81e-5)	**3.98e-3(4.60e-5)**	25.27(0.04)	**24.07(0.05)**	0.79(0.04)	**0.80(0.03)**

[a] ℓ_2-Autoencoder represents ℓ_2 loss
[b] ℓ_1-Autoencoder represents ℓ_1 loss
[c] GAN + ℓ_1 represents adding ℓ_1 regularization in GAN generator loss

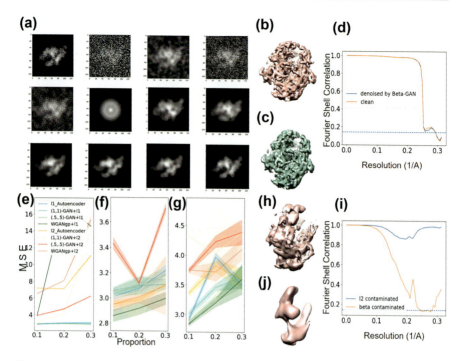

Fig. 5 Results for RNAP dataset. (**a**) is denoised images in different denoised methods (from left to right, top to bottom): clean, noisy, BM3D, KSVD, nonlocal means, CWF, ℓ_1-autoencoder, ℓ_2-autoencoder, (1,1)-GAN + ℓ_1, (0, 0)-GAN + ℓ_1, (.5, .5)-GAN + ℓ_1, and WGANgp + ℓ_1. (**b**) and (**c**) are reconstruction of clean images and (.5, .5)-GAN + ℓ_1 denoised images. (**d**) is FSC curve of (**b**) and (**c**). (**e**), (**f**), and (**g**) are robustness tests of various methods under $\epsilon \in \{0.1, 0.2, 0.3\}$-proportion contamination in three types of contamination: (**e**) type A, replacing the reference images with random noise; (**f**) type B, replacing the noisy images with random noise; (**g**) type C, replacing both with random noise. (**h**) and (**j**) are reconstructions of images with (.5, .5)-GAN + ℓ_1 and ℓ_2-autoencoder under type A contamination, respectively, where ℓ_2-autoencoder totally fails but (.5, .5)-GAN + ℓ_1 is robust. (**i**) shows FSC curves of (**h**) and (**j**)

conformation of Fig. 3) to present, and the performances show that WGANgp and (α, β)-GAN can grasp the "open" shape completely and derive the more explicit pictures than other methods.

What's more, in order to test the denoised results of β-GAN, we reconstruct the 3D volume by RELION in 200,000 images of SNR 0.1, which are denoised by (.5, .5)-GAN + ℓ_1. The value of pixel size, amplitude contrast, spherical aberration, and voltage are 1.6, 2.26, 0.1, and 300. For the other terms, retain the default settings in RELION software. Figure 5b and c separately shows the 3D volume recovered by clean images and denoised images. Also, the related FSC curves are shown in Fig. 5d. Specifically, the blue curve, which represents the denoised images in (.5, .5)-GAN + ℓ_1, is closed to red curves representing the clean images. We use the 0.143 cutoff criterion in literature (the resolution as Fourier shell correlation reaches 0.143, shown by dash lines in Fig. 5d) to choose the final resolution: $3.39 \mathring{A}$.

The structure recovered by (.5, .5)-GAN + ℓ_1 and FSC curve are as good as the original structure, which illustrates that the denoised result of β-GAN can identify the details of image and be helpful in 3D reconstruction.

In addition, Appendix "Clustering to Solve the Conformational Heterogeneity" also shows an example that GAN with ℓ_1-autoencoder helps heterogeneous conformation clustering.

Robustness Under Contamination

We also consider the contamination model $\epsilon \neq 0$ and Q from purely noisy images. We randomly replace partial samples of our training dataset of RNAP by noise to test whether our model is robust or not. There are three types of contaminations to test: (A) only replacing the clean reference images (it implies the reference images are wrong or missing, such that we do not have the reference images to compare; this is the worst contamination case), (B) only replacing the noisy images (it means the cryo-EM images which the machine produces are broken), and (C) replacing both, which indicates both A and B happen. The latter two are mild contamination cases, especially C that replaces both reference and noisy images by Gaussian noise whose ℓ_1 or ℓ_2 loss is thus well-controlled.

Here we test our robustness of various deep learning-based methods using the RNAP data of SNR 0.1, and the former three types of contamination are applied to randomly replace the samples in the proportion of $\epsilon \in \{0.1, 0.2, 0.3\}$ of the whole dataset.

Figure 5e, f, and g compares the robustness of different methods. In all the cases, some β-GANs ((.5, .5)- and (1, 1)-) with ℓ_1-autoencoder exhibit relatively universal robustness. Particularly, (1) the MSE with ℓ_1 loss is less than the MSE with ℓ_2 loss, which represents the ℓ_1 loss is more robust than ℓ_2 as desired. (2) The autoencoder method in ℓ_2 loss and WGANgp show certain robustness in cases B and C but are largely influenced by contamination in case A (shown in Fig. 5e), indicating the most serious damage arising from type A, merely replacing only the reference image by Gaussian noise. The reason is that the ℓ_2-autoencoder and WGANgp method are confused by the wrong reference images so that they cannot learn the mapping from data distribution to reference distribution accurately. (3) In the type C, the standard deviations of the five best models are larger compared the other two types. The contamination of both noisy y and clean x images influence the stability of model more than the other two types.

Furthermore, we take an example in type A contamination with $\epsilon = 0.1$ for 3D reconstruction. The 3D reconstructions in denoised images with (.5, .5)-GAN + ℓ_1 and l_2-autoencoder are shown in Fig. 5h and j, and related FSC curve is Fig. 5i. Specifically, on the one hand, the blue FSC curve of ℓ_2-autoencoder doesn't drop, which leads to the worse reconstruction; on the other hand, the red FSC curve of (.5, .5)-GAN + ℓ_1 drops quickly but begins to rise again, whose reason is that some unclear detail of structure mixed angular information in reconstruction. When applying 0.143 cutoff criterion (dashed line in FSC curve), the resolution of (.5, .5)-GAN + ℓ_1 is about $4\mathring{A}$. Although reconstruction of images and final resolution is not better than the clean images, it is much clearer than ℓ_2-autoencoder which totally

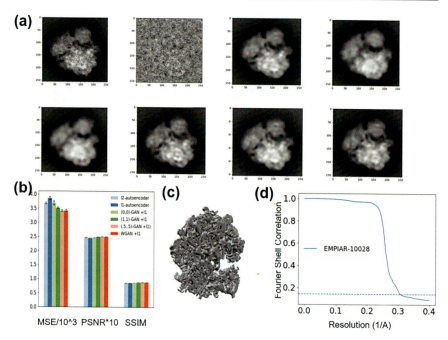

Fig. 6 Results for EMPIAR-10028. (a) Comparison in EMPIAR-10028 dataset in different deep learning methods (from left to right, top to bottom): clean image, noisy image, ℓ_1-autoencoder, ℓ_2-autoencoder, (0, 0)-GAN + ℓ_1, (1, 1)-GAN + ℓ_1, (.5, .5)-GAN + ℓ_1, WGANgp + ℓ_1. (b) is the MSE, PSNR, and SSIM in different denoised methods. (c) and (d) are the 3D reconstruction of denoised images by (.5, .5)-GAN + ℓ_1 and the FSC curve, respectively. The resolution of reconstruction from (.5, .5)-GAN + ℓ_1 denoised images is 3.20Å, which is as good as the original resolution

fails in the contamination case. The outcome of the reconstruction demonstrates that (.5, .5)-GAN + ℓ_1 is relatively robust, whose 3D result is consistent with the clean image reconstruction.

In summary, some (α, β)-GAN methods, such as the ((.5, .5)-GAN and (1, 1)-GAN, with ℓ_1-autoencoder are more resistant to sample contamination, which are better to be applied into the denoising of cryo-EM data.

Results for EMPIAR-10028

The following Fig. 6a and b shows the denoising results by different deep learning methods in experimental data, ℓ_1- or ℓ_2-autoencoders, JS-GAN ((0, 0)-GAN), WGANgp, and (α, β)-GAN, where we add ℓ_1 loss in all of the GAN-based structures. Although the autoencoder can grasp the shape of macromolecules, it is a little blur in some parts. What is more, WGANgp and (.5, .5)-GAN perform better than other deep learning methods according to MSE and PSNR, which is largely consistent with the result of the RNAP dataset. The improvements of such GANs over pure autoencoders lie in their ability of utilizing structural information among similar images to learn the data distribution better.

Finally, we implement reconstruction via RELION of 100,000 images, which are denoised by (.5, .5)-GAN +ℓ_1. The parameters are the same as the ones set in the paper (Wong et al. 2014). The reconstruction results are shown in Fig. 6c. It is demonstrated that the final resolution is $3.20\mathring{A}$, which is derived by FSC curve in Fig. 6d using the same 0.143 cutoff (dashed line) to choose the final resolution. We note that the final resolution by RELION after denoising is as good as the original resolution $3.20\mathring{A}$ reported in Wong et al. (2014).

Conclusion

In this chapter, we set a connection between the traditional image forward model and Huber contamination model in solving the complex contamination in the cryo-EM dataset. The joint training of autoencoder and GAN has been proved to substantially improve the performance in cryo-EM image denoising. In this joint training scheme, the reconstruction loss of autoencoder helps GAN to avoid mode collapse and stabilize training. GAN further helps autoencoder in denoising by utilizing the highly correlated cryo-EM images since they are 2D projections of one or a few 3D molecular conformations. To overcome the low signal-to-noise ratio challenge in cryo-EM images, joint training of ℓ_1-autoencoder combined with (.5, .5)-GAN, (1, 1)-GAN, and WGAN with gradient penalty is often among the best performances in terms of MSE, PSNR, and SSIM when the data is contamination-free. However, when a portion of data is contaminated, especially when the reference data is contaminated, WGAN with ℓ_1-autoencoder may suffer from the significant deterioration of reconstruction accuracy. Therefore, robust ℓ_1-autoencoder combined with robust GANs ((.5, .5)-GAN and (1, 1)-GAN) is the overall best choice for robust denoising with contaminated and high-noise datasets.

Part of the results in this chapter is based on a technical report (Gu et al. 2020). Most of the deep learning-based techniques in image denoising need reference data, limiting themselves in the application of cryo-EM denoising. For example, in our experimental dataset EMPIAR-10028, the reference data is generated by the cryoSPARC, which itself becomes problematic in highly heterogeneous conformations. Therefore, the reference image we learn may follow a fake distribution. How to denoise without the reference image thus becomes a significant problem. It is still open how to adapt to different experiments and those without reference images. In order to overcome this drawback, an idea called "image-blind denoising" was offered by the literature (Lehtinen et al. 2018; Krull et al. 2019), which viewed the noisy image or void image as the reference image to denoise. Besides, Chen et al. (2018) tried to extract the noise distribution from the noisy image and gain denoised images through removing the noise for noisy data; Quan et al. (2020) augmented the data by Bernoulli sampling and denoise image with dropout. Nevertheless, all of the methods need noise is independent of the elements themselves. Thus it is hard to remove noise in cryo-EM because the noise from ice and machine is related to the particles.

In addition, for reconstruction problems in cryo-EM, Zhong et al. (2020) proposed an end-to-end 3D reconstruction approach based on the network from cryo-EM images, where they attempt to borrow the variational autoencoder (VAE) to approximate the forward reconstruction model and recover the 3D structure directly by combining the angle information and image information learned from data. This is one future direction to pursue.

Appendix

Influence of Parameter(α, β) Brings in β-GAN

In this part, we have applied β-GAN into denoising problem. How to pick up a good parameter: (α, β) in the β-GAN becomes an important issue. Therefore, we investigate the impact of the parameter (α, β) on the outcome of denoising. We choose eight significant groups of α, β. Our result is shown in Table 2. It is demonstrated that the effect of these groups in different parameters is not large. The best result appears in $\alpha = 1, \beta = 1$ and $\alpha = 0.5, \beta = 0.5$

Clustering to Solve the Conformational Heterogeneity

In this part, we try to analyze whether the denoised result is good in solving conformation heterogeneity in simulated RNAP dataset. Specifically, for heterogeneous conformations in simulation data, we mainly choose the following two typical conformations: *open* and *close* conformations (the leftmost and rightmost conformations in Fig. 3) as our testing data. Our goal is to distinguish these two classes of conformations. However, different from the paper (Xian et al. 2018), we do not have the template images to calculate the distance matrix, so what we try is unsupervised learning – clustering. Our clustering method is firstly using manifold learning, Isomap (Tenenbaum et al. 2000), to reduce the dimension of the denoised images and then making use of k-means ($k = 2$) to group the different conformations.

Figure 7a displays the 2D visualizations of two conformations about the clustering effect in different denoised methods. Here the SNR of noisy data is 0.05. In correspondence to those visualizations, the accuracy of competitive methods is reported: (1, 1)-GAN+ℓ_1, 54/60 (54 clustering correctly in 60); WGANgp+ℓ_1, 54/60; ℓ_2-autoencoder, 44/60; BM3D, 34/60; and KSVD, 36/60. This result shows that clean images separate well; (α, β)-GAN and WGANgp with l_1-autoencoder can distinguish the open and close structure partially, although there exist several wrong points; ℓ_2-autoencoder and traditional techniques have poor performance because it is hard to detect the clamp shape.

Table 2 The result of β-GANs with ResNet architecture: MSE, PSNR, and SSIM of different (α, β) in β-GAN under various levels of Gaussian noise corruption in RNAP dataset

Parameter/SNR	MSE		PSNR		SSIM	
	0.1	0.05	0.1	0.05	0.1	0.05
$\alpha=1, \beta=1$	**2.99e-3(3.51e-5)**	4.01e-3(1.54e-4)	**25.30(0.05)**	**24.07(0.16)**	**0.82(0.03)**	0.79(0.03)
$\alpha=0.5, \beta=0.5$	3.01e-3(2.81e-5)	**3.98e-3(4.60e-5)**	25.27(0.04)	**24.07(0.05)**	0.79(0.04)	**0.80(0.03)**
$\alpha=-0.5, \beta=-0.5$	3.02e-3(1.69e-5)	4.15e-3(5.05e-5)	25.27(0.02)	23.91(0.05)	0.80(0.03)	**0.80(0.03)**
$\alpha=-1, \beta=-1$	3.05e-3(3.54e-5)	4.12e-3(8.30e-5)	25.23(0.05)	23.93(0.08)	0.80(0.05)	0.77(0.04)
$\alpha=1, \beta=-1$	3.05e-3(4.30e-5)	4.10e-3(5.80e-5)	25.24(0.06)	23.96(0.06)	**0.82(0.02)**	0.76(0.03)
$\alpha=0.5, \beta=-0.5$	3.09e-3(6.79e-5)	4.05e-3(6.10e-5)	25.17(0.04)	24.01(0.06)	0.79(0.04)	0.77(0.05)
$\alpha=0, \beta=0$	3.06e-3(5.76e-5)	4.02e-3(5.67e-4)	25.23(0.04)	24.00(0.06)	0.78(0.03)	0.78(0.03)
$\alpha=0.1, \beta=-0.1$	3.07e-3(5.62e-5)	4.05e-3(8.55e-5)	25.23(0.08)	23.98(0.04)	0.78(0.02)	0.79(0.03)

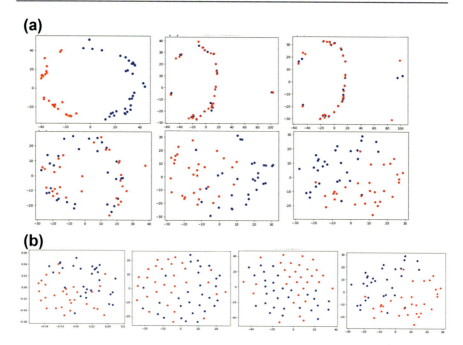

Fig. 7 2D visualization of two-conformation images in manifold learning. Red point and blue point separately represent the open and closed conformation. (**a**) is 2D visualization of two-conformation image by ISOMAP in different methods (from the left and top to the right and bottom): clean image, BM3D, KSVD, ℓ_2-autoencoder, (1, 1)-GAN+ ℓ_1, WGANgp+ ℓ_1. (**b**) is 2D visualization of two-conformation image in different manifold learning methods (from left to right): spectral methods, MDS, TSNE, and ISOMAP

Furthermore, the reason we use Isomap is it performs the best in our case, and comparisons of different manifold learning methods are shown in Fig. 7b. It demonstrates that blue and red points separate most in the graph of ISOMAP. Specifically, the accuracy of these four methods are 50/60 (spectral method), 46/50 (MDS), 46/50 (TSNE), and 54/60 (ISOMAP). It is shown that Isomap can distinguish best in the two structures' images compared to other methods, such as the spectral method (Ng et al. 2002), MDS (Cox and Cox 2008), and TSNE (Maaten and Hinton 2008).

Convolution Network

We present the result of simple deep convolution network (remove the ResNet block); the performances in all of criterion are worse than performances of the residue's architecture work. Table 3 compares the MSE and PSNR performance of various methods in the RNAP dataset with SNR 0.1 and 0.05. And Fig. 8a displays the denoised image of different methods in the RNAP dataset with SNR 0.05.

Table 3 MSE and PSNR of different models under various levels of Gaussian noise corruption in RNAP dataset, where the architectures of GANs or autoencoders are simply convolution network

	MSE		PSNR	
Method/SNR	0.1	0.05	0.1	0.05
BM3D	3.5e-2(7.8e-3)	5.9e-2(9.9e-3)	14.535(0.1452)	12.134(0.1369)
KSVD	1.8e-2(6.6e-3)	3.5e-2(7.6e-3)	17.570(0.1578)	14.609(0.1414)
Nonlocal means	5.0e-2(5.5e-3)	5.8e-2(8.9e-3)	13.040(0.4935)	12.404(0.6498)
CWF	2.5e-2(2.0e-3)	9.3e-3(8.8e-4)	16.059(0.3253)	20.314(0.4129)
ℓ_2-Autoencoder	4.0e-3(6.0e-4)	6.7e-3(9.0e-4)	24.202(0.6414)	21.739(0.7219)
$(0, 0)$-GAN + ℓ_1	3.8e-3(6.0e-4)	5.6e-3(8.0e-4)	24.265(0.6537)	22.594(0.6314)
WGANgp + ℓ_1	**3.1e-3(5.0e-4)**	5.0e-3(8.0e-4)	**25.086(0.6458)**	23.010(0.6977)
$(1, -1)$-GAN + ℓ_1	3.4e-3(5.0e-4)	**4.9e-3(9.0e-4)**	24.748(0.7233)	**23.116(0.7399)**
$(.5, -.5)$-GAN + ℓ_1	3.5e-3(5.0e-4)	5.6e-3(9.0e-4)	24.556(0.6272)	22.575(0.6441)

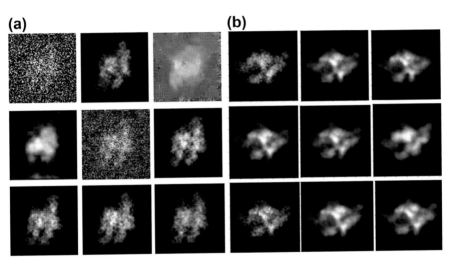

Fig. 8 (a) Denoised images with convolution network without ResNet structure in different methods in RNAP dataset with SNR 0.05 (from left to right, top to bottom): clean, noisy, BM3D, ℓ_2-autoencoder, KSVD, JS-GAN + ℓ_1, WGANgp + ℓ_1, $(1, -1)$-GAN + ℓ_1, $(.5, -.5)$-GAN + ℓ_1. (b) Denoised and reference images in different regularization λ (we use $(.5, .5)$-GAN +λ ℓ_1 as an example) in corresponding to Table 4. From left to right, top to bottom, the image is clean image, $\lambda = 0.1, \lambda = 1, \lambda = 5, \lambda = 10, \lambda = 50, \lambda = 100, \lambda = 500, \lambda = 10,000$

It shows the advantage of residue structure in our GAN-based denoising cryo-EM problem.

Test RNAP Dataset with PGGAN Strategy

PGGAN (Karras et al. 2018) is a popular method to generate high-resolution images from low-resolution ones by gradually adding layers of generator and discriminator.

It accelerates and stabilizes the model training. Since cryo-EM images are in large pixel size that fits well the PGGAN method, here we choose its structure[2] instead of the ResNet and convolution structures above to denoise cryo-EM images. Our experiments partially demonstrate two things: (1) the denoised images sharpen more, though the MSE changes to be higher; (2) we do not need to add ℓ_1 regularization to make model training stable; it can also detect the outlier of images for both real and simulated data without regularization.

In detail, based on the PGGAN architecture and parameters, we test the following two objective functions developed in the section "Robust Denoising Method": WGANgp and WGANgp + ℓ_1, in the RNAP simulated dataset with SNR 0.05 as an example to explain. The denoised images are presented in Fig. 9; it is noted that the model is hard to collapse regardless of adding ℓ_1 regularization. The MSE of adding regularization is 8.09e-3(1.46e-3), which is less than 1.01e-2(1.81e-3)

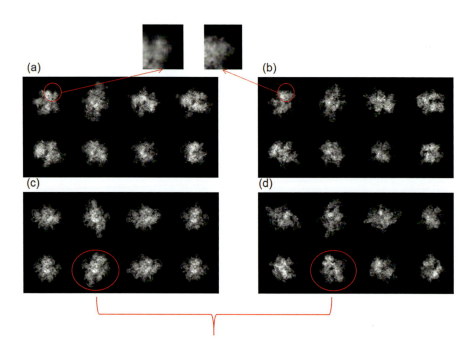

Fig. 9 Denoised and reference images by PGGAN instead of simple ResNet and convolution structure in RNAP dataset with SNR 0.05. The PGGAN strategy is tested in two objective functions: WGANgp + ℓ_1 and WGANgp. (**a**) and (**b**) are denoised and reference images using PGGAN with WGANgp + ℓ_1; (**c**) and (**d**) are denoised and reference images using PGGAN in WGANgp, respectively. Specifically, the images highlighted in red color show the structural difference between denoised images and reference images. It demonstrates that denoised images are different from reference images using PGGAN strategy

---

[2]We set the same architecture and parameters as https://github.com/nashory/pggan-pytorch and the input image size is 128×128.

Table 4 MSE, PSNR, and SSIM of different λ in $(.5,.5)$-GAN $+ \lambda l_1$ in RNAP dataset

λ/criterion	MSE	PSNR	SSIM
0.1	3.06e-3(4.50e-5)	25.22(0.07)	**0.82(0.06)**
1	3.05e-3(4.49e-5)	25.24(0.06)	0.81(0.05)
5	3.03e-3(2.80e-5)	25.26(0.04)	0.80(0.04)
10	**3.01e-3(2.81e-5)**	**25.27(0.04)**	0.79(0.04)
50	3.07e-3(3.95e-5)	25.20(0.06)	0.79(0.02)
100	3.11e-3(5.96e-5)	25.15(0.06)	0.80(0.02)
500	3.17e-3(5.83e-5)	25.01(0.07)	0.78(0.04)
10,000	3.17e-3(2.90e-5)	25.03(0.04)	0.79(0.04)

without adding regularization. Nevertheless, both of them don't exceed the results based on the ResNet structure above. This shows that PGGAN architecture does not have more power than the ResNet structure. But an advantage of PGGAN lies in its efficiency in training. So it is an interesting problem to improve PGGAN toward the accuracy of ResNet structure.

Another thing that needs to be highlighted is MSE may not be a good criterion because denoised images by PGGAN are clearer in some details than the front methods we propose. This phenomenon is also shown in Appendix "Influence of the Regularization Parameter: λ." So how to find a better criterion to evaluate the model and combine two strengths of ResNet-GAN and PGGAN await us to explore.

Influence of the Regularization Parameter: λ

In this chapter, we add ℓ_1 regularization to make model stable, but how to choose λ of ℓ_1 regularization becomes a significant problem. Here we take $(.5, .5)$-GAN to denoise in RNAP dataset with SNR 0.1. According to some results in different λ in Table 4, we find as the λ tends to infinity, the MSE results tend to ℓ_1-autoencoder, which is reasonable. Also, the MSE result becomes the smallest as the $\lambda = 10$.

What's more, an interesting phenomenon is found that a much clearer result could be obtained at $\lambda = 100$ than that at $\lambda = 10$, although the MSE is not the best (shown in Fig. 8b).

References

Agostinelli, F., Anderson, M., Lee, H.: Adaptive multi-column deep neural networks with application to robust image denoising. In: Advances in Neural Information Processing Systems, pp. 1493–1501 (2013)

Aharon, M., Elad, M., Bruckstein, A.: K-SVD: an algorithm for designing overcomplete dictionaries for sparse representation. IEEE Trans. Sig. Process. **54**(11), 4311–4322 (2006)

Arjovsky, M., Chintala, S., Bottou, L.: Wasserstein generative adversarial networks. In: Proceeding of the International Conference on Machine Learning, pp. 214–223 (2017)

Bae, B., Feklistov, A., Lass-Napiorkowska, A., Landick, R., Darst, S.: Structure of a bacterial RNA polymerase holoenzyme open promoter complex. Elife **4**, e08504 (2015)

Bai, X.C., McMullan, G., Scheres, S.: How Cryo-EM is revolutionizing structural biology. Trends Biochem. Sci. **40**(1), 49–57 (2015)

Baldi, P.: Autoencoders, unsupervised learning, and deep architectures. In: Proceedings of ICML Workshop on Unsupervised and Transfer Learning, JMLR Workshop and Conference Proceedings, pp. 37–49 (2012)

Bau, D., Zhu, J.Y., Wulff, J., Peebles, W., Strobelt, H., Zhou, B., Torralba, A.: Seeing what a gan cannot generate. In: Proceedings of the IEEE/CVF International Conference on Computer Vision, pp. 4502–4511 (2019)

Bhamre, T., Zhang, T., Singer, A.: Denoising and covariance estimation of single particle Cryo-EM images. J. Struct. Biol. **195**(1), 72–81 (2016)

Browning, D., Busby, S.: The regulation of bacterial transcription initiation. Nat. Rev. Microbiol. **2**(1), 57–65 (2004)

Chen, J., Chen, J., Chao, H., Yang, M.: Image blind denoising with generative adversarial network based noise modeling. In: Proceedings of the IEEE Conference on Computer Vision and Pattern Recognition, pp. 3155–3164 (2018)

Chen, J., Chiu, C., Gopalkrishnan, S., Chen, A., Olinares, P., Saecker, R., Winkelman, J., Maloney, M., Chait, B., Ross, W. et al.: Stepwise promoter melting by bacterial RNA polymerase. Mol. Cell **78**, 275–288.e6 (2020)

Cox, M., Cox, T.: Multidimensional scaling. In: Handbook of Data Visualization. Springer, Berlin, pp. 315–347 (2008)

Dabov, K., Foi, A., Katkovnik, V., Egiazarian, K.: Image denoising by sparse 3-D transform-domain collaborative filtering. IEEE Trans. Image Process. **16**(8), 2080–2095 (2007)

Dai, Z., Yang, Z., Yang, F., Cohen, W.W., Salakhutdinov, R.: Good semi-supervised learning that requires a bad gan. In: Proceedings of the 31st International Conference on Neural Information Processing Systems, pp. 6513–6523 (2017)

Dong, Z., Liu, G., Ni, G., Jerwick, J., Duan, L., Zhou, C.: Optical coherence tomography image denoising using a generative adversarial network with speckle modulation. J. Biophotonics **13**(4), e201960135 (2020)

Frank, J.: Three-dimensional electron microscopy of macromolecular assemblies: visualization of biological molecules in their native state. Oxford University Press, New York (2006)

Gao, C., Liu, J., Yao, Y., Zhu, W.: Robust estimation and generative adversarial nets. In: Interational Conference on Learning Representation, New Orleans (2019)

Gao, C., Yao, Y., Zhu, W.: Generative adversarial nets for robust scatter estimation: a proper scoring rule perspective. J. Mach. Learn. Res. **21**, 160–161 (2020)

Goodfellow, I., Pouget-Abadie, J., Mirza, M., Xu, B., Warde-Farley, D., Ozair, S., Courville, A., Bengio, Y.: Generative adversarial nets. In: Advances in Neural Information Processing Systems, pp. 2672–2680 (2014)

Gu, H., Unarta, I.C., Huang, X., Yao, Y.: Robust autoencoder gan for cryo-em image denoising (2020)

Gulrajani, I., Ahmed, F., Arjovsky, M., Dumoulin, V., Courville, A.: Improved training of Wasserstein GANs. In: Advances in Neural Information Processing Systems, pp. 5767–5777 (2017)

Hua, Y., Li, R., Zhao, Z., Chen, X., Zhang, H.: Gan-powered deep distributional reinforcement learning for resource management in network slicing. IEEE J. Sel. Areas Commun. **38**(2), 334–349 (2019)

Huber, P.: Robust estimation of a location parameter. In: Breakthroughs in Statistics. Springer, New York, pp. 492–518 (1992)

Karras, T., Aila, T., Laine, S., Lehtinen, J.: Progressive growing of GANs for improved quality, stability, and variation. In: International Conference on Learning Representation, Vancouver (2018)

Kenzaki, H., Koga, N., Hori, N., Kanada, R., Li, W., Okazaki, K., Yao, X.Q., Takada, S.: CafeMol: a coarse-grained biomolecular simulator for simulating proteins at work. J. Chem. Theory Comput. **7**(6), 1979–1989 (2011)

Kingma, D.P., Ba, J.: Adam: a method for stochastic optimization. In: International Conference on Learning Representation, San Diego (2015)

Krivov, G., Shapovalov, M., Dunbrack R.L. Jr.: Improved prediction of protein side-chain conformations with SCWRL4. Proteins: Struct. Funct. Bioinform. **77**(4), 778–795 (2009)

Krull, A., Buchholz, T.O., Jug, F.: Noise2void-learning denoising from single noisy images. In: Proceedings of the IEEE Conference on Computer Vision and Pattern Recognition, pp. 2129–2137 (2019)

Kühlbrandt, W.: The resolution revolution. Science **343**(6178), 1443–1444 (2014)

Lehtinen, J., Munkberg, J., Hasselgren, J., Laine, S., Karras, T., Aittala, M., Aila, T.: Noise2noise: learning image restoration without clean data. In: Proceeding of the International Conference on Machine Learning, pp. 2965–2974 (2018)

Maaten, L., Hinton, G.: Visualizing data using t-SNE. J. Mach. Learn. Res. **9**(11), 2579–2605 (2008)

Marabini, R., Masegosa, I., San Martın, M., Marco, S., Fernandez, J., De la Fraga, L., Vaquerizo, C., Carazo, J.: Xmipp: an image processing package for electron microscopy. J. Struct. Biol. **116**(1), 237–240 (1996)

Moore, B., Kelley, L., Barber, J., Murray, J., MacDonald, J.: High–quality protein backbone reconstruction from alpha carbons using Gaussian mixture models. J. Comput. Chem. **34**(22), 1881–1889 (2013)

Murakami, K., Masuda, S., Darst, S.: Structural basis of transcription initiation: Rna polymerase holoenzyme at 4 å resolution. Science **296**(5571), 1280–1284 (2002)

Ng, A., Jordan, M., Weiss, Y.: On spectral clustering: analysis and an algorithm. In: Advances in Neural Information Processing Systems, pp. 849–856 (2002)

Okazaki, K., Koga, N., Takada, S., Onuchic, J., Wolynes, P.: Multiple-basin energy landscapes for large-amplitude conformational motions of proteins: structure-based molecular dynamics simulations. Proc. Natl. Acad. Sci. **103**(32), 11844–11849 (2006)

Punjani, A., Rubinstein, J.L., Fleet, D.J., Brubaker, M.A.: CryoSPARC: algorithms for rapid unsupervised Cryo-EM structure determination. Nat. Methods **14**(3), 290 (2017)

Quan, Y., Chen, M., Pang, T., Ji, H.: Self2self with dropout: Learning self-supervised denoising from single image. In: Proceedings of the IEEE/CVF Conference on Computer Vision and Pattern Recognition, pp. 1890–1898 (2020)

Sarmad, M., Lee, H.J., Kim, Y.M.: Rl-gan-net: a reinforcement learning agent controlled gan network for real-time point cloud shape completion. In: Proceedings of the IEEE/CVF Conference on Computer Vision and Pattern Recognition, pp. 5898–5907 (2019)

Scheres, S.: Processing of structurally heterogeneous Cryo-EM data in RELION. In: Methods in Enzymology. Elsevier, Academic Press, vol. 579, pp. 125–157 (2016)

Shen, P.: The 2017 Nobel Prize in Chemistry: Cryo-EM comes of age. Anal. Bioanal. Chem. **410**(8), 2053–2057 (2018)

Su, M., Zhang, H., Schawinski, K., Zhang, C., Cianfrocco, M.: Generative adversarial networks as a tool to recover structural information from cryo-electron microscopy data. BioRxiv, p. 256792 (2018)

Tenenbaum, J., De Silva, V., Langford, J.: A global geometric framework for nonlinear dimensionality reduction. Science **290**(5500), 2319–2323 (2000)

Tran, L., Nguyen, S.M., Arai, M.: GAN-based noise model for denoising real images. In: Proceedings of the Asian Conference on Computer Vision (2020)

Tripathi, S., Lipton, Z.C., Nguyen, T.Q.: Correction by projection: denoising images with generative adversarial networks (2018)

Vincent, P., Larochelle, H., Bengio, Y., Manzagol, P.A.: Extracting and composing robust features with denoising autoencoders. In: Proceeding of the International Conference on Machine Learning, pp. 1096–1103 (2008)

Wang, J., Yin, C.C (2013) A zernike-moment-based non-local denoising filter for cryo-em images. Sci. China Life Sci. **56**(4), 384–390

Wang, Z., Bovik, A., Sheikh, H., Simoncelli, E.: Image quality assessment: from error visibility to structural similarity. IEEE Trans. Image Process. **13**(4), 600–612 (2004)

Wang, F., Gong, H., Liu, G., Li, M., Yan, C., Xia, T., Li, X., Zeng, J.: DeepPicker: a deep learning approach for fully automated particle picking in Cryo-EM. J. Struct. Biol. **195**(3), 325–336 (2016)

Warren, B.E.: X-Ray Diffraction. Courier Corporation. Dover Publications; Reprint Edition (1990)

Wei, D.Y., Yin, C.C.: An optimized locally adaptive non-local means denoising filter for cryo-electron microscopy data. J. Struct. Biol. **172**(3), 211–218 (2010)

Wong, W., Bai, X.C., Brown, A., Fernandez, I., Hanssen, E., Condron, M., Tan, Y.H., Baum, J., Scheres, S.: Cryo-EM structure of the Plasmodium falciparum 80s ribosome bound to the anti-protozoan drug emetine. Elife **3**, e03080 (2014)

Wüthrich, K.: NMR with proteins and nucleic acids. Europhys. News **17**(1), 11–13 (1986)

Xian, Y., Gu, H., Wang, W., Huang, X., Yao, Y., Wang, Y., Cai, J.F.: Data-driven tight frame for cryo-em image denoising and conformational classification. In: Proceeding of the IEEE Global Conference on Signal and Information Processing (GlobalSIP), pp. 544–548 (2018)

Xie, J., Xu, L., Chen, E.: Image denoising and inpainting with deep neural networks. In: Advances in Neural Information Processing Systems, pp. 341–349 (2012)

Yang, Q., Yan, P., Zhang, Y., Yu, H., Shi, Y., Mou, X., Kalra, M., Zhang, Y., Sun, L., Wang, G.: Low-dose CT image denoising using a generative adversarial network with wasserstein distance and perceptual loss. IEEE Trans. Med. Imaging **37**(6), 1348–1357 (2018)

Zhang, K., Zuo, W., Chen, Y., Meng, D., Zhang, L.: Beyond a gaussian denoiser: Residual learning of deep CNN for image denoising. IEEE Trans. Image Process. **26**(7), 3142–3155 (2017)

Zhong, E., Bepler, T., Davis, J., Berger, B.: Reconstructing continuous distributions of 3D protein structure from Cryo-EM images. In: International Conference on Learning Representation, Addis Ababa (2020)

27. Variational Models and Their Combinations with Deep Learning in Medical Image Segmentation: A Survey

Luying Gui, Jun Ma, and Xiaoping Yang

Contents

Introduction	1002
Conventional Algorithms Based on Variational Methods	1003
The Data Term	1004
The Regularization Term	1006
Variational Models Meet Deep Learning in Medical Image Segmentation	1011
Variational Models Guided Deep Learning	1011
Deep Learning-Driven Variational Models	1015
Conclusion	1017
References	1017

Abstract

Image segmentation means to partition an image into separate meaningful regions. Segmentation in medical images can extract different organs, lesions, and other regions of interest, which helps in subsequent disease diagnosis, surgery planning, and efficacy assessment. However, medical images have many unavoidable interference factors, such as imaging noise, artificial artifacts, and mutual occlusion of organs, which make accurate segmentation highly difficult. Incorporating prior knowledge and image information into segmentation model based on variational methods has proven efficient for more accurate segmentation. In recent years, segmentation based on deep learning has been significantly developed, and the combination of classical variational method-

L. Gui · J. Ma
Department of Mathematics, Nanjing University of Science and Technology, Nanjing, China
e-mail: ly.gui@njust.edu.cn; junma@njust.edu.cn

X. Yang (✉)
Department of Mathematics, Nanjing University, Nanjing, China
e-mail: xpyang@nju.edu.cn

© Springer Nature Switzerland AG 2023
K. Chen et al. (eds.), *Handbook of Mathematical Models and Algorithms in Computer Vision and Imaging*, https://doi.org/10.1007/978-3-030-98661-2_109

based models with deep learning is a hot topic. In this survey, we briefly review the segmentation methods based on a variational method making use of image information and regularity information. Subsequently, we clarify how the integration of variational methods into the deep learning framework leads to more precise segmentation results.

Keywords

Medical image segmentation · Variational models · Deep learning

Introduction

Medical image segmentation plays an important role in clinical practices, such as quantitative analysis of lesions, radiotherapy planning, pre-operative planning, intra-operative navigation, and post-operative evaluation. A large number of segmentation methods have been proposed in the past few decades such as graph cut-based (Boykov et al. 2001; Boykov and Funka-Lea 2006), atlas-based methods (Iglesias and Sabuncu 2015), etc. Variational model-based methods are one of the most widely used approaches in medical image segmentation.

The key idea behind the variational model is to make the contour reach the object boundary and minimize the energy functional, which is usually related to information such as intensity, gradient, and texture of the image itself, and also usually includes the desired properties in order to achieve a better segmentation result.

Variational model-based segmentation methods have many desired features, for example, they have transparent and explainable mathematical formulations. Customized constraints and priors can be easily and naturally incorporated into the energy functionals. Moreover, they do not rely on large training data. However, variational models still in general suffer from several shortages:

– The final segmentation results rely on good and reasonable initializations
– The hyperparameters need to be tuned for each testing case
– They lack the ability to learn efficient representations from labeled data

During the past 5 years, fully supervised deep learning methods have revolutionized medical image segmentation (Litjens et al. 2017), and many convolutional neural networks (CNNs) (Long et al. 2015; Shelhamer et al. 2017; Ronneberger et al. 2015; Isensee et al. 2021) have achieved unprecedented performance, such as liver segmentation (Bilic et al. 2019; Kavur et al. 2021), cardiac segmentation (Bernard et al. 2018), kidney segmentation (Heller et al. 2020), and so on. CNN-based segmentation methods directly build the end-to-end mapping between images and annotations by automatically learning object feature representations from a number of training data. The learned models can be directly applied to testing images without any hyperparameter tuning. However, these methods lack interpretability and rely on the large training sets. In this paper, we mainly focus on fully supervised deep learning methods, while there are also weakly supervised methods (Cheplygina et al. 2019) for medical image segmentation.

Thanks to the complementary roles between classical variational models and modern deep learning approaches, a natural trend is to combine the advantages of the two types of approaches to design more accurate, data-efficient, and transparent segmentation methods.

This paper aims to present an overview of classical variational models and their extensions in deep learning era, especially in medical image segmentation. The remainder of this article is organized as follows. First, we introduce the conventional variational models with typical data terms and regularization terms. Then, we present the different combination mechanisms between variational models and deep learning: variational model-guided deep learning and deep learning-driven variational models. Finally, we draw a brief conclusion.

Conventional Algorithms Based on Variational Methods

In 1989 Mumford and Shah proposed a famous image segmentation model, named Mumford-Shah (MS) model (Mumford and Shah 1989), which assumes the image I as a piece-wise smooth function u with the following energy functional:

$$E_{MS}(C, u) = \underbrace{\int_{\Omega} |I - u|^2 dx}_{E_{\text{fidelity}}} + \underbrace{\nu \int_{\Omega \setminus C} |\nabla u|^2 dx + \gamma \mathcal{H}^1(C)}_{E_{\text{regularization}}}, \quad (1)$$

where C is a closed subset of image domain Ω and represents the boundary of the object, and $\mathcal{H}^1$ is the one-dimensional Hausdorff measure.

The solution of the functional (1) is formed by smooth regions R_i, which is represented by u and with sharp boundaries C. A reduced form of this problem is to simplify the restriction of E_{MS} to piecewise constant functions u, that is, $u = c_i$ on each R_i. The reduced case is proposed by Chan and Vese (2001); the energy functional of Chan-Vese (CV) model is as follows:

$$E_{CV}(C, c_1, c_2) = \underbrace{\lambda_1 \int_{\text{inside}(C)} |I - c_1|^2 dx + \lambda_2 \int_{\text{outside}(C)} |I - c_2|^2, dx}_{E_{\text{fidelity}}}$$

$$+ \underbrace{\mu |C|}_{E_{\text{regularization}}}, \quad (2)$$

where c_1 and c_2 are two constants, respectively.

The MS and CV models are based on the assumptions of the segmented regions. Differing from the above models, the "snakes" model focus on boundary detection, and these kinds of models have been extensively studied since the original work of Kass et al. (1988). The main idea is based on deforming the initial contour so that it is oriented towards the boundary of the object to be detected. The classical snakes model relates the parametrized planar curve $C(q) : [0, 1] \to \mathcal{R}^2$ to an energy which is given by

$$E_{\text{snakes}}(C) = \underbrace{-\int_0^1 |\nabla I(C(q))|^2 dq}_{E_{\text{fidelity}}} + \underbrace{\int_0^1 \alpha |C'(q)|^2 + \beta |C''(p)|^2 dq}_{E_{\text{regularization}}}. \quad (3)$$

The above three most typical methods are based on regional information and boundary information, respectively. Many researchers also classify the variational model-based segmentation methods into two categories: region information-based and boundary information-based. This classification method is mainly according to the usage of different types of information in data terms. However, in the actual segmentation of medical images, there are inevitably disturbing factors such as imaging noise, artifacts, and occlusions, which can easily mislead the segmentation algorithm and lead to imprecise segmentation results. In this case, it has become a current inevitable trend to impose proper features or constraints on the segmentation models. The energy term to achieve this function is called the regularization term. The functional of the above three classical methods also consists of two types of energy, the fidelity term and the regularization term, as labeled in these energy functionals. One is the term driven by image information, which guarantees the correspondence between segmentation results and image data and is called the fidelity term. The other guarantees specific properties of the contour or region. This category is called the regularization term.

The Data Term

In image segmentation, the fidelity term is also called the data term for two main reasons. First, the energy of this term usually originates from the image itself, such as E_{fidelity} in the snakes model, which utilizes the gradient information of the image, and E_{fidelity} in the CV model, which utilizes the mean values of the intensity of the different regions of the image. In addition, segmentation models also usually make assumptions about the image, such as the MS model, in which a piecewise smooth function u is used to approximate the image. The fidelity term $\int_\Omega |I-u|^2 dx$ ensures that the function u does not deviate too far from the actual image I. According to the different types of image information utilized by the fidelity, we classify them into two categories, boundary information-based, and regional information-based.

The Boundary Information

Boundary and edge information usually includes important image features that are often used to delineate the object of interest in an image. In image segmentation, the actual boundary of the object is usually considered to be where the pixel changes most dramatically, so the boundary information can be obtained by applying edge detectors, which typically involve first- or second-order spatial differential operators.

One of the most popular segmentation models using edge information is the GAC model (Caselles et al. 1997), which uses image gradient to construct a monotonically decreasing function as a stopping function to control the contour evolution. Since the object boundary is usually expressed as the maximum gradient in the image, this method enables the contour to stop at the desired object boundary.

Since segmentation algorithms aim to find the boundaries of the objects, the detection of boundaries and boundary-based segmentation algorithms is a very intuitive idea and has very accurate segmentation results on better-quality images. However, since interferences such as noise and pseudo-boundaries are often present on medical images and segmentation targets often show weak or missing boundaries, in these cases, boundary-dependent algorithms are often fragile. Therefore, some researchers have also emphasized the importance of integrating regional information for accurate segmentation (Haddon and Boyce 1990; Falah et al. 1994; Chan et al. 1996; Muñoz et al. 2003).

The Regional Information

Although boundaries of the objects provide a natural data-fitting target, it is commonly believed that region-based formulations exhibit less local minima than approaches that solely rely on gradient information of the objects (Cremers et al. 2007). In region-based methods, the intensity/gray value of the image is usually used, such as in CV model, where the gray values of the target and background are assumed to be close to two different constants, respectively. Under this circumstance, the intensity on each pixel is considered to be spatially independent. However, for textured images, the gray value of a pixel is considered to be correlated with its surroundings. The texture is a special attribute of an image for which there is no formal scientific definition (Tuceryan and Jain 1998), and local correlations of intensities usually characterize the textures. Although texture can be visually recognized, it is difficult to define one mathematically, so it is difficult to segment images with texture by general methods. The texture features have been proposed to capture these local correlations. Common representations of texture properties are gray-level co-occurrence matrices (Reska et al. 2015; Wu et al. 2015; Haddon and Boyce 1990; Boonnuk et al. 2015; Lu et al. 2017; Pons et al. 2008), Gabor filters (Gui et al. 2017c), local binary patterns (LBP) (Gui and Yang 2018), sparse texture dictionaries, variational image decompositions, and rapidly developing deep learning based on convolutional neural networks(CNN) in recent years.In addition, since the Gaussian mixture model (GMM) is theoretically capable of fitting any distribution of pixels, it is also commonly used as a regional term in the segmentation of medical images (Martinez-Uso et al. 2010; Balafar 2014; Ji et al. 2012).

However, in medical images, irregular intensity distributions are often presented. Intensity inhomogeneity on medical images is a common phenomenon that can be caused by many factors, such as complex noise, unavoidable artifacts produced by the imaging equipment, and the nature of the imaging object itself. For different situations, researchers have proposed various schemes to solve these problems. For example, to suppress the effects of noise, researchers used a combination of a local denoising term and a local fidelity term to ensure segmentation accuracy (Ali et al. 2018); in Niu et al. (2017), researchers used a local similarity factor to resist the influence of noise. To fitting the unevenly distributed intensity, researchers Yu et al. (2019) generated an adaptive perturbation factor to integrate the external energy functional of the curve evolution. In Li et al. (2008), researchers investigated two fitting functions that locally approximate the image intensities on the two sides of the contour, respectively. In addition, different methods are proposed to correct the bias due to uneven illumination and imaging artifacts (Zhou et al. 2017; Li et al. 2009). Some researchers have also proposed a quantitative assessment of the degree of inhomogeneity of the regions themselves, so as to find the boundaries of different regions and thus segment the desired objects (Li et al. 2011, 2020b; Gui et al. 2017a).

In practical medical image segmentation, the boundary and region information are usually used in combination to achieve a better discriminative object description. For example, the corner detection detects the boundary points, but the critical points inside the object are extracted. On the other hand, the active shape model (Cootes et al. 2000) creates a position-dependent statistical model of all critical points at the object boundary and interior (Cootes et al. 1994). Based on this, statistical texture information is incorporated to form an active appearance model (Cootes et al. 2001; Beichel et al. 2005). Figure 1 shows the different segmentation results given by the two methods, which use intensity information (Chan and Vese 2001) of the image and intensity combined with texture information (Gui et al. 2017c), respectively. The differences between the two segmentation results can be observed by zooming in on the region.

The Regularization Term

In segmentation, the regularization term, also known as constraint, keeps the model from overfitting or imposing some restrictions so that the segmentation curve or segmented region has specific desired properties. Based on the purpose of these regularization terms, they can be divided into two categories: generic regularization terms, which are not related to the segment objects, and specific regularization terms, which are related to the segment objects. Furthermore, they constrain and guide the segmentation model according to some characteristics of the objects.

Generic Regularization

The constraints imposed on the curve are usually independent of the specific segmentation target, by which the smoothness or other characteristics of the curve are guaranteed.

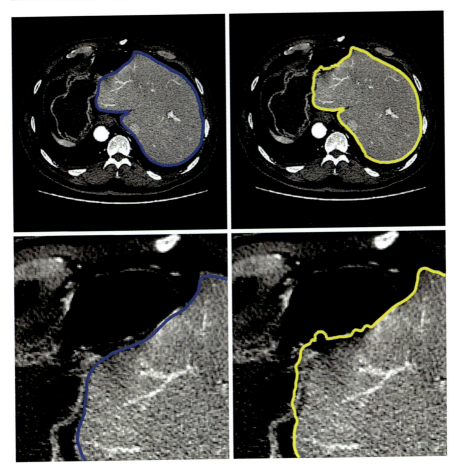

Fig. 1 The 1st row: liver segmentation results, from left to right: by CV model (Chan and Vese 2001) and method from Gui et al. (2017c); the 2nd row: zoomed regions of the segmentation results

The penalty for length is one of the most famous regularization terms in the segmentation model, such as the MS model (1) and CV model (2). Although the constraint on the length of the contour helps cope with problems such as a certain amount of noise in the image, it also brings a bias towards smaller-length contour lines, which leads to isotropic smooth segmentation curves, and small/shortened objects.

The total variation regularization can smooth only the tangent direction of each level line

$$R_{TV}(\phi) = sup\{\int_\Omega u \, div\phi \; : \phi \in C'_c, \|\phi\| \leq \infty\} \qquad (4)$$

and the H^1 regularization

$$R_{H^1}(\phi) = \mu \int_\Omega |\nabla \phi(x)|^2 dx, \qquad (5)$$

applies a purely isotropic smoothing at every pixel x.

Curvature-based regularity is another valuable type of regularization. In psychophysical experiments on contour completion (Kanizsa 1974), the curvature is considered to be an important part of human perception. So curvature regularity (Osher and Sethian 1988) is often used to segment obscured targets (Esedoglu and March 2003) and some thin and elongated targets. Comparative experiments in Schoenemann and Cremers (2007) show that the length-based regularity term usually converges to a small curve enclosing a few pixels only due to intensity inhomogeneity, low contrast, initial positions, etc. In contrast, the curvature-based regularity term usually provides a more meaningful area, i.e., the region of the entire objects.

Another famous curvature-based regularization is the elasticity regularity (Tai et al. 2011); the standard Euler's elastic energy of the curve γ can be written as follows:

$$R_{\text{elastic}}(\gamma) = \int_\gamma (a + b\kappa^2) ds, \qquad (6)$$

where κ is the curvature γ, and two parameters $a, b > 0$. The most remarkable feature of elastic regularity is that it promotes convex contours. It may therefore be used for some particular task of segmenting objects with a convex shape (Bae et al. 2017). And in the snake model (3), the regularization term then consists of two components, the bending energy and the elastic energy, where the bending energy is defined as the sum of the squared curvature of the curve, generating the bending force. In contrast, the elastic energy prevents the stretching of the curve by introducing tension.

In addition to restriction on the nature of the curve itself, regularization terms on the curve have also been proposed as a guarantee of stability and speed of evolution. For instance, Li et al. (2010) avoided re-initialization of the level set by imposing restriction on the gradient of the high-dimensional surface ϕ while ensuring evolutionary stability, making larger steps and faster speeds possible. Yu et al. (2019) performed a restriction on a small neighborhood of zero-level set functions by adding a perturbation factor, thus breaking the pseudo-balance due to heavy noise and then reaching the global optimum.

Targeted Regularization Terms Arising from Object Properties

This type of constraint is usually derived from the nature of the segmentation target itself and is therefore also commonly referred to as prior information. The shape, geometric, and topological properties of the target are widely applied to promote segmentation efficiency.

The segmentation task in medical images is usually to segment out some organs, tissues, or lesions. Fortunately, some organs and tissues have generally similar morphological features. Although the images are subject to imaging errors and individual differences, the shape prior is a robust semantic descriptor for specifying targeted objects. In our categorization, shape prior can be modeled in two ways: building statistical templates and representing by analytical expressions.

Some simple shapes, such as circles or ellipses, can be expressed analytically, and by optimizing the parameters of these analytic expressions, the shape constraints of this analytic representation can be adapted to different variations of the segmented objects, including scale, rotation, and translation (Ray and Acton 2004).

For complex shapes that are difficult to express analytically, an alternative approach is to use a prior shape representation in the form of templates. Template-based shape priors are usually obtained by training on a set of similar shapes. Some researchers have studied the distribution of points on significant positions of the object, also called landmark points, to build a shape template for the object (Cootes et al. 1995), and some researchers employed boundary points as the shape templates (Grenander et al. 2012; Mardia et al. 1991). Subsequently, this kind of parametric point distribution shape prior was also extended into a hybrid segmentation model incorporating intensities (Grenander and Miller 1994) or both gradient and region-homogeneity information (Chakraborty et al. 1994). In the level-set-based approaches, shape constraint is represented as a zero level set of a higher-dimensional surface. Any deviation from the shape can be penalized (Leventon et al. 2002); a simple way to calculate the dissimilarity between them is given by $\int_{\Omega} (\phi_1 - \phi_2)^2 dx$, where ϕ_1 and ϕ_2 are shape constraint and segmented contour, respectively. Usually, to fit the unknown segmentation target, parameters of position, scale, orientation, and other information are also included in the shape energy term (Chen et al. 2002; Pluempitiwiriyawej et al. 2005).

In addition to specific shapes, segmentation targets on medical images may have other more general morphological properties that allow researchers to add them as high-level information to the energy functional as effective constraints. For example, many objects have convex characteristics. As mentioned above, the curvature-based elastic energy term can maintain the convexity of the target. In addition, the limitation of the region can also provide the convexity of the segmentation target (Li et al. 2019; Yan et al. 2020; Luo et al. 2019). In medical images, the left ventricle segmentation is a representative example of the need to preserve the convexity of the object (Feng et al. 2016; Shi and Li 2021; Hajiaghayi et al. 2016). Segmentation of the left ventricle (LV) is critical for the diagnosis of cardiovascular disease. Accurate assessment of crucial clinical parameters such as ejection fraction, myocardial mass, and beat volume depends on the segmentation of the LV, that is, the precise segmentation of the endocardial border. According to the anatomy of the left ventricle, the left ventricle includes the cardiac chambers, trabeculae, and papillary muscles surrounded by the myocardium. Although there is good contrast between myocardium and blood flow on MR images, there are still difficulties in segmentation. This problem is mainly due to the presence of papillary muscles and trabeculae (irregular walls) within the ventricles. They have the same intensity distribution as the surrounding myocardial tissue. Therefore, they can easily mislead the segmentation algorithm and prevent the walls from being clearly depicted, causing critical difficulties in endocardial segmentation.

In addition to the above geometric features, many other regularization terms proposed for segmented object characteristics can also facilitate segmentation. For example, some segmentation objects have a tendency to cluster together, which is defined as compactness. This characteristic can be used as constraint in segmentation organs, such as liver, prostate, as well as cysts and most hepatocellular carcinoma (Gui et al. 2017b). Considering that segmented objects in medical images may present deformation due to lesions, researchers used low-order moment as regularity to constrain the size/volume (Ayed et al. 2008) or location (Klodt and Cremers 2011) of the objects. Figure 2 shows the different segmentation results given by the two methods, one using the classical GAC method (Caselles et al. 1997) without any prior and the other using the intensity information of the image and the isoperimetric shape prior (Gui et al. 2017b). The differences between the two segmentation results can be observed by zooming in on the region.

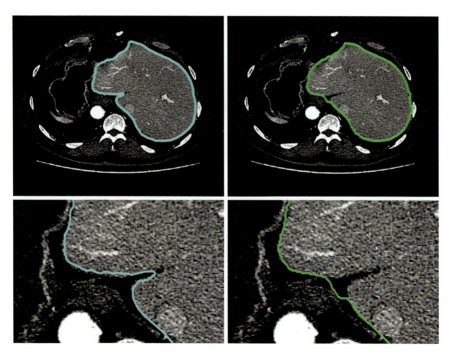

Fig. 2 The 1st row: liver segmentation results, from left to right: by geodesic active contours (GAC) (Caselles et al. 1997) and method from Gui et al. (2017b) ; the 2nd row: zoomed regions of the segmentation results

Variational Models Meet Deep Learning in Medical Image Segmentation

Since 2015, deep learning has gradually dominated medical image segmentation methods. A typical segmentation network is composed of an encoder network followed by a decoder network. The encoder network aims to extract and aggregate features from input images, and the decoder network is to project the features onto the pixel space to get dense predictions. In this way, the deep learning network can directly generate pixel-wise segmentation results with input images. Thus, a natural problem is that could one combine the advantages of deep learning networks and variational models. In this section, we will summarize the progress in this direction.

Variational Models Guided Deep Learning

Variational Model-Inspired Network Modules

Variational models lack learning ability that cannot obtain discrimination ability from the labeled dataset[1]. On the other hand, deep learning methods have poor interpretabilities. In order to formulate the variational model in a learnable framework and increase the interpretability of deep learning, Le et al. (2018b) reformulated the level set (Chan and Vese 2001) evolution as a deep recurrent neural network (Cho et al. 2014) because both of them are time sequence process. In general, the level set function ϕ was updated by

$$\phi_{t+1} = \phi_t + \eta \frac{\partial \phi_t}{\partial t}, \tag{7}$$

where η is the step size (or the learning rate in deep learning). Then, sequence data $\{x_t\}$ for recurrent network input are generated based on the level set evolution:

$$x_{t+1} = \kappa(\phi_t) - F_\theta(I - c_1)^2 + B_\theta(I - c_2)^2, \tag{8}$$

where $\kappa(\phi) = -\mathbf{div}(\frac{\nabla \phi}{|\nabla \phi|})$ is the curvature, and F_θ and B_θ are the learnable parameters that control the force of foreground and background, respectively. This procedure corresponds to the minimization of Chan-Vese energy functional (Chan and Vese 2001) composed of the data fitting term and the contour length term. The final network layer output is computed from the hidden state ϕ_t followed by a Softmax layer to obtain foreground and background segmentation probability maps. The variational level set and this deep learning level set have the same input, including the image and the initial level set function. However, they have

---

[1] The network module is a combination of several network layers, which is part of the network. For example, the well-known U-Net consists of multiple Convolution-Batch Normalization-ReLU modules.

different update rules and outputs. Specifically, the variational level set is updated by the gradient flow of the energy functional, and the output is still a level set function, while deep learning level set is updated by network layers with learnable hyperparameters, and the output is the Softmax probability map.

This network module can be directly connected to existing segmentation networks with convolutional layers and deconvolutional layers for medical image segmentation. For example, Le et al. proposed deep recurrent level set network for brain tumor segmentation (Le et al. 2018a), which achieved less computational time during inference and improved the Dice Similarity Coefficient (DSC) by 1–2%.

In addition to unrolling the level set evolution as network modules, regularizers or priors in classical variational models can also be incorporated into segmentation networks for end-to-end learning. The main challenge is to formulate the non-smooth constraints as differentiable network modules. Typical segmentation CNNs (Ronneberger et al. 2015; Çiçek et al. 2016; Shelhamer et al. 2017) predict each pixel independently and do not explicitly consider the dependency between pixels, which could lead to isolated or scattered small segmentation errors, especially when only few training data is available. To embed spatial regularity in segmentation CNNs, Jia et al. proposed total variation (TV) regularized segmentation CNNs (Jia et al. 2021) to add spatial regularization to the segmented networks, which can produce smooth edges and eliminate isolated segmentation errors. This approach was further applied to pancreas segmentation (Fan and Tai 2019) by unfolding the primal-dual block of TV regularizer and embedding in 2D U-Net (Ronneberger et al. 2015). This type of method has two main benefits. On the one hand, it can produce smooth segmentation edges and eliminate isolated segmentation errors. On the other hand, it is more efficient than the commonly used post-processing methods (Kamnitsas et al. 2017). In order to explicitly add non-local priors to CNNs, Jia et al. (2020) introduced graph total variation to the Softmax function by a primal-dual hybrid gradient method, which can capture long-range information.

Some common shape priors were embedded in segmentation CNNs by reformulating the Softmax layer. Liu et al. (2020b) proposed a Soft Threshold Dynamics framework to integrate many spatial priors of the classical variational models into segmentation CNNs, including spatial regularization, volume, and star-shape priors. The key idea to interpret the Softmax function s is to consider it as a solution of the following variational problem:

$$\min_s - <s, o> + <s, \ln s>, \qquad (9)$$

where o is the network output in the last layer and $\sum_{i=1}^{N} s_i = 1$ (N is the number of classes). In this way, many spatial priors can be imposed on the Softmax results by adding corresponding terms on the energy functional (9). Furthermore, a Soft Threshold Dynamics algorithm was designed to solve the regularized variation problems, which enable stable and fast convergence during forward and backward propagation. Similarly, the convex shape prior (Liu et al. 2020a) and volume-preserving regularization (Li et al. 2020a) were also imposed on segmentation

CNNs. In addition, different priors can be used in combination. For example, using both special regularization and the convex prior can make the segmentation boundary simultaneously smooth and convex.

Variational Model-Inspired Loss Functions

The energy functional of variational models can be directly used as loss functions to guide the learning procedure of segmentation CNNs.

The Mumford-Shah model-inspired loss function (Kim and Ye 2019) This loss function is based on the observation that the characteristic function in the Mumford-Shah model has a striking similarity to the Softmax function in segmentation CNNs. Thus, Kim et al. proposed the following loss function by replacing the characteristic function with Softmax function:

$$L_{MS}(\Theta; I) = \sum_{i=1}^{N} \int_{\Omega} |I(\mathbf{x}) - c_i|^2 S_i(I(x); \Theta) d\mathbf{x} + \lambda \sum_{i=1}^{N} \int_{\Omega} |\nabla S_i(I(\mathbf{x}); \Theta)| d\mathbf{x}, \tag{10}$$

where Θ is the trainable network parameters and

$$c_i = \frac{\int_{\Omega} I(\mathbf{x} S_i(\mathbf{x}; \Theta))}{\int_{\Omega} S_i(\mathbf{x}; \Theta) d\mathbf{x}} \tag{11}$$

is the average intensity value of the i-th class. This loss function enables semi-supervised and unsupervised segmentation, which only requires limited labeled data.

Chan-Vese model-inspired loss function Kim et al. introduced level set loss (Kim et al. 2019) by using the region term of Chan-Vese model, which is defined by

$$L_{LevelSet} = \int_{\Omega} |I_{GT} - c_1|^2 H_\epsilon(\phi_\Theta) d\mathbf{x} + \int_{\Omega} |I_{GT} - c_2|^2 (1 - H_\epsilon(\phi_\Theta)) d\mathbf{x}, \tag{12}$$

where ϕ_Θ is the predicted level set function by the network with parameters Θ and $H_\epsilon(\phi_\Theta) = \frac{1}{2}(1 + \tanh(\frac{\phi_\Theta}{\epsilon}))$. c_1 and c_2 denote the average values of the interior and exterior of the contour, which are defined by

$$c_1 = \frac{\int_{\Omega} I_{GT} H_\epsilon(\phi_\Theta) d\mathbf{x}}{\int_{\Omega} H_\epsilon(\phi_\Theta) d\mathbf{x}}$$

and

$$c_2 = \frac{\int_{\Omega} I_{GT}(1 - H_\epsilon(\phi_\Theta)) d\mathbf{x}}{\int_{\Omega} 1 - H_\epsilon(\phi_\Theta) d\mathbf{x}},$$

respectively.

Chen et al. proposed an active contour loss (Chen et al. 2019) to consider the area inside and outside objects as well as the size of boundaries during learning. In particular, it introduces total variation to approximate the boundary length and membership functions to compute the region area, which is defined by

$$L_{ActiveContour} = Length + \lambda Region$$

$$= \int_{\Omega} |\nabla S_{\Theta}| d\mathbf{x} + \int_{\Omega} |I_{GT} - c_1|^2 S_{\Theta} + |I_{GT} - c_2|^2 (1 - S_{\Theta}) d\mathbf{x}, \quad (13)$$

where S_{Θ} is the predicted Softmax probability map.

Both level set loss (Kim et al. 2019) and active contour loss (Chen et al. 2019) were derived from the Chan-Vese model (Chan and Vese 2001). The main difference is that the mean intensity values of the interior and exterior of the contour are fixed to 1 (foreground) and 0 (background), respectively, in the active contour loss, while the values are iteratively updated in the level set loss.

In addition to fully supervised segmentation tasks, Gur et al. (2019) introduced a new loss term for unsupervised micro-vascular image segmentation. The loss term was based on the morphological optimization method of Chan-Vese model (Marquez-Neila et al. 2013), which is defined by

$$L_{morph-AC} = ||\nabla S_{\Theta}||_1 ((I - c_1)^2 - 2(I - c_2)^2), \quad (14)$$

where ∇S_{Θ} is the intermediate segmentation derivative, computed by the central differences.

Geodesic active contour inspired loss (Ma et al. 2021b) To explicitly embed object global information in segmentation CNNS, Ma et al. proposed a level set regression network with the geodesic active contour loss function:

$$L_{GAC} = \int_{\Omega} g_I \delta_{\epsilon}(\phi_{\Theta}) |\nabla \phi_{\Theta}| d\mathbf{x}, \quad (15)$$

where $g_I = \frac{1}{1+|\nabla I|}$ is the edge indicator function. Different from the level set loss and active contour loss that only used the groundtruth information, the geodesic active contour loss explicitly introduced the image gradient information, which can guide the CNNs to capture detailed boundary information.

Figure 3 presents the visualized segmentation results of different methods on left atrial MRI and pancreas CT images (Fig. 3-a). Commonly used Dice loss (Milletari et al. 2016) (Fig. 3-b) may have obvious segmentation errors because it does not have any global constraint. Level set loss (Kim et al. 2019) (Fig. 3-c) and active contour loss (Chen et al. 2019) (Fig. 3-d) generate similar results that are better than the Dice loss. However, there are still some isolated outliers in the segmentation results. In contrast, the learning GAC (Ma et al. 2021b) (Fig. 3-e) significantly reduces the isolated segmentation masses, and the boundaries are closer to the

27 Variational Models and Their Combinations with Deep Learning in... 1015

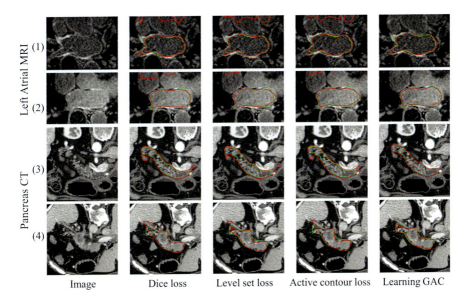

Fig. 3 Qualitative comparisons between commonly used Dice loss (Milletari et al. 2016), Chan-Vese model-inspired level set loss (Kim et al. 2019), active contour loss (Chen et al. 2019), and geodesic active contours inspired learning GAC method (Ma et al. 2021b). The green and red contours denote groundtruth and segmentation results, respectively. (**a**) Image. (**b**) Dice loss. (**c**) Level set loss. (**d**) Active contour loss. (**e**) Learning GAC

ground truth. The is because the learning GAC explicitly considers the image boundary information and geodesic geometry constraint, which can guide the network outputs to achieve lower-energy state of geodesic active contour model and then lead to more accurate results in boundary regions. In addition, it should be noted that the above variational model-inspired loss functions should be added to the Dice loss in a supervised learning framework.

Deep Learning-Driven Variational Models

Classical variational models are sensitive to initializations and hyperparameters settings. To address this limitation, many researches use deep learning to directly generate initial segmentation contours and learn hyperparameters. On the other hand, variational models can help deep learning methods to obtain more accurate boundaries. The learning paradigm can be classified into two categories: two-stage framework and end-to-end framework.

Learning Hyperparameters in Two-Stage Framework

Hoogi et al. (2017) used CNN to estimate the hyperparameters of the mean separation model (Yezzi et al. 2002), and the energy functional was defined by

$$\min_{\phi,c_1,c_2} \int_\Omega \delta(\phi)|\nabla\phi|d\mathbf{x} + \lambda_1 \int_\Omega \frac{(I-c_1)^2}{A_1} H(\phi)d\mathbf{x} + \lambda_2 \int_\Omega \frac{(I-c_2)^2}{A_2}(1-H(\phi))d\mathbf{x} \quad (16)$$

where $A_1 = \int_\Omega H(\phi)dx$ and $A_2 = \int_\Omega (1-H(\phi))dx$ are the area of the local interior and exterior regions surrounding the contour. To adaptively estimate the region term weights λ_1 and λ_2 separately for each case during contour evolution, a CNN was employed to predict the location of the zero level set contour relative to the segmentation target (e.g., lesions), and the output was a probability for each of three classes: inside the lesion and far from its boundaries ($p1$), close to the boundaries of the lesion ($p2$), or outside the lesion and far away from its boundaries ($p3$). The weight parameters were set as follows:

$$\lambda_1 = \exp(\frac{1+p_2+p_3}{1+p_1+p_2}), \quad \lambda_2 = \exp(\frac{1+p_1+p_2}{1+p_2+p_3}). \quad (17)$$

If $p_1 > p_3$, then $\lambda_2 > \lambda_1$ and the contour will expand. Conversely, if $p_3 > p_1$, then $\lambda_1 > \lambda_2$ and the contour tend to shrink. In this way, the contour can be adaptively expanded or shrinked towards the object boundary without any manual tuning.

Instead of predicting the contour location, Hatamizadeh et al. (2019) used an encoder-decoder network to predict the segmentation probability map S_θ. The weights was set as follows:

$$\lambda_1 = \exp(\frac{2-S_\theta}{1+S_\theta}), \quad \lambda_2 = \exp(\frac{1+S_\theta}{2-S_\theta}). \quad (18)$$

Experiments on various lesion segmentation tasks (e.g., brain lesion, liver lesion, lung lesion) and image modalities (CT and MR) show that the proposed method can produce more accurate and detailed boundaries compared with only using CNNs.

Learning Hyperparameters in End-to-End Framework

In order to avoid manual hyperparameter tuning, Zhang et al. (2020) proposed a deep active contour network (DACN) by integrating the convexified Chan-Vese model (Chan et al. 2006) into the DenseUNet (Huang et al. 2017; Ronneberger et al. 2015). The original Chan-Vese model is reduced to a convex minimization problem:

$$\min_{0 \leq u \leq 1} |\nabla u|_1 + \lambda(u, (I-c_1)^2 - (I-c_2)^2). \quad (19)$$

This minimization problem can be solved by the split Bregman algorithm (Goldstein et al. 2010). In the forward propagation, the DenseU-Net generated initial contours and pixel-wise hyperparameter maps of Eq. (19). Then, the contours, maps, and input images were transmitted to the active contour model that was solved by the split Bregman algorithm (Goldstein et al. 2010). The whole network was trained by comparing the final output to the ground truth with cross-entropy loss function.

Ali et al. proposed Trainable Deep Active Contours (TDACs) (Hatamizadeh et al. 2020) based on a standard encoder-decoder CNN and localized Chan-Vese model (Lankton and Tannenbaum 2008), which can explicitly capture local image information. The network also directly predicted pixel-wise hyperparameter maps and the initialization map that were used by the localized Chan-Vese model to update the segmentation results. The network and active contour modules of TDAC was simultaneously trained in an end-to-end manner. Both the initialization map and the active contour model output were passed to a Sigmoid function to generate final segmentation predictions. The loss function is the combination between cross entropy and Dice loss (Milletari et al. 2016) because the compound loss has been proved to be robust in segmentation tasks (Ma et al. 2021a).

Conclusion

In this paper, we have introduced the typical variational models and their combinations with modern deep learning methods, which have many applications in medical image segmentation. We have witnessed several different strategies to fuse the merits of variational models and deep learning methods. However, there is still a lack of the public segmentation benchmark to evaluate and compare these methods in a common and fair platform. We hope this survey can reach broad audiences with diverse backgrounds and inspire more inter-crossing researches between variational models and deep learning.

References

Ali, H., Rada, L., Badshah, N.: Image segmentation for intensity inhomogeneity in presence of high noise. IEEE Trans. Image Process. **27**(8), 3729–3738 (2018)

Ayed, I.B., Li, S., Islam, A., Garvin, G., Chhem, R.: Area prior constrained level set evolution for medical image segmentation. In: Medical Imaging 2008: Image Processing, vol. 6914, p. 691402. International Society for Optics and Photonics (2008)

Bae, E., Tai, X.C., Wei, Z.: Augmented lagrangian method for an Euler's elastica based segmentation model that promotes convex contours (2017)

Balafar, M.: Gaussian mixture model based segmentation methods for brain MRI images. Artif. Intell. Rev. **41**(3), 429–439 (2014)

Beichel, R., Bischof, H., Leberl, F., Sonka, M.: Rosbust active appearance models and their application to medical image analysis. IEEE Trans. Med. Imaging **24**(9), 1151–1169 (2005)

Bernard, O., Lalande, A., Zotti, C., Cervenansky, F., Yang, X., Heng, P.A., Cetin, I., Lekadir, K., Camara, O., Ballester, M.A.G., et al.: Deep learning techniques for automatic mri cardiac multi-structures segmentation and diagnosis: is the problem solved? IEEE Trans. Med. Imaging **37**(11), 2514–2525 (2018)

Bilic, P., Christ, P.F., Vorontsov, E., Chlebus, G., Chen, H., Dou, Q., Fu, C.W., Han, X., Heng, P.A., Hesser, J., et al.: The liver tumor segmentation benchmark (lits). arXiv preprint arXiv:1901.04056 (2019)

Boonnuk, T., Srisuk, S., Sripramong, T.: Texture segmentation using active contour model with edge flow vector. Int. J. Inf. Electron. Eng. **5**(2), 107 (2015)

Boykov, Y., Funka-Lea, G.: Graph cuts and efficient ND image segmentation. Int. J. Comput. Vis. **70**(2), 109–131 (2006)

Boykov, Y., Veksler, O., Zabih, R.: Fast approximate energy minimization via graph cuts. IEEE Trans. Pattern Anal. Mach. Intell. **23**(11), 1222–1239 (2001)

Caselles, V., Kimmel, R., Sapiro, G.: Geodesic active contours. Int. J. Comput. Vis. **22**(1), 61–79 (1997)

Çiçek, Ö., Abdulkadir, A., Lienkamp, S.S., Brox, T., Ronneberger, O.: 3D U-Net: learning dense volumetric segmentation from sparse annotation. In: International Conference on Medical Image Computing and Computer-Assisted Intervention, pp. 424–432 (2016)

Chakraborty, A., Staib, L.H., Duncan, J.S.: An integrated approach to boundary finding in medical images. In: Proceedings of IEEE Workshop on Biomedical Image Analysis, pp. 13–22. IEEE (1994)

Chan, T.F., Vese, L.A.: Active contours without edges. IEEE Trans. Image Process. **10**(2), 266–277 (2001)

Chan, F., Lam, F., Poon, P., Zhu, H., Chan, K.: Object boundary location by region and contour deformation. IEE Proc.-Vis. Image Sig. Process. **143**(6), 353–360 (1996)

Chan, T.F., Esedoglu, S., Nikolova, M.: Algorithms for finding global minimizers of image segmentation and denoising models. SIAM J. Appl. Math. **66**(5), 1632–1648 (2006)

Chen, Y., Tagare, H.D., Thiruvenkadam, S., Huang, F., Wilson, D., Gopinath, K.S., Briggs, R.W., Geiser, E.A.: Using prior shapes in geometric active contours in a variational framework. Int. J. Comput. Vis. **50**(3), 315–328 (2002)

Chen, X., Williams, B.M., Vallabhaneni, S.R., Czanner, G., Williams, R., Zheng, Y.: Learning active contour models for medical image segmentation. In: Proceedings of the IEEE Conference on Computer Vision and Pattern Recognition, pp. 11632–11640 (2019)

Cheplygina, V., de Bruijne, M., Pluim, J.P.: Not-so-supervised: a survey of semi-supervised, multi-instance, and transfer learning in medical image analysis. Med. Image Anal. **54**, 280–296 (2019)

Cho, K., van Merrienboer, B., Gülçehre, Ç., Bahdanau, D., Bougares, F., Schwenk, H., Bengio, Y.: Learning phrase representations using rnn encoder-decoder for statistical machine translation. In: Empirical Methods in Natural Language Processing (EMNLP) (2014)

Cootes, T.F., Hill, A., Taylor, C.J., Haslam, J.: Use of active shape models for locating structures in medical images. Image Vis. Comput. **12**(6), 355–365 (1994)

Cootes, T.F., Taylor, C.J., Cooper, D.H., Graham, J.: Active shape models-their training and application. Comput. Vis. Image Underst. **61**(1), 38–59 (1995)

Cootes, T., Baldock, E., Graham, J.: An introduction to active shape models. Image Process. Anal. **328**, 223–248 (2000)

Cootes, T.F., Edwards, G.J., Taylor, C.J.: Active appearance models. IEEE Trans. Pattern Anal. Mach. Intell. **23**(6), 681–685 (2001)

Cremers, D., Rousson, M., Deriche, R.: A review of statistical approaches to level set segmentation: integrating color, texture, motion and shape. Int. J. Comput. Vis. **72**(2), 195–215 (2007)

Esedoglu, S., March, R.: Segmentation with depth but without detecting junctions. J. Math. Imaging Vis. **18**(1), 7–15 (2003)

Falah, R.K., Bolon, P., Cocquerez, J.P.: A region-region and region-edge cooperative approach of image segmentation. In: Proceedings of 1st International Conference on Image Processing, vol. 3, pp. 470–474. IEEE (1994)

Fan, J., Tai, X.c.: Regularized unet for automated pancreas segmentation. In: Proceedings of the Third International Symposium on Image Computing and Digital Medicine, pp. 113–117 (2019)

Feng, C., Zhang, S., Zhao, D., Li, C.: Simultaneous extraction of endocardial and epicardial contours of the left ventricle by distance regularized level sets. Med. Phys. **43**(6Part1), 2741–2755 (2016)

Goldstein, T., Bresson, X., Osher, S.: Geometric applications of the split bregman method: segmentation and surface reconstruction. J. Sci. Comput. **45**(1), 272–293 (2010)

Grenander, U., Miller, M.I.: Representations of knowledge in complex systems. J. R. Stat. Soc.: Ser. B (Methodological) **56**(4), 549–581 (1994)

Grenander, U., Chow, Y.-S., Keenan, D.M.: Hands: A pattern theoretic study of biological shapes, vol. 2. Springer Science & Business Media, New York (2012)

Gui, L., Yang, X.: Automatic renal lesion segmentation in ultrasound images based on saliency features, improved lbp, and an edge indicator under level set framework. Med. Phys. **45**(1), 223–235 (2018)

Gui, L., He, J., Qiu, Y., Yang, X.: Integrating compact constraint and distance regularization with level set for hepatocellular carcinoma (HCC) segmentation on computed tomography (CT) images. Sens. Imaging **18**(1), 4 (2017a)

Gui, L., Li, C., Yang, X.P.: Medical image segmentation based on level set and isoperimetric constraint. Phys. Med. **42**, 162–173 (2017b)

Gui, L., Yang, X., Cremers, A.B., Chen, Y.: Dempster-shafer evidence theory-based CV model for renal lesion segmentation of medical ultrasound images. J. Med. Imaging Health Inform. **7**(3), 595–606 (2017c)

Gur, S., Wolf, L., Golgher, L., Blinder, P.: Unsupervised microvascular image segmentation using an active contours mimicking neural network. In: Proceedings of the IEEE International Conference on Computer Vision, pp. 10722–10731 (2019)

Haddon, J.F., Boyce, J.F.: Image segmentation by unifying region and boundary information. IEEE Trans. Pattern Anal. Mach. Intell. **12**(10), 929–948 (1990)

Hajiaghayi, M., Groves, E.M., Jafarkhani, H., Kheradvar, A.: A 3-D active contour method for automated segmentation of the left ventricle from magnetic resonance images. IEEE Trans. Biomed. Eng. **64**(1), 134–144 (2016)

Hatamizadeh, A., Hoogi, A., Sengupta, D., Lu, W., Wilcox, B., Rubin, D., Terzopoulos, D.: Deep active lesion segmentation. In: International Workshop on Machine Learning in Medical Imaging, pp. 98–105 (2019)

Hatamizadeh, A., Sengupta, D., Terzopoulos, D.: End-to-end trainable deep active contour models for automated image segmentation: delineating buildings in aerial imagery. In: European Conference on Computer Vision, pp. 730–746 (2020)

Heller, N., Isensee, F., Maier-Hein, K.H., Hou, X., Xie, C., Li, F., Nan, Y., Mu, G., Lin, Z., Han, M., Yao, G., Gao, Y., Zhang, Y., Wang, Y., Hou, F., Yang, J., Xiong, G., Tian, J., Zhong, C., Ma, J., Rickman, J., Dean, J., Stai, B., Tejpaul, R., Oestreich, M., Blake, P., Kaluzniak, H., Raza, S., Rosenberg, J., Moore, K., Walczak, E., Rengel, Z., Edgerton, Z., Vasdev, R., Peterson, M., McSweeney, S., Peterson, S., Kalapara, A., Sathianathen, N., Papanikolopoulos, N., Weight, C.: The state of the art in kidney and kidney tumor segmentation in contrast-enhanced ct imaging: results of the kits19 challenge. Med. Image Anal. **67**, 101821 (2020)

Hoogi, A., Subramaniam, A., Veerapaneni, R., Rubin, D.L.: Adaptive estimation of active contour parameters using convolutional neural networks and texture analysis. IEEE Trans. Med. Imaging **36**(3), 781–791 (2017)

Huang, G., Liu, Z., Van Der Maaten, L., Weinberger, K.Q.: Densely connected convolutional networks. In: Proceedings of the IEEE Conference on Computer Vision and Pattern Recognition, pp. 4700–4708 (2017)

Iglesias, J.E., Sabuncu, M.R.: Multi-atlas segmentation of biomedical images: a survey. Med. Image Anal. **24**(1), 205–219 (2015)

Isensee, F., Jäeger, P.F., Kohl, S.A.A., Petersen, J., Maier-Hein, K.H.: nnU-Net: a self-configuring method for deep learning-based biomedical image segmentation. Nat. Methods **18**(2), 203–211 (2021)

Jia, F., Tai, X.C., Liu, J.: Nonlocal regularized cnn for image segmentation. Inverse Probl. Imaging **14**(5), 891 (2020)

Jia, F., Liu, J., Tai, X.C.: A regularized convolutional neural network for semantic image segmentation. Anal. Appl. **19**(01), 147–165 (2021)

Ji, Z., Xia, Y., Sun, Q., Chen, Q., Xia, D., Feng, D.D.: Fuzzy local Gaussian mixture model for brain MR image segmentation. IEEE Trans. Inf. Technol. Biomed. **16**(3), 339–347 (2012)

Kamnitsas, K., Ledig, C., Newcombe, V.F., Simpson, J.P., Kane, A.D., Menon, D.K., Rueckert, D., Glocker, B.: Efficient multi-scale 3D CNN with fully connected crf for accurate brain lesion segmentation. Med. Image Anal. **36**, 61–78 (2017)

Kanizsa, G.: Contours without gradients or cognitive contours? Giornale Italiano di Psicologia (1974)

Kass, M., Witkin, A., Terzopoulos, D.: Snakes: active contour models. Int. J. Comput. Vis. **1**(4), 321–331 (1988)

Kavur, A.E., Gezer, N.S., Barış, M., Aslan, S., Conze, P.H., Groza, V., Pham, D.D., Chatterjee, S., Ernst, P., Özkan, S., Baydar, B., Lachinov, D., Han, S., Pauli, J., Isensee, F., Perkonigg, M., Sathish, R., Rajan, R., Sheet, D., Dovletov, G., Speck, O., Nürnberger, A., Maier-Hein, K.H., Bozdağı Akar, G., Ünal, G., Dicle, O., Selver, M.A.: Chaos challenge – combined (CT-MR) healthy abdominal organ segmentation. Med. Image Anal. **69**, 101950 (2021)

Kim, B., Ye, J.C.: Mumford–Shah loss functional for image segmentation with deep learning. IEEE Trans. Image Process. **29**, 1856–1866 (2019)

Kim, Y., Kim, S., Kim, T., Kim, C.: CNN-based semantic segmentation using level set loss. In: 2019 IEEE Winter Conference on Applications of Computer Vision (WACV), pp. 1752–1760 (2019)

Klodt, M., Cremers, D.: A convex framework for image segmentation with moment constraints. In: 2011 International Conference on Computer Vision, pp. 2236–2243. IEEE (2011)

Lankton, S., Tannenbaum, A.: Localizing region-based active contours. IEEE Trans. Image Process. **17**(11), 2029–2039 (2008)

Le, T.H.N., Gummadi, R., Savvides, M.: Deep recurrent level set for segmenting brain tumors. In: International Conference on Medical Image Computing and Computer-Assisted Intervention, pp. 646–653 (2018a)

Le, T.H.N., Quach, K.G., Luu, K., Duong, C.N., Savvides, M.: Reformulating level sets as deep recurrent neural network approach to semantic segmentation. IEEE Trans. Image Process. **27**(5), 2393–2407 (2018b)

Leventon, M.E., Grimson, W.E.L., Faugeras, O.: Statistical shape influence in geodesic active contours. In: 5th IEEE EMBS International Summer School on Biomedical Imaging, 2002, p. 8. IEEE (2002)

Li, C., Kao, C.Y., Gore, J.C., Ding, Z.: Minimization of region-scalable fitting energy for image segmentation. IEEE Trans. Image Process. **17**(10), 1940–1949 (2008)

Li, C., Xu, C., Anderson, A.W., Gore, J.C.: MRI tissue classification and bias field estimation based on coherent local intensity clustering: a unified energy minimization framework. In: International Conference on Information Processing in Medical Imaging, pp. 288–299. Springer (2009)

Li, C., Xu, C., Gui, C., Fox, M.D.: Distance regularized level set evolution and its application to image segmentation. IEEE Trans. Image Process. **19**(12), 3243–3254 (2010)

Li, C., Huang, R., Ding, Z., Gatenby, J.C., Metaxas, D.N., Gore, J.C.: A level set method for image segmentation in the presence of intensity inhomogeneities with application to MRI. IEEE Trans. Image Process. **20**(7), 2007–2016 (2011)

Li, L., Luo, S., Tai, X.C., Yang, J.: Convex hull algorithms based on some variational models. arXiv preprint arXiv:1908.03323 (2019)

Li, H., Liu, J., Cui, L., Huang, H., Tai, X.C.: Volume preserving image segmentation with entropy regularized optimal transport and its applications in deep learning. J. Vis. Commun. Image Rep. **71**, 102845 (2020a)

Li, X., Yang, X., Zeng, T.: A three-stage variational image segmentation framework incorporating intensity inhomogeneity information. SIAM J. Imaging Sci. **13**(3), 1692–1715 (2020b)

Litjens, G., Kooi, T., Bejnordi, B.E., Setio, A.A.A., Ciompi, F., Ghafoorian, M., Van Der Laak, J.A., Van Ginneken, B., Sánchez, C.I.: A survey on deep learning in medical image analysis. Med. Image Anal. **42**, 60–88 (2017)

Liu, J., Tai, X.C., Luo, S.: Convex shape prior for deep neural convolution network based eye fundus images segmentation. arXiv preprint arXiv:2005.07476 (2020a)

Liu, J., Wang, X., Tai, X.C.: Deep convolutional neural networks with spatial regularization, volume and star-shape prior for image segmentation. arXiv preprint arXiv:2002.03989 (2020b)

Long, J., Shelhamer, E., Darrell, T.: Fully convolutional networks for semantic segmentation. In: Proceedings of the IEEE Conference on Computer Vision and Pattern Recognition, pp. 3431–3440 (2015)

Lu, J., Wang, G., Pan, Z.: Nonlocal active contour model for texture segmentation. Multimedia Tools Appl. **76**(8), 10991–11001 (2017)

Luo, S., Tai, X.C., Huo, L., Wang, Y., Glowinski, R.: Convex shape prior for multi-object segmentation using a single level set function. In: Proceedings of the IEEE/CVF International Conference on Computer Vision, pp. 613–621 (2019)

Ma, J., Chen, J., Ng, M., Huang, R., Li, Y., Li, C., Yang, X., Martel, A.: Loss odyssey in medical image segmentation. Med. Image Anal. **71**, 102035 (2021a)

Ma, J., He, J., Yang, X.: Learning geodesic active contours for embedding object global information in segmentation CNNs. IEEE Trans. Med. Imaging **40**(1), 93–104 (2021b)

Mardia, K., Kent, J., Walder, A.: Statistical shape models in image analysis. In: Proceedings of the 23rd Symposium on the Interface, Seattle, pp. 550–557 (1991)

Marquez-Neila, P., Baumela, L., Alvarez, L.: A morphological approach to curvature-based evolution of curves and surfaces. IEEE Trans. Pattern Anal. Mach. Intell. **36**(1), 2–17 (2013)

Martinez-Uso, A., Pla, F., Sotoca, J.M.: A semi-supervised Gaussian mixture model for image segmentation. In: 2010 20th International Conference on Pattern Recognition, pp. 2941–2944. IEEE (2010)

Milletari, F., Navab, N., Ahmadi, S.A.: V-Net: Fully convolutional neural networks for volumetric medical image segmentation. In: 2016 Fourth International Conference on 3D Vision (3DV), pp. 565–571 (2016)

Mumford, D., Shah, J.: Optimal approximations by piecewise smooth functions and associated variational problems. Commun. Pure Appl. Math. **42**(5), 577–685 (1989)

Muñoz, X., Freixenet, J., Cufı, X., Martı, J.: Strategies for image segmentation combining region and boundary information. Pattern Recogn. Lett. **24**(1–3), 375–392 (2003)

Niu, S., Chen, Q., De Sisternes, L., Ji, Z., Zhou, Z., Rubin, D.L.: Robust noise region-based active contour model via local similarity factor for image segmentation. Pattern Recogn. **61**, 104–119 (2017)

Osher, S., Sethian, J.A.: Fronts propagating with curvature-dependent speed: algorithms based on Hamilton-Jacobi formulations. J. Comput. Phys. **79**(1), 12–49 (1988)

Pluempitiwiriyawej, C., Moura, J.M., Wu, Y.J.L., Ho, C.: Stacs: new active contour scheme for cardiac MR image segmentation. IEEE Trans. Med. Imaging **24**(5), 593–603 (2005)

Pons, S.V., Rodríguez, J.L.G., Pérez, O.L.V.: Active contour algorithm for texture segmentation using a texture feature set. In: 2008 19th International Conference on Pattern Recognition, pp. 1–4. IEEE (2008)

Ray, N., Acton, S.T.: Motion gradient vector flow: an external force for tracking rolling leukocytes with shape and size constrained active contours. IEEE Trans. Med. Imaging **23**(12), 1466–1478 (2004)

Reska, D., Boldak, C., Kretowski, M.: A texture-based energy for active contour image segmentation. In: Image Processing & Communications Challenges, vol. 6, pp. 187–194. Springer (2015)

Ronneberger, O., Fischer, P., Brox, T.: U-Net: Convolutional networks for biomedical image segmentation. In: International Conference on Medical Image Computing and Computer-Assisted Intervention, pp. 234–241 (2015)

Schoenemann, T., Cremers, D.: Introducing curvature into globally optimal image segmentation: minimum ratio cycles on product graphs. In: 2007 IEEE 11th International Conference on Computer Vision, pp. 1–6. IEEE (2007)

Shelhamer, E., Long, J., Darrell, T.: Fully convolutional networks for semantic segmentation. IEEE Trans. Pattern Anal. Mach. Intell. **39**(4), 640–651 (2017)

Shi, X., Li, C.: Convexity preserving level set for left ventricle segmentation. Magn. Reson. Imaging **78**, 109–118 (2021)

Tai, X.C., Hahn, J., Chung, G.J.: A fast algorithm for Euler's elastica model using augmented lagrangian method. SIAM J. Imaging Sci. **4**(1), 313–344 (2011)

Tuceryan, M., Jain, A.K.: Texture analysis. In: Chen, CH, Pau, LF, Wang, PSP (eds) The Handbook of Pattern Recognition and Computer Vision, 2nd Edn., pp. 207–248. World Scientific (1998)

Wu, Q., Gan, Y., Lin, B., Zhang, Q., Chang, H.: An active contour model based on fused texture features for image segmentation. Neurocomputing **151**, 1133–1141 (2015)

Yan, S., Tai, X.C., Liu, J., Huang, H.Y.: Convexity shape prior for level set-based image segmentation method. IEEE Trans. Image Process. **29**, 7141–7152 (2020)

Yezzi Jr, A., Tsai, A., Willsky, A.: A fully global approach to image segmentation via coupled curve evolution equations. J. Vis. Commun. Image Rep. **13**(1–2), 195–216 (2002)

Yu, H., He, F., Pan, Y.: A novel segmentation model for medical images with intensity inhomogeneity based on adaptive perturbation. Multimedia Tools Appl. **78**(9), 11779–11798 (2019)

Zhang, M., Dong, B., Li, Q.: Deep active contour network for medical image segmentation. In: International Conference on Medical Image Computing and Computer-Assisted Intervention, pp. 321–331 (2020)

Zhou, S., Wang, J., Zhang, M., Cai, Q., Gong, Y.: Correntropy-based level set method for medical image segmentation and bias correction. Neurocomputing **234**, 216–229 (2017)

Bidirectional Texture Function Modeling

28

Michal Haindl

Contents

Introduction	1025
Visual Texture	1027
Bidirectional Texture Function	1027
BTF Measurement	1029
Compound Markov Model	1030
Principal Markov Model	1031
Principal Single Model Markov Random Field	1032
Non-parametric Markov Random Field	1032
Non-parametric Markov Random Field with Iterative Synthesis	1033
Non-parametric Markov Random Field with Fast Iterative Synthesis	1035
Potts Markov Random Field	1037
Potts-Voronoi Markov Random Field	1038
Bernoulli Distribution Mixture Model	1040
Gaussian Mixture Model	1041
Local Markov and Mixture Models	1042
3D Causal Simultaneous Autoregressive Model	1042
3D Moving Average Model	1046
Spatial 3D Gaussian Mixture Model	1047
Applications	1049
Texture Synthesis and Enlargement	1050
Texture Compression	1053
Texture Editing	1053
Illumination Invariants	1053
(Un)supervised Image Recognition	1054
Multispectral/Multi-channel Image Restoration	1056
Conclusion	1057
References	1058

M. Haindl (✉)
Institute of Information Theory and Automation, Czech Academy of Sciences, Prague, Czechia
e-mail: haindl@utia.cas.cz

© Springer Nature Switzerland AG 2023
K. Chen et al. (eds.), *Handbook of Mathematical Models and Algorithms in Computer Vision and Imaging*, https://doi.org/10.1007/978-3-030-98661-2_103

Abstract

An authentic material's surface reflectance function is a complex function of over 16 physical variables, which are unfeasible both to measure and to mathematically model. The best simplified measurable material texture representation and approximation of this general surface reflectance function is the seven-dimensional bidirectional texture function (BTF). BTF can be simultaneously measured and modeled using state-of-the-art measurement devices and computers and the most advanced mathematical models of visual data. However, such an enormous amount of visual BTF data, measured on the single material sample, inevitably requires state-of-the-art storage, compression, modeling, visualization, and quality verification. Storage technology is still the weak part of computer technology, which lags behind recent data sensing technologies; thus, even for virtual reality correct materials modeling, it is infeasible to use BTF measurements directly. Hence, for visual texture synthesis or analysis applications, efficient mathematical BTF models cannot be avoided. The probabilistic BTF models allow unlimited seamless material texture enlargement, texture restoration, tremendous unbeatable appearance data compression (up to 1:1000 000), and even editing or creating new material appearance data. Simultaneously, they require neither storing actual measurements nor any pixel-wise parametric representation. Unfortunately, there is no single universal BTF model applicable for physically correct modeling of visual properties of all possible BTF textures. Every presented model is better suited for some subspace of possible BTF textures, either natural or artificial. In this contribution, we intend to survey existing mathematical BTF models which allow physically correct modeling and enlargement measured texture under any illumination and viewing conditions while simultaneously offering huge compression ratio relative to natural surface materials optical measurements. Exceptional 3D Markovian or mixture models, which can be either solved analytically or iteratively and quickly synthesized, are presented. Illumination invariants can be derived from some of its recursive statistics and exploited in content-based image retrieval, supervised or unsupervised image recognition. Although our primary goal is physically correct texture synthesis of any unlimited size, the presented models are equally helpful for various texture analytical applications. Their modeling efficiency is demonstrated in several analytical and modeling image applications, in particular, on a (un)supervised image segmentation, bidirectional texture function (BTF) synthesis and compression, and adaptive multispectral and multi-channel image and video restoration.

Keywords

Bidirectional texture function · Texture modeling · Markov random fields · Discrete distribution mixtures · Expectation-Maximization algorithm

Introduction

Multidimensional data modeling or understanding (or set of spatially related objects) is more accurate and efficient if we respect all interdependencies between single objects. Objects to be processed, for example, multispectral pixels, in a digitized image are often mutually dependent (e.g., correlated) with a dependency degree related to a distance between two objects in their corresponding data space. These relations can be incorporated into a pattern recognition or visualization process through an appropriate multidimensional data model. If such a model is probabilistic, we can benefit from a consistent Bayesian framework for solving many related visual or pattern recognition tasks.

Features derived from multidimensional data models are information preserving in the sense that they can be used to synthesize data spaces closely resembling original measurement data space as can be illustrated on the recent best visual representation of real material surfaces in the form of seven-dimensional bidirectional texture function (Haindl and Filip 2007; Filip and Haindl 2009). Virtual or augmented reality systems require object surfaces covered with physically correct nature-like color textures to enhance realism in visual scenes applied in computer games, CAD systems, or other computer graphics applications. Surface material appearance modeling thus aims to generate and enlarge a synthetic texture visually indiscernible from the visual properties of measured material, whatever the observation conditions might be.

While simple color textures can be either digitized measured natural textures or textures synthesized from an appropriate mathematical model, realistic 7D BTF textures require mathematical modeling. Measured BTF textures are far less convenient alternative, because of extreme virtual system memory demands, limited size measurements, visible discontinuities (if we apply some usual computer graphics sampling approach for texture enlargement (De Bonet 1997; Efros and Freeman 2001; Praun et al. 2000; Xu et al. 2000; Wei and Levoy 2000, 2001; Liang et al. 2001; Soler et al. 2002; Dong and Chantler 2002; Zelinka and Garland 2002; Haindl and Hatka 2005a,b; Ngan and Durand 2006)), or several other drawbacks (Haindl 1991). Some of these methods are based on per-pixel sampling (Wei and Levoy 2001; Tong et al. 2002; Zelinka and Garland 2003; Zhang et al. 2003) while other are patch-based sampling methods (Praun et al. 2000; Xu et al. 2000; Efros and Freeman 2001; Liang et al. 2001; Soler et al. 2002; Kwatra et al. 2003; Dong et al. 2010). Texture synthesis algorithms (Heeger and Bergen 1995; Liu and Picard 1996; Efros and Leung 1999; Portilla and Simoncelli 2000) view surface texture as a stochastic process and aim to produce new realizations that resemble an input exemplar by either copying pixels (non-parametric methods) or matching image statistics (parametric techniques). Some of these simple gray scale/color texture modeling methods, which also allow texture enlargement, could be formally applied independently for each BTF material space. However, this is infeasible for all about a thousand measurements for a single BTF material due to their enormous computing time and memory constraints. Furthermore, for example, a car interior usually has about 20 different materials to synthesize.

Principle component analysis (PCA)-based BTF approximation (Müller et al. 2003; Sattler et al. 2003; Ruiters et al. 2013) allows BTF lossy compression but not enlargement. Furthermore, projecting the measured data onto a linear space constructed by statistical analysis such as PCA results in low-quality data compression. Another compression method (Tsai and Shih 2012) is based on K-clustered tensor approximation or the polynomial wavelet tree (Baril et al. 2008).

BTF data can be approximated using separate texel models, i.e., spatially varying bidirectional reflectance distribution function (SVBRDF) models that combine texture mapping and BRDF models but sacrifice some spatial dependency information. A linear combination of multivariate spherical radial basis functions is used to model BTF as a set of texelwise BRDFs (SVBRDF) in Tsai et al. (2011). Another SVBRDF method (Wu et al. 2011) uses a parametric mixture model with a basis analytical BRDF function for texel modeling. Several SVBRDF models use multilayer perceptron neural networks (Aittala et al. 2016; Deschaintre et al. 2018; Rainer et al. 2020). A deep convolutional neural network VGG-19 is used in Aittala et al. (2016), while the convolutional neural network recovers SVBRDF from estimated normal, diffuse albedo, specular albedo, and specular roughness from a single image lit by a handheld flash in Deschaintre et al. (2018). A learned SVBRDF decoder in a multilayer perceptron neural model approximates BRDF values in Rainer et al. (2020). The SVBRDF methods approximate BTF quality, are computationally expensive due to the nonlinear optimization, allow only moderate compression ratio, require several manually tuned parameters, and do not allow BTF space enlargement.

Mathematical multidimensional data models are useful for describing many of the multidimensional data types provided that we can assume some data homogeneity, so some data characteristics are a translation invariant. While the 1D models like time series (Anderson 1971; Broemeling 1985) are relatively well researched, and they have a rich application history in control theory, econometrics, medicine, meteorology, and many other data mining or machine learning applications, multidimensional models are much less known (e.g., more than three-dimensional MRF), and their applications are still limited. The reason is not only unsolved theory difficulties but mainly their vast computing power demands, which prevented their more extensive use until recently.

Visual data models need nonstandard multidimensional (three-dimensional for static color textures, four-dimensional for videos, or even seven-dimensional for static BTFs) models. However, if such a nD data space can be factorized, then these data can also be approximated using a set of lower-dimensional probabilistic models. Although full visual nD models allow unrestricted spatial-spectral-temporal-angular correlation modeling, their main drawback is many parameters to be estimated, which require a correspondingly large learning set. In some models (e.g., Markov models), the necessity is to estimate all these parameters simultaneously.

We introduced (Haindl and Havlíček 1998, 2000, 2010, 2016, 2017b, 2018a,b; Haindl et al. 2012, 2015b), several efficient fast multiresolution Markov random field (MRF)-based models which exploit BTF space factorization. Our methods avoid the time-consuming Markov chain Monte Carlo simulation (MCMC) so typical for Markov models applications with one exception of the Potts MRF. Our models avoid some problems of alternative options (see Haindl 1991 for details), but they are also easy to analyze as well as to synthesize, and last but not least, they are still flexible enough to correctly imitate a broad set of natural and artificial textures or other spatial data.

We can categorize the model's applications into synthesis and analysis. Analytical applications include static or dynamic data un-/semi-/supervised recognition, scene understanding, data space analysis, motion detection, and numerous others. Typical synthesis applications are missing data reconstruction, restoration, image compression, and static or dynamic texture synthesis.

Visual Texture

The visual texture notion is closely tied to the human semantic meaning of surface material appearance, and texture analysis is an essential and frequently published area of image processing. However, there is still no mathematically rigorous definition of the texture that would be accepted throughout the computer vision community.

We understand a textured image or the *visual texture* (Haindl and Filip 2013) to be a realization of a random field, and our effort is to find its parameterizations in such a way that the real texture representing the specific material appearance measurements will be visually indiscernible from the corresponding random field's realization, whatever the observation conditions might be. Some work distinguishes between texture and color. We regard such separation between spatial structure and spectral information to be artificial and principally wrong because there is no bijective mapping between gray scale and multispectral textures. Thus, our random field model is always multispectral.

Bidirectional Texture Function

A natural material's surface general reflectance function (GRF), representing physically correct visual properties of surface materials and their variations under any observation conditions, is a complex function of 16 physical variables. It is currently unfeasible to measure or to model such a function mathematically. Practical applications thus require significant simplification, namely, using additional assumptions. These approximative assumptions neglect the most less significant variables to achieve a solvable problem, with the solution still far more realistic

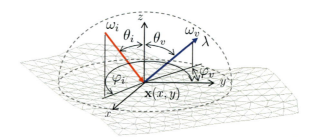

Fig. 1 BTF reflectance model

than the traditional three-dimensional static color texture representation. BTF can model complex lighting effects such as self-shadows, masking, foreshortening, interreflections, and multiple subsurface light scattering due to material surface microgeometry.

The seven-dimensional bidirectional texture function (BTF) reflectance model Fig. 1 is the best recent visual texture representation, which can still be simultaneously measured and modeled using state-of-the-art measurement devices and computers as well as the most advanced mathematical models of visual data. Thus, it is the most important representation for the high-end and physically correct surface materials appearance modeling. Nevertheless, BTF requires the most advanced modeling as well as high-end hardware support. The BTF reflectance model

$$Y_r^{BTF} = BTF(\lambda, x, y, \theta_i, \varphi_i, \theta_v, \varphi_v), \tag{1}$$

where Y_r^{BTF} is a random spectral reflectance vector at location r, r is a multiindex, and Y_r^{BTF} accepts six simplifying assumptions from GRF – light transport in material takes zero time ($t_i = t_v$ (incident time is equal to the reflection time) and $t_v = \emptyset$), reflectance behavior of the surface is time invariant ($t_v = t_i = const.$, $t_v = t_i = \emptyset$); interaction with the material does not change wavelength ($\lambda_i = \lambda_v$), i.e., $\lambda_v = \emptyset$), constant radiance along light rays ($z_i = z_v = \emptyset$), no transmittance ($\theta_t = \varphi_t = \emptyset$), and incident light leaves at the same point.

Multispectral BTF is a seven-dimensional random function, which considers measurement dependency on color spectrum and planar material position, as well as its dependence on illumination incident light (lower index i) and viewing reflection light (lower index v) angles $BTF(r, \theta_i, \phi_i, \theta_v, \phi_v)$, where the multiindex $r = [r_1, r_2, r_3]$ specifies planar horizontal and vertical position in material sample image, r_3 is the spectral index, and θ, ϕ are elevation and azimuthal angles of the illumination and view direction vectors. The BTF measurements comprise a whole the hemisphere of light and camera positions in observed material sample coordinates according to selected quantization steps, and this is the main difference compared to the standard three-dimensional static color texture. This difference significantly improves the visual quality and realism of BTF representation and simultaneously complicates its measurement and modeling.

BTF Measurement

Accurate and reliable BTF acquisition is not a trivial task; only a few BTF measurement systems currently exist (for details see Haindl and Filip 2013; Schwartz et al. 2014; Dana et al. 1997; Koudelka et al. 2003; Sattler et al. 2003; Han and Perlin 2003; Müller et al. 2004; Wang and Dana 2006; Ngan and Durand 2006; Debevec et al. 2000; Marschner et al. 2005; Holroyd et al. 2010; Ren et al. 2011; Aittala et al. 2013, 2015). However, their number increases every year in response to the growing demand for photorealistic virtual representations of real-world materials. These systems are (similar to bidirectional reflectance distribution function (BRDF) measurement systems) based on the light source, video/still camera, and material sample. The main difference between individual BTF measurement systems is in the type of measurement setup allowing four degrees of freedom for camera/light, the type of measurement sensor (CCD, video, and some other), and light.

In some systems, the camera is moving, and the light is fixed (Dana et al. 1997; Sattler et al. 2003; Neubeck et al. 2005), while in others, e.g., Koudelka et al. (2003), it is just the opposite. There are also systems where both camera and light source remain fixed (Han and Perlin 2003; Müller et al. 2004).

The UTIA gonioreflectometer setup Fig. 2 consists of independently controlled arms with a camera and light. Its parameters, such as angular precision 0.03 degree, spatial resolution 1000 DPI, or selective spatial measurement, classify this

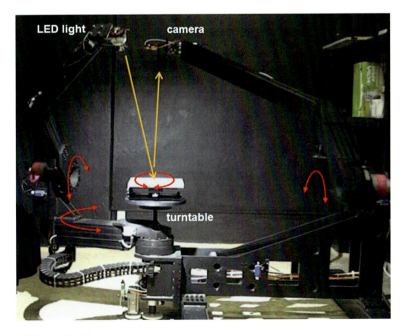

Fig. 2 UTIA gonioreflectometer

gonioreflectometer to the state-of-the-art devices. The typical resolution of the area of interest is around 2000 × 2000 pixels, sample size 7 × 7 [cm], and sensor distance ≈2 [m] with a field of view angle of 8.25°, and each of them is represented using at least 16-bit floating-point value for a reasonable representation of high-dynamic-range visual information. Illumination source is 11 LED arrays, each having a flux of 280 lm at 0.7 A, spectral wavelength 450 − 700 [nm], and its optics. The memory requirements for storage of a single material sample amount to 360 gigabytes per color channel but can be much more for a more precise spectral measurement.

We measure each material sample mostly in 81 viewing positions n_v and 81 illumination positions n_i, resulting in 6561 images per sample (4 terabytes of data).

Compound Markov Model

BTF data space is seven-dimensional, and thus it also requires seven-dimensional probabilistic models for physically correct BTF modeling, data compression, and enlargement with all related problems needed for robust estimation of all their numerous parameters. A practical alternative is to factorize a seven-dimensional problem into a set of lower-dimensional models with fewer parameters dedicated to model subparts of a BTF texture combined into a compound BTF model.

We exploit the compound Markov model for physically correct BTF modeling for either synthesis or analytical applications. Let us denote a multiindex $r = (r_1, r_2)$, $r \in I$, where I is a discrete two-dimensional rectangular lattice and r_1 is the row and r_2 the column index, respectively. The principal field pixel $X_r \in \mathcal{K}$ where $\mathcal{K}$ is the index set of K distinguished sub-models, i.e., $X_r \in \{1, 2, \ldots, K\}$ is a random variable with natural number value (a positive integer). Y_r is the multispectral pixel at location r and $Y_{r,j} \in \mathcal{R}$ is its j-th spectral plane component. Both random fields (X, Y) are indexed on the same $M \times N$ lattice I.

Let us assume that each multispectral observed texture $\tilde{Y}$ (composed of d spectral planes, e.g., $d = 3$ for color textures) and indexed on the $\tilde{M} \times \tilde{N}$ lattice $\tilde{I}$ (usually $\tilde{I} \subseteq I$ and $\tilde{M}, \tilde{N}$ are number of rows and columns of the measured BTF texture) can be modeled by a compound Markov random field model (CMRF), where the principal Markov random field (MRF) X controls switching to a regional local MRF model $^i Y$ where $Y = \bigcup_{i=1}^{K} {^i Y}$. Single K regional random field sub-models $^i Y$ are defined on their corresponding lattice subsets $^i I$, $^i I \cap {^j I} = \emptyset$ $\forall i \neq j$, $I = \bigcup_{i=1}^{K} {^i I}$ ($X_r = X_s$ $\forall r, s \in {^i I}$) and they are of the same MRF type. These models differ only in their contextual support set $^i I_r$ and corresponding parameter sets $^i \theta$ (a set of all i-th local random field parameters). The same type of sub-models are assumed only for simplicity and can be omitted without any problems if needed. The BTF-CMRF model has a posterior probability

$$P(X, Y \mid \tilde{Y}) = P(Y \mid X, \tilde{Y}) P(X \mid \tilde{Y}) \qquad (2)$$

and the corresponding optimal maximum a posteriori (MAP) solution is

$$(\hat{X}, \hat{Y}) = \arg \max_{X \in \Omega_X, Y \in \Omega_Y} P(Y \mid X, \tilde{Y}) \, P(X \mid \tilde{Y}),$$

where Ω_X, Ω_Y are the corresponding configuration spaces for both random fields (X, Y). To avoid an iterative MCMC MAP solution for parameter estimation, we proposed the following two-step approximation $\check{X}, \check{Y}$ (Haindl and Havlíček 2010):

$$(\check{X}) = \arg \max_{X \in \Omega_X} P(X \mid \tilde{Y}), \tag{3}$$

$$(\check{Y}) = \arg \max_{Y \in \Omega_Y} P(Y \mid \check{X}, \tilde{Y}). \tag{4}$$

This approximation significantly simplifies the BTF-CMRF estimation without compromising random sampling for its synthesis because it allows us to take advantage of the possible analytical estimation of all regional MRF models $^i Y$ in (4). We randomly sample the required enlarged texture in the same order, i.e., at first (3) and, consequently, based on this principal random field realization, the local random fields (4). Furthermore, there is no need to have a unique solution of the (3), (4) approximation because the aim is to obtain a visually indiscernible result or results from the target observation. The subsequent Markovian/mixture compound models use the notation BTF-CMRF$^{principal\_model \; local\_model}$ where the upper indices indicate the principal as well as the local model families.

Principal Markov Model

The principal part (X) of the BTF compound Markov models (BTF-CMRF) is assumed to be independent on illumination and observation angles, i.e., it is identical for all possible combinations $\phi_i, \phi_v, \theta_i, \theta_v$ azimuthal and elevation illumination/viewing angles, respectively. This assumption does not compromise the resulting BTF space quality because it influences only a material texture macrostructure independent of these angles for static BTF textures.

The principal random field $\check{X}$ is estimated using simple K-means clustering of $\tilde{Y}$ in the RGB color space into a predefined number of K classes, where cluster indices are $\check{X}_r \quad \forall r \in I$ estimates. We further use for simplicity the RGB color space, but any other color space can be used as well. The number of classes K can be estimated using the Kullback-Leibler divergence and considering a sufficient amount of data necessary to estimate all local Markovian models reliably. If the BTF texture contains subparts with distinct texture but similar colors, any more sophisticated texture segmenter (e.g., Haindl and Mikeš 2007; Haindl et al. 2009a,b, 2015a) can be used.

Principal Single Model Markov Random Field

The simplest principal model is a constant field that contains only one model BTF-CMRF$^{c\cdots}$ $P(X \mid \tilde{Y}) = const.$, i.e., $P(X_r \mid \tilde{Y}) = P(X_s \mid \tilde{Y})$ $\forall r, s$. Then there is no need to use the MAP approximation (3), (4), and the compound Markov model simplifies into a single random field BTF-MRF model, and the BTF-MRF model can be any of the following local MRF models.

Non-parametric Markov Random Field

If we do not assume any specific principal control field parametric model, but rather we seamlessly and directly enlarge its realization from measured data (Fig. 3), we get several non-parametric principal control field approaches. The non-parametric principal field BTF-CMRF$^{NProl\cdots}$ (NProl... – a non-parametric roller-based principal field with any local random fields denoted as ...; see Figs. 3, 4, 16) can be modeled using the roller method (Haindl and Havlíček 2010) for optimal $\check{X}$ compression and speedy enlargement to any required field size. The roller method (Haindl and Hatka 2005a,b) principle is the overlapping tiling and subsequent minimum error boundary cut. One or several optimal double toroidal data patches are seamlessly and randomly repeated during the synthesis step. This fully automatic method starts with minimal tile size detection, which is limited by the size of the principal field, the number of toroidal tiles we are looking for, and the sample spatial frequency content.

Fig. 3 Measured brick principal field (upper left), its optimal double toroidal patch (bottom left), and enlarged synthetic principal field (right, $K = 8$)

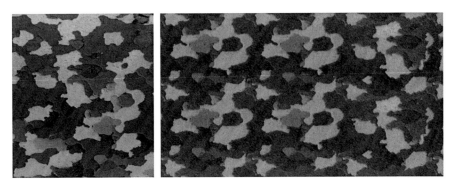

Fig. 4 Synthetic BTF-CMRF^{NProl3DCAR} enlarged color bark (right) estimated from their natural measurements (left)

Non-parametric Markov Random Field with Iterative Synthesis

The non-parametric principal random field $\check{X}$ is estimated using simple K-means clustering of $\tilde{Y}$ in the RGB color space into a predefined number of K classes, where cluster indices ω_i are $\check{X}_r$ $\forall r \in I$ estimates. The clustering resulting thematic map is used to compute region size histograms $\tilde{h}_i$ for all $i = 1, \ldots, K$ classes. Let us order classes according to the decreasing number of pixels $\tilde{n}_i$ belonging to each class, i.e., $\tilde{n}_1 \geq \tilde{n}_2 \geq \ldots \geq \tilde{n}_K$. Histograms $\tilde{h}_i$ are the only parameters required to store for the principal field.

Iterative Principal Field Synthesis

The iterative algorithm (Haindl and Havlíček 2018b) (Figs. 5 and 6) uses a data structure that describes membership in the region for each pixel. This data structure for each region additionally contains the class membership, size of the region and the requested number of regions of its size, all border pixels from both sides of the border, possibility to decrease or increase the region, and, for all classes, the histogram and regions, which can be increased or decreased. After any change in a pixel class assignment, this structure has to be updated.

0. The synthesized $M \times N$ required principal field is initialized to the largest class, and all histograms cells are rescaled using the scaling factor $\frac{MN}{\tilde{M}\tilde{N}}$, where $\tilde{M} \times \tilde{N}$ is the target (measured) texture size, i.e., $X_r^{(0)} = \omega_1$ $\forall r \in I$ and $\tilde{h}_i \to h_i$ for $i = 1, \ldots, K$. A lattice multiindex r is randomly generated starting from the second-largest class ω_2 till the smallest size class ω_K. Class index X_r is changed to new value $X_r = \omega_i$ only if its previous value was $X_r = \omega_1$ and the total number of principal field pixels with class indicator ω_i is smaller than its final value n_i. After this initialization step, all classes have their correct required number of pixels but not yet their correct region size histograms.

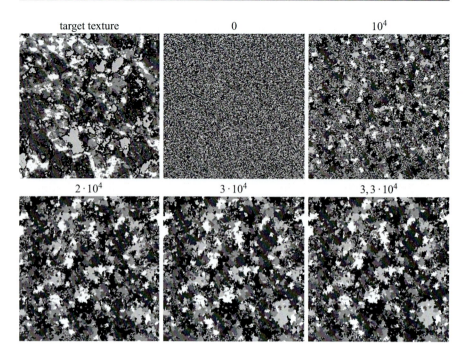

Fig. 5 The granite (Fig. 6) principal field synthesis. The target texture principal field, initialization, and selected iteration steps rightwards

Fig. 6 The granite measurement and its synthetic enlargement (BTF-CMRF^{NPi3AR})

1. Pixels r and s are randomly selected with the following properties: The pixel r from the class ω_i is on the border between region $\downarrow \omega_i^A$ (a region A which can be decreased) and region $\uparrow \omega_j^B$ (a region B which can be increased). The pixel s from the class ω_j is on the border between region $\downarrow \omega_j^C$ (a region C which can be decreased) and region $\uparrow \omega_i^D$ (a region D which can be increased). These regions have to be distinct, i.e., $A \cap D = \emptyset$ and $B \cap C = \emptyset$. If such pixels r, s exist, go to step 5. If not repeat this step once more.

2. Gradually check all class couples starting from $\omega_1, \omega_2, \ldots, \omega_K$ to find pixels r, s which meet conditions in step 1. All regions corresponding to the chosen classes, ω_i and ω_j, are selected randomly. If such pixels r, s exist, go to step 5.
3. Randomly select a region from class ω_i, which has two neighboring regions of class ω_j such as one can be decreased and another increased. If there exist two border pixels r, s in the region ω_j, where r is a border pixel with a region to be increased and s with a region to be decreased, go to step 5.
4. Gradually check all classes with incorrect histogram, starting from $\omega_1, \omega_2, \ldots, \omega_K$; for every class ω_i gradually check all its regions $\uparrow \omega_i^A$ which can be increased; for each region $\uparrow \omega_i^A$, check every region neighboring border pixel r from class ω_j and region $\downarrow \omega_j^B$ (a region B which can be decreased), and find pixel s with the following properties: pixel s is from the class ω_i and region $\downarrow \omega_i^C$ (a region C which can be decreased), and pixel s is on the boarder of the region $\uparrow \omega_j^D$ from class ω_j (a region which can be increased). These regions have to be distinct, i.e., $A \cap C = \emptyset$ and $B \cap D = \emptyset$. If such pixels do not exist, go to step 7.
5. $X_r = \omega_j, X_s = \omega_i$ update the data structure.
6. If the number of iterations is less than a selected limit, go to 1.
7. Store the resulting principal field and stop.

Steps 1 and 2 allow simultaneous improvement of four regions, while step 3 improves two regions only. The algorithm converges to the correct class histograms $h_i \; i = 1, \ldots, K$.

Non-parametric Markov Random Field with Fast Iterative Synthesis

The non-parametric principal field (Haindl and Havlíček 2018a) BTF-CMRF$^{NPfi\ldots}$ is estimated as in the previous section, and its synthesis is modified to be significantly faster at the cost of slightly compromised principal field variability. The fast algorithm compromise is its preference for convex regions instead of their general shapes but profits with faster convergency.

The median speed up between this method and the approach for the non-parametric principal field synthesis in section "Non-parametric Markov Random Field with Iterative Synthesis" is one-fifth of the required cycles to converge. Some textures (e.g., granite; Fig. 7) have sufficiently similar statistics of the synthesized regions with the principal target field already in the initialization step. Hence, the principal field synthesis even does not need any iterations. The lichen Fig. 8 principal target field (512×512) requires 29 137 iterations, while the previous iterative method needs nearly 5 times more (140 146) iterations to converge.

Iterative Principal Field Synthesis

The iterative algorithm is based on a similar data structure, which describes membership in the region for each pixel, as in the previous section. Both iterative algorithms differ only in their initialization steps.

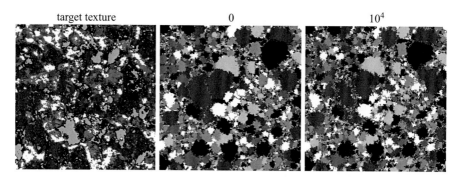

Fig. 7 The granite principal field synthesis. The target texture principal field, initialization, and a similar 10^4-th iteration step result

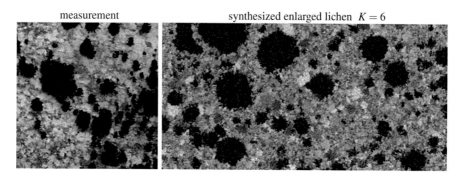

Fig. 8 The lichen measurement and its synthetic enlargement (BTF-CMRF^{NPfi3DCAR})

0. The synthesized $M \times N$ required principal field is initialized to the value ω_0 it means that pixel was not assigned to any class ω_i for $i = 1, \ldots, K$. All histogram cells are rescaled using the scaling factor $\frac{MN}{\tilde{M}\tilde{N}}$, i.e., $X_r^{(0)} = \omega_1 \; \forall r \in I$ and $\tilde{h}_i \to h_i$ for $i = 1, \ldots, K$. All regions from all classes $i = 1, \ldots, K$ are sorted by region size. Starting from the biggest region A_1 till the smallest region A_M, where M the is number of all regions, a lattice multiindex r is randomly generated. The first pixel X_r of the region A_j where $j = 1, \ldots, M$ and class ω_i is randomly selected and is changed to new value $X_r = \omega_i$ only if its previous value was $X_r = \omega_0$ All neighbors X_s of the pixel X_r which fulfil conditions $X_s = \omega_0$ and pixel X_s that has no neighbor from the class ω_i are added to the queue Q. Till the size of region A_j is higher than the number of actually added pixels, the next pixel X_r is randomly selected from the queue Q, the values are changed to $X_r = \omega_i$ and its neighbors are added to the queue Q if they meet the mentioned conditions. If the queue Q is empty and the size of the region A_j is higher than the number of actually assigned pixels, the rest of the pixels is randomly assigned to the class ω_i after the initialization of the last region A_M. After this initialization step,

all classes have their correct required number of pixels but not their correct region size histograms.

1.–7. Identical with the corresponding items in section "Iterative Principal Field Synthesis".

Steps 1 and 2 allow simultaneous improvement of four regions, while step 3 improves two regions only. The algorithm converges to the correct class histograms $h_i \; i = 1, \ldots, K$.

Potts Markov Random Field

The resulting thematic principal map $\check{X}$ BTF-CMRF$^{2P\cdots}$ is represented by the hierarchical two-scale Potts model (Haindl et al. 2012)

$$\check{X}^{(a)} = \frac{1}{Z^{(a)}} \exp\left\{-\beta^{(a)} \sum_{s \in I_r} \delta_{X_r^{(a)} X_s^{(a)}}\right\} \quad (5)$$

where Z is the appropriate normalizing constant and $\delta()$ is the Kronecker delta function. The rough-scale-upper- level Potts model ($a = 1$) regions are further elaborated with the detailed fine-scale-level ($a = 2$) Potts model which models the corresponding subregions in each upper-level region. The parameter $\beta^{(a)}$ for both level models is estimated using an iterative estimator which starts from the upper β limit ($\beta_{\max}$) and adjusts (decreases or increases) its value until the Potts model regions have similar parameters (average inscribed squared region size and/or the region's perimeter) with the target texture switching field. This iterative estimator gives more resembling results with the target texture than the alternative maximum pseudo-likelihood method (Levada et al. 2008). The corresponding Potts models are synthesized (Fig. 9 – middle) using the fast Swendsen-Wang sampling method (Swendsen and Wang 1987).

Fig. 9 The rusty plate texture measurement, its principal synthetic field, and the final synthetic CMRF^{P3AR} model texture

Potts-Voronoi Markov Random Field

The principal field (X) of the CMRF BTF-CMRF$^{PV\cdots}$ model (Haindl et al. 2015b) is a mosaic represented as a Voronoi diagram (Aurenhammer 1991), and the distribution of the particular colors (texture classes) of the mosaic is modeled as a Potts random field which is built on top of the adjacency graph (G) of the mosaic. Figure 10 illustrates this model applied to the floor mosaic, while Fig. 11 shows this model applied to a glass mosaic synthesis in St. Vitus Cathedral in Prague Castle. The algorithm requires input in the form of a segmented mosaic with distinguishable regions of the same texture type.

After that follows the identification of the mosaic field centers and the estimation of the parameters of the 2D discrete point process, which samples the control points of the newly synthesized Voronoi mosaic. This sampling is done using a 2D histogram, which has shown to be sufficient for the good quality estimate. The only other parameter is the number of points to be sampled, which grows linearly with the required area of the synthetic image in the case of texture enlargement applications.

With the control points for the Voronoi mosaic cells having been sampled, we compute the Voronoi diagram, and optionally mark the delimiting edges between adjacent cells. The assignment of a regional texture model to each mosaic cell (the principal MRF ($P(X \mid \tilde{Y})$)) is then mapped by the flexible K−state Potts random field (Potts and Domb 1952; Wu 1982).

Let us denote $G = (V, E)$ the adjacency graph of the mosaic areas and

$$N_u = \{\forall v \in V : (u, v) \in E\}, \ u \in V \quad (6)$$

the 1st-order neighborhood, where V, E are the vertex and edge sets. Vertexes correspond to the particular areas in the mosaic, and there is an edge between two vertexes if their corresponding areas are directly next to each other.

The resulting thematic principal map $\check{X}$ is represented by the Potts model for a general graph

Fig. 10 The floor mosaic measurement and its synthesis (BTF-CMRF^{PV3DCAR})

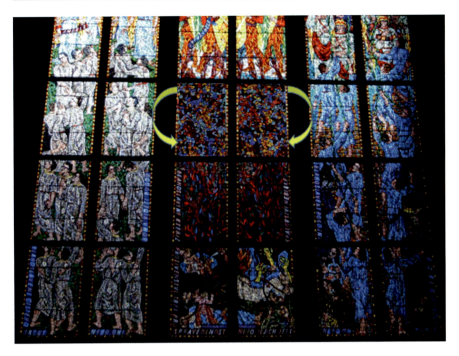

Fig. 11 An example of St. Vitus Cathedral in Prague Castle stained glass window with two original panels (yellow arrows) replaced with synthetic images (BTF-CMRF^{PV3DCAR})

$$p(\check{X}|\beta) = \frac{1}{Z} \exp\left\{-\beta \sum_{u \in V, v \in N_u} \delta(X_u, X_v)\right\} \quad (7)$$

where Z is the appropriate normalizing constant and $\delta()$ is the Kronecker delta function. The parameter β is estimated from the K-means clustered input mosaic using the maximum pseudo-likelihood method described by Levada et al. (2008). The local density of the Potts field can be expressed as

$$p(X_u = q | X_{v \in N_u}, \beta) = \frac{\exp\left\{\beta \sum_{s \in N_u} \delta(q, X_v)\right\}}{\sum_{k=1}^{K} \exp\left\{\beta \sum_{v \in N_u} \delta(k, X_v)\right\}} \quad (8)$$

for which the pseudo-likelihood approximation is

$$PL(\beta) = \prod_{u \in V} p(X_u = q | X_{v \in N_u}, \beta). \quad (9)$$

Calculating the logarithm, differentiating, and setting the result equal to 0, we get the maximum pseudo-likelihood equation (10) for the β estimate:

$$\Psi(\beta) = -\sum_{u \in V} \frac{\sum_{k=1}^{K} \left(\sum_{v \in N_u} \delta(X_u, X_v) \right) \exp \left\{ \beta \sum_{v \in N_u} \delta(k, X_v) \right\}}{\sum_{k=1}^{K} \exp \left\{ \beta \sum_{v \in N_u} \delta(k, X_v) \right\}}$$

$$+ \sum_{u \in V} \sum_{v \in N_u} \delta(X_u, X_v) = 0. \tag{10}$$

The corresponding Potts models are synthesized using the fast Swendsen-Wang sampling method (Swendsen and Wang 1987), although for smaller fields, which the mosaics undoubtedly are, other sampling MCMC methods such as the Gibbs sampler (Geman and Geman 1984) can be used. Alternatively, the Metropolis algorithm (Metropolis et al. 1953) should also work sufficiently fast enough.

Bernoulli Distribution Mixture Model

The distribution $P(X_{\{r\}})$ is assumed to be multivariable Bernoulli mixture (BM) (Haindl and Havlíček 2017b). The mixture distribution $P(X_{\{r\}})$ has the form

$$P(X_{\{r\}}) = \sum_{m \in \mathcal{M}} P(X_{\{r\}} | m) \, p(m) = \sum_{m \in \mathcal{M}} \prod_{s \in I_r} p_s(Y_s | m) \, p(m), \tag{11}$$

where $\mathcal{M}$ is set of all mixture components, m a mixture component index, $\{r\}$ is a set of indices from I_r, and the principal field BTF-CMRF$^{BM\cdots}$ is further decomposed into separate binary bit planes of binary variables $\xi \in \mathcal{B}$, $\mathcal{B} = \{0, 1\}$ which are separately modeled and can be learned from much smaller training texture than a multi-level discrete mixture model (see examples in Fig. 14). We suppose that a bit factor of a principal field can be fully characterized by a marginal probability distribution of binary levels on pixels within the scope of a window centered around the location r and specified by the index set $I_r \subset I$, i.e., $X_{\{r\}} \in \mathcal{B}^\eta$ and $P(X_{\{r\}})$ is the corresponding marginal distribution of $P(X | \tilde{Y})$. The component distributions $P(\cdot | m)$ are factorizable, and multivariable Bernoulli

$$P(X_{\{r\}} | m) = \prod_{s \in I_r} \dot{\theta}_{m,s}^{X_s} (1 - \dot{\theta}_{m,s})^{1-X_s} \qquad X_s \in X_{\{r\}}. \tag{12}$$

The mixture model parameters (11), (12) include component weights $p(m)$ and the univariate discrete distributions of binary levels. They are defined by one parameter $\dot{\theta}_{m,s}$ as a vector of probabilities:

$$p_s(\cdot | m) = (\dot{\theta}_{m,s}, 1 - \dot{\theta}_{m,s}). \tag{13}$$

The EM solution is (14), (15):

$$q^{(t)}(m \mid X_{\{r\}}) = \frac{p^{(t)}(m)\, P^{(t)}(X_{\{r\}} \mid m)}{\sum_{j \in \mathcal{M}} p^{(t)}(j) P^{(t)}(X_{\{r\}} \mid j)}, \qquad (14)$$

$$p^{(t+1)}(m) = \frac{1}{|\mathcal{S}|} \sum_{X_{\{r\}} \in \mathcal{S}} q^{(t)}(m \mid X_{\{r\}}), \qquad (15)$$

and

$$p_s^{(t+1)}(\xi \mid m) = \frac{1}{|\mathcal{S}|\, p^{(t+1)}(m)} \sum_{X_{\{r\}} \in \mathcal{S}} \delta(\xi, X_s)\, q^{(t)}(m \mid X_{\{r\}}), \quad \xi \in \mathcal{B}. \qquad (16)$$

The total number of mixture (11), (13) parameters is thus $\dot{M}(1+\eta)$ $\dot{M} \in \mathcal{M}$ – confined to the appropriate norming conditions. The advantage of the multivariable Bernoulli model (13) is a simple switchover to any marginal distribution by deleting superfluous terms in the products $P(X_{\{r\}} \mid m)$.

Gaussian Mixture Model

The discrete principal field can be alternatively modeled (Haindl and Havlíček 2017b) by a continuous RF BTF-CMRF$^{GM\cdots}$ if we map single indices into continuous random variables with uniformly separated mean values and small variance. The synthesis results are subsequently inversely mapped back into a corresponding synthetic discrete principal field. We assume the joint probability distribution $P(X_{\{r\}})$, $X_{\{r\}} \in \mathcal{K}^\eta$ in the form of a normal mixture, and the mixture components are defined as products of univariate Gaussian densities

$$P(X_{\{r\}} \mid \mu_m, \sigma_m) = \prod_{s \in I_r} p_s(X_s \mid \mu_{ms}, \sigma_{ms}), \qquad (17)$$

$$p_s(X_s \mid \mu_{ms}, \sigma_{ms}) = \frac{1}{\sqrt{2\pi}\, \sigma_{ms}} \exp\left\{ -\frac{(X_s - \mu_{ms})^2}{2\sigma_{ms}^2} \right\},$$

i.e., the components are multivariate Gaussian densities with diagonal covariance matrices. The maximum-likelihood estimates of the parameters $p(m)$, μ_{ms}, σ_{ms} can be computed by the expectation-maximization (EM) algorithm (Dempster et al. 1977; Grim and Haindl 2003). Anew we use a data set $\mathcal{S}$ obtained by pixel-wise shifting the observation window within the original texture image $\mathcal{S} = \{X_{\{r\}}^{(1)}, \ldots, X_{\{r\}}^{(K)}\}$, $X_{\{r\}}^{(k)} \subset X$. The corresponding log-likelihood function is maximized by the EM algorithm ($m \in \mathcal{M}, n \in \mathcal{N}, X_{\{r\}} \in \mathcal{S}$), and the iterations are (14), (15) and

$$\mu_{m,n}^{(t+1)} = \frac{1}{\sum_{X_{\{r\}} \in \mathscr{S}} q^{(t)}(m \mid X_{\{r\}})} \sum_{X_{\{r\}} \in \mathscr{S}} X_n \, q(m \mid X_{\{r\}}), \tag{18}$$

$$(\sigma_{m,n}^{(t+1)})^2 = -(\mu_{m,n}^{(t+1)})^2 + \frac{\sum_{X_{\{r\}} \in \mathscr{S}} X_n^2 \, q^{(t)}(m \mid X_{\{r\}})}{\sum_{X_{\{r\}} \in \mathscr{S}} q(m \mid X_{\{r\}})}. \tag{19}$$

Local Markov and Mixture Models

While the principal models control the overall large-scale low-frequency textural structure, the local models synthesize the detail, regional and fine-granularity spatial-spectral BTF information. Once we have synthesized the required size's principal random field, using some of the previously described models, we use it to synthesize the local random part (3) of the BTF compound random model Y. This local model is a mosaic of K random field sub-models. These sub-models are assumed to be of the same type, but they differ in parameters and contextual support sets. This assumption is for simplicity only and is not restrictive because every sub-model is estimated and synthesized independently; thus, the Y mosaic can be easily composed of different types of random field models.

Local i-th texture region (not necessarily continuous) models are view and illumination dependent; thus, they need to be ideally represented by models which can be analytically estimated as well as easily non-iteratively synthesized (BTF-CMRF^{NProl3DCAR} (Haindl and Havlíček 2010), BTF-CMRF^{2P3DCAR} (Haindl et al. 2012), BTF-CMRF^{PV3DCAR} (Haindl et al. 2015b), BTF-CMRF^{c3DGM} (Haindl and Havlíček 2016), BTF-CMRF^{BM3DCAR} (Haindl and Havlíček 2017b), BTF-CMRF^{GM3DCAR}, BTF-CMRF^{NProl3DMA} (Haindl and Havlíček 2017a), BTF-CMRF^{NPi3DCAR} (Haindl and Havlíček 2018b), BTF-CMRF^{NPfi3DCAR} (Haindl and Havlíček 2018a)).

3D Causal Simultaneous Autoregressive Model

The 3D causal simultaneous autoregressive model (3DCAR) is an exceptional model because all its statistics can be solved analytically, and it can be utilized to build much more complex nD data models. For example, the 7D BTF models illustrated in Fig. 4 are composed from up to one hundred 3DCARs.

A digitized image Y is assumed to be defined on a finite rectangular $N \times M \times d$ lattice I, and $r = (r_1, r_2, r_3) \in I$ denotes a pixel multiindex with the row, columns, and spectral indices, respectively. The notation $I_r^c \subset I$ is a causal or unilateral neighborhood of pixel r, i.e.,

$$I_r^c \subset I_r^C = \{s : 1 \leq s_1 \leq r_1, 1 \leq s_2 \leq r_2, s \neq r\}.$$

The 3D causal simultaneous autoregressive model (3DCAR) is the wide-sense Markov model that can be written in the following regression equation form:

$$\tilde{Y}_r = \sum_{s \in I_r^c} A_s \tilde{Y}_{r-s} + e_r \qquad \forall r \in I \qquad (20)$$

where A_s are matrices (21) and the zero mean white Gaussian noise vector e_r has uncorrelated components with data indexed from I_r^c but noise vector components can be mutually correlated with a constant covariance matrix Σ.

$$A_{s_1,s_2} = \begin{pmatrix} a_{1,1}^{s_1,s_2}, \ldots, a_{1,d}^{s_1,s_2} \\ \vdots, \ddots, \vdots \\ a_{d,1}^{s_1,s_2}, \ldots, a_{d,d}^{s_1,s_2} \end{pmatrix} \qquad (21)$$

where $d \times d$ are parameter matrices. The model can be expressed in the matrix form

$$Y_r = \gamma Z_r + e_r, \qquad (22)$$

where

$$Z_r = [\tilde{Y}_{r-s}^T : \forall s \in I_r^c], \qquad (23)$$

Z_r is a $d\eta \times 1$ vector, $\eta = card(I_r^c)$ and γ

$$\gamma = [A_1, \ldots, A_\eta] \qquad (24)$$

is a $d \times d\eta$ parameter matrix. To simplify notation the multiindexes $r, s, \ldots$ have only two components further on in this section.

An optimal support can be selected as the most probable model given past data

$$Y^{(r-1)} = \{Y_{r-1}, Y_{r-2}, \ldots, Y_1, Z_r, Z_{r-1}, \ldots, Z_1\},$$

i.e., $\max_j \{p(\mathcal{M}_j \mid Y^{(r-1)})\}$. Simultaneous conditional density can be evaluated analytically from

$$p(Y^{(r-1)} \mid \mathcal{M}_j) = \int \int p(Y^{(r-1)} \mid \gamma, \Sigma^{-1}) p(\gamma, \Sigma^{-1} \mid \mathcal{M}_j) d\gamma d\Sigma^{-1} \qquad (25)$$

, and for the implemented uniform priors start, we get a decision rule (Haindl and Šimberová 1992):

The most probable AR model given past data $Y^{(r-1)}$, the normal-Wishart parameter prior and the uniform model prior is the model $\mathcal{M}_i$ (Haindl 1983) for which

$$i = \arg\max_j \{D_j\}$$

$$D_j = -\frac{d}{2}\ln|V_{x(r-1)}| - \frac{\beta(r) - d\eta + d + 1}{2}\ln|\lambda_{(r-1)}| + \frac{d^2\eta}{2}\ln\pi \qquad (26)$$

$$+ \sum_{i=1}^{d}\left[\ln\Gamma\left(\frac{\beta(r) - d\eta + d + 2 - i}{2}\right) - \ln\Gamma\left(\frac{\beta(0) - d\eta + d + 2 - i}{2}\right)\right]$$

where $V_{z(r-1)} = \tilde{V}_{z(r-1)} + V_{z(0)}$ with $\tilde{V}_{z(r-1)}$ defined in (31), $V_{z(0)}$ is an appropriate part of V_0 (31), $\beta(r)$ is defined in (27), (28) and $\lambda_{(r-1)}$ is (29).

The statistics (26) uses the following notation (27), (28), (29), (30) and (31):

$$\beta(r) = \beta(0) + r - 1 = \beta(r-1) + 1, \qquad (27)$$

$$\beta(0) > \eta - 2, \qquad (28)$$

and

$$\lambda_{(r)} = V_{y(r)} - V_{zy(r)}^T V_{z(r)}^{-1} V_{zy(r)}. \qquad (29)$$

$$V_{r-1} = \tilde{V}_{r-1} + V_0, \qquad (30)$$

$$\tilde{V}_{r-1} = \begin{pmatrix} \sum_{k=1}^{r-1}\tilde{Y}_k\tilde{Y}_k^T & \sum_{k=1}^{r-1}\tilde{Y}_k\tilde{Z}_u^T \\ \sum_{k=1}^{r-1}\tilde{Z}_k\tilde{Y}_k^T & \sum_{k=1}^{r-1}\tilde{Z}_k\tilde{Z}_k^T \end{pmatrix} = \begin{pmatrix} \tilde{V}_{y(r-1)} & \tilde{V}_{zy(r-1)}^T \\ \tilde{V}_{zy(r-1)} & \tilde{V}_{z(r-1)} \end{pmatrix}. \qquad (31)$$

Marginal densities $p(\gamma \mid Y^{(r-1)})$ and $p(\Sigma^{-1} \mid Y^{(r-1)})$ can be evaluated from (32), (33), respectively.

$$p(\gamma \mid Y^{(r-1)}) = \int p(\gamma, \Sigma^{-1} \mid Y^{(r-1)}) d\Sigma^{-1} \qquad (32)$$

$$p(\Sigma^{-1} \mid Y^{(r-1)}) = \int p(\gamma, \Sigma^{-1} \mid Y^{(r-1)}) d\gamma \qquad (33)$$

The marginal density $p(\Sigma^{-1} \mid Y^{(r-1)})$ is the Wishart distribution density (Haindl 1983)

$$p(\Sigma^{-1} \mid Y^{(r-1)}) = \frac{\pi^{\frac{d(1-d)}{4}} |\Sigma^{-1}|^{\frac{\beta(r) - d\eta}{2}}}{2^{\frac{d(\beta(r) - d\eta + d + 1)}{2}} \prod_{i=1}^{d}\Gamma(\frac{\beta(r) - d\eta + 2 + d - i}{2})} |\lambda_{(r-1)}|^{\frac{\beta(r) - d\eta + d + 1}{2}}$$

$$\exp\left\{-\frac{1}{2}tr\{\Sigma^{-1}\lambda_{(r-1)}\}\right\} \qquad (34)$$

with

$$E\left\{\Sigma^{-1} \mid Y^{(r-1)}\right\} = (\beta(r) - d\eta + d + 1)\lambda_{(r-1)}^{-1} \qquad (35)$$

$$E\left\{(\Sigma^{-1} - E\{\Sigma^{-1} \mid Y^{(r-1)}\})^T (\Sigma^{-1} - E\{\Sigma^{-1} \mid Y^{(r-1)}\}) \mid Y^{(r-1)}\right\} =$$

$$\frac{2(\beta(r) - d\eta + 1)}{\lambda_{(r-1)}\lambda_{(r-1)}^T}. \qquad (36)$$

The marginal density $p(\gamma \mid Y^{(r-1)})$ is matrix t distribution density (Haindl 1983)

$$p(\gamma \mid Y^{(r-1)}) = \frac{\prod_{i=1}^d \Gamma(\frac{\beta(r)+d+2-i}{2})}{\prod_{i=1}^d \Gamma(\frac{\beta(r)-d\eta+d+2-i}{2})} \pi^{-\frac{d^2\eta}{2}} |\lambda_{(r-1)}|^{-\frac{d\eta}{2}} |V_{x(r-1)}|^{\frac{d}{2}}$$

$$\left| I + \lambda_{(r-1)}^{-1} (\gamma - \hat{\gamma}_{r-1}) V_{z(r-1)} (\gamma - \hat{\gamma}_{r-1})^T \right|^{-\frac{\beta(r)+d+1}{2}} \qquad (37)$$

with the mean value

$$E\left\{\gamma \mid Y^{(r-1)}\right\} = \hat{\gamma}_{r-1} \qquad (38)$$

and covariance matrix

$$E\left\{(\gamma - \hat{\gamma}_{r-1})^T (\gamma - \hat{\gamma}_{r-1}) \mid Y^{(r-1)}\right\} = \frac{V_{z(r-1)}^{-1} \lambda_{(r-1)}}{\beta(r) - d\eta}. \qquad (39)$$

Similar statistics can be easily derived (Haindl 1983) for the alternative Jeffreys non-informative parameter prior. Similar to other model statistics, also the predictive density can be analytically derived.

The one-step-ahead predictive posterior density for the normal-Wishart parameter prior has the form of d-dimensional Student's probability density (40) (Haindl 1983)

$$p(Y_r \mid Y^{(r-1)}) = \frac{\Gamma(\frac{\beta(r)-d\eta+d+2}{2})}{\Gamma(\frac{\beta(r)-d\eta+2}{2}) \pi^{\frac{d}{2}} (1 + Z_r^T V_{z(r-1)}^{-1} Z_r)^{\frac{d}{2}} |\lambda_{(r-1)}|^{\frac{1}{2}}}$$

$$\left(1 + \frac{(Y_r - \hat{\gamma}_{r-1} Z_r)^T \lambda_{(r-1)}^{-1} (Y_r - \hat{\gamma}_{r-1} Z_r)}{1 + Z_r^T V_{z(r-1)}^{-1} Z_r}\right)^{-\frac{\beta(r)-d\eta+d+2}{2}}, \qquad (40)$$

with $\beta(r) - d\eta + 2$ degrees of freedom; if $\beta(r) > d\eta$ then the conditional mean value is

$$E\left\{Y_r \mid Y^{(r-1)}\right\} = \hat{\gamma}_{r-1} Z_r, \qquad (41)$$

and

$$E\left\{(Y_r - \hat{\gamma}_{r-1} Z_r)(Y_r - \hat{\gamma}_{r-1} Z_r)^T \mid Y^{(r-1)}\right\} = \frac{1 + Z_r V_{z(r-1)}^{-1} Z_r^T}{(\beta(r) - d\eta)} \lambda_{(r-1)}. \qquad (42)$$

The 3DCAR model can be made adaptive if we modify its recursive statistics using an exponential forgetting factor, i.e., a constant $\varphi \approx 0.99$. This forgetting factor smaller than 1 is used to weigh the influence of older data. The numerical stability of 3DCAR can be guaranteed if all its recursive statistics use the square root factor updating applying either the Cholesky or LDL^T decomposition (Haindl 2000), respectively.

The 3DCAR (analogously also the 2DCAR model) model has advantages in analytical solutions (Bayes, ML, or LS estimates) for I_r, $\hat{\gamma}$, $\hat{\sigma}^2$, $\hat{Y}_r$ statistics. It allows straightforward, fast synthesis, adaptivity, and building efficient recursive application algorithms.

3D Moving Average Model

Single multispectral texture factors Y are modeled using the extended version ($3D\,MA$) of the moving average model (Li et al. 1992; Haindl and Havlíček 2017a). A stochastic multispectral texture can be considered to be a sample from a 3D random field defined on an infinite 2D lattice. The model assumes that each factor is the output of an underlying system, which completely characterizes it in response to a 3D uncorrelated random input. This system can be represented by the impulse response of a linear 3D filter. The intensity values of the most significant pixels, together with their neighbors, are collected and averaged. The resultant 3D kernel is used as an estimate of the impulse response of the underlying system. A synthetic mono-spectral factor can be generated by convolving an uncorrelated 3D random field with this estimate. Suppose a stochastic multispectral texture denoted by Y is the response of an underlying linear system that completely characterizes the texture in response to a 3D uncorrelated random input E_r; then, Y_r is determined by the difference equation

$$Y_r = \sum_{s \in I_r} B_s E_{r-s} \qquad (43)$$

where B_s are constant matrix coefficients and $I_r \subset I$.

Hence, Y_r can be represented as $Y_r = h(r) * E_r$ where the convolution filter $h(r)$ contains all parameters B_s. In this equation, the underlying system behaves as a 3D filter, where we restrict the system impulse response to have significant values only within a finite region. The geometry of I_r determines the causality or non-causality of the model.

The parameter estimation can be based on the modified random decrement technique (RDT) (Cole Jr 1973; Asmussen 1997). RDT assumes that the input is an uncorrelated random field. If every pixel component is higher than its corresponding threshold vector component and simultaneously at least one of its four neighbors is less than this threshold, the pixel is saved in the data accumulator. The procedure begins by selecting thresholds usually chosen as some percentage of the standard deviation of each spectral plane's intensities separately. In addition to that, a 3D MA model also requires to estimate the noise spectral correlation, i.e.,

$$E\{E_r E_s\} = 0 \qquad \forall r_1 \neq s_1 \vee r_2 \neq s_2,$$
$$E\{E_{r_1,r_2,r_3} E_{r_1,r_2,\bar{r}_3}\} \neq 0 \qquad \forall r_3 \neq \bar{r}_3.$$

The synthetic factor can be generated simply by convolving an uncorrelated 3D RF E with the estimate of B according to (43). All generated factors form a new Gaussian pyramid. Fine resolution synthetic smooth texture is obtained by the collapse of the pyramid, i.e., an inverse procedure of that one creating the pyramid. This model can be used for materials which consist of several types of relatively small regions with fine-granular inner structure such as sand, grit, cork, lichen, or plaster. Figure 12 illustrates the visual quality of this simple model if the regional textures violate this fine-granularity assumption.

Spatial 3D Gaussian Mixture Model

A static homogeneous three-dimensional textural factor Y is assumed to be defined on a finite rectangular $M \times N \times d$ lattice I, $r = (r_1, r_2) \in I$ denotes a pixel multiindex with the row, columns, and indices, respectively. Let us suppose that Y represents a realization of a random vector with a probability distribution $P(Y)$. The statistical properties of interior pixels of the moving window on Y are translation invariant due to assumed textural homogeneity. They can be represented by a joint probability distribution, and the properties of the texture can be fully characterized

Fig. 12 The stone measurement and its synthesis (BTF-CMRF^{NP3DMA})

by statistical dependencies on a sub-field, i.e., by a marginal probability distribution of spectral levels on pixels within the scope of a window centered around the location r and specified by the index set:

$$I_r = \{r + s : |r_1 - s_1| \leq \alpha \wedge |r_2 - s_2| \leq \beta\} \subset I. \tag{44}$$

The index set I_r depends on modeled visual data and can have any other than this rectangular shape. $Y_{\{r\}}$ denotes the corresponding matrix containing all $d \times 1$ vectors Y_s in some fixed order arrangement such that $s \in I_r$, $Y_{\{r\}} = [Y_s \ \forall s \in I_r]$, $Y_{\{r\}} \subset Y$, $\eta = $ cardinality$\{I_r\}$, and $P(Y_{\{r\}})$ is the corresponding marginal distribution of $P(Y)$.

If we assume the joint probability distribution $P(Y_{\{r\}})$, in the form of a normal mixture (Haindl and Havlíček 2016)

$$P(Y_{\{r\}}) = \sum_{m \in \mathcal{M}} p(m) \, P(Y_{\{r\}} \mid \mu_m, \Sigma_m) \qquad Y_{\{r\}} \subset Y,$$

$$= \sum_{m \in \mathcal{M}} p(m) \prod_{s \in I_r} p_s(Y_s \mid \mu_{m,s}, \Sigma_{m,s}) \tag{45}$$

where $Y_{\{r\}} \in \mathfrak{R}^{d \times \eta}$ is $d \times \eta$ matrix, μ_m is $d \times \eta$ mean matrix, Σ_m is $d \times d \times \eta$ a covariance tensor, and $p(m)$ are probability weights and the mixture components are defined as products of multivariate Gaussian densities

$$P(Y_{\{r\}} \mid \mu_m, \Sigma_m) = \prod_{s \in I_{\{r\}}} p_s(Y_s \mid \mu_{ms}, \Sigma_{ms}), \tag{46}$$

$$p_s(Y_s \mid \mu_{ms}, \Sigma_{ms}) = \frac{1}{(2\pi)^{\frac{d}{2}} |\Sigma_{m,s}|^{\frac{1}{2}}} \exp\left\{-\frac{1}{2}(Y_r - \mu_{m,s})^T \Sigma_{m,s}^{-1}(Y_r - \mu_{m,s})\right\}, \tag{47}$$

i.e., the components are multivariate Gaussian densities with covariance matrices (53).

The underlying structural model of conditional independence is estimated from a data set $\mathscr{S}$ obtained by the step-wise shifting of the contextual window I_r within the original textural image, i.e., for each location r one realization of $Y_{\{r\}}$.

$$\mathscr{S} = \{Y_{\{r\}} \ \forall r \in I, \ I_r \subset I\} \quad Y_{\{r\}} \in \mathfrak{R}^{d \times \eta}. \tag{48}$$

Parameter Estimation

The unknown parameters of the approximating mixture can be estimated using the iterative EM algorithm (Dempster et al. 1977). In order to estimate the unknown distributions $p_s(\cdot \mid m)$ and the component weights $p(m)$ we maximize the likelihood function (49) corresponding to the training set (48):

$$L = \frac{1}{|\mathscr{S}|} \sum_{Y_{\{r\}} \in \mathscr{S}} \log \left[\sum_{m \in \mathscr{M}} P(Y_{\{r\}} \mid \mu_m, \Sigma_m) \, p(m) \right]. \quad (49)$$

The likelihood is maximized using the iterative EM algorithm (with non-diagonal covariance matrices):

E:

$$q^{(t)}(m \mid Y_{\{r\}}) = \frac{\tilde{P}^{(t)}(Y_{\{r\}} \mid \mu_m, \Sigma_m) \, p^{(t)}(m)}{\sum_{j \in \mathscr{M}} P^{(t)}(Y_{\{r\}} \mid \mu_j, \Sigma_j) \, p^{(t)}(j)}, \quad (50)$$

M:

$$p^{(t+1)}(m) = \frac{1}{|\mathscr{S}|} \sum_{Y_{\{r\}} \in \mathscr{S}} q^{(t)}(m \mid Y_{\{r\}}), \quad (51)$$

$$\mu_{m,s}^{(t+1)} = \frac{1}{\sum_{Y_{\{r\}} \in \mathscr{S}} q^{(t)}(m \mid Y_{\{r\}})} \sum_{Y_{\{r\}} \in \mathscr{S}} Y_s q^{(t)}(m \mid Y_{\{r\}}). \quad (52)$$

The covariance matrices are

$$\Sigma_{m,s}^{(t+1)} = \frac{\sum_{Y_{\{r\}} \in \mathscr{S}, Y_s \in Y_{\{r\}}} q^{(t)}(m \mid Y_{\{r\}})}{\sum_{Y_r \in \mathscr{S}} q^{(t)}(m \mid Y_{\{r\}})} (Y_s - \mu_{m,s}^{(t+1)})(Y_s - \mu_{m,s}^{(t+1)})^T \quad (53)$$

$$= \frac{\sum_{Y_{\{r\}} \in \mathscr{S}, Y_s \in Y_{\{r\}}} q^{(t)}(m \mid Y_{\{r\}}) \, Y_s Y_s^T}{\sum_{Y_r \in \mathscr{S}} q^{(t)}(m \mid Y_{\{r\}})} - \frac{p^{(t+1)}(m) \, |\mathscr{S}| \, \mu_{m,s}^{(t+1)} \left(\mu_{m,s}^{(t+1)} \right)^T}{\sum_{Y_r \in \mathscr{S}} q^{(t)}(m \mid Y_{\{r\}})}.$$

The iteration process stops when the criterion increments are sufficiently small. The EM algorithm iteration scheme has the monotonic property $L^{(t+1)} \geq L^{(t)}$, $t = 0, 1, 2, \ldots$ which implies the convergence of the sequence $\{L^{(t)}\}_0^\infty$ to a stationary point of the EM algorithm (local maximum or a saddle point of L). Figure 13 illustrates the usefulness of the BTF-CMRF^{3DGM} model for textile material modeling, while Fig. 18 shows this model applied to scratch restoration.

Applications

Numerous modeling applications can exploit the BTF models. The synthesis is beneficial not only for physically correct appearance modeling of surface materials under realistic and variable observation conditions (Figs. 15 and 17, upper row)

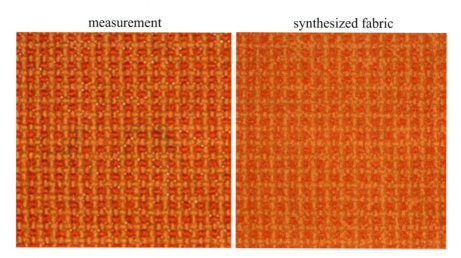

Fig. 13 The fabric measurement and its synthesis (BTF-CMRF^{3DGM})

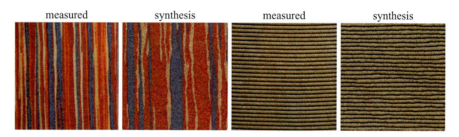

Fig. 14 Measured original cloth and corduroy materials and their synthesis using the $CRF^{BM-3CAR}$ model

but also for texture editing (Fig. 16), texture compression, or texture inpainting and restoration (Fig. 18). Various state-of-the-art unsupervised, semi-supervised, or supervised visual scene classification and understanding under variable observation conditions is the primary application for BTF analysis.

Texture Synthesis and Enlargement

Texture synthesis methods may be divided primarily into intelligent sampling and model-based methods (Fig. 14). They differ in need to store (sampling) or not (modeling) some actual texture measurements for new texture synthesis. Thus, even some methods which view texture as a stochastic process (Heeger and Bergen 1995; Efros and Leung 1999) still require to store an input exemplar. Sampling approaches De Bonet (1997), Efros and Leung (1999), Efros and Freeman (2001), Heeger and Bergen (1995), Xu et al. (2000), Dong and Chantler (2002), and Zelinka and Garland

(2002) rely on sophisticated sampling from real texture measurements, while the model-based techniques (Kashyap 1981; Haindl 1991; Haindl and Havlíček 1998, 2000; Bennett and Khotanzad 1998, 1999; Gimelfarb 1999; Paget and Longstaff 1998; Zhu et al. 2000) describe texture data using multidimensional mathematical models, and their synthesis is based on the estimated model parameters only. The mathematical model-based synthesis has an advantage in the possibility of seamless texture enlargement to any size (e.g., Fig. 6). The enlargement of a restricted texture measurement is always required in any application but cannot be achieved with sampling approaches without visible seams or repetitions.

The BTF modeling's ultimate aim is to create a visual impression of the same material without a pixel-wise correspondence to the finding condition model of the original measurements. Figure 15 shows the finding condition model of the beautiful gothic style relief (around 1370) of the Christ in Gethsemane (Prague) in the right and restored condition to a possible original appearance in the left.

The cornerstone of our BTF compression and modeling methods is the replacement of a vast number of original BTF measurements by their efficient parametric estimates derived from an underlying set of spatial probabilistic models and thus to allow a huge BTF compression ratio unattainable by any alternative sampling-based BTF synthesis method. Simultaneously these models can be used to reconstruct missing parts of the BTF measurement space or the controlled BTF space editing (Haindl and Havlíček 2009, 2012; Haindl et al. 2015b) by changing some of the model's parameters.

Textures without significant low frequencies such as Fig. 14-corduroy or Fig. 13-fabric can be modeled using simple local models only, either Markovian or mixtures such as 3DCAR, 3DMA, 3DBM, 3DGM, etc. Textures with substantial low frequencies (Figs. 4, 9, 14-cloth) will benefit from a compound version of the BTF model. Non-BTF textures can approximate low frequencies using a multiscale version of these models, e.g., pyramidal model (Haindl and Filip 2013).

Fig. 15 3D model of the beautiful gothic style relief of the Christ in Gethsemane, Prague (finding condition model right, restored condition to a possible original appearance left) mapped with the BTF synthetic plaener using the $CMRF^{3CAR}$ model

Fig. 16 Synthetic BTF-CMRF^{NProl3DCAR} edited and enlarged maple bark texture (second and fourth rows) with single sub-models estimated from their natural measurements (maple bark first and flowers third row)

The 3DCAR model is synthesized directly from its predictor (41) and Gaussian noise generator (22), (39). The advantage of a mixture model is its simple synthesis based on the marginals:

$$p_{n|\rho}(Y_n \mid Y_{\{\rho\}}) = \sum_{m=1}^{\dot{M}} W_m(Y_{\{\rho\}}) \, p_n(Y_n \mid m), \tag{54}$$

where $W_m(Y_{\{\{\rho\}\}})$ are the a posteriori component weights corresponding to the given sub-matrix $Y_{\{\rho\}} \subset Y_{\{r\}}$:

$$W_m(Y_{\{\rho\}}) = \frac{p(m) P_\rho(Y_{\{\rho\}} \mid m)}{\sum_{j=1}^{\dot{M}} p(j) P_\rho(Y_{\{\rho\}} \mid j)}, \tag{55}$$

$$P_\rho(Y_{\{\rho\}} \mid m) = \prod_{n \in \rho} p_n(Y_n \mid m). \tag{56}$$

There are several alternatives for the 3DGM model synthesis (Haindl et al. 2011) (Fig. 13). The unknown multivariate vector-levels Y_n can be synthesized by random sampling from the conditional density (54), or the mixture RF can be approximated using the GM mixture prediction.

Texture Compression

BTF – the best current measurable representation of a material appearance – requires tens of thousands of images using a sophisticated high-precision automatic measuring device. Such measurements result in a massive amount of data that can easily reach tens of terabytes for a single measured material. Nevertheless, these data have still insufficient spatial extent for any real virtual reality applications and have to be further enlarged using advanced modeling techniques. The resulting BTF size excludes its direct rendering in graphical applications, and compression of these huge BTF data spaces is inevitable. The usual car interior model requires more than 20 of such demanding BTF material measurements, and a similar problem holds for other applications of the physically correct appearance modeling such as computer games or film animations. A related problem is measurement data storage because storage technology is still the weak link, lagging behind recent developments in data sensing technologies. The apparent solution is mathematical modeling which allows replacing massive measured data with few thousand parameters and thus to reach tremendous unbeatable appearance data compression apart from unlimited seamless material texture enlargement. For example, the compression ratio relative to our BTF measurements is up to 1 : 1000000.

Texture Editing

Material-appearance editing is a practical approach with vast potential for significant speedup and cost reduction in industrial virtual prototyping or various design applications. An editing process can simulate materials for which no direct measurements are available or not existing in Nature (Fig. 16). Another example of the edited texture is two panels with the artificial but fitting glass mosaic synthesis in St. Vitus Cathedral in Prague Castle stained glass window on Fig. 11. Such edited artifacts allow an artist to test several possible design alternatives or model defunct monuments.

Illumination Invariants

Textures are essential clues to specify objects present in a visual scene. However, the appearance of natural textures is highly illumination and view angle-dependent. As a consequence, the most recent realistic texture-based classification or segmentation methods require multiple training images (Varma and Zisserman 2005) captured under all possible illumination and viewing conditions for each class. Such learning is clumsy, probably expensive, and very often even impossible if required measurements are not available.

If we assume fixed positions of viewpoint and illumination sources, uniform illumination sources, and Lambertian surface reflectance, then two images $\tilde{Y}, Y$ acquired with different illumination spectra can be linearly transformed to each other:

$$\tilde{Y}_r = B Y_r \quad \forall r. \tag{57}$$

It is possible to show (Vacha and Haindl 2007) that assuming (57) the following 3DCAR model-derived features are illumination invariant:

1. trace: $trace\ A_m,\ m = 1, \ldots, \eta K$
2. eigenvalues: $v_{m,j}$ of $A_m,\ m = 1, \ldots, \eta K,\ j = 1, \ldots, C$
3. $1 + X_r^T V_x^{-1} X_r$,
4. $\sqrt{\sum_r (Y_r - \hat{\gamma} X_r)^T \lambda^{-1} (Y_r - \hat{\gamma} X_r)}$,
5. $\sqrt{\sum_r (Y_r - \mu)^T \lambda^{-1} (Y_r - \mu)}$,

where μ is the mean value of the vector Y_r.

Above textural features derived from the 3DCAR model are robust to illumination direction changes, invariant to illumination brightness and spectrum changes, and simultaneously also robust to Gaussian noise degradation. We extensively verified this property on the BTF texture measurements, where illumination sources are spanned over 75% of possible illumination half-sphere. Figure 17 illustrates the application of 3DCAR model-derived features are illumination invariants to the unsupervised wood mosaic segmentation.

(Un)supervised Image Recognition

Unsupervised or supervised texture segmentation is the prerequisite for successful content-based image retrieval, scene analysis, automatic acquisition of virtual models, quality control, security, medical applications, and many others.

Similarly, robust surface material recognition requires the BTF data learning set. We classified 65 wood species measured in the BTF representation in the study Mikeš and Haindl (2019) using the state-of-the-art convolutional neural network (TensorFlow library (Google 2019; Krizhevsky 2009; Krizhevsky et al. 2012; Pattanayak 2017)). We documented (Mikeš and Haindl 2019) sharp classification accuracy decrease when using standard texture recognition approach, i.e., small

Fig. 17 BTF wood mosaic and the MW3-AR8^i model-based (Haindl et al. 2015a) unsupervised segmentation results

learning set size and the vertical viewing and illumination angle, which is a very inadequate representation of the enormous material appearance variability.

Although plentiful different methods were already published (Zhang 1997), the image recognition problem is still far from being solved. This situation is among others due to missing reliable performance comparison between different techniques. Only limited results were published (Martin et al. 2001; Sharma and Singh 2001; Ojala et al. 2002; Haindl and Mikeš 2008) on suitable quantitative measures that allow us to evaluate and compare the quality of segmentation algorithms.

Spatial interaction models and especially Markov random field-based models are increasingly popular for texture representation (Kashyap 1986; Reed and du Buf 1993; Haindl 1991), etc. Several researchers dealt with the difficult problem of unsupervised segmentation using these models, see for example Panjwani and Healey (1995), Manjunath and Chellapa (1991), Andrey and Tarroux (1998), Haindl (1999), and Matuszak and Schreiber (2009).

Our unsupervised segmenters (Haindl and Mikeš 2004, 2005, 2006; Haindl et al. 2015a) assume the multispectral or multi-channel textures to be locally represented by the parameters (Θ_r) of the multidimensional random field models possibly recursively evaluated for each pixel and several scales. The segmentation part of the algorithm is then based on the underlying Gaussian mixture model ($p(\Theta_r) = \sum_{i=1}^{K} p_i \, p(\Theta_r \mid \nu_i, \Sigma_i)$) representing the Markovian parametric space and starts with an over-segmented initial estimation, which is adaptively modified until the optimal number of homogeneous mammogram segments is reached. The corresponding mixture model equations ($p(\Theta_r), p(\Theta_r \mid \nu_i, \Sigma_i)$) are solved using a modified EM algorithm (Haindl and Mikeš 2007).

The concept of decision fusion for high-performance pattern recognition is well known and widely accepted in the area of supervised classification, where (often very diverse) classification technologies, each providing complementary sources of information about class membership, can be integrated to provide more accurate, robust, and reliable classification decisions than single-classifier applications. Our method (Haindl and Mikeš 2007) circumvents the problem of multiple unsupervised segmenter combination by fusing multiple-processed measurements into a single segmenter feature vector.

Multispectral/Multi-channel Image Restoration

Physical imaging, processing or transmission systems, and a recording medium are imperfect, and thus a recorded image represents a degraded version of the original scene.

The image restoration task is to recover an unobservable image given the observed corrupted image $\ddot{Y}$ with respect to some statistical criterion. Image restoration is a busy research area for already several decades, and many restoration algorithms have been proposed (Andrews and Hunt 1977; Geman and Geman 1984; Acton and Bovik 1999; Loubes and Rochet 2009; Felsberg 2009; Burgeth et al. 2009; Polzehl and Tabelow 2009).

The image degradation is often supposed to be approximated by the linear degradation model:

$$\ddot{Y}_r = \sum_{s \in I_r} f_s Y_{r-s} + e_r \tag{58}$$

where f is a discrete representation of the unknown point-spread function. The point-spread function can be non-homogeneous, but we assume its slow changes relative to the size of an image. I_r is some contextual support set, and the degradation noise e is uncorrelated with the unobservable image. The point-spread function is unknown but such that we can assume the unobservable image Y to be reasonably well approximated by the expectation of the corrupted image

$$\hat{Y} = E\{\ddot{Y}\} \tag{59}$$

in regions with gradual pixel value changes.

Let us approximate after having observed $\ddot{Y}^{(j-1)} = \{\ddot{Y}_{j-1}, \ldots, \ddot{Y}_1\}$ the mean value $\hat{Y}_j = E\{\ddot{Y}_j\}$ by the $E\{\ddot{Y}_j \mid \ddot{Y}^{(j-1)} = \ddot{y}^{(j-1)}\}$ where $\ddot{y}^{(j-1)}$ are known past realization for j. Thus, we suppose that all other possible realizations $\ddot{y}^{(j-1)}$ than the true past pixel values have negligible probabilities. This assumption implies conditional expectations approximately equal to unconditional ones, i.e.,

$$E\{\ddot{Y}_j\} \approx E\{\ddot{Y}_j \mid \ddot{Y}^{(j-1)}\}, \tag{60}$$

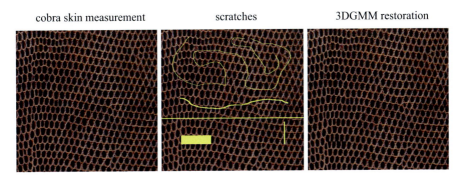

Fig. 18 Cobra skin scratch restoration using the spatial 3D Gaussian mixture model

and assuming the noisy image $\ddot{Y}$ can be represented by a 3DCAR model, then the restoration model as well as the local estimation of the point-spread function leads to a fast analytical solution (Haindl 2002). A similar restoration approach can also be derived for a multi-channel (Haindl and Šimberová 2002) or multitemporal (Haindl and Šimberová 2005) image restoration problems typically caused by random fluctuations originating mostly in the Earth's atmosphere during ground-based telescope observations.

A difficult restoration problem is to restore missing parts of an image or a spatially correlated data field. For example, every movie deteriorates with usage and time irrespective of any care it gets. Movies (on both optical and magnetic materials) suffer from blotches, dirt, sparkles, noise, scratches (Fig. 18), missing or heavily corrupted frames, mold, flickering, jittering, image vibrations, and other problems. For each kind of defect, usually a different kind of restoration algorithm is needed. The scratch notion means every coherent region with missing data (simultaneously in all spectral bands) in a color movie frame (Haindl and Filip 2002), static image, range map, radio-spectrograph (Haindl and Šimberová 1996), radar observation, color textures (Haindl and Havlíček 2015), etc. These missing data restoration methods (inpainting) exploit correlations in the spatial/spectral/temporal data space and benefit from the discussed Markovian or mixture (Fig. 18) random field models.

Conclusion

There is no single universal BTF model applicable for physically correct modeling of visual properties of all possible BTF textures. Every presented model is better suited for some subspace of possible BTF textures, either natural or artificial. Their selection depends primarily on their spectral and spatial frequency content as well as on available learning data. We present exceptional adaptive 3D Markovian or mixture models, either solved analytically or iteratively and quickly synthesized.

The presented compound Markovian models are rare exceptions in the Markovian model family that allow deriving extraordinarily efficient and fast data processing algorithms. All their statistics can be either evaluated recursively, and they either do not need any Monte Carlo sampling typical for other Markovian models or can use a fast form of such sampling (Potts random field). The 3DCAR models have an advantage over non-causal (3DAR) in their analytical treatment. It is possible to find the analytical solution of model parameters, optimal model support, model predictor, etc. Similarly, the 3DCAR model synthesis is straightforward, and this model can be directly generated from the model equation.

The mixture models are capable of reducing additive noise and restore missing textural parts simultaneously. They produce high-quality results, especially of regular or near-regular color textures. Their typical drawback the extensive learning date set requirement is lessened by the ample available BTF measurement space using a transfer learning approach.

The BTF-CMRF models offer a large data compression ratio (only tens of parameters per BTF), easy simulation, and fast, seamless synthesis of any required texture size. The methods have no restriction to the number of spectral channels; thus, they can be easily applied to hyperspectral BTFs. The methods can be easily generalized for color or BTF texture editing by estimating some local models from different target materials or for image restoration or inpainting.

The Markovian models can be used for image enhancement, e.g., utterly automatic mammogram enhancement, multispectral and multiresolution texture qualitative measures development, or image or video segmentation. Some of these models also allow robust textural features for texture classification when learning and classified textures differ in scale. The classifiers based on Markovian features can exploit illumination or geometric invariance properties and often outperform the state-of-the-art alternative methods on tested public databases (e.g., eye, bark, needles, textures).

Acknowledgments The Czech Science Foundation Project GAČR 19-12340S supported this research.

References

Acton, S., Bovik, A.: Piecewise and local image models for regularized image restoration using cross-validation. IEEE Trans. Image Process. **8**(5), 652–665 (1999)

Aittala, M., Weyrich, T., Lehtinen, J.: Practical SVBRDF capture in the frequency domain. ACM Trans. Graph. (Proc. SIGGRAPH) **32**(4), 110:1– 110:13 (2013)

Aittala, M., Weyrich, T., Lehtinen, J.: Two-shot SVBRDF capture for stationary materials. ACM Trans. Graph. **34**(4), 110:1–110:13 (2015). https://doi.org/10.1145/2766967

Aittala, M., Aila, T., Lehtinen, J.: Reflectance modeling by neural texture synthesis. ACM Trans. Graph. **35**(4), 65:1–65:13 (2016). https://doi.org/10.1145/2897824.2925917

Anderson, T.W.: The Statistical Analysis of Time Series. Wiley, New York (1971)

Andrews, H.C., Hunt, B.: Digital Image Restoration. Prentice-Hall, Englewood Cliffs (1977)

Andrey, P., Tarroux, P.: Unsupervised segmentation of markov random field modeled textured images using selectionist relaxation. IEEE Trans. Pattern Anal. Mach. Intell. **20**(3), 252–262 (1998)

Asmussen, J.C.: Modal analysis based on the random decrement technique: application to civil engineering structures. PhD thesis, University of Aalborg (1997)

Aurenhammer, F.: Voronoi diagrams-a survey of a fundamental geometric data structure. ACM Comput. Surv. (CSUR) **23**(3), 345–405 (1991)

Baril, J., Boubekeur, T., Gioia, P., Schlick, C.: Polynomial wavelet trees for bidirectional texture functions. In: SIGGRAPH'08: ACM SIGGRAPH 2008 talks, p. 1. ACM, New York (2008). https://doi.org/10.1145/1401032.1401072

Bennett, J., Khotanzad, A.: Multispectral random field models for synthesis and analysis of color images. IEEE Trans. Pattern Anal. Mach. Intell. **20**(3), 327–332 (1998)

Bennett, J., Khotanzad, A.: Maximum likelihood estimation methods for multispectral random field image models. IEEE Trans. Pattern Anal. Mach. Intell. **21**(6), 537–543 (1999)

Broemeling, L.D.: Bayesian Analysis of Linear Models. Marcel Dekker, New York (1985)

Burgeth, B., Pizarro, L., Didas, S., Weickert, J.: Coherence-enhancing diffusion filtering for matrix fields. In: Locally Adaptive Filtering in Signal and Image Processing. Springer, Berlin (2009)

Cole Jr, H.A.: On-line failure detection and damping measurement of aerospace structures by random decrement signatures. Technical Report TMX-62.041, NASA (1973)

Dana, K.J., Nayar, S.K., van Ginneken, B., Koenderink, J.J.: Reflectance and texture of real-world surfaces. In: CVPR, pp. 151–157. IEEE Computer Society (1997)

De Bonet, J.: Multiresolution sampling procedure for analysis and synthesis of textured images. In: ACM SIGGRAPH 97, pp. 361–368. ACM Press (1997)

Debevec, P., Hawkins, T., Tchou, C., Duiker, H.P., Sarokin, W., Sagar, M.: Acquiring the reflectance field of a human face. In: Proceedings of ACM SIGGRAPH 2000, Computer Graphics Proceedings, Annual Conference Series, pp. 145–156 (2000)

Dempster, A., Laird, N., Rubin, D.: Maximum likelihood from incomplete data via the em algorithm. J. R. Stat. Soc. B **39**(1), 1–38 (1977)

Deschaintre, V., Aittala, M., Durand, F., Drettakis, G., Bousseau, A.: Single-image svbrdf capture with a rendering-aware deep network. ACM Trans. Graph. **37**(4), 1–15 (2018). https://doi.org/10.1145/3197517.3201378

Dong, J., Chantler, M.: Capture and synthesis of 3D surface texture. In: Texture 2002, vol. 1, pp. 41–45. Heriot-Watt University (2002)

Dong, J., Wang, R., Dong, X.: Texture synthesis based on multiple seed-blocks and support vector machines. In: 2010 3rd International Congress on Image and Signal Processing (CISP), vol. 6, pp. 2861–2864 (2010). https://doi.org/10.1109/CISP.2010.5646815

Efros, A.A., Freeman, W.T.: Image quilting for texture synthesis and transfer. In: Fiume, E. (ed.) ACM SIGGRAPH 2001, pp. 341–346. ACM Press (2001). citeseer.nj.nec.com/efros01image.html

Efros, A.A., Leung, T.K.: Texture synthesis by non-parametric sampling. In: Proceedings of International Conference on Computer Vision (2), Corfu, pp. 1033–1038 (1999). citeseer.nj.nec.com/efros99texture.html

Felsberg, M.: Adaptive filtering using channel representations. In: Locally Adaptive Filtering in Signal and Image Processing. Springer, Berlin (2009)

Filip, J., Haindl, M.: Bidirectional texture function modeling: a state of the art survey. IEEE Trans. Pattern Anal. Mach. Intell. **31**(11), 1921–1940 (2009). https://doi.org/10.1109/TPAMI.2008.246

Geman, S., Geman, D.: Stochastic relaxation, gibbs distributions and bayesian restoration of images. IEEE Trans. Pattern Anal. Mach. Intel. **6**(11), 721–741 (1984)

Gimelfarb, G.: Image Textures and Gibbs Random Fields. Kluwer Academic Publishers, Dordrecht (1999)

Google (2019) Tensorflow. Technical report, Google AI, http://www.tensorflow.org/

Grim, J., Haindl, M.: Texture modelling by discrete distribution mixtures. Comput. Stat. Data Anal. **41**(3–4), 603–615 (2003)

Haindl, M.: Identification of the stochastic differential equation of the type arma. PhD thesis, ÚTIA Czechoslovak Academy of Sciences, Prague (1983)

Haindl, M.: Texture synthesis. CWI Q. **4**(4), 305–331 (1991)

Haindl, M.: Texture segmentation using recursive Markov random field parameter estimation. In: Bjarne, K.E., Peter, J. (eds.) Proceedings of the 11th Scandinavian Conference on Image Analysis, Pattern Recognition Society of Denmark, Lyngby, pp. 771–776 (1999). http://citeseer.ist.psu.edu/305262.html; http://www.ee.surrey.ac.uk/Research/VSSP/3DVision/virtuous/Publications/Haindl-SCIA99.ps.gz

Haindl, M.: Recursive square-root filters. In: Sanfeliu, A., Villanueva, J., Vanrell, M., Alquezar, R., Jain, A., Kittler, J. (eds.) Proceedings of the 15th IAPR International Conference on Pattern Recognition, vol. II, pp. 1018–1021. IEEE Press, Los Alamitos (2000). https://doi.org/10.1109/ICPR.2000.906246

Haindl, M.: Recursive model-based colour image restoration. Lect. Notes Comput. Sci. (2396), 617–626 (2002)

Haindl, M., Filip, J.: Fast restoration of colour movie scratches. In: Kasturi, R., Laurendeau, D., Suen, C. (eds.) Proceedings of the 16th International Conference on Pattern Recognition, vol. III, pp. 269–272. IEEE Computer Society, Los Alamitos (2002). https://doi.org/10.1109/ICPR.2002.1047846

Haindl, M., Filip, J.: Extreme compression and modeling of bidirectional texture function. IEEE Trans. Pattern Anal. Mach. Intell. **29**(10), 1859–1865 (2007). https://doi.org/10.1109/TPAMI.2007.1139

Haindl, M., Filip, J.: Visual Texture. Advances in Computer Vision and Pattern Recognition. Springer, London (2013). https://doi.org/10.1007/978-1-4471-4902-6

Haindl, M., Hatka, M.: BTF Roller. In: Chantler, M., Drbohlav, O. (eds.) Texture 2005. Proceedings of the 4th International Workshop on Texture Analysis, pp. 89–94. IEEE, Los Alamitos (2005a)

Haindl, M., Hatka, M.: A roller – fast sampling-based texture synthesis algorithm. In: Skala, V. (ed.) Proceedings of the 13th International Conference in Central Europe on Computer Graphics, Visualization and Computer Vision, pp. 93–96. UNION Agency – Science Press, Plzen (2005b)

Haindl, M., Havlíček, V.: Multiresolution colour texture synthesis. In: Dobrovodský, K. (ed.) Proceedings of the 7th International Workshop on Robotics in Alpe-Adria-Danube Region, pp. 297–302. ASCO Art, Bratislava (1998)

Haindl, M., Havlíček, V.: A multiresolution causal colour texture model. Lect. Notes Comput. Sci. (1876), 114–122 (2000)

Haindl, M., Havlíček, V.: Texture editing using frequency swap strategy. In: Jiang, X., Petkov, N. (eds.) Computer Analysis of Images and Patterns. Lecture Notes in Computer Science, vol. 5702, pp. 1146–1153. Springer (2009). https://doi.org/10.1007/978-3-642-03767-2_139

Haindl, M., Havlíček, V.: A compound MRF texture model. In: Proceedings of the 20th International Conference on Pattern Recognition, ICPR 2010, pp. 1792–1795. IEEE Computer Society CPS, Los Alamitos (2010). https://doi.org/10.1109/ICPR.2010.442

Haindl, M., Havlíček, V.: A plausible texture enlargement and editing compound markovian model. In: Salerno, E., Cetin, A., Salvetti, O. (eds.) Computational Intelligence for Multimedia Understanding. Lecture Notes in Computer Science, vol. 7252, pp. 138–148. Springer, Berlin/Heidelberg (2012). https://doi.org/10.1007/978-3-642-32436-9_12, http://www.springerlink.com/content/047124j43073m202/

Haindl, M., Havlíček, V.: Color Texture Restoration, pp. 13–18. IEEE, Piscataway (2015). https://doi.org/10.1109/ICCIS.2015.7274540

Haindl, M., Havlíček, V.: Three-dimensional gaussian mixture texture model. In: The 23rd International Conference on Pattern Recognition (ICPR), pp. 2026–2031. IEEE (2016). https://doi.org/978-1-5090-4846-5/16/\protect\T1\textdollar31.0, http://www.icpr2016.org/site/

Haindl, M., Havlíček, M.: A compound moving average bidirectional texture function model. In: Zgrzynowa, A., Choros, K., Sieminski, A. (eds.) Multimedia and Network Information Systems, Advances in Intelligent Systems and Computing, vol. 506, pp. 89–98. Springer International Publishing (2017a). https://doi.org/10.1007/978-3-319-43982-2_8

Haindl, M., Havlíček, V.: Two compound random field texture models. In: Beltrán-Castañón, C., Nyström, I., Famili, F. (eds.) 2016 the 21st IberoAmerican Congress on Pattern Recognition (CIARP 2016). Lecture Notes in Computer Science, vol. 10125, pp. 44–51. Springer International Publishing AG, Cham (2017b). https://doi.org/10.1007/978-3-319-52277-7_6

Haindl, M., Havlíček, V.: BTF compound texture model with fast iterative non-parametric control field synthesis. In: di Baja, G.S., Gallo, L., Yetongnon, K., Dipanda, A., Castrillon-Santana, M., Chbeir, R. (eds.) Proceedings of the 14th International Conference on Signal-Image Technology & Internet-Based Systems (SITIS 2018), pp. 98–105. IEEE Computer Society CPS, Los Alamitos (2018a). https://doi.org/10.1109/SITIS.2018.00025

Haindl, M., Havlíček, V.: BTF compound texture model with non-parametric control field. In: The 24th International Conference on Pattern Recognition (ICPR 2018), pp. 1151–1156. IEEE (2018b). http://www.icpr2018.org/

Haindl, M., Mikeš, S.: Model-based texture segmentation. Lect. Notes Comput. Sci. (3212), 306–313 (2004)

Haindl, M., Mikeš, S.: Colour texture segmentation using modelling approach. Lect. Notes Comput. Sci. (3687), 484–491 (2005)

Haindl, M., Mikeš, S.: Unsupervised texture segmentation using multispectral modelling approach. In: Tang, Y., Wang, S., Yeung, D., Yan, H., Lorette, G. (eds.) Proceedings of the 18th International Conference on Pattern Recognition, ICPR 2006, vol. II, pp. 203–206. IEEE Computer Society, Los Alamitos (2006). https://doi.org/10.1109/ICPR.2006.1148

Haindl, M., Mikeš, S.: Unsupervised texture segmentation using multiple segmenters strategy. In: Haindl, M., Kittler, J., Roli, F. (eds.) MCS 2007. Lecture Notes in Computer Science, vol. 4472, pp. 210–219. Springer (2007). https://doi.org/10.1007/978-3-540-72523-7_22

Haindl, M., Mikeš, S.: Texture segmentation benchmark. In: Lovell, B., Laurendeau, D., Duin, R. (eds.) Proceedings of the 19th International Conference on Pattern Recognition, ICPR 2008, pp. 1–4. IEEE Computer Society, Los Alamitos (2008). https://doi.org/10.1109/ICPR.2008.4761118

Haindl, M., Šimberová, S.: A multispectral image line reconstruction method. In: Theory & Applications of Image Analysis. Series in Machine Perception and Artificial Intelligence, pp. 306–315. World Scientific, Singapore (1992). https://doi.org/10.1142/9789812797896_0028

Haindl, M., Šimberová, S.: A high – resolution radiospectrograph image reconstruction method. Astron. Astrophys. **115**(1), 189–193 (1996)

Haindl, M., Šimberová, S.: Model-based restoration of short-exposure solar images. In: Abraham, A., Ruiz-del Solar, J., Koppen, M. (eds.) Soft Computing Systems Design, Management and Applications, pp. 697–706. IOS Press, Amsterdam (2002)

Haindl, M., Šimberová, S.: Restoration of multitemporal short-exposure astronomical images. Lect. Notes Comput. Sci. (3540), 1037–1046 (2005)

Haindl, M., Mikeš, S., Pudil, P.: Unsupervised hierarchical weighted multi-segmenter. In: Benediktsson, J., Kittler, J., Roli, F. (eds.) Lecture Notes in Computer Science. MCS 2009, vol. 5519, pp. 272–282. Springer (2009a). https://doi.org/10.1007/978-3-642-02326-2_28

Haindl, M., Mikeš, S., Vácha, P.: Illumination invariant unsupervised segmenter. In: Bayoumi, M. (ed.) IEEE 16th International Conference on Image Processing – ICIP 2009, pp. 4025–4028. IEEE (2009b). https://doi.org/10.1109/ICIP.2009.5413753

Haindl, M., Havlíček, V., Grim, J.: Probabilistic mixture-based image modelling. Kybernetika **46**(3), 482–500 (2011). http://www.kybernetika.cz/content/2011/3/482/paper.pdf

Haindl, M., Remeš, V., Havlíček, V.: Potts compound markovian texture model. In: Proceedings of the 21st International Conference on Pattern Recognition, ICPR 2012, pp. 29–32. IEEE Computer Society CPS, Los Alamitos (2012)

Haindl, M., Mikeš, S., Kudo, M.: Unsupervised surface reflectance field multi-segmenter. In: Azzopardi, G., Petkov, N. (eds.) Computer Analysis of Images and Patterns. Lecture Notes in Computer Science, vol. 9256, pp. 261–273. Springer International Publishing (2015a). https://doi.org/10.1007/978-3-319-23192-1_22

Haindl, M., Remeš, V., Havlíček, V.: BTF Potts Compound Texture Model, vol. 9398, pp. 939807-1–939807-11. SPIE, Bellingham (2015b). https://doi.org/10.1117/12.2077481

Han, J.Y., Perlin, K.: Measuring bidirectional texture reflectance with a kaleidoscope. ACM Trans. Graph. **22**(3), 741–748 (2003)

Heeger, D., Bergen, J.: Pyramid based texture analysis/synthesis. In: ACM SIGGRAPH 95, pp. 229–238. ACM Press (1995)

Holroyd, M., Lawrence, J., Zickler, T.: A coaxial optical scanner for synchronous acquisition of 3D geometry and surface reflectance. ACM Trans. Graph. (Proc. SIGGRAPH 2010) (2010). http://www.cs.virginia.edu/~mjh7v/Holroyd10.php

Kashyap, R.: Analysis and synthesis of image patterns by spatial interaction models. In: Kanal, L., Rosenfeld, A. (eds.) Progress in Pattern Recognition 1. Elsevier, North-Holland (1981)

Kashyap, R.: Image models. In: Young, T.Y., Fu, K.S. (eds.) Handbook of Pattern Recognition and Image Processing. Academic, New York (1986)

Koudelka, M.L., Magda, S., Belhumeur, P.N., Kriegman, D.J.: Acquisition, compression, and synthesis of bidirectional texture functions. In: Texture 2003: Third International Workshop on Texture Analysis and Synthesis, Nice, pp. 59–64 (2003)

Krizhevsky, A.: Learning multiple layers of features from tiny images. Master's thesis, University of Toronto (2009)

Krizhevsky, A., Sutskever, I., Hinton, G.E.: Imagenet classification with deep convolutional neural networks. In: Advances in Neural Information Processing Systems, pp. 1097–1105 (2012)

Kwatra, V., Schodl, A., Essa, I., Turk, G., Bobick, A.: Graphcut textures: image and video synthesis using graph cuts. ACM Trans. Graph. **22**(3), 277–286 (2003)

Levada, A., Mascarenhas, N., Tannus, A.: Pseudolikelihood equations for potts mrf model parameter estimation on higher order neighborhood systems. Geosci. Remote Sens. Lett. IEEE **5**(3), 522–526 (2008). https://doi.org/10.1109/LGRS.2008.920909

Li, X., Cadzow, J., Wilkes, D., Peters, R., Bodruzzaman II, M.: An efficient two dimensional moving average model for texture analysis and synthesis. In: Proceedings IEEE Southeastcon'92, vol. 1, pp. 392–395. IEEE (1992)

Liang, L., Liu, C., Xu, Y.Q., Guo, B., Shum, H.Y.: Real-time texture synthesis by patch-based sampling. ACM Trans. Graph. (TOG) **20**(3), 127–150 (2001)

Liu, F., Picard, R.: Periodicity, directionality, and randomness: wold features for image modeling and retrieval. IEEE Trans. Pattern Anal. Mach. Intell. **18**(7), 722–733 (1996). https://doi.org/10.1109/34.506794

Loubes, J., Rochet, P.: Regularization with approximated L^2 maximum entropy method. In: Locally Adaptive Filtering in Signal and Image Processing. Springer, Berlin (2009)

Manjunath, B., Chellapa, R.: Unsupervised texture segmentation using Markov random field models. IEEE Trans. Pattern Anal. Mach. Intell. **13**, 478–482 (1991)

Marschner, S.R., Westin, S.H., Arbree, A., Moon, J.T.: Measuring and modeling the appearance of finished wood. ACM Trans. Graph. **24**(3), 727–734 (2005)

Martin, D., Fowlkes, C., Tal, D., Malik, J.: A database of human segmented natural images and its application to evaluating segmentation algorithms and measuring ecological statistics. In: Proceedings of 8th International Conference on Computer Vision, vol. 2, pp. 416–423 (2001). http://www.cs.berkeley.edu/projects/vision/grouping/segbench/

Matuszak, M., Schreiber, T.: Locally specified polygonal Markov fields for image segmentation. In: Locally Adaptive Filtering in Signal and Image Processing. Springer, Berlin (2009)

Metropolis, N., Rosenbluth, A.W., Rosenbluth, M.N., Teller, A.H., Teller, E.: Equation of state calculations by fast computing machines. J. Chem. Phys. **21**, 1087–1092 (1953)

Mikeš, S., Haindl, M.: View dependent surface material recognition. In: Bebis, G., Boyle, R., Parvin, B., Koračin, D., Ushizima, D., Chai, S., Sueda, S., Lin, X., Lu, A., Thalmann, D., Wang, C., Xu, P. (eds.) 14th International Symposium on Visual Computing (ISVC 2019). Lecture Notes in Computer Science, vol. 11844, pp. 156–167. Springer Nature Switzerland AG (2019). https://doi.org/10.1007/978-3-030-33720-9_12, https://www.isvc.net/

Müller, G., Meseth, J., Klein, R.: Compression and real-time rendering of measured BTFs using local PCA. In: Vision, Modeling and Visualisation 2003, pp. 271–280 (2003)

Müller, G., Meseth, J., Sattler, M., Sarlette, R., Klein, R.: Acquisition, synthesis and rendering of bidirectional texture functions. In: Eurographics 2004, STAR – State of The Art Report, Eurographics Association, pp. 69–94 (2004)

Neubeck, A., Zalesny, A., Gool, L.: 3D texture reconstruction from extensive BTF data. In: Chantler, M., Drbohlav, O. (eds.) Texture 2005. Heriot-Watt University, Edinburgh (2005)

Ngan, A., Durand, F.: Statistical acquisition of texture appearance. In: Eurographics Symposium on Rendering, Eurographics (2006)

Ojala, T., Maenpaa, T., Pietikainen, M., Viertola, J., Kyllonen, J., Huovinen, S.: Outex: new framework for empirical evaluation of texture analysis algorithms. In: International Conference on Pattern Recognition, pp. I:701–706 (2002)

Paget, R., Longstaff, I.D.: Texture synthesis via a noncausal nonparametric multiscale markov random field. IEEE Trans. Image Process. **7**(8), 925–932 (1998)

Panjwani, D., Healey, G.: Markov random field models for unsupervised segmentation of textured color images. IEEE Trans. Pattern Anal. Mach. Intell. **17**(10), 939–954 (1995)

Pattanayak, S.: Pro Deep Learning with TensorFlow. Apress (2017). https://doi.org/10.1007/978-1-4842-3096-1

Polzehl, J., Tabelow, K.: Structural adaptive smoothing: principles and applications in imaging. In: Locally Adaptive Filtering in Signal and Image Processing. Springer, Berlin (2009)

Portilla, J., Simoncelli, E.: A parametric texture model based on joint statistics of complex wavelet coefficients. Int. J. Comput. Vis. **40**(1), 49–71 (2000)

Potts, R., Domb, C.: Some generalized order-disorder transformations. In: Proceedings of the Cambridge Philosophical Society, vol. 48, pp. 106–109 (1952)

Praun, E., Finkelstein, A., Hoppe, H.: Lapped textures. In: ACM SIGGRAPH 2000, pp. 465–470 (2000)

Rainer, G., Ghosh, A., Jakob, W., Weyrich, T.: Unified neural encoding of BTFs. In: Computer Graphics Forum, vol. 39, pp. 167–178. Wiley Online Library (2020)

Reed, T.R., du Buf, J.M.H.: A review of recent texture segmentation and feature extraction techniques. CVGIP–Image Underst. **57**(3), 359–372 (1993)

Ren, P., Wang, J., Snyder, J., Tong, X., Guo, B.: Pocket reflectometry. ACM Trans. Graph. (Proc. SIGGRAPH) **30**(4) (2011). https://doi.org/10.1145/2010324.1964940

Ruiters, R., Schwartz, C., Klein, R.: Example-based interpolation and synthesis of bidirectional texture functions. In: Computer Graphics Forum, vol. 32, pp. 361–370. Wiley Online Library (2013)

Sattler, M., Sarlette, R., Klein, R.: Efficient and realistic visualization of cloth. In: Eurographics Symposium on Rendering (2003)

Schwartz, C., Sarlette, R., Weinmann, M., Rump, M., Klein, R.: Design and implementation of practical bidirectional texture function measurement devices focusing on the developments at the university of bonn. Sensors **14**(5), 7753–7819 (2014). https://doi.org/10.3390/s140507753. http://www.mdpi.com/1424-8220/14/5/7753

Sharma, M., Singh, S.: Minerva scene analysis benchmark. In: Seventh Australian and New Zealand Intelligent Information Systems Conference, pp. 231–235. IEEE (2001)

Soler, C., Cani, M., Angelidis, A.: Hierarchical pattern mapping. ACM Trans. Graph. **21**(3), 673–680 (2002)

Swendsen, R.H., Wang, J.S.: Nonuniversal critical dynamics in Monte Carlo simulations. Phys. Rev. Lett. **58**(2), 86–88 (1987). https://doi.org/10.1103/PhysRevLett.58.86

Tong, X., Zhang, J., Liu, L., Wang, X., Guo, B., Shum, H.Y.: Synthesis of bidirectional texture functions on arbitrary surfaces. ACM Trans. Graph. (TOG) **21**(3), 665–672 (2002)

Tsai, Y.T., Shih, Z.C.: K-clustered tensor approximation: a sparse multilinear model for real-time rendering. ACM Trans. Graph. **31**(3), 19:1–19:17 (2012). https://doi.org/10.1145/2167076.2167077

Tsai, Y.T., Fang, K.L., Lin, W.C., Shih, Z.C.: Modeling bidirectional texture functions with multivariate spherical radial basis functions. Pattern Anal. Mach. Intell. IEEE Trans. **33**(7), 1356–1369 (2011). https://doi.org/10.1109/TPAMI.2010.211

Vacha, P., Haindl, M.: Image retrieval measures based on illumination invariant textural mrf features. In: CIVR'07: Proceedings of the 6th ACM International Conference on Image and Video Retrieval, pp. 448–454. ACM Press, New York (2007). https://doi.org/10.1145/1282280.1282346

Varma, M., Zisserman, A.: A statistical approach to texture classification from single images. Int. J. Comput. Vis. **62**(1–2), 61–81 (2005)

Wang, J., Dana, K.: Relief texture from specularities. IEEE Trans. Pattern Anal. Mach. Intell. **28**(3), 446–457 (2006)

Wei, L., Levoy, M.: Texture synthesis using tree-structure vector quantization. In: ACM SIGGRAPH 2000, pp. 479–488. ACM Press/Addison Wesley/Longman (2000). citeseer.nj.nec.com/wei01texture.html

Wei, L., Levoy, M.: Texture synthesis over arbitrary manifold surfaces. In: SIGGRAPH 2001, pp. 355–360. ACM (2001)

Wu, F.: (1982) The Potts model. Rev. Modern Phys. **54**(1), 235–268

Wu, H., Dorsey, J., Rushmeier, H.: A sparse parametric mixture model for BTF compression, editing and rendering. Comput. Graph. Forum **30**(2), 465–473 (2011)

Xu, Y., Guo, B., Shum, H.: Chaos mosaic: fast and memory efficient texture synthesis. Technical Report MSR-TR-2000-32, Redmont (2000)

Zelinka, S., Garland, M.: Towards real-time texture synthesis with the jump map. In: 13th European Workshop on Rendering, p. 99104 (2002)

Zelinka, S., Garland, M.: Interactive texture synthesis on surfaces using jump maps. In: Christensen, P., Cohen-Or, D. (eds.) 14th European Workshop on Rendering, Eurographics (2003)

Zhang, Y.J.: Evaluation and comparison of different segmentation algorithms. Pattern Recogn. Lett. **18**, 963–974 (1997)

Zhang, J.D., Zhou, K., Velho ea, L.: Synthesis of progressively-variant textures on arbitrary surfaces. ACM Trans. Graph. **22**(3), 295–302 (2003)

Zhu, S., Liu, X., Wu, Y.: Exploring texture ensembles by efficient Markov Chain Monte Carlo – toward a "trichromacy" theory of texture. IEEE Trans. Pattern Anal. Mach. Intell. **22**(6), 554–569 (2000)

Regularization of Inverse Problems by Neural Networks

29

Markus Haltmeier and Linh Nguyen

Contents

Introduction	1066
Ill-Posedness	1067
Data-Driven Reconstruction	1068
Outline	1069
Preliminaries	1069
Right Inverses	1070
Regularization Methods	1072
Deep Learning	1074
Regularizing Networks	1075
Null-Space Networks	1076
Convergence Analysis	1078
Extensions	1079
The NETT Approach	1080
Learned Regularization Functionals	1080
Convergence Analysis	1082
Related Methods	1088
Conclusion and Outlook	1090
References	1091

Abstract

Inverse problems arise in a variety of imaging applications, including computed tomography, non-destructive testing, and remote sensing. Characteristic features of inverse problems are the non-uniqueness and instability of their solutions.

M. Haltmeie (✉)
Department of Mathematics, University of Innsbruck, Innsbruck, Austria
e-mail: markus.haltmeier@uibk.ac.at

L. Nguyen
Department of Mathematics, University of Idaho, Moscow, ID, USA
e-mail: lnguyen@uidaho.edu

Therefore, any reasonable solution method requires the use of regularization tools that select specific solutions and, at the same time, stabilize the inversion process. Recently, data-driven methods using deep learning techniques and neural networks showed to significantly outperform classical solution methods for inverse problems. In this chapter, we give an overview of inverse problems and demonstrate the necessity of regularization concepts for their solution. We show that neural networks can be used for the data-driven solution of inverse problems and review existing deep learning methods for inverse problems. In particular, we view these deep learning methods from the perspective of regularization theory, the mathematical foundation of stable solution methods for inverse problems. This chapter is more than just a review as many of the presented theoretical results extend existing ones.

Keywords

Inverse problems · Deep learning · Neural networks · Regularization theory · Ill-posedness · Stability · Theoretical foundation

Introduction

The solution of inverse problems arises in a variety of practically important applications, including medical imaging, computer vision, geophysics, as well as many other branches of pure and applied sciences. Inverse problems are most efficiently formulated as an estimation problem of the form

$$\text{recover} \quad \mathbf{x}^* \in \mathbb{X} \quad \text{from data} \quad \mathbf{y} = \mathbf{A}(\mathbf{x}^*) + \xi \in \mathbb{Y}. \tag{1}$$

Here, $\mathbf{A} \colon \mathbb{X} \to \mathbb{Y}$ is a mapping between normed spaces, $\mathbf{x}^* \in \mathbb{X}$ is the true unknown solution, $\mathbf{y}$ represents the given data, and ξ is an unknown data perturbation. In this context, the application of the operator $\mathbf{A}$ is referred to as the forward operator or forward problem, and solving (1) is the corresponding inverse problem. In the absence of noise where $\xi = 0$, we refer to $\mathbf{y} = \mathbf{A}(\mathbf{x}^*)$ as exact data, and in the case where $\xi \neq 0$, we refer to $\mathbf{y}$ as noisy data.

One of the prime examples of inverse problems are image reconstruction problems, where the forward operator describes the data generation process depending on the image reconstruction modality. For example, in X-ray computed tomography (CT), the forward operator is the sampled Radon transform, whereas in magnetic resonance imaging (MRI), the forward operator is the sampled Fourier transform. Reconstructing the diagnostic image from experimentally collected data leads to solving an inverse problem of the form (1). In these and other applications, the underlying forward operator is naturally formulated on infinite-dimensional spaces, because the object to be reconstructed is a function of a continuous spatial variable. Even though the numerical solution is performed in a finite dimensional discretization, the mathematical properties of the continuous formulation are crucial for understanding and improving image formation algorithms.

Ill-Posedness

The inherent character of inverse problems is their ill-posedness. This means that even in the case of exact data, the solution of (1) is either not unique, not existent, or does not stably depend on the given data. More formally, for an inverse problem, at least one of the following three unfavorable properties holds:

(I1) NON-UNIQUENESS: For some $\mathbf{x}_1^* \neq \mathbf{x}_2^* \in \mathbb{X}$, we have $\mathbf{A}(\mathbf{x}_1^*) = \mathbf{A}(\mathbf{x}_2^*)$.
(I2) NON-EXISTENCE: For some $\mathbf{y} \in \mathbb{Y}$, the equation $\mathbf{A}(\mathbf{x}) = \mathbf{y}$ has no solution.
(I3) INSTABILITY: Smallness of $\|\mathbf{A}(\mathbf{x}_1^*) - \mathbf{A}(\mathbf{x}_2^*)\|$ does not imply smallness of $\|\mathbf{x}_1^* - \mathbf{x}_2^*\|$.

These conditions imply that the forward operator does not have a continuous inverse, which could be used to directly solve (1). Instead, regularization methods have to be applied, which result in stable methods for solving inverse problem.

Regularization methods approach the ill-posedness by two steps. First, to address non-uniqueness and non-existence issues (I1), (I2), one restricts the image and pre-image space of the forward operator to sets $\mathbb{M} \subseteq \mathbb{X}$ and $\text{ran}(\mathbf{A}) \subseteq \mathbb{Y}$, such that the restricted forward operator $\mathbf{A}_{\text{res}} : \mathbb{M} \to \text{ran}(\mathbf{A})$ becomes bijective. For any exact data, the equation $\mathbf{A}(\mathbf{x}) = \mathbf{y}$ then has a unique solution in $\mathbb{M}$, which is given by the inverse of the restricted forward operator applied to $\mathbf{y}$. Second, in order to address the instability issue (I3), in a second step, one considers a family of continuous operators $\mathbf{B}_\alpha : \mathbb{Y} \to \mathbb{X}$ for $\alpha > 0$ that converge to $\mathbf{A}_{\text{res}}^{-1}$ in a suitable sense; see section "Preliminaries" for precise definitions.

Note that the choice of the set $\mathbb{M}$ is crucial as it represents the class of desired reconstructions and acts as selection criteria for picking a particular solution of the given inverse problem. The main challenge is that this class is actually unknown or at least it cannot be described properly. For example, in CT for medical imaging, the set of desired solutions represents the set of all functions corresponding to spatially attenuation inside patients, a function class that is clearly challenging, if not impossible, to describe in simple mathematical terms.

Variational regularization and variants (Scherzer et al. 2009) have been the most successful class of regularization methods for solving inverse problems. Here, $\mathbb{M}$ is defined as solutions having a small value of a certain regularization functional that can be interpreted as a measure for the deviation from the desired solutions. Various regularization functionals have been analyzed for inverse problems, including Hilbert space norms (Engl et al. 1996), total variation (Acar and Vogel 1994), and sparse ℓ^q-penalties (Daubechies et al. 2004; Grasmair et al. 2008). Such handcrafted regularization functionals have limited complexity and are unlikely to accurately model complex signal classes arising in applications such as medical imaging. On the other hand, their regularization effects are well understood, efficient numerical algorithms have been developed for their realization, they work reasonably well in practice, and they have been rigorously analyzed mathematically.

Data-Driven Reconstruction

Recently data-driven methods based on neural networks and deep learning demonstrated to significantly outperform existing variational and iterative reconstruction algorithms for solving inverse problems. The essential idea is to use neural networks to define a class $(\mathbf{R}_\theta)_{\theta \in \Theta}$ of reconstruction networks $\mathbf{R}_\theta \colon \mathbb{Y} \to \mathbb{X}$ and to select the parameter vector $\theta \in \Theta$ of the network in a data-driven manner. The selection is based on a set of training data $(\mathbf{x}_1, \mathbf{y}_1), (\mathbf{x}_2, \mathbf{y}_2), \ldots, (\mathbf{x}_N, \mathbf{y}_N)$, where $\mathbf{x}_i \in \mathbb{M}$ are desired reconstructions and $\mathbf{y}_i = \mathbf{A}(\mathbf{x}_i^*) + \xi_i \in \mathbb{Y}$ are corresponding data. Even if the set $\mathbb{M}$ of desired reconstructions is unknown, the available samples $\mathbf{x}_1, \ldots, \mathbf{x}_N$ can be used to select the particular reconstruction method. A typical selection strategy is to minimize a penalized least-squares functional having the form

$$\theta^* \in \arg\min_\theta \left\{ \frac{1}{N} \sum_{i=1}^N \|\mathbf{x}_i - \mathbf{R}_\theta(\mathbf{y}_i)\|^2 + P(\theta) \right\}. \tag{2}$$

The final neural network-based reconstruction method is then given by $\mathbf{R}_{\theta^*} \colon \mathbb{Y} \to \mathbb{X}$ and is such that in average, it performs well on the given training dataset.

Existing deep learning-based methods include post-processing networks (Han et al. 2016; Jin et al. 2017), null-space networks (Schwab et al. 2019, 2020), variational networks (Kobler et al. 2017), iterative networks (Yang et al. 2016; Adler and Öktem 2017; Aggarwal et al. 2018), network cascades (Kofler et al. 2018; Schlemper et al. 2017), and learned regularization functional (Li et al. 2020; Lunz et al. 2018; Obmann et al. 2020b). We refer to the review Arridge et al. (2019) for other data-driven reconstruction methods such as GANs (Bora et al. 2017; Mardani et al. 2018), dictionary learning, deep basis pursuit (Sulam et al. 2019), or deep image priors (Ulyanov et al. 2018; Van Veen et al. 2018; Dittmer et al. 2020), which we do not touch in this chapter. Post-processing networks and null-space networks are explicit, where the reconstruction network is given explicitly and its parameters are trained to fit the given training data. Methods using learned regularizers are implicit, and the reconstruction network $\mathbf{R}_\theta(\mathbf{y}) = \arg\min \mathcal{T}_{\theta,\mathbf{y}}$ is defined by minimizing a properly trained Tikhonov functional $\mathcal{T}_{\theta,\mathbf{y}} \colon \mathbb{X} \to [0, \infty]$. Variational networks and iterative networks are in between, where $\arg\min \mathcal{T}_{\theta,\mathbf{y}}$ is approximated via an iterative scheme using L steps.

Any reasonable method for solving an inverse problem, including all learned reconstruction schemes, has to include some form of regularization. However, regularization may be imposed implicitly, even without noticing by the researcher developing the algorithm. Partially, this is the case because discretization, early stopping, or other techniques to numerically stabilizing an optimization algorithm at the same time have a regularization effect on the underlying inverse problem. Needless to say, understanding and analyzing where exactly the regularization effect comes from will increase the reliability of any algorithm and allows its further improvement. In conclusion, any data-driven reconstruction method has to include either explicitly or implicitly a form of regularization. In this chapter, we will analyze the regularization properties of various deep learning methods for solving inverse problems.

Outline

The outline of this chapter is as follows. In section "Preliminaries", we present the background of inverse problems and deep learning. In section "Regularizing Networks", we analyze direct neural network-based reconstructions, whereas in section "The NETT Approach", we study variational and iterative reconstruction methods based on neural networks. The chapter concludes with a discussion and some final remarks given in section "Conclusion and Outlook". While the concepts presented in the subsequent sections are known, most of the presented results extend existing ones. Therefore, this chapter is much more than just a review over existing results.

For the sake of clarity, in this chapter, we focus on linear inverse problems; even several results can be extended non-linear problems as well. We will provide remarks pointing to such results. Throughout the study, we allow an infinite-dimensional setting, because in many applications, the unknowns to be recovered as well as the data are most naturally modeled as functions that lie in infinite-dimensional spaces $\mathbb{X}$ and $\mathbb{Y}$. However, everything said in this chapter applies to finite dimensional spaces as well. In limited data problems, such as sparse-view CT, the finite dimension of the data space $\mathbb{Y}$ is even an intrinsic part of the forward model. Therefore, the reader not familiar with infinite-dimensional vector space can think of $\mathbb{X}$ and $\mathbb{Y}$ as finite-dimensional vector spaces each equipped with a standard vector norm.

Preliminaries

In this section, we provide necessary background on linear inverse problems, their regularization, and their solution with neural networks.

Throughout the following, $\mathbb{X}$ and $\mathbb{Y}$ are Banach spaces. We study solving inverse problems of the form (1) in a deterministic setting with a bounded linear forward operator $\mathbf{A} \colon \mathbb{X} \to \mathbb{Y}$. Hence we aim to estimate the unknown signal $\mathbf{x}^* \in \mathbb{X}$ from the available data $\mathbf{y} = \mathbf{A}(\mathbf{x}^*) + \xi$, where $\xi \in \mathbb{Y}$ is the noise that is assumed to satisfy an estimate of the form $\|\xi\| \leq \delta$. Here, $\delta \geq 0$ is called the noise level, and in the case $\delta = 0$, we call $\mathbf{y} = \mathbf{A}(\mathbf{x}^*)$ the exact data.

Right Inverses

As we have explained in the Introduction, the main feature of inverse problems is their ill-posedness. Regularization methods approach the ill-posedness by two steps. In the first step, they address (I1) and (I2) by restricting the image and the pre-image spaces, which gives a certain right inverse defined on ran(**A**). In order to address the instability issue (I3), in a second step, regularization methods are applied for stabilization. We first consider right inverse and their instability and consider the regularization in the following subsection.

Definition 1 (Right inverse). A possibly non-linear mapping $\mathbf{B}\colon \mathrm{ran}(\mathbf{A}) \subseteq \mathbb{Y} \to \mathbb{X}$ is called right inverse of **A** if $\mathbf{A}(\mathbf{B}(\mathbf{y})) = \mathbf{y}$ for all $\mathbf{y} \in \mathrm{ran}(\mathbf{A})$.

Clearly, a right inverse always exists because for any $\mathbf{y} \in \mathrm{ran}(\mathbf{A})$, there exists an element $\mathbf{B}\mathbf{y} := \mathbf{x}$, such that $\mathbf{A}\mathbf{x} = \mathbf{y}$. However, in general, no continuous right inverse exists. More precisely, we have the following result (compare Nashed 1987).

Proposition 1 (Continuous right inverses). *Let* $\mathbf{B}\colon \mathrm{ran}(\mathbf{A}) \to \mathbb{X}$ *be a continuous right inverse. Then,* $\mathrm{ran}(\mathbf{A})$ *is closed.*

Proof. By continuity, **B** can be extended in a unique way to a mapping $\mathbf{H}\colon \overline{\mathrm{ran}(\mathbf{A})} \to \mathbb{X}$. The continuity of **H** and **A** implies $\mathbf{A} \circ \mathbf{H} = \mathrm{Id}_{\overline{\mathrm{ran}(\mathbf{A})}}$. Therefore $\overline{\mathrm{ran}(\mathbf{A})} = \mathrm{ran}(\mathbf{A} \circ \mathbf{H}) \subseteq \mathrm{ran}(\mathbf{A}) \subseteq \overline{\mathrm{ran}(\mathbf{A})}$ which shows that $\mathrm{ran}(\mathbf{A})$ is closed.

Proposition 1 implies that whenever ran(**A**) is non-closed, **A** does not have a continuous right inverse.

The next question we study is the existence of a linear right inverse. For that purpose recall that a mapping $\mathbf{P}\colon \mathbb{X} \to \mathbb{X}$ is called projection if $\mathbf{P}^2 = \mathbf{P}$. If **P** is a linear bounded projection, then ran(**P**) and ker(**P**) are closed subspaces and $\mathbb{X} = \mathrm{ran}(\mathbf{P}) \oplus \mathrm{ker}(\mathbf{P})$.

Definition 2 (Complemented subspace). A closed (linear) subspace $\mathbb{V}$ of $\mathbb{X}$ is called complemented in $\mathbb{X}$ if there exists a bounded linear projection **P** with $\mathrm{ran}(\mathbf{P}) = \mathbb{V}$

A closed subspace $\mathbb{V} \subseteq \mathbb{X}$ is complemented if and only if there is another closed subspace $\mathbb{U} \subseteq \mathbb{X}$ with $\mathbb{X} = \mathbb{U} \oplus \mathbb{V}$. In a Hilbert space, any closed subspace is complemented, and $\mathbb{X} = \mathbb{V}^\perp \oplus \mathbb{V}$ with the orthogonal complement $\mathbb{V}^\perp := \{u \in \mathbb{X} \mid \forall v \in \mathbb{V}\colon \langle u, v \rangle = 0\}$. However, as shown in Lindenstrauss and Tzafriri (1971), in every Banach space that is not isomorphic to a Hilbert space, there exist closed subspaces which are not complemented.

29 Regularization of Inverse Problems by Neural Networks

Proposition 2 (Linear right inverses).

(a) $\mathbf{A}$ has a linear right inverse with bounded $\mathbf{B} \circ \mathbf{A}$ if and only if $\ker(\mathbf{A})$ is complemented.

(b) A linear right inverse as in (a) is continuous if and only $\operatorname{ran}(\mathbf{A})$ is closed.

Proof. (a) First, suppose that $\mathbf{A}$ has a linear right inverse $\mathbf{B} \colon \operatorname{ran}(\mathbf{A}) \to \mathbb{X}$ such that $\mathbf{B} \circ \mathbf{A}$ is bounded. For any $\mathbf{x} \in \mathbb{X}$, we have $(\mathbf{B} \circ \mathbf{A})^2(\mathbf{x}) = \mathbf{B} \circ (\mathbf{A} \circ \mathbf{B})(\mathbf{A}(\mathbf{x})) = (\mathbf{B} \circ \mathbf{A})(\mathbf{x})$. Hence, $\mathbf{B} \circ \mathbf{A}$ is a linear bounded projection. This implies the topological decomposition $\mathbb{X} = \operatorname{ran}(\mathbf{B} \circ \mathbf{A}) \oplus \ker(\mathbf{B} \circ \mathbf{A})$ with closed subspaces $\operatorname{ran}(\mathbf{B} \circ \mathbf{A})$ and $\ker(\mathbf{B} \circ \mathbf{A})$. It holds $\ker(\mathbf{B} \circ \mathbf{A}) \supseteq \ker(\mathbf{A}) = \ker(\mathbf{A} \circ \mathbf{B} \circ \mathbf{A}) \supseteq \operatorname{ran}(\mathbf{B} \circ \mathbf{A})$, which shows that $\ker(\mathbf{A}) = \operatorname{ran}(\mathbf{B} \circ \mathbf{A})$ is complemented. Conversely let $\ker(\mathbf{A})$ be complemented, and write $\mathbb{X} = \mathbb{X}_1 \oplus \ker(\mathbf{A})$. Then $\mathbf{A}_{\text{res}} \colon \mathbb{X}_1 \to \operatorname{ran}(\mathbf{A})$ is bijective and therefore has a linear inverse $\mathbf{A}_{\text{res}}^{-1}$ defining a desired right inverse for $\mathbf{A}$.

(b) For any continuous right inverse, $\operatorname{ran}(\mathbf{A})$ is closed according to Proposition 1. Conversely, let $\mathbf{B} \colon \operatorname{ran}(\mathbf{A}) \to \mathbb{X}$ be linear right inverse such that $\mathbf{B} \circ \mathbf{A}$ is bounded and $\operatorname{ran}(\mathbf{A})$ closed. In particular, $\ker(\mathbf{A})$ is complemented, and we can write $\mathbb{X} = \mathbb{X}_1 \oplus \ker(\mathbf{A})$. The restricted mapping $\mathbf{A}_{\text{res}} \colon \mathbb{X}_1 \to \overline{\operatorname{ran}(\mathbf{A})}$ is bijective, therefore bounded according to the bounded inverse theorem. This implies that $\mathbf{B}$ is bounded, too. □

In a Hilbert space $\mathbb{X}$, the kernel $\ker(\mathbf{A})$ of a bounded linear operator is complemented, as any other closed subspace of $\mathbb{X}$. Therefore, according to Proposition 2, any bounded linear operator defined on a Hilbert space has a linear right inverse. However, in a general Banach space, this is not the case, as the following example shows.

Example 1 (Bounded linear operator without linear right inverse). Consider the set $c_0(\mathbb{N})$ of all sequences converging to zero as a subspace of the space $\ell^\infty(\mathbb{N})$ of all bounded sequences $\mathbf{x} \colon \mathbb{N} \to \mathbb{R}$ with the supremum norm $\|\mathbf{x}\|_\infty := \sup_{n \in \mathbb{N}} |\mathbf{x}(n)|$. Note that $c_0(\mathbb{N}) \subseteq \ell^\infty(\mathbb{N})$ is a classic example for a closed subspace that is not complemented in a Banach space, as first shown in Phillips (1940). Now consider the quotient space $\mathbb{Y} = \ell^\infty(\mathbb{N})/c_0(\mathbb{N})$, where elements in $\ell^\infty(\mathbb{N})$ are identified if their difference is contained in $c_0(\mathbb{N})$. Then the quotient map $\mathbf{A} \colon \ell^\infty(\mathbb{N}) \to \mathbb{Y} \colon \mathbf{x} \mapsto [\mathbf{x}]$ is clearly linear, bounded, and onto with $\ker(\mathbf{A}) = c_0(\mathbb{N})$. It is clear that a right inverse of $\mathbf{A}$ exists, which can be constructed by simply choosing any representative in $[\mathbf{x}]$. However, because $c_0(\mathbb{N})$ is not complemented, the kernel of $\mathbf{A}$ is not complemented, and according to Proposition 2, no linear right inverse $\mathbf{B}$ of $\mathbf{A}$ such that $\mathbf{B} \circ \mathbf{A}$ is bounded.

At first glance it might be surprising that bounded linear forward operators do not always have suitable linear right inverses. However, following Example 1, one constructs bounded linear operators without linear right inverses for every Banach space that is not isomorphic to a Hilbert space. This in particular includes the function spaces $L^p(\Omega)$ with $p \neq 2$, where inverse problems are often formulated on.

Proposition 3 (Right inverses in Hilbert spaces). *Let $\mathbb{X}$ be a Hilbert space and let $\mathbf{P}_{\ker(\mathbf{A})} \colon \mathbb{X} \to \mathbb{X}$ denote the orthogonal projection onto* $\ker(\mathbf{A})$.

(a) $\mathbf{A}$ has a unique linear right inverse $\mathbf{A}^\dagger \colon \operatorname{ran}(\mathbf{A}) \to \mathbb{X}$ with $\mathbf{A} \circ \mathbf{A}^\dagger = \operatorname{Id} - \mathbf{P}_{\ker(\mathbf{A})}$.
(b) $\forall \mathbf{y} \in \operatorname{ran}(\mathbf{A}) \colon \mathbf{A}^\dagger(\mathbf{y}) = \arg\min\{\|\mathbf{x}\| \mid \mathbf{A}(\mathbf{x}) = \mathbf{y}\}$.
(c) $\mathbf{A}^\dagger$ is continuous if and only if $\operatorname{ran}(\mathbf{A})$ is closed.
(d) If $\operatorname{ran}(\mathbf{A})$ is non-closed, then any right inverse is discontinuous.

Proof. In a Hilbert space, the orthogonal complement $\ker(\mathbf{A})^\perp$ defines a complement of $\ker(\mathbf{A})$, and therefore, (a), (c), (d) follow from Propositions 1 and 2. Item (b) holds because any solution of the equation $\mathbf{A}(\mathbf{x}) = \mathbf{y}$ has the form $\mathbf{x} = \mathbf{x}_1 + \mathbf{x}_2 \in \ker(\mathbf{A})^\perp \oplus \ker(\mathbf{A})$, and we have $\|\mathbf{x}\|^2 = \|\mathbf{x}_1\|^2 + \|\mathbf{x}_2\|^2$ according to Pythagoras theorem. □

In the case that $\mathbb{X}$ and $\mathbb{Y}$ are both Hilbert spaces, there is a unique extension $\mathbf{A}^\dagger \colon \operatorname{ran}(\mathbf{A}) \oplus \operatorname{ran}(\mathbf{A})^\perp \to \mathbb{X}$, such that $\mathbf{A}^\dagger(\mathbf{y}_1 \oplus \mathbf{y}_2) = \mathbf{A}^\dagger(\mathbf{y}_1)$ for all $\mathbf{y}_1 \oplus \mathbf{y}_2 \in \operatorname{ran}(\mathbf{A}) \oplus \operatorname{ran}(\mathbf{A})^\perp$. The operator $\mathbf{A}^\dagger$ is referred to as the Moore-Penrose inverse of $\mathbf{A}$. For more background on generalized in inverses in Hilbert and Banach spaces, see Nashed (1987).

Regularization Methods

Let $\mathbf{B} \colon \operatorname{ran}(\mathbf{A}) \subseteq \mathbb{Y} \to \mathbb{X}$ be a right inverse of $\mathbf{A}$, set $\mathbb{M} := \operatorname{ran}(\mathbf{B})$ and suppose $\mathbb{M}^* \subseteq \mathbb{M}$. Moreover, let $\mathcal{D} \colon \mathbb{Y} \times \mathbb{Y} \to [0, \infty]$ be some functional measuring closeness in the data space. The standard choice is the norm distance $\mathbf{d}_\mathbb{Y}(\mathbf{y}, \mathbf{y}^\delta) := \|\mathbf{y} - \mathbf{y}^\delta\|$ but also other choices will be considered in this chapter.

Definition 3 (Regularization method). A function $\mathbf{R} \colon (0, \infty) \times \mathbb{Y} \to \mathbb{X}$ with

$$\forall \mathbf{x} \in \mathbb{M}^* \colon \lim_{\delta \to 0} \sup \left\{ \|\mathbf{x} - \mathbf{R}(\delta, \mathbf{y}^\delta)\| \mid \mathbf{y}^\delta \in \mathbb{Y} \wedge \mathcal{D}(\mathbf{A}(\mathbf{x}), \mathbf{y}^\delta) \le \delta \right\} = 0 \quad (3)$$

is called (convergent) regularization method for (1) on the signal class $\mathbb{M}^* \subseteq \mathbb{M}$ with respect to the similarity measure $\mathcal{D}$. We also write $(\mathbf{R}_\delta)_{\delta > 0}$ instead of $\mathbf{R}$.

The following lemma gives a useful guideline for creating regularization methods based on point-wise approximations of $\mathbf{B}$.

Proposition 4 (Point-wise approximations are regularizations). *Let $(\mathbf{B}_\alpha)_{\alpha > 0}$ be a family of continuous operators $\mathbf{B}_\alpha \colon \mathbb{Y} \to \mathbb{X}$ that converge uniformly to $\mathbf{B}$ on $\mathbf{A}(\mathbb{M}^*)$ as $\alpha \to 0$. Then, there is a function $\alpha_0 \colon (0, \infty) \to (0, \infty)$ such that*

$$\mathbf{R} \colon (0, \infty) \times \mathbb{Y} \to \mathbb{X} \colon (\delta, \mathbf{y}^\delta) \mapsto \mathbf{R}(\delta, \mathbf{y}^\delta) := \mathbf{B}_{\alpha_0(\delta)}(\mathbf{y}^\delta) \quad (4)$$

is a regularization method for (1) on the signal class $\mathbb{M}^*$ with respect to the norm distance $\mathbf{d}_\mathbb{Y}$. One calls α_0 an a-prior parameter choice over the set $\mathbb{M}^*$.

Proof. For any $\epsilon > 0$, choose $\alpha(\epsilon)$ such $\|\mathbf{B}_{\alpha(\epsilon)}(\mathbf{y}) - \mathbf{x}\| \leq \epsilon/2$ for all $\mathbf{x} \in \mathbb{M}^*$. Moreover, choose $\tau(\epsilon)$ such that for all $\mathbf{z} \in \mathbb{Y}$ with $\|\mathbf{y} - \mathbf{z}\| \leq \tau(\epsilon)$, we have $\|\mathbf{B}_{\alpha(\epsilon)}(\mathbf{y}) - \mathbf{B}_{\alpha(\epsilon)}(\mathbf{z})\| \leq \epsilon/2$. Without loss of generality, we can assume that $\tau(\epsilon)$ is strictly increasing and continuous with $\tau(0+) = 0$. We define $\alpha_0 := \alpha \circ \tau^{-1}$. Then, for every $\delta > 0$ and $\|\mathbf{y} - \mathbf{y}^\delta\| \leq \delta$,

$$\|\mathbf{B}_{\alpha_0(\delta)}(\mathbf{y}^\delta) - \mathbf{x}\| \leq \|\mathbf{B}_{\alpha_0(\delta)}(\mathbf{y}) - \mathbf{x}\| + \|\mathbf{B}_{\alpha_0(\delta)}(\mathbf{y}) - \mathbf{B}_{\alpha_0(\delta)}(\mathbf{y}^\delta)\|$$
$$= \|\mathbf{B}_{\alpha \circ \tau^{-1}(\delta)}(\mathbf{y}) - \mathbf{x}\| + \|\mathbf{B}_{\alpha \circ \tau^{-1}(\delta)}(\mathbf{y}) - \mathbf{B}_{\alpha \circ \tau^{-1}(\delta)}(\mathbf{y}^\delta)\|$$
$$\leq \tau^{-1}(\delta)/2 + \tau^{-1}(\delta)/2 = \tau^{-1}(\delta).$$

Because $\tau^{-1}(\delta) \to 0$ as $\delta \to 0$ this completes the proof. □

A popular class of regularization methods is convex variational regularization defined by a convex functional $\mathcal{R}: \mathbb{X} \to [0, \infty]$. These methods approximate right inverses, given by the $\mathcal{R}$-minimizing solutions of $\mathbf{A}(\mathbf{x}) = \mathbf{y}$. Such solutions are elements in $\arg\min\{\mathcal{R}(\mathbf{x}) \mid \mathbf{x} \in \mathbb{X} \wedge \mathbf{A}(\mathbf{x}) = \mathbf{y}\}$. Note that an $\mathcal{R}$-minimizing solution exists whenever $\mathbb{X}$ is reflexive, $\mathcal{R}$ is coercive and weakly lower semi-continuous, and the equation $\mathbf{A}(\mathbf{x}) = \mathbf{y}$ has at least one solution in the domain of $\mathcal{R}$. Moreover, the $\mathcal{R}$-minimizing solution is unique if $\mathcal{R}$ is strictly convex. In this case this immediately defines a right inverse for $\mathbf{A}$. Convex variational regularization is defined by minimizing the Tikhonov functional $\mathcal{T}_{\mathbf{y}^\delta, \alpha}: \mathbb{X} \to [0, \infty]: \mathbf{x} \mapsto \mathcal{D}(\mathbf{A}(\mathbf{x}), \mathbf{y}^\delta) + \alpha\mathcal{R}(\mathbf{x})$ for data $\mathbf{y}^\delta \in \mathbb{Y}$ and regularization parameter $\alpha > 0$. In section "The NETT Approach", we will study a more general form, including non-convex regularizers defined by a neural network. At this point, we only state one result on convex variational regularization.

Theorem 1 (Tikhonov regularization in Banach spaces). *Let $\mathbb{X}$ be reflexive, strictly convex, and $p, q > 1$. Moreover, suppose that $\mathbb{X}$ satisfies the Radon–Riesz property; that is, for any sequence $(\mathbf{x}_k)_{k \in \mathbb{N}} \in \mathbb{X}^\mathbb{N}$, the weak convergence $\mathbf{x}_k \rightharpoonup \mathbf{x} \in \mathbb{X}$ together with the convergence in the norm $\|\mathbf{x}_k\| \to \|\mathbf{x}\|$ implies $\lim_{k \to \infty} \mathbf{x}_k = \mathbf{x}$ in the norm topology. Then the following holds:*

(a) $\mathbf{A}^\dagger: \operatorname{ran}(\mathbf{A}) \to \mathbb{X}: \mathbf{y} \mapsto \arg\min\{\|\mathbf{x}\| \mid \mathbf{x} \in \mathbb{X} \wedge \mathbf{A}(\mathbf{x}) = \mathbf{y}\}$ is well defined.
(b) For all, $\alpha > 0$ the mapping $\mathbf{B}_\alpha: \mathbb{Y} \to \mathbb{X}: \mathbf{y}^\delta \mapsto \arg\min\{\|\mathbf{A}(\mathbf{x}) - \mathbf{y}^\delta\|^p + \|\mathbf{x}\|^q \mid \mathbf{x} \in \mathbb{X}\}$ is well defined and continuous.
(c) For any $\alpha_0: (0, \infty) \to (0, \infty)$ with $\alpha_0 \to 0$ and $\delta^p/\alpha_0(\delta) \to 0$ as $\delta \to 0$, the mapping defined by (4) and (b) is a regularization method for (1) on $\mathbf{A}^\dagger(\mathbb{X})$ with respect to the norm distance $\mathbf{d}_\mathbb{X}$

Proof. See Ivanov et al. (2002) and Scherzer et al. (2009).

In the Hilbert space setting, the mapping $\mathbf{A}^\dagger$ defined In Theorem 1 is given by the Moore-Penrose inverse; see Proposition 3 and the text below this Proposition.

Deep Learning

In this subsection, we give a brief review of neural networks and deep learning. Deep learning can be characterized as the field where deep neural networks are used to solve various learning problems (LeCun et al. 2015; Goodfellow et al. 2016). Several such methods recently appeared as a new paradigm for solving inverse problems. In deep learning literature, neural networks are often formulated in a finite dimensional setting. To allow a unified treatment, we consider here a general setting, including the finite dimensional as well as the infinite-dimensional setting.

Problem 1 (The supervised learning problem). Suppose the aim is to find an unknown function $\boldsymbol{\Phi} \colon \mathbb{Y} \to \mathbb{X}$ between two Banach spaces. Similar to classical regression, we are given data $(\mathbf{y}_i, \mathbf{x}_i) \in \mathbb{Y} \times \mathbb{X}$ with $\boldsymbol{\Phi}(\mathbf{y}_i) \simeq \mathbf{x}_i$ for $i = 1, \ldots, N$. From this data, we aim to estimate the function $\boldsymbol{\Phi}$. For that purpose, one chooses a certain class $(\boldsymbol{\Phi}_\theta)_{\theta \in \Theta}$ of functions $\boldsymbol{\Phi}_\theta \colon \mathbb{Y} \to \mathbb{X}$ and defines $\boldsymbol{\Phi} := \boldsymbol{\Phi}_{\theta^*}$ where θ^* minimizes the penalized empirical risk functional

$$R_N \colon \Theta \to [0, \infty] \colon \theta \mapsto \frac{1}{N} \sum_{i=1}^{N} L(\boldsymbol{\Phi}_\theta(\mathbf{y}_i), \mathbf{x}_i) + P(\theta). \tag{5}$$

Here $L \colon \mathbb{X} \times \mathbb{X} \to \mathbb{R}$ is the so-called loss function, which is a measure for the error made by the function $\boldsymbol{\Phi}_\theta$ on the training examples, and P is a penalty that prevents overfitting of the network and also stabilizes the training process.

Both the numerical minimization of the functional (5) and investigating properties of θ^* as $N \to \infty$ are of interest in its own (Glorot and Bengio 2010; Chen et al. 2018) but not subject of our analysis. Instead, most theory in this chapter is developed under the assumption of suitable trained prediction function.

Definition 4 (Neural network). Let Θ be a parameter set and $H_{\ell,\theta} \colon \mathbb{X}_0 \times \cdots \times \mathbb{X}_{\ell-1} \to \mathbb{X}_\ell$, for $\ell = 1, \ldots, L$ and $\theta \in \Theta$ be mappings between Banach spaces with $\mathbb{X}_0 = \mathbb{Y}$ and $\mathbb{X}_L = \mathbb{X}$. We call a family $(\boldsymbol{\Phi}_\theta)_{\theta \in \Theta}$ of recursively defined mappings

$$\boldsymbol{\Phi}_\theta := a_{L,\theta} \colon \mathbb{Y} \to \mathbb{X} \quad \text{where } \forall \ell \in \{1, \ldots, L\} \colon a_{\ell,\theta} = H_{\ell,\theta}(\mathrm{Id}, a_{1,\theta}, \ldots, a_{\ell-1,\theta}) \tag{6}$$

a neural network. In that context, $\mathbb{X}_1, \ldots, \mathbb{X}_{L-1}$ are called the hidden spaces. We refer to the individual members $\boldsymbol{\Phi}_\theta$ of a neural network as neural network functions.

A neural network in finite dimension can be seen as discretization of $(\boldsymbol{\Phi}_\theta)_{\theta \in \Theta}$, where $\mathbb{Y}$ and $\mathbb{X}$ are discretized using any standard discretization approach.

Example 2 (Layered neural network). As a typical example for a neural network, consider a layered neural network $(\mathbf{\Phi}_\theta)_{\theta \in \Theta}$ with L layers between finite dimensional spaces. In this case, each network function has the form $\mathbf{\Phi}_\theta : \mathbb{R}^p \to \mathbb{R}^q : \mathbf{y} \mapsto (\sigma_L \circ \mathcal{V}_L^\theta) \circ \cdots \circ (\sigma_1 \circ \mathcal{V}_1^\theta)(\mathbf{y})$, where $\mathcal{V}_\ell^\theta : \mathbb{R}^{d(\ell-1)} \to \mathbb{R}^{d(\ell)}$ are affine mappings and $\sigma_\ell : \mathbb{R}^{d(\ell)} \to \mathbb{R}^{d(\ell)}$ are nonlinear mappings with $d(0) = p$ and $d(L) = q$. The notion indicates that the affine mappings depend on the parameters $\theta \in \Theta$, while the nonlinear mappings are taken fixed. Although this is standard in neural networks, modifications where the nonlinearities contain trainable parameters have been proposed (Agostinelli et al. 2014; Ramachandran et al. 2017). The affine parts $\mathcal{V}_\ell^\theta$, which are the learned parts in the neural network, can be represented by a $d(\ell) \times d(\ell - 1)$ matrix for the linear part and a vector of size $1 \times d(\ell)$ for the translation part.

In standard neural networks, the entries of the matrix and the bias vector are taken as independent parameters. For typical inverse problems where the dimensions p and q are large, learning all these numbers is challenging and perhaps an impossible task. For example, the matrix describing the linear part of a layer mapping a 200×200 image to an image of the same size already contains 1.6 billion parameters. Learning these parameters from data seems challenging. Recent neural networks and, in particular, convolutional neural networks (CNNs) use the concepts of sparsity and weight sharing to significantly reduce the number of parameters to be learned.

Example 3 (CNNs using sparsity and weight sharing). In order to reduce the number of free parameter between a linear mapping between images, say of sizes $q = n \times n$ and $p = n \times n$, CNNs implement sparsity and weight sharing via convolution operators. In fact, a convolution operation $\mathbf{K} \colon \mathbb{R}^{n \times n} \to \mathbb{R}^{n \times n}$ with kernel size $k \times k$ is represented by k^2 numbers, which clearly enormously reduces the number n^4 of parameters required to represent a general linear mapping on $\mathbb{R}^{n \times n}$. To enrich the expressive power of the neural network, actual CNN architectures use multiple-input multiple-output convolutions $\mathbf{K} \colon \mathbb{R}^{n \times n \times c} \to \mathbb{R}^{n \times n \times d}$, which uses one convolution kernel for each pair in $\{1, \ldots, c\} \times \{1, \ldots, d\}$ formed between each input channel and each output channel. This now increases the number of learnable parameters to cdk^2, but overall the number of parameters remains much smaller than for a full dense layer between large images. Moreover, the use of multiple-input multiple-output convolutions in combination with typical nonlinearities introduces a flexible and complex structure, which demonstrated to give state-of-the art results in various imaging tasks.

Regularizing Networks

Throughout, this section let $\mathbf{A} \colon \mathbb{X} \to \mathbb{Y}$ be a linear forward operator between Banach spaces and $\mathbf{B} \colon \mathrm{ran}(\mathbf{A}) \to \mathbb{X}$ a linear right inverse with $\mathbb{U} := \mathrm{ran}(\mathbf{B})$. In particular, the kernel of $\mathbf{A}$ is complemented, and we can write $\mathbb{X} = \mathbb{U} \oplus \ker(\mathbf{A})$.

The results in this section generalize the methods and some of the results of Schwab et al. (2019) from the Hilbert case to the Banach space case.

Null-Space Networks

The idea of post-processing networks is to improve a given right inverse by applying a network. Standard networks, however, will destroy data consistency of the initial reconstruction. Null-space networks are the natural class of neural networks restoring data consistency.

Definition 5 (Null-space network). We call the family $(\mathrm{Id}_{\mathbb{X}} + \mathbf{P}_{\ker(\mathbf{A})} \circ \mathbf{N}_\theta)_{\theta \in \Theta}$ a null-space network if $(\mathbf{N}_\theta)_{\theta \in \Theta}$ is any network of Lipschitz continuous functions $\mathbf{N}_\theta \colon \mathbb{X} \to \mathbb{X}$. We will also refer to individual functions $\mathbf{\Phi}_\theta = \mathrm{Id}_{\mathbb{X}} + \mathbf{P}_{\ker(\mathbf{A})} \circ \mathbf{N}_\theta$ as null-space networks.

Any null-space network $\mathbf{\Phi}_\theta = \mathrm{Id}_{\mathbb{X}} + \mathbf{P}_{\ker(\mathbf{A})} \circ \mathbf{N}_\theta$ preserves data consistency in the sense that $\mathbf{A}(\mathbf{x}) = \mathbf{y}$ implies $\mathbf{A}(\mathbf{\Phi}_\theta(\mathbf{x})) = \mathbf{y}$, which can be seen from

$$\mathbf{A} \circ \bigl(\mathrm{Id}_{\mathbb{X}} + \mathbf{P}_{\ker(\mathbf{A})} \circ \mathbf{N}_\theta\bigr)(\mathbf{x}) = \mathbf{A}(\mathbf{x}) + \mathbf{A} \circ \mathbf{P}_{\ker(\mathbf{A})} \circ \mathbf{N}_\theta(\mathbf{x}) = \mathbf{y}. \tag{7}$$

A standard residual network $\mathrm{Id}_{\mathbb{X}} + \mathbf{N}_\theta$ often used as post-processing network in general does not satisfy this such a data consistency property.

Remark 1 (Computation of the projection layer). One of the main ingredient in the null-space network is the computation of the projection layer $\mathbf{P}_{\ker(\mathbf{A})}$. In some cases, it can be computed explicitly. For example, if $\mathbf{A} = \mathbf{S}_I \circ \mathbf{F}$ is the subsampled Fourier transform, then $\mathbf{P}_{\ker(\mathbf{A})} = \mathbf{F}^* \circ \mathbf{S}_I \circ \mathbf{F}$. For a general forward operator between Hilbert spaces, the projection $\mathbf{P}_{\ker(\mathbf{A})}\mathbf{z}$ can be implemented via standard methods for solving linear equation. For example, using the starting value $\mathbf{z}$ and solving the equation $\mathbf{A}(\mathbf{x}) = 0$ with the CG (conjugate gradient) method for the normal equation or Landwebers methods gives a sequence that converges to the projection $\mathbf{P}_{\ker(\mathbf{A})}\mathbf{z} = \arg\min\{\|\mathbf{x} - \mathbf{z}\| \mid \mathbf{A}(\mathbf{x}) = 0\}$.

An example comparing a standard residual network $\mathrm{Id}_{\mathbb{X}} + \mathbf{N}_\theta$ and a null-space network $\mathrm{Id} + \mathbf{P}_{\ker(\mathbf{A})} \circ \mathbf{N}_\theta$ both with two weight layers are shown in Fig. 1.

Proposition 5 (Right inverses defined by null-space networks). *Let* $\mathbf{B} \colon ran(A) \to \mathbb{X}$ *be a given linear right inverse such that* $\mathbf{B} \circ \mathbf{A}$ *is bounded and* $\mathbf{\Phi}_\theta = \mathrm{Id}_{\mathbb{X}} + \mathbf{P}_{\ker(\mathbf{A})} \circ \mathbf{N}_\theta$ *be a null-space network. Then the composition*

$$\mathbf{\Phi}_\theta \circ \mathbf{B} \colon ran(\mathbf{A}) \to \mathbb{X} \colon \mathbf{y} \mapsto (\mathrm{Id}_{\mathbb{X}} + \mathbf{P}_{\ker(\mathbf{A})} \circ \mathbf{N}_\theta)(\mathbf{B}\mathbf{y}) \tag{8}$$

is right inverse of $\mathbf{A}$*. Moreover, the following assertions are equivalent:*

29 Regularization of Inverse Problems by Neural Networks

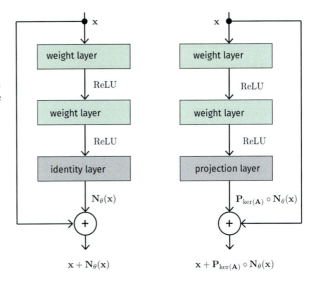

Fig. 1 Residual network Id $+\mathbf{N}_\theta$ (left) versus null-space network Id $+\mathbf{P}_{\ker(\mathbf{A})} \circ \mathbf{N}_\theta$ (right). The difference between the two architectures is the projection layer $\mathbf{P}_{\ker(\mathbf{A})}$ in the null-space network after the last weight layer

(i) $\mathbf{\Phi}_\theta \circ \mathbf{B}$ is continuous
(ii) $\mathbf{B}$ is continuous
(iii) $\mathrm{ran}(\mathbf{A})$ is closed.

Proof. Because $\mathbf{B}$ is a right inverse, we have $(\mathbf{A} \circ \mathbf{B})(\mathbf{x}) = \mathbf{y}$ for all $\mathbf{y} \in \mathrm{ran}(\mathbf{A})$. Hence, the data consistency property (7) implies $\mathbf{A}(((\mathrm{Id}_\mathbb{X} + \mathbf{P}_{\ker(\mathbf{A})} \circ \mathbf{N}_\theta) \circ \mathbf{B})(\mathbf{x})) = \mathbf{y}$, showing that $\mathbf{\Phi}_\theta \circ \mathbf{B}$ is a right inverse of $\mathbf{A}$. The implication (i) $\Rightarrow$ (ii) follows from the identity $\mathbf{P}_\mathbb{U} \circ \mathbf{\Phi}_\theta \circ \mathbf{B} = \mathbf{B}$ and the continuity of the projection. The implication (ii) $\Rightarrow$ (iii) follows from the continuity of $\mathbf{\Phi}_\theta$. Finally, the equivalence (ii) $\Leftrightarrow$ (iii) has been established in Proposition 2. □

The benefit of non-linear right inverses defined by null-space networks is that they can be adjusted to a given image class. A possible network training is given as follows:

Remark 2 (Possible training strategy). The null-space network $\mathbf{\Phi}_\theta = \mathrm{Id} + \mathbf{P}_{\ker(\mathbf{A})} \circ \mathbf{N}_\theta$ can be trained to map elements in $\mathbb{M}$ to the elements from the desired class of images. For that purpose, select training data pairs $(\mathbf{x}_1, \mathbf{z}_1), \ldots, (\mathbf{x}_N, \mathbf{z}_N)$ with $\mathbf{z}_i = \mathbf{B} \circ \mathbf{A}(\mathbf{x}_i)$ and minimize the regularized empirical risk,

$$R_N : \Theta \to [0, \infty] : \theta \mapsto \frac{1}{N} \sum_{n=1}^{N} \|\mathbf{x}_i - \mathbf{\Phi}_\theta(\mathbf{z}_i)\|^2 + \beta \|\theta\|_2^2. \tag{9}$$

Note that for our analysis, it is not required that (9) is exactly minimized. Any null-space network $\mathbf{\Phi}_\theta$ where $\sum_{n=1}^{N} \|\mathbf{x}_i - \mathbf{\Phi}_\theta(\mathbf{z}_i)\|^2$ is small yields a right inverse $\mathbf{\Phi}_\theta \mathbf{B}$ that does a better job in estimating $\mathbf{x}_n$ from data $\mathbf{A}\mathbf{x}_n$ than the original right inverse $\mathbf{B}$.

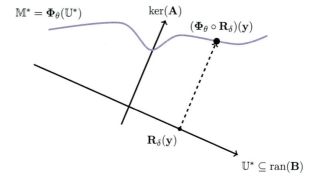

Fig. 2 Linear Regularization $(\mathbf{R}_\delta)_{\delta>0}$ combined with a null-space network $\mathbf{\Phi}_\theta = \mathrm{Id} + \mathbf{P}_{\ker(\mathbf{A})} \circ \mathbf{N}_\theta$. We start with a linear regularization $\mathbf{R}_\delta \mathbf{y}$ and the null-space network $\mathbf{\Phi}_\theta = \mathrm{Id} + \mathbf{P}_{\ker(\mathbf{A})} \circ \mathbf{N}_\theta$ adds missing parts along the null space ker(**A**)

Proposition 5 implies that the solution of ill-posed problems by null-space networks requires the use of stabilization methods similar to the case of classical methods. In the following subsection, we show that the combination of null-space network with a regularization of **B** in fact yields a regularization method on a signal class related to the null-space network.

Convergence Analysis

Throughout the following, let $\mathbf{\Phi}_\theta = \mathrm{Id} + \mathbf{P}_{\ker(\mathbf{A})} \circ \mathbf{N}_\theta$ be null-space network and $(\mathbf{R}_\delta)_{\delta>0}$ be a regularization method for (1) on the signal class $\mathbb{U}^* \subseteq \mathbb{U}$ with respect to the similarity measure $\mathcal{D}$ as introduced in Definition 3. As illustrated in Fig. 2, we consider the family $(\mathbf{\Phi}_\theta \circ \mathbf{R}_\delta)_{\delta>0}$ of compositions of the regularization method with the null-space network.

Theorem 2 (Regularizing null-space network). *For a given null-space network $\mathbf{\Phi}_\theta = \mathrm{Id} + \mathbf{P}_{\ker(\mathbf{A})} \circ \mathbf{N}_\theta$ and a given regularization method $(\mathbf{R}_\delta)_{\delta>0}$ on the signal class $\mathbb{U}^*$, the family $(\mathbf{\Phi}_\theta \circ \mathbf{R}_\delta)_{\delta>0}$ is regularization method for (1) on the signal class $\mathbf{\Phi}_\theta(\mathbb{U}^*)$ with respect to the similarity measure $\mathcal{D}$. We call $(\mathbf{\Phi}_\theta \circ \mathbf{R}_\delta)_{\delta>0}$ a regularizing null-space network.*

Proof. Let L be a Lipschitz constant of $\mathbf{\Phi}_\theta$ and recall $\mathbf{\Phi}_\theta = \mathrm{Id} + \mathbf{P}_{\ker(\mathbf{A})} \circ \mathbf{N}_\theta$. For any $\mathbf{x} \in \mathbf{\Phi}_\theta(\mathbb{U}^*)$ and $\mathbf{y}^\delta \in \mathbb{Y}$, we have

$$\|\mathbf{x} - \mathbf{\Phi}_\theta \circ \mathbf{R}_\delta(\mathbf{y}^\delta)\| = \|(\mathrm{Id} + \mathbf{P}_{\ker(\mathbf{A})} \circ \mathbf{N}_\theta)(\mathbf{B} \circ \mathbf{A}(\mathbf{x})) - (\mathrm{Id} + \mathbf{P}_{\ker(\mathbf{A})} \circ \mathbf{N}_\theta)(\mathbf{R}_\delta(\mathbf{y}^\delta))\|$$

$$\leq L\|(\mathbf{B} \circ \mathbf{A})(\mathbf{x}) - \mathbf{R}_\delta(\mathbf{y}^\delta)\|.$$

Here, we have used the identity $\mathbf{x} = (\mathrm{Id} + \mathbf{P}_{\ker(\mathbf{A})} \circ \mathbf{N}_\theta)((\mathbf{B} \circ \mathbf{A})(\mathbf{x}))$ for $\mathbf{x} \in \mathrm{ran}(\mathbf{\Phi}_\theta)$. Consequently

$$\sup\{\|\mathbf{x} - (\mathbf{\Phi}_\theta \circ \mathbf{R}_\delta)(\mathbf{y}^\delta)\| \mid \mathbf{y}^\delta \in \mathbb{Y} \wedge \mathcal{D}(\mathbf{y}^\delta, \mathbf{y}) \leq \delta\}$$
$$\leq L \sup\{\|(\mathbf{B} \circ \mathbf{A})(\mathbf{x}) - \mathbf{R}_\delta(\mathbf{y}^\delta)\| \mid \mathbf{y}^\delta \in \mathbb{Y} \wedge \mathcal{D}(\mathbf{y}^\delta, \mathbf{y}) \leq \delta\} \to 0.$$

In particular, $(\mathbf{\Phi}_\theta \circ \mathbf{R}_\delta)_{\delta>0}$ is a regularization method for (1) on $\mathbf{\Phi}_\theta(\mathbb{U}^*)$ with respect to the similarity measure $\mathcal{D}$. □

In Hilbert spaces, a wide class of regularizing reconstruction networks can be defined by regularizing filters.

Example 4 (Regularizations defined by filters). Let $\mathbb{X}$ and $\mathbb{Y}$ be Hilbert spaces. A family $(g_\alpha)_{\alpha>0}$ of functions $g_\alpha \colon [0, \|\mathbf{A}^* \circ \mathbf{A}\|] \to \mathbb{R}$ is called a regularizing filter if it satisfies

- For all $\alpha > 0$, g_α is piecewise-continuous and boudned.
- $\exists C > 0 \colon \sup\{|\lambda g_\alpha(\lambda)| \mid \alpha > 0 \wedge \lambda \in [0, \|\mathbf{A}^* \circ \mathbf{A}\|]\} \leq C$.
- $\forall \lambda \in (0, \|\mathbf{A}^* \circ \mathbf{A}\|] \colon \lim_{\alpha \to 0} g_\alpha(\lambda) = 1/\lambda$.

For a given regularizing filter $(g_\alpha)_{\alpha>0}$, define $\mathbf{B}_\alpha := g_\alpha(\mathbf{A}^* \circ \mathbf{A})\mathbf{A}^*$. Then for a suitable parameter choice $\alpha = \alpha(\delta, \mathbf{y})$, the family $(\mathbf{B}_{\alpha(\delta, \cdot)})_{\delta>0}$ is a regularization method on $\operatorname{ran}(\mathbf{A}^\dagger)$. Therefore, according to Theorem 2, the family $(\mathbf{\Phi}_\theta \circ \mathbf{B}_{\alpha(\delta, \cdot)})_{\delta>0}$ is a regularization method on $\mathbf{\Phi}_\theta(\operatorname{ran}(\mathbf{A}^\dagger))$. Note that in this setting, one can derive quantitative error estimates (convergence rates); we refer the interested reader to the original paper Schwab et al. (2019).

Extensions

The regularizing null-space networks defined in Theorem 2 are of the form $\mathbf{\Phi}_\theta \circ \mathbf{R}_\delta$, where $\mathbf{R}_\delta$ is a classical regularization and $\mathbf{\Phi}_\theta$ only acts in the null space of $\mathbf{A}$. In order to better account for noise, it is beneficial to allow the networks to modify $\mathbf{R}_\delta$ also on the complement $\mathbb{U}$.

Definition 6 (Regularizing family of networks). Let $(\mathbf{R}_\delta)_{\delta>0}$ be a regularization method for (1) on the signal class $\mathbb{U}^* \subseteq \mathbb{U}$ with respect to the similarity measure $\mathcal{D}$ as introduced in Definition 3. A family $(\mathbf{\Phi}_{\theta(\delta)} \circ \mathbf{R}_\delta)_{\delta>0}$ is called regularizing family of networks if $(\mathbf{\Phi}_\theta)_{\theta \in \Theta}$ is a neural network, such that the network functions $\mathbf{\Phi}_{\theta(\delta)} \colon \mathbb{X} \to \mathbb{X}$, for $\delta > 0$, are uniformly Lipschitz continuous and

$$\forall \mathbf{z} \in \operatorname{ran}(\mathbf{B}) \colon \lim_{\delta \to 0} \mathbf{\Phi}_{\theta(\delta)}(\mathbf{R}_\delta \circ \mathbf{A}(\mathbf{z})) = \mathbf{N}(\mathbf{z}),$$

for some null-space network $\mathbf{N}$.

Regularizing families of networks have been introduced in Schwab et al. (2020), where it has been shown that a regularizing family of networks defines a regularization method together with convergence rates. Moreover, an example in the form of a data-driven extension of truncated SVD regularization has been given. In a finite dimensional setting, related extension of null-space networks named deep decomposition learning has been introduced in Chen and Davies (2019). A combination of null-space learning with shearlet reconstruction for limited angle tomography has been introduced in Bubba et al. (2019). In Dittmer and Maass (2019), a neural network-based projection approach based on approximate data consistency sets has been studied. Relaxed versions of null-space networks, where approximate data consistency is incorporated via a confidence region or a soft penalty, are proposed in Huang et al. (2020) and Kofler et al. (2020). Finally, extensions of the null-space approach to non-linear problems are studied in Boink et al. (2020).

The NETT Approach

Let us recall that convex variational regularization of the inverse problem (1) consists in minimizing the generalized Tikhonov functional $\mathcal{D}(\mathbf{A}(\mathbf{x}), \mathbf{y}^\delta) + \alpha \mathcal{R}(\mathbf{x})$, where $\mathcal{R}$ is a convex functional and $\mathcal{D}$ a similarity measure (see section "Regularization Methods"). The regularization term $\mathcal{R}$ is traditionally a semi-norm defined on a dense subspace of $\mathbb{X}$. In this section, we will extend this setup by using deep learning techniques with learned regularization functionals.

Learned Regularization Functionals

We assume that the regularizer takes the form

$$\forall \mathbf{x} \in \mathbb{X}: \quad \mathcal{R}(\mathbf{x}) = \mathcal{R}_\theta(\mathbf{x}) := \psi_\theta(\mathbf{\Phi}_\theta(\mathbf{x})). \tag{10}$$

Here $\psi_\theta \colon \Xi \to [0, \infty]$ is a scalar functional, and $\mathbf{\Phi}_\theta(\cdot) \colon \mathbb{X} \to \Xi$ a neural network where $\theta \in \Theta$, for some vector space Θ containing free parameters that can be adjusted by available training data. From the representation learning point of view (Bengio et al. 2013), $\mathbf{\Phi}_\theta(\mathbf{x})$ can be interpreted as a learned representation of $\mathbf{x}$. It could be constructed in such a way that $\psi_\theta \circ \mathbf{\Phi}_\theta$ is minimal for a low-dimensional manifold where the true signals $\mathbf{x}$ are clustered around. Finding such manifold for biomedical images has been an active research topic on manifold learning (Georg et al. 2008; Wachinger et al. 2012). Deep learning has also been used for this purpose (Brosch et al. 2013). A learned regularizer $\mathcal{R}_\theta = \psi_\theta \circ \mathbf{\Phi}_\theta$ reflects the statistics of the signal space, which penalizes those who deviate from the data manifold.

The similarity measure is taken as $\mathcal{D}_\theta \colon \mathbf{C} \times \mathbf{C} \to [0, \infty]$, where $\mathbf{C}$ is a conic closed subset in $\mathbb{Y}$. It is not necessarily symmetric in its arguments. One may take $\mathcal{D}_\theta(\mathbf{A}(\mathbf{x}), \mathbf{y})$ to be a common hard-coded consistency measure such as $\mathcal{D}(\mathbf{A}(\mathbf{x}), \mathbf{y}) =$

$\|\mathbf{A}(\mathbf{x}) - \mathbf{y}\|^2$ or the Kullback-Leibler divergence (which, among others, is used in emission tomography). On the other hand, it can be a learned measure, defined via a neural network. A learned consistency measure $\mathcal{D}_\theta$ reflects the statistics in the data (measurement) space. It can learn to reduce uncertainty in the data measurement process, e.g., by identifying non-functional transducers. It can also learn to reduce the error in the forward model (Aljadaany et al. 2019). Finally, it may encode the range description of the forward operator, which has not been successfully exploited in inverse problems by traditional methods. In summary, it can be said that learned consistency measures have potentially high impact in solving inverse problems.

Using the neural network-based learned regularizer (10) and a learned discrepancy measure as discussed above results in the following optimization problem:

$$\arg\min_{\mathbf{x} \in \mathbb{D}} \mathcal{T}_\theta(\mathbf{x}) := \mathcal{D}_\theta(\mathbf{A}(\mathbf{x}), \mathbf{y}) + \alpha \mathcal{R}_\theta(\mathbf{x}). \tag{11}$$

Solving (11) can be seen as a neural network-based variant of generalized Tikhonov regularization for solving (1). Following Li et al. (2020) we therefore call (11) the network Tikhonov (NETT) approach for solving inverse problems. Currently, there are two main approaches for integrating neural networks in the NETT approach (11): (T1) training the neural networks simultaneously with solving the optimization problem and (T2) training the network independently before solving the optimization problem.

Approach (T1) fuses the data with a solution method of the optimization problem (11). The resulting neural networks, therefore, depend on the method to solve the optimization. This approach enforces the neural networks to learn particular representations that are useful for the chosen optimization technique. These representations will be called *solver-dependent*. The biggest advantage of this end-to-end approach is to provide a direct and relatively fast solution $\mathbf{x}$ for given new data $\mathbf{y}$. It is commonly realized by unrolling an iterative process (Arridge et al. 2019). The resulting neural network is a cascade of relatively small neural networks; each of them is possibly a variant of those appearing in the data consistency or regularization term. It is worth noting that the neural network does not aim for representation learning. Each layer or block serves to move the approximate solution closer to the exact solution. In contrast to typical iterative methods, each block in an unrolled neural network can be different from others. This is explained to speed up the convergence of the learned iterative method. The success of this approach is an interesting phenomenon that needs further investigation. The use of neural networks to implement and accelerate iterative methods to solve traditional regularization methods has been intensively studied. We refer the reader to Arridge et al. (2019) and the references contained therein.

The approach (T2) is more modular (Li et al. 2020; Lunz et al. 2018) and results in smaller training problems and is closer to the meaning of representation learning. The training of the regularizer may or may not depend on the forward operator $\mathbf{A}$. In the former case, the resulting representation is called *model-dependent*, while the latter is *model-independent*. Model-dependent representation seems to be crucial in inverse problem for two reasons. The first reason is that it aligns with the inverse problem (and better serves any solution approach). Secondly, in medical imaging applications, the training signals are often not the groundtruth signals. They are normally obtained with a reconstruction method from high-quality data. Therefore, while training the regularizer, one should also keep in mind the reconstruction mechanism of the training data. A possible approach is to first train a baseline neural network to learn model-independent representation. Then an additional block is added on top to train for model-dependent representation. This has been shown in Obmann et al. (2020a) to be a very efficient strategy.

Let us mention that approach (T1) has richer literature than (T2) but less (convergence) analysis. In this section, we focus more on (T2), where we establish the convergence analysis and convergence rate in section "Convergence Analysis". This is an extension of our works Haltmeier et al. (2019) and Obmann et al. (2020a). In section "Related Methods", we review a few existing methods that are most relevant to our discussion, including some works in approach (T1). We also propose INDIE, which can be regarded as an operator inversion-free variant of the MODL technique (Aggarwal et al. 2018) and can make better use of parallel computation.

Convergence Analysis

Analysis for regularization with neural networks has been studied in Li et al. (2020) and Haltmeier et al. (2019). In this section, we further investigate the issue. To this end, we consider the approach (T2), where the neural networks are trained independently of the optimization problem (11). That is, $\theta = \theta^*$ is already fixed a priori. For the sake of simplicity, we will drop θ from the notation of $\mathcal{R}_\theta$ and $\mathcal{D}_\theta$. We focus on how the problem depends on the regularization parameter α and noise level δ in the data. Such analysis in standard situations is well studied; see, e.g., Scherzer et al. (2009). However, we need to extend the analysis to more general cases to accommodate the fact that $\mathcal{R}$ comes from a neural network and is likely non-convex.

Let us make several assumptions on the regularizer and fidelity term.

Condition 3.

(A1) Network regularizer $\mathcal{R}: \mathbb{X} \to [0, \infty]$ satisfies
    (a) $0 \in \operatorname{dom}(\mathcal{R}) := \{\mathbf{x} \mid \mathcal{R}(\mathbf{x}) < \infty\}$;
    (b) $\mathcal{R}$ is lower semi-continuous;
    (c) $\mathcal{R}(\cdot)$ is coercive, that is $\mathcal{R}(\mathbf{x}) \to \infty$ as $\|\mathbf{x}\| \to \infty$.
(A2) Data consistency term $\mathcal{D}: \mathbf{C} \times \mathbf{C} \to [0, \infty]$ satisfies
    (a) $\operatorname{dom}(\mathcal{D}(0, \cdot)) = \mathbf{C}$;
    (b) If $\mathcal{D}(\mathbf{y}_0, \mathbf{y}_1) < \infty$ and $\mathcal{D}(\mathbf{y}_1, \mathbf{y}_2) < \infty$ then $\mathcal{D}(\mathbf{y}_0, \mathbf{y}_2) < \infty$;
    (c) $\mathcal{D}(\mathbf{y}, \mathbf{y}') = 0 \iff \mathbf{y} = \mathbf{y}'$;
    (d) $\mathcal{D}(\mathbf{y}, \mathbf{y}') \geq C\|\mathbf{y} - \mathbf{y}'\|^2$ holds in any bounded subset of $\operatorname{dom}(\mathcal{D})$;
    (e) For any $\mathbf{y}$, the function $\mathcal{D}(\mathbf{y}, \cdot)$ is continuous and coercive on its domain;

(f) The functional $(\mathbf{x}, \mathbf{y}) \mapsto \mathcal{D}(\mathbf{A}(\mathbf{x}), \mathbf{y})$ is sequentially lower semi-continuous in the weak topology of $\mathbb{X}$ and strong topology of $\mathbb{Y}$.

For (A1), the coercivity condition (c) is the most restrictive. However, it can be accommodated. One such regularizer is proposed in our recent work Haltmeier et al. (2019) as follows:

$$\mathcal{R}(\mathbf{x}) = \phi(\mathbf{E}(\mathbf{x})) + \frac{\beta}{2} \|\mathbf{x} - (\mathbf{D} \circ \mathbf{E})(\mathbf{x})\|_2^2. \tag{12}$$

Here, $\mathbf{D} \circ \mathbf{E} \colon \mathbb{X} \to \mathbb{X}$ is an encoder-decoder network. The regularizer $\mathcal{R}$ is to enforce that a reasonable solution $\mathbf{x}$ satisfies $\mathbf{x} \simeq (\mathbf{D} \circ \mathbf{E})(\mathbf{x})$ and $\phi(\mathbf{E}(\mathbf{x}))$ is small. The term $\phi(\mathbf{E}(\mathbf{x}))$ implements learned prior knowledge, which is normally a sparsity measure in a non-linear basis. The second term $\|\mathbf{x} - (\mathbf{D} \circ \mathbf{E})(\mathbf{x})\|_2^2$ forces $\mathbf{x}$ to be close to data manifold $\mathcal{M}$. Their combination also guarantees the coercivity of the regularization functional $\mathcal{R}$. Another choice for $\mathcal{R}$ was suggested in Li et al. (2020).

For (A2), $\mathbf{C}$ is a conic set in $\mathbb{Y}$. For any $\mathbf{y} \in \mathbf{C}$, we define $\mathrm{dom}(\mathcal{D}(\mathbf{y}, \cdot)) = \{\mathbf{y}' \mid \mathcal{D}(\mathbf{y}, \mathbf{y}') < \infty\}$. The data consistency conditions in (A2) are flexible enough to be satisfied by a few interesting cases. The first example is that $\mathcal{D}(\mathbf{y}, \mathbf{y}') = \|\mathbf{y} - \mathbf{y}'\|^2$, which is probably the most popular data consistency measure. Another case is the Kullback-Leibler divergence, which reads as follows. Let $\mathbb{Y} = \mathbb{R}^n$, and $\mathbf{A} \colon \mathbb{X} \to \mathbb{Y}$ is a bounded linear positive operator.[1] Consider nonnegative cone $\mathbf{C} = \{(\mathbf{y}_1, \ldots, \mathbf{y}_n) \mid \forall i \colon \mathbf{y}_i \geq 0\}$. We define $\mathcal{D} \colon \mathbf{C} \times \mathbf{C} \to [0, \infty]$ by

$$\mathcal{D}(\mathbf{y}, \mathbf{y}') = \sum_{i=1}^{n} \mathbf{y}_i \log \frac{\mathbf{y}_i}{\mathbf{y}'_i} + \mathbf{y}'_i - \mathbf{y}_i.$$

It is straight forward to check that Condition (A2) is satisfied in this case. In particular, item (d) has been verified in Resmerita and Anderssen (2007, Equation (13)).

To emphasize the fact that our data is the noisy version $\mathbf{y}^\delta$ of $\mathbf{y}$, we rewrite (11) as follows:

$$\arg \min_{\mathbf{x} \in \mathbb{D}} \mathcal{T}_{\mathbf{y}^\delta, \alpha}(\mathbf{x}) := \mathcal{D}(\mathbf{A}(\mathbf{x}), \mathbf{y}^\delta) + \alpha \mathcal{R}(\mathbf{x}). \tag{13}$$

Here, $\mathbb{D}$ is a weakly closed conic set in $\mathbb{X}$ such that $\mathbf{A}(\mathbb{D}) \subseteq \mathbf{C}$.

Theorem 4 (Well-posedness and convergence). *Let Condition 3 be satisfied. Then the following assertions hold true:*

(a) *Existence: For all $\mathbf{y} \in \mathbf{C}$ and $\alpha > 0$, there exists a minimizer of $\mathcal{T}_{\mathbf{y}, \alpha}$ in $\mathbb{D}$.*

---

[1] $\mathbf{A}$ is positive if: $\mathbf{y} \geq 0 \Rightarrow \mathbf{A}\mathbf{y} \geq 0$.

(b) *Stability:* If $\mathbf{y}_k \to \mathbf{y}$, $\mathcal{D}(\mathbf{y}, \mathbf{y}_k) < \infty$ and $\mathbf{x}_k \in \arg\min \mathcal{T}_{\alpha; \mathbf{y}_k}$, then weak accumulation points of $(\mathbf{x}_k)_{k \in \mathbb{N}}$ exist and are minimizers of $\mathcal{T}_{\alpha; \mathbf{y}}$.

(c) *Convergence:* Let $\mathbf{y} \in \operatorname{ran}(\mathbf{A}) \cap \mathbf{C}$ and $(\mathbf{y}_k)_{k \in \mathbb{N}}$ satisfy $\mathcal{D}(\mathbf{y}, \mathbf{y}_k) \leq \delta_k$ for some sequence $(\delta_k)_{k \in \mathbb{N}} \in (0, \infty)^{\mathbb{N}}$ with $\delta_k \to 0$. Suppose $\mathbf{x}_k \in \arg\min_{\mathbf{x}} \mathcal{T}_{\mathbf{y}_k, \alpha(\delta_k)}(\mathbf{x})$, and let the parameter choice $\alpha : (0, \infty) \to (0, \infty)$ satisfy

$$\lim_{\delta \to 0} \alpha(\delta) = \lim_{\delta \to 0} \frac{\delta}{\alpha(\delta)} = 0. \tag{14}$$

Then the following holds:

(1) All weak accumulation points of $(\mathbf{x}_k)_{k \in \mathbb{N}}$ are $\mathcal{R}$-minimizing solutions of the equation $\mathbf{A}(\mathbf{x}) = \mathbf{y}$;

(2) $(\mathbf{x}_k)_{k \in \mathbb{N}}$ has at least one weak accumulation point $\mathbf{x}^\dagger$;

(3) Every subsequence $(\mathbf{x}_{k(n)})_{n \in \mathbb{N}}$ that weakly converges to $\mathbf{x}^\dagger$ satisfies $\mathcal{R}(\mathbf{x}_{k(n)}) \to \mathcal{R}(\mathbf{x}^\dagger)$;

(4) If the $\mathcal{R}$-minimizing solution of $\mathbf{A}(\mathbf{x}) = \mathbf{y}$ is unique, then $\mathbf{x}_k \rightharpoonup \mathbf{x}^\dagger$.

Before starting the proof, we recall that $\mathbf{x}^\dagger$ is an $\mathcal{R}$-minimizing solution of the equation $\mathbf{A}\mathbf{x} = \mathbf{y}$ if $\mathbf{x}^\dagger \in \arg\min \{\mathcal{R}(\mathbf{x}) \mid \mathbf{x} \in \mathbb{D} \wedge \mathbf{A}\mathbf{x} = \mathbf{y}\}$.

Proof. (a) First, we observe that $c := \inf_{\mathbf{x}} \mathcal{T}_{\mathbf{y}, \alpha}(\mathbf{x}) \leq \mathcal{T}_{\mathbf{y}, \alpha}(0) < \infty$. Let $(\mathbf{x}_k)_k$ be a sequence such that $\mathcal{T}_{\mathbf{y}, \alpha}(\mathbf{x}_k) \to c$. There exists $M > 0$ such that $\mathcal{T}_{\mathbf{y}, \alpha}(\mathbf{x}_k) \leq M$, which implies $\alpha \mathcal{R}(\mathbf{x}_k) \leq M$. Since $\mathcal{R}$ is coercive, we obtain that $(\mathbf{x}_k)_k$ is bounded. By passing into a subsequence, $\mathbf{x}_{k_i} \rightharpoonup \mathbf{x}^* \in \mathbb{D}$. Due to the lower semi-continuity of $\mathcal{T}_{\alpha, \cdot}(\cdot)$, we have $\mathbf{x}^* \in \arg\min \mathcal{T}_{\mathbf{y}, \alpha}$.

(b) Since $\mathbf{x}_k \in \arg\min \mathcal{T}_{\mathbf{y}, \alpha}$, it holds $\mathcal{T}_{\mathbf{y}_k, \alpha}(\mathbf{x}_k) \leq \mathcal{T}_{\mathbf{y}_k, \alpha}(0) = \mathcal{D}(0, \mathbf{y}_k) + \alpha \mathcal{R}(0)$. Thanks to the continuity of $\mathcal{D}(0, \cdot)$ on $\mathbf{C}$, $(\mathcal{D}(0, \mathbf{y}_k))_k$ is a bounded sequence. Therefore, $\alpha \mathcal{R}(\mathbf{x}_k) \leq \mathcal{T}_{\mathbf{y}_k, \alpha}(\mathbf{x}_k) \leq M$, for a constant M independent of k. Since $\mathcal{R}$ is coercive, $(\mathbf{x}_k)_k$ is bounded and hence has a weakly convergent subsequence $\mathbf{x}_{k_i} \rightharpoonup \mathbf{x}^\dagger$.

Let us now prove that $\mathbf{x}^\dagger$ is a minimizer of $\mathcal{T}_{\mathbf{y}, \alpha}$. Since $\mathcal{T}_{\mathbf{y}, \alpha}(\mathbf{x})$ is lower semi-continuous in $\mathbf{x}$ and $\mathbf{y}$,

$$\liminf_{k_i \to \infty} \mathcal{T}_{\mathbf{y}_{k_i}, \alpha}(\mathbf{x}_{k_i}) \geq \mathcal{T}_{\mathbf{y}, \alpha}(\mathbf{x}^\dagger). \tag{15}$$

On the other hand, let $\mathbf{x} \in \mathbb{D}$ be such that $\mathcal{T}_{\mathbf{y}, \alpha}(\mathbf{x}) < \infty$. We obtain $\mathcal{D}(\mathbf{A}(\mathbf{x}), \mathbf{y}) < \infty$ and $\mathcal{R}(\mathbf{x}) < \infty$. Condition (A2)(d) and $\mathcal{D}(\mathbf{y}, \mathbf{y}_k) < \infty$ give $\mathcal{D}(\mathbf{A}(\mathbf{x}), \mathbf{y}_k) < \infty$. That is, $\mathbf{y}_k \in \operatorname{dom}(\mathcal{D}(\mathbf{A}(\mathbf{x}), \cdot)$. The continuity of $\mathcal{D}(\mathbf{A}(\mathbf{x}), \cdot)$ on its domain implies $\mathcal{D}(\mathbf{A}(\mathbf{x}), \mathbf{y}_k) \to \mathcal{D}(\mathbf{A}(\mathbf{x}), \mathbf{y})$. Since $\mathbf{x}_k$ is the minimizer of $\mathcal{T}_{\mathbf{y}_k, \alpha}$, $\mathcal{T}_{\mathbf{y}_k, \alpha}(\mathbf{x}_k) \leq \mathcal{T}_{\mathbf{y}_k, \alpha}(\mathbf{x})$. Taking the limit, we obtain $\limsup_k \mathcal{T}_{\mathbf{y}_k, \alpha}(\mathbf{x}_k) \leq \mathcal{T}_{\mathbf{y}, \alpha}(\mathbf{x})$. From (15), $\mathcal{T}_{\mathbf{y}, \alpha}(\mathbf{x}^\dagger) \leq \mathcal{T}_{\mathbf{y}, \alpha}(\mathbf{x})$ for any $\mathbf{x} \in \mathbb{D}$. We conclude that $\mathbf{x}^\dagger$ is a minimizer of $\mathcal{T}_{\mathbf{y}, \alpha}$.

(c) We prove the properties item by item.

(1) Since $\mathbf{y} \in \mathbb{R}(\mathbf{A})$, we can pick solution $\bar{\mathbf{x}}$ of $\mathbf{A}(\mathbf{x}) = \mathbf{y}$. We have

$$\mathcal{D}(\mathbf{A}(\mathbf{x}_k), \mathbf{y}_k) + \alpha_k \mathcal{R}(\mathbf{x}_k) \leq \mathcal{D}(\mathbf{y}, \mathbf{y}_k) + \alpha_k \mathcal{R}(\bar{\mathbf{x}}) \leq \delta_k + \alpha_k \mathcal{R}(\bar{\mathbf{x}}). \tag{16}$$

Assume that $\mathbf{x}^\dagger$ is a weak accumulation point of $\mathbf{x}_k$, then

$$\mathcal{D}(\mathbf{A}(\mathbf{x}^\dagger), \mathbf{y}) \leq \liminf_{k \to \infty} \mathcal{D}(\mathbf{A}(\mathbf{x}_k), \mathbf{y}_k) \leq \liminf_{k \to \infty} (\delta_k + \alpha_k \mathcal{R}(\bar{\mathbf{x}})) = 0.$$

Therefore, $\mathcal{D}(\mathbf{A}(\mathbf{x}^\dagger), \mathbf{y}) = 0$ or $\mathbf{A}(\mathbf{x}^\dagger) = \mathbf{y}$. Moreover, $\mathcal{R}(\mathbf{x}_k) \leq \delta_k/\alpha_k + \mathcal{R}(\bar{\mathbf{x}})$, which implies $\mathcal{R}(\mathbf{x}^\dagger) \leq \liminf \mathcal{R}(\mathbf{x}_k) \leq \mathcal{R}(\bar{\mathbf{x}})$. Since this holds for all possible solution $\bar{\mathbf{x}}$ of $\mathbf{A}(\mathbf{x}) = \mathbf{y}$, we conclude that $\mathbf{x}^\dagger$ is a $\mathcal{R}$-minimizing solution of $\mathbf{A}(\mathbf{x}) = \mathbf{y}$.
(2) Using again the inequality $\mathcal{R}(\mathbf{x}_k) \leq \delta_k/\alpha_k + \mathcal{R}(\bar{\mathbf{x}})$ and $\mathcal{R}$ is coercive, we obtain that $\{\mathbf{x}_k\}$ is bounded. Therefore, $\{\mathbf{x}_k\}$ has a weak accumulation point $\mathbf{x}^\dagger$.
(3) Using (16) again for $\bar{\mathbf{x}} = \mathbf{x}^\dagger$, we obtain $\mathcal{R}(\mathbf{x}_k) \leq \delta_k/\alpha_k + \mathcal{R}(\mathbf{x}^\dagger)$, which gives $\limsup_k \mathcal{R}(\mathbf{x}_k) \leq \mathcal{R}(\mathbf{x}^\dagger)$. This together with the fact that $\mathcal{R}$ is lower semi-continuous gives $\mathcal{R}(\mathbf{x}_{k(n)}) \to \mathcal{R}(\mathbf{x}^\dagger)$.
(4) The last conclusion follows straightforwardly from the above three.

$\square$

Let us proceed to obtain some convergence results in the norm. Following Li et al. (2020), we introduce the absolute Bregman distance.

Definition 7 (Absolute Bregman distance). Let $\mathbb{F}: \mathbb{D} \subseteq \mathbb{X} \to \mathbb{R}$ be Gâteaux differentiable at $\mathbf{x} \in \mathbb{X}$. The *absolute Bregman distance* $\mathbf{\Delta}_\mathbb{F}(\cdot, \mathbf{x}): \mathbb{D} \to [0, \infty]$ with respect to $\mathbb{F}$ at $\mathbf{x}$ is defined by

$$\forall \tilde{\mathbf{x}} \in \mathbb{X}: \quad \mathbf{\Delta}_\mathbb{F}(\tilde{\mathbf{x}}, \mathbf{x}) := \left| \mathbb{F}(\tilde{\mathbf{x}}) - \mathbb{F}(\mathbf{x}) - \mathbb{F}'(\mathbf{x})(\tilde{\mathbf{x}} - \mathbf{x}) \right|. \tag{17}$$

Here $\mathbb{F}'(\mathbf{x})$ denotes the Gâteaux derivative of $\mathbb{F}$ at $\mathbf{x}$.

From Theorem 4, we can conclude convergence of $\mathbf{x}_\alpha^\delta$ to the exact solution in the absolute Bregman distance $\mathbf{\Delta}_\mathcal{R}$. Below we show that this implies strong convergence under some additional assumption on the regularization functional. For this purpose, we define the concept of total non-linearity, which was introduced in Li et al. (2020).

Definition 8 (Total non-linearity). Let $\mathbb{F}: \mathbb{D} \subseteq \mathbb{X} \to \mathbb{R}$ be Gâteaux differentiable at $\mathbf{x} \in \mathbb{D}$. We define the *modulus of total non-linearity* of $\mathbb{F}$ at $\mathbf{x}$ as $\nu_\mathbb{F}(\mathbf{x}, \cdot): [0, \infty) \to [0, \infty]$,

$$\forall t > 0: \quad \nu_\mathbb{F}(x, t) := \inf \{ \mathbf{\Delta}_\mathbb{F}(\tilde{\mathbf{x}}, \mathbf{x}) \mid \tilde{\mathbf{x}} \in \mathbb{D} \wedge \|\tilde{\mathbf{x}} - \mathbf{x}\| = t \}. \tag{18}$$

The function $\mathbb{F}$ is called *totally non-linear* at $\mathbf{x}$ if $\nu_\mathbb{F}(\mathbf{x}, t) > 0$ for all $t \in (0, \infty)$.

The following result, due to Li et al. (2020), connects the convergence in absolute Bregman distance and in norm

Proposition 6. *For* $\mathbf{F}\colon D \subseteq \mathbb{X} \to \mathbb{R}$ *and and any* $\mathbf{x} \in D$, *the followings are equivalent:*

(i) *The function* $\mathbf{F}$ *is totally nonlinear at at* $\mathbf{x}$;
(ii) $\forall (\mathbf{x}_n)\colon (\lim_{n\to\infty} \mathcal{B}_\mathbf{F}(\mathbf{x}_n, \mathbf{x}) = 0 \wedge (\mathbf{x}_n)$ *bounded*$) \Rightarrow \lim_{n\to\infty} \|\mathbf{x}_n - \mathbf{x}\| = 0$.

As a consequence, we have the following convergence result in the norm topology.

Theorem 5 (Strong convergence). *Assume that* $\mathbf{A}(\mathbf{x}) = \mathbf{y}$ *has a solution; let* $\mathcal{R}_\theta$ *be totally nonlinear at all* $\mathcal{R}_\theta$-*minimizing solutions of* $\mathbf{A}(\mathbf{x}) = \mathbf{y}$, *and let* $(\mathbf{x}_k)_{k\in\mathbb{N}}, (y_k)_{k\in\mathbb{N}}, (\alpha_k)_{k\in\mathbb{N}}, (\delta_k)_{k\in\mathbb{N}}$ *be as in Theorem 4. Then there is a subsequence* $(\mathbf{x}_{k(\ell)})_{\ell\in\mathbb{N}}$ *of* $(\mathbf{x}_k)_{k\in\mathbb{N}}$ *and an* $\mathcal{R}_\theta$-*minimizing solution* $\mathbf{x}^\dagger$ *of* $\mathbf{A}(\mathbf{x}) = \mathbf{y}$, *such that* $\lim_{\ell\to\infty} \|\mathbf{x}_{k(\ell)} - \mathbf{x}^\dagger\| = 0$. *Moreover, if the* $\mathcal{R}_\theta$-*minimizing solution of* $\mathbf{A}(\mathbf{x}) = \mathbf{y}$ *is unique, then* $\mathbf{x} \to \mathbf{x}^\dagger$ *in the norm topology.*

We now focus on the convergence rate. To this end, we make the following assumptions:

(B1) $\mathbb{Y}$ is a finite dimensional space;
(B2) $\mathcal{R}$ is coercive and weakly sequentially lower semi-continuous;
(B3) $\mathcal{R}$ is Lipschitz;
(B4) $\mathcal{R}$ is Gâteaux differentiable.

The most restrictive condition in the above list is that $\mathbf{A}$ has finite-dimensional range. However, this assumption holds true in practical applications such as sparse data tomography, which is the main focus of deep learning techniques for inverse problems. For infinite-dimensional space result, see Li et al. (2020).

We start our analysis with the following result.

Proposition 7. *Let* (B1)–(B4) *be satisfied and assume that* $\mathbf{x}^\dagger$ *is an* $\mathcal{R}$-*minimizing solution of* $\mathbf{A}(\mathbf{x}) = \mathbf{y}$. *Then there exists a constant* $C > 0$ *such that*

$$\forall \mathbf{x} \in \mathbb{X}\colon \quad \Delta_\mathcal{R}(\mathbf{x}, \mathbf{x}^\dagger) \leq \mathcal{R}(\mathbf{x}) - \mathcal{R}(\mathbf{x}^\dagger) + C\|\mathbf{A}(\mathbf{x}) - \mathbf{A}(\mathbf{x}^\dagger)\|.$$

The proof follows Obmann et al. (2020a). We present it here for the sake of completeness.

Proof. Let us first prove that for some constant $\gamma \in (0, \infty)$, it holds

$$\forall \mathbf{x} \in \mathbb{X}\colon \quad \mathcal{R}(\mathbf{x}^\dagger) - \mathcal{R}(\mathbf{x}) \leq \gamma \|\mathbf{A}(\mathbf{x}^\dagger) - \mathbf{A}(\mathbf{x})\|. \tag{19}$$

Indeed, let $\mathbf{P}$ be the orthogonal projection onto $\ker(\mathbf{A})$ and define $\mathbf{x}_0 = (\mathbf{x}^\dagger - \mathbf{P}(\mathbf{x}^\dagger)) + \mathbf{P}(\mathbf{x})$. Then, we have $\mathbf{A}(\mathbf{x}_0) = \mathbf{A}(\mathbf{x}^\dagger)$ and $\mathbf{x} - \mathbf{x}_0 \in \ker(\mathbf{A})^\perp$. Since the restricted operator $\mathbf{A}|_{\ker(\mathbf{A})^\perp} : \ker(\mathbf{A})^\perp \to \mathbb{Y}$ is injective and has finite-dimensional range, it is bounded from below by a constant γ_0. Therefore,

$$\|\mathbf{A}(\mathbf{x}^\dagger) - \mathbf{A}(\mathbf{x})\| = \|\mathbf{A}(\mathbf{x}_0) - \mathbf{A}(\mathbf{x})\| = \|\mathbf{A}(\mathbf{x}_0 - \mathbf{x})\| \geq \gamma_0 \|\mathbf{x}_0 - \mathbf{x}\|. \tag{20}$$

On the other hand, since $\mathbf{x}^\dagger$ is the $\mathcal{R}$-minimizing solution of $\mathbf{A}(\mathbf{x}) = \mathbf{y}$ and $\mathcal{R}$ is Lipschitz, we have $\mathcal{R}(\mathbf{x}^\dagger) - \mathcal{R}(\mathbf{x}) \leq \mathcal{R}(\mathbf{x}_0) - \mathcal{R}(\mathbf{x}) \leq L\|\mathbf{x}_0 - \mathbf{x}\|$. Together with (20) we obtain (19).

Next we prove that there is a constant γ_1 such that

$$\langle \mathcal{R}'(\mathbf{x}^\dagger), \mathbf{x}^\dagger - \mathbf{x} \rangle \leq \gamma_1 \|\mathbf{A}(\mathbf{x}^\dagger) - \mathbf{A}(\mathbf{x})\|. \tag{21}$$

Indeed, since $\mathbf{x}^\dagger$ is an $\mathcal{R}$-minimizing solution of $\mathbf{A}(\mathbf{x}) = \mathbf{y}$, we obtain $\langle \mathcal{R}'(\mathbf{x}^\dagger), \mathbf{x}^\dagger - \mathbf{x}_0 \rangle \leq 0$. Therefore,

$$\langle \mathcal{R}'(\mathbf{x}^\dagger), \mathbf{x}^\dagger - \mathbf{x} \rangle = \langle \mathcal{R}'(\mathbf{x}^\dagger), \mathbf{x}^\dagger - \mathbf{x}_0 \rangle + \langle \mathcal{R}'(\mathbf{x}^\dagger), \mathbf{x}_0 - \mathbf{x} \rangle$$
$$\leq \langle \mathcal{R}'(\mathbf{x}^\dagger), \mathbf{x}_0 - \mathbf{x} \rangle \leq \|\mathcal{R}'(\mathbf{x}^\dagger)\| \|\mathbf{x}_0 - \mathbf{x}\|.$$

Using (20), again we obtain (21).

To finish the proof, we consider two cases:

- $\mathcal{R}(\mathbf{x}^\dagger) \leq \mathcal{R}(\mathbf{x}) \Rightarrow |\mathcal{R}(\mathbf{x}^\dagger) - \mathcal{R}(\mathbf{x})| = \mathcal{R}(\mathbf{x}) - \mathcal{R}(\mathbf{x}^\dagger)$
- $\mathcal{R}(\mathbf{x}^\dagger) \geq \mathcal{R}(\mathbf{x}) \Rightarrow |\mathcal{R}(\mathbf{x}^\dagger) - \mathcal{R}(\mathbf{x})| = \mathcal{R}(\mathbf{x}) - \mathcal{R}(\mathbf{x}^\dagger) + 2(\mathcal{R}(\mathbf{x}^\dagger) - \mathcal{R}(\mathbf{x}))$.

Therefore, using (19) and (21), we obtain

$$\boldsymbol{\Delta}_{\mathcal{R}}(\mathbf{x}, \tilde{\mathbf{x}}) \leq |\mathcal{R}(\mathbf{x}^\dagger) - \mathcal{R}(\mathbf{x})| + |\langle \mathcal{R}'(\mathbf{x}^\dagger), \mathbf{x} - \mathbf{x}^\dagger \rangle|$$
$$\leq \mathcal{R}(\mathbf{x}) - \mathcal{R}(\mathbf{x}^\dagger) + (2\gamma + \gamma_1)\|\mathbf{A}(\mathbf{x}) - \mathbf{A}(\mathbf{x}^\dagger)\|,$$

which concludes our proof with $C := 2\gamma_0 + \gamma_1$. $\square$

Here is our convergence rates result, which is an extension of Obmann et al. (2020a, Theorem 3.1).

Theorem 6 (Convergence rates results). *Let* (B1)–(B4) *be satisfied, and suppose* $\alpha \sim \delta$. *Then* $\boldsymbol{\Delta}_{\mathcal{R}}(\mathbf{x}_\alpha^\delta, \mathbf{x}^\dagger) = \mathcal{O}(\delta)$ *as* $\delta \to 0$.

Proof. From Proposition 7, we obtain

$$\alpha \Delta_F(\mathbf{x}_\alpha^\delta, \mathbf{x}^\dagger) \leq \alpha \mathcal{R}(\mathbf{x}_\alpha^\delta) - \alpha \mathcal{R}(\mathbf{x}^\dagger) + C\alpha \|\mathbf{A}(\mathbf{x}_\alpha^\delta) - \mathbf{A}(\mathbf{x}^\dagger)\|$$
$$= \mathcal{T}_{\alpha,\delta}(\mathbf{x}_\alpha^\delta) - \mathcal{D}(\mathbf{A}(\mathbf{x}_\alpha^\delta), \mathbf{y}^\delta) - \left(\mathcal{T}_{\alpha,\delta}(\mathbf{x}^\dagger) - \mathcal{D}(\mathbf{A}(\mathbf{x}^\dagger), \mathbf{y}^\delta)\right)$$
$$+ C\alpha\|\mathbf{A}(\mathbf{x}_\alpha^\delta) - \mathbf{A}(\mathbf{x}^\dagger)\|$$
$$\leq \delta^2 + C\alpha\delta - \mathcal{D}(\mathbf{A}(\mathbf{x}_\alpha^\delta), \mathbf{y}^\delta) + C\alpha\|\mathbf{A}(\mathbf{x}_\alpha^\delta) - \mathbf{y}^\delta\|$$
$$\leq \delta^2 + C\alpha\delta - \mathcal{D}(\mathbf{A}(\mathbf{x}_\alpha^\delta), \mathbf{y}^\delta) + C\alpha\sqrt{\mathcal{D}(\mathbf{A}(\mathbf{x}_\alpha^\delta), \mathbf{y}^\delta)}.$$

Cauchy's inequality gives $\alpha \Delta_\mathcal{R}(\mathbf{x}_\alpha^\delta, \mathbf{x}^\dagger) \leq \delta^2 + C\alpha\delta + C^2\alpha^2/4$. For $\alpha \sim \delta$, we easily conclude $\Delta_\mathcal{R}(\mathbf{x}_\alpha^\delta, \mathbf{x}^\dagger) = \mathcal{O}(\delta)$. □

Related Methods

The use of neural networks as regularizers or similarity measures is an active research direction. Many interesting works have been done. We briefly review several techniques: variational networks (Kobler et al. 2017), deep cascaded networks (Kofler et al. 2018; Schlemper et al. 2017), and the MODL approach (Aggarwal et al. 2018). Further, we propose INDIE as a new operator-inversion-free variant of MODL. As opposed to the discussion in section "Convergence Analysis", these works make use of the approach (T1): employing solver-dependent training. Finally, we will discuss a synthesis variant of the NETT framework.

Variational networks: Variational networks (Kobler et al. 2017) connect variational methods and deep learning. They are based on the fields of experts model (Roth and Black 2005) and consider the Tikhonov functional

$$\mathcal{T}_{\mathbf{y},\alpha}(\mathbf{x}) = \sum_{c=1}^{N_c} \mathcal{T}_c(\mathbf{x}) := \sum_{c=1}^{N_c}\left(\sum_j\sum_i \phi_i^c((\bar{K}_i^c \mathbf{x})_j) + \alpha \sum_j\sum_i \psi_i^c((K_i^c(\mathbf{A}(\mathbf{x})-\mathbf{y}))_j)\right),$$

where $\bar{K}_i^c$ and, K_i^c are learnable convolutional operators, and ϕ_i, ψ_i are learnable functionals. Alternating gradient descent method for minimizing $\mathcal{T}_{\mathbf{y},\alpha}$ provides the update formula

$$\mathbf{x}_{n+1} = \mathbf{x}_n - \eta_n \nabla_\theta \mathcal{T}_{c(n)}(\mathbf{x}_n) \quad \text{where } c(n) = 1 + (n \mod N_c). \tag{22}$$

Direct calculations show $\nabla_\theta \mathcal{T}_c(\mathbf{x}) = \sum_i (\bar{K}_i^c)^T (\phi_i^c)'(\bar{K}_i^c \mathbf{x}) + \mathbf{A}^T \sum_i (K_i^c)^T (\psi_i^c)'(K_i^c(\mathbf{A}(\mathbf{x})-\mathbf{y}))$. Minimizing the $\mathcal{T}_{\mathbf{y},\alpha}$ is then replaced by training the neural network that consists of a L blocks realizing the iterative update (22).

Network cascades: Deep network cascades (Kofler et al. 2018; Schlemper et al. 2017) alternate between the application of post-processing networks and so-called data consistency layers. The data consistency condition proposed in Kofler et al. (2018) for sparse data problems $\mathbf{A} = \mathbf{S} \circ \mathbf{A}_\mathrm{F}$, where $\mathbf{S}$ is a sampling operator and $\mathbf{A}_\mathrm{F}$ a full data forward operator (such as the fully sampled Radon transform), takes the form

$$\mathbf{x}_{n+1} = \mathbf{B}_\mathrm{F}\left(\arg\min_{\mathbf{z}} \|\mathbf{z} - \mathbf{A}_\mathrm{F}(\mathcal{N}_{\theta(n)}(\mathbf{x}_n))\|_2^2 + \alpha \|\mathbf{y} - \mathbf{S}(\mathbf{z})\|_2^2\right), \quad (23)$$

with initial reconstruction $\mathbf{x}_0 = (\mathbf{B}_\mathrm{F} \circ \mathbf{S}^*)(\mathbf{y})$, where $\mathbf{B}_\mathrm{F} \colon \mathbb{Y} \to \mathbb{X}$ is a reconstruction method for the full data forward operator and $\mathcal{N}_{\theta(n)}$ are networks. For example, in MRT the operator $\mathbf{B}_\mathrm{F}$ is the inverse Fourier transform (Schlemper et al. 2017), and in CT, the operator $\mathbf{B}_\mathrm{F}$ can be implemented by the filtered backprojection (Kofler et al. 2018). The resulting neural network consists of L steps of (23) that can be trained end-to-end.

MODL approach: The model-based deep learning (MODL) approach of Aggarwal et al. (2018) starts with the Tikhonov functional $\mathcal{T}_{\mathbf{y},\alpha}(\mathbf{x}) = \|\mathbf{A}(\mathbf{x}) - \mathbf{y}\|_2^2 + \alpha \|\mathbf{x} - \mathcal{N}_\theta(\mathbf{x})\|_2^2$, where $\mathcal{N}_\theta(\mathbf{x})$ is interpreted as denoising network. By designing $\mathcal{N}_\theta$ as a convolutional block, then $\mathbf{x} - \mathcal{N}_\theta(\mathbf{x})$ is a small residual network (He et al. 2016). The authors of Aggarwal et al. (2018) proposed the following heuristic iterative scheme $\mathbf{x}_{n+1} = \arg\min_{\mathbf{x}} \|\mathbf{A}(\mathbf{x}) - \mathbf{y}\|^2 + \alpha \|\mathbf{x} - \mathcal{N}_\theta(\mathbf{x}_n)\|_2^2$ based on $\mathcal{T}_{\mathbf{y},\alpha}$ whose closed-form solution is

$$\mathbf{x}_{n+1} = (\mathbf{A}^\mathsf{T}\mathbf{A} + \alpha\,\mathrm{Id})^{-1}(\mathbf{A}^\mathsf{T}\mathbf{y} + \alpha\,\mathcal{N}_\theta(\mathbf{x}_n)). \quad (24)$$

Concatenating these steps together, one arrives at a deep neural network. Similar to network cascades, each block (24) consists of a trainable layer $\mathbf{z}_n = \mathbf{A}^\mathsf{T}\mathbf{y} + \alpha\mathcal{N}_\theta(\mathbf{x}_n)$ and a non-trainable data consistency layer $\mathbf{x}_{n+1} = (\mathbf{A}^\mathsf{T}\mathbf{A} + \lambda\,\mathrm{Id})^{-1}(\mathbf{z}_n)$.

INDIE approach: Let us present an alternative to the above procedures, inspired by Daubechies et al. (2004). Namely, we propose the iterative update

$$\mathbf{x}_{n+1} = \arg\min \mathcal{L}_n(\mathbf{x})$$
$$\mathcal{L}_n(\mathbf{x}) := \|\mathbf{A}(\mathbf{x}) - \mathbf{y}\|^2 + \alpha \|\mathbf{x} - \mathcal{N}_\theta(\mathbf{x}_n)\|^2 + C\|\mathbf{x} - \mathbf{x}_n\|^2 - \|\mathbf{A}(\mathbf{x} - \mathbf{x}_n)\|^2.$$

Here the constant $C > 0$ is an upper bound for the operator norm $\|\mathbf{A}\|$. Elementary manipulations show the identity

$$\mathcal{L}_n(\mathbf{x}) = -2\langle \mathbf{A}^\mathsf{T}(\mathbf{y} - \mathbf{A}(\mathbf{x}_n)) + \alpha\mathcal{N}_\theta(\mathbf{x}_n) + C\mathbf{x}_n, \mathbf{x}\rangle$$
$$+ (\alpha + C)\|\mathbf{x}\|^2 - (\alpha\|\mathcal{N}_\theta(\mathbf{x}_n)\| + C\|\mathbf{x}_n\|^2) - \|\mathbf{A}(\mathbf{x}_n)\|^2 + \|\mathbf{y}\|^2.$$

The minimizer of $\mathcal{L}_n$ can therefore be computed explicitly by setting the gradient of the latter expression to zero. This results in the proposed network block

$$\mathbf{x}_{n+1} = \frac{1}{\alpha + C} \left(\mathbf{A}^\mathsf{T}(\mathbf{y} - \mathbf{A}(\mathbf{x}_n)) + \alpha \mathcal{N}_\theta(\mathbf{x}_n) + C\mathbf{x}_n \right). \tag{25}$$

This results at a deep neural network similar to the MODL iteration. However, each block in (25) is clearly simpler than the blocks in (24). In fact, as opposed to MODL, our proposed learned iterative scheme does not require costly matrix inversion. We name the resulting iteration INDIE (for **in**version-free **d**eep **ite**rative) cascades. We consider the numerical comparison of MODL and INDIE as well as the theoretical analysis of both architectures to be interesting lines of future research.

Learned synthesis regularization: Let us finish this section by pointing out that regularization by neural network is not restricted to the form (11). For example, one can consider the synthesis version, which reads (Obmann et al. 2020b)

$$\mathbf{x}^{\mathrm{syn}} = \mathbf{D}_\theta \left(\arg\min_\xi \|\mathbf{A} \circ \mathbf{D}_\theta(\xi) - \mathbf{y}\|^2 + \alpha \sum_{\lambda \in \Lambda} \omega_\lambda |\xi_\lambda|^p \right), \tag{26}$$

where Λ is a countable set, $1 \leq p < 2$, and $\mathcal{D}_\theta \colon \ell^2(\Lambda) \to \mathbb{X}$ is a learned operator that performs nonlinear synthesis of $\mathbf{x}$. Rigorous analysis of the above formulation was derived in Obmann et al. (2020b).

Finally, note that one can generalize the frameworks (11) and (26) by allowing the involved neural networks to depend on the regularization parameter α or the noise-level δ. The dependence on α has been studied in, for example, Obmann et al. (2020b). The dependence on δ can be realized by mimicking the Morozov's stopping criteria, when training the neural networks, either independently or together with the optimization problem. In the later case, δ can help decide the depth of the unrolled neural network.

Conclusion and Outlook

Inverse problems are central to solving a wide range of important practical problems within and outside of imaging and computer vision. Inverse problems are characterized by the ambiguity and instability of their solution. Therefore, stabilizing solution methods based on regularization techniques is necessary to solve them in a reasonable way. In recent years, neural networks and deep learning have emerged as the rising stars for the solution of inverse problems. In this chapter, we have developed the mathematical foundations for solving inverse problems with deep learning. In addition, we have shown stability and convergence for selected neural networks to solve inverse problems. The investigated methods, which combine the strengths of both worlds, are regularizing null-space networks and the NETT (Network-Tikhonov) approach for inverse problems.

References

Acar, R., Vogel, C.R.: Analysis of bounded variation penalty methods for ill-posed problems. Inverse Probl. **10**(6), 1217–1229 (1994)

Adler, J., Öktem, O.: Solving ill-posed inverse problems using iterative deep neural networks. Inverse Probl. **33**(12), 124007 (2017)

Aggarwal, H.K., Mani, M.P., Jacob, M.: MoDL: model-based deep learning architecture for inverse problems. IEEE Trans. Med. Imaging **38**(2), 394–405 (2018)

Agostinelli, F., Hoffman, M., Sadowski, P., Baldi, P.: Learning activation functions to improve deep neural networks. arXiv:1412.6830 (2014)

Aljadaany, R., Pal, D.K., Savvides, M.: Douglas-rachford networks: learning both the image prior and data fidelity terms for blind image deconvolution. In: Proceedings of the IEEE Computer Society Conference on Computer Vision and Pattern Recognition, pp. 10235–10244 (2019)

Arridge, S., Maass, P., Öktem, O., Schönlieb C.: Solving inverse problems using data-driven models. Acta Numer. **28**, 1–174 (2019)

Bengio, Y., Courville, A., Vincent, P.: Representation learning: a review and new perspectives. IEEE Trans. Pattern Anal. **35**(8), 1798–1828 (2013)

Boink, Y.E., Haltmeier, M., Holman, S., Schwab, J.: Data-consistent neural networks for solving nonlinear inverse problems. arXiv:2003.11253 (2020), to apper in Inverse Probl. Imaging

Bora, A., Jalal, A., Price, E., Dimakis, A.G.: Compressed sensing using generative models. In: Proceedings of the 34th International Conference on Machine Learning, vol. 70, pp. 537–546 (2017)

Brosch, T., Tam, R., et al.: Manifold learning of brain MRIs by deep learning. In: International Conference on Medical Image Computing and Computer-Assisted Intervention, pp. 633–640. Springer (2013)

Bubba, T.A., Kutyniok, G., Lassas, M., Maerz, M., Samek, W., Siltanen, S., Srinivasan, V.: Learning the invisible: a hybrid deep learning-shearlet framework for limited angle computed tomography. Inverse Probl. **35**(6), 064002 (2019)

Chen, D., Davies, M.E.: Deep decomposition learning for inverse imaging problems. In *European Conference on Computer Vision*, pp. 510–526). Springer, Cham (2020)

Chen, T.Q., Rubanova, Y., Bettencourt, J., Duvenaud, D.K.: Neural ordinary differential equations. In: Advances in Neural Information Processing Systems, pp. 6571–6583 (2018)

Daubechies, I., Defrise, M., De Mol, C.: An iterative thresholding algorithm for linear inverse problems with a sparsity constraint. Commun. Pure Appl. Math. **57**(11), 1413–1457 (2004)

Dittmer, S., Maass, P.: A projectional ansatz to reconstruction. arXiv:1907.04675 (2019)

Dittmer, S., Kluth, T., Maass, P., Baguer, D.O.: Regularization by architecture: a deep prior approach for inverse problems. J. Math. Imaging Vis. **62**, 456–470 (2020)

Engl, H.W., Hanke, M., Neubauer, A.: Regularization of Inverse Problems, vol. 375. Kluwer Academic Publishers Group, Dordrecht (1996)

Georg, M., Souvenir, R., Hope, A., Pless, R.: Manifold learning for 4D CT reconstruction of the lung. In: IEEE Computer Society Conference on Computer Vision and Pattern Recognition, pp. 1–8. IEEE (2008)

Glorot, X., Bengio, Y.: Understanding the difficulty of training deep feedforward neural networks. In: Proceedings of 13th International Conference on Artificial Intelligence and Statistics, pp. 249–256 (2010)

Goodfellow, I., Bengio, Y., Courville, A.: Deep Learning. MIT Press, London (2016)

Grasmair, M., Haltmeier, M., Scherzer, O.: Sparse regularization with l^q penalty term. Inverse Probl. **24**(5), 055020 (2008)

Han, Y., Yoo, J.J., Ye, J.C.: Deep residual learning for compressed sensing CT reconstruction via persistent homology analysis (2016). http://arxiv.org/abs/1611.06391

He, K., Zhang, X., Ren, S., Sun, J.: Deep residual learning for image recognition. In: IEEE Computer Society Conference on Computer Vision and Pattern Recognition, pp. 770–778 (2016)

Huang, Y., Preuhs, A., Manhart, M., Lauritsch, G., Maier, A.: Data consistent ct reconstruction from insufficient data with learned prior images. arXiv:2005.10034 (2020)

Ivanov, V.K., Vasin, V.V., Tanana, V.P.: Theory of Linear Ill-Posed Problems and Its Applications. Inverse and Ill-Posed Problems Series, 2nd edn. VSP, Utrecht, (2002). Translated and revised from the 1978 Russian original

Jin, K.H., McCann, M.T., Froustey, E., Unser, M.: Deep convolutional neural network for inverse problems in imaging. IEEE Trans. Image Process. **26**(9), 4509–4522 (2017)

Kobler, E., Klatzer, T., Hammernik, K., Pock, T.: Variational networks: connecting variational methods and deep learning. In: German Conference on Pattern Recognition, pp. 281–293. Springer (2017)

Kofler, A., Haltmeier, M., Kolbitsch, C., Kachelrieß, M., Dewey, M.: A U-Nets cascade for sparse view computed tomography. In: Proceedings of 1st Workshop on Machine Learning for Medical Image Reconstruction, pp. 91–99. Springer (2018)

Kofler, A., Haltmeier, M., Schaeffter, T., Kachelrieß, M., Dewey, M., Wald, C., Kolbitsch, C.: Neural networks-based regularization of large-scale inverse problems in medical imaging. Phys. Med. Biol. **65**, 135003 (2020)

LeCun, Y., Bengio, Y., Hinton, G.: Deep learning. Nature **521**(7553), 436–444 (2015)

Li, H., Schwab, J., Antholzer, S., Haltmeier, M.: NETT: solving inverse problems with deep neural networks. Inverse Probl. **36**, 065005 (2020)

Lindenstrauss, J., Tzafriri, L.: On the complemented subspaces problem. Israel J. Math. **9**(2), 263–269 (1971)

Lunz, S., Öktem, O., Schönlieb, C.: Adversarial regularizers in inverse problems. In: Advances in Neural Information Processing Systems, vol. 31, pp. 8507–8516 (2018)

Mardani, M., Gong, E., Cheng, J.Y., Vasanawala, S.S., Zaharchuk, G., Xing, L., Pauly, J.M.: Deep generative adversarial neural networks for compressive sensing MRI. IEEE Trans. Med. Imag. **38**(1), 167–179 (2018)

Nashed, M.Z.: Inner, outer, and generalized inverses in banach and hilbert spaces. Numer. Func. Anal. Opt. **9**(3–4), 261–325 (1987)

Obmann, D., Nguyen, L., Schwab, J., Haltmeier, M.: Sparse aNETT for solving inverse problems with deep learning. In *2020 IEEE 17th International Symposium on Biomedical Imaging Workshops (ISBI Workshops)* (pp. 1–4). IEEE (2020a)

Obmann, D., Schwab, J., Haltmeier, M.: Deep synthesis network for regularizing inverse problems. *Inverse Problems*, **37**(1), 015005 (2020b)

Obmann, D., Nguyen, L., Schwab, J., Haltmeier, M.: Augmented NETT regularization of inverse problems. J. Phys. Commun. **5**(10), 105002 (2021)

Phillips, R.S.: On linear transformations. Trans. Am. Math. Soc. **48**(3), 516–541 (1940)

Ramachandran, P., Zoph, B., Le, Q.V.: Searching for activation functions. arXiv:1710.05941 (2017)

Resmerita, E., Anderssen, R.S.: Joint additive Kullback–Leibler residual minimization and regularization for linear inverse problems. Math. Methods Appl. Sci. **30**(13), 1527–1544 (2007)

Roth, S., Black, M.J.: Fields of experts: a framework for learning image priors. In: Proceedings of the IEEE Computer Society Conference on Computer Vision and Pattern Recognition, vol. 2, pp. 860–867. IEEE (2005)

Scherzer, O., Grasmair, M., Grossauer, H., Haltmeier, M., Lenzen, F.: Variational Methods in Imaging. Applied Mathematical Sciences, vol. 167. Springer, New York (2009)

Schlemper, J., Caballero, J., Hajnal, J.V., Price, A., Rueckert, D.: A deep cascade of convolutional neural networks for MR image reconstruction. In: Proceedings of Information Processing in Medical Imaging, pp. 647–658. Springer (2017)

Schwab, J., Antholzer, S., Haltmeier, M.: Deep null-space learning for inverse problems: convergence analysis and rates. Inverse Probl. **35**(2), 025008 (2019)

Schwab, J., Antholzer, S., Haltmeier, M.: Big in Japan: regularizing networks for solving inverse problems. J. Math. Imaging Vis. **62**, 445–455 (2020)

Sulam, J., Aberdam, A., Beck, A., Elad, M.: On multi-layer basis pursuit, efficient algorithms and convolutional neural networks. IEEE Trans. Pattern Anal. Mach. Intell. **42**(8), 1968–1980 (2019)

Ulyanov, D., Vedaldi, A., Lempitsky, V.: Deep image prior. In: Proceedings of the IEEE Computer Society Conference on Computer Vision and Pattern Recognition, pp. 9446–9454 (2018)

Van Veen, D., Jalal, A., Soltanolkotabi, M., Price, E., Vishwanath, S., Dimakis, A.G.: Compressed sensing with deep image prior and learned regularization. arXiv:1806.06438 (2018)

Wachinger, C., Yigitsoy, M., Rijkhorst, E., Navab, N.: Manifold learning for image-based breathing gating in ultrasound and MRI. Med. Image Anal. **16**(4), 806–818 (2012)

Yang, Y., Sun, J., Li, H., Xu, Z.: Deep ADMM-net for compressive sensing MRI. In: Proceedings of 30th International Conference on Neural Information Processing Systems, pp. 10–18 (2016)

Shearlets: From Theory to Deep Learning

30

Gitta Kutyniok

Contents

Introduction	1096
The Applied Harmonic Analysis Viewpoint	1096
Frame Theory Comes into Play	1097
Wavelets	1098
From Wavelets to Shearlets	1098
From Inverse Problems to Deep Learning	1099
Outline	1100
Continuous Shearlet Systems	1100
Classical Continuous Shearlet Systems	1101
Cone-Adapted Continuous Shearlet Systems	1103
Resolution of the Wavefront Set	1104
Discrete Shearlet Systems	1105
Cone-Adapted Discrete Shearlet Systems	1105
Frame Properties	1106
Sparse Approximation	1109
Extensions of Shearlets	1111
Higher Dimensions	1111
α-Molecules	1113
Universal Shearlets	1114
Digital Shearlet Systems	1116
Digital 2D Shearlet Transform	1116
Extensions of the Digital 2D Shearlet Transform and ShearLab3D	1118
Applications of Shearlets	1119
Sparse Regularization Using Shearlets	1120
Shearlets Meet Deep Learning	1124
Conclusion	1129
References	1129

G. Kutyniok (✉)
Ludwig-Maximilians-Universität München, Mathematisches Institut, München, Germany
e-mail: kutyniok@math.lmu.de

© Springer Nature Switzerland AG 2023
K. Chen et al. (eds.), *Handbook of Mathematical Models and Algorithms in Computer Vision and Imaging*, https://doi.org/10.1007/978-3-030-98661-2_80

Abstract

Many important problem classes are governed by anisotropic features, which typically appear as singularities concentrated on lower-dimensional embedded manifolds. Examples include edges in images or shock fronts in solutions of transport-dominated equations. Shearlets are the first representation system which exhibits optimal sparse approximation properties in combination with a unified treatment of the continuum and digital realm, leading to faithful implementations. A prominent class of applications are inverse problems, foremost in imaging science, where shearlets are utilized for sparse regularization. Recently, shearlet systems have also been used in combination with data-driven approaches, predominately deep neural networks. This chapter shall serve as an introduction to and a survey about the theory of shearlets and their applications.

Keywords

Deep neural networks · Frames · Shearlets · Sparse approximation · Wavelets

Introduction

In the twenty-first century, technological advances have generated an unprecedented deluge of highly complex data sets, posing enormous challenges to provide efficient methodologies for acquisition and analysis. While there exists a huge variety of different types of data, the majority of it falls into the category of images and videos. Prominent examples of areas in science producing massive data sets of this type are astronomy, medicine, or seismology. One key problem to tackle is the question of suitable representations of such data. This led to an intense study in the research community of applied harmonic analysis aiming to provide highly efficient multivariate encoding methodologies.

The Applied Harmonic Analysis Viewpoint

The viewpoint of applied harmonic analysis concerning the application of representation systems in data processing can be summarized as follows: Let C be a class of data in a Hilbert space $\mathcal{H}$, and assume $(\psi_\lambda)_{\lambda \in \Lambda} \subset \mathcal{H}$ is a carefully constructed collection of vectors with Λ being a countable indexing set.

On the one hand, $(\psi_\lambda)_{\lambda \in \Lambda}$ can then be utilized to *decompose* the data by

$$C \ni f \mapsto (\langle \cdot, \psi_\lambda \rangle)_{\lambda \in \Lambda}. \tag{1}$$

This can be regarded as an encoding step, often aiming to reveal important features of the data f such as singularities by analyzing the associated coefficient sequence. On the other hand, $(\psi_\lambda)_{\lambda \in \Lambda}$ can also serve as a means to *expand* the data by representing it as

$$f = \sum_{\lambda \in \Lambda} c(f)_\lambda \psi_\lambda \quad \text{for all } f \in C. \tag{2}$$

Since efficient expansions are typically desirable, one usually aims for the coefficient sequence $(c(f)_\lambda)_{\lambda \in \Lambda}$ to be sparse in the sense of rapid decay to allow efficient encoding of the data f.

In case that $(\psi_\lambda)_{\lambda \in \Lambda}$ forms an orthonormal basis, it is well-known that $c(f)_\lambda = \langle \cdot, \psi_\lambda \rangle$ for all $\lambda \in \Lambda$. However, it might not be possible to design an orthonormal basis with the desirable properties, or redundancy is for other reasons such as robustness required. This then leads to the notion of a frame, in which case (1) and (2) cannot be that easily linked, but requires methods from frame theory.

Frame Theory Comes into Play

The area of frame theory focuses on redundant representation systems in the sense of nonunique expansions, thereby going beyond the concept of orthonormal bases. It provides a general framework for redundant systems $(\psi_\lambda)_{\lambda \in \Lambda}$ while allowing to control their stability.

A system $(\psi_\lambda)_{\lambda \in \Lambda}$ is called a *frame* for $\mathcal{H}$, if there exist constants $0 < A \leq B < \infty$ such that

$$A \|f\|^2 \leq \sum_{\lambda \in \Lambda} |\langle f, \psi_\lambda \rangle|^2 \leq B \|f\|^2 \quad \text{for all } f \in \mathcal{H}.$$

In case $A = B = 1$, it is coined a *Parseval frame*. In fact, referring to section "The Applied Harmonic Analysis Viewpoint", Parseval frames are the most general systems which can satisfy $c(f)_\lambda = \langle \cdot, \psi_\lambda \rangle$ for all $\lambda \in \Lambda$.

The associated *frame operator* is defined by

$$S : \mathcal{H} \to \mathcal{H}, \quad f \mapsto \sum_{\lambda \in \Lambda} \langle f, \psi_\lambda \rangle \psi_\lambda,$$

which is self-adjoint with spectrum $\sigma(S) \subset [A, B]$. The sequence $(\tilde{\psi}_\lambda)_{\lambda \in \Lambda} := (S^{-1}\psi_\lambda)_{\lambda \in \Lambda}$ is then referred to as the *canonical dual frame*. It allows reconstruction of some $f \in \mathcal{H}$ from the decomposition (1) and the construction of an explicit coefficient sequence in the expansion (2) by considering

$$f = \sum_{\lambda \in \Lambda} \langle f, \psi_\lambda \rangle \tilde{\psi}_\lambda \quad \text{and} \quad f = \sum_{\lambda \in \Lambda} \langle f, \tilde{\psi}_\lambda \rangle \psi_\lambda \quad \text{for all } f \in \mathcal{H},$$

respectively. The coefficient sequence $(\langle f, \tilde{\psi}_\lambda \rangle)_{\lambda \in \Lambda}$ can even be shown to be the smallest in ℓ_2 norm among all possible ones.

For further information on frame theory, we refer to Casazza et al. (2012) and Christensen (2003).

Wavelets

One first highlight in applied harmonic analysis was the development of the system of wavelets, based on translation $(x \mapsto x-m)$ and dilation $(x \mapsto 2^j x)$ leading to the representation of functions in $L^2(\mathbb{R}^d)$ at different locations and different resolution levels.

Definition 1. For $\psi^1, \ldots, \psi^L \in L^2(\mathbb{R}^d)$, the associated *(discrete) wavelet system* is defined by

$$\{\psi^\ell_{j,m} = 2^{\frac{dj}{2}} \psi^\ell(2^j \cdot -m) : j \in \mathbb{Z}, m \in \mathbb{Z}^d, \ell = 1, \ldots, L\}. \tag{3}$$

The generating functions $\psi^1, \ldots, \psi^L$ can be chosen such that the associated wavelet system forms an orthonormal basis (more generally, a frame) for $L^2(\mathbb{R}^d)$. The functions $\psi^1, \ldots, \psi^L$ are then typically referred to as *wavelets* with the parameter j serving as *scale* and m as *position*. In fact, one key aspect of wavelet theory which has significantly contributed to its success is its rich mathematical structure. This allows to design families of wavelets with various desirable properties expressed in terms of regularity, decay, or vanishing moments. On the application side, wavelets have revolutionized various areas such as imaging science, for instance, for compression tasks by developing JPEG2000, and numerical analysis of partial differential equations.

The literature on wavelets is very rich, and for the sake of brevity, we here just refer to the books Cohen (2003), Daubechies (1992), Mallat (1998), and references therein.

From Wavelets to Shearlets

Multivariate functions are distinctively different from univariate functions, since they are, in particular, typically governed by anisotropic (i.e., directional) singularities. Let us exemplary mention that indeed edges are prominent features in images similar to shock fronts in the solutions of transport-dominated equations. More generally, in high-dimensional data information is often contained in lower-dimensional embedded manifolds. Thus, it is fair to say that a system which aims for efficient encoding of such data should, in particular, be able to efficiently encode anisotropic features.

Although wavelets can be shown to optimally encode functions governed by point singularities in the sense of decay rates of the error of best N-term approximation, it is evident that due to their isotropic structure, they are not capable to efficiently encode anisotropic features. Indeed the isotropic scaling matrix with a dyadic scaling factor 2^j (see (3)) prevents a wavelet system from delivering optimal approximation rates of such data.

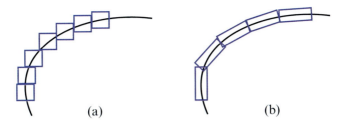

Fig. 1 (**a**) Approximation of a curvilinear structure by isotropic elements. (**b**) Approximation of a curvilinear structure by anisotropic elements

This argumentation shows the need to develop anisotropic representation systems, by going beyond systems consisting of translation and dilation. Figure 1 depicts the problem of an isotropic system such as a wavelet system as opposed to the advantage of anisotropically shaped elements. A list of desirable properties for an anisotropic representation system can be summarized as follows:

(1) Underlying group structure for availability of deep mathematical tools.
(2) Provably optimal sparse approximations of anisotropic features.
(3) Compactly supported analyzing elements for high spatial localization.
(4) Uniform treatment of the continuum and digital realm.
(5) Fast implementation of the associated decomposition.

This problem has led to the development of various novel anisotropic representation systems within the area of applied harmonic analysis. Some of the key contributions are *steerable pyramid* by Simoncelli et al. (1992), *directional filter banks* by Bamberger and Smith (1992), *2D directional wavelets* by Antoine et al. (1993), *curvelets* by Candès and Donoho (2004), *contourlets* by Do and Vetterli (2005), *bandelets* by Le Pennec and Mallat (2005), and *shearlets* (Guo et al. 2006; Labate et al. 2005). Shearlet systems indeed satisfy all desiderata one commonly requires from an anisotropic system as stated before.

From Inverse Problems to Deep Learning

The main application areas of shearlets are inverse problems, foremost in imaging. A common approach to solve an ill-posed inverse problem $Tf = g$ for a linear, bounded operator $T : \mathcal{H} \to \mathcal{H}$ is by Tikhonov regularization. A generalization of this conceptual approach to sparse regularization was suggested in Daubechies et al. (2004). Given a representation system $(\psi_\lambda)_{\lambda \in \Lambda}$, an approximation of the solution can be computed by minimizing the functional

$$\|Tf - g\|^2 + \beta \cdot \|(\langle \cdot, \psi_\lambda \rangle)_{\lambda \in \Lambda}\|_{\ell_1}, \qquad (4)$$

with β being the regularization parameter. This approach exploits the fact that, when carefully designing the system $(\psi_\lambda)_{\lambda \in \Lambda}$, the solution of $Tf = g$ exhibits a sparse coefficient sequence $(\langle \cdot, \psi_\lambda \rangle)_{\lambda \in \Lambda}$. Exemplary general inverse problems are inpainting (Genzel and Kutyniok 2014; King et al. 2014), morphological component analysis (Donoho and Kutyniok 2013; Kutyniok and Lim 2012) and segmentation (Häuser and Steidl 2013) or inverse problems from medical diagnosis such as magnetic resonance imaging (Kutyniok and Lim 2018).

Recently, deep learning has swept the area of imaging science with deep neural network-based approaches often outperforming the to-date state-of-the-art algorithms. The last years though have shown that in fact hybrid methods, i.e., combinations of model-based and data-driven approaches, typically lead to the best results by taking the best out of both worlds. Since the shearlet representation is particularly well suited to analyze anisotropic features, several hybrid approaches were suggested which combine the shearlet transform with deep neural networks such as for limited-angle computed tomography (Bubba et al. 2019) as well as for wavefront set and semantic edge detection (Andrade-Loarca et al. 2020, 2019).

In the following, we will provide an introduction to and a survey about the theory and applications of shearlets. For additional information, we refer to Kutyniok and Labate (2012).

Outline

We start by discussing continuous shearlet systems and their associated transforms in section "Continuous Shearlet Systems", including their ability to resolve the wavefront set. This is followed by the introduction of their discrete counterparts with a presentation of their optimal sparse approximation properties for anisotropic features (see section "Discrete Shearlet Systems"). Section "Extensions of Shearlets" is devoted to extensions of shearlet systems such as extensions to higher dimensions, α-molecules, and universal shearlets. The faithful digitalization as also implemented in www.ShearLab.org is then presented in section "Digital Shearlet Systems". Finally, in section "Applications of Shearlets" applications of shearlets to inverse problems, also in combination with deep learning, are discussed.

Continuous Shearlet Systems

We start by introducing the main notation and the definition of continuous shearlets. Shearlet systems are composed of three operators, namely, scaling, shearing, and translation, applied to a generating function, related to different resolution levels, orientations, and positions, respectively. The term "continuous" indicates that continuous parameter sets are considered. Notice that also the continuous shearlet system and associated transform can be generalized in a canonical way to $L^2(\mathbb{R}^n)$ for $n \geq 3$ with the results from sections "Classical Continuous Shearlet Systems" and "Cone-Adapted Continuous Shearlet Systems" holding in a similar manner (Dahlke et al. 2008, 2009, 2010, 2013).

Classical Continuous Shearlet Systems

We will first present the classical version of continuous shearlet systems. For this, let the *parabolic scaling matrix* A_a, $a \in \mathbb{R}^* := \mathbb{R} \setminus \{0\}$ and the *shearing matrix* S_s, $s \in \mathbb{R}$, be given by

$$A_a = \begin{pmatrix} a & 0 \\ 0 & |a|^{1/2} \end{pmatrix} \quad \text{and} \quad S_s = \begin{pmatrix} 1 & s \\ 0 & 1 \end{pmatrix}, \tag{5}$$

respectively. Letting now the *dilation operator* $D_M : L^2(\mathbb{R}^2) \to L^2(\mathbb{R}^2)$, $M \in \mathbb{R}^{2 \times 2}$, be defined by

$$(D_M f)(x) \mapsto |\det(M)|^{-1/2} f(M^{-1} x)$$

and the *translation operator* $T_t : L^2(\mathbb{R}^2) \to L^2(\mathbb{R}^2)$, $t \in \mathbb{R}^2$, by $(T_t f)(x) \mapsto f(x - t)$ yields the definition of continuous shearlet systems.

Definition 2. For $\psi \in L^2(\mathbb{R}^2)$, the *continuous shearlet system* $\mathcal{SH}(\psi)$ is defined by

$$\mathcal{SH}(\psi) = \{\psi_{a,s,t} := T_t D_{A_a} D_{S_s} \psi = a^{-3/4} \psi(A_a^{-1} S_s^{-1}(x - t)) : a \in \mathbb{R}^*, s \in \mathbb{R}, t \in \mathbb{R}^2\},$$

and the associated *continuous shearlet transform* of $f \in L^2(\mathbb{R}^2)$ is given by

$$SH_\psi f(a, s, t) := \langle f, \psi_{a,s,t} \rangle, \quad (a, s, t) \in \mathbb{R}^* \times \mathbb{R} \times \mathbb{R}^2.$$

This transform is invertible provided ψ satisfies an admissibility condition, whose definition requires to take a group theoretic viewpoint. We now endow $\mathbb{R}^* \times \mathbb{R} \times \mathbb{R}^2$ with a group structure, namely, the *(full) shearlet group* $\mathbb{S} := \mathbb{R}^* \times \mathbb{R} \times \mathbb{R}^2$ with group operation given by

$$(a, s, t) \circ (a', s', t') = (aa', s + \sqrt{|a|}s', t + S_s A_a t').$$

This is a locally compact group with left Haar measure $d_\mu(a, s, t) = da/|a|^3 ds dt$ (Dahlke et al. 2009). The map from $\mathbb{S}$ into the group of unitary operators on $L^2(\mathbb{R}^2)$, $\mathcal{U}(L^2(\mathbb{R}^2))$, given by $(a, s, t) \mapsto \psi_{a,s,t}$ can now be regarded as a unitary representation of the shearlet group. This allows to analyze square-integrability of this mapping, i.e., irreducibility and the existence of a nontrivial *admissible* function $\psi \in L^2(\mathbb{R}^2)$ which, for all $f \in L^2(\mathbb{R}^2)$, satisfies the *admissibility condition*

$$\int_{\mathbb{S}} |\langle f, \psi_{a,s,t} \rangle|^2 \, d\mu(a, s, t) < \infty.$$

A function is then defined to be a shearlet, if a condition equivalent to the admissibility condition is fulfilled.

Definition 3. A function $\psi \in L^2(\mathbb{R}^2)$ is called a *shearlet*, if

$$\int_{\mathbb{R}^2} \frac{|\hat{\psi}(\xi)|^2}{|\xi_1|^2} \, d\xi < \infty,$$

where $\xi = (\xi_1, \xi_2)$ and $\hat{\psi}$ denote the Fourier transform of ψ.

This leads to the following result, which heavily relies on group theoretic arguments:

Theorem 1 (Dahlke et al. 2008). *Let $\psi \in L^2(\mathbb{R}^2)$ be a shearlet. Then*

$$SH_\psi : L^2(\mathbb{R}^2) \to L^2(\mathbb{S}), \quad f \mapsto SH_\psi f(a, s, t)$$

is an isometry.

Let us now consider some examples of shearlets. The first and most extensively studied shearlet is the so-called classical shearlet, which is a band-limited function introduced in Labate et al. (2005). For an illustration of the support of the associated Fourier transform, we refer to Fig. 2a.

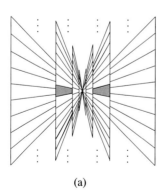
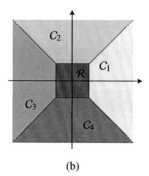

Fig. 2 (**a**) Partitioning of Fourier domain by supports of several elements of the classical shearlet system, with the support of the Fourier transform of the classical shearlet itself being highlighted. (**b**) The partition of Fourier domain into four conic regions $C_1 - C_4$ and a centered rectangle $\mathcal{R} = \{(\xi_1, \xi_2) : |\xi_1|, |\xi_2| \leq 1\}$ as the low-frequency regime

Example 1. A *classical shearlet* $\psi \in L^2(\mathbb{R}^2)$ is defined by

$$\hat{\psi}(\xi) = \hat{\psi}(\xi_1, \xi_2) = \hat{\psi}_1(\xi_1)\, \hat{\psi}_2(\tfrac{\xi_2}{\xi_1}),$$

where $\psi_1 \in L^2(\mathbb{R})$ is a wavelet, i.e., it satisfies the discrete Calderón condition given by

$$\sum_{j \in \mathbb{Z}} |\hat{\psi}_1(2^{-j}\xi)|^2 = 1 \quad \text{for a.e. } \xi \in \mathbb{R},$$

with $\hat{\psi}_1 \in C^\infty(\mathbb{R})$ and supp $\hat{\psi}_1 \subseteq [-\tfrac{5}{4}, -\tfrac{1}{4}] \cup [\tfrac{1}{4}, \tfrac{5}{4}]$, and $\psi_2 \in L^2(\mathbb{R})$ is a "bump function," namely,

$$\sum_{k=-1}^{1} |\hat{\psi}_2(\xi + k)|^2 = 1 \quad \text{for a.e. } \xi \in [-1, 1],$$

satisfying $\hat{\psi}_2 \in C^\infty(\mathbb{R})$ and supp $\hat{\psi}_2 \subseteq [-1, 1]$.

In general, shearlets of both band-limited and compactly supported type have been constructed and analyzed (see, e.g., Dahlke et al. 2008, 2011 and Grohs 2011b). We would also like to remark, that by using coorbit space theory, associated smoothness spaces can be derived together with their atomic decompositions and (Banach) frames for these spaces, (see, e.g., Dahlke et al. (2009, 2011) and Labate et al. (2013)).

Cone-Adapted Continuous Shearlet Systems

The group-theoretic approach leading to continuous shearlet systems allows to directly apply various results and methodologies from abstract harmonic analysis. This approach is however problematic, since it leads to a directional bias of the system in the sense of an imbalance of the directional sensitivity; for an illustration, we refer to Fig. 2a. This creates a problem when shearlet systems are, for instance, applied to resolve the wavefront set.

To circumvent this issue, cone-adapted continuous shearlet systems were constructed. The key idea is to decompose the Fourier domain in a suitable way which enforces a more balanced decomposition of the different directions as depicted in Fig. 2b. In the following definition, notice that the system $\Psi(\psi)$ is associated with the horizontal cones $C_1 \cup C_3$, whereas choosing the shearlet $\tilde{\psi}$ with the roles of ξ_1 and ξ_2 reversed, i.e., $\tilde{\psi}(\xi_1, \xi_2) = \psi(\xi_2, \xi_1)$, the system $\tilde{\Psi}(\tilde{\psi})$ is then associated with the vertical cones $C_2 \cup C_4$.

Definition 4. For $\phi, \psi, \tilde{\psi} \in L^2(\mathbb{R}^2)$, the *cone-adapted continuous shearlet system* is defined by $\mathcal{SH}(\phi, \psi, \tilde{\psi}) = \Phi(\phi) \cup \Psi(\psi) \cup \tilde{\Psi}(\tilde{\psi})$, where

$$\Phi(\phi) = \{\phi_t = \phi(\cdot - t) : t \in \mathbb{R}^2\},$$

$$\Psi(\psi) = \{\psi_{a,s,t} = a^{-\frac{3}{4}}\psi(A_a^{-1}S_s^{-1}(\cdot - t)) : a \in (0, 1], |s| \le 1 + a^{1/2}, t \in \mathbb{R}^2\},$$

$$\tilde{\Psi}(\tilde{\psi}) = \{\tilde{\psi}_{a,s,t} = a^{-\frac{3}{4}}\tilde{\psi}(\tilde{A}_a^{-1}S_s^{-T}(\cdot - t)) : a \in (0, 1], |s| \le 1 + a^{1/2}, t \in \mathbb{R}^2\},$$

and $\tilde{A}_a = \mathrm{diag}(a^{1/2}, a)$.

The associated transform can be defined in a similar manner as before in the pure group-theoretic approach.

Definition 5. For $\phi, \psi, \tilde{\psi} \in L^2(\mathbb{R}^2)$, let $SH(\phi, \psi, \tilde{\psi})$ be the associated cone-adapted continuous shearlet system. Then the associated *cone-adapted continuous shearlet transform* of $f \in L^2(\mathbb{R}^2)$ is given by

$$SH_{\phi,\psi,\tilde{\psi}} f(a, s, t, \iota) := \begin{cases} \langle f, \psi_{a,s,t}\rangle : \iota = -1, \\ \langle f, \phi_t\rangle : \iota = 0, \\ \langle f, \tilde{\psi}_{a,s,t}\rangle : \iota = 1. \end{cases}$$

where $(a, s, t) \in \mathbb{R}^* \times \mathbb{R} \times \mathbb{R}^2$.

We mention that this transform satisfies similar isometry properties as the continuous shearlet transform (cf. Kutyniok and Labate 2009).

Resolution of the Wavefront Set

The ability of a (cone-adapted) continuous shearlet system to resolve different directions can be analyzed using the notion of a wavefront set from microlocal analysis. Coarsely speaking, a wavefront set consists of the elements of the singular support of a distribution together with the directions in which the singularity propagates. For more details on microlocal analysis and wavefront sets, we refer to Hörmander (2003).

Definition 6. Let f be a distribution. Then a point $(x, \lambda) \in \mathbb{R}^2 \times \mathbb{S}^1$ is a *regular directed point* of f, if there exist open neighborhoods U_x and U_λ of x and λ, respectively, and a smooth function $\phi \in C^\infty(\mathbb{R}^2)$ with $\mathrm{supp}\phi \subset U_x$ and $\phi(x) = 1$ such that

$$|\widehat{\phi f}(\xi)| \le C_k(1 + |\xi|)^{-k} \quad \text{for all } \xi \in \mathbb{R}^2 \setminus \{0\} \text{ such that } \xi/|\xi| \in V_\lambda$$

holds for some $C_k > 0$. The *wavefront set* $WF(f)$ is then defined as the complement of the set of all regular directed points.

The notion of wavefront allows us to derive a precise statement about the resolution of different directions by cone-adapted continuous shearlet systems.

Theorem 2 (Kutyniok and Labate 2009). *Let $\psi \in L^2(\mathbb{R}^2)$ be a shearlet, and $f \in L^2(\mathbb{R}^2)$. Let $\mathcal{D} = \mathcal{D}_1 \cup \mathcal{D}_2$, where $\mathcal{D}_1 = \{(t_0, s_0) \in \mathbb{R}^2 \times [-1, 1] : \text{for } (s, t) \text{ in a neighborhood } U \text{ of } (s_0, t_0), |SH_{\phi,\psi,\tilde{\psi}} f(a, s, t, -1)| = O(a^k) \text{ as } a \to 0, \text{ for all } k \in \mathbb{N}, \text{ with the } O(\cdot)\text{-term uniform over } (s, t) \in U\}$ and $\mathcal{D}_2 = \{(t_0, s_0) \in \mathbb{R}^2 \times [1, \infty) : \text{for } (\frac{1}{s}, t) \text{ in a neighborhood } U \text{ of } (s_0, t_0), |SH_{\phi,\psi,\tilde{\psi}} f(a, s, t, 1)| = O(a^k) \text{ as } a \to 0, \text{ for all } k \in \mathbb{N}, \text{ with the } O(\cdot)\text{-term uniform over } (\frac{1}{s}, t) \in U\}$. Then*

$$WF(f)^c = \mathcal{D}.$$

An extension of this result to a more general class of shearlets $\psi, \tilde{\psi} \in L^2(\mathbb{R}^2)$ was derived in Grohs (2011a). Stronger results in the sense of more precise decay estimates can be found in Guo et al. (2009) for the band-limited case and in Kutyniok and Petersen (2017) for the compactly supported case.

Discrete Shearlet Systems

Discrete shearlet systems are derived by sampling the parameter set of continuous shearlet systems. Thus, similar to continuous shearlet systems, both a "classical" and a cone-adapted variant are available. Due to the fact that the first variant in the discrete setting not only is incapable of detecting the horizontal direction precise – only asymptotically – but also faces numerical instabilities due to the occurrence of arbitrarily small support sets, we will focus in the sequel only on the cone-adapted variant.

Cone-Adapted Discrete Shearlet Systems

The discretization of the parameter sets of parabolic scaling and shearing as defined in (5) is typically performed by choosing A_{2j} and S_k with $j, k \in \mathbb{Z}$. Coorbit theory (cf. section "Classical Continuous Shearlet Systems") then yields the discretization (for $c \in (\mathbb{R}_+)^2$ to add flexibility)

$$(a, s, t) \mapsto (2^{-j}, -k2^{-j/2}, A_{2j}^{-1} S_k^{-1} cm),$$

which when applied to Definition 4 leads to the following definition of a cone-adapted discrete shearlet system: The association of the different subsystems with the conic regions from Fig. 2b evidently carries over the discrete situation.

Definition 7. Let $c = (c_1, c_2) \in (\mathbb{R}_+)^2$. For $\phi, \psi, \tilde{\psi} \in L^2(\mathbb{R}^2)$, the *cone-adapted discrete shearlet system* $\mathcal{SH}(\phi, \psi, \tilde{\psi}; c) = \Phi(\phi; c_1) \cup \Psi(\psi; c) \cup \tilde{\Psi}(\tilde{\psi}; c)$ is defined by

$$\Phi(\phi; c_1) = \{\phi_m := \phi(\cdot - m) : m \in c_1 \mathbb{Z}^2\},$$

$$\Psi(\psi; c) = \{\psi_{j,k,m} := 2^{\frac{3}{4}j} \psi(S_k A_{2^j} \cdot - m) : j \geq 0, |k| \leq \lceil 2^{j/2} \rceil, m \in M_c \mathbb{Z}^2\},$$

$$\tilde{\Psi}(\tilde{\psi}; c) = \{\tilde{\psi}_{j,k,m} := 2^{\frac{3}{4}j} \tilde{\psi}(S_k^T \tilde{A}_{2^j} \cdot - m) : j \geq 0, |k| \leq \lceil 2^{j/2} \rceil, m \in \tilde{M}_c \mathbb{Z}^2\},$$

where $\tilde{A}_{2^j} = \operatorname{diag}(2^{j/2}, 2^j)$, $M_c = \operatorname{diag}(c_1, c_2)$ and $\tilde{M}_c = \operatorname{diag}(c_2, c_1)$. If $c = (1, 1)$, we also use the notions $\Phi(\phi)$, $\Psi(\psi)$, and $\tilde{\Psi}(\tilde{\psi})$.

One often refers to ϕ as a *scaling function* and to the functions ψ and $\tilde{\psi}$ as *(discrete) shearlets*.

As in the continuous setting, we can also define an associated transform, which arises as a discretization of the continuous version.

Definition 8. For $\phi, \psi, \tilde{\psi} \in L^2(\mathbb{R}^2)$ and $c = (c_1, c_2) \in (\mathbb{R}_+)^2$, let $SH(\phi, \psi, \tilde{\psi}; c)$ be the associated cone-adapted discrete shearlet system. Then the associated *cone-adapted discrete shearlet transform* of $f \in L^2(\mathbb{R}^2)$ is given by

$$SH_{\phi,\psi,\tilde{\psi}} f(j, k, m, \iota) := \begin{cases} \langle f, \psi_{j,k,m} \rangle : \iota = -1, \ j \geq 0, |k| \leq \lceil 2^{j/2} \rceil, m \in M_c \mathbb{Z}^2, \\ \langle f, \phi_m \rangle : \iota = 0, \ m \in c_1 \mathbb{Z}^2, \\ \langle f, \tilde{\psi}_{j,k,m} \rangle : \iota = 1, \ j \geq 0, |k| \leq \lceil 2^{j/2} \rceil, m \in \tilde{M}_c \mathbb{Z}^2. \end{cases}$$

The tiling of Fourier domain provided by the cone-adapted discrete shearlet system with classical shearlets as generating functions is depicted in Fig. 3a. Figure 3b shows a classical shearlet in spatial domain. One notices the "needlelike" structure of the function, which is of size $2^{-j} \times 2^{-j/2}$, hence becoming even more anisotropic shaped as $j \to \infty$.

Frame Properties

It is evident that the frame properties of a cone-adapted discrete shearlet system are closely linked to the chosen shearlets. One can identify band-limited shearlets and compactly supported shearlets as the two main classes of shearlets. Some applications such as seismology have a natural band-limited structure which then makes the first type of shearlets preferable, whereas other applications might require high spatial localization, which requires the second type.

30 Shearlets: From Theory to Deep Learning

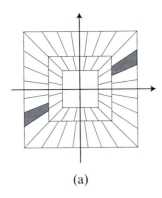

(a)   (b)

Fig. 3 (**a**) Partitioning of Fourier domain by the cone-adapted discrete shearlet system with classical shearlets as generating functions. (**b**) One shearlet in spatial domain

Band-Limited Shearlets

Classical shearlets as introduced in Example 1 are the most well-known type of band-limited shearlets. With slight modifications, the associated cone-adapted discrete shearlet system forms a Parseval frame for $L^2(\mathbb{R}^2)$.

Theorem 3 (Guo et al. 2006). *Let $C = \{(\xi_1, \xi_2) \in \mathbb{R}^2 : |\xi_2/\xi_1| \leq 1\}$ and $\widetilde{C} = \mathbb{R}^2 \setminus C$ with P_C and $P_{\widetilde{C}}$ denoting the associated orthogonal projections in $L^2(\mathbb{R}^2)$. Further, let $\psi \in L^2(\mathbb{R}^2)$ be a classical shearlet, let $\tilde{\psi}(\xi_1, \xi_2) = \psi(\xi_2, \xi_1)$, and let $\phi \in L^2(\mathbb{R}^2)$ be chosen so that, for a.e. $\xi \in \mathbb{R}^2$,*

$$|\hat{\phi}(\xi)|^2 + \sum_{j \geq 0} \sum_{|k| \leq \lceil 2^{j/2} \rceil} |\hat{\psi}(S_{-k}^T A_{2^{-j}} \xi)|^2 \chi_C + \sum_{j \geq 0} \sum_{|k| \leq \lceil 2^{j/2} \rceil} |\hat{\tilde{\psi}}(S_{-k} \tilde{A}_{2^{-j}} \xi)|^2 \chi_{\widetilde{C}} = 1.$$

Then the modified cone-adapted discrete shearlet system $\Phi(\phi) \cup P_C \Psi(\psi) \cup P_{\widetilde{C}} \tilde{\Psi}(\tilde{\psi})$ is a Parseval frame for $L^2(\mathbb{R}^2)$.

Refinements of this result leading to a smooth Parseval frame were derived in Guo and Labate (2013) and Bodmann et al. (2019). Moreover, in Grohs (2013), constructions of band-limited shearlet frames with dual frames such that both frames possess distinctive time-frequency localization properties were provided.

Compactly Supported Shearlets

Despite the advantage of high spatial localization, the construction of a Parseval frame is not as straightforward as in the band-limited case. In fact, it is still not clear whether a cone-adapted discrete shearlet system associated with compactly supported shearlets can be introduced, which forms a Parseval frame for $L^2(\mathbb{R}^2)$. Despite this obstacle, various constructions of compactly supported shearlets were suggested which yield cone-adapted discrete shearlet frames for $L^2(\mathbb{R}^2)$ with numerically proven ratio of the frame bounds of approximately 4, hence sufficiently stable from a numerical standpoint.

We now describe the general framework introduced in Kittipoom et al. (2012) for deriving sufficient conditions for cone-adapted discrete shearlet systems to form a frame alongside with theoretical estimates for the associated frame bounds. We start by defining the rectangle Ω_0 and the conic region Ω_1 by

$$\Omega_0 = \{\xi \in \mathbb{R}^2 : \max\{|\xi_1|, |\xi_2|\} \leq \tfrac{1}{2}\}, \quad \Omega_1 = \{\xi \in \mathbb{R}^2 : \tfrac{1}{2} < |\xi_2| < 1, |\xi_2|/|\xi_1| < 1\}.$$

Letting $\psi \in L^2(\mathbb{R}^2)$ and $\phi \in L^2(\mathbb{R}^2)$, we assume that

$$\operatorname*{ess\,inf}_{\xi \in \Omega_0} |\hat{\phi}(\xi)| > 0 \quad \text{and} \quad \operatorname*{ess\,inf}_{\xi \in \Omega_1} |\hat{\psi}(\xi)| > 0. \tag{6}$$

Setting $\tilde{\psi}(x_1, x_2) = \psi(x_2, x_1)$, it can be shown that those conditions ensure

$$\operatorname*{ess\,inf}_{\xi \in \mathbb{R}^2} |\hat{\phi}(\xi)|^2 + \sum_{j \geq 0} \sum_{|k| \leq \lceil 2^{j/2} \rceil} \left(|\hat{\psi}(S_k^T A_{2^{-j}} \xi)|^2 + |\hat{\tilde{\psi}}(\tilde{S}_k^T \tilde{A}_{2^{-j}} \xi)|^2 \right) > 0.$$

The following result then proves that, provided the Fourier transforms of the scaling function and shearlets decay fast enough with sufficient vanishing moments and satisfy (6), we obtain a shearlet frame $\mathcal{SH}(\phi, \psi, \tilde{\psi}; c)$.

Theorem 4 (Kittipoom et al. 2012). *Let $\phi, \psi \in L^2(\mathbb{R}^2)$ be functions such that*

$$\hat{\phi}(\xi_1, \xi_2) \leq C_1 \cdot \min\{1, |\xi_1|^{-\gamma}\} \cdot \min\{1, |\xi_2|^{-\gamma}\} \quad \text{and}$$

$$|\hat{\psi}(\xi_1, \xi_2)| \leq C_2 \cdot \min\{1, |\xi_1|^{\alpha}\} \cdot \min\{1, |\xi_1|^{-\gamma}\} \cdot \min\{1, |\xi_2|^{-\gamma}\}, \tag{7}$$

for some positive constants $C_1, C_2 < \infty$ and $\alpha > \gamma > 3$. Define $\tilde{\psi}(x_1, x_2) = \psi(x_2, x_1)$ and assume that ϕ, ψ satisfy (6). Then, there exists some positive constant c^* such that $\mathcal{SH}(\phi, \psi, \tilde{\psi}, c)$ forms a frame for $L^2(\mathbb{R}^2)$ for any $c = (c_1, c_2)$ with $\max\{c_1, c_2\} \leq c^*$.

For various explicit constructions of compactly supported shearlets leading to numerically stable cone-adapted discrete shearlet systems, we refer to Kittipoom et al. (2012). Let us further remark that this theorem can also be applied to band-limited cone-adapted discrete shearlet systems, since band-limited shearlets trivially satisfy condition (7).

Sparse Approximation

Recalling the goal to derive suitable decompositions (1) and efficient representations (2) of data, we will now show that within a certain model setting, shearlets can be proven to serve for both tasks in an optimal way.

For this, we first focus on the approximation properties of shearlets and introduce the related basic notions of approximation theory. Given a class of functions and a representation system, one main goal of approximation theory is to analyze the suitability of this system for uniformly approximating functions from this class. This leads to the notion of best N-term approximation.

Definition 9. Letting $N \in \mathbb{N}$, $C \subseteq L^2(\mathbb{R}^2)$ be a class of functions and $(\psi_\lambda)_{\lambda \in \Lambda} \subset L^2(\mathbb{R}^2)$ be a representation system, we call $f_N \in L^2(\mathbb{R}^2)$ *best N-term approximation of* f, if

$$\|f - f_N\|_{L^2} \leq \|f - g\|_{L^2} \quad \text{for all } g = \sum_{\lambda \in \Lambda_N} c_\lambda \psi_\lambda, \quad \text{where } \#\Lambda_N = N, \ \Lambda_N \subseteq \Lambda.$$

The *error of best N-term approximation* of some $f \in C$ is then given by $\|f - f_N\|_{L^2}$. The largest $\gamma > 0$ such that

$$\sup_{f \in C} \|f - f_N\|_{L^2} = O(N^{-\gamma}) \quad \text{as } N \to \infty$$

determines the *optimal (sparse) approximation rate* of C by $(\psi_\lambda)_{\lambda \in \Lambda}$.

Thus, the optimal (sparse) approximation rate relates approximation accuracy with the complexity of the approximating system in terms of sparsity.

We discussed earlier that one key aspect of multivariate functions is the fact that they are typically governed by anisotropic features. The model class of cartoonlike functions introduced by Donoho in (2001) makes this mathematically precise. We refer to Fig. 4 for an illustration.

Definition 10. For fixed $\nu > 0$, the *class* $\mathcal{E}_\nu^2$ *of cartoonlike functions* is the set of functions $f : \mathbb{R}^2 \to \mathbb{C}$ of the form

$$f = f_0 + f_1 \chi_B,$$

Fig. 4 Example of a cartoonlike function

where $B \subset [0, 1]^2$ with ∂B being a closed C^2-curve with curvature bounded by ν as well as $f_i \in C^2(\mathbb{R}^2)$ with supp $f_i \subset [0, 1]^2$ and $\|f_i\|_{C^2} \leq 1$ for each $i = 0, 1$.

The optimal (sparse) approximation rate for cartoonlike functions was proven by Donoho as well and can be stated in the situation of frames as follows. We wish to emphasize that the original result is proven for more general function systems.

Theorem 5 (Donoho 2001). *Let $(\psi_\lambda)_{\lambda \in \Lambda}$ be a frame for $L^2(\mathbb{R}^2)$. Then the optimal asymptotic approximation error of $f \in \mathcal{E}_\nu^2$ is given by*

$$\|f - f_N\|_{L^2} \leq C \cdot N^{-1} \quad \text{as } N \to \infty,$$

with f_N being a best N-term approximation of f and $C > 0$.

This benchmark result allows to make the phrase "optimal sparse approximations of cartoonlike functions" mathematically precise, namely, being justified in case a representation system does satisfy this rate. Indeed, it can be proven that, under weak assumptions on the generating functions, cone-adapted discrete shearlet systems associated with compactly supported shearlets provide this optimal rate up to a log-factor, which is typically assumed to be negligible.

Theorem 6 (Kutyniok and Lim 2011). *Let $c > 0$, and let $\phi, \psi, \tilde{\psi} \in L^2(\mathbb{R}^2)$ be compactly supported. Suppose that, in addition, for all $\xi = (\xi_1, \xi_2) \in \mathbb{R}^2$, the shearlet ψ satisfies*

(i) $|\hat{\psi}(\xi)| \leq C_1 \cdot \min\{1, |\xi_1|^\alpha\} \cdot \min\{1, |\xi_1|^{-\gamma}\} \cdot \min\{1, |\xi_2|^{-\gamma}\}$ *and*
(ii) $\left|\frac{\partial}{\partial \xi_2} \hat{\psi}(\xi)\right| \leq |h(\xi_1)| \cdot \left(1 + \frac{|\xi_2|}{|\xi_1|}\right)^{-\gamma}$,

where $\alpha > 5$, $\gamma \geq 4$, $h \in L^1(\mathbb{R})$, and C_1 are constant, and suppose that the shearlet $\tilde{\psi}$ satisfies (i) and (ii) with the roles of ξ_1 and ξ_2 reversed. Further, suppose that $\mathcal{SH}(\phi, \psi, \tilde{\psi}; c)$ forms a frame for $L^2(\mathbb{R}^2)$. Then, for any $\nu > 0$, the shearlet frame $\mathcal{SH}(\phi, \psi, \tilde{\psi}; c)$ provides (almost) optimal sparse approximations of functions $f \in \mathcal{E}_\nu^2$ in the sense that there exists some $C > 0$ such that

$$\|f - f_N\|_{L^2} \leq C \cdot N^{-1} \cdot (\log N)^{\frac{3}{2}} \quad \text{as } N \to \infty,$$

where f_N is the nonlinear N-term approximation obtained by choosing the N largest shearlet coefficients of f.

A similar result can also be derived in the setting of band-limited shearlets (Guo and Labate 2007).

Extensions of Shearlets

Several extensions of the described discrete shearlet systems were developed. In the sequel, we will discuss shearlet systems for arbitrary dimensions, α-molecules, and universal shearlets. Besides those, other generalizations of shearlets include irregular discrete shearlet frames arising from a different type of sampling of continuous shearlet systems (Kittipoom et al. 2011) and bendlets, which can be regarded as a second-order shearlet system (Lessig et al. 2019).

Higher Dimensions

We will first describe the extension to the three-dimensional situation, i.e., to derive a frame for $L^2(\mathbb{R}^3)$. In this situation, the four cones will be replaced by six pyramids again leading to a uniform way to treat the different directions. Accordingly, we define *paraboloidal scaling matrices* $A_{2^j}, \tilde{A}_{2^j}$ and $\check{A}_{2^j}, j \in \mathbb{Z}$ by

$$A_{2^j} = \text{diag}(2^j, 2^{j/2}, 2^{j/2}), \quad \tilde{A}_{2^j} = \text{diag}(2^{j/2}, 2^j, 2^{j/2}), \quad \text{and} \quad \check{A}_{2^j} = \text{diag}(2^{j/2}, 2^{j/2}, 2^j)$$

as well as *shear matrices* $S_k, \tilde{S}_k$, and $\check{S}_k, k = (k_1, k_2) \in \mathbb{Z}^2$ by

$$S_k = \begin{pmatrix} 1 & k_1 & k_2 \\ 0 & 1 & 0 \\ 0 & 0 & 1 \end{pmatrix}, \quad \tilde{S}_k = \begin{pmatrix} 1 & 0 & 0 \\ k_1 & 1 & k_2 \\ 0 & 0 & 1 \end{pmatrix}, \quad \text{and} \quad \check{S}_k = \begin{pmatrix} 1 & 0 & 0 \\ 0 & 1 & 0 \\ k_1 & k_2 & 1 \end{pmatrix}.$$

The lattices in $\mathbb{R}^3$ for defining translation are chosen as $M_c = \text{diag}(c_1, c_2, c_2)$, $\tilde{M}_c = \text{diag}(c_2, c_1, c_2)$, and $\check{M}_c = \text{diag}(c_2, c_2, c_1)$, where $c_1, c_2 > 0$.

The discrete shearlet system is then defined according to a partition of the Fourier domain into a rectangular region and six pyramids (see Fig. 5) similar to the conic regions of cone-adapted discrete shearlet systems.

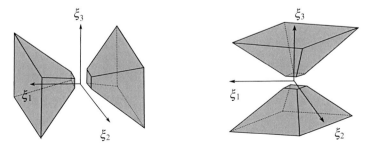

Fig. 5 The partition of Fourier domain by four of the six pyramids

The definition of pyramid-adapted discrete shearlet systems can now be stated as follows: Each part of the system is responsible for covering one set of pyramids, similar to covering the set of cones in Definition 7.

Definition 11. For $c = (c_1, c_2) \in (\mathbb{R}_+)^2$, the *pyramid-adapted discrete shearlet system* $\mathcal{SH}(\phi, \psi, \tilde{\psi}, \check{\psi}; c)$ generated by $\phi, \psi, \tilde{\psi}, \check{\psi} \in L^2(\mathbb{R}^3)$ is defined by

$$\mathcal{SH}(\phi, \psi, \tilde{\psi}, \check{\psi}; c) = \Phi(\phi; c_1) \cup \Psi(\psi; c) \cup \tilde{\Psi}(\tilde{\psi}; c) \cup \check{\Psi}(\check{\psi}; c),$$

where

$$\Phi(\phi; c_1) = \{\phi_m := \phi(\cdot - m) : m \in c_1 \mathbb{Z}^3\},$$
$$\Psi(\psi; c) = \{\psi_{j,k,m} := 2^j \psi(S_k A_{2^j} \cdot -m) : j \geq 0, \|k\|_\infty \leq \lceil 2^{j/2} \rceil, m \in M_c \mathbb{Z}^3\},$$
$$\tilde{\Psi}(\tilde{\psi}; c) = \{\tilde{\psi}_{j,k,m} := 2^j \tilde{\psi}(\tilde{S}_k \tilde{A}_{2^j} \cdot -m) : j \geq 0, \|k\|_\infty \leq \lceil 2^{j/2} \rceil, m \in \tilde{M}_c \mathbb{Z}^3\},$$
$$\check{\Psi}(\check{\psi}; c) = \{\check{\psi}_{j,k,m} := 2^j \check{\psi}(\check{S}_k \check{A}_{2^j} \cdot -m) : j \geq 0, \|k\|_\infty \leq \lceil 2^{j/2} \rceil, m \in \check{M}_c \mathbb{Z}^3\}.$$

The sufficient conditions for a cone-adapted discrete shearlet system to form a frame for $L^2(\mathbb{R}^2)$ as stated in Theorem 4 can be extended to the shearlet systems from Definition 11 (see Kutyniok et al. (2012)). The notion of cartoonlike functions (Definition 10) was in Kutyniok et al. (2012) suitably extended as well by considering a 3D body with a surface singularity, leading to a benchmark result as Theorem 5 now with the rate $N^{-\frac{1}{2}}$. In Kutyniok et al. (2012), it could then be proven that, similar to Theorem 6, also pyramid-adapted discrete shearlet systems lead to (almost) optimal sparse approximations of cartoonlike functions.

We also wish to mention that in fact a generalization of the definition of a discrete shearlet system, the frame properties, and the sparse approximation result to $L^2(\mathbb{R}^d)$, for $d \in \mathbb{N}$ is similarly possible, in this case the optimal rate being $N^{-\frac{1}{d-1}}$ (Kutyniok et al. 2012). In fact, the crucial step is from dimension 2 to 3, since this is the first time that anisotropic features of different dimensions can occur, namely, in this situation filament-like and sheetlike structures. Why do we then have just one type of shearlets for $L^2(\mathbb{R}^3)$? In fact, the shearlet elements we defined are in the spatial domain of size $2^{-j} \times 2^{-j/2} \times 2^{-j/2}$, making them "platelike" as $j \to \infty$. A different, seemingly also valid strategy would be to consider the scaling matrix $A_{2^j} = \text{diag}(2^j, 2^j, 2^{j/2})$ with similar changes for $\tilde{A}_{2^j}$ and $\check{A}_{2^j}$, leading to "needlelike" shearlet elements (of size $2^{-j} \times 2^{-j} \times 2^{-j/2}$) as $j \to \infty$. However, a shearlet system consisting of such "needlelike" shearlet elements lacks frame properties, making them unattractive for image processing. Moreover, the sparse approximation results show that even when introducing one-dimensional singularities as well, the rate is always determined by the singularities of the largest dimension.

α-Molecules

Several anisotropic representation systems based on parabolic scaling such as band-limited shearlets, compactly supported shearlets, and also second-generation curvelets provide (almost) optimal sparse approximations of cartoonlike functions (cf. Theorem 6), proven on a case-by-case basis. This raises the question whether such approximation results hold for a much more general class of anisotropic systems. In fact, the unified framework of *parabolic molecules* introduced in Grohs and Kutyniok (2014) provides such a general class, encompassing all known anisotropic frame constructions based on parabolic scaling. It allows to transfer approximation results from one system to another, thereby enabling that all the desirable approximation properties of shearlets can be deduced for virtually any other system based on parabolic scaling (see Grohs and Kutyniok (2014)).

The framework of parabolic molecules was even further generalized to α-*molecules* (Grohs et al. 2016a), which, for instance, also includes ridgelets and wavelets. In this approach, the parameter α measures the degree of anisotropy. The conceptual idea relies on the introduction of a general *parameter space*

$$\mathbb{P} := \mathbb{R}_+ \times \mathbb{T} \times \mathbb{R}^2,$$

where $(s, \theta, x) \in \mathbb{P}$ describes scale 2^s, orientation θ, and location x, and a flexibly applicable *parametrization* defined as a pair (Λ, Φ_Λ), where Λ is a discrete index set and Φ_Λ is a mapping

$$\Phi_\Lambda : \begin{cases} \Lambda \to \mathbb{P}, \\ \lambda \mapsto (s_\lambda, \theta_\lambda, x_\lambda). \end{cases}$$

This allows the definition of α-molecules, which includes a variety of anisotropic systems such as ridgelets for $\alpha = 0$, curvelets and shearlets for $\alpha = \frac{1}{2}$, and wavelets for $\alpha = 1$. It also includes α-*shearlets*, which are defined as a cone-adapted discrete shearlet system with the scaling depending on α and suitable adaption of shearing (Kutyniok et al. 2012).

Definition 12. Let $\alpha \in [0, 1]$, and let (Λ, Φ_Λ) be a parametrization. Then $(m_\lambda)_{\lambda \in \Lambda}$ is a *system of α-molecules of order* $(L, M, N_1, N_2) \in (\mathbb{Z}_+ \cup \{\infty\})^2 \times \mathbb{Z}_+^2$, if, for all $\lambda \in \Lambda$,

$$m_\lambda(x) = s_\lambda^{(1+\alpha)/2} a^{(\lambda)} \left(A_{\alpha, s_\lambda} R_{\theta_\lambda} (x - x_\lambda) \right), \quad \Phi_\Lambda(\lambda) = (s_\lambda, \theta_\lambda, x_\lambda),$$

such that, for all $|\beta| \leq L$,

$$\left| \partial^\beta \hat{a}^{(\lambda)}(\xi) \right| \lesssim \min\left(1, s_\lambda^{-1} + |\xi_1| + s_\lambda^{-(1-\alpha)} |\xi_2| \right)^M \left(1 + |\xi|^2 \right)^{-\frac{N_1}{2}} \left(1 + \xi_2^2 \right)^{-\frac{N_2}{2}}.$$

We next state the key result enabling the transfer of sparse approximation results from one system to all other systems within this framework for the same α. It provides an estimate for the decay of the entries of the cross-Gramian matrix away from the main diagonal, which requires an appropriate notion of distance. For this, let (Λ, Φ_Λ) and $(\tilde{\Lambda}, \Phi_{\tilde{\Lambda}})$ be parametrizations. For $\lambda \in \Lambda$ and $\mu \in \tilde{\Lambda}$, we then define the *index distance* by

$$\omega(\lambda, \mu) := \omega(\Phi_\Lambda(\lambda), \Phi_{\tilde{\Lambda}}(\mu)) := 2^{|s_\lambda - s_\mu|}\left(1 + 2^{\operatorname{argmin}(s_\lambda, s_\mu)}d(\lambda, \mu)\right),$$

where

$$d(\lambda, \mu) := |\theta_\lambda - \theta_\mu|^2 + |x_\lambda - x_\mu|^2 + |\langle (\cos(\theta_\lambda), \sin(\theta_\lambda))^\top, x_\lambda - x_\mu \rangle|.$$

This allows us to now formulate the result.

Theorem 7 (Grohs and Kutyniok 2014; Grohs et al. 2016a). *Let $\alpha \in [0, 1]$, $N > 0$, and let $(m_\lambda)_{\lambda \in \Lambda}$, $(p_\mu)_{\mu \in \tilde{\Lambda}}$ be systems of α-molecules of order (L, M, N_1, N_2) with*

$$L \geq 2N, \quad M > 3N - \frac{3-\alpha}{2}, \quad N_1 \geq N + \frac{1+\alpha}{2}, \quad N_2 \geq 2N.$$

Then, for all $\lambda \in \Lambda$ and $\mu \in \tilde{\Lambda}$,

$$\left|\langle m_\lambda, p_\mu \rangle\right| \lesssim \omega(\lambda, \mu)^{-N}.$$

For detailed results concerning frame properties of parabolic molecules as well as the more general α-molecules and sufficient conditions for those to provide optimal sparse approximation properties up to a log-factor, we refer to Grohs and Kutyniok (2014) and Grohs et al. (2016a,b).

Universal Shearlets

Another extension of cone-adapted discrete shearlet systems are *universal shearlets* introduced in Genzel and Kutyniok (2014), which provide even more flexibility in the type of scaling than α-molecules. In fact, universal shearlets allow a different type of scaling at each scaling level of α-shearlets by setting $\alpha = (\alpha_j)_j$ with j being the scale and $\alpha_j \in (0, 2)$. The generalized dilation matrices $A_{\alpha_j, 2^j}$ and $\tilde{A}_{\alpha_j, 2^j}$ are then defined by

$$A_{\alpha_j, 2^j} := \begin{pmatrix} 2^j & 0 \\ 0 & 2^{\frac{\alpha_j}{2}j} \end{pmatrix} \quad \text{and} \quad \tilde{A}_{\alpha_j, 2^j} := \begin{pmatrix} 2^{\frac{\alpha_j}{2}j} & 0 \\ 0 & 2^j \end{pmatrix}.$$

Based on these, universal shearlet systems are defined as follows (Genzel and Kutyniok 2014):

Definition 13. For $\phi, \psi, \tilde{\psi} \in L^2(\mathbb{R}^2)$, $\alpha = (\alpha_j)_j$, $\alpha_j \in (0, 2)$, and $c = (c^j)_j$ with $c^j = (c_1^j, c_2^j) \in (\mathbb{R}_+)^2$ for each scale j, the *universal shearlet system* $\mathcal{SH}(\phi, \psi, \tilde{\psi}; \alpha, c)$ is defined by

$$\mathcal{SH}(\phi, \psi, \tilde{\psi}; \alpha, c) := \Phi(\phi; c_1^0) \cup \Psi(\psi; \alpha, c) \cup \tilde{\Psi}(\tilde{\psi}; \alpha, c),$$

where

$$\Phi(\phi; c_1^0) := \{\phi_m = \phi(\cdot - c_1^0 m) : m \in \mathbb{Z}^3\},$$

$$\Psi(\psi; \alpha, c) := \{\psi_{j,k,m} = 2^{\frac{\alpha_j+2}{4}j} \psi(S_k A_{\alpha_j, 2^j} \cdot - M_{c^j} m) : j \geq 0, |k| \leq \lceil 2^{\frac{j(2-\alpha_j)}{2}} \rceil, m \in \mathbb{Z}^2\},$$

$$\tilde{\Psi}(\tilde{\psi}; \alpha, c) := \{\tilde{\psi}_{j,k,m} = 2^{\frac{\alpha_j+2}{4}j} \tilde{\psi}(S_k^T \tilde{A}_{\alpha_j, 2^j} \cdot - \tilde{M}_{c^j} m) : j \geq 0, |k| \leq \lceil 2^{\frac{j(2-\alpha_j)}{2}} \rceil, m \in \mathbb{Z}^2\}.$$

In the special situation when all α_j and c^j coincide, i.e., $\alpha_j = \alpha_0$ and $(c_1^j, c_2^j) = (c_1, c_2)$ for all scales j, and $\alpha_0 = 1$, the system reduces to cone-adapted discrete shearlet systems in the sense that $\mathcal{SH}(\phi, \psi, \tilde{\psi}; \alpha, c) = \mathcal{SH}(\phi, \psi, \tilde{\psi}; c)$. If in this situation $\alpha_0 = 2$, then the universal shearlet systems reduce to isotropic wavelet systems. Finally, for $\alpha_0 \to 0$, the system of ridgelets is approached.

Since the implementation of ShearLab3D in www.ShearLab.org relies on universal shearlets, we also state the associated transform explicitly.

Definition 14. Retain the notions from Definition 13, and let $SH(\phi, \psi, \tilde{\psi}; \alpha, c)$ be a universal shearlet system. Then the associated *universal shearlet transform* of $f \in L^2(\mathbb{R}^2)$ is given by

$$SH_{\phi,\psi,\tilde{\psi}} f(j, k, m, \iota) := \begin{cases} \langle f, \psi_{j,k,m} \rangle : \iota = -1, \ j \geq 0, |k| \leq \lceil 2^{\frac{j(2-\alpha_j)}{2}} \rceil, m \in \mathbb{Z}^2, \\ \langle f, \phi_m \rangle : \iota = 0, \ m \in \mathbb{Z}^2, \\ \langle f, \tilde{\psi}_{j,k,m} \rangle : \iota = 1, \ j \geq 0, |k| \leq \lceil 2^{\frac{j(2-\alpha_j)}{2}} \rceil, m \in \mathbb{Z}^2. \end{cases}$$

On the theoretical side, this approach has so far been only analyzed for band-limited generators concerning their frame properties. More precisely, in Genzel and Kutyniok (2014), it has been shown that there exists a large class of scaling sequences $\alpha = (\alpha_j)_j$ such that, using classical shearlets with small modifications, the system $SH(\phi, \psi, \tilde{\psi}; \alpha, c)$ forms a Parseval frame for $L^2(\mathbb{R}^2)$.

Digital Shearlet Systems

One main advantage of shearlets is the fact that they admit a faithful digitalization and hence a consistent implementation, mainly due to the fact that directional sensitivity is incorporated by a shearing operator (instead of, for instance, a rotation operator, which would change the digital grid). The first digital version was introduced in Easley et al. (2008) as the nonsubsampled shearlet transform in 2D and 3D, which digitalized the cone-adapted discrete shearlet transform based on band-limited shearlets. The first faithful digital shearlet transform using compactly supported shearlets was suggested in Lim (2010). It utilizes separable shearlets to achieve low complexity. This approach was later improved in Lim (2013) by an implementation called nonseparable shearlet transform. It uses the fact that nonseparable compactly supported shearlet generators can much better approximate classical band-limited shearlets, which in turn can be designed to form Parseval frames.

Digital 2D Shearlet Transform

In the sequel, we will describe the concept of digital shearlet systems and associated transforms as developed in Lim (2013). In fact, these are also the basis for the software package ShearLab3D provided on the webpage www.ShearLab.org (see also Kutyniok et al. (2016)), which extends this concept to both universal shearlets and the 3D situation.

The digital shearlet systems we will introduce are a faithful digitalization of cone-adapted discrete shearlet systems $\mathcal{SH}(\phi, \psi, \tilde{\psi}; c) = \Phi(\phi; c_1) \cup \Psi(\psi; c) \cup \tilde{\Psi}(\tilde{\psi}; c)$ as in Definition 7. Since the component $\Phi(\phi; c_1)$ is just the scaling part coinciding with a wavelet scaling part, we refer for its digitalization to the common wavelet literature (Daubechies 1992; Mallat 1998). Furthermore, we restrict to discussing $\Psi(\psi; c)$, since $\tilde{\Psi}(\tilde{\psi}; c)$ can be digitalized similarly except for switching the order of variables.

We first define a separable shearlet $\psi^{\text{sep}} \in L^2(\mathbb{R}^2)$, which will be the basis for defining a nonseparable variant. For this, let ψ^1 and $\phi^1 \in L^2(\mathbb{R})$ be a compactly supported 1D wavelet and an associated (orthonormal) scaling function, respectively, satisfying the two scale relations

$$\phi^1(x_1) = \sum_{n_1 \in \mathbb{Z}} h(n_1) \sqrt{2} \phi^1(2x_1 - n_1)$$

and

$$\psi^1(x_1) = \sum_{n_1 \in \mathbb{Z}} g(n_1) \sqrt{2} \phi^1(2x_1 - n_1),$$

with some appropriately chosen filters g and h in the sense that both ψ^1 and ϕ^1 are sufficiently smooth and ψ^1 has sufficient vanishing moments. For later use, we also define

$$H_j(\xi_1) := \prod_{k=0}^{j-1} H(2^k \xi_1) \quad \text{and} \quad G_j(\xi_1) := G(2^{j-1}\xi_1) H_{j-1}(\xi_1),$$

where $H(\xi) := \sum_{n \in \mathbb{Z}^d} h_n e^{-2\pi i \langle n, \xi \rangle}$. Then the separable shearlet generator is chosen as $\psi^{\text{sep}} := \psi^1 \otimes \phi^1$.

Based on this, we define the nonseparable generator ψ such as

$$\hat{\psi}(\xi) = P(\tfrac{\xi_1}{2}, \xi_2) \hat{\psi}^{\text{sep}}(\xi), \tag{8}$$

where the trigonometric polynomial P is a 2D fan filter (cf. Do and Vetterli 2005). With a suitable choice for P, we indeed have

$$P(\tfrac{\xi_1}{2}, \xi_2) \hat{\psi}^1(\xi_1) \hat{\phi}^1(\xi_2) \approx \hat{\psi}^1(\xi_1) \hat{\psi}^2(\tfrac{\xi_2}{\xi_1}),$$

where $\hat{\psi}^1(\xi_1) \hat{\psi}^2(\tfrac{\xi_2}{\xi_1})$ is a classical shearlet as introduced in Example 1. The functions P, ψ^1, and ϕ^1 can be chosen in such a way that the sufficient conditions (cf. Theorem 4) for the resulting shearlet system to form a frame are satisfied.

The second step consists of digitalizing the associated shearlet coefficients $\langle f, \psi_{j,k,m} \rangle$ for $j = 0, \ldots, J-1$, of a function f given as

$$f(x) = \sum_{m \in \mathbb{Z}^2} f_J(m) 2^J \phi^1(2^J x_1 - m_1) \phi^1(2^J x_2 - m_2), \tag{9}$$

where

$$\psi_{j,k,m}(x) := 2^{\frac{3}{4}j} \psi(S_k A_{2^j} x - M_c m) = \psi_{j,0,m}(S_{k 2^{-j/2}} x) \tag{10}$$

with the sampling matrix given by $M_c = \text{diag}(c_1, c_2)$. Without loss of generality, we will from now on assume that $j/2$ is integer; otherwise, we take either $\lceil j/2 \rceil$ or $\lfloor j/2 \rfloor$.

To obtain a faithful discretization of $\psi_{j,0,m}$ in (10) by using the structure of the multiresolution analysis associated with (8), we let $p_j := (p_{j,n})_{n \in \mathbb{Z}^2}$ and g_j be the Fourier coefficients of $P(2^{J-j-1}\xi_1, 2^{J-j/2}\xi_2)$ and G_j, respectively. Then we have

$$\langle f, \psi_{j,0,m} \rangle = (f_J * \overline{(p_j * g_j)})_{A_{2^j}^{-1} 2^J M_c m}, \tag{11}$$

assuming that the sampling matrix M_{c_j} satisfies $A_{2^j}^{-1} 2^J M_c m \in \mathbb{Z}^2$. The associated discrete filter coefficients for $\psi_{j,0,m}$ can be shown to equal $p_j * g_j$. Digitizing (10) also requires a digital shearing operator. Since the shear matrix $S_{k 2^{-j/2}}$ does not preserve the regular grid $\mathbb{Z}^2$, this problem is resolved by refining the regular grid $\mathbb{Z}^2$ along the horizontal axis x_1 by a factor of $2^{-j/2}$, leading to the new grid $2^{-j/2}\mathbb{Z} \times \mathbb{Z}$. Now, let $\uparrow 2^{j/2}$, $\downarrow 2^{j/2}$, and $*_1$ be the 1D upsampling, downsampling, and convolution operator along the horizontal axis x_1 by a factor of $2^{j/2}$, respectively. For a 2D discrete signal $f^d = (f_n^d)_{n \in \mathbb{Z}^2} \in \ell^2(\mathbb{Z}^2)$, the shear operator $S_{k2^{-j/2}}$ can then be digitalized by

$$S_{k2^{-j/2}}^d(f^d) := \left(((\tilde{f}^d)_{S_k(\cdot)} *_1 \overline{h_{j/2}})\right)_{\downarrow 2^{j/2}}, \tag{12}$$

where $\tilde{f}^d$ given by $\tilde{f}^d := ((f^d)_{\uparrow 2^{j/2}} *_1 h_{j/2})$ is resampled by S_k.

The discussed digitalization of (10) leads to the following definition of a faithful digital shearlet transform:

Definition 15. Let $f_J \in \ell^2(\mathbb{Z}^2)$ be the scaling coefficients given in (9), and retain the notions from this subsection. Then the *digital shearlet transform* associated with $\Psi(\psi; c)$ is defined by

$$DST_\psi^{2D} f(j, k, m) := (f_J * \overline{\psi_{j,k}^d})(2^J A_{2^j}^{-1} M_c m) \quad \text{for } j = 0, \ldots, J - 1,$$

where

$$\psi_{j,k}^d := S_{k2^{-j/2}}^d(p_j * g_j),$$

with the shearing operator defined by (12) and the sampling matrix M_c chosen so that $2^J A_{2^j}^{-1} M_c m \in \mathbb{Z}^2$.

Considering the full shearlet system and not only $\Psi(\psi; c)$ then leads in a canonical manner to the digital shearlet transform $DST_{\phi, \psi, \tilde{\psi}}^{2D} f(j, k, m, \iota)$ with ι playing a similar role as in Definition 7.

Extensions of the Digital 2D Shearlet Transform and ShearLab3D

We now discuss the extension of the digital shearlet transform to both universal shearlets and the 3D situation, as it is implemented in ShearLab3D (www.ShearLab.org). For details of the implementation, we refer to Kutyniok et al. (2016).

For this, recall the notion from the definition of a universal shearlet system $SH(\phi, \psi, \tilde{\psi}; \alpha, c)$ (Definition 13). The nonseparable shearlet in $L^2(\mathbb{R}^3)$ is now chosen as

$$\hat{\psi}(\xi) = \left(P(\tfrac{\xi_1}{2}, \xi_2)\hat{\psi}^1(\xi_1)\hat{\phi}^1(\xi_2)\right)\left(P(\tfrac{\xi_1}{2}, \xi_3)\hat{\phi}^1(\xi_3)\right).$$

Canonically extending the arguments in section "Digital 2D Shearlet Transform" and as before only focusing on $\Psi(\psi; \alpha, c)$, we can digitalize the shearlet coefficients $\langle f, \psi_{j,k,m}\rangle$ for a function $f \in L^2(\mathbb{R}^3)$ given by

$$f(x) = \sum_{m \in \mathbb{Z}^3} f_{J,m} 2^{J \cdot 3/2} (\phi^1 \otimes \phi^1 \otimes \phi^1)(2^J x - m) \tag{13}$$

as follows:

Definition 16. Let $f_J \in \ell^2(\mathbb{Z}^3)$ be the scaling coefficients given in (13), and retain the notions from this section. Then the *digital shearlet transform* associated with $\Psi(\psi; \alpha, c)$ is defined by

$$DST_\psi^{3D} f(j, k, m) := (f_J * \overline{\psi_{j,k}^d})(\tilde{m}) \quad \text{for } j = 0, \ldots, J-1,$$

where the sampling constants c_1^j and c_2^j are chosen so that

$$\tilde{m} := (2^{J-j} c_1^j m_1, 2^{J-\frac{\alpha_j}{2}j} c_2^j m_2, 2^{J-\frac{\alpha_j}{2}j} c_2^j m_3) \in \mathbb{Z}^3,$$

and the discrete-time Fourier transforms of the 3D digital shearlet filters $\psi_{j,k}^d$ are defined by

$$\psi_{j,k}^d(\xi) := G_{J-j}(\xi_1) \Phi_{j,k_1}^d(\xi_1, \xi_2) \Phi_{j,k_2}^d(\xi_1, \xi_3)$$

with Φ_{j,k_1}^d and Φ_{j,k_2}^d being the discrete-time Fourier transforms of

$$\phi_{j,k_1,(n_1,n_2)}^d := \left(S_{k_1 2^{-d\alpha_j}}^d (h_{J-\frac{\alpha_j}{2}j} *_{x_2} p_j) \right)_{(n_1,n_2)}$$

and

$$\phi_{j,k_2,(n_1,n_3)}^d := \left(S_{k_2 2^{-d\alpha_j}}^d (h_{J-\frac{\alpha_j}{2}j} *_{x_3} p_j) \right)_{(n_1,n_3)},$$

respectively.

Similar to the 2D situation, the definition of the full 3D digital shearlet transform $DST_{\phi,\psi,\tilde{\psi},\check{\psi}}^{3D} f(j, k, m, \iota)$ is then canonical.

Applications of Shearlets

This section is devoted to applications of shearlet systems and the associated transforms. We will foremost exploit the fact that shearlets provide optimal sparse approximations of functions which are governed by anisotropic features (section "Sparse Approximation"), alongside with a faithful implementation (section "Digital Shearlet Systems"). Due to their high spatial localization and their equal treatment of different directions, we will focus on compactly supported cone-adapted discrete shearlet systems (section "Compactly Supported Shearlets"). We wish to remark that the problem settings we present in this section such as image inpainting can also be handled in the 3D setting, i.e., video inpainting, by similar means.

The main areas of application of shearlets are inverse problems from imaging sciences. Indeed, images are typically governed by anisotropic features such as edges, which also the human visual cortex is particularly sensitive to recognize. The traditional (model-based) approach to solving an inverse problem using the fact that the original image is sparsely approximated by a representation system, here shearlets, is by sparse regularization. In the sequel, we will discuss the inpainting (Grohs and Kutyniok 2014; King et al. 2014) and the separation problem (Kutyniok and Lim 2012; Donoho and Kutyniok 2013; Kutyniok 2014) as exemplary problem instances. For a more extensive survey about applications of shearlets using pure model-based approaches, we refer to Easley and Labate (2012) and Kutyniok et al. (2016).

Due to the increasing complexity of problems in imaging, pure model-based methods are often today not sufficient anymore. At the same time, we witness the tremendous success of data-driven methodologies such as deep neural networks for various problem classes, in particular, in imaging sciences. However, entirely replacing physical knowledge about a problem by learned insights is usually not a sensible strategy. The type of approaches which intuitively lead to an optimal combination of the model-based and data-driven realm pursues the strategy to use model-based methods as far as they are reliable and data-driven methods where it is necessary. This concept also circumvents the problem that as of now methodologies such as deep learning act as a black box without any comprehensive theoretical underpinning. In the sequel, the approach "Learning the Invisible" to the limited-angle computed tomography problem (Bubba et al. 2019) shall serve as an example. The classical problem of edge detection, even wavefront set detection, will show another possibility to optimally combine shearlets with deep learning approaches as it is done in DeNSE (Deep Network Shearlet Edge Extractor), leading to superior performance over model-based methods (Andrade-Loarca et al. 2019, 2020).

Sparse Regularization Using Shearlets

Given an ill-posed inverse problem $Tf = g$, where $T : \mathcal{H} \to \mathcal{H}$ is a linear, bounded operator and $g \in \mathcal{H}$, classical Tikhonov regularization aims to solve this problem by minimizing the functional

$$\|Tf - g\|^2 + \beta \cdot \|f\|^2,$$

with β being the regularization parameter. However, the regularization term $\|f\|^2$ might not be appropriate for each inverse problem, and other prior information of f is known and should be incorporated. For functions in $L^2(\mathbb{R}^2)$ governed by anisotropic features such as images, shearlet systems provide optimal sparse approximations. The generalization of Tikhonov regularization introduced in Daubechies et al. (2004) exploits such information by suggesting to minimize

$$\|Tf - g\|^2 + \beta \cdot \|(\langle \cdot, \psi_\lambda \rangle)_{\lambda \in \Lambda}\|_{\ell_1},$$

with $(\psi_\lambda)_{\lambda \in \Lambda}$ being a shearlet system, instead. We remark that the concept of sparse regularization is closely related to, and in fact might also be seen as belonging to, the area of compressed sensing (Davenport et al. 2012).

We now discuss two different special situations in which this conceptual approach can be applied.

Image Separation

Images are typically a composition of morphologically distinct components. The problem of image separation, which is a highly ill-posed inverse problem, aims to decompose the image into those components. To be mathematically precise, assuming just two components, the problem can be modeled as follows: Let $f_1, f_2 \in L^2(\mathbb{R}^2)$ and $g = f_1 + f_2$; we aim to recover f_1 and f_2 from g. One possible setting is the separation of curve-like and point-like objects, which, for example, appears in neurobiological imaging in the form of spines (point-like objects) and dendrites (curve-like objects) or astronomical imaging in the form of stars (point-like objects) and filaments (curve-like objects). For further examples, we refer to Starck et al. (2010).

This problem can only be solved by assuming prior information on the components. The approach of sparse regularization assumes that each component f_1 and f_2 can be sparsified by a representation system $(\psi_\lambda^1)_{\lambda \in \Lambda}$ and $(\psi_\lambda^2)_{\lambda \in \Lambda}$, respectively. This leads to the following minimization problem:

$$\arg\min_{u_1,u_2} \|(\langle u_1, \psi_\lambda^1 \rangle)_{\lambda \in \Lambda}\|_{\ell_1} + \|(\langle u_2, \psi_\lambda^2 \rangle)_{\lambda \in \Lambda}\|_{\ell_1} \quad \text{subject to} \quad g = u_1 + u_2, \tag{14}$$

where we chose the constrained form of the optimization problem, for which also the theoretical results are formulated. Let us now consider the situation that f_1 are point-like features and f_2 are curve-like features. In this case, we would choose $(\psi_\lambda^1)_{\lambda \in \Lambda}$ to be a wavelet system and $(\psi_\lambda^2)_{\lambda \in \Lambda}$ to be a shearlet system. For an illustration, we refer to Fig. 6.

To explain the associated theoretical results, assume for f_1 and f_2 models for point-like and curve-like features, namely,

$$f_1 := \sum_{i=1}^{P} |x - x_i|^{-3/2} \quad \text{and} \quad f_2 := \int \delta_{\tau(t)} dt,$$

where $x_i \in \mathbb{R}^2$ and $\tau : [0,1] \to \mathbb{R}^2$ are closed curves. To aim for an asymptotic analysis, let $(F_j)_j$ be a sequence of filters such as wavelet filters satisfying

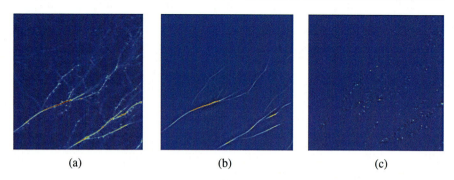

Fig. 6 Separation of spines and dendrites in neurobiological imaging (Kutyniok and Lim 2012) using ShearLab3D to solve (14). (**a**) Original image. (**b**) Extracted dendrites (curve-like objects). (**c**) Extracted spines (point-like objects)

$$g = \sum_j F_j * (F_j * g), \quad \text{for all } g \in L^2(\mathbb{R}^2). \tag{15}$$

This leads to a scale-dependent decomposition. Consider then the accordingly filtered components $(f_{i,j})_j := (f_i * F_j)_j$, $i = 1, 2$ as well as the image at scale j, i.e., $g_j := f_{1,j} + f_{2,j}$. The following result analyzes the microlocal structure of the problem and shows that at all sufficiently fine scales, nearly perfect separation is achieved. The key reason for the success of the separation approach is the morphological difference between the point and curve structures, which is mirrored in the difference between the associated sparsifying systems.

Theorem 8 (Donoho and Kutyniok 2009, 2013). *Retaining the notation from this subsection and letting $\widetilde{f_{1,j}}$, $\widetilde{f_{2,j}}$ denote the solution of (14) for the separation problem $g_j = f_{1,j} + f_{2,j}$, we have*

$$\frac{\|f_{1,j} - \widetilde{f_{1,j}}\|_{L^2} + \|f_{2,j} - \widetilde{f_{2,j}}\|_{L^2}}{\|f_{1,j}\|_{L^2} + \|f_{2,j}\|_{L^2}} \to 0, \quad j \to \infty.$$

A stronger result concerning recovery of the wavefront sets of the models for point-like and curve-like features using a thresholding algorithm was derived in Kutyniok (2013, 2014) studies the separation of cartoon and texture using as sparsifying systems a shearlet and a Gabor system. Finally, for similar results in the general Hilbert space setting, we refer to Donoho and Kutyniok (2013).

Image Inpainting

Image inpainting aims to recover missing or deteriorated parts of an image. It is thus a special case of a data recovery problem; and the approach we discuss can be generalized to this setting as well. The problem can be formulated as follows:

Let $f \in L^2(\mathbb{R}^2)$ and a (measurable) mask $M \subset \mathbb{R}^2$; we aim to recover f from $g := f \cdot 1_{\mathbb{R}^2 \setminus M}$.

Let now $(\psi_\lambda)_{\lambda \in \Lambda}$ be a shearlet system. Sparse regularization using shearlets assumes the following model for the solution, where – similar as in the previous subsection – we choose the constrained form of the optimization problem:

$$\min_u \|(\langle u, \psi_\lambda \rangle)_{\lambda \in \Lambda}\|_{\ell_1} \quad \text{subject to} \quad g = u \cdot 1_{\mathbb{R}^2 \setminus M}. \tag{16}$$

Figure 7 shows some numerical experiments. For further examples as well as comparison to other state-of-the-art approaches, we refer to Kutyniok et al. (2016).

Theoretical results have been achieved in the case that f is a distribution with a curvilinear singularity, i.e.:

$$f := \int_{-\rho}^{\rho} w(t) \delta_{\tau(t)} dt,$$

where $\tau : [-1, 1] \to \mathbb{R}^2$ is a C^2-curve, $\rho < 1$, and $w : [-\rho, \rho] \to \mathbb{R}_0^+$ is a "bump" function. The mask is then defined as a vertical strip intersecting the curve, with a flexible width:

$$M_h = \{(x_1, x_2) \in \mathbb{R}^2 : |x_1| \leq h\}, \quad h > 0.$$

Again aiming for an asymptotic analysis, let $(F_j)_j$ be a sequence of filters (cf. (15)), which leads to a scale-dependent decomposition, and consider the filtered image $f_j := f * F_j$ as well as the filtered observed image $g_j := (f \cdot 1_{\mathbb{R}^2 \setminus M_{h_j}}) * F_j$, where we also make the width of the mask dependent on the scale j. The following result shows that at all sufficiently fine scales, nearly perfect inpainting is achieved in case the shearlets are asymptotically larger than the width of the mask.

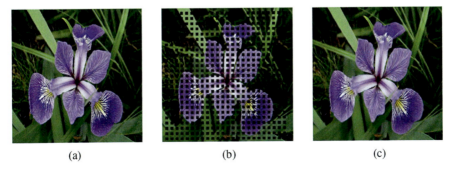

Fig. 7 Numerical experiments using ShearLab3D to solve (16). (**a**) Original image. (**b**) Masked image. (**c**) Inpainted image

Theorem 9 (King et al. 2014). *Retaining the notation from this subsection and letting $\tilde{f}_j$ denote the solution of (16) for the inpainting problem $g_j = (f \cdot 1_{\mathbb{R}^2 \setminus M_{h_j}}) * F_j$, if $h_j = o(2^{-j/2})$ as $j \to \infty$, we have*

$$\frac{\|\tilde{f}_j - f_j\|_{L^2}}{\|f_j\|_{L^2}} \to 0, \quad j \to \infty.$$

A similar result holds for wavelet inpainting, then with the sufficient condition that $h_j = o(2^{-j})$ as $j \to \infty$ according to the smaller width of a wavelet element. An extension to inpainting using universal shearlet systems can be found in Genzel and Kutyniok (2014). For similar results in the general Hilbert space setting, we refer to Donoho and Kutyniok (2013) and Genzel and Kutyniok (2014).

Shearlets Meet Deep Learning

Deep learning approaches have recently swept the area of inverse problems, predominantly from imaging, the main reason being that no physical model for images exists, consequently making data-driven methods very effective. A standard feed-forward *deep neural network* consists of affine-linear maps $W_\ell : \mathbb{R}^{N_{\ell-1}} \to \mathbb{R}^{N_\ell}$, $\ell = 1, \ldots, L$, i.e., $W_\ell(x) = A_\ell x + b_\ell$, where $A_\ell \in \mathbb{R}^{N_\ell \times N_{\ell-1}}$ and $b_\ell \in \mathbb{R}^{N_\ell}$, as well as a (nonlinear) univariate function $\sigma : \mathbb{R} \to \mathbb{R}$ called *activation function*, and realizes the map $\mathcal{NN}_\theta : \mathbb{R}^d \to \mathbb{R}^{N_L}$

$$\mathcal{NN}_\theta(x) = W_L \sigma(W_{L-1}\sigma(\ldots \sigma(W_1(x)))),$$

with σ being applied componentwise and θ denoting all parameters of the neural network, i.e., the weight matrices A_ℓ and biases b_ℓ. In applications, the activation function is typically chosen as the ReLU (Rectified Linear Unit) given by $\sigma(x) := \max\{0, x\}$. Corresponding to the depiction as a graph, L is referred to as the number of layers. Given samples $(x_i, f(x_i))_{i=1}^m$ of a function $f : \mathbb{R}^d \to \mathbb{R}^{N_L}$, learning algorithms such as stochastic gradient descent learn θ according to minimizing a certain empirical risk functional.

For an introduction and overview, also concerning the various types of neural networks, we refer to Goodfellow et al. (2017).

Convolutional neural networks, in which convolutions are performed in each layer, are the state-of-the-art for imaging applications. The network architecture typically utilized for solving inverse problems is the *U-Net* as introduced in Ronneberger et al. (2015), which can be regarded as an autoencoder with additional skip connections to allow the transportation of additional information across the compressed layers.

The most basic approach to solving an inverse problem $Tf = g$ by deep learning is to train a neural network Φ on samples $(Tf_i, f_i)_{i=1}^m$, thereby pursuing a pure data-driven method while entirely discarding physical knowledge. Another elementary approach, which was suggested in Jin et al. (2017), first recovers an approximation of f from g by standard model-based approaches followed by a convolutional neural network, which acts as a denoiser. More sophisticated types of approaches aim to insert deep neural networks in iterative reconstruction schemes, for instance, by replacing certain steps such as a denoising step by a neural network, which was pioneered in Gregor and LeCun (2010), or replacing some of the proximal operators by networks (see, e.g., Meinhardt et al. (2017) and Adler and Öktem (2018)). For an overview of deep learning approaches to inverse problems, we refer to Adler and Öktem (2017) and McCann et al. (2017).

In contrast to the previously discussed approaches, we will now present two exemplary algorithms which combine the model-based realm represented by shearlets with the data-driven realm of deep neural networks following the philosophy of using model-based methods as far as they are reliable and data-driven methods where it is necessary. This conceptual type of approach not only avoids that deep neural networks affect the entire data set during inversion, which presumably causes instabilities (Gottschling et al. 2020), but also allows a better interpretation of the results.

Limited-Angle Computed Tomography

Computed tomography (CT) is one of the main imaging technologies for medical diagnosis. A CT scanner samples the *Radon transform*

$$\mathcal{R}f(\phi, s) = \int_{L(\phi,s)} f(x) dS(x),$$

where $L(\phi, s) = \{x \in \mathbb{R}^2 : x_1 \cos(\phi) + x_2 \sin(\phi) = s\}$, $\phi \in [-\pi/2, \pi/2)$, and $s \in \mathbb{R}$ (Natterer 2001). The inverse problem of reconstructing f from its Radon transform $g := \mathcal{R}f$ becomes even more challenging when only partial data is available. One instance of this problem complex is *limited-angle computed tomography*, where $\mathcal{R}f(\cdot, s)$ is only sampled on $[-\phi, \phi] \subset [-\pi/2, \pi/2)$. Examples include breast tomosynthesis, dental CT, and electron tomography. Due to the large missing part in the measured data – in contrast to, for instance, low-dose CT – model-based approaches only provide crude reconstructions, since no model-based priors exist which model a human body sufficiently accurately.

Depending on the missing angle, it is known which information about the wavefront set of the original image is contained in the measured data, hence in this sense what is "visible" (Quinto 1993). This allows to view the problem of limited-angle computed tomography as an inpainting problem of the wavefront set. Due to the sensitivity of shearlets to the wavefront set (Theorem 2), it is suggestive to exploit this system in this problem setting.

The approach "Learning the Invisible" (Bubba et al. 2019) pursues this strategy, by first reconstructing the image using sparse regularization with shearlets as sparsifying system, followed by surgically precisely learning the invisible data corresponding to the missing part of the wavefront set by a deep learning approach. The algorithm can be outlined as follows:

- *Step 1: Reconstruct the Visible.* Solve

$$f^* := \arg\min_{f \geq 0} \|\mathcal{R}f - g\|_2^2 + \|SH_{\phi,\psi,\tilde{\psi}} f\|_{1,w},$$

with $SH_{\phi,\psi,\tilde{\psi}}$ being a shearlet transform and $\|\cdot\|_{1,w}$ a suitably chosen weighted ℓ_1 norm. The wavefront set can then be approximately assessed via a sparsity prior on shearlets in the following sense, where $\mathcal{I}_{\texttt{inv}}$ corresponds to the "invisible" shearlet coefficients and $\mathcal{I}_{\texttt{vis}}$ to the "visible" coefficients:
  - For $(j,k,m,\iota) \in \mathcal{I}_{\texttt{inv}}$: $SH_{\phi,\psi,\tilde{\psi}} f^*(j,k,m,\iota) \approx 0$.
  - For $(j,k,m,\iota) \in \mathcal{I}_{\texttt{vis}}$: $SH_{\phi,\psi,\tilde{\psi}} f^*(j,k,m,\iota)$ is reliable and near perfect.
- *Step 2: Learn the Invisible.* Apply a neural network $\mathcal{NN}_\theta$ with a U-Net-like CNN architecture of 40 layers coined *PhantomNet* (Bubba et al. 2019), which is trained using training data $(f_i^*, f_i^{\texttt{gt}})_{i=1}^m$ ("gt" = "groundtruth") by minimizing

$$\min_\theta \frac{1}{m} \sum_{i=1}^m \|\mathcal{NN}_\theta(SH_{\phi,\psi,\tilde{\psi}} f_i^*) - SH_{\phi,\psi,\tilde{\psi}} f_i^{\texttt{gt}}|_{\mathcal{I}_{\texttt{inv}}}\|_{w,2}^2,$$

and compute

$$\mathcal{NN}_\theta : SH_{\phi,\psi,\tilde{\psi}} f^*|_{\mathcal{I}_{\texttt{vis}}} \longrightarrow F \left(\stackrel{!}{\approx} SH_{\phi,\psi,\tilde{\psi}} f^{\texttt{gt}}|_{\mathcal{I}_{\texttt{inv}}} \right).$$

- *Step 3: Combine.* Compute the reconstruction

$$f_{\texttt{LtI}} = SH_{\phi,\psi,\tilde{\psi}}^{-1} \left(SH_{\phi,\psi,\tilde{\psi}} f^*|_{\mathcal{I}_{\texttt{vis}}} + F \right).$$

Figure 8 shows numerical results, which prove superiority not only over the model-based approach but even over the pure deep learning approach from Gu and Ye (2017).

Wavefront Set Detection

Edge detection is a widely studied problem, which aims to detect singularity points in an image. As argued before, edges carry most of the information of an image; in addition, it is believed that rough sketching involving edge detection is actually the first of the operations of the human visual cortex. Various approaches to edge detection have been suggested with maybe the most famous one being the Canny edge detector (Canny 1986).

However, sometimes not only the detection of the edge but also its directionality in the sense of detecting the wavefront set is required. One example is – also related to the previous subsection – tomographic imaging. In fact, the wavefront set of an

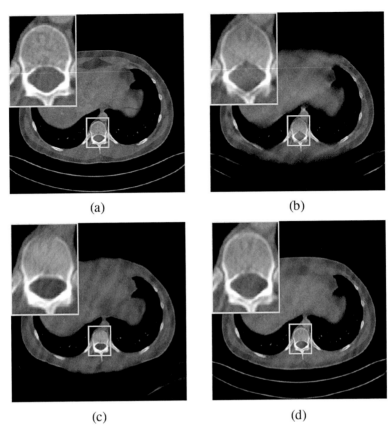

Fig. 8 Numerical experiments from Bubba et al. (2019) using data from Mayo-60° with a missing wedge of 60°, where RE stands for relative error and HaarPSI is the Haar wavelet-based perceptual similarity index for image quality assessment (Reisenhofer et al. 2018). (**a**) Original image. (**b**) f^* (RE: 0.19, HaarPSI: 0.43). (**c**) Result from Gu and Ye (2017) (RE: 0.22, HaarPSI: 0.40). (**d**) f_{LtI} (RE: 0.09, HaarPSI: 0.76)

image can be related to the wavefront set of its transformed version such as its Radon transform by (microlocal) canonical relations. Being able to detect the wavefront set of the Radon transform, say, allows to compute an approximation of the wavefront set of the original image by a (microlocal) canonical relation and use it as a prior for reconstruction (Andrade-Loarca et al. 2020).

Cone-adapted continuous shearlet systems are able to resolve wavefront sets (Theorem 2). But algorithms following this model such as Yi et al. (2009) and Reisenhofer et al. (2015) often suffer from the fact that real-world scenarios are highly complex and the theoretical analysis only provides an asymptotic estimate.

In the sequel, we will discuss an approach coined DeNSE (Deep Network Shearlet Edge Extractor) (Andrade-Loarca et al. 2019), which again follows the philosophy to use a model-based approach as far as it is reliable and use a deep

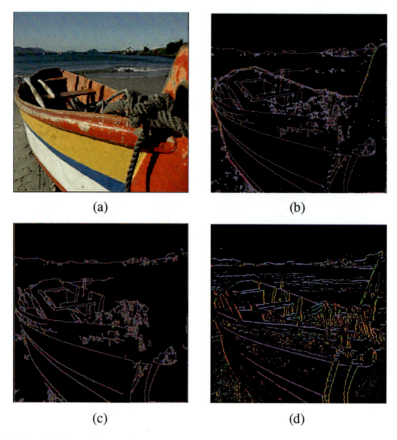

Fig. 9 Numerical experiments from Andrade-Loarca et al. (2019), where the color coding indicates the detected direction. (**a**) Original image. (**b**) Result from Yi et al. (2009). (**c**) Result from Reisenhofer et al. (2015). (**d**) Result using DeNSE. Copyright ©2019 Society for Industrial and Applied Mathematics. Reprinted with permission. All rights reserved

neural network where it is necessary. More precisely, it first computes a shearlet transform whose ability to detect wavefront sets is subsequently improved by deep learning when operating in shearlet domain. The algorithm can be outlined as follows:

- *Step 1: Reveal Directionality in the Shearlet Domain.* For a given test image $f \in \mathbb{R}^{M \times M}$, compute the digital shearlet transform of f with 49 shearlet generators, i.e.,

$$\left(DST^{2D}_{\phi,\psi,\tilde{\psi}} f(j,k,m,\iota)\right)_{j,k,m \in [1,M]^2, \iota \in \{-1,0,1\}}.$$

- *Step 2: Shearlet Transform.* For every location $m^* = (m_1^*, m_2^*) \in [11, M-10]^2$, apply a neural network classifier consisting of four convolutional layers plus one fully connected layer to the associated patch

$$\left(DST^{2D}_{\phi,\psi,\tilde{\psi}} f(j,k,m,\iota)\right)_{j,k,m \in [m_1^*-10, m_1^*+10] \times [m_2^*-10, m_2^*+10], \iota \in \{-1,0,1\}}.$$

If the network predicts the presence of an edge with direction ϑ, then (m^*, ϑ) is detected as an element of the wavefront set of f.

For an example of the effectiveness of this hybrid approach, we refer to Fig. 9.

Conclusion

The area of applied harmonic analysis provides representation systems for data processing, aiming for both decomposition and expansion of data/functions. Shearlet systems are specifically designed for the setting of multivariate functions and exist as continuous, discrete, and digital systems. While the continuous version allows a precise resolution of wavelet fronts, the discrete version provides optimally sparse approximations of cartoon-like functions as a model class of functions being governed by anisotropic features, and the digital version yields faithful implementations. Shearlet systems can be extended to higher dimensions as well as also to more general universal shearlets and a-molecules. Shearlet systems are typically used for sparse regularization of inverse problems such as feature extraction and inpainting, for which both theoretical and numerical results are available. Recent applications combine the shearlet transform with deep neural networks in a smart way targeting problems such as limited-angle computed tomography and wavefront set detection.

Acknowledgments G.K. would like to thank Hector Andrade-Loarca for producing several of the figures.

References

Adler, J., Öktem, O.: Solving ill-posed inverse problems using iterative deep neural networks. Inverse Probl. **33**, 124007 (2017)

Adler, J., Öktem, O.: Learned primal-dual reconstruction. IEEE T. Med. Imaging **37**, 1322–1332 (2018)

Andrade-Loarca, H., Kutyniok, G., Öktem, O.: Shearlets as feature extractor for semantic edge detection: the model-based and data-driven realm. Proc. R. Soc. A. **476**(2243), 20190841 (2020). https://royalsocietypublishing.org/toc/rspa/2020/476/2243

Andrade-Loarca, H., Kutyniok, G., Öktem, O., Petersen, P.: Extraction of digital wavefront sets using applied harmonic analysis and deep neural networks. SIAM J. Imaging Sci. **12**, 1936–1966 (2019)

Antoine, J.P., Carrette, P., Murenzi, R., Piette, B.: Image analysis with two-dimensional continuous wavelet transform. Sig. Process. **31**, 241–272 (1993)

Bamberger, R.H., Smith, M.J.T.: A filter bank for the directional decomposition of images: theory and design. IEEE Trans. Sig. Process. **40**, 882–893 (1992)

Bodmann, B.G., Labate, D., Pahari, B.R.: Smooth projections and the construction of smooth Parseval frames of shearlets. Adv. Comput. Math. **45**, 3241–3264 (2019)

Bubba, T.A., Kutyniok, G., Lassas, M., März, M., Samek, W., Siltanen, S., Srinivasan, V.: Learning the invisible: a hybrid deep learning-shearlet framework for limited angle computed tomography. Inverse Probl. **35**, 064002 (2019). https://iopscience.iop.org/article/10.1088/1361-6420/ab10ca

Candès, E.J., Donoho, D.L.: New tight frames of curvelets and optimal representations of objects with piecewise C^2 singularities. Commun. Pure Appl. Math. **56**, 216–266 (2004)

Canny, J.: A computational approach to edge detection. IEEE Trans. Pattern Anal. Mach. Intell. **8**, 679–698 (1986)

Casazza, P.G., Kutyniok, G., Philipp, F.: Introduction to finite frame theory. In: Finite Frames: Theory and Applications, pp. 1–53. Birkhäuser, Boston (2012)

Christensen, O.: An Introduction to Frames and Riesz Bases. Birkhäuser, Boston (2003)

Cohen, A.: Numerical Analysis of Wavelet Methods. Studies in Mathematics and Its Applications, vol. 32. JAI Press, Greenwich (2003)

Dahlke, S., Kutyniok, G., Maass, P., Sagiv, C., Stark, H.-G., Teschke, G.: The uncertainty principle associated with the continuous shearlet transform. Int. J. Wavelets Multiresolut. Inf. Process. **6**, 157–181 (2008)

Dahlke, S., Kutyniok, G., Steidl, G., Teschke, G.: Shearlet coorbit spaces and associated banach frames. Appl. Comput. Harmon. Anal. **27**, 195–214 (2009)

Dahlke, S., Steidl, G., Teschke, G.: The continuous shearlet transform in arbitrary space dimensions. J. Fourier Anal. Appl. **16**, 340–354 (2010)

Dahlke, S., Steidl, G., Teschke, G.: Shearlet coorbit spaces: compactly supported analyzing shearlets, traces and embeddings. J. Fourier Anal. Appl. **17**, 1232–1255 (2011)

Dahlke, S., Häuser, S., Steidl, G., Teschke, G.: Shearlet coorbit spaces: traces and embeddings in higher dimensions. Monatsh. Math. **169**, 15–32 (2013)

Daubechies, I.: Ten Lectures on Wavelets. SIAM, Philadelphia (1992)

Daubechies, I., Defrise, M., De Mo, C.: An iterative thresholding algorithm for linear inverse problems with a sparsity constraint. Commun. Pure Appl. Math. **57**, 1413–1457 (2004)

Davenport, M., Duarte, M., Eldar, Y., Kutyniok, G.: Introduction to compressed sensing. In: Compressed Sensing: Theory and Applications, pp. 1–64. Cambridge University Press (2012)

Do, M.N., Vetterli, M.: The contourlet transform: an efficient directional multiresolution image representation. IEEE Trans. Image Process. **14**, 2091–2106 (2005)

Donoho, D.L.: Sparse components of images and optimal atomic decomposition. Constr. Approx. **17**, 353–382 (2001)

Donoho, D.L., Kutyniok, G.: Geometric separation using a wavelet-shearlet dictionary. SampTA'09, Marseille. Proceedings (2009)

Donoho, D.L., Kutyniok, G.: Microlocal analysis of the geometric separation problem. Commun. Pure Appl. Math. **66**, 1–47 (2013)

Easley, G., Labate, D.: Image processing using shearlets. In: Shearlets: Multiscale Analysis for Multivariate Data, pp. 283–325. Birkhäuser, Boston (2012)

Easley, G., Labate, D., Lim, W.-Q.: Sparse directional image representation using the discrete shearlet transform. Appl. Comput. Harmon. Anal. **25**, 25–46 (2008)

Genzel, M., Kutyniok, G.: Asymptotic analysis of inpainting via universal shearlet systems. SIAM J. Imaging Sci. **7**, 2301–2339 (2014)

Goodfellow, I., Bengio, Y., Courville, A.: Deep Learning. Adaptive Computation and Machine Learning. MIT Press, Cambridge (2017)

Gottschling, N., Antun, V., Adcock, B., Hansen, A.C.: The troublesome kernel: why deep learning for inverse problems is typically unstable. preprint, arXiv:2001.01258 (2020)

Gregor, K., LeCun, Y.: Learning fast approximations of sparse coding. In: International Conference on Machine Learning (ICML), pp. 399–406 (2010)

Grohs, P.: Continuous Shearlet frames and Resolution of the Wavefront Set. Monatsh. Math. **164**, 393–426 (2011a)

Grohs, P.: Continuous shearlet tight frames. J. Fourier Anal. Appl. **17**, 506–518 (2011b)

Grohs, P.: Bandlimited shearlet frames with nice duals. J. Comput. Appl. Math. **142**, 139–151 (2013)

Grohs, P., Kutyniok, G.: Parabolic molecules. Found. Comput. Math. **14**, 299–337 (2014)

Grohs, P., Keiper, S., Kutyniok, G., Schäfer, M.: α-molecules. Appl. Comput. Harmon. Anal. **42**, 297–336 (2016a)

Grohs, P., Keiper, S., Kutyniok, G., Schäfer, M.: Cartoon approximation with α-curvelets. J. Fourier Anal. Appl. **22**, 1235–1293 (2016b)

Gu, J., Ye, J.C.: Multi-scale wavelet domain residual learning for limited-angle CT reconstruction. In: Procs Fully3D, pp. 443–447 (2017)

Guo, K., Labate, D.: Optimally sparse multidimensional representation using shearlets. SIAM J Math. Anal. **39**, 298–318 (2007)

Guo, K., Labate, D.: The construction of smooth parseval frames of shearlets. Math. Model. Nat. Phenom. **8**, 82–105 (2013)

Guo, K., Kutyniok, G., Labate, D.: Sparse multidimensional representations using anisotropic dilation and shear operators. In: Wavelets and Splines, Athens, 2005, pp. 189–201. Nashboro Press, Nashville (2006)

Guo, K., Labate, D., Lim, W.-Q.: Edge analysis and identification using the continuous shearlet transform. Appl. Comput. Harmon. Anal. **27**, 24–46 (2009)

Häuser, S., Steidl, G.: Convex multiclass segmentation with shearlet regularization. Int. J. Comput. Math. **90**, 62–81 (2013)

Hörmander, L.: The analysis of linear partial differential operators. I. Distribution theory and Fourier analysis. Springer, Berlin (2003)

Jin, K.H., McCann, M.T., Froustey, E., Unser, M.: Deep convolutional neural network for inverse problems in imaging. IEEE Trans. Image Proc. **26**, 4509–4522 (2017)

King, E.J., Kutyniok, G., Zhuang, X.: Analysis of inpainting via clustered sparsity and microlocal analysis. J. Math. Imaging Vis. **48**, 205–234 (2014)

Kittipoom, P., Kutyniok, G., Lim, W.-Q.: Irregular shearlet frames: geometry and approximation properties. J. Fourier Anal. Appl. **17**, 604–639 (2011)

Kittipoom, P., Kutyniok, G., Lim, W.-Q.: Construction of compactly supported shearlet frames. Constr. Approx. **35**, 21–72 (2012)

Kutyniok, G.: Clustered sparsity and separation of cartoon and texture. SIAM J. Imaging Sci. **6**, 848–874 (2013)

Kutyniok, G.: Geometric separation by single-pass alternating thresholding. Appl. Comput. Harmon. Anal. **36**, 23–50 (2014)

Kutyniok, G., Labate, D.: Resolution of the wavefront set using continuous shearlets. Trans. Am. Math. Soc. **361**, 2719–2754 (2009)

Kutyniok, G., Labate, D.: Introduction to shearlets. In: Shearlets: Multiscale Analysis for Multivariate Data, pp. 1–3. Birkhäuser, Boston (2012)

Kutyniok, G., Lim, W.-Q.: Compactly supported shearlets are optimally sparse. J. Approx. Theory **163**, 1564–1589 (2011)

Kutyniok, G., Lim, W.-Q.: Image separation using wavelets and shearlets. In: Curves and Surfaces, Avignon, 2010). Lecture Notes in Computer Science, vol. 6920, pp. 416–430. Springer (2012)

Kutyniok, G., Lim, W.-Q.: Optimal compressive imaging of Fourier data. SIAM J. Imaging Sci. **11**, 507–546 (2018)

Kutyniok, G., Petersen, P.: Classification of edges using compactly supported shearlets. Appl. Comput. Harmon. Anal. **42**, 245–293 (2017)

Kutyniok, G., Lemvig, J., Lim, W.-Q.: Optimally sparse approximations of 3D functions by compactly supported shearlet frames. SIAM J. Math. Anal. **44**, 2962–3017 (2012)

Kutyniok, G., Lim, W.-Q., Reisenhofer, R.: ShearLab 3D: faithful digital shearlet transforms based on compactly supported shearlets. ACM Trans. Math. Softw. **42**, 5 (2016)

Labate, D., Lim, W.-Q., Kutyniok, G., Weiss, G.: Sparse multidimensional representation using shearlets. In: Wavelets XI, Proceedings of SPIE, Bellingham, vol. 5914, pp. 254–262 (2005)

Labate, D., Mantovani, L., Negi, P.S.: Shearlet smoothness spaces. J. Fourier Anal. Appl. **19**, 577–611 (2013)

Le Pennec, E.L., Mallat, S.: Sparse geometric image representations with bandelets. IEEE Trans. Image Process. **14**, 423–438 (2005)

Lessig, C., Petersen, P., Schäfer, M.: Bendlets: a second-order shearlet transform with bent elements. Appl. Comput. Harmon. Anal. **46**, 384–399 (2019)

Lim, W.-Q.: The discrete shearlet transform: a new directional transform and compactly supported shearlet frames. IEEE Trans. Image Proc. **19**, 1166–1180 (2010)

Lim, W.-Q.: Nonseparable shearlet transform. IEEE Trans. Image Proc. **22**, 2056–2065 (2013)

Mallat, S.: A Wavelet Tour of Signal Processing. Academic Press, San Diego (1998)

McCann, M.T., Jin, K.H., Unser, M.: Convolutional neural networks for inverse problems in imaging: a review. IEEE Signal Proc. Mag. **34**, 85–95 (2017)

Meinhardt, T., Möller, M., Hazirbas, C., Cremers, D.: Learning proximal operators: using denoising networks for regularizing inverse imaging problems. In: International Conference on Computer Vision (ICCV) (2017)

Natterer, F.: The Mathematics of Computerized Tomography. Society for Industrial and Applied Mathematics (SIAM), Philadelphia (2001)

Quinto, E.T.: Singularities of the X-ray transform and limited data tomography in $\mathbb{R}^2$ and $\mathbb{R}^3$. SIAM J. Math. Anal. **24**, 1215–1225 (1993)

Reisenhofer, R., Kiefer, J., King, E.J.: Shearlet-based detection of flame fronts. Exp. Fluids **57**, 11 (2015)

Reisenhofer, R., Bosse, S., Kutyniok, G., Wiegand, T.: A Haar wavelet-based perceptual similarity index for image quality assessment. Sig. Process. Image **61**, 33–43 (2018)

Ronneberger, O., Fischer, P., Brox, T.: U-Net: convolutional networks for biomedical image segmentation. In: Medical Image Computing and Computer-Assisted Intervention (MICCAI). LNCS, vol. 9351, pp. 234–241. Springer (2015)

Simoncelli, E.P., Freeman, W.T., Adelson, E.H., Heeger, D.J.: Shiftable multiscale transforms. IEEE Trans. Inform. Theory **38**, 587–607 (1992)

Starck, J.-L., Murtagh, F., Fadili, J.: Sparse Image and Signal Processing: Wavelets, Curvelets, Morphological Diversity. Cambridge University Press, Cambridge (2010)

Yi, S., Labate, D., Easley, G.R., Krim, H.: A shearlet approach to edge analysis and detection. IEEE Trans. Image Process. **18**, 929–941 (2009)

Learned Regularizers for Inverse Problems 31

Sebastian Lunz

Contents

Introduction ... 1134
Shallow Learned Regularizers ... 1136
    Bilevel Learning ... 1136
    Dictionary Learning ... 1137
Deep Regularizers ... 1138
    Regularization Properties of Learned Regularizers ... 1138
    Adversarial Regularization ... 1141
    Total Deep Variation ... 1145
Summary and Outlook ... 1149
    Conclusion ... 1149
    Outlook ... 1151
References ... 1152

Abstract

In the past years, there has been a surge of interest in methods to solve inverse problems that are based on neural networks and deep learning. A variety of approaches have been proposed, showing improvements in reconstruction quality over existing methods. Among those, a class of algorithms builds on the well-established variational framework, training a neural network as a regularization functional. Those approaches come with the advantage of a theoretical understanding and a stability theory that is built on existing results for variational regularization. We discuss various approaches for learning a

S. Lunz (✉)
Department of Applied Mathematics and Theoretical Physics, University of Cambridge, Cambridge, UK
e-mail: sl767@cam.ac.uk

© Springer Nature Switzerland AG 2023
K. Chen et al. (eds.), *Handbook of Mathematical Models and Algorithms in Computer Vision and Imaging*, https://doi.org/10.1007/978-3-030-98661-2_68

regularization functional, aiming at giving an overview at the multiple directions investigated by the research community.

Keywords

Inverse problems · Variational regularization · Deep learning

Introduction

We consider an inverse problem of the form

$$y = Ax + \epsilon, \qquad (1)$$

where $x \in \mathcal{X}$ is an image we wish to reconstruct from measurements $y \in \mathcal{Y}$, the operator $A : \mathcal{X} \to \mathcal{Y}$ is linear, and $\epsilon \in \mathcal{Y}$ is random noise. A well-established framework for recovering x is via solving a variational problem of the form

$$\arg\min_x \mathcal{D}(Ax, y) + \lambda \mathcal{R}(x), \qquad (2)$$

where $\mathcal{D} : \mathcal{Y} \times \mathcal{Y} \to \mathbb{R}$ is a distance functional, typically chosen to be the ℓ^2 distance if the noise ϵ is Gaussian. The regularization functional $\mathcal{R}$ is chosen such that minimization is well-posed despite the pseudo-inverse $A^\dagger$ possibly being unbounded. A classical choice of $\mathcal{R}$ is the Tikhonov regularization functional $\mathcal{R}(x) := \|x\|_2^2$. This allows deducing various stability and convergence results on the reconstruction (see, e.g., Engl et al. 1996).

Taking the viewpoint of Bayesian statistics, we can interpret a solution to (2) as a maximum a posteriori likelihood estimator via

$$\arg\max_x \log p(x|y) = \arg\min_x -\log p(x|y) - \log p(x). \qquad (3)$$

The expression $\log p(x|y)$ is captured by the data term $\mathcal{D}(Ax, y)$, whereas the regularization functional can be viewed as an approximation to the log prior. This viewpoint motivates investigating priors beyond their ability to stabilize reconstruction, explaining the success of wildly used handcrafted priors such as total variation (TV) that capture the distinct properties of the distribution of images, such as sharp edges, more closely than Tikhonov-type regularization.

While TV has enjoyed great success in the past decades, its representation of the behavior of images remains limited, assuming them to be piecewise constant. As this is not true for many images, TV-based regularization is known to introduce staircasing artefacts into reconstructions. To overcome these drawbacks, the research community has shifted their focus on learning priors from data directly, with the goal of obtaining a more realistic and detailed image representation. More precisely, one aims at utilizing a training set $\{(\tilde{x}^i, y^i)\}$ of ground truth images $\tilde{x}^i$

and associated measurements y^i to learn powerful characterization of images from data directly. We want to note at this point that the setting $\{(\widetilde{x}^i, y^i)\}$ corresponds to a supervised training setting and that some algorithms require less structure in the training data, as, for example, dictionary learning (section "Dictionary Learning") or the adversarial regularizers we discuss later (section "Adversarial Regularization"). Well-established approaches for learning priors from data include dictionary learning and bilevel learning, as outlined in section "Shallow Learned Regularizers". Recently, attention has shifted to methods based on deep neural networks (Kobler et al. 2017; Adler and Öktem 2017, 2018; Jin et al. 2017; Li et al. 2020; Lunz et al. 2018; Kobler et al. 2020). The majority of approaches is based on a direct parametrization of the reconstruction operator $\Psi_\Theta(\cdot, A) : \mathcal{Y} \to \mathcal{X}$ that is trained using a loss function $\ell : \mathcal{X} \times \mathcal{X} \to \mathbb{R}$ and empirical risk minimization

$$\min_\Theta \sum_i \ell(\Psi_\Theta(y^i, A), \widetilde{x}^i). \qquad (4)$$

For those methods, the trained network $\Psi_\Theta(\cdot, A)$ can be applied directly to new measurements at inference. On the other hand, approaches based on learning a regularization functional $\mathcal{R}_\Theta$ typically separate between the training procedure of $\mathcal{R}_\Theta$ and the reconstruction step, using a variational functional of the form (2) or a similar functional for reconstruction. While those methods in general perform slightly worse than methods based on a direct parametrization that are trained *end-to-end* (Adler and Öktem 2018), they often allow for stability and convergence guarantees and enable a statistical interpretation of the learned functional. In this survey, we will in particular discuss Network Tikhnonov (NETT) in section "Regularization Properties of Learned Regularizers", adversarial regularizers in section "Adversarial Regularization", and total deep variation in section "Total Deep Variation".

Some hybrid approaches invoke a variational problem (or an early stopped version of it) but aim at parametrizing the gradient of the regularization functional instead of the functional directly (Kobler et al. 2017; Romano et al. 2017). While these methods have shown very good reconstruction results, we will omit them in our summary, focusing instead on methods that parametrize a regularization functional directly. In particular, approaches like regularization by denoising (RED) cannot always guarantee that the learned gradient is in fact the gradient of some functional.

Finally, deep image priors (Ulyanov et al. 2018) use the network architecture itself, without prior training, as a regularization term. These methods however are crucially reliant on early stopping, and we will not discuss them in detail here, but instead, refer to Ulyanov et al. (2018) for details.

Outline In this summary, we first give a brief overview over classical approaches at learning regularization functionals that do not make use of deep neural networks in section "Shallow Learned Regularizers". We then discuss three approaches for using neural networks as regularization functionals in detail in section "Deep Regularizers": Network Tikhonov in section "Regularization Properties of Learned

Regularizers", adversarial regularizers in section "Adversarial Regularization", and total deep variation in section "Total Deep Variation". We finish this review by giving a short summary and outline of potential for future research in section "Summary and Outlook".

Shallow Learned Regularizers

In this section, we review some approaches for learning a regularization functional that do not make use of neural networks. We in particular discuss bilevel learning as a technique for parameter optimization in regularization functionals and dictionary learning as a prominent unsupervised approach.

Bilevel Learning

Given a training set of the form $\{(\tilde{x}^i, y^i)\}$ of some images $\tilde{x}^i$ and associated measurements y^i, the bilevel problem of finding the optimal parameters Θ is given by

$$\begin{cases} \hat{\Theta} \in \arg\min_{\Theta} \sum_i [\ell(x_\Theta^i, \tilde{x}^i)] \\ x_\Theta^i := \arg\min_{x^i} \mathcal{D}(Ax^i, y^i) + \mathcal{R}_\Theta(x^i). \end{cases} \quad (5)$$

The generic framework of (5) has been used in various contexts to learn a regularization functional $\mathcal{R}_\Theta$. A prominent example is learning TV-type regularizers that consist of one or multiple regularization functionals based on the ℓ^1 norm of the gradient or smoothed versions thereof (Kunisch and Pock 2013; Calatroni et al. 2012). More complex regularization functionals, such as the field of experts (FoE) model (Roth and Black 2005), have also been trained using bilevel learning (Chen et al. 2013). In this setting, a linear combination of filters is learned from data.

Deriving sharp optimality conditions for bilevel learning generally requires the lower-level problem in (5) to be sufficiently regular. Under sufficient smoothness assumptions on the inner problem, optimality conditions can be established, and the problem (5) can be solved utilizing suitable techniques from PDE-constrained optimization.

In general, solving (5) is hard, with the problem being non-convex in Θ even in simple scenarios such as the Operator $A = Id$ being the identity (Arridge et al. 2019), making it challenging to scale bilevel techniques to highly parametric regularization functionals such as those given by neural networks. However, the concept of empirical risk minimization, i.e., of using a term of the form

$$\hat{\Theta} \in \arg\min_{\Theta} \sum_i [\ell(x_\Theta^i, \tilde{x}^i)]$$

is wildly used to train neural networks, and we will see an approach that utilizes a term of this form to train a deep regularization functional in the chapter on total deep variation (Kobler et al. 2020).

Dictionary Learning

Dictionary learning is based on the concept that the model parameter has a sparse representation in a some dictionary D. Approaches for dictionary learning (Aharon et al. 2006; Dabov et al. 2007; Xu et al. 2012) can be classified by the strategy taken to learn the dictionary D, which can be defined a priori in an analytical form, can be learned before reconstruction from data, or can be generated at reconstruction time, where the latter is mostly used in patch-based approaches.

A common approach in this context is sparse dictionary learning, aiming at learning a dictionary S from a collection of samples $\tilde{x}_i \in \mathcal{X}$ by minimizing the functional

$$\arg\min_{D,\xi} \sum_i \ell_{\mathcal{X}}(\tilde{x}_i, D\xi_i) + \mu \|\xi\|_1, \tag{6}$$

where $D : \Xi \to \mathcal{X}$ is a matrix containing the atoms of the dictionary in its columns and ξ_i is the representation of $\tilde{x}_i$ in the dictionary D. The distance on image space $\ell_{\mathcal{X}} : \mathcal{X} \times \mathcal{X} \to \mathbb{R}$ can, for example, be chosen to be ℓ_2. The ℓ_1 penalty term $\|\xi\|_1$ is chosen as a convex relaxation of a sparsity constraint on the representation ξ_i that limits $\|\xi_i\|_0 < s$ in its non-relaxed form. This formulation allows to learn a dictionary D that can represent each image x_i sparsely. Once learned, it can be used as a sparsity penalty during reconstruction, for example, by solving the problem

$$\arg\min_x \|AD\xi - y\| + \lambda \|\xi\|_1 \tag{7}$$

at reconstruction time, leading to the reconstruction $x = D\xi$. A drawback of this approach is that the sparsity level s, parametrized by μ in the relaxed formulation, needs to be chosen beforehand. This can be challenging as a too low sparsity level will not allow the dictionary to capture details of the images $\tilde{x}_i$, while a too high sparsity level will lead to a that does not act as an efficient regularizer at reconstruction time.

Finally, we note that when learning a dictionary of sparse representations from a large body of training samples, only *unsupervised* training data samples are required. To be more precise, we require access to a collection of $\tilde{x}^i$ of true images only, without requiring any pairing to corresponding measurements y^i. This makes training data more readily accessible in this context. We will later see how other learned distribution-based approaches, in particular the adversarial regularizer discussed in section "Adversarial Regularization", inherit this property.

Deep Regularizers

In this section, we move from shallow to deep regularization functionals, presenting three recent approaches for training a deep neural network as a regularization functional. These works are motivated by the established theory for variational regularization outlined above, some building on well-posedness and stability results, some on the statistical viewpoint of inverse problems. In terms of training strategy, some approaches build on the paradigm of empirical risk minimization previously seen in the context of bilevel learning, while some take an unsupervised approach to the problem, much like dictionary learning. We highlight that these approaches put an emphasis on statistical understanding, cross-modality flexibility, stability, and convergence results, separating them from the majority of deep learning approaches that directly parametrize a reconstruction operator. While those obtain state-of-the-art results, they offer little room for theoretical understanding.

Regularization Properties of Learned Regularizers

In the network Tikhonov (NETT) paper (Li et al. 2020), the authors propose one of the earliest approaches at learning a regularization functional from data using tools from deep learning. The authors put a strong emphasis on deducing stability results for the resulting algorithm that resemble the classical theory of variational regularization (Engl et al. 1996).

The authors study the inverse problem associated with (1) in the general setting of $(\mathcal{X}, \|\cdot\|)$ and $(\mathcal{Y}, \|\cdot\|)$ being reflexive Banach spaces with domain D. We denote by δ the noise level such that the noise ϵ satisfies $\|\epsilon\| \leq \delta$. The authors restrict their study to regularization functional of the form

$$\mathcal{R}_\Theta(x) = \phi(\Psi_\Theta(x)), \tag{8}$$

where $\phi : \mathbb{X}_L \to [0, \infty]$ is a scalar functional and $\Psi_\Theta : \mathcal{X} \to \mathbb{X}_L$ is a neural network of depth L, with parameters Θ. An example of a regularization functional of this form is given by a neural network $\Psi_\Theta : \mathcal{X} \to \mathcal{X}$ that maps an input image to some other element in image space, which is then mapped to a scalar via the ℓ_2 norm, $\Phi = \|\cdot\|_2$. The network Ψ_Θ is as usual given by a concatenation of affine functions and pointwise nonlinear activation functions that we denote by σ. Given this regularization function, an image x can be reconstructed form measurements y by minimizing the variational functional

$$\mathcal{T}_{\lambda, y_\delta}(x) := \mathcal{D}(A(x), y_\delta) + \lambda \mathcal{R}_\Theta(x) \to \min_{x \in D} \tag{9}$$

where $\mathcal{D} : Y \times Y \to [0, \infty]$ is the data consistency term.

A key contribution of the authors is the result that, under certain assumptions, reconstructions via (9) provide a stable solution scheme for (1). In addition to the well-posedness and weak convergence, the authors provide a complete analysis of norm-convergence and various convergence rates results, introducing the absolute Bregman distance as a new generalization of the standard Bregman distance from the convex to the non-convex setting. In the following, we report their key results.

To start, we discuss convergence of NETT regularization. To this end, the authors make the following assumptions.

Assumption 1.

- Network regularizer $\mathcal{R}$:
    - the regularizer is defined by (8);
    - The linear part of the affine layers in Ψ_Θ is bounded;
    - The activation functions σ are weakly continuous;
    - The functional ϕ is weakly lower semi-continuous.
- Data consistency term $\mathcal{D}$:
    - For some $\tau > 1$ we have $\forall y_0, y_1, y_2 \in Y : \mathcal{D}(y_0, y_1) \leq \tau \mathcal{D}(y_0, y_2) + \tau \mathcal{D}(y_2, y_1)$;
    - $\forall y_0, y_1 \in Y : \mathcal{D}(y_0, y_1) = 0 \iff y_0 = y_1$;
    - $\forall (y_k)_{k \in \mathbb{N}} \in Y^{\mathbb{N}} : y_k \to y \implies \mathcal{D}(y_k, y) \to 0$;
    - The functional $(x, y) \mapsto \mathcal{D}(A(x), y)$ is sequentially lower semi-continuous.
- Coercivity condition:
    - $\mathcal{R}_\Theta(\cdot)$ is coercive, that is $\mathcal{R}_\Theta(x) \to \infty$ as $\|x\| \to \infty$.

The conditions on the network regularizer guarantee the lower semicontinuity of the regularizer. Both those conditions and the assumptions on the data consistency term are not very restrictive and are satisfied by most natural choices of consistency term (such as the ℓ^2 distance) and network architectures. We hence point the reader's attention to the coercivity condition, which is not straightforward and will be violated by standard network architectures without introducing further restrictions. In particular, in the following chapter on adversarial regularizers, we will see that the authors allow a class of networks that can violate this coercivity assumption and they hence rely on the data term for providing the coercivity that is crucial for theoretical stability guarantees. The authors of Li et al. (2020) describe several ways to obtain coercivity by tuning the architecture of the network. The proposed approaches include skip and residual connections as well as layer-wise coercivity constraints using, for example, leaky ReLU or max-pooling.

We now state a key result of the paper that can be deduced under the above assumptions, demonstrating that they are sufficiently powerful to obtain results similar to the classical stability theory for variational problems with convex regularization functionals.

Theorem 1 (Well-posedness of CNN-regularization (Thm 2.6 Li et al. 2020)). *Let Assumption 1 be satisfied. Then the following assertions hold true:*

- *Existence:* For all $y \in Y$ and $\lambda > 0$, there exists a minimizer of $\mathcal{T}_{\lambda,y}$;
- *Stability:* If $y_k \to y$ and $x_k \in \arg\min \mathcal{T}_{\lambda,y_k}$, then weak accumulation points of $(x_k)_{k\in\mathbb{N}}$ exist and are minimizers of $\mathcal{T}_{\lambda,y}$;
- *Convergence:* Let $x \in \mathcal{X}$, $y := A(x)$, $(y_k)_{k\in\mathbb{N}}$ satisfy $\mathcal{D}(y_k, y), \mathcal{D}(y, y_k) \le \delta_k$ for some sequence $(\delta_k)_{k\in\mathbb{N}} \in (0, \infty)^{\mathbb{N}}$ with $\delta_k \to 0$, suppose $x_k \in \arg\min_x \mathcal{T}_{\lambda(\delta_k),y_k}(x)$, and let the parameter choice $\lambda : (0, \infty) \to (0, \infty)$ satisfy

$$\lim_{\delta \to 0} \lambda(\delta) = \lim_{\delta \to 0} \frac{\delta}{\lambda(\delta)} = 0 \tag{10}$$

Then the following holds:
- Weak accumulation points of $(x_k)_{k\in\mathbb{N}}$ are $\mathcal{R}_\Theta(\cdot)$-minimizing solutions of $A(x) = y$;
- $(x_k)_{k\in\mathbb{N}}$ has at least one weak accumulation point x_+;
- Any weakly convergent subsequence $(x_{k(n)})_{n\in\mathbb{N}}$ satisfies $\mathcal{R}_\Theta(x_{k(n)}) \to \mathcal{R}_\Theta(x_+)$;
- If the $\mathcal{R}_\Theta(\cdot)$-minimizing solution of $A(x) = y$ is unique, then $x_k \rightharpoonup x_+$ (weak convergence).

This theorem establishes the classical results of existence and stability of solutions along with convergence of solutions to the true image as the noise level $\delta \to 0$ for reconstructions via the variational problem (9). The authors strengthen the convergence results by introducing the notion of total nonlinearity as an additional assumption. We refer to Theorem 2.11 in the paper for details as well as proof of the theorem.

Finally, the authors also establish convergence rates in the absolute Bregmann distance. We refer the reader to Section 3 in Li et al. (2020) for further details on the resulting theorems as well as on the conditions necessary to obtain convergence rates in the absolute Bregman distance.

To summarize, we note that the combination of theoretical results deduced in Li et al. (2020) forms the most extensive theoretical analysis of a learned regularization functional conducted so far, including stability and convergence results as well as convergence rates. However, in order for the theorems to apply, one requires various constraints on the network architecture that need to be imposed either by network design or during training. Enforcing those might potentially be harmful in terms of model performance, but comes with the benefit of guaranteed stability and convergence as shown above.

Training scheme and results While the main emphasis of the paper is on an extensive stability and convergence theory, the authors also propose an algorithm for training a regularization functional. In particular, they choose a parametrization of the form

$$\mathcal{R}_\Theta(x) = \sum_i \|\Psi_{\Theta,i}(x)\|_q^q,$$

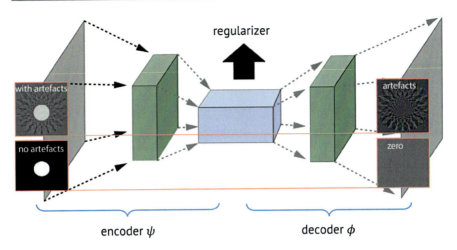

Fig. 1 Training setup in NETT. (Taken from Li et al. 2020)

where $\Psi_{\Theta,i}(x)$ denotes the i the component of $\Psi_{\Theta}(x)$. In order to train Ψ_{Θ}, the authors propose an encoder-decoder-based architecture that invokes a decoder network Φ in addition to the encoder Ψ_{Θ}. The joint architecture is trained to detect the characteristic artefacts in unregularized reconstructions, as shown in Fig. 1. The heuristic motivation behinds is that the resulting network is able to decompose the parts of a given reconstruction that are part of the underlying images and the ones that are reconstruction artefacts only. By penalizing the ℓ_q norm of the noise part only, typical noise patterns are suppressed during reconstruction without introducing artefacts in the underlying image. Note the similarity to adversarial training as discussed in the next section on adversarial regularizers (Lunz et al. 2018).

The authors employ subgradient descent for solving the minimization problem (9) and show results for photoacoustic tomography (PAT), as seen in Fig. 2.

Note that the authors and further researchers have published a variety of extension papers based on the NETT theory discussed here. These papers include discussions on improved training schemes and architectures as well as on further fields of applications (Obmann et al. 2020a,b). The NETT paper (Li et al. 2020) can be viewed as the theoretical foundation and first result in this direction.

Adversarial Regularization

The paper "adversarial regularizers" (Lunz et al. 2018) introduces a regime for learning regularization functionals, training the functional to reduce the *distributional* distance between reconstructions and true images. While there are similarities between the training regimes in this paper and in the previously discussed NETT (Li et al. 2020) approach, the authors of the adversarial regularizer paper focus their

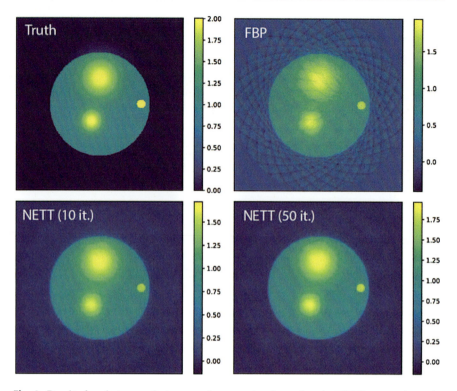

Fig. 2 Results for photoacoustic tomography reconstruction using the NETT approach on the Shepp-Logan phantom. (Taken from Li et al. 2020)

viewpoint on a statistical and distributional understanding of the learned regularization functional in contrast to the clear emphasis on convergence results in Li et al. (2020). The main contribution of the paper is a training technique the authors entitle adversarial training. Using this approach, the authors are able to train very complex regularization functionals. While this training regime is not necessarily limited to regularization functionals given by neural networks, it is particularly appealing when training complex functionals such as those parametrized by neural networks.

The authors rely on two distributions, given by their empirical counterparts of $\tilde{x}_i \in \mathcal{X}$ independent samples from the distribution of ground truth images $\mathbb{P}_r$ and by $y_i \in \mathcal{Y}$ independent samples from the distribution of measurements $\mathbb{P}_Y$.

The authors then consider the mapping of the distribution $\mathbb{P}_Y$ to a distribution on image space by applying via a pseudo-inverse $A_\delta^\dagger$, yielding the distribution $\mathbb{P}_n = (A_\delta^\dagger)_\# \mathbb{P}_Y$ of distorted reconstructions. Here, # denotes the push-forward of measures, i.e., $A_\delta^\dagger Y \sim (A_\delta^\dagger)_\# \mathbb{P}_Y$ for $Y \sim \mathbb{P}_Y$. Samples drawn from $\mathbb{P}_n$ are corrupted with noise that depends both on the noise model e and on the operator A.

The authors argue that a good regularization functional $\mathcal{R}_\Theta$ is able to tell apart the distributions $\mathbb{P}_r$ and $\mathbb{P}_n$. The authors use this as a motivation to choose the loss functional for training a neural network Ψ_Θ that directly parametrizes the regularization functional $\mathcal{R}_\Theta = \Psi_\Theta$ as

$$\mathbb{E}_{X \sim \mathbb{P}_r}\left[\Psi_\Theta(X)\right] - \mathbb{E}_{X \sim \mathbb{P}_n}\left[\Psi_\Theta(X)\right] + \mu \cdot \mathbb{E}\left[\left(\|\nabla_x \Psi_\Theta(X)\| - 1\right)_+^2\right], \tag{11}$$

where the last term in the loss functional serves to enforce the trained network Ψ_Θ to be Lipschitz continuous with constant one.

Written using the empirical distributions instead, the training loss (11) reads as

$$\sum_i \Psi_\Theta(\tilde{x}_i) - \sum_i \Psi_\Theta(A^\dagger y_i) + \mu \sum_i \left(\|\nabla_x \Psi_\Theta(\xi_i)\| - 1\right)_+^2,$$

where the points ξ are chosen randomly on the straigt line between $\tilde{x}_i$ and $A^\dagger y$.

The authors make this choice of penalty term for its connection to the Wasserstein distance between the distributions $\mathbb{P}_r$ and $\mathbb{P}_n$ that allows them to deduce the following theorem on the gradient flow over a perfectly trained regularization functional. Here, perfectly trained refers to the functional being 1-Lipschitz and perfectly minimizing the Wasserstein distance in the Kontorovich duality formulation

$$\text{Wass}(\mathbb{P}_r, \mathbb{P}_n) = \sup_{f \in 1-Lip} \mathbb{E}_{X \sim \mathbb{P}_n}\left[f(X)\right] - \mathbb{E}_{X \sim \mathbb{P}_r}\left[f(X)\right]. \tag{12}$$

Consider the distribution $\mathbb{P}_\eta := (g_\eta)_\# \mathbb{P}_n$ of samples obtained after a single gradient descent over Ψ_Θ of step of size η, starting from noisy reconstructions.

$$g_\eta(x) := x - \eta \cdot \nabla_x \Psi_\Theta(x). \tag{13}$$

The authors show the following theorem.

Theorem 2 (Wasserstein distance descent (Thm 1 Lunz et al. 2018)). *Assume that $\eta \mapsto \text{Wass}(\mathbb{P}_r, \mathbb{P}_\eta)$ admits a left and a right derivative at $\eta = 0$ and that they are equal. Then*

$$\frac{d}{d\eta} \text{Wass}(\mathbb{P}_r, \mathbb{P}_\eta)|_{\eta=0} = -\mathbb{E}_{X \sim \mathbb{P}_n}\left[\|\nabla_x \Psi_\Theta(X)\|^2\right].$$

The authors strengthen this result to

$$\frac{d}{d\eta}[\Psi_\Theta(g_\eta(X))]|_{\eta=0} = -1 \tag{14}$$

under weak assumptions.

The authors are hence able to show that the regularization functional trained via (11) can in fact optimally reduce the Wasserstein distance between reconstructions and ground truth images, at least at the initial step of the gradient descent scheme. The authors extend their analysis by deducing an explicit form of the regularization functional in the specific scenario of the true distribution being concentrated along a manifold $m \subset \mathcal{X}$.

Assumption 2. Denote by

$$P_m : D \to m, \quad x \to \arg\min_{y \in m} \|x - y\| \tag{15}$$

the data manifold projection, where D denotes the set of points for which such a projection exists. We assume $\mathbb{P}_n(D) = 1$. This can be guaranteed under weak assumptions on m and $\mathbb{P}_n$. We make the assumption that the measures $\mathbb{P}_r$ and $\mathbb{P}_n$ satisfy

$$(P_m)_\#(\mathbb{P}_n) = \mathbb{P}_r \tag{16}$$

i.e., for every measurable set $A \subset \mathcal{X}$, we have $\mathbb{P}_n(P_m^{-1}(A)) = \mathbb{P}_r(A)$

The authors motivate this as a low-noise assumption under which it is guaranteed that the distortions of the true data present in the distribution of pseudo-inverses $\mathbb{P}_n$ are sufficiently well behaved to recover the distribution of true images from noisy ones by projecting back onto the manifold. Under this assumption, the authors prove the following theorem.

Theorem 3 (Data Manifold Distance (Thm 2 Lunz et al. 2018)). *Under Assumption 2, a maximizer to the functional*

$$\sup_{f \in 1-Lip} \mathbb{E}_{X \sim \mathbb{P}_n} f(X) - \mathbb{E}_{X \sim \mathbb{P}_r} f(X) \tag{17}$$

is given by the distance function to the data manifold

$$d_m(x) := \min_{y \in m} \|x - y\| \tag{18}$$

The authors motivate the theorem as a consistency result, demonstrating that the approach yields reasonable regularization functionals in the particular setting of the theorem.

The paper also contains stability result with a similar flavor, the NETT paper in Theorem 1. The analysis is however less exhaustive and requires stronger assumptions on the operator A, making it less readily applicable to all inverse problems than Theorem 1. On a technical level, the key difference is that the NETT paper develops assumptions that ensure that the learned regularization functional is itself coercive, whereas the adversarial regularizer relies on the coercivity of the data term. The latter makes use of the additional 1-Lipschitz property of the regularization functional $\mathcal{R}_\Theta$ to ensure that a coercive regularization functional yields a coercive variational functional even if the regularization functional is not bounded from below. In the following, we state the results shown for adversarial regularizers in Lunz et al. (2018) and refer to the paper for the proof and further details.

Theorem 4 (Stability (Thm 3 Lunz et al. 2018)). *Let y_n be a sequence in Y with $y_n \to y$ in the norm topology and denote by x_n a sequence of minimizers of the functional*

$$\arg\min_{x \in X} \|Ax - y_n\|^2 + \lambda \mathcal{R}_\Theta(x)$$

Under appropriate assumptions on the operator A (see Appendix of Lunz et al. 2018), x_n has a weakly convergent subsequence, and the limit x is a minimizer of $\|Ax - y\|^2 + \lambda \mathcal{R}_\Theta(x)$.

Computational Results The authors show results for the discussed algorithm for denoising and computed tomography reconstruction. They show improved results compared to classical approaches such as total variation (Engl et al. 1996), but do not match results obtained with end-to-end trained algorithms such as post-processing approach for computed tomography (Jin et al. 2017) or a DnCNN (Zhang et al. 2017) for denoising. The results in Fig. 3 show results on denoising, whereas Fig. 4 contains results for computed tomography reconstruction.

Total Deep Variation

The recent paper Kobler et al. (2020) follows the paradigm of *end-to-end* learning in order to obtain a regularization functional, using a distance functional between reconstruction and ground truth as training objective. In general, unrolling methods

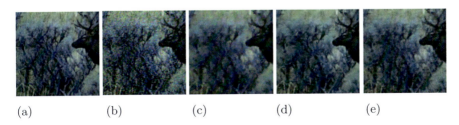

Fig. 3 Denoising results for the adversarial regularizer on the Berkeley Segmentation dataset (BSDS500). (Taken from Lunz et al. 2018). (**a**) Ground Truth. (**b**) Noisy Image. (**c**) TV. (**d**) Denoising N.N. (**e**) Adversarial Reg.

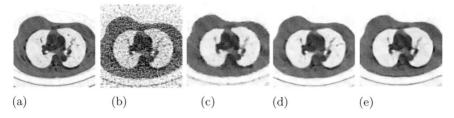

Fig. 4 Reconstruction from simulated CT measurements on the LIDC dataset using adversarial regularizers. (Taken from Lunz et al. 2018). (**a**) Ground Truth. (**b**) FBP. (**c**) TV. (**d**) Post-Processing. (**e**) Adversarial Reg.

such as Adler and Öktem (2017), Meinhardt et al. (2017), and Kobler et al. (2017) recover an image x_T from measurements y by applying

$$x_{n+1} = x_n - \lambda \Psi_\Theta(A^t(Ax_n - y), x_n), \qquad (19)$$

where the iteration is typically initialized with a pseudo-inverse x_0 and stopped after a fixed predefined number of steps N. The parameters Θ are trained by minimizing a loss functional

$$\sum_i \ell(x_N^i, x_T^i) \qquad (20)$$

over the parameters Θ for a collection of samples $\{x_T^i, y^i\}$ and a notion of distance ℓ that is typically chosen to be the ℓ^2 distance. Various approaches differ in their choice of parametrization of Ψ_Θ, ranging from architectures that do not further restrict the mapping properties of Ψ_Θ to those that explicitly separate out a gradient terms obtained from the data term and the image prior, leading to the form

$$\Psi_\Theta(A^t(Ax_n - y), x_n) = A^t(Ax_n - y) + \mu \Phi_\Theta(x_n). \qquad (21)$$

While these methods have shown to yield high-quality reconstructions, they cannot readily be understood using the viewpoint of variational regularization, as the regularization or image prior is implicitly contained in the mapping properties of the network Ψ_Θ. Even if parameterized as in (21), the network parametrizes the gradients of an implicit regularization functional rather than the functional directly.

An additional challenge in bridging the gap between unrolling based methods and a variational methods lies in the fixed choice of iterations N that is typically small and prohibits viewing x_N as the result of a minimization of a variational problem.

The authors of Kobler et al. (2020) bridge these problems by introducing two novel contributions: firstly, instead of parametrizing the gradient of the regularization functional, the functional itself is parametrized directly. While this makes training slightly more challenging, requiring double backpropagation for minimization, it yields a true regularization functional that can be interpreted as a prior on image distribution. Secondly, instead of fixing the number of gradient steps a priori, the authors introduce an optimal stopping time that allows for a flexible number of gradient descent iterations on the variational functional. While this still leaves a gap to classical regularization functionals that do not necessarily require a stopping criterion, the flexibility in the number of iterations makes the *total deep variation* approach the closest candidate for a generic method to yield a deep regularization functional that is trained by differentiating through the minimization of the corresponding variational functional.

Architecture For an image $x \in \mathbb{R}^{nC}$, where n denotes the number of pixels and C the number of channels, the authors parametrize a regularization functional $\mathcal{R}$ of the form

$$\mathcal{R}_\Theta(x) = \sum_{i=1}^{n} r(x,\Theta)_i, \quad r(x,\Theta) = \omega^t \mathcal{N}(Kx) \in \mathbb{R}^n. \tag{22}$$

Here, K denotes a zero-mean convolution kernel, ω is a learned weight vector contracting over channels but not over the spatial component, and $\mathcal{N}$ is a multi-scale neural network that is inspired by a UNet (Ronneberger et al. 2015) architecture. The authors employ a smooth log-student-t-distribution of the form $\Phi(x) = \frac{1}{2\mu} \log(1 + \mu x^2)$ as activation function, leading to a smooth regularization functional. This is advantageous for the double backpropagation used for minimization as discussed later (Fig. 5).

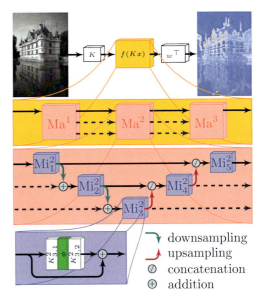

Fig. 5 Architecture of the regularization functional for TDV. (Taken from Kobler et al. 2020)

Training procedure We assume to be given a training set $\{(\tilde{x}^i, y^i, x_0^i)\}$ of ground truth image $\tilde{x}^i \in \mathcal{X}$, measurement $y^i \in \mathcal{Y}c$ and an initial guess, such as a pseudo-inverse of y_i, $x_0^i \in \mathcal{X}$. The authors cast the training process as an optimal control problem, introducing an optimal time horizon T. Using a fixed time discretization level $S \in \mathbb{N}$, their sampled objective function on the training set $\{(\tilde{x}^i, y^i, x_0^i)\}$ reads as

$$\inf_{T \in [0, T_{Max}]} \left\{ \frac{1}{N} \sum_{i=1}^{N} l(x_S^i - \tilde{x}^i) \right\}, \tag{23}$$

subject to the state equation

$$x_s^{i+1} = x_s^i - \frac{T}{S} A^t \left(A x_{s+1}^i - y^i \right) - \frac{T}{S} \nabla \mathcal{R}_\Theta(x_s^i), \tag{24}$$

where $l : \mathcal{X} \times \mathcal{X} \to \mathbb{R}$ denotes a loss functional in (23), which is typically chosen as either the ℓ^2 or ℓ^1 loss. An equivalent formulation of the state equation that is solved for x_s^{i+1} reads as

$$x_s^{i+1} = \left(\mathrm{Id} + \frac{T}{S} A^t A \right)^{-1} \left(x_s^i + \frac{T}{S} \left(A^t y^i - \nabla \mathcal{R}_\Theta(x_s^i) \right) \right) \tag{25}$$

The training objective is simultaneously minimized for the time horizon T and the parameters Θ that determine the form of the regularization functional. The stochastic ADAM optimizer is used for minimization. The zero-mean constraint on the regularization functional is enforced projection after every minimization step. Note that differentiating (23) with respect to (T, Θ) involves derivatives of the regularization functional $\mathcal{R}_\Theta(x)$ with respect to both Θ and x. These terms are handled in a numerically efficient way using the *double backpropagation* algorithm. The algorithm can be applied in this context as the architecture and activation functions used have been chosen to be C^2. The application of double backpropagation separates this work from earlier attempts at learning a regularization functional with a loss functional of the form (20).

The authors also derive various theorems to characterize the solutions of (23).

Theorem 5 (Existence of a solution (Thm 2.1 Kobler et al. 2020)). *The time continuous version of* (23) *alongside its corresponding state equation* (24) *admits a solution in the sense that the infimum is attained.*

The authors provide a characterization of the optimal solution in terms of the adjoint state in Theorem 3.1 in Kobler et al. (2020). They also include a sensitivity analysis of the results with respect to changes in the model parameters (T, Θ), bounding changes in the reconstruction by the differences in model parameters and some experiment specific quantities like the Lipschitz norm of the regularization

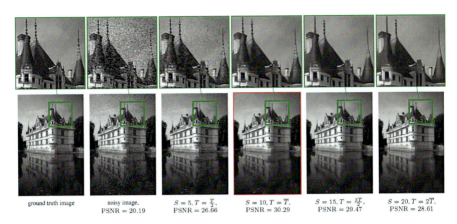

Fig. 6 TDV results for denoising with various choices for the time horizon. (Taken from Kobler et al. 2020)

functional $\mathcal{R}_\Theta$, the norm of the gradient of the regularization functional, and various others. Details can be found in Theorem 3.2 in the paper.

Results The authors show results for their TDV approach on a variety of inverse problems. For denoising, the approach is able to outperform approaches like BM3D (Dabov et al. 2007) as well as some end-to-end trained approaches like DnCNN (Zhang et al. 2017), but slightly underperforms compared to FOCNet (Jia et al. 2019). The latter has roughly one hundred times more parameter than the TDV approach. Results for denoising are shown in Fig. 6 for various choices of time discretization S and time horizon T. As expected, choosing the time horizon lower than the learned optimal parameter leads to under-regularization, while choosing it higher leads to over-regularization.

This chapter also discusses applications of the approach to medical imaging, demonstrating that a prior trained for computed tomography reconstruction of abdominal CT images can be readily applied for MRI reconstruction of knee images – a task that differs both in the imaging modality and in the characteristics of the images occurring. This shows that TDV generalizes well between different tasks and image characteristics. Results for MRI reconstruction can be seen in Fig. 7.

Summary and Outlook

Conclusion

We have discussed various approaches for training neural networks as regularization functionals that have been proposed in the past years. Network Tychonov (NETT) focuses on deriving stability and convergence results for regularization functionals

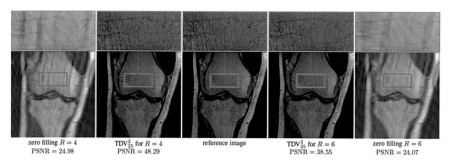

Fig. 7 TDV results for MRI reconstruction. (Taken from Kobler et al. 2020)

Table 1 Comparison of learned regularization approaches discussed in this survey

	Stability theory	Data structure	Performance
NETT	Fully developed, applies to wide variety of inverse problems	Paired, unpaired extenstions	No comparison to SOTA in original publication, follow-up work (Obmann et al. 2020b)
Adv. Reg.	Stability theory only applicable to some inverse problems	Unpaired training data suffices	Strong improvements on classical approaches (TV), slightly under SOTA for supervised reconstruction
TDV	No equivalent results for classical variational stability derived in paper	Fundamentally reliant on paired training data	Essentially SOTA performance, generalizability between tasks demonstrated

based on neural networks, allowing to deduce guarantees on the behavior of the resulting algorithm. The main contribution of adversarial regularizers and total deep variation lies in the proposal of novel schemes for training regularization functionals. The first introduces an approach that is based on training the network to tell apart ground truth images from noisy reconstructions, yielding an algorithm that can be trained in an unsupervised manner. The latter investigates the idea of supervised training of regularization functionals, made possible by the use of double backpropagation and the introduction of an optimal stopping time. We now turn to comparing the algorithms presented in terms of their results on stability, the structure of training data needed, and the performance demonstrated. This discussion is summarized in Table 1.

Stability Results The NETT paper contains an extensive stability analysis that is applicable to a wide variety of inverse problems. On a technical level, the theory does not make any assumptions on the data term being coercive and ensures coercivity by discussing sufficient conditions for the learned regularization functional to be coercive. This in particular allows the application of the theory to ill-posed inverse problems. The adversarial regularizer paper on the other hand makes strong assumptions on the properties of the forward operator, which can be violated in the context of ill-posed inverse problems. Most of the theoretical analysis in the paper focuses on discussing the effects of the learned regularization functional on the distribution of reconstructions instead of focusing on an instance-level stability theory. For the total deep variation approach, the authors include a discussion in terms of optimal control theory as well as stability with respect to changes in the training dataset, but do not derive stability results that are equivalent to the classical stability theory for inverse problems.

Training Data Both the NETT and the TDV approach rely on paired training data consisting of measurements and their corresponding ground truth images. While the first one can be extended to an unpaired setting when changing the training scheme (Obmann et al. 2020b), TDV is fundamentally dependent on paired data. Looking at marginals of distributions only, the adversarial regularizer approach can naturally handle unpaired training data.

Performance The authors of the NETT paper compare to backprojection only; an assessment on how the method compares to the state of the art is hence difficult. More extensive comparisons are included in the authors' more recent follow-up publications (Obmann et al. 2020a,b). The adversarial regularizer has been demonstrated to clearly outperform classical regularization techniques like total variation regularization, but does not quite reach the performance of state-of-the-art reconstruction methods that are trained with supervised data. The authors of TDV report results that are essentially state of the art and also demonstrate generalizability of the learned regularization functional between different imaging tasks, a property not yet investigated in the other papers discussed.

Outlook

In future work, combining the viewpoints of NETT and adversarial regularizers or NETT and total deep variation could be an interesting direction to explore. This could yield an algorithm that is provably stable while still being built on the training heuristics proposed in adversarial regularizers and total deep variation, respectively. As an example, we are recently working on introducing convexity constraints on the adversarial regularizer, resulting in an algorithm with better stability and convergence guarantees.

Finally, building a regularization functionals that approximate the prior on the image distribution more directly and more closely than the approaches discussed in this survey is another possible line of research. Notable algorithms based on generative models and in particular flow-based probabilistic models are being discussed within the research community for their potential to learn the image prior distribution without the need for any information on the operator or the specific noise distribution used.

References

Adler, J., Öktem, O.: Solving ill-posed inverse problems using iterative deep neural networks. Inverse Probl. **33**(12), 124007 (2017)

Adler, J., Öktem, O.: Learned primal-dual reconstruction. IEEE Trans. Med. Imaging **37**(6), 1322–1332 (2018)

Aharon, M., Elad, M., Bruckstein, A.: K-SVD: an algorithm for designing overcomplete dictionaries for sparse representation. IEEE Trans. Signal Process. **54**(11), 4311–4322 (2006)

Arridge, S., Maass, P., Öktem, O., Schönlieb, C.-B.: Solving inverse problems using data-driven models. Acta Numer. **28**, 1–174 (2019)

Calatroni, L., Cao, C., De Los Reyes, J.C., Schönlieb, C.-B., Valkonen, T.: Bilevel approaches for learning of variational imaging models. RADON Book Series **18** (2012)

Chen, Y., Pock, T., Ranftl, R., Bischof, H.: Revisiting loss-specific training of filter-based MRFs for image restoration. In: German Conference on Pattern Recognition, pp. 271–281. Springer (2013)

Dabov, K., Foi, A., Katkovnik, V., Egiazarian, K.: Image denoising by sparse 3-D transform-domain collaborative filtering. IEEE Trans. Image Process. **16**(8), 2080–2095 (2007)

Engl, H.W., Hanke, M., Neubauer, A.: Regularization of Inverse Problems, vol. 375. Springer Science & Business Media, Dordrecht (1996)

Jia, X., Liu, S., Feng, X., Zhang, L.: Focnet: a fractional optimal control network for image denoising. In: Proceedings of the IEEE Conference on Computer Vision and Pattern Recognition, pp. 6054–6063 (2019)

Jin, K.H., McCann, M., Froustey, E., Unser, M.: Deep convolutional neural network for inverse problems in imaging. IEEE Trans. Image Process. **26**(9), 4509–4522 (2017)

Kobler, E., Klatzer, T., Hammernik, K., Pock, T.: Variational networks: connecting variational methods and deep learning. In: German Conference on Pattern Recognition, pp. 281–293. Springer (2017)

Kobler, E., Effland, A., Kunisch, K., Pock, T.: Total deep variation for linear inverse problems (2020)

Kunisch, K., Pock, T.: A bilevel optimization approach for parameter learning in variational models. SIAM J. Imaging Sci. **6**(2), 938–983 (2013)

Li, H., Schwab, J., Antholzer, S., Haltmeier, M.: NETT: solving inverse problems with deep neural networks. Inverse Probl. **36**, 065005 (2020)

Lunz, S., Öktem, O., Schönlieb, C.-B.: Adversarial regularizers in inverse problems. In: Bengio, S., Wallach, H., Larochelle, H., Grauman, K., Cesa-Bianchi, N., Garnett, R. (eds.) Advances in Neural Information Processing Systems 31, pp. 8507–8516. Curran Associates, Inc., Red Hook (2018)

Meinhardt, T., Moller, M., Hazirbas, C., Cremers, D.: Learning proximal operators: using denoising networks for regularizing inverse imaging problems. In: Proceedings of the IEEE International Conference on Computer Vision, pp. 1781–1790 (2017)

Obmann, D., Schwab, J., Haltmeier, M.: Deep synthesis regularization of inverse problems. arXiv preprint arXiv:2002.00155 (2020a)

Obmann, D., Nguyen, L., Schwab, J., Haltmeier, M.: Sparse aNETT for solving inverse problems with deep learning. arXiv preprint arXiv:2004.09565 (2020b)

Romano, Y., Elad, M., Milanfar, P.: The little engine that could: regularization by denoising (red). SIAM J. Imaging Sci. **10**(4), 1804–1844 (2017)

Ronneberger, O., Fischer, P., Brox, T.: U-net: convolutional networks for biomedical image segmentation. In: International Conference on Medical Image Computing and Computer-Assisted Intervention, pp. 234–241. Springer (2015)

Roth, S., Black, M.J.: Fields of experts: a framework for learning image priors. In: 2005 IEEE Computer Society Conference on Computer Vision and Pattern Recognition (CVPR'05), vol. 2, pp. 860–867. IEEE (2005)

Ulyanov, D., Vedaldi, A., Lempitsky, V.: Deep image prior. In: Proceedings of the IEEE Conference on Computer Vision and Pattern Recognition, pp. 9446–9454 (2018)

Xu, Q., Yu, H., Mou, X., Zhang, L., Hsieh, J., Wang, G.: Low-dose x-ray CT reconstruction via dictionary learning. IEEE Trans. Med. Imaging **31**(9), 1682–1697 (2012)

Zhang, K., Zuo, W., Chen, Y., Meng, D., Zhang, L.: Beyond a gaussian denoiser: residual learning of deep cnn for image denoising. IEEE Trans. Image Process. **26**(7), 3142–3155 (2017)

Filter Design for Image Decomposition and Applications to Forensics

32

Robin Richter, Duy H. Thai, Carsten Gottschlich, and Stephan F. Huckemann

Contents

Introduction	1156
Applications and Challenges for Automated Image Decomposition	1157
Diffusion Methods	1160
Fourier and Wavelet Methods	1160
Variational Problems	1162
Non-linear Spectral Decompositions	1164
Texture Information	1165
Machine Learning	1166
Adaptive Balancing	1167
Adapting the Data-Fidelity-Norm	1168
Connection to the G-Norm	1168
Other Choices of M	1169
Connections with Machine Learning	1169
Solving via the ADMM/AL-Algorithm	1170
Interpretation via a Feasibility Problem	1171
A General Learning Problem	1175
Filter Design Using Factor Families	1177
Conclusion	1178
References	1179

R. Richter (✉) · S. F. Huckemann (✉)
Felix-Bernstein-Institute for Mathematical Statistics in the Biosciences, University of Göttingen, Göttingen, Germany
e-mail: robin.richter@mathematik.uni-goettingen.de; huckeman@math.uni-goettingen.de

D. H. Thai (✉)
Department of Mathematics, Colorado State University, Fort Collins, CO, USA
e-mail: duy.hoang-thai@mathematik.uni-goettingen.de

C. Gottschlich (✉)
Institute for Mathematical Stochastics, University of Göttingen, Göttingen, Germany
e-mail: gottschl@math.uni-goettingen.de

© Springer Nature Switzerland AG 2023
K. Chen et al. (eds.), *Handbook of Mathematical Models and Algorithms in Computer Vision and Imaging*, https://doi.org/10.1007/978-3-030-98661-2_92

Abstract

Employing image filters in image processing applications, essentially matrix convolution operators, has been an active field of research since a long time, and it is so very much still today. In the first part, we give a brief overview of imaging methods with emphasis on applications in fingerprint recognition and shoeprint forensics. In the second part, we propose a generalized discrete scheme for image decomposition that encompasses many of the existing methods. Due to its generality, it has the potential to learn, for specific use cases, a highly flexible set of imaging filters that are related to one another by rather general conditions.

Keywords

Image decomposition · Variational methods · Texture · Forensics · Fingerprint recognition

Introduction

Image decomposition is one of the first and crucial steps in image analysis, be it decomposition into *signal* and *noise*, *foreground* and *background*, or more refined, such as decompositions into *cartoon*, *texture*, and *noise*. Often, under theoretical technical assumptions, precise objective functions are employed to this end; however, for specific applications, they are not a priori available. The latter is the case, for instance, when at crime scenes, latent fingerprints or shoeprints are to be compared to print scans taken from suspects at hand who are released immediately after. For expert comparison taking place afterwards, the quality of the scanned prints is decisive. This quality, however, can only be defined indirectly, for example, that improved quality is proportional to improved (lowered) error rates. In this application scenario, other image processing steps surface as well, namely, *image enhancement*, for example, of latent prints, and *image compression* to significant features, for example, in large databases.

In this chapter we give, guided by examples from forensics, a brief overview of image decomposition methods from the past to the present with emphasis on a unified viewpoint for some current challenges.

In acoustic signal processing, digital filter design has first been inspired by analog electric filtering circuits, and this has also inspired filter design for images. Images, however, have fundamentally different features than acoustic signals. While for the former Fourier decomposition was highly effective, image analysis required different types of analysis, for example, *Haar wavelet frames* for sharp edge modeling (Daubechies 1992; Mallat 2008). Other popular approaches are given by diffusion equations (Perona and Malik 1990; Weickert 1998) or minimization problems (Mumford and Shah 1989; Scherzer et al. 2009). This has led to the development of entirely new mathematical frameworks, often connected with one another (Steidl et al. 2004; Burger et al. 2016).

In this context, Chambolle and Pock (2016) give an overview of a multitude of optimization algorithms for a multitude of proposed minimization problems. Such reconstruction methods often *balance* between *data fidelity* and, possibly, several *reconstruction regularity* objectives. In this context, the assertion of unique optima and the development of convergent algorithms has spurred an abundance of publications, in particular when, as is often the case in realistic applications, linearity is relaxed and modeled by additional constraints. This has led to the development/application of iterative algorithms for *saddle points* of associated Lagrange functionals which are *augmented* to obtain strict convexity, which results in additional robustness (e.g., Bertsekas 1982; Eckstein and Bertsekas 1992; Wu and Tai 2010).

The advent of *machine learning* allowed to train modified regularization filters in view of specific application tasks, given larger databases for training and testing. When only moderately sized databases are available, as is the case, for example, in academic forensic sciences, learning methods improve by drawing on prior structure information at hand. For instance, in view of fingerprint analysis, it is a priori known that the object of interest comprises a fringe pattern, sometimes forking, of nearly constant frequency that follows a smooth orientation field, featuring only three types of singularities (Maltoni et al. 2009).

Considering minimization problems with a global balancing parameter, as a learning model, however, comes at a price such as the well-known *loss of contrast* dilemma: Removing highly oscillating structures while preserving steps of small intensity differences between otherwise flat structures cannot be simultaneously achieved (e.g., Figure 2 of Strong and Chan 2003). As a workaround, adaptive balancing filters have been introduced, localizing in the spatial or frequency domain (Osher et al. 2003; Buades et al. 2010; Bredies et al. 2013). In conclusion of this chapter, in generalization, we introduce a flexible, discrete learning model featuring a general *alternating direction method of multipliers* (ADMM) inspired algorithm based on a *feasibility problem*. This framework draws flexibility from decoupling involved families of filters from one another only requiring rather general regularity conditions.

It includes several of the abovementioned methods as special cases. For specific application scenarios at hand, suitable filter families can be learned. In application to forensics, we illustrate how to employ the new model for shoeprint decompositions. For shoeprint analysis, as detailed, challenges are much higher than for fingerprint analysis (which are still high), and scientifically based shoeprint image analysis is still in its very beginnings.

Applications and Challenges for Automated Image Decomposition

Very often, images contain an object of interest (or several) within a *region of interest* (ROI), for example, the area covered by a latent fingerprint or shoeprint in *forensics* (cf. Fig. 1), a tumor within an organ in *medical imaging*, faces observed

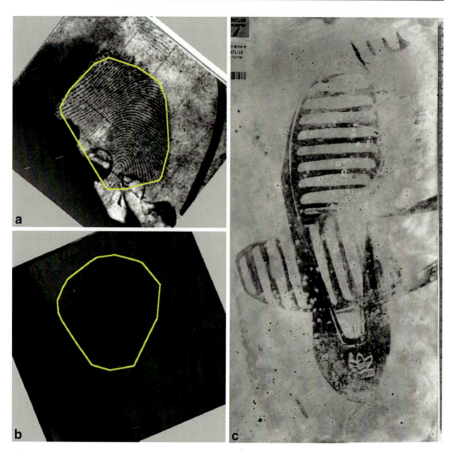

Fig. 1 Latent fingerprint images from the NIST special database 27 (left) (cf. Garris and McCabe (2000) with boundary (drawn in yellow) of the estimated region of interest by the DG3PD of Thai and Gottschlich (2016a)) and from Wiesner et al. (2020b) two overlapping shoeprints with similar shoe pattern elements (right). A natural question is: Are those from the same shoe?

by *surveillance cameras* or by *web searches*, or structures of buildings in *satellite images*, to name just a few. In many applications, upon closer inspection, certain parts or features of these objects are of concern, e.g., texture information of doting material in the *material sciences*, connectivity structure in *brain imaging*, or minutiae loci in fingerprints (see Fig. 2) for smartphone user *authentication* and *identification*. Extracting this kind of information out of often high-dimensional input images $F \in \mathbb{R}^{n \times m}$ is especially challenging when the inputs can consist of heterogeneous images, such as fingerprint images taken at a crime scene that are hard to model.

Notably, since all images are based on individual pixels, in this chapter, we consider only the discrete case, viewing images as matrices.

32 Filter Design for Image Decomposition and Applications to Forensics 1159

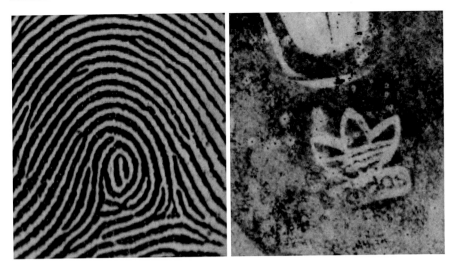

Fig. 2 Fingerprint ridge lines of very good quality following an *orientation field*, ending or forking at *minutiae* (left, from Turroni et al. 2011). Shoeprint (detail from Fig. 1) with sole *pattern* and pattern damages called *accidentals* (right). Here the black dots with circular white halo due to sand grains and the dark black clusters due to dirt need to be discriminated from true wear effects, for instance, on the left side of the brand's logo

These challenges have led to a surge of decomposition methods aimed at automatic removal of *noise* and/or *texture*, returning a piecewise constant or smooth *cartoon* component $U \in \mathbb{R}^{n \times m}$, possibly a second *texture* component $V \in \mathbb{R}^{n \times m}$, and a *noise* component $\varepsilon \in \mathbb{R}^{n \times m}$. Such decompositions can simplify the extraction of information as in applications one is often interested either in large-scale information (e.g., edges of buildings in aerial photographs) or small-scale information (e.g., fringe patterns in fingerprints). Well-known side effects of such methods are *loss of contrast* or artifacts such as *ringing* and *straircasing*.

In addition to image enhancement, decomposition of overlapping structures in the ROI is a frequent challenge in forensics; see Fig. 1. Moreover, structure at different scale is of high importance, e.g., shoe pattern *elements* identifying a shoe brand (cf. Fig. 1) and damages to the pattern due to wear or other damaging effects, called *accidentals*, identifying an individual shoe; cf. Fig. 2. In fingerprint analysis, the ridge line structure with its *orientation field* is the coarse structure to be identified, and *minutiae* (ending or forking ridge lines) convey the microstructure identifying individuals; cf. Fig. 2.

Automated fingerprint comparison utilizes minutiae loci and possibly the ridge line structure (orientation field); cf. (Maltoni et al. 2009). These are extracted by identifying a ROI. Bad quality images can be enhanced, or, while fingerprint scans are taken, bad quality scans can be rejected; cf. (NFI 2015; Yao et al. 2016; Richter et al. 2019).

For shoeprint analysis, due to the larger challenges given by the huge diversity of shoe element patterns and accidental structures, automated comparison is still in its very beginnings, e.g., Wiesner et al. (2020a,b).

Diffusion Methods

Solving the heat equation with initial conditions given by the image at hand, and following it over time, is one of the oldest smoothing methods. Over time, first smaller structures are smoothed, and then also bigger structures disappear, until, after infinite time, no information remains. This calls for smart choices of stopping times, and, in order to preserve specific structures for a longer time, alterations of the diffusion differential equation. For instance, Perona and Malik (1990), and subsequently Alvarez et al. (1992), impede diffusion along image gradients by *anisotropic nonlinear* diffusion, thus steering diffusion along rather constant image intensity regions.

In fingerprint images, among others, as detailed above, estimation of orientation fields is of high importance. Due to small interridge distances in fingerprints, in low-quality fingerprint images, however, image gradients are heavily influenced by noise and cannot be relied on. To this end, Perona (1998) applied orientation diffusion to estimate a smooth orientation field. Such separately estimated orientation fields (for alternate methods, e.g., Bazen and Gerez 2002) have been used by Gottschlich and Schönlieb (2012) for fingerprint enhancement (cf. Fig. 3 for this and related methods):

(1) Orientation field (OF) estimation
(2) *Oriented diffusion*
(3) Contrast enhancement

An overview of structure tensor-based diffusion methods is given in Weickert (1998); for more broad structure-based image analysis with application in face and fingerprint recognition, see Bigun (2006).

Notably, solving the heat equation can be viewed as applying a low-pass Gauss filter, and anisotropic diffusion has been shown to be strongly connected to TV-ℓ^2 minimization and Haar-wavelet soft-thresholding, e.g., Steidl et al. (2004), linking to spectral and wavelet methods briefly discussed in the next section and minimization methods in the next but one section.

Fourier and Wavelet Methods

In the context of image processing, Fourier, wavelet, curvelet, and similar transformations map an image from an image domain into a spectral, wavelet, etc. domain, apply some form of thresholding, and map the result under the inverse transformation back to the image domain, giving a *filtered* image. Such methods may serve all ends of noise removal, cartoon and texture identification, image

32 Filter Design for Image Decomposition and Applications to Forensics

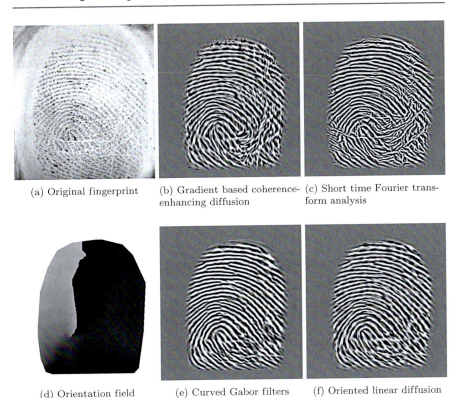

(a) Original fingerprint (b) Gradient based coherence-enhancing diffusion (c) Short time Fourier transform analysis

(d) Orientation field (e) Curved Gabor filters (f) Oriented linear diffusion

Fig. 3 A low-quality fingerprint (**a**) from Turroni et al. (2011) and the corresponding orientation field (**d**), where orientations in degrees are encoded as gray values between 0 and 179, with 0 denoting the x-axis' direction angle and angles increase clock-wise. Compared enhancement methods are (**b**) gradients-based coherence-enhancing diffusion filtering according to Weickert (1999), (**c**) STFT analysis by Chikkerur et al. (2007), (**e**) curved Gabor filters by Gottschlich (2012), and (**f**) oriented diffusion filtering by Gottschlich and Schönlieb (2012)

enhancement, and image compression. In extension of the Fourier transform, wavelet transforms also include local information, and thus draw strength from multiresolution analysis. Very popular is the Haar wavelet which is a special case of the Daubechies wavelet, for which many mulitresolution filter banks are available. For an overview, cf. Daubechies (1992); Chui (1992); Mallat (2008). Curvelets have been introduced by Candès et al. (2006), and Ma and Plonka (2010) give a concise survey.

Particularly fingerprint images, due to their periodic fringe pattern, and also shoeprint images with repeating element patterns, are well suited to Fourier and wavelet methods. *Wavelet scalar quantization* (WSQ) has been used for fingerprint image compression by Hopper et al. (1993). Chikkerur et al. (2007) and Bartůněk et al. (2013) use the *short time Fourier transform* (STFT) for image enhancement; cf. Fig. 3. Gragnaniello et al. (2014) apply a three-level wavelet transform to fingerprint images from which in subsequent processing steps, features are derived

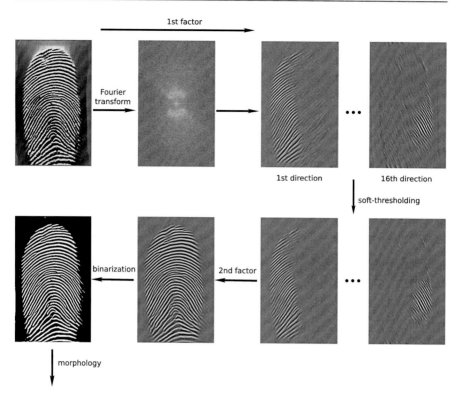

Fig. 4 Factorized directional bandpass (FDB) method from Thai et al. (2016): Soft-thresholding the result of 16 directional filters in the Fourier domain (first factor), binarizing the reconstructing (second factor) in the image domain, and morphological operations lead to identification of the ROI used in Fig. 5

to discriminate between real and *spoof* fingers. Spoof fingerprints are artificial fingerprints created from gelatin or latex; say, cf. Maltoni et al. (2009). *Factorized directional bandpass* (FDB) filters have been built by Thai et al. (2016) using the directional Hilbert transform of a Butterworth bandpass (DHBB) filter and soft-thresholding; cf. Fig. 4. Curiously, thresholding can be viewed as testing with statistical significance for the presence of non-zero filter response coefficients; cf. (Donoho and Johnstone 1994; Frick et al. 2012). The FDB filters have been optimized for texture extraction from fingerprint images with the purpose of segmentation; see Fig. 5.

Variational Problems

Variational problems have played an important role in imaging over the last decades; see, for example, Scherzer et al. (2009) and Aubert and Kornprobst (2006). As for anisotropic diffusion the emphasis lies on computing image approximations

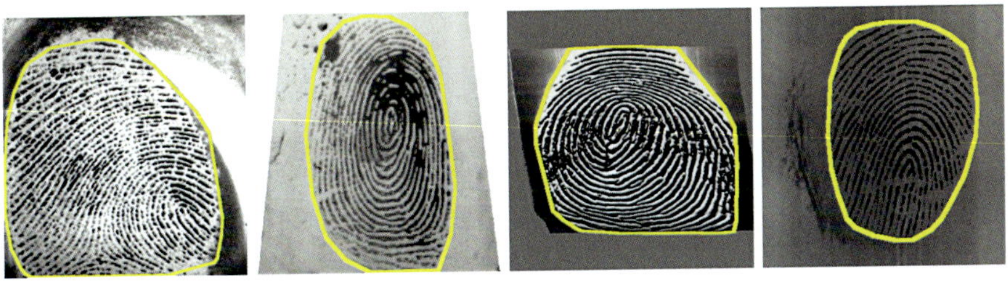

Fig. 5 Four examples of estimated fingerprint segmentation by FDB from Thai et al. (2016)

that keep sharp edges (discontinuities) while removing uninformative noise and/or texture. The, possibly, most influential model is the *Rudin-Osher-Fatemi* (ROF) model of Rudin et al. (1992). In its unconstrained form it is below given as Problem 1 (cf. Chambolle and Lions 1997; they use Neumann boundary conditions, however, as discussed below). It can be seen as a convex recast of the more intrigued *Mumford-Shah* (MS) model (Mumford and Shah 1989) that involves a non-convex term (Hausdorff measure of discontinuities); and for this the MS model is difficult to minimize numerically. Instead, the ROF model includes a total-variation term as regularizer that, when discretized, is given by an ℓ^1-norm of the discrete gradient. Note that instead of the Neumann boundary conditions that are often enforced in the continuous domain (e.g., in the classical Rudin et al. 1992), we, being in the discrete domain, prefer periodic boundary conditions. Then, the discrete gradient comes in handy as a circular convolution operator denoted by $\underline{\mathfrak{C}_D}$; cf. Definition 1 in section "Adaptive Balancing".

Problem 1 (Discrete ROF Model). Let $F \in \mathbb{R}^{n \times m}$ be an input image and $\mu \in \mathbb{R}_+$. The *(isotropic)* $TV - \ell^2$-*model* is given by

$$\text{minimize} \quad \mathcal{J}_{\text{ROF}}(U) := \left|\underline{\mathfrak{C}_D}(U)\right|_{1,2} + \frac{\mu}{2}\|U - F\|^2, \quad (1)$$
$$\text{over} \quad U \in \mathbb{R}^{n \times m},$$

where $|\cdot|_{1,2}$ is the ℓ^1-norm of the ℓ^2-norms of the gradients at each pixel and $\|\cdot\|$ is the usual Euclidean norm.

Solving (1) via steepest descent has led Andreu et al. (2001) to consider a corresponding partial differential equation (PDE) with weak solutions coined as *total variation* (TV) *flow*.

Meanwhile, alleviating for the systematic loss of contrast in the classical ROF-model, Osher et al. (2005) propose iterative Bregman iterations beginning with the ROF solution, passing near a putative noise free version and eventually converging in an *inverse scale-space flow* to the original noisy image.

An extension using higher order derivatives has led to the *total generalized variation* (TGV) model in Bredies et al. (2010) with more detail in Papafitsoros and

Bredies (2015). For a detailed overview of total variation in imaging, see Caselles et al. (2015) and the chapter in this book. In the context of relating different imaging techniques to one another, (Steidl et al. 2004), among others, link the balancing parameter μ of (1) to the stopping time in the anisotropic diffusion model discussed in section "Diffusion Methods". In the following we also consider the general regularization problem given for an input image $F \in \mathbb{R}^{n \times m}$ by

$$\begin{aligned} & \text{minimize} \quad \mathcal{R}(U) + \mu \mathcal{D}(F, U), \\ & \text{over} \quad U \in \mathbb{R}^{n \times m}, \end{aligned} \qquad (2)$$

with $\mathcal{R} : \mathbb{R}^{n \times m} \to \mathbb{R}$ the *regularization term*, $\mathcal{D} : \mathbb{R}^{n \times m} \times \mathbb{R}^{n \times m} \to \mathbb{R}$ the *data-fidelity term*, and $\mu \in \mathbb{R}_+$ a *balancing parameter*.

Non-linear Spectral Decompositions

In analogy to filtering of (linearly) transformed coefficients – discussed in section "Fourier and Wavelet Methods" – a non-linear scale-space approach, the *TV transform*, has been developed by Gilboa (2014). The basis of the TV transform (the rescaled second derivative in the distributional sense of the TV flow) is the definition of so-called eigenfunctions for the TV flow – corresponding to functions f such that αf minimizes the continuous analog of (1) for some $\alpha \in \mathbb{R}$. The TV transform of a general image is then obtained by decomposing into such eigenfuctions (atoms), which for the TV flow are simply disks. Upon observing that the phenomenon of loss of contrast is rooted in the fact that no stopped TV flow is an ideal low-pass filter because the disks loose height, Gilboa (2014) proposes genuine low- and band-pass filters with resepect to his TV transform.

The concept of the non-linear TV transform has been generalized to non-linear spectral decompositions for one-homogeneous functionals in Burger et al. (2016). To this end, the notion of the "eigenfunction" introduced above was extended to (2) with any one-homogeneous convex regularization term $\mathcal{R}$ and $\mathcal{D}$ being ℓ^2-fidelity. It is not clear, however, whether finite linear combinations of eigenfunctions (also called singular vectors) are decomposable into their respective atoms; corresponding circumstances are addressed in Schmidt et al. (2018). Moreover, Burger et al. (2016) defined the non-linear spectral decomposition not only on the basis of the "forward" scale-space flow (as Gilboa 2014 with the TV flow) but also on the basis of the regularization model (2) – with ℓ^2-fidelity – and the inverse scale-space flow, as introduced for the ROF model by Osher et al. (2005) and discussed above.

On the application side, the TV transform has been used for color image denoising in Moeller et al. (2015), texture decomposition into different scales in Horesh and Gilboa (2016), segmentation in Zeune et al. (2017), and image manipulation and image fusion in Hait and Gilboa (2018).

Texture Information

A typical example for texture is the fringe pattern in fingerprint applications. Decomposition methods as the ROF model into a cartoon U and a texture/noise part $V := U - F$ are well suited to obtain a binary ROI that segments an image into *foreground* (e.g., fingerprint) and *background*. Among the many descendants of the ROF model, there are the *decompositions into three parts*: cartoon, texture, and noise (e.g., Aujol and Chambolle 2005; Shen 2005), which are particularly useful in fingerprint analysis. One decisive step is introducing the theory of the *G-space* from Meyer (2001), a space particularly designed to feature small corresponding *G-norms* for oscillating functions. In general, for a function $f : \Omega \to \mathbb{R}$ from a bounded image domain Ω, the G-norm is given by

$$\|f\|_G := \inf\{\|g\|_{L^\infty(\Omega,\mathbb{R}^2)} : g = (g_1, g_2); f = \partial_1 g_1(x) + \partial_2 g_2\}, \qquad (3)$$

where $\|g\|_{L^\infty(\Omega,\mathbb{R}^2)} = \operatorname{ess\,sup}_{x \in \Omega} \left(x \mapsto \sqrt{g_1(x)^2 + g_2(x)^2} \right)$. Due to its indirect definition via the g's, solving a minimization problem involving the G-norm is rather hard. There is quite a body of literature devoted to analyzing the G-space (and its related E- and F-space also introduced in Meyer 2001) and proposing approximations or simplifications to a $TV - G$ model (see, e.g., Vese and Osher 2003; Le and Vese 2005; Aujol et al. 2005). In section "Adaptive Balancing" a more detailed inspection of some of these approaches will be given.

For ROI extraction in fingerprint images, the *global three parts decomposition* (G3PD) model has been proposed by Thai and Gottschlich (2016b). It decomposes an image into cartoon, texture, and noise, using an anisotropic total variation regularizer for the cartoon U, a curvelet-ℓ^1-norm plus an ℓ^1-norm on the texture part V, while the ℓ^∞-norm of the curvelet coefficients of $\varepsilon = F - U - V$ is bounded. The model involves the following objective function

$$\mathcal{J}_{\text{G3PD}}(U, V) := \left|\underline{\mathfrak{C}}_{\underline{D}}(U)\right|_{1,1} + \mu_1 |\mathcal{C}(V)|_1 + \mu_2 |V|_1 \;,$$

for $U, V \in \mathbb{R}^{n \times m}$ and $\mathcal{C}(V)$ being the curvelet decomposition (Candès et al. 2006) of V, which is to be minimized under the constraints

$$|\mathcal{C}(\varepsilon)|_\infty \leq \delta, \qquad F = U + V + \varepsilon,$$

where $\mathcal{C}$ is the same curvelet transform of Candès et al. (2006) and $\mu_1, \mu_2, \delta \in \mathbb{R}$. Due to orientation sensitivity of the curvelet transform, G3PD is well suited to capture the fringe pattern of a fingerprint in the texture component; see Fig. 6. In automated practice, when applied to images, not containing other small-scale information featuring similar frequencies as the fingerprint pattern, parameters can be well tuned to specific sensors, such that ROIs are reliably extracted. On crime

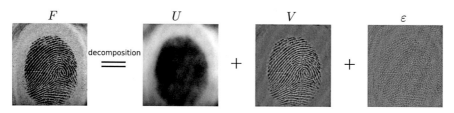

Fig. 6 Decomposition by G3PD of a fingerprint image F from Thai and Gottschlich (2016b) into three parts: cartoon (U), texture (V), noise (ε)

scene images, however, ideal parameter choices often vary substantially over images with different background, calling for more flexibility of the model and specific learning methods.

Machine Learning

We have thus far reported how for specific tasks at hand (e.g., segmentation, enhancement) specific tools have been designed, often using elaborate parameter tuning. In fact such ideal parameters often vary over varying use cases (e.g., G3PD requires different parameter choices when large regions contain small-scale patterns not related to the fingerprint at hand, as is often the case in real crime scene images). This calls for designing more flexible models and learning methods to incorporate heterogeneous use cases. Notably, when abundant data are available, nearly any machine learning method off the shelf usually works well. The less data are available, however, the more a priori structure must be built into learning methods. This is, for instance, the case in academic forensic research. For example, supervised learning models involving second order minutiae structure have resulted in a highly discriminatory test for separating real fingerprints from synthetic images where training set, validation set, and test set have summed up only 110 fingers (and 8 impressions per finger) per class; cf. Gottschlich and Huckemann (2014).

The very small size of data sets in fingerprint recognition and forensic applications stands in stark contrast to databases for image classification and visual object recognition like ImageNet which contains more than 14 million images (http://image-net.org/about-stats). Very large data sets enable fully automatic end-to-end learning by neural networks (Bengio 2009), whereas a very small number of training examples pose a huge additional machine learning challenge for biometric and forensic research. For the task of fingerprint quality estimation using image decomposition, Richter et al. (2019) proposed a new robust biometric quality validation scheme (RBQ VS) based on repeated random subsampling cross-validation to deal with problematic lack of a preferable number of training and test images. For fingerprint alteration detection, even fewer examples are available for training and testing and Gottschlich et al. (2015) also resort to cross-validation in order to compare different approaches. Biometric and forensic applications could profit immensely from research on, e.g., deep learning (LeCun et al. 2015), evolutionary algorithms (Kennedy and Shi 2001), Bayesian learning (Neal 2012), support vector machines (Schoelkopf and Smola 2002), or random forests (Breiman 2001) if only larger data sets were available.

Regarding the aforementioned imaging approaches (cf. sections "Diffusion Methods", "Fourier and Wavelet Methods", "Variational Problems"), there have been a multitude of machine learning extensions proposed in the literature. For example, anisotropic diffusion has been learned by De los Reyes and Schönlieb (2013), and Chen and Pock (2017) have learned reaction diffusion models, while (Grossmann et al. 2020) learn TV transform filters. Arridge et al. (2019) give a survey on solving ill-posed inverse problems based on deep learning, with domain-specific knowledge contained in physical–analytical models.

Adaptive Balancing

Augmenting the ℓ^1-regularization model in (1), obtaining a more general training model can be achieved by making the balancing parameter μ in the spatial or in the frequency domain adaptive. In fact, the former corresponds to bilevel minimization problems that choose the balancing parameter (or a more general balancing function) via its own minimization (e.g., Bredies et al. 2013; Calatroni et al. 2017), while the latter relates to various approaches to model texture following intuition from Meyer (2001). In the following we report on this connection for the discrete case and propose a way of extending the model class even beyond.

Definition 1. Define the *matrix-family convolution* in the following circular way by

$$\underline{\mathfrak{C}}_{\underline{B}}(U) := (B_1 * U, B_2 * U, \ldots, B_p * U),$$

where $B * U$ is the usual circular convolution of matrices (e.g., Mallat 2008) with components given by

$$(B * U)[r, s] = \sum_{k=0}^{n-1} \sum_{\ell=0}^{m-1} B[k, \ell] U[r - k, s - \ell],$$

where k is taken modulo n and ℓ is taken modulo m. Moreover, let us denote by Γ_P the space of matrix-families $(\mathbb{R}^{n \times m})^P$.

Then the (forward) discrete gradient (with periodic boundary conditions) is given by the matrix-family convolution

$$\underline{\mathfrak{C}}_{\underline{D}} : \mathbb{R}^{n \times m} \to \Gamma_2, \quad U \mapsto \left(\mathfrak{C}_{D_1}(U), \mathfrak{C}_{D_2}(U)\right)^T,$$

where

$$D_1[k,\ell] := \begin{cases} -1 & \text{if } [k,\ell] = [0,0] \\ 1 & \text{if } [k,\ell] = [1,0] \\ 0 & \text{else} \end{cases}, \quad D_2[k,\ell] := \begin{cases} -1 & \text{if } [k,\ell] = [0,0] \\ 1 & \text{if } [k,\ell] = [0,1] \\ 0 & \text{else} \end{cases}.$$

Adapting the Data-Fidelity-Norm

As mentioned in section "Variational Problems", following Meyer (2001) there has been much research devoted to change the data-fidelity norm towards making it more adaptive to capture oscillating patterns. In the discrete setting, many of these models can be brought into the form of the general *TV-Hilbert model* proposed in Aujol and Gilboa (2006). Absorbing the balancing parameter in M, one can interpret the *TV*-Hilbert model as *adaptive TV-regularization* minimizing the functional

$$\mathcal{J}_M(U) := \left|\underline{\mathfrak{C}}_{\underline{D}}(U)\right|_{1,2} + \frac{1}{2}\left\|\mathfrak{C}_M(F-U)\right\|^2, \tag{4}$$

with the discrete gradient $\underline{\mathfrak{C}}_{\underline{D}}$ and the new *balancing filter* $M \in \mathbb{R}^{n \times m}$ featuring $\widehat{M} \in \mathbb{R}_+^{n \times m}$ (where $\widehat{M}$ is the discrete Fourier transform). Notably, Aujol and Gilboa (2006) also allow operators more general than the circular convolution operator $\mathfrak{C}_M$ above. Of course, the ROF model is a special case of (4) by choosing $\mathfrak{C}_M$ as a multiplication with the balancing parameter μ. Let us ponder first on a connection of (4) with the G-norm given in (3) and secondly on some literature considering (4).

Connection to the G-Norm

The *Osher-Solè-Vese* (OSV) *model* considered in Osher et al. (2003) serves as an example of the connection between the $TV - G$ model and minimizing the functional in (4). To sketch the underlying ideas, let u, f be functions defined on a bounded domain $\Omega \subset \mathbb{R}^2$. Recall that the G-norm of $f - u$ is defined as the infimum of L^∞-norms over all $g \in L^\infty(\Omega, \mathbb{R}^2)$ such that $\text{div}(g) = f - u$. In Vese and Osher (2003) the G-norm is approximated by introducing g as a variable and replacing the G-norm of $f-u$ by an L^2 penalization $f-u-\text{div}(g)$ plus an L^p-norm on g with $1 \le p < \infty$. Building on this model in Vese and Osher (2003), the OSV model in Osher et al. (2003) simplifies by assuming the existence of $g \in L^2(\Omega^2)$ with $\text{div}(g) = f - u$ and $g = \nabla P$ for some scalar-valued function $P \in H^1(\Omega)$. Hence,

$$f - u = \text{div}(g) = \Delta P.$$

Plugging the above into the model of Vese and Osher (2003), one obtains (cf. Equation (2.1) in Osher et al. 2003) for $\lambda \in \mathbb{R}_+$ the objective function

$$\int_\Omega |\nabla u| + \lambda \int_\Omega \left|\nabla(\Delta^{-1})(f-u)\right|^2.$$

Assuming that $\mathfrak{C}_{\widetilde{M}}$ is an appropriate discrete version of the pseudo-differential operator Δ^{-1}, then discretizing the OSV model leads to the matrix-convolution operator $\mathfrak{C}_D \mathfrak{C}_{\widetilde{M}}$. Since $(\mathfrak{C}_D \mathfrak{C}_{\widetilde{M}})^* \mathfrak{C}_D \mathfrak{C}_{\widetilde{M}}$ is self-adjoint and thus has real and nonnegative eigenvalues, it allows for a unique positive-semidefinite square-root $\mathfrak{C}_M$ which is the one from (4) and λ is set to 1, as it can be absorbed by M.

Other Choices of M

In Aujol et al. (2006), for a matrix-family convolution $\mathfrak{C}_B$ with Gabor wavelet frames, *Gabor wavelet filters* of form $\mathfrak{C}_M = \mathfrak{C}_B^* \mathfrak{C}_B$ have been proposed for (4).

Garnett et al. (2007) use Besov norms of the Besov spaces $\dot{B}_{p,q}^\alpha$ for $1 \leq p, 1 \leq q < \infty$ and $\alpha \in \mathbb{R}$ to approximate the G-norm. For $p = q = 2$ specific filters K_α (see Definition 5 of Garnett et al. 2007) are associated to the Besov spaces $\dot{B}_{2,2}^\alpha$. Then a discrete version of the Besov space norm on $F - U$ is given by

$$\left\|\mathfrak{C}_{K_\alpha}(U - F)\right\|^2,$$

cf. (56) in Garnett et al. (2007).

A model proposed by Buades et al. (2010) considers a special $\mathfrak{C}_M$ in (4), defined in the frequency domain by the continuous filter

$$\widehat{L}_\sigma(\xi) := \frac{1}{1 + (2\pi\sigma|\xi|)^4}.$$

Discretization of the above then yields $\mathfrak{C}_M$.

Connections with Machine Learning

Yang et al. (2016) consider ℓ^1-regularization in the spirit of machine learning approaches: They implement a general learning problem based on (2) with ℓ^2-data-fidelity term. Upon closer inspection one can show that the learning architecture is constructed in such a way that it also learns over adaptive balancing parameters (Richter et al. 2020). In the remainder of this chapter we ponder on the connection of adaptive balancing and intersection point problems arising from an ADMM/AL algorithm solving (5) on which (Yang et al. 2016) build. To this end, denote the larger class of *adaptive ℓ^1-regularizations* by

$$\mathcal{J}_{B,M}(U) := \left|\mathfrak{C}_B(U)\right|_{1,\kappa} + \frac{\mu}{2}\left\|\mathfrak{C}_M(F - U)\right\|^2. \tag{5}$$

Here $\underline{B} \in \Gamma_P$ is a suitable matrix-family convolution, $\kappa \in \{1, 2\}$, $\mu \in \mathbb{R}_+$ and $M \in \mathbb{R}^{n \times m}$ is the balancing filter.

Solving via the ADMM/AL-Algorithm

The advantage of the functional $\mathcal{J}_{B,M}$ given in (5) lies in its convexity, in the smoothness of its data-fidelity term, and in the norm of its regularizer, being well understood. In the following we focus on the *alternating directions method of multipliers* (ADMM) in the context of *augmented Lagrangian* (AL) approaches. While the convergence of ADMM/AL to the exact solution is often slower when compared to other methods, its convergence to a neighborhood, when given bad starting values, is rather satisfactory. The method of multipliers has been introduced by Powell (1969) and Hestenes (1969). For a general result on the setup and convergence of ADMM/AL algorithms in the context of minimization via the augmented Lagrangian, see Theorem 8 of Eckstein and Bertsekas (1992) and references therein.

There have been various other algorithms proposed for minimizing functionals such as $\mathcal{J}_{B,M}$ from (5). The original ROF model, a special case of (4), was solved by Rudin et al. (1992) via a rather slow gradient descent algorithm. Popular later approaches include projection algorithms (Chambolle 2004; Aujol and Gilboa 2006), the use of Bregman distances (Goldstein and Osher 2009), graph-cut methods (Darbon and Sigelle 2006a,b), and forward-backward splitting (Chambolle and Pock 2011). For an in-depth overview, we refer to Chambolle and Pock (2016) and Goldstein et al. (2014).

The functional $\mathcal{J}_{B,M}$ of (5) contains the non-linear regularization term $|\mathfrak{C}_{\underline{B}}(U)|_{1,\kappa}$ which cannot be minimized simply by differentiation with respect to $\overline{U}$. For this reason a new additional variable $\underline{W}$ is introduced, taking the place of $\mathfrak{C}_{\underline{B}}(U)$. This yields the constrained problem

$$\text{minimize } \widetilde{\mathcal{J}}_{B,M}(U, \underline{W}) := |\underline{W}|_{1,\kappa} + \frac{\mu}{2} \left\| \mathfrak{C}_M(F - U) \right\|^2, \quad (6)$$

$$\text{such that } \mathfrak{C}_{\underline{B}}(U) = \underline{W}, \quad \text{over } U \in \mathbb{R}^{n \times m} \text{ and } \underline{W} \in \Gamma_P,$$

which is equivalent to minimizing $\mathcal{J}_{B,M}$. Problem (6) can now be solved by computing the saddle point of the augmented Lagrangian functional $\mathcal{J}_{\text{AL}}$ given below for $\beta \in \mathbb{R}_+$ and Lagrangian multiplier $\underline{\lambda} \in \Gamma_P$ (e.g., Bertsekas 1982),

$$\begin{aligned}\mathcal{J}_{\text{AL}}(U, \underline{W}, \underline{\lambda}) := |\underline{W}|_{1,\kappa} &+ \frac{\mu}{2} \left\| \mathfrak{C}_M(F - U) \right\|^2 \\ &+ \frac{\beta}{2} \left\| \underline{W} - \mathfrak{C}_{\underline{B}}(U) \right\|^2 + \left\langle \underline{\lambda}, \underline{W} - \mathfrak{C}_{\underline{B}}(U) \right\rangle.\end{aligned} \quad (7)$$

Algorithm 1 ADMM/AL for adaptive ℓ^1-regularization (one step)

Input: $F \in \mathbb{R}^{n \times m}$.
Input Filters: $M \in \mathbb{R}^{n \times m}, \underline{B} \in \Gamma_P$.
Customizable Parameters: $\mu \in \mathbb{R}_+, \kappa \in \{1, 2\}$.
Initialization: $U^{(0)} = F \in \mathbb{R}^{n \times m}, \underline{\lambda}^{(1)} = \underline{0} \in \Gamma_{P,\cdot}$
for $\tau = 1, 2, \ldots$ **do**

$$\underline{W}^{(\tau)} = \underset{\underline{W} \in \Gamma_P}{\arg\min} \left(\mathcal{J}_{AL} \left(U^{(\tau-1)}, \underline{W}; \underline{\lambda}^{(\tau)} \right) \right),$$

$$U^{(\tau)} = \underset{U \in \mathbb{R}^{n \times m}}{\arg\min} \left(\mathcal{J}_{AL} \left(U, \underline{W}^{(\tau)}; \underline{\lambda}^{(\tau)} \right) \right),$$

$$\underline{\lambda}^{(\tau+1)} = \underline{\lambda}^{(\tau)} + \beta \left(\underline{W}^{(\tau)} - \underline{\mathfrak{C}}_B \left(U^{(\tau)} \right) \right).$$

end for

Notably, a saddle-point of (7) does not depend on the choice of β. To solve for the saddle-point, Algorithm 1 alternates between minimizing $\mathcal{J}_{AL}$ for $\underline{W}$ and U (one iteration of an ADMM algorithm), while updating in each iteration the Lagrangian multiplier $\underline{\lambda}$ via a gradient step.

Interpretation via a Feasibility Problem

We now show that Algorithm 1, which converges to the saddle-point of $\mathcal{J}_{AL}$, solves a special case of a broader feasibility problem (Problem 3). Before, we state a (seemingly) different feasibility problem directly derived from the above updating rules.

Problem 2. Given $F, M \in \mathbb{R}^{n \times m}, \underline{B} \in \Gamma_P, \mu \in \mathbb{R}_+, \kappa \in \{1, 2\}$, and $\beta \in \mathbb{R}_+$, with discrete Fourier transform $\widehat{M} \in \mathbb{R}_+^{n \times m}$, find a point $(U^\dagger, \underline{W}^\dagger, \underline{\lambda}^\dagger) \in \Gamma_{1+2P}$ in the intersection of the following three sets

$$\Omega_1^\kappa := \left\{ (U, \underline{W}, \underline{\lambda}) \in \Gamma_{1+2P} : \underline{W} = \underset{\underline{\widetilde{W}} \in \Gamma_P}{\arg\min} \mathcal{J}_{AL} \left(U, \underline{\widetilde{W}}, \underline{\lambda} \right) \right\},$$

$$\Omega_2 := \left\{ (U, \underline{W}, \underline{\lambda}) \in \Gamma_{1+2P} : U = \underset{\widetilde{U} \in \mathbb{R}^{n \times m}}{\arg\min} \mathcal{J}_{AL} \left(\widetilde{U}, \underline{W}, \underline{\lambda} \right) \right\}, \quad (8)$$

$$\Omega_C := \left\{ (U, \underline{W}, \underline{\lambda}) \in \Gamma_{1+2P} : \underline{\mathfrak{C}}_B (U) = \underline{W} \right\}.$$

To prepare for the proof of equivalence of the above feasibility problem and the one stated further below (Problem 3), let us first compute the minimizers of $\mathcal{J}_{AL}$

with respect to $\underline{W}$ and U, using standard variational calculus (from, e.g., Boyd and Vandenberghe 2004; Bauschke and Combettes 2011).

- For given $U \in \mathbb{R}^{n \times m}$, $\underline{B}, \underline{\lambda} \in \Gamma_P, \kappa \in \{1, 2\}$, as well as $\beta \in \mathbb{R}_+$, the unique minimizer of

$$\mathcal{J}_1(\underline{W}) := |\underline{W}|_{1,\kappa} + \frac{\beta}{2}\|\underline{W} - \underline{\mathfrak{C}}_{\underline{B}}(U)\|^2 + \langle \underline{\lambda}, \underline{W} \rangle,$$

is given by

$$\underline{W}^\dagger = \mathsf{S}_\kappa \left(\underline{\mathfrak{C}}_{\underline{B}}(U) - \frac{1}{\beta}\underline{\lambda}; \frac{1}{\beta} \right), \qquad (9)$$

where $\mathsf{S}_\kappa : \Gamma_P \to \Gamma_P$ is the isotropic ($\kappa = 2$) or anisotropic ($\kappa = 1$) *soft-shrinkage function*.

- For given $F \in \mathbb{R}^{n \times m}$, $\underline{B}, \underline{W}, \underline{\lambda} \in \Gamma_P$, and $\beta \in \mathbb{R}_+$, the unique minimizer of

$$\mathcal{J}_2(U) := \frac{\mu}{2}\|\mathfrak{C}_M(F - U)\|^2 + \frac{\beta}{2}\|\underline{W} - \underline{\mathfrak{C}}_{\underline{B}}(U)\|^2 - \langle \underline{\lambda}, \underline{\mathfrak{C}}_{\underline{B}}(U) \rangle,$$

is given by

$$U^\dagger = \mu \left(\mu \mathfrak{C}_M^* \mathfrak{C}_M + \beta \underline{\mathfrak{C}}_{\underline{B}}^* \underline{\mathfrak{C}}_{\underline{B}} \right)^{-1} \mathfrak{C}_M^* \mathfrak{C}_M(F) \\ + \beta \left(\mu \mathfrak{C}_M^* \mathfrak{C}_M + \beta \underline{\mathfrak{C}}_{\underline{B}}^* \underline{\mathfrak{C}}_{\underline{B}} \right)^{-1} \underline{\mathfrak{C}}_{\underline{B}}^* \left(\underline{W} + \frac{1}{\beta}\underline{\lambda} \right), \qquad (10)$$

given that $\mu \mathfrak{C}_M^* \mathfrak{C}_M + \beta \underline{\mathfrak{C}}_{\underline{B}}^* \underline{\mathfrak{C}}_{\underline{B}}$ is invertible, which is the case because by $\widehat{M} \in \mathbb{R}_+^{n \times m}$ we have $\ker(\mathfrak{C}_M) = \{0\}$.

Abbreviating the two operators in (10), we introduce $A \in \mathbb{R}^{n \times m}$ and $\underline{\tilde{B}} \in \Gamma_P$ such that

$$\mathfrak{C}_A := \mu \left(\mu \mathfrak{C}_M^* \mathfrak{C}_M + \beta \underline{\mathfrak{C}}_{\underline{B}}^* \underline{\mathfrak{C}}_{\underline{B}} \right)^{-1} \mathfrak{C}_M^* \mathfrak{C}_M, \qquad (11)$$

and

$$\underline{\mathfrak{C}}_{\underline{\tilde{B}}}^* := \beta \left(\mu \mathfrak{C}_M^* \mathfrak{C}_M + \beta \underline{\mathfrak{C}}_{\underline{B}}^* \underline{\mathfrak{C}}_{\underline{B}} \right)^{-1} \underline{\mathfrak{C}}_{\underline{B}}^*.$$

This gives the above anticipated feasibility problem. As before, for any matrix A, $\widehat{A}$ denotes its discrete Fourier transform.

Problem 3. Consider arbitrary $F, Y \in \mathbb{R}^{n \times m}$, $\underline{B} \in \Gamma_P$, $\kappa \in \{1, 2\}$, and $\nu \in \mathbb{R}_+$, such that the following two conditions are satisfied

1. $\widehat{Y} \in \mathbb{R}_+^{n \times m}$,
2. $\mathfrak{C}_Y \mathfrak{C}_{\underline{B}}^* \mathfrak{C}_{\underline{B}}$ has all eigenvalues in $[0, 1)$.

Moreover, define $\underline{\widetilde{B}} \in \Gamma_P$ via

$$\widetilde{B}_p[k, \ell] = \widehat{Y}[k, \ell]\widehat{B}_p[k, \ell],$$

for all $0 \leq k \leq n-1$ and $0 \leq \ell \leq m-1$ and $1 \leq p \leq P$ and $A \in \mathbb{R}^{n \times m}$ as the matrix corresponding to the matrix-convolution given by

$$\mathfrak{C}_A = \mathfrak{E} - \mathfrak{C}_{\underline{B}}^* \mathfrak{C}_{\underline{B}},$$

where $\mathfrak{E}$ is the identity operator on $\mathbb{R}^{n \times m}$. Find a point $(\underline{U}^\dagger, \underline{W}^\dagger, \underline{\lambda}^\dagger) \in \Gamma_{1+2P}$ in the intersection of the following three sets

$$\Omega_1'^\kappa := \left\{ (\underline{U}, \underline{W}, \underline{\lambda}) \in \Gamma_{1+2P} : \underline{W} = \mathsf{S}_\kappa \left(\mathfrak{C}_{\underline{B}}(\underline{U}) - \frac{1}{\nu}\underline{\lambda}; \frac{1}{\nu} \right) \right\},$$

$$\Omega_2^G := \left\{ (\underline{U}, \underline{W}, \underline{\lambda}) \in \Gamma_{1+2P} : \underline{U} = \mathfrak{C}_A(F) + \mathfrak{C}_{\underline{\widetilde{B}}}^* \left(\underline{W} + \frac{1}{\nu}\underline{\lambda} \right) \right\}, \quad (12)$$

$$\Omega_C := \left\{ (\underline{U}, \underline{W}, \underline{\lambda}) \in \Gamma_{1+2P} : \mathfrak{C}_{\underline{B}}(\underline{U}) = \underline{W} \right\}.$$

Replacing M, we have introduced in Problem 3 a new matrix Y balancing now the interplay of the matrix-families $\underline{B}$ and $\underline{\widetilde{B}}$. It turns out that this balancing Y corresponds in the following way to the adaptive balancing filter M of (5), guaranteeing the equivalence of Problems 2 and 3.

Theorem 1. Let $F \in \mathbb{R}^{n \times m}$, $\underline{B} \in \Gamma_P$ and $\kappa \in \{1, 2\}$. For given $M \in \mathbb{R}^{n \times m}$ such that $\widehat{M} \in \mathbb{R}_+^{n \times m}$, let $\mu \in \mathbb{R}_+$ and let $(\underline{U}^\dagger, \underline{W}^\dagger, \underline{\lambda}^\dagger) \in \Gamma_{1+2P}$ be a solution of Problem 2. Then, letting $\nu = \mu = \beta$, and defining $Y \in \mathbb{R}^{n \times m}$ via

$$\mathfrak{C}_Y = \beta \left(\mu \mathfrak{C}_M^* \mathfrak{C}_M + \beta \mathfrak{C}_{\underline{B}}^* \mathfrak{C}_{\underline{B}} \right)^{-1},$$

we have that $\widehat{Y} \in \mathbb{R}_+^{n \times m}$, that $\mathfrak{C}_Y \mathfrak{C}_{\underline{B}}^* \mathfrak{C}_{\underline{B}}$ has all eigenvalues in $[0, 1)$, and that $(\underline{U}^\dagger, \underline{W}^\dagger, \underline{\lambda}^\dagger)$ is a solution of Problem 3.

Vice versa, let $Y \in \mathbb{R}^{n \times m}$, such that $\widehat{Y} \in \mathbb{R}_+^{n \times m}$ and $\mathfrak{C}_Y \mathfrak{C}_{\underline{B}}^* \mathfrak{C}_{\underline{B}}$ has all eigenvalues in $[0, 1)$, let $\nu \in \mathbb{R}_+$, and let $(\underline{U}^\dagger, \underline{W}^\dagger, \underline{\lambda}^\dagger) \in \Gamma_{1+2P}$ be a solution of Problem 3. Then, defining $\mu = 1$ and $\mathfrak{C}_M$ as the unique positive semi-definite square root of

$$\mathfrak{C}_{\widetilde{M}} = \nu \mathfrak{C}_Y^{-1}\left(\mathfrak{E} - \mathfrak{C}_Y \underline{\mathfrak{C}}_B^* \underline{\mathfrak{C}}_B\right) = \nu \mathfrak{C}_Y^{-1} - \nu \underline{\mathfrak{C}}_B^* \underline{\mathfrak{C}}_B$$

(existing due to the eigenvalues of $\mathfrak{C}_Y \underline{\mathfrak{C}}_B^* \underline{\mathfrak{C}}_B$ being strictly less than 1), then $\widehat{M} \in \mathbb{R}_+^{n\times m}$ and $(U^\dagger, \underline{W}^\dagger, \underline{\lambda}^\dagger)$ is a solution of Problem 2.

Proof. Let, as in the assertion, $F, M \in \mathbb{R}^{n\times m}$, $B \in \Gamma_P$, $\mu \in \mathbb{R}_+$ and $\kappa \in \{1, 2\}$ be given with $\widehat{M} \in \mathbb{R}_+^{n\times m}$, and let $(U^\dagger, \underline{W}^\dagger, \underline{\lambda}^\dagger) \in \Gamma_{1+2P}$ be a solution of Problem 2. Recall that the solution does not depend on the choice of $\beta \in \mathbb{R}_+$ for $\mathcal{J}_{AL}$. Hence, w.l.o.g., we can set $\beta = \mu$. Moreover setting $\nu = \mu$ we have at once by (9) that the definitions of Ω_1^κ of Problem 2 and of $\Omega_1^{'\kappa}$ Problem 3 coincide. Since Ω_C is the same for both problems, we are left to show that $(U^\dagger, \underline{W}^\dagger, \underline{\lambda}^\dagger) \in \Omega_2^G$.

Defining $Y \in \mathbb{R}^{n\times m}$ as in the assertion, we have at once that $\widehat{Y} \in \mathbb{R}_+^{n\times m}$. Moreover, since matrix convolution operators are diagonalized by the discrete Fourier transform, the eigenvalues of $\mathfrak{C}_Y \underline{\mathfrak{C}}_B^* \underline{\mathfrak{C}}_B$ are given by

$$\beta\left(\mu\widehat{M}[k, \ell]^2 + \beta\sum_{p=1}^P \left|\widehat{B}_p[k, \ell]\right|^2\right)^{-1} \sum_{p=1}^P \left|\widehat{B}_p[k, \ell]\right|^2 \in [0, 1),$$

because $\widehat{M} \in \mathbb{R}_+^{n\times m}$.

Last, by (10) we have that

$$U^\dagger = \mu\left(\mu\mathfrak{C}_M^*\mathfrak{C}_M + \beta\underline{\mathfrak{C}}_B^*\underline{\mathfrak{C}}_B\right)^{-1} \mathfrak{C}_M^*\mathfrak{C}_M(F)$$

$$+ \beta\left(\mu\mathfrak{C}_M^*\mathfrak{C}_M + \beta\underline{\mathfrak{C}}_B^*\underline{\mathfrak{C}}_B\right)^{-1} \underline{\mathfrak{C}}_B^*\left(\underline{W}^\dagger + \frac{1}{\beta}\underline{\lambda}^\dagger\right)$$

$$= \frac{\mu}{\beta}\mathfrak{C}_Y\mathfrak{C}_M^*\mathfrak{C}_M(F) + \mathfrak{C}_Y\underline{\mathfrak{C}}_B^*\left(\underline{W}^\dagger + \frac{1}{\beta}\underline{\lambda}^\dagger\right)$$

$$= \mathfrak{C}_A(F) + \underline{\mathfrak{C}}_{\widetilde{B}}^*\left(\underline{W}^\dagger + \frac{1}{\nu}\underline{\lambda}^\dagger\right),$$

where the last equality holds true due to $\nu = \mu = \beta$ and

$$\mathfrak{C}_A = \mathfrak{E} - \underline{\mathfrak{C}}_{\widetilde{B}}^*\underline{\mathfrak{C}}_{\widetilde{B}} = \mathfrak{E} - \mathfrak{C}_Y\underline{\mathfrak{C}}_B^*\underline{\mathfrak{C}}_B = \mathfrak{C}_Y(\mathfrak{C}_Y^{-1} - \underline{\mathfrak{C}}_B^*\underline{\mathfrak{C}}_B)$$

$$= \mathfrak{C}_Y\left(\frac{\mu}{\beta}\mathfrak{C}_M^*\mathfrak{C}_M + \underline{\mathfrak{C}}_B^*\underline{\mathfrak{C}}_B - \underline{\mathfrak{C}}_B^*\underline{\mathfrak{C}}_B\right) = \mathfrak{C}_Y\mathfrak{C}_M^*\mathfrak{C}_M.$$

Hence, $(U^\dagger, \underline{W}^\dagger, \underline{\lambda}^\dagger) \in \Omega_2^G$ yielding that $(U^\dagger, \underline{W}^\dagger, \underline{\lambda}^\dagger)$ is a solution of Problem 3.

Vice versa, let now $F, Y \in \mathbb{R}^{n\times m}$, $B \in \Gamma_P$, $\nu \in \mathbb{R}_+$, and $\kappa \in \{1, 2\}$, with $\widehat{Y} \in \mathbb{R}_+^{n\times m}$, and suppose that $\mathfrak{C}_Y\underline{\mathfrak{C}}_B\underline{\mathfrak{C}}_{\widetilde{B}}$ has all eigenvalues in $[0, 1)$. Further, let $(U^\dagger, \underline{W}^\dagger, \underline{\lambda}^\dagger) \in \Gamma_{1+2P}$ be a solution of Problem 3.

Choose $\mu = 1$ and $M \in \mathbb{R}^{n \times m}$ as in the assertion. Since $\mathfrak{C}_Y \underline{\mathfrak{C}}_B^* \underline{\mathfrak{C}}_B$ has all eigenvalues in $[0, 1)$ and $\widehat{Y} \in \mathbb{R}_+^{n \times m}$, we have that $\mathfrak{C}_M^* \mathfrak{C}_M = \beta \mathfrak{C}_Y^{-1}(\mathfrak{E} - \mathfrak{C}_Y \underline{\mathfrak{C}}_B^* \underline{\mathfrak{C}}_B)$ has all eigenvalues in $\mathbb{R}_+$. In consequence, by definition of the unique positive semi-definite square root, all eigenvalues of $\mathfrak{C}_M$ are positive, yielding $\widehat{M} \in \mathbb{R}_+^{n \times m}$.

Next, let $\mathcal{J}_{\mathrm{AL}}$ be defined via $\beta = \nu$, then again via (9) the spaces Ω_1^K and $\Omega_1'^K$ defined in the two Problems 2 and 3 coincide, and the space Ω_C is the same anyway. Moreover, we have

$$U^\dagger = \mathfrak{C}_A(F) + \underline{\mathfrak{C}}_{\widetilde{B}}^* \left(W^\dagger + \frac{1}{\nu} \underline{\lambda}^\dagger \right)$$

$$= (\mathfrak{E} - \mathfrak{C}_Y \underline{\mathfrak{C}}_B^* \underline{\mathfrak{C}}_B)(F) + \mathfrak{C}_Y \underline{\mathfrak{C}}_B^* \left(W^\dagger + \frac{1}{\beta} \underline{\lambda}^\dagger \right)$$

$$= \left(\mathfrak{E} - \beta \left(\mathfrak{C}_M^* \mathfrak{C}_M + \beta \underline{\mathfrak{C}}_B^* \underline{\mathfrak{C}}_B \right)^{-1} \underline{\mathfrak{C}}_B^* \underline{\mathfrak{C}}_B \right)(F)$$

$$+ \beta \left(\mathfrak{C}_M^* \mathfrak{C}_M + \beta \underline{\mathfrak{C}}_B^* \underline{\mathfrak{C}}_B \right)^{-1} \underline{\mathfrak{C}}_B^* \left(W^\dagger + \frac{1}{\beta} \underline{\lambda}^\dagger \right)$$

$$= \mu \left(\mu \mathfrak{C}_M^* \mathfrak{C}_M + \beta \underline{\mathfrak{C}}_B^* \underline{\mathfrak{C}}_B \right)^{-1} \mathfrak{C}_M^* \mathfrak{C}_M(F)$$

$$+ \beta \left(\mu \mathfrak{C}_M^* \mathfrak{C}_M + \beta \underline{\mathfrak{C}}_B^* \underline{\mathfrak{C}}_B \right)^{-1} \underline{\mathfrak{C}}_B^* \left(W^\dagger + \frac{1}{\beta} \underline{\lambda}^\dagger \right).$$

Hence $(U^\dagger, W^\dagger, \underline{\lambda}^\dagger)$ is by (10) a minimizer of $\mathcal{J}_{\mathrm{AL}}$ for fixed $W^\dagger$ and $\underline{\lambda}^\dagger$ over $U \in \mathbb{R}^{n \times m}$, i.e., $(U^\dagger, W^\dagger, \underline{\lambda}^\dagger) \in \Omega_2$. Thus, $(U^\dagger, W^\dagger, \underline{\lambda}^\dagger)$ solves Problem 2 for M and μ, as defined. □

A General Learning Problem

The filter $Y \in \mathbb{R}^{n \times m}$ introduced in Problem 3 had to satisfy two properties. In order to generalize beyond these, we formalize them as relations between $\underline{B}$ and $\underline{\widetilde{B}}$ and add already a relaxed version, which comes first.

Definition 2. Let $(\underline{B}, \underline{\widetilde{B}}) \in \Gamma_{2P}$.

- We say that $(\underline{B}, \underline{\widetilde{B}})$ *factor weakly* if for all $1 \leq p \leq P$ and $0 \leq k \leq n-1$ and $0 \leq \ell \leq m-1$ we have

$$\widehat{\widetilde{B}}_p[k, \ell] = \widehat{Y}_p[k, \ell] \widehat{B}_p[k, \ell],$$

for some $\underline{Y} = (Y_p)_{p=1}^P \in \Gamma_P$ with $\widehat{Y}_p \in \mathbb{R}_+^{n \times m}$ for all $1 \leq p \leq P$, called *factor matrix-family*.

- We say that $(\underline{B}, \underline{\widetilde{B}})$ *factor strongly* if for all $1 \leq p \leq P$ and $0 \leq k \leq n-1$ and $0 \leq \ell \leq m-1$ we have

$$\widetilde{\widehat{B}}_p[k, \ell] = \widehat{Y}[k, \ell] \widehat{B}_p[k, \ell],$$

for some $Y \in \mathbb{R}^{n \times m}$ with $\widehat{Y} \in \mathbb{R}_+^{n \times m}$, called *factor matrix*.
- We say that $(\underline{B}, \underline{\widetilde{B}})$ satisfy the *contraction and positive semidefinite condition* (CPC) if

$$0 \leq \sum_{p=1}^{P} \overline{\widetilde{\widehat{B}}_p[k, \ell]} \widehat{B}_p[k, \ell] < 1, \quad \text{for all } 0 \leq k \leq n-1 \text{ and } 0 \leq \ell \leq m-1.$$

Relaxing the feasibility Problem 3 from strongly factoring to weakly factoring, we obtain a more general problem. Moreover, we let the filter $A \in \mathbb{R}^{n \times m}$ be flexible as well.

Problem 4. Given $F \in \mathbb{R}^{n \times m}, \kappa \in \{1, 2\}$ and $\beta \in \mathbb{R}_+$, as well as input filters $(A, \underline{B}, \underline{\widetilde{B}}) \in \Gamma_{1+2P}$, find a point $(\underline{U}^\dagger, \underline{W}^\dagger, \underline{\lambda}^\dagger) \in \Gamma_{1+2P}$ in the intersection of the following three sets

$$\Omega_1''^\kappa := \left\{ (\underline{U}, \underline{W}, \underline{\lambda}) \in \Gamma_{1+2P} : \underline{W} = S_\kappa \left(\underline{\mathfrak{C}}_{\underline{B}} (\underline{U}) - \frac{1}{\beta} \underline{\lambda}; \frac{1}{\beta} \right) \right\},$$

$$\Omega_2''^G := \left\{ (\underline{U}, \underline{W}, \underline{\lambda}) \in \Gamma_{1+2P} : \underline{U} = \mathfrak{C}_A (F) + \underline{\mathfrak{C}}_{\underline{B}}^* \left(\underline{W} + \frac{1}{\beta} \underline{\lambda} \right) \right\}, \quad (13)$$

$$\Omega_C := \left\{ (\underline{U}, \underline{W}, \underline{\lambda}) \in \Gamma_{1+2P} : \underline{\mathfrak{C}}_{\underline{B}} (\underline{U}) = \underline{W} \right\}.$$

Generalizing in a similar manner Algorithm 1, we obtain the following Algorithm 2.

Notably, weakly factoring families allow for at least $(P-1)\lceil \frac{mn}{2} \rceil$ new trainable parameters while keeping the eigenvalues of $\underline{\mathfrak{C}}_{\underline{B}}^* \underline{\mathfrak{C}}_{\underline{B}}$ real and positive. We have the following result on existence of a solution.

Theorem 2 (Richter 2019; Richter et al. 2020). *Let $F \in \mathbb{R}^{n \times m}, \kappa \in \{1, 2\}, \beta \in \mathbb{R}_+$ and let $(A, \underline{B}, \underline{\widetilde{B}}) \in \Gamma_{1+2P}$ be input filters, with weakly factoring $(\underline{B}, \underline{\widetilde{B}})$ satisfying the (CPC). Then Problem 4 has a solution.*

Uniqueness of the solution and convergence of Algorithm 2 to it, say, by showing that Algorithm 2 is again an ADMM/AL algorithm for Problem 4 remains an open problem. In practice, in all numerical experiments conducted by Richter (2019) and Richter et al. (2020) convergence has been observed.

Algorithm 2

Input: $F \in \mathbb{R}^{n \times m}$.
Input Filters: $(A, \underline{B}, \underline{\widetilde{B}}) \in \Gamma_{1+2P}$.
Customizable Parameters: $\beta \in \mathbb{R}_+, \kappa \in \{1, 2\}$.
Initialization: $U^{(0)} = F \in \mathbb{R}^{n \times m}, \underline{\lambda}^{(1)} = \underline{0} \in \Gamma_P$.
for $\tau = 1, 2, \ldots$ **do**

$$\underline{W}^{(\tau)} = S_\kappa \left(\underline{\mathfrak{C}}_{\underline{B}} \left(U^{(\tau-1)} \right) - \frac{1}{\beta} \underline{\lambda}^{(\tau)}; \frac{1}{\beta} \right),$$

$$U^{(\tau)} = \mathfrak{C}_A(F) + \underline{\mathfrak{C}}_{\underline{\widetilde{B}}}^* \left(\underline{W}^{(\tau)} + \frac{1}{\beta} \underline{\lambda}^{(\tau)} \right),$$

$$\underline{\lambda}^{(\tau+1)} = \underline{\lambda}^{(\tau)} + \beta \left(\underline{W}^{(\tau)} - \underline{\mathfrak{C}}_{\underline{B}} (U^{(\tau)}) \right).$$

end for

Filter Design Using Factor Families

We conclude by reporting on weakly factoring filters proposed for Problem 4 in Richter (2019) and Richter et al. (2020). These filters lead to cartoon texture separation with desirable properties (keeping edges, removing texture and no blurring, caveat: mosaic pattern appearing); see Fig. 7. The construction is based heuristically on the filter A of the ROF model derived via (11) given by

$$\mathfrak{C}_A = \mu (\mu + \beta \underline{\mathfrak{C}}_{\underline{D}}^* \underline{\mathfrak{C}}_{\underline{D}})^{-1}.$$

As $\underline{\mathfrak{C}}_{\underline{D}}$ is a discrete gradient, the operator $\underline{\mathfrak{C}}_{\underline{D}}^* \underline{\mathfrak{C}}_{\underline{D}}$ is a discrete Laplace operator. The filter A can now be recast by the Laplacian B-spline ϕ defined in Van De Ville et al. (2005) given in the frequency domain by

$$\widehat{\phi}(x, y) := \left(\frac{4 \left(\sin^2 \left(\frac{x}{2} \right) + \sin^2 \left(\frac{y}{2} \right) \right) - \frac{8}{3} \left(\sin \left(\frac{x}{2} \right) \sin \left(\frac{y}{2} \right) \right)}{(x^2 + y^2)} \right)^{\frac{\gamma}{2}}. \quad (14)$$

In Van De Ville et al. (2005) the function ϕ served as a scaling function to construct bi-orthogonal wavelets. Doing a similar construction $(\underline{B}, \underline{\widetilde{B}})$ can be obtained by a bi-orthogonal, directional wavelet frames construction (for the exact construction, see Richter et al. 2020, Appendix C of Richter 2019 and also Mallat 2008; Unser and Ville 2010). Note that if one were to use orthogonal wavelet frames we would be in the realm of strongly factoring, which is exactly the case for the Gabor wavelet frames proposed by Aujol and Gilboa (2006). The heuristic derivation of the new filter A, elaborated above, draws on a similar connection as in Cai et al. (2012), where the discrete gradient $\underline{\mathfrak{C}}_{\underline{D}}$ is recast as a Haar wavelet frame, the first order cardinal B-spline (e.g., Chui 1992).

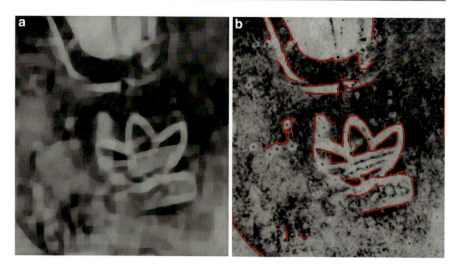

Fig. 7 Applying Algorithm 2 with filter families based on (14) in Problem 4 to the shoeprint detail (from Fig. 2), featuring sharp edges, little blurring, and minimal loss of contrast (left). From this cartoon picture, shoeprint elements are detected by a classical edge detection (Canny 1986) filter (right). For instance, the wear effect (called *accidental*, cf. section "Applications and Challenges for Automated Image Decomposition") on the left of the brand's logo is no longer part of the corresponding element's edge

Conclusion

With advanced computational power and increased numbers of training images, learning methods have entered the field of image analysis and image decomposition. While in fingerprint recognition, automated methods have been around for decades, for forensics applications (latent shoeprint or fingerprint images of bad quality from crime scenes) such methods are far more difficult to design, due to the great heterogeneity of real life use case images. This calls for the development of

(1) Highly flexible families of filters
(2) Corresponding minimization/feasibility problems with solution guarantees
(3) Corresponding algorithms with convergence guarantees

Additionally, since the use case is often defined only indirectly (e.g., improved quality results by improved matching rates, as in Richter et al. 2019), this calls for the development of

(4) Learning methods based on objective functions, only indirectly available

In this chapter we have given a short survey on current research with emphasis on a recent development that seems promising in view of the above-stated goals (1)–(4).

Acknowledgments The authors thank the anonimous referee for the valuable comments and the first and last author gratefully acknowledge funding by the DFG within the RTG 2088.

References

Nist fingerprint quality (NFIQ) (2015) https://www.nist.gov/services-resources/software/nist-biometric-image-software-nbis. Accessed: 2017-12-04

Alvarez, L., Lions, P.-L., Morel, J.-M.: Image selective smoothing and edge detection by nonlinear diffusion. II. SIAM J. Numer. Anal. **29**(3), 845–866 (1992)

Andreu, F., Ballester, C., Caselles, V., Mazón, J.M.: Minimizing total variation flow. Differ. Integral Equ. **14**(3), 321–360 (2001)

Arridge, S., Maass, P., Öktem, O., Schönlieb, C.-B.: Solving inverse problems using data-driven models. Acta Numerica **28**, 1–174 (2019)

Aubert, G., Kornprobst, P.: Mathematical problems in image processing, volume 147 of *Applied Mathematical Sciences*. Springer, New York, 2nd edn. Partial differential equations and the calculus of variations, With a foreword by Olivier Faugeras (2006)

Aujol, J.-F., Aubert, G., Blanc-Féraud, L., Chambolle, A.: Image decomposition into a bounded variation component and an oscillating component. J. Math. Imaging Vision **22**(1), 71–88 (2005)

Aujol, J.-F., Chambolle, A.: Dual norms and image decomposition models. Int. J. Comput. Vis. **63**(1), 85–104 (2005)

Aujol, J.-F., Gilboa, G.: Constrained and SNR-based solutions for TV-Hilbert space image denoising. J. Math. Imaging Vision **26**(1–2), 217–237 (2006)

Aujol, J.-F., Gilboa, G., Chan, T., Osher, S.: Structure-texture image decomposition—modeling, algorithms, and parameter selection. Int. J. Comput. Vis. **67**(1), 111–136 (2006)

Bartůněk, J., Nilsson, M., Sällberg, B., Claesson, I.: Adaptive fingerprint image enhancement with emphasis on preprocessing of data. IEEE Trans. Image Process. **22**(2), 644–656 (2013)

Bauschke, H.H., Combettes, P.L.: Convex analysis and monotone operator theory in Hilbert spaces. CMS Books in Mathematics/Ouvrages de Mathématiques de la SMC. Springer, New York. With a foreword by Hédy Attouch (2011)

Bazen, A., Gerez, S.: Systematic methods for the computation of the directional fields and singular points of fingerprints. IEEE Trans. Pattern Anal. Mach. Intell. **24**(7), 905–919 (2002)

Bengio, Y.: Learning deep architectures for AI. Found. Trends Mach. Learn. **2**(1), 1–127 (2009)

Bertsekas, D.P.: Constrained Optimization and Lagrange Multiplier Methods. Computer Science and Applied Mathematics. Academic Press, Inc. [Harcourt Brace Jovanovich, Publishers], New York/London (1982)

Bigun, J.: Vision with Direction. Springer, Berlin/Germany (2006)

Boyd, S., Vandenberghe, L.: Convex Optimization. Cambridge University Press, Cambridge (2004)

Bredies, K., Dong, Y., Hintermüller, M.: Spatially dependent regularization parameter selection in total generalized variation models for image restoration. Int. J. Comput. Math. **90**(1):109–123 (2013)

Bredies, K., Kunisch, K., Pock, T.: Total generalized variation. SIAM J. Imag. Sci. **3**(3), 492–526 (2010)

Breiman, L.: Random forests. Mach. Learn. **45**, 5–32 (2001)

Buades, A., Le, T., Morel, J.-M., Vese, L.: Fast cartoon + texture image filters. IEEE Trans. Image Process. **19**(8), 1978–1986 (2010)

Burger, M., Gilboa, G., Moeller, M., Eckardt, L., Cremers, D.: Spectral decompositions using one-homogeneous functionals. SIAM J. Imag. Sci. **9**(3), 1374–1408 (2016)

Cai, J.-F., Dong, B., Osher, S., Shen, Z.: Image restoration: total variation, wavelet frames, and beyond. J. Am. Math. Soc. **25**(4), 1033–1089 (2012)

Calatroni, L., Cao, C., De Los Reyes, J.C., Schönlieb, C.-B., Valkonen, T.: Bilevel approaches for learning of variational imaging models. Variational Meth Imaging Geometric Control **18**(252), 2 (2017)

Candès, E., Demanet, L., Donoho, D., Ying, L.: Fast discrete curvelet transforms. Multiscale Model. Simul. **5**(3), 861–899 (2006)

Canny, J.: A computational approach to edge detection. IEEE Trans. Pattern Anal. Mach. Intell. **8**(6), 679–698 (1986)

Caselles, V., Chambolle, A., Novaga, M.: Total variation in imaging. In Handbook of Mathematical Methods in Imaging. Vol. 1, 2, 3. Springer, New York (2015), pp. 1455–1499

Chambolle, A.: An algorithm for total variation minimization and applications. J. Math. Imaging Vision **20**(1–2), 89–97 (2004)

Chambolle, A., Lions, P.-L.: Image recovery via total variation minimization and related problems. Numerische Mathematik **76**(2), 167–188 (1997)

Chambolle, A., Pock, T.: A first-order primal-dual algorithm for convex problems with applications to imaging. J. Math. Imaging Vision **40**(1), 120–145 (2011)

Chambolle, A., Pock, T.: An introduction to continuous optimization for imaging. Acta Numerica **25**, 161–319 (2016)

Chen, Y., Pock, T.: Trainable nonlinear reaction diffusion: A flexible framework for fast and effective image restoration. IEEE Trans. Pattern Anal. Mach. Intell. **39**(6), 1256–1272 (2017)

Chikkerur, S., Cartwright, A., Govindaraju, V.: Fingerprint image enhancement using STFT analysis. Pattern Recogn. **40**(1), 198–211 (2007)

Chui, C.K.: An introduction to Wavelets, Volume 1 of Wavelet Analysis and its Applications. Academic Press, Inc., Boston (1992)

Darbon, J., Sigelle, M.: Image restoration with discrete constrained total variation. I. Fast and exact optimization. J. Math. Imaging Vision **26**(3), 261–276 (2006a)

Darbon, J., Sigelle, M.: Image restoration with discrete constrained total variation. II. Levelable functions, convex priors and non-convex cases. J. Math. Imaging Vision **26**(3), 277–291 (2006b)

Daubechies, I.: Ten lectures on wavelets, volume 61 of CBMS-NSF Regional Conference Series in Applied Mathematics. Society for Industrial and Applied Mathematics (SIAM), Philadelphia (1992)

De los Reyes, J.C., Schönlieb, C.-B.: Image denoising: learning the noise model via nonsmooth PDE-constrained optimization. Inverse Probl. Imaging **7**(4), 1183–1214 (2013)

Donoho, D.L., Johnstone, J.M.: Ideal spatial adaptation by wavelet shrinkage. Biometrika **81**(3), 425–455 (1994)

Eckstein, J., Bertsekas, D.P.: On the Douglas-Rachford splitting method and the proximal point algorithm for maximal monotone operators. Math. Program. **55**(3, Ser. A), 293–318 (1992)

Frick, K., Marnitz, P., Munk, A., et al. Statistical multiresolution dantzig estimation in imaging: fundamental concepts and algorithmic framework. Electron. J. Stat. **6**, 231–268 (2012)

Garnett, J.B., Le, T.M., Meyer, Y., Vese, L.A.: Image decompositions using bounded variation and generalized homogeneous Besov spaces. Appl. Comput. Harmon. Anal. **23**(1), 25–56 (2007)

Garris, M.D., McCabe, R.M.: Nist special database 27: Fingerprint minutiae from latent and matching tenprint images. Technical Report 6534, National Institute of Standards and Technology, Gaithersburg (2000)

Gilboa, G.: A total variation spectral framework for scale and texture analysis. SIAM J. Imag. Sci. **7**(4), 1937–1961 (2014)

Goldstein, T., O'Donoghue, B., Setzer, S., Baraniuk, R.: Fast alternating direction optimization methods. SIAM J. Imag. Sci. **7**(3), 1588–1623 (2014)

Goldstein, T., Osher, S.: The split Bregman method for $L1$-regularized problems. SIAM J. Imag. Sci. **2**(2), 323–343 (2009)

Gottschlich, C.: Curved-region-based ridge frequency estimation and curved Gabor filters for fingerprint image enhancement. IEEE Trans. Image Process. **21**(4), 2220–2227 (2012)

Gottschlich, C., Huckemann, S.: Separating the real from the synthetic: Minutiae histograms as fingerprints of fingerprints. IET Biom. **3**(4), 291–301 (2014)

Gottschlich, C., Mikaelyan, A., Olsen, M., Bigun, J., Busch, C.: Improving fingerprint alteration detection. In: Proceedings of 9th International Symposium on Image and Signal Processing and Analysis (ISPA 2015), pp. 83–86, Zagreb (2015)

Gottschlich, C., Schönlieb, C.-B.: Oriented diffusion filtering for enhancing low-quality fingerprint images. IET Biom. **1**(2), 105–113 (2012)

Gragnaniello, D., Poggi, G., Sansone, C., Verdoliva, L.: Wavelet-Markov local descriptor for detecting fake fingerprints. Electron. Lett. **50**(6), 439–441 (2014)

Grossmann, T.G., Korolev, Y., Gilboa, G., Schönlieb, C.-B.: Deeply learned spectral total variation decomposition. arXiv preprint arXiv:2006.10004 (2020)

Hait, E., Gilboa, G.: Spectral total-variation local scale signatures for image manipulation and fusion. IEEE Trans. Image Process. **28**(2), 880–895 (2018)

Hestenes, M.R.: Multiplier and gradient methods. J. Optim. Theory Appl. **4**, 303–320 (1969)

Hopper, T., Brislawn, C., Bradley, J.: WSQ gray-scale fingerprint image compression specification. Technical report, Federal Bureau of Investigation (1993)

Horesh, D., Gilboa, G.: Separation surfaces in the spectral tv domain for texture decomposition. IEEE Trans. Image Process. **25**(9), 4260–4270 (2016)

Kennedy, J.R.E., Shi, Y.: Swarm Intelligence. Academic, San Diego (2001)

Le, T.M., Vese, L.A.: Image decomposition using total variation and div(BMO). Multiscale Model. Simul. **4**(2), 390–423 (2005)

LeCun, Y., Bengio, Y., Hinton, G.: Deep learning. Nature **521**(7553), 436–444 (2015)

Ma, J., Plonka, G.: The curvelet transform. IEEE Signal Process. Mag. **27**(2), 118–133 (2010)

Mallat, S.: A Wavelet Tour of Signal Processing. Academic, San Diego (2008)

Maltoni, D., Maio, D., Jain, A.K., Prabhakar, S.: Handbook of Fingerprint Recognition. Springer, London (2009)

Meyer, Y.: Oscillating Patterns in Image Processing and Nonlinear Evolution Equations. American Mathematical Society, Boston (2001)

Moeller, M., Diebold, J., Gilboa, G., Cremers, D.: Learning nonlinear spectral filters for color image reconstruction. In: Proceedings of the IEEE International Conference on Computer Vision, pp. 289–297 (2015)

Mumford, D., Shah, J.: Optimal approximations by piecewise smooth functions and associated variational problems. Commun. Pure Appl. Math. **42**(5), 577–685 (1989)

Neal, R.M.: Bayesian Learning for Neural Networks, Vol. 118. Springer Science & Business Media (2012)

Osher, S., Burger, M., Goldfarb, D., Xu, J., Yin, W.: An iterative regularization method for total variation-based image restoration. Multiscale Model. Simul. **4**(2), 460–489 (2005)

Osher, S., Solé, A., Vese, L.: Image decomposition and restoration using total variation minimization and the H^{-1} norm. Multiscale Model. Simul. **1**(3), 349–370 (2003)

Papafitsoros, K., Bredies, K.: A study of the one dimensional total generalised variation regularisation problem. Inverse Prob. Imaging **9**(2), 511 (2015)

Perona, P.: Orientation diffusions. IEEE Trans. Image Process. **7**(3), 457–467 (1998)

Perona, P., Malik, J.: Scale-space and edge detection using anisotropic diffusion. IEEE Trans. Pattern Anal. Mach. Intell. **12**(7), 629–639 (1990)

Powell, M.J.D.: A method for nonlinear constraints in minimization problems. In: Optimization (Symposium, University of Keele, Keele, 1968), pp. 283–298. Academic, London (1969)

Richter, R.: Cartoon-Residual Image Decompositions with Application in Fingerprint Recognition. Ph.D. thesis, Georg-August-University of Goettingen (2019)

Richter, R., Gottschlich, C., Mentch, L., Thai, D., Huckemann, S.: Smudge noise for quality estimation of fingerprints and its validation. IEEE Trans. Inf. Forensics Secur. **14**(8), 1963–1974 (2019)

Richter, R., Thai, D.H., Huckemann, S.: Generalized intersection algorithms with fixpoints for image decomposition learning. arXiv preprint arXiv:2010.08661 (2020)

Rudin, L., Osher, S., Fatemi, E.: Nonlinear total variation based noise removal algorithms. Physica D **60**(1–4), 259–268 (1992)

Scherzer, O., Grasmair, M., Grossauer, H., Haltmeier, M., Lenzen, F.: Variational Methods in Imaging. Springer (2009)

Schmidt, M.F., Benning, M., Schönlieb, C.-B.: Inverse scale space decomposition. Inverse Prob. **34**(4), 1–34 (2018)

Schoelkopf, B., Smola, A.: Learning with Kernels. MIT Press, Cambridge (2002)

Shen, J.: Piecewise $H^{-1} + H^0 + H^1$ images and the Mumford-Shah-Sobolev model for segmented image decomposition. AMRX Appl. Math. Res. Express (4), 143–167 (2005)

Steidl, G., Weickert, J., Brox, T., Mrázek, P., Welk, M.: On the equivalence of soft wavelet shrinkage, total variation diffusion, total variation regularization, and sides. SIAM J. Numer. Anal. **42**(2), 686–713 (2004)

Strong, D., Chan, T.: Edge-preserving and scale-dependent properties of total variation regularization. Inverse Prob. **19**(6), S165–S187 (2003). Special section on imaging

Thai, D., Gottschlich, C.: Directional global three-part image decomposition. EURASIP J. Image Video Process. **2016**(12), 1–20 (2016a)

Thai, D., Gottschlich, C.: Global variational method for fingerprint segmentation by three-part decomposition. IET Biom. **5**(2), 120–130 (2016b)

Thai, D., Huckemann, S., Gottschlich, C.: Filter design and performance evaluation for fingerprint image segmentation. PLoS ONE **11**(5), e0154160 (2016)

Turroni, F., Maltoni, D., Cappelli, R., Maio, D.: Improving fingerprint orientation extraction. IEEE Trans. Inf. Forensics Secur. **6**(3), 1002–1013 (2011)

Unser, M., Ville, D.V.D.: Wavelet steerability and the higher-order Riesz transform. IEEE Trans. Image Process. **19**(3), 636–652 (2010)

Van De Ville, D., Blu, T., Unser, M.: Isotropic polyharmonic B-splines: scaling functions and wavelets. IEEE Trans. Image Process. **14**(11), 1798–1813 (2005)

Vese, L., Osher, S.: Modeling textures with total variation minimization and oscillatory patterns in image processing. J. Sci. Comput. **19**(1–3), 553–572 (2003)

Weickert, J.: Anisotropic Diffusion in Image Processing. Teubner, Stuttgart (1998)

Weickert, J.: Coherence-enhancing diffusion filtering. Int. J. Comput. Vis. **31**(2/3), 111–127 (1999)

Wiesner, S., Kaplan-Damary, N., Eltzner, B., Huckemann, S.F.: Shoe prints: The path from practice to science. In: Banks, D., Kafadar, K., Kaye, D. (eds.) Handbook of Forensic Statistics, pp. 391–410. Springer (2020a)

Wiesner, S., Shor, Y., Tsach, T., Kaplan-Damary, N., Yekutieli, Y.: Dataset of digitized racs and their rarity score analysis for strengthening shoeprint evidence. J. Forensic Sci. **65**(3), 762–774 (2020b)

Wu, C., Tai, X.-C.: Augmented Lagrangian method, dual methods, and split Bregman iteration for ROF, vectorial TV, and high order models. SIAM J. Imag. Sci. **3**(3), 300–339 (2010)

Yang, Y., Sun, J., Li, H., Xu, Z.: Deep ADMM-net for compressive sensing MRI. In: 30th Conference on Neutral Information Processing Systems (NIPS 2016), pp. 10–18 (2016)

Yao, Z., Le Bars, J.-M., Charrier, C., Rosenberger, C.: A literature review of fingerprint quality assessment and its evaluation. IET J. Biom. **5**(3), 243–251 (2016)

Zeune, L., van Dalum, G., Terstappen, L.W., van Gils, S.A., Brune, C.: Multiscale segmentation via bregman distances and nonlinear spectral analysis. SIAM J. Imaging Sci. **10**(1), 111–146 (2017)

ns in X-Ray Tomography

Johannes Schwab

Contents

Introduction	1184
Background	1185
Tomographic Image Reconstruction	1186
Deep Learning	1187
Case Examples in X-Ray CT	1190
Limited Angle Computed Tomography	1192
Reduction of Metal Artefacts	1196
Low-Dose Computed Tomography	1197
Further Methods	1198
Conclusion	1199
References	1200

Abstract

Successful medical diagnosis heavily relies on the reconstruction and analysis of images showing organs, bones, and other structures in the interior of the human body. In the last couple of years, the stored image data has increased tremendously, and also the computing power of modern GPUs experienced huge progress. Machine learning methods, and in particular deep learning methods, are on the rise to tackle advanced image reconstruction and image analysis tasks to support medical doctors in their diagnostic routines. In this chapter, we focus on the reconstruction task; especially consider tomographic imaging problems with incomplete, corrupted, or noisy data; and demonstrate how deep learning methods enable us to solve such tasks in a unified manner. We present the basic ideas of these methods assuming paired training data (supervised learning) and

J. Schwab (✉)
Department of Mathematics, University of Innsbruck, Innsbruck, Austria
e-mail: schwab@mrc-lmb.cam.ac.uk

utilizing only feed-forward networks. In particular, we illustrate the underlying concepts for missing data problems in classical computed tomography (CT), noting that most of the concepts can be transferred to other inverse imaging problems.

Keywords

Computed tomography · Deep learning · Inverse problem · Limited Data · Regularization

Introduction

Most modern medical imaging methods rely on the solution of an inverse problem, meaning that for given measured data $g \in \mathcal{Y}$ and physics-based forward model $\mathbf{R} \colon \mathcal{X} \to \mathcal{Y}$, the task is to estimate the cause $f \in \mathcal{X}$ for the observed measurements under the model $\mathbf{R}$. In an ideal setting, this amounts in solving the following task:

$$\text{Find } f \text{ from measurements } \quad g = \mathbf{R}(f). \tag{1}$$

In tomographic imaging, the space $\mathcal{X}$ is typically a space of functions $f \colon \Omega \to \mathbb{C}$, where the domain Ω denotes a subset of $\mathbb{R}^2$ (slice) or $\mathbb{R}^3$. The corresponding model $\mathbf{R}$ is an operator modelling the physical effects used for the tomographic modality. In computed tomography, $\mathbf{R}$ describes the absorption of X-ray radiation in the investigated tissue (Hounsfield 1973), whereas in magnetic resonance imaging, $\mathbf{R}$ describes the excitation and detection of radio-frequency signals of hydrogen atoms in the human tissue (Purcell et al. 1946). Tomographic imaging includes a great variety of applications in different fields, for example, electrical impedance tomography, optical tomography, positron emission tomography, seismic tomography, ultrasound tomography, and many more. In most of these applications, the forward model can be described by Radon type transforms, which use integrals over different families of one-dimensional manifolds. This can be integrals along lines as it is the case in X-ray transmission tomography (Natterer 2001), or integrals over circles in photoacoustic tomography (Beard 2011). In the following, we will present data-driven reconstruction methods based on three typical examples of ill-posedness in classical computed tomography. We assume that paired training data are available and, to make the article more readable, restrict ourselves to feed-forward neural networks. In principle, however, more complex network architectures, which are constantly being developed and improved, could be employed as well. Also the ideas illustrated on the example problems in this articles can be adopted to missing data problems in different inverse imaging applications.

Given the operator $\mathbf{R} \colon \mathcal{X} \to \mathcal{Y}$, in an ideal world, data would be given by $g = \mathbf{R}(f)$. However, in the real world, this is not the case, and $\mathbf{R}(f)$ is corrupted and modified by several sources. In this chapter, we consider and review three different

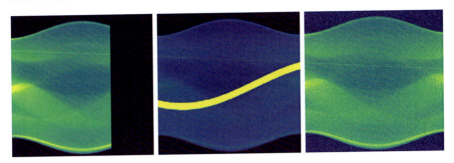

Fig. 1 Different sources of imperfect data in tomographic imaging. LEFT, incomplete data (e.g., limited angle CT); MIDDLE, corrupted data due to high-intensity region (e.g., a metal artifact); RIGHT, noisy data (e.g., low-dose CT)

frequently occurring problems in tomographic imaging, which can essentially be formulated as follows:

- Only **incomplete data** is available, meaning that only parts of the complete measurement data are given (Fig. 1).
- Partially **corrupted data** is measured. Here parts of the measurements are affected by physical effects not modelled by **R** (Fig. 1).
- Presence of strong **noise in the data**. Physical measurements are inevitably affected by statistical uncertainty; therefore, the measured data cannot be fully described by the model **R** (Fig. 1).

All of these scenarios typically lead to ill-posed inverse problems, where the reconstruction is either non-unique, the reconstruction process unstable, or the data not in the range of the operator **R**. These issues can be analyzed by mathematical regularization theory (Engl et al. 1996). Incomplete data and partially corrupted data can lead to severe artifacts in the reconstruction. The noise in the data is propagating to the reconstructed image and can be severely amplified in the reconstruction process if the inverse of the forward operator is discontinuous. In all of these problems, exact direct reconstruction methods are either unavailable or lead to strong degradation of the reconstructed images. Iterative methods are extremely flexible and show good performance in all three cases, but come with very high computational cost. Deep Learning offers an alternative approach that can achieve good performance while being computationally efficient (Wang 2016).

Background

We begin with a brief description of the inverse problem in computed tomography and the three limited data problems mentioned earlier. Subsequently, we present the very basics of deep learning as well as a definition of feed-forward neural networks.

Tomographic Image Reconstruction

Analytic Reconstruction Methods

Common analytic reconstruction methods for tomographic imaging refer to numerical implementations of analytic inversion formulas and are of particular interest in application because they can be efficiently implemented. Most explicit inversion formulas for an operator $\mathbf{R}$ are based on its adjoint operator $\mathbf{R}^*$ or defined by some infinite series expansion. For many tomographic imaging problems, exact inversion formulas exist under the assumption that full, perfect data is available. Nevertheless, these inversion formulas only hold for specific scenarios. If the data is incomplete or not in the range of $\mathbf{R}$, the inversion formulas are not valid. As a consequence, the reconstructions are bad, if the data deviate from the mathematical model from which the inversion formula has been obtained. In addition, it is often challenging to incorporate existing prior knowledge into direct methods. To address these issues, iterative reconstruction methods can be used. As it turned out, deep learning constitutes a great opportunity to improve analytic image reconstruction methods (Fig. 2).

Iterative Reconstruction Methods

In contrast to direct methods, iterative methods rely on optimization tools for finding the minimizer f^* of a functional depending on the data g

$$f^* = \arg\min_{f \in \mathcal{X}} \|g - \mathbf{R}(f)\|_{\mathcal{Y}}^2.$$

Minimization problems of this type can be solved by various iterative methods. For example, assuming that $\mathbf{R}$ is a linear operator between Hilbert spaces $\mathcal{X}$ and $\mathcal{Y}$ an iterative solution method is Landwebers algorithm (Landweber 1951). This algorithm is defined by the update formula

$$f_{k+1} = f_k + \mathbf{R}^*(g - \mathbf{R}(f_k)).$$

If g is in the domain of the Moore-Penrose inverse $\mathbf{R}^+$ defined by

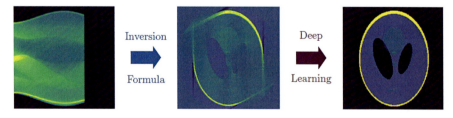

Fig. 2 Basic deep learning approach to improve analytic image reconstruction. First an analytic inversion method (derived for ideal data) is applied. In a second step, a deep learning algorithm is used to improve the initial reconstruction

$$\mathbf{R}^+ : \mathcal{Y} \supset \operatorname{range}(\mathbf{R}) \oplus \operatorname{range}(\mathbf{R})^\perp \to \mathcal{X}$$
$$y \mapsto \arg\min\{\|x\|_{\mathcal{X}} \mid x \in \mathcal{X} \wedge \mathbf{R}^*\mathbf{R}x = \mathbf{R}^*y\}$$

the sequence f_k converges to the minimum norm least squares solution $\mathbf{R}^+(g)$ of the inverse problem (1) (Engl et al. 1996).

One advantage of indirect methods is that they are very flexible and one can easily add a penalty term $\mathscr{P} : \mathcal{X} \to [0, \infty]$ to obtain solutions that have specific characteristics (prior knowledge) by finding

$$f^* \in \arg\min_{f \in \mathcal{X}} \|g - \mathbf{R}(f)\|_{\mathcal{Y}}^2 + \mathscr{P}(f). \tag{2}$$

Such an approach for solving inverse problems is called variational regularization (Scherzer et al. 2009) or generalized Tikhonov regularization.

A popular choice of penalty term, also called regularizer, is the total variation (TV) of a function f, or some functional that enforces sparsity in a given basis or frame (Acar and Vogel 1994; Daubechies et al. 2004). Also penalty terms adapted to a data set of known solutions have been considered to describe signal characteristics for the class of desired solutions. For example, learning a basis or dictionary for signals to be recovered in which the reconstruction should have a spars representation was proposed (Elad 2010). Further regularizers that are represented by deep neural networks have been proposed as well. A mathematical analysis of methods using learned regularizers has been developed in Lunz et al. (2018), Mukherjee et al. (2020), Li et al. (2020), and Obmann et al. (2020). Recently also, data-driven iterative algorithms serving to minimize (2) were introduced (Adler and Öktem 2017, 2018) and applied to various types of inverse problems (Wu et al. 2019; Guazzo 2020; Boink et al. 2019).

In the next subsection, we will provide a brief introduction to deep learning.

Deep Learning

In machine learning, the goal is to solve a given problem based on available observations. Analogous to physicists trying to explain the universe, for given observational data, one wants to find a model (or a theory in physics) that explains this data. But explaining data alone is not the most difficult challenge, since this can always be achieved with a model of sufficient complexity. For a good model, the real demand consists in enabling it to generate to new, unseen data and to make predictions. In recent years, a lot of research has been done on how such models can be calculated. An overview of common methods can be found, for example, in Goodfellow et al. (2016), Hastie et al. (2009), and LeCun et al. (2015).

Roughly speaking, machine learning tasks can be classified in Goodfellow et al. (2016).

- **Supervised learning:** Here the input to the task and the corresponding solution are known for the training set. Therefore, the training set consists of a subset of $\mathcal{A} \times \mathcal{B}$ where the training pairs are coupled by the problem to be solved.
- **Unsupervised learning:** No paired data set is available, and the training set only consists of the inputs; the solutions or even the concrete task is unknown. The training set consist of a subset of $\mathcal{A}$ assumed to have some particular property which is to be discovered.

In this chapter, we exclusively focus on supervised learning tasks, which are described in the following. A model for solving a problem can be interpreted as an operator $\Phi: \mathcal{A} \to \mathcal{B}$. If the given data consists of input instances $(a_i)_{i=1}^N$ and the corresponding solutions $(b_i)_{i=1}^N$ fitting the data means finding and operator

$$\Phi^* = \underset{\Phi: \mathcal{A} \to \mathcal{B}}{\arg \min} \frac{1}{N} \sum_{i=1}^N \mathcal{D}(\Phi(a_i), b_i),$$

where $\mathcal{D}$ is some similarity measure in the space $\mathcal{B}$. However, a model Φ^* also has to predict meaningful solutions for data different from the data used for the fitting. A model, which is unable to make predictions, is more or less useless. To achieve this, the class of admissible operators is restricted to a subset $\mathcal{C}$ of all mappings $\Phi: \mathcal{A} \to \mathcal{B}$. In practice, additional strategies are adopted in the optimization procedure in order to restrict the class of possible solution operators.

The ultimate goal for the application is to implement a computer program, which finds a good approximation of the operator that is able to solve some specific task, as, for example, image analysis and image reconstruction tasks in medical imaging (Wang 2016). This model optimization is also termed learning of the model. For this purpose, the user has to feed the computer with experience, called training data, for example, images or measurement data.

We now introduce a popular approach of setting up such a task-solving machinery for supervised learning problems. The approach consists in parametrizing the function, which maps a given input to the solution of the problem. A particular class of such functions is called artificial neural networks (Werbos 1974).

After discretization of the spaces $\mathcal{A} := \mathbb{R}^L$ and $\mathcal{B} := \mathbb{R}^M$, a feed-forward artificial neural network is given by

$$\Phi_\mathcal{W}: \mathbb{R}^L \to \mathbb{R}^M$$
$$a \mapsto \Phi_\mathcal{W}(a) := (\sigma_K \circ \mathbf{W}_K \circ \sigma_{K-1} \circ \mathbf{W}_{K-1} \circ \ldots \circ \sigma_1 \circ \mathbf{W}_1)(a), \qquad (3)$$

where $\mathbf{W}_i: \mathbb{R}^{n_i} \to \mathbb{R}^{n_{i+1}}$ and $\sigma_i: \mathbb{R}^{n_i} \to \mathbb{R}^{n_i}$ for $i \in \{1, \ldots, K\}$ are affine linear operators and point-wise nonlinear mappings, respectively. Further $\mathcal{W}$ denotes the dependence of the function $\Phi_\mathcal{W}$ on the operators $\mathbf{W}_i$. We consider the real vector spaces $\mathbb{R}^{n_1} = \mathbb{R}^L$ and $\mathbb{R}^{n_K} = \mathbb{R}^M$ as input and output spaces. Networks of the form (3) are called feed-forward networks as they have a sequential, forward directed structure. We note that a great variety of more complex network architectures exist that, for example, also allow cycles or loops. In all of what follows, the network architecture is not essential, and everything can equally be applied to more sophisticated network designs. In a feed-forward neural network, fixing the depth K, the dimension of the intermediate spaces $\mathbb{R}^{n_i}$, and the functions σ_i gives a class of operators only depending on the parameters of the affine linear functions $\mathbf{W}_i$. These parameters are the entries of the matrices and are called weights of the artificial neural network. A particular choice of the linear operators is discrete convolution operators. One of the main advantages of convolutions is that the corresponding matrices only contain a small number of nonzero weights, which is computationally much more efficient than using full matrices (fully connected layers). Networks consisting of such discrete convolutions are called convolutional neural networks and are of particular interest for imaging tasks since they are able to detect local correlations. Further if K is not very small, a neural network is called deep, although there is no strict definition of when a network is considered to be deep. A typical choice for the nonlinear mappings σ_i is the rectified linear unit (ReLU)

$$\sigma(x) = \mathrm{ReLU}(x) := \max\{0, x\},$$

or sigmoid functions.

Given a set of training data $(a_i, b_i)_{i=1}^{N}$, the goal now is to find good linear operators, such that the neural network fits the training data and is able to generalize the learned expertise. If we denote the vector of weights by $\mathcal{W} := (\mathbf{W}_1, \ldots, \mathbf{W}_N)$, the corresponding minimization problem can be formulated by

$$\text{Find } \Phi_\mathcal{W} \text{ minimizing} \quad \mathcal{L}(\mathcal{W}) := \frac{1}{N} \sum_{i=1}^{N} \mathbf{D}(\Phi_\mathcal{W}(a_i), b_i), \qquad (4)$$

where $\mathbf{D}: \mathbb{R}^M \times \mathbb{R}^M \to [0, \infty)$ is some distance measure on the output space and $\mathcal{L}$ represents the cost function. Assuming that $\mathcal{L}$ admits calculation of a (sub-)gradient, minimization of $\mathcal{L}$ is typically done by gradient descent methods. In these procedures, the parameters are iteratively updated by

$$\mathcal{W} \to \mathcal{W} - \eta \nabla \mathcal{L}(\mathcal{W}),$$

where η is a parameter determining the step size, also called learning rate in machine learning. In practice a much cheaper alternative is deployed which only takes into account the gradient of the cost function $\mathcal{L}$ corresponding to a subset of the training data. Typically these subsets of training instances are randomly selected, resulting in stochastic gradient descent methods. The partial derivatives of the gradient are computed by the backpropagation algorithm (Hecht-Nielsen 1992; Higham and Higham 2019).

Optimization of (4) is challenging, since the cost function is non-convex. Various techniques to an improvement of this optimization process as well as the generalization properties of an artificial neural network have been proposed. These techniques include:

- Evaluating the model with a data set not contained in the training set during training to estimate the generalization capability of the network; this set is typically called **validation set**.
- Including other operations (layers) in the network architecture; some examples are **pooling layers**, which reduce the dimension by taking the maximum or average over a small region of an intermediate output. Further possibilities to improve generalization properties and optimization are **dropout layers** and **batch normalization layers** and also **residual connections** and other skip connections that add or concatenate outputs obtained earlier in the network to inputs in later stages. Detailed explanation of these building blocks can be found in Goodfellow et al. (2016) and Lundervold and Lundervold (2019).
- More sophisticated variants of gradient descent algorithms including **momentum** or **Nesterov updates**; a summary and explanation of popular optimization algorithms are given in Ruder (2016).
- Including a penalty term $\mathscr{P}$ for the weights in the cost function and minimizing

$$\mathscr{L}(\mathcal{W}) = \frac{1}{N} \sum_{i=1}^{N} \mathbf{D}(\Phi_{\mathcal{W}}(a_i), b_i) + \mathscr{P}(\mathcal{W}).$$

The choice of the particular network and optimizer is very important for obtaining the best possible results and depends on the specific task to be solved. Likewise, the choice of the loss function **D** plays an important role to obtain a valuable model. Depending on the specific task, a huge amount of different loss functions have been proposed, ℓ^2, ℓ^1, structural similarity (SSIM) and Wasserstein distance being the most popular, when working with images. In the following, however, we concentrate on illustrating the conceivable application of neural networks rather than on the concrete network design and optimization strategy.

Case Examples in X-Ray CT

To illustrate deep learning methods for tomographic image reconstruction, in the following, we consider the parallel beam geometry for X-ray computed tomography. In this imaging method, the particular mathematical model **R** is the Radon transform, which evaluates the integrals over all lines across the radiative absorption coefficient of the tissue. In this case, the sought-after function f is the spatially depending absorption coefficient, and the measured data follows the physical model.

$$g(\theta, s) = \mathbf{R} f(\theta, s) = \int_{L(\theta, s)} f(x) \, d\sigma(x). \tag{5}$$

Here $\mathbb{S}^{d-1}$ denotes the $(d-1)$–dimensional unit sphere, which declares the direction of the line, $s \in \mathbb{R}$ determines the distance of the line to the origin, and σ denotes the surface measure on $L(\theta, s)$. The data consists of all line integrals along lines $L(\theta, s) := s\theta^\perp + \mathbb{R}\theta$ where $(\theta, s) \in \mathbb{S}^{d-1} \times \mathbb{R}$ (Fig. 3). This operator $\mathbf{R}$ is called Radon transform, and in theory exact inversion of the transform is possible. An extensive overview of the mathematical formulation of X-ray tomography and solution methods can be found, among others, in Natterer (2001), Deans (2007), and Scherzer et al. (2009).

In the following, we will shortly describe common reconstruction methods for X-ray computed tomography.

Analytic Reconstruction

For the Radon transform (5) for $d = 2$, such an exact reconstruction formula (Natterer 2001) is given by

$$f(x) = \frac{1}{4\pi^2} \int_{\mathbb{S}^1} \left(\int_{\mathbb{R}} \frac{\partial_s \mathbf{R} f(\theta, s)}{\theta \cdot x - s} \, ds \right) d\theta, \tag{6}$$

where ∂_s denotes the partial derivative in the s component and $\cdot$ the standard inner product in $\mathbb{R}^2$. Using the Hilbert transform

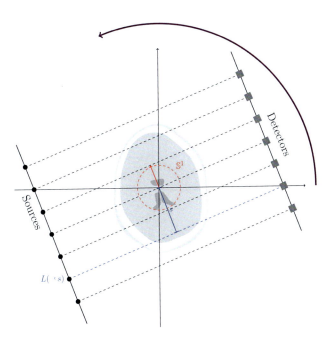

Fig. 3 Illustration of parallel beam CT. The sources and detectors rotate around the object. The vector $\theta \in \mathbb{S}^1$ determines the angle of the parallel lines and the scalar $s \in \mathbb{R}$ the distance of the line to the origin

$$H_s g(\theta, s) = \frac{1}{\pi} \int_{\mathbb{R}} \frac{g(\theta, t)}{s - t} \, dt$$

the inversion formula above can be written as

$$f(x) = \frac{1}{4\pi} \mathbf{R}^* \left(H_s \partial_s \mathbf{R}(f) \right)(x), \tag{7}$$

where $\mathbf{R}^*$ denotes the adjoint of the Radon transform defined by

$$\mathbf{R}^* g(x) = \int_{\mathbb{S}^1} g(\theta, \theta \cdot x) \, d\theta.$$

Here the improper integral arising in the Hilbert transform is understood in the sense of Cauchy. The operation $H_s \partial_s$ is called filtering, whereas the adjoint of the Radon transform is also called backprojection operator. Such inversion formulas of filtered backprojection type are available for different variants of the Radon transform as well, occurring in various tomographic imaging problems.

Analytic reconstruction in practice then consists of implementing a discretized version of the inversion formula. The inversion formula (7) can, for example, be implemented by:

- Approximation of the filtering operation $H_s \partial_s$ by discrete convolution in Fourier domain by a non-singular filter (e.g., Ram-Lak filter, Shepp-Logan filter)
- Interpolation to compute the values of the filtered data at the points $(\theta, \theta \cdot x)$.
- Numerical integration methods to compute the integral over $\mathbb{S}^1$ (backprojection).

Limited Angle Computed Tomography

For some applications, it is favorable to only measure line integrals along a limited range of angles, to reduce scanning time or being able to reduce the scanning area to a smaller region. In some applications, it is even impossible to measure all line integrals around the object under investigation due to physical constraints. Therefore, the data is limited to certain areas which make high-quality reconstructions with simple inversion formulas an unsolvable task. Recently, however, deep learning algorithms have made a huge advance that has made it possible to get a good reconstruction despite the limited data. In the case of limited angle computed tomography, we consider subsets of the form $\Lambda := \Gamma \times \mathbb{R} \subset \mathbb{S}^1 \times \mathbb{R}$, where Γ is a connected subset of $\mathbb{S}^1$ denoting the set of directions in which measurements are available. The set Γ corresponds to the range of measurement angles not covering the full range of 180° required for exact reconstruction. Thus, for example, the set covering only 90° is given by $\Gamma := \{(\sin(\alpha), \cos(\alpha)) \mid \alpha \in [0, \pi/2]\} \subset \mathbb{S}^1$. The restriction of the data can be formulated by $\chi_\Lambda g$, where

$$\chi_\Lambda(\theta, s) = \begin{cases} 1 & \theta \in \Gamma \\ 0 & \theta \notin \Gamma. \end{cases}$$

A reconstruction can then be found by finding a solution f minimizing the penalty term $\mathcal{P}$ and matching the available data

$$f^* = \arg\min_{f \in \mathcal{X}} \|\chi_\Lambda g - \chi_\Lambda \mathbf{R}(f)\|_{\mathcal{Y}} + \mathcal{P}(f). \tag{8}$$

Iterative reconstruction methods employing a penalty term of the image gradient ($\mathcal{P}$ total variation functional) yield satisfying results, but are computationally expensive. Therefore, deep learning methods that can be trained prior to reconstruction are a good option, as they can be employed very quickly to make predictions once their training process is complete.

The two most prominent deep learning methods for improving limited angle tomography can be assigned to two classes: methods that work in data domain and approaches that already use some initial reconstruction.

Learning in Data Domain
Deep learning methods working in data domain do not only aim for minimizing the data fidelity on the restricted data $\|\chi_\Lambda g - \chi_\Lambda \mathbf{R}(f)\|_{\mathcal{Y}}$ but also for finding an extension of the data to the set $\Lambda^c := \left(\mathbb{S}^1 \setminus \Gamma\right) \times \mathbb{R}$.

Given a set of N pairs of data $(\chi_\Lambda g_i, g_i)_{i=1}^N \subset \mathcal{Y} \times \mathcal{Y}$, the goal is to find some data extension operator $\Phi: \mathcal{Y} \to \mathcal{Y}$ in a certain operator class $\mathcal{C}$ that maps $\chi_\Lambda g_i$ to g_i for every training sample. For natural images, this task would consist of image completion. This can be formulated by finding the operator

$$\Phi^* := \arg\min_{\Phi \in \mathcal{C}} \frac{1}{N} \sum_{i=1}^{N} \|\Phi(\chi_\Lambda g_i) - g_i\|_{\mathcal{Y}}, \tag{9}$$

where the norm $\|\cdot\|_{\mathcal{Y}}$ can be replaced by any distance measure $\mathcal{D}_{\mathcal{Y}}$ on $\mathcal{Y}$. Subsequently, given this extension operator Φ^*, one can obtain a reconstruction either by solving

$$f^* = \arg\min_{f \in \mathcal{X}} \|\Phi^*(\chi_\Lambda g) - \mathbf{R}f\|_{\mathcal{Y}} + \mathcal{P}(f),$$

or using a direct reconstruction method for the full operator $\mathbf{R}$. Many algorithms have been proposed to extend the data to the full range of 180°, as, for example, sinogram restoration based on Helgason-Ludwig consistency conditions (Huang et al. 2017) and other consistency conditions. A popular approach is to approximate the extension operator by a fully convolutional network or a generative adversarial network. In Anirudh et al. (2018), the authors propose a 1D convolutional network to generate a latent code from the partial sinogram. Subsequently this latent code is fed to a 2D convolutional generator network, which is optimized together with a discriminator network, rating the generated image. Applying the full-view Radon transform to the generated image yields the projection data for all angles, and the missing values in the original sinogram are replaced by the new data. Typically one wants the extension operator Φ^* to be consistent with the given data, meaning that it does not change the available measured values of the data in Γ.

Learning in Image Domain

A second approach consists in using the limited data for an initial reconstruction (8) which is then refined by a learned operator. Artifacts occurring in limited angle computed tomography have been studied and characterized (Quinto 1988; Frikel and Quinto 2013) for a long time. Since these artefacts are deterministic and have a directional property, deep convolutional networks, which have proven very successful in detecting signal features and patterns, also seem to be suitable for removing limited angle artifacts.

Given a set of N functions in the manifold of desired solutions $\mathcal{M} \subset \mathcal{X}$, one can obtain pairs $(f_i, g_i)_{i=1}^N \subset \mathcal{M} \times \mathbf{R}(\mathcal{M})$ by computing the Radon transform and generate a set of training data $(f_i, \chi_\Lambda g_i)$ by restricting the Radon data to the set Λ. The goal in this approach is to find some operator that removes artefacts in the reconstructions f_i^* obtained by some iterative reconstruction algorithm for finding

$$f_i^* = \arg\min_{f \in \mathcal{X}} \|\chi_\Lambda g_i - \chi_\Lambda \mathbf{R}(f)\|_Y + \mathcal{P}(f).$$

or by some direct reconstruction algorithm (e.g., (6)). The refinement operator can then be calculated by

$$\Phi^* := \arg\min_{\Phi \in \mathcal{C}} \frac{1}{N} \sum_{i=1}^N \mathcal{D}_\mathcal{X}(\Phi(f_i^*), f_i), \qquad (10)$$

for some distance measure $\mathcal{D}_\mathcal{X}$ on $\mathcal{X}$. Here $\mathcal{C}$ again is a class of operators which can be defined by neural networks after discretization. One shortcoming of these post-processing networks is that they depend heavily on the set of training data and are vulnerable to adversarial examples or changes in the noise characteristics (Huang et al. 2018b). Including knowledge of the operator $\mathbf{R}$ within the deep learning can potentially remedy these problems and are discussed in the following.

Using Knowledge of the Operator

For the Radon transform, the missing information in the projection data can be characterized in Fourier domain (Frikel and Quinto 2013). For the reconstructed image from incomplete data, the frequency components in a double wedge are missing. This characterization in frequency domain has been exploited by Gu and Ye (2017) where they apply a directional wavelet transform (conourlets) partitioning the frequency domain and in a further step train a convolutional neural network that estimates the artifacts and finally adds the missing frequencies. A similar approach was proposed by the authors in Bubba et al. (2019). They use a shearlet frame for $\mathcal{X}$ which can be split in visible and invisible coefficients. The shearlet frame is adapted to the operator $\mathbf{R}$ and the set Λ, such that the visible coefficients carry reliable information about the unknown f, whereas the invisible coefficients do not contain relevant information. In a first step, the visible coefficients are obtained by an initial nonlinear reconstruction

$$f^+ \in \arg\min_{f \geq 0} \|\mathcal{S}(f)\|_{1,w} + \frac{1}{2}\|\chi_\Lambda g - \chi_\Lambda \mathbf{R}(f)\|_y.$$

Here $\mathcal{S}$ denotes the analysis operator for the shearlet frame and $\|\cdot\|_{1,w}$ a weighted ℓ^1 norm. In a second step, a neural network Φ^* is trained to estimate the invisible coefficients from the visible ones.

$$r = \Phi^*(\mathcal{S}(f^+)). \tag{11}$$

The final reconstruction is obtained by applying the synthesis operator $\mathcal{S}^*$ of the frame to the reliable visible coefficients combined with the learned invisible coefficients (11)

$$f^* = \mathcal{S}^*\left(\mathcal{S}(f^+) + r\right).$$

Other data-consistent deep learning approaches exploiting the knowledge of the operator were proposed in Schwab et al. (2019a,b) and Boink et al. (2020).

Learned Backprojection
Although some works for fully learned reconstructions for tomographic imaging $\Phi: \mathcal{Y} \to \mathcal{X}$ exist (Zhu et al. 2018; Boink and Brune 2019), they are strongly limited by the size of the images and the data, and for a known forward operator, a fully learned scheme seems inadequate. Nevertheless, it is possible to improve direct inversion methods by deep neural networks. In Würfl et al. (2018), the authors propose a reconstruction framework based on a filtered backprojection algorithm for limited angle tomography. Their framework consists in a weighting layer $\mathbf{W}$, which performs a pixel-wise independent weighting of the projection data, a 1D convolutional layer Φ with a single convolution mimicking the filtering operation in (7), and a backprojection step. The reconstruction is obtained by

$$f^* = \mathbf{R}^*\left(\Phi(\mathbf{W}(g))\right).$$

A similar approach of using the backprojection algorithm as basis for the network for photoacoustic tomographic imaging was studied in Schwab et al. (2018, 2019c), where compensation weights were learned in order to improve reconstruction for limited data problems.

Reduction of Metal Artefacts

In the presence of metal in the investigated tissue located in the region $\Omega_m \subset \mathbb{R}^2$, its radiative attenuation coefficient can be modelled by

$$f = \chi_{\Omega_m^c} f + \chi_{\Omega_m} f.$$

Due to the linearity of the Radon transform, this leads to a composition of the data

$$\mathbf{R} f = \mathbf{R}(\chi_{\Omega_m^c} f) + f_m \mathbf{R}(\chi_{\Omega_m}),$$

where f_m denotes the radiative absorption coefficient of the metal. Most methods for artifact reduction in the presence of metal now aim at finding the set

$$\mathbf{M} := \left\{ (\theta, s) \in \mathbb{S}^1 \times \mathbb{R} \mid \mathbf{R}(\chi_{\Omega_m})(\theta, s) = 0 \right\},$$

consisting of the data which is responsible for the artifacts in the reconstruction. The set $\mathbf{M}$ contains the reliable information of the measured data coming from the non-metal region; therefore, knowledge of this set would give the opportunity to remove the corrupted data in this region and apply a data extension operator. One possible approach to identify this set consists in three steps:

(1) Reconstruction of an image from the raw measurements
(2) Segmentation of metal in the reconstructed image
(3) Application of the forward operator to the segmented image to obtain $\mathbf{M}$.

If the set $\mathbf{M}$ is found, similar to (9), a training set can be generated by computing the Radon transform of the training examples f_i s and the corresponding corrupted data by setting $g_i(\theta, s) = 0$ for $(\theta, s) \notin \mathbf{M}_i$. Here the sets $\mathbf{M}_i$ denote the region of corrupted data for the ith training instance. The extension operator $\Phi^* \colon \mathcal{Y} \to \mathcal{Y}$ should then satisfy

$$\Phi^* := \underset{\Phi \in \mathcal{C}}{\arg\min} \frac{1}{N} \sum_{i=1}^{N} \mathcal{D}_\mathcal{Y} \left(\Phi(\chi_{\mathbf{M}_i} g_i), g_i \right),$$

for the training data $(\chi_{\mathbf{M}_i} g_i, g_i)_{i=1}^{N} \subset \mathcal{Y} \times \mathcal{Y}$ and some distance measure $\mathcal{D}_\mathcal{Y}$ on $\mathcal{Y}$. Convolutional neural networks are best suited to learn such an extension operator between the discretized spaces $\mathbf{Y} \to \mathbf{Y}$; in particular, multi-scale residual networks like the U-net (Park et al. 2017) or generative adversarial networks (Ghani and Karl 2019) are a popular choice for the sinogram correction task. In contrast to the analogue approach for limited angle tomography, here the mask $\mathbf{M}_i$ depends on the particular instance in the training data. Therefore, it is necessary to extract the metal mask before the application of the data extension network. The design of a network that takes the linear interpolated masked data as well as the mask itself as an input was shown to be beneficial (Zhang and Yu 2018). After extending the data by the network, in a second step, any reconstruction method can be used to obtain an image with reduced metal artifacts. The authors in Bayaraa et al. (2020) tackle both, the problem of missing data due to detector offset and artifact reduction from high-density objects in dental cone-beam computerized tomography. They first apply a sinogram correction algorithm to extend the data to the missing region and then apply a complementary deep convolutional network to further improve the reconstruction quality.

A second class of deep learning methods utilizes a reconstruction from raw data and targets the removal of generated artifacts in a post-processing step (cf. Fig. 2). In this case, the training strategy is similar to (10), with some works proposing a patch-wise artifact removal (Gjesteby et al. 2018; Huang et al. 2018a). In Zhang and Yu (2018), the authors employ a convolutional network to obtain a prior image from an initial reconstruction, which is further utilized to generate a full sinogram. The full sinogram is then used for the purpose of replacing the metal trace in the original data and creating a corrected sinogram for the final reconstruction.

Low-Dose Computed Tomography

Since X-ray radiation creates a potential risk for the patient, it is desired to lower the radiation dose. There are two main strategies to achieve a reduction of the X-ray radiation for computed tomography, namely, limiting the X-ray flux by reducing the operating current and minimizing the number of measurements.

Lowering the radiation dose results in a noisy data and consequently a noisy reconstructed image with a low signal-to-noise ratio. This can potentially make medical diagnosis more difficult, and therefore a great amount of algorithms were proposed for improving image reconstruction for low-dose computed tomography. Especially with the availability of large datasets, such as the Low-Dose Parallel Beam (LoDoPaB)-CT data set (Leuschner et al. 2021), there has been a large body of work aimed at improving the reconstruction of low-dose data (Kang et al. 2017). Generally these methods can be categorized into (Chen et al. 2017):

- Filtering in data domain
- Iterative reconstruction
- Image processing

Algorithms of the first and third category have the advantage that they can be efficiently combined with classical reconstruction methods, whereas iterative reconstruction algorithms tend to suffer from numerical complexity. Deep learning methods have proven to be particularly suitable for tackling the reconstruction problem, as they are able to achieve image quality, either favorable or comparable to commercial iterative reconstructions, while at the same time being computationally more efficient (Shan et al. 2019).

Deep learning offers the possibility to account for filtering in the sinogram domain and image processing after some initial reconstruction. Given full dose measurements $g \in \mathcal{Y}$, the low-dose measurements can be defined by $\sigma(g)$, where $\sigma: \mathcal{Y} \to \mathcal{Y}$ denotes the transformation mapping from full-dose to low-dose data. A neural network Φ^* can then be trained to approximate the inverse of σ (Ghani and Karl 2018). Denoting the training data by $(\sigma(g_i), g_i)_{i=1}^{N} \subset \mathcal{Y} \times \mathcal{Y}$ and a suitable operator class $\mathcal{C} \subset \{\Phi: \mathcal{Y} \to \mathcal{Y}\}$, we want to find

$$\Phi^* = \arg\min_{\Phi \in \mathcal{C}} \frac{1}{N} \sum_{i=1}^{N} \mathcal{D}_{\mathcal{Y}}\left(\Phi(\sigma(g_i)), g_i\right).$$

The reconstruction can then be carried out by a classical method after the application of Φ^* to the sinogram.

In contrast to this approach, if we denote the reconstruction from low-dose measurements with a classical method by f^σ, then the learning task in image domain can be formulated by

$$\Phi^* = \arg\min_{\Phi \in \mathcal{C}} \frac{1}{N} \sum_{i=1}^{N} \mathcal{D}_{\mathcal{X}}\left(\Phi(f_i^\sigma), f_i\right),$$

where $\mathcal{C} \subset \{\Phi: \mathcal{X} \to \mathcal{X}\}$ is a class of possible denoising operators. Convolutional encoder-decoder networks have been found to be particularly well suited for this task (Chen et al. 2017), especially networks denoising not the whole reconstructed image but only image patches.

For low-dose CT reconstructions with undersampled data (small numbers of measurements), approaches in the same line have been proposed. In this application area, the learning task in the data domain entails upsampling the sinogram, whereas the learned image processing should remove the streak artifacts that occur in reconstructions from undersampled data.

Further Methods

The field of data-driven image reconstruction is a very dynamic one and is constantly evolving. We now describe another very important class of methods, namely, the learned iterative methods (Adler and Öktem 2017, 2018; Hauptmann et al. 2020). They also give the opportunity to incorporate knowledge about the forward model into the reconstruction process and yield impressive results

(Wu et al. 2017; Chen et al. 2020; Guazzo 2020). The structure of these learned iterative schemes is as follows. For an existing iterative procedure, a number of iterations is chosen, and in all iterations up to this number, the update process is augmented by a neural network. For given data g and an initial guess f_0, the final reconstruction $\hat{f}$ in its simplest form are given by

$$f^* = \left(\Phi^N_{W_N} \circ \mathbf{G}^N(g) \circ \ldots \circ \Phi^1_{W_1} \circ \mathbf{G}^1(g) \right)(f_0),$$

where $\Phi^i_{W_i}$ denote the augmentation networks and $\mathbf{G}^i(g)$ iterative updates of the reconstruction. These updates depend on the current iteration but also on the data g, which results in a final reconstruction f^* that is consistent to the given data. This class of algorithms can also be utilized for every inverse imaging problem, where the forward operator can be modeled, including the three example problems discussed above.

Other approaches that will continue to play a very important role in the future are unsupervised methods that do not rely on a paired dataset. Recently a variety of methods have been published in this field of research as well (Kwon and Ye 2021; Lee et al. 2020; Kuanar et al. 2019), just to name a few. The great advantage of these methods is that no paired training data need to be collected, which is very difficult or even impossible in many experimental applications.

Completeness in such a rapidly developing field of research is impossible; nevertheless, a more complete and detailed survey of deep learning methods for inverse imaging is given in the nice review article (Arridge et al. 2019).

Conclusion

Deep learning methods show excellent results for tomographic image reconstruction and represent a promising framework for obtaining good image quality for different measurement cases that create incomplete, corrupted, or noisy data. Various deep learning-based methods have meanwhile been designed in order to optimize tomographic image reconstruction. Among them are learned iterative schemes, network cascades, learned regularizers, and two-step approaches; we presented two-stage strategies of deploying data-driven methods to improve image reconstruction in frequently occurring imperfect data situations in X-ray CT. Most of these approaches can be similarly adapted to other tomographic imaging modalities as well. Nevertheless, it is important to consciously harness the power of deep learning to ensure robustness and guarantee meaningful images for diagnosis. In my opinion, knowledge of the physics (the modelling operator $\mathbf{R}$) and consistency constraints such as data consistency can help overcome these issues and should be incorporated in the design of deep learning approaches in tomographic imaging. Furthermore, careful and extensive validation and evaluation of these methods including experts' opinions from radiologists and medical doctors are necessary to exploit the indisputable power of deep learning for medical imaging.

References

Acar, R., Vogel, C.R.: Analysis of bounded variation penalty methods for ill-posed problems. Inverse Probl. **10**(6), 1217 (1994)

Adler, J., Öktem, O.: Solving ill-posed inverse problems using iterative deep neural networks. Inverse Probl. **33**(12), 124007 (2017)

Adler, J., Öktem O.: Learned primal-dual reconstruction. IEEE Trans. Med. Imaging **37**(6), 1322–1332 (2018)

Anirudh, R., Kim, H., Thiagarajan, J.J., Mohan, K.A., Champley, K., Bremer, T.: Lose the views: limited angle CT reconstruction via implicit sinogram completion. In: Proceedings of the IEEE Conference on Computer Vision and Pattern Recognition, pp. 6343–6352 (2018)

Arridge, S., Maass, P., Öktem, O., Schönlieb, C.-B.: Solving inverse problems using data-driven models. Acta Numer. **28**, 1–174 (2019)

Bayaraa, T., Hyun, C.M., Jang, T.J., Lee, S.M., Seo, J.K.: A two-stage approach for beam hardening artifact reduction in low-dose dental CBCT. IEEE Access **8**, 225981–225994 (2020)

Beard, P.: Biomedical photoacoustic imaging. Interface Focus **1**(4), 602–631 (2011)

Boink, Y.E., Brune, C.: Learned SVD: solving inverse problems via hybrid autoencoding. arXiv preprint arXiv:1912.10840 (2019)

Boink, Y.E., Manohar, S., Brune, C.: A partially-learned algorithm for joint photo-acoustic reconstruction and segmentation. IEEE Trans. Medi. Imaging **39**(1), 129–139 (2019)

Boink, Y.E., Haltmeier, M., Holman, S., Schwab, J.: Data-consistent neural networks for solving nonlinear inverse problems. arXiv preprint arXiv:2003.11253 (2020)

Bubba, T.A., Kutyniok, G., Lassas, M., Maerz, M., Samek, W., Siltanen, S., Srinivasan, V.: Learning the invisible: a hybrid deep learning-shearlet framework for limited angle computed tomography. Inverse Probl. **35**(6), 064002 (2019)

Chen, H., Zhang, Y., Kalra, M.K., Lin, F., Chen, Y., Liao, P., Zhou, J., Wang, G.: Low-dose CT with a residual encoder-decoder convolutional neural network. IEEE Trans. Med. Imaging **36**(12), 2524–2535 (2017)

Chen, G., Hong, X., Ding, Q., Zhang, Y., Chen, H., Fu, S., Zhao, Y., Zhang, X., Ji, H., Wang, G. et al.: Airnet: fused analytical and iterative reconstruction with deep neural network regularization for sparse-data CT. Med. Phys. **47**(7), 2916–2930 (2020)

Daubechies, I., Defrise, M., De Mol, C.: An iterative thresholding algorithm for linear inverse problems with a sparsity constraint. Commun. Pure Appl. Math. J. Issued Courant Inst. Math. Sci. **57**(11), 1413–1457 (2004)

Deans, S.R.: The Radon Transform and Some of Its Applications. Courier Corporation. Dover Publications, INC., Mineola, New York (2007)

Elad, M.: Sparse and Redundant Representations: From Theory to Applications in Signal and Image Processing. Springer Science & Business Media, New York (2010)

Engl, H.W., Hanke, M., Neubauer, A.: Regularization of Inverse Problems, vol. 375. Springer Science & Business Media, Dordrecht (1996)

Frikel, J., Quinto, E.T.: Characterization and reduction of artifacts in limited angle tomography. Inverse Probl. **29**(12), 125007 (2013)

Ghani, M.U., Karl, W.C.: CNN based sinogram denoising for low-dose CT. In: Mathematics in Imaging, pp. MM2D–5. Optical Society of America, Optical Society of America, Orlando, Florida (2018)

Ghani, M.U., Karl, W.C.: Fast enhanced CT metal artifact reduction using data domain deep learning. IEEE Trans. Comput. Imaging, IEEE Trans. Comput. Imaging, vol. 6, 181–193 (2019)

Gjesteby, L., Shan, H., Yang, Q., Xi, Y., Claus, B., Jin, Y., De Man, B., Wang, G.: Deep neural network for CT metal artifact reduction with a perceptual loss function. In: Proceedings of the Fifth International Conference on Image Formation in X-Ray Computed Tomography, vol. 1 (2018)

Goodfellow, I., Bengio, Y., Courville, A.: Deep Learning. MIT Press, Cambridge, MA (2016)

Gu, J., Ye, J.C.: Multi-scale wavelet domain residual learning for limited-angle CT reconstruction. arXiv preprint arXiv:1703.01382 (2017)

Guazzo, A.: Deep learning for PET imaging: from denoising to learned primal-dual reconstruction (2020)

Hastie, T., Tibshirani, R., Friedman, J.: The Elements of Statistical Learning: Data Mining, Inference, and Prediction. Springer Science & Business Media (2009)

Hauptmann, A., Adler, J., Arridge, S.R., Öktem, O.: Multi-scale learned iterative reconstruction. IEEE Trans. Comput. Imaging, vol. 6, 843–856 (2020)

Hecht-Nielsen, R.: Theory of the backpropagation neural network. In: Neural Networks for Perception, pp. 65–93. Elsevier (1992)

Higham, C.F., Higham, D.J.: Deep learning: an introduction for applied mathematicians. SIAM Rev. **61**(4), 860–891 (2019)

Hounsfield, G.N.: Computerized transverse axial scanning (tomography): part 1. description of system. Br. J. Radiol. **46**(552), 1016–1022 (1973)

Huang, Y., Huang, X., Taubmann, O., Xia, Y., Haase, V., Hornegger, J., Lauritsch, G., Maier, A.: Restoration of missing data in limited angle tomography based on Helgason–Ludwig consistency conditions. Biomed. Phys. Eng. Express **3**(3), 035015 (2017)

Huang, X., Wang, J., Tang, F., Zhong, T., Zhang, Y.: Metal artifact reduction on cervical CT images by deep residual learning. Biomed. Eng. Online **17**(1), 175 (2018a)

Huang, Y., Würfl, T., Breininger, K., Liu, L., Lauritsch, G., Maier, A.: Some investigations on robustness of deep learning in limited angle tomography. In: International Conference on Medical Image Computing and Computer-Assisted Intervention, pp. 145–153. Springer (2018b)

Kang, E., Min, J., Ye, J.C.: A deep convolutional neural network using directional wavelets for low-dose x-ray CT reconstruction. Med. Phys. **44**(10), e360–e375 (2017)

Kuanar, S., Athitsos, V., Mahapatra, D., Rao, K.R., Akhtar, Z., Dasgupta, D.: Low dose abdominal CT image reconstruction: an unsupervised learning based approach. In: 2019 IEEE International Conference on Image Processing (ICIP), pp. 1351–1355. IEEE (2019)

Kwon, T., Ye, J.C.: Cycle-free cyclegan using invertible generator for unsupervised low-dose CT denoising. arXiv preprint arXiv:2104.08538 (2021)

Landweber, L.: An iteration formula for Fredholm integral equations of the first kind. Am. J. Math. **73**(3), 615–624 (1951)

LeCun, Y., Bengio, Y., Hinton, G.: Deep learning. Nature **521**(7553), 436–444 (2015)

Lee, J., Gu, J., Ye, J.C.: Unsupervised CT metal artifact learning using attention-guided beta-cyclegan. arXiv preprint arXiv:2007.03480 (2020)

Leuschner, J., Schmidt, M., Baguer, D.O., Maass, P.: LoDoPab-CT, a benchmark dataset for low-dose computed tomography reconstruction. Sci. Data **8**(1), 1–12 (2021)

Li, H., Schwab, J., Antholzer, S., Haltmeier, M.: Nett: solving inverse problems with deep neural networks. Inverse Probl. **36**(6), 065005 (2020)

Lundervold, A.S., Lundervold, A.: An overview of deep learning in medical imaging focusing on mri. Zeitschrift für Medizinische Physik **29**(2), 102–127 (2019)

Lunz, S., Öktem, O., Schönlieb, C.-B.: Adversarial regularizers in inverse problems. arXiv preprint arXiv:1805.11572 (2018)

Mukherjee, S., Dittmer, S., Shumaylov, Z., Lunz, S., Öktem, O., Schönlieb, C.-B.: Learned convex regularizers for inverse problems. arXiv preprint arXiv:2008.02839 (2020)

Natterer, F.: The Mathematics of Computerized Tomography. SIAM, Philadelphia (2001)

Obmann, D., Nguyen, L., Schwab, J., Haltmeier, M.: Sparse anett for solving inverse problems with deep learning. In: 2020 IEEE 17th International Symposium on Biomedical Imaging Workshops (ISBI Workshops), pp. 1–4. IEEE (2020)

Park, H.S., Chung, Y.E., Lee, S.M., Kim, H.P., Seo, J.K.: Sinogram-consistency learning in CT for metal artifact reduction. arXiv preprint arXiv:1708.00607, 1 (2017)

Purcell, E.M., Torrey, H.C., Pound, R.V.: Resonance absorption by nuclear magnetic moments in a solid. Phys. Rev. **69**(1–2), 37 (1946)

Quinto, E.T.: Tomographic reconstructions from incomplete data-numerical inversion of the exterior radon transform. Inverse Probl. **4**(3), 867 (1988)

Ruder, S.: An overview of gradient descent optimization algorithms. arXiv preprint arXiv:1609.04747 (2016)

Scherzer, O., Grasmair, M., Grossauer, H., Haltmeier, M., Lenzen, F.: Variational Methods in Imaging. Springer, New York (2009)

Schwab, J., Antholzer, S., Nuster, R., Haltmeier, M.: Real-time photoacoustic projection imaging using deep learning. arXiv preprint arXiv:1801.06693 (2018)

Schwab, J., Antholzer, S., Haltmeier, M.: Big in Japan: regularizing networks for solving inverse problems. J. Math. Imaging Vis., vol. 62, 445–455 (2019a)

Schwab, J., Antholzer, S., Haltmeier, M.: Deep null space learning for inverse problems: convergence analysis and rates. Inverse Probl. **35**(2), 025008 (2019b)

Schwab, J., Antholzer, S., Haltmeier, M.: Learned backprojection for sparse and limited view photoacoustic tomography. In: Photons Plus Ultrasound: Imaging and Sensing 2019, vol. 10878, p. 1087837. International Society for Optics and Photonics, SPIE BiOS, San Francisco, California (2019c)

Shan, H., Padole, A., Homayounieh, F., Kruger, U., Khera, R.D., Nitiwarangkul, C., Kalra, M.K., Wang, G.: Competitive performance of a modularized deep neural network compared to commercial algorithms for low-dose CT image reconstruction. Nat. Mach. Intell. **1**(6), 269–276 (2019)

Wang, G.: A perspective on deep imaging. IEEE Access **4**, 8914–8924 (2016)

Werbos, P.: Beyond regression: new tools for prediction and analysis in the behavioral sciences. Ph. D. dissertation, Harvard University (1974)

Wu, D., Kim, K., El Fakhri, G., Li, Q.: Iterative low-dose CT reconstruction with priors trained by artificial neural network. IEEE Trans. Med. Imaging **36**(12), 2479–2486 (2017)

Wu, D., Kim, K., Kalra, M.K., De Man, B., Li, Q.: Learned primal-dual reconstruction for dual energy computed tomography with reduced dose. In: 15th International Meeting on Fully Three-Dimensional Image Reconstruction in Radiology and Nuclear Medicine, vol. 11072, p. 1107206. International Society for Optics and Photonics (2019)

Würfl, T., Hoffmann, M., Christlein, V., Breininger, K., Huang, Y., Unberath, M., Maier, A.K.: Deep learning computed tomography: learning projection-domain weights from image domain in limited angle problems. IEEE Trans. Med. Imaging **37**(6), 1454–1463 (2018)

Zhang, Y., Yu, H.: Convolutional neural network based metal artifact reduction in x-ray computed tomography. IEEE Trans. Med. Imaging **37**(6), 1370–1381 (2018)

Zhu, B., Liu, J.Z., Cauley, S.F., Rosen, B.R., Rosen, M.S.: Image reconstruction by domain-transform manifold learning. Nature **555**(7697), 487–492 (2018)

MRI Bias Field Estimation and Tissue Segmentation Using Multiplicative Intrinsic Component Optimization and Its Extensions

34

Samad Wali, Chunming Li, and Lingyan Zhang

Contents

Introduction	1204
Multiplicative Intrinsic Component Optimization	1207
Decomposition of MR Images into Multiplicative Intrinsic Components	1207
Mathematical Description of Multiplicative Intrinsic Components	1208
Energy Formulation for MICO	1209
Optimization of Energy Function and Algorithm	1211
Numerical Stability Using Matrix Analysis	1213
Execution of MICO	1215
Some Extensions	1217
Introduction of Spatial Regularization in MICO	1217
The Proposed TV-Based MICO Model and Its Solver	1217
Spatiotemporal Regularization for 4D Segmentation	1222
Modified MICO Formulation with Weighting Coefficients for Different Tissues	1224
Results and Discussions	1224
Conclusion	1232
References	1232

S. Wali
School of Information and Communication Engineering, University of Electronic Science and Technology of China, Chengdu, China

Department of Mathematics, Namal Univeristy, Mianwali, Pakistan
e-mail: samad.walikhan@outlook.com

C. Li (✉) · L. Zhang
School of Information and Communication Engineering, University of Electronic Science and Technology of China, Chengdu, China
e-mail: chunming.li@uestc.edu.cn; 1799031768@qq.com

© Springer Nature Switzerland AG 2023
K. Chen et al. (eds.), *Handbook of Mathematical Models and Algorithms in Computer Vision and Imaging*, https://doi.org/10.1007/978-3-030-98661-2_110

Abstract

In medical image analysis, energy minimization-based optimization approaches are invaluable. This chapter presents a joint optimization method called multiplicative intrinsic component optimization (MICO) for magnetic resonance (MR) images in bias field estimation and segmentation. Due to the intensity inhomogeneity in MR images, there are overlaps between the ranges of the intensities of different tissues, which often causes misclassification of tissues. To overcome this problem, our proposed method MICO can estimate bias field without avoiding intensity inhomogeneity and can benefit to achieve superior tissue segmentation results. We extended MICO formulation by connecting total variation (TV) as a convex regularization. In addition, for the TV-based MICO model, we implemented the alternating direction method of multipliers (ADMM), which can solve the model efficiently and guarantee its convergence. Quantitative evaluations and comparisons with other popular software have shown that MICO and TVMICO outperform them in terms of robustness and accuracy.

Keywords

MRI · Brain segmentation · Intensity inhomogeneity · Bias field estimation · Bias field correction · Energy minimization · Multiplicative intrinsic component optimization · 4D segmentation · Total variation · ADMM

Introduction

Image segmentation is a fundamental task in image processing in which an image is divided into numerous disjoint parts so that pixels in the same region have certain consistent properties such as intensity, color, and texture (Stockman and Shapiro 2001). Due to an inherent artifact known as intensity inhomogeneity, segmentation in magnetic resonance imaging (MRI) is a challenging task. It appears as slow intensity variations in the same tissue across the image domain (Li et al. 2008; Vovk et al. 2007). In MRI, intensity inhomogeneity can be caused by a variety of factors, including B0 and B1 field inhomogeneities and patient-centered interactions. Because of intensity inhomogeneity, there are overlaps between the ranges of intensities of various tissues, which frequently leads to tissue misclassification. Intensity inhomogeneities can also mislead other image analysis methods, such as image registration. As a result, before doing a quantitative analysis of MRI data, it is typically necessary to eliminate intensity inhomogeneity using a process known as bias field correction. Bias field correction is typically achieved by estimating the bias field that accounts for the intensity inhomogeneity in the MR image and then dividing the image by the estimated bias field to obtain a bias field corrected image.

Traditional segmentation techniques, such as the K-means clustering algorithm, frequently fail in the presence of image intensity inhomogeneities (Zheng et al. 2018). To use these techniques, bias field correction must be performed as a separate preprocessing step to eliminate the intensity inhomogeneity (Juntu et al. 2005; Tustison et al. 2010). Because some modern image segmentation algorithms feature an inherent mechanism for dealing with intensity inhomogeneities, they may be used immediately for segmentation without the necessity for bias field correction in a subsequent preprocessing phase. In an iterative procedure, these algorithms often interleave bias field estimation and image segmentation (Li et al. 2014; Guillemaud and Brady 1997).

Wells et al. established a strategy for interleaved bias field estimation and segmentation based on an expectation-maximization (EM) algorithm (Wells et al. 1996). Guillemaud and Brady improved on this strategy in Guillemaud and Brady (1997). However, appropriate initialization is required for either the bias field or the classification estimate in these EM-based approaches (Styner et al. 2000). To accomplish initialization in MRI, these techniques often require manual choices of representative locations for each tissue class. Such initializations are often imprecise and irreproducible (Leemput et al. 1999). Furthermore, the outcome of bias field correction and segmentation is sensitive to the initial condition selections (Vovk et al. 2007).

Pham and Prince introduced an energy minimization strategy for segmentation and bias field estimation in (1999), which employed a fuzzy c-means (FCM) algorithm for image segmentation. Their technique, known as adaptive fuzzy c-means (AFCM), is an extension of FCM that includes a bias field as a component in the cluster centers. A smoothing factor was incorporated in their energy function to assure the smoothness of the bias field. The coefficient of the smoothing component, on the other hand, is sometimes challenging to adjust (Vovk et al. 2007), limiting the algorithm's effectiveness. Pham expanded AFCM to an improved formulation named FANTASM in a subsequent article Pham (2001) by including a spatial regularization procedure on the tissue membership functions. Although spatial regularization reduces the influence of noise, FANTASM suffers from the same issue as AFCM in terms of the smoothing term for the bias field.

The correction of bias fields is a crucial challenge in medical image processing. Over the last two decades, many bias field correction techniques have been presented. Prospective approaches Condon et al. (1987), Simmons et al. (1991), Wicks et al. (1993), Tincher et al. (1993), Axel et al. (1987), McVeigh et al. (1986), Narayana et al. (1988) and retrospective methods (Wells et al. 1996; Johnston et al. 1996; Dawant et al. 1993; Sled et al. 1998; Pham and Prince 1999; Leemput et al. 1999; Styner et al. 2000; Ahmed et al. 2002; Salvado et al. 2005; Li et al. 2008) are the two primary categories of existing bias correction methods. Prospective methods use special hardware or particular sequences to avoid intensity inhomogeneity throughout the sampling process. These approaches can correct some of the intensity inhomogeneities induced by the MR scanner, but they cannot address patient-dependent inhomogeneities, making them of limited utility in practical applications (Likar et al. 2001). Retrospective approaches, in contrast to prospective methods, focus only on the information contained within the collected image and can be used to reduce intensity inhomogeneities induced by patient-dependent effects. Vovk et al. (2007) provides a current survey of bias correcting approaches.

Homomorphic filtering (Johnston et al. 1996) is one of the earliest retrospective approaches for bias field elimination. This approach posits that intensity inhomogeneity is a low-frequency signal that can be smothered by using high-pass filtering. However, because the imaged objects typically contain low frequencies as well, filtering approaches frequently fail to achieve sufficient bias field corrections (Vovk et al. 2007). Dawant et al. (1993) presented a method for estimating the inhomogeneity field using splines fitting to the intensities of chosen points. Their approach is based on manually picking reference points inside white tissue. For bias field correction, in Sled et al. (1998), authors suggested an iterative approach named N3 that is based on intensity histograms. It seeks to generate the smooth bias field that sharpens the image's intensity histogram optimum. In Tustison et al. (2010), the N3 algorithm's implementation was enhanced by employing a quicker and more robust B-spline approximation to construct the bias field.

Variational models using total variation (TV) have been widely employed in a wide range of image applications, including bias field estimation and segmentation (He et al. 2012; Tu et al. 2016; Li et al. 2010). Because of its edge preservation property, convexity, and L1 norm sparsity behavior, total variation is quite beneficial (Li et al. 2016). It was initially employed as a regularization for image denoising (Rudin et al. 1992), and it has since been studied and is still useful for a variety of image-processing applications (Chen 2013). The non-smoothness of the total variation semi-norm, on the other hand, poses a barrier to its minimization. To address this issue, the most popular approach is to replace total variation in image restoration models with smoothed versions of the total variation (Liu et al. 2015). To tackle the non-smoothness issue in total variation, alternating direction method of multiplier (ADMM) (Gabay and Mercier 1976; Glowinski and Marroco 1975) is used, which is similar to split Bregman (Goldstein and Osher 2009) and proved to be particularly beneficial for L1 and TV-type optimization problems. We develop an efficient method by introducing two sets of auxiliary variables with closed-form solutions to all subproblems.

In this chapter, firstly, we propose a novel technique for bias field estimation and tissue segmentation in an energy minimization setting. In an energy minimization technique, bias field estimation and tissue membership functions are performed simultaneously. The proposed method optimizes two multiplicative intrinsic components of an MR image: the bias field, which compensates for intensity inhomogeneity, and the true image, which represents a physical property of the tissues. The spatial features of these two components are completely incorporated in their physical representations with the help of the proposed energy minimization approach. Secondly, we have extended the proposed MICO to total variational-based MICO, which we called TVMICO. We use an alternating direction method of multiplier (ADMM) to solve the TVMICO model. By introducing two new constraints, we have closed-form solutions to each sub-variational problem. Because of the convexity of the energy function in each of its variables, our technique, which we term multiplicative intrinsic component optimization (MICO) and TVMICO, both are robust. The proposed MICO formulation can be naturally extended to 3D/4D segmentation with spatial and temporal regularization.

Multiplicative Intrinsic Component Optimization

The formulation of MICO for bias field estimation and tissue segmentation based on the decomposition of an MRI into two multiplicative components is presented in this section. The proposed energy minimization technique leads to the MICO algorithm for combined bias field estimation and tissue segmentation. We follow Li et al. (2014) for most mathematical formulation and notations.

Decomposition of MR Images into Multiplicative Intrinsic Components

Consider $I(x)$ to be the intensity of an observed MR image at voxel x. In most cases, an MR image can be modeled as follows:

$$I(x) = b(x)J(x) + n(x), \qquad (1)$$

where $J(x)$ is the clean image, $b(x)$ is the bias field that accounts for the observed image's intensity inhomogeneity, and $n(x)$ is zero-mean additive noise. The widely accepted assumptions in the literature for both J and b are given in Wells et al. (1996), Leemput et al. (1999), and Pham and Prince (1999). The bias field b is supposed to vary smoothly. The true image J describes a physical characteristic of the tissues being imaged, which should ideally take a specific value for voxels of the same tissue type. As a result, for all point x in the i-th tissue, we assume that $J(x)$ is approximately a constant c_i.

In this chapter, we consider Eq. (1) decomposes the MR image I into two multiplicative components b and J, as well as additive zero-mean noise n. From this aspect, we specify systematically biased field estimation and tissue segmentation as a variational-based problem, which is seeking accurate decomposition of given MRI I into two multiplicative components b and J. It is important to mention here that the bias field b and the true image J are intrinsic components of the observed MR image I. In this chapter, we consider an observed image I as a function $I : \Omega \to \mathcal{R}$ on a continuous domain Ω.

In computer vision, a given observed image I can be decomposed into reflectance image R and the illumination image S that can be shown in multiplicative form as $I = RS$. These multiplicative components of an observed image are similar to Eq. (1). The terminologies intrinsic images were introduced by Barrow and Tenenbaum in (1978) to express these two multiplicative components. In computer vision, estimating intrinsic images from an observed scene image has been a significant challenge. Several methods for estimating the intrinsic images from a scene image based on different assumptions on the two intrinsic images have been presented (Tappen et al. 2005; Weiss 2001; Kimmel et al. 2003).

The bias field b and the real image J are considered as multiplicative intrinsic components of an observed MR image in this study. From an observed MR image, we present a unique approach for estimating these two components. We should point out that the method proposed in this chapter differs from those used in computer vision to estimate reflectance and illumination images. In fact, due to a lack of knowledge about the unknown intrinsic images R and S, estimation of intrinsic images is an ill-posed problem.

If no prior knowledge of the multiplicative components b and J of the observed MR image I is used, estimation of these components is an underdetermined or ill-posed problem. To solve the problem, we have to gain some knowledge about the bias field b and true image J. The piecewise constant property of the true image J and the smoothly varying property of the bias field b are used in this study to present a strategy that uses the basic properties of the true image and bias field. In the development of our proposed technique, the decomposition of the MR image I into two multiplicative intrinsic components b and J with their respective spatial properties is completely exploited.

Mathematical Description of Multiplicative Intrinsic Components

We could use a suitable mathematical representation and description of the bias field b and true image J to appropriately utilize their features. Assume we have a collection of functions $g_1, \cdots, g_M$ that ensures the bias field's smoothly varying property. The bias field in our method is a linear combination of a series of smooth basis functions. It has been studied that for a given sufficiently large number of M basis, a function can be approximated by a linear combination of several basis functions to an arbitrary degree of accuracy (Powell 1981). We use 20 polynomials of the first three degrees as the basis functions in MICO applications to 1.5T and 3T MRI images. The optimal coefficients $w_1, \cdots, w_M$ in the linear combination $b(x) = \sum_{k=1}^{M} w_k g_k$ are needed to determine and used to estimate the bias field. The coefficients $w_1, \cdots, w_M$ are represented as a column vector $= (w_1, \cdots, w_M)^T$, where $(\cdot)^T$ is the transpose operator. A column vector-valued function $G(x) = (g_1(x), \cdots, g_M(x))^T$ represents the basis functions $g_1(x), \cdots, g_M(x)$. Therefore, the bias field $b()$ can be expressed in the vector form shown below.

$$b(x) = \mathbf{w}^T G(x). \tag{2}$$

In our proposed variational-based minimization approach for bias field estimation, Eq. (2) will be utilized as a vector representation. It enables us to calculate the optimal bias field obtained from the energy minimization problem using efficient vector and matrix calculations, as will be explained in section "Optimization of Energy Function and Algorithm".

More formally, the true image J has piecewise approximately constant property, and it can be expressed as follows. We suppose that there are N different types of tissues in the image domain Ω. For x in the i-th tissue, the true image $J(x)$ is approximately a constant c_i. The location where the i-th tissue is located is denoted as Ω_i. The membership function u_i may be used to represent each Ω_i region (tissue).

The membership function u_i is a binary membership function in the ideal case when each voxel contains just one kind of tissue, with $u_i(x) = 1$ for $x \in \Omega_1$ and $u_i(x) = 0$ for $x \notin \Omega_i$. Because of the partial volume effect, one voxel may include more than one type of tissue, especially at the interface between adjacent tissues. In this scenario, the N tissues are represented by fuzzy membership functions $u_i(x)$ with values ranging from 0 to 1 and satisfying $\sum_{i=1}^{N} u_i = 1$. The fuzzy membership function $u_i(x)$ value can be construed as the proportion of the i-th tissue within the voxel x. A column vector-valued function $\mathbf{u} = (u_1, \cdots u_N)^T$, where T is the transpose operator, can be used to express such membership functions $u_1, \cdots u_N$. The space of all such vector-valued functions is denoted as $\mathcal{U}$.

$$\mathcal{U} \triangleq \{\mathbf{u} = (u_1, \cdots, u_N)^T : 0 \leq u_i(x) \leq 1, i = 1, \cdots, N, \text{ and }$$
$$\sum_{i=1}^{N} u_i(x) = 1, \text{ for all } x \in \Omega\} \tag{3}$$

The true image J can be approximated by the following combination of membership functions u_i and constants c_i.

$$J(x) = \sum_{i=1}^{N} c_i u_i(x). \tag{4}$$

The function in Eq. (4) is a piecewise constant function when the membership functions u_i are binary functions, with $J(x) = c_i$ for $x \in_i = \{x : u_i(x) = 1\}$. If $u_1, \cdots, u_N$ are the binary membership functions, the segmentation is called the hard segmentation, while the corresponding regions $\Omega_1, \cdots, \Omega_N$ show an image domain Ω partition, with the conditions as $\cup_{i=1}^{N} \Omega_i = \Omega$ and $\Omega_i \cap \Omega_j = \emptyset$. On the other hand, the functions $u_1, \ldots u_N$ are fuzzy membership functions with values between 0 and 1 representing a soft segmentation result.

We propose an energy minimization approach for simultaneous bias field estimation and tissue segmentation based on the image model Eq. (1). The membership function $\mathbf{u} = (u_1, \cdots, u_N)$ gives the outcome of tissue segmentation. The estimated bias field b is used to compute the bias field corrected image, which is expressed as the reciprocal, i.e., $\frac{I}{b}$.

Energy Formulation for MICO

Based on the image model Eq. (1) and the intrinsic features of the bias field and the true image as mentioned in section "Decomposition of MR Images into Multiplicative Intrinsic Components", we present an energy minimization formulation for bias field estimation and tissue segmentation. In light of the image model (1), we address the problem of determining the multiplicative intrinsic components b and J of an observed MR image I to minimize the following energy.

$$F(b, J) = \int_{\Omega} |I(x) - b(x)J(x)|^2 dx. \tag{5}$$

Minimization of energy problem Eq. (5) is obviously an ill-posed problem if the variables b and J are not constrained. Indeed, in the absence of constraints, every nonzero function b and $J = I/B$ optimizes the energy $F(b, J)$. To solve the problem, we must limit the search spaces of b and J by utilizing some information about the unknowns b and J. The characteristics of the bias field b and the true image J described in section "Decomposition of MR Images into Multiplicative Intrinsic Components" are the information that may be used to limit the search spaces of b and J to specific search subspaces that reflect these properties. Using binary membership functions $u_1, \cdots u_N$ and the knowledge that the true image J is piecewise approximately constant, we can confine the true image J's search space to the subspace of piecewise constant functions as in Eq. (4) $J(x) = \sum_{i=1}^{N} c_i u_i(x)$. The search space of the bias field b, on the other hand, is constrained to the subspace of all functions of the type $b(x) = \mathbf{w}^T G(x)$, as shown in Eq. (2). The energy $F(b, J)$ may be written in terms of three variables, $\mathbf{u} = (u_1, \cdots, u_N)^T$, $\mathbf{c} = (c_1, \cdots, c_N)^T$, and $= (w_1, \cdots, w_M)^T$, i.e.:

$$F(b, J) = F(\mathbf{u}, \mathbf{c}, \mathbf{w}) = \int_{\Omega} \left| I(x) - \mathbf{w}^T G(x) \sum_{i=1}^{N} c_i u_i(x) \right|^2 dx, \tag{6}$$

Thus, optimizing b and J involves minimizing the energy F with respect to $\mathbf{u}$, $\mathbf{c}$, and $\mathbf{w}$. Because u_i is the binary membership function of the region Ω_i, we obtain the following:

$$u_i(x) = \begin{cases} 1, & x \in \Omega_i; \\ 0, & x \notin \Omega_i. \end{cases}$$

Therefore, we have as follows:

$$\sum_{i=1}^{N} c_i u_i(x) = c_i \text{ for } x \in \Omega_i$$

As a result, the energy F may be stated as follows:

$$F(\mathbf{u}, \mathbf{c}, \mathbf{w}) = \int_{\Omega} \left| I(x) - \mathbf{w}^T G(x) \sum_{i=1}^{N} c_i u_i(x) \right|^2 dx$$

$$= \sum_{i=1}^{N} \int_{\Omega_i} \left| I(x) - \mathbf{w}^T G(x) c_i \right|^2 dx$$

$$= \sum_{i=1}^{N} \int_{\Omega} \left| I(x) - \mathbf{w}^T G(x) c_i \right|^2 u_i(x) dx \tag{7}$$

We obtain by rearranging the order of summation and integration in Eq. (7) as follows:

$$F(\mathbf{u}, \mathbf{c}, \mathbf{w}) = \int_{\Omega} \sum_{i=1}^{N} \left| I(x) - \mathbf{w}^T G(x) c_i \right|^2 u_i(x) dx. \tag{8}$$

The formulation of the energy F in Eq. (8) allows us to construct an efficient energy minimization technique, which is discussed in section "Optimization of Energy Function and Algorithm". We derive the optimal membership function $\hat{\mathbf{u}} = (\hat{u}_1, \cdots, \hat{u}_N)^T$ as the segmentation result by minimizing the energy $F(\mathbf{u}, \mathbf{c}, \mathbf{w})$, as well as the optimal vector $\hat{\mathbf{w}}$, from which the estimated bias field is defined by $b(x) = \hat{\mathbf{w}}^T G(x)$.

The ideal membership functions $u_1, \cdots, u_N$ that minimize the energy given in Eq. (8) are binary functions with values of 0 or 1, leading to a hard segmentation conclusion, as will be demonstrated in section "Optimization of Energy Function and Algorithm". Many applications prefer fuzzy (or soft) segmentation results, which are provided by fuzzy membership functions with values ranging from 0 to 1, as in the fuzzy C-means (FCM) clustering approach (Bezdek et al. 1984). To accomplish fuzzy segmentation, we change the energy function F in Eq. (8) by adding a fuzzifier $q \geq 1$ to generate the following energy:

$$F_q(\mathbf{u}, \mathbf{c}, \mathbf{w}) = \int_{\Omega} \sum_{i=1}^{N} \left| I(x) - \mathbf{w}^T G(x) c_i \right|^2 u_i^q(x) dx. \tag{9}$$

The optimal membership functions that minimize the energy $F(\mathbf{u}, \mathbf{c}, \mathbf{w})$ for the scenario $q > 1$ are fuzzy membership functions with values between 0 and 1. By minimizing the energy $F(\mathbf{u}, \mathbf{c}, \mathbf{w})$ in Eq. (8) or $F_q(\mathbf{u}, \mathbf{c}, \mathbf{w})$ in Eq. (9), our technique accomplishes image segmentation and bias field estimation, subject to the constraints $\mathbf{u} \in \mathcal{U}$. The fact that the energy $F_q(\mathbf{u}, \mathbf{c}, \mathbf{w})$ is convex in each variable, $\mathbf{u}$, $\mathbf{c}$, or $\mathbf{w}$, is a desired characteristic (Li et al. 2009). This characteristic guarantees that the energy $F_q(\mathbf{u}, \mathbf{c}, \mathbf{w})$ has a unique minimum point for each of its variables.

Optimization of Energy Function and Algorithm

We used alternating minimization technique in which one can achieve the minimum and independent solution of $F_q(\mathbf{u}, \mathbf{c}, \mathbf{w})$ with respect to each of its variables given the other two fixed. The alternating minimization of $F_q(\mathbf{u}, \mathbf{c}, \mathbf{w})$ with respect to each of its variables is described below.

Optimization of c

The energy $F_q(\mathbf{u}, \mathbf{c}, \mathbf{w})$ is optimized with respect to the variable $\mathbf{c}$ for fixed $\mathbf{w}$ and $\mathbf{u} = (u_1, \cdots, u_N)^T$. It is simple to present that $F_q(\mathbf{u}, \mathbf{c}, \mathbf{w})$ is minimized by $\mathbf{c} = \hat{\mathbf{c}} = (\hat{c}_1, \cdots, \hat{c}_N)^T$ with the following:

$$\hat{c}_i = \frac{\int_\Omega I(x)b(x)u_i^q(x)dx}{\int_\Omega b^2(x)u_i^q(x)dx}, \quad i = 1, \cdots, N. \tag{10}$$

Optimization of w and Bias Field Estimation $\hat{b}$

We minimize the energy $F(\mathbf{u}, \mathbf{c}, \mathbf{w})$ with respect to the variable $\mathbf{w}$ for fixed $\mathbf{u}$ and $\mathbf{c}$. This may be accomplished by solving the equation $\frac{\partial F}{\partial \mathbf{w}} = 0$. It is simple to demonstrate that:

$$\frac{\partial F}{\partial \mathbf{w}} = -2\mathbf{v} + 2A\mathbf{w}$$

where $\mathbf{v}$ is a column vector with M dimensions and here A is an $M \times M$ matrix provided by the following:

$$\mathbf{v} = \int_\Omega G(x)I(x)\left(\sum_{i=1}^N c_i u_i^q(x)\right) dx, \tag{11}$$

$$A = \int_\Omega G(x)G^T(x)\left(\sum_{i=1}^N c_i^2 u_i^q(x)\right) dx. \tag{12}$$

The equation $\frac{\partial F}{\partial \mathbf{w}} = 0$ can be represented as a linear equation:

$$A\mathbf{w} = \mathbf{v} \tag{13}$$

We compute the estimated bias field as $\hat{b}(x) = \hat{\mathbf{w}}^T G(x)$ given the solution to this equation, $\hat{\mathbf{w}} = A^{-1}\mathbf{v}$. The non-singularity of matrix A is demonstrated in section "Numerical Stability Using Matrix Analysis".

As a result, the linear equation $\frac{\partial F}{\partial \mathbf{w}} = -2\mathbf{v} + 2A\mathbf{w} = 0$ has a unique solution $\hat{\mathbf{w}} = A^{-1}\mathbf{v}$. The vector $\hat{\mathbf{w}}$ can be represented explicitly by using Eq. (12) as follows:

$$\hat{\mathbf{w}} = \left(\int_\Omega G(x)G^T(x)\left(\sum_{i=1}^N c_i^2 u_i^q(x)\right) dx\right)^{-1} \int_\Omega G(x)I(x)\left(\sum_{i=1}^N c_i u_i^q(x)\right) dx. \tag{14}$$

The estimated bias field is obtained using the optimum vector $\hat{\mathbf{w}}$ provided by Eq. (14).

$$\hat{b}(x) = \hat{\mathbf{w}}^T G(x) \tag{15}$$

We will verify the non-singularity of the matrix A, as well as the numerical stability of the foregoing calculation for solving the linear system (13) in section "Numerical Stability Using Matrix Analysis". These are two critical concerns in the implementation of our proposed technique.

Optimization of u

We begin with the scenario where $q > 1$ and minimize the energy $F(\mathbf{u}, \mathbf{c}, \mathbf{w})$ for fixed $\mathbf{c}$ and $\mathbf{w}$, subject to the constraint that $\mathbf{u} \in \mathcal{U}$. It can be demonstrated that $F(\mathbf{u}, \mathbf{c}, \mathbf{w})$ is minimized at $\mathbf{u} = \hat{\mathbf{u}} = (\hat{u}_1, \cdots, \hat{u}_N)^T$, obtained by the following:

$$\hat{u}_i(x) = \frac{(\delta_i(x))^{\frac{1}{1-q}}}{\sum_{j=1}^{N}(\delta_j(x))^{\frac{1}{1-q}}}, \quad i = 1, \cdots, N, \tag{16}$$

where:

$$\delta_i(x) = |I(x) - \mathbf{w}^T G(x) c_i|^2. \tag{17}$$

For $q = 1$, it can be presented that the minimizer $\hat{\mathbf{u}} = (\hat{u}_1, \cdots, \hat{u}_N)^T$ is provided by the following:

$$\hat{u}_i(x) = \begin{cases} 1, i = i_{\min}(x); \\ 0, i \neq i_{\min}(x), \end{cases} \tag{18}$$

where:

$$i_{\min}(x) = \arg\min_i\{\delta_i(I(x))\}.$$

Numerical Stability Using Matrix Analysis

The bias field estimate computation comprises calculating the vector $\mathbf{v}$ in (11), the matrix A in (12), and the inverse matrix A^{-1} in (14). The matrix A is an $M \times M$ matrix, where M is the number of basis functions. We use $M = 20$ basis functions in this chapter; hence the dimension of matrix A is a 20×20. The non-singularity of the matrix A assures that the inverse matrix A^{-1} exists and that the Eq. (13) has a unique solution. We will also demonstrate that the numerical calculation of the inverse matrix A^{-1} is stable.

The non-singularity of matrix A stated in Eq. (12) is demonstrated in the following way. We begin by defining $h_m(x) \triangleq g_m(x)\sqrt{\sum_{i=1}^{N} c_i^2 u_i^q(x)}$. Thus, the (m, k) entry of the matrix A can be represented as the inner product of h_m and h_k provided by the following:

$$\langle h_m, h_k \rangle = \int_\Omega h_m(x) h_k(x) dx.$$

As a result, the matrix A is the *Gramian matrix* of $h_1, \cdots, h_M$. The Gramian matrix of $h_1, \cdots, h_M$ is non-singular according to linear algebra (Horn and Johnson 1985) if and only if they are linearly independent. It is clear that the above-defined functions $h_1, \cdots, h_M$ are linearly independent, implying that A is non-singular.

The importance of numerical stability in solving the Eq. (13) cannot be overstated. The *condition number* of the matrix A characterizes the numerical stability of solving the Eq. (13); for more details see Golub and Loan (1996). A positive-definite matrix A's condition number is given by the following:

$$\kappa(A) = \lambda_{\max}(A)/\lambda_{\min}(A),$$

where $\lambda_{\min}(A)$ and $\lambda_{\max}(A)$ are the minimal and maximal eigenvalues of matrix A, respectively. For very large value of the condition number $\kappa(A)$, minor variations in the matrix A or the vector $\mathbf{v}$, which are most likely caused by image noise and accumulating intermediate rounding errors, can cause very large variation of the solution $\hat{b}w$ to the Eq. (13). As a result, it is vital to guarantee that the condition number $kappa(A)$ is not huge, as shown below, to ensure the robustness of the bias field computation.

The matrix analysis that follows is predicated on the orthogonality of the basis functions, that is:

$$\int_\Omega g_m(x) g_k(x) dx = \delta_{mk}, \qquad (19)$$

here $\delta_{mk} = 0$ for $m \neq k$ and $\delta_{mk} = 1$ for $m = k$.

It can be demonstrated that for the above-specified matrix A in Eq. (12) with the basis functions $g_1, \cdots, g_M$ satisfying the orthogonality criterion in Eq. (19):

$$0 < \min_i\{c_i^2\} \leq \lambda_{\min}(A) \leq \lambda_{\max}(A) \leq \max_i\{c_i^2\}$$

As a result, A's condition number is determined by the following:

$$\kappa(A) \leq \frac{\max_i\{c_i^2\}}{\min_i\{c_i^2\}}. \qquad (20)$$

For instance, if $\max_i\{c_i\} = 250$ and $\min_i\{c_i\} = 50$, by the inequality (20), we have $\kappa(A) \leq \frac{250^2}{50^2} = 25$. We observed that the condition numbers of the matrix A are at this level in the implementations of our approach to actual MRI data, which is small enough to assure the numerical stability of the inversion operation.

Execution of MICO

We summarize the technique for minimizing the energy $F_q(\mathbf{u}, \mathbf{c}, \mathbf{w})$ for $q \geq 1$ as the following iteration process from section "Optimization of Energy Function and Algorithm":

- Step-1. Initialize $\mathbf{u}$ and $\mathbf{c}$.
- Step-2. Update b as $\hat{b}$ in Eq. (15).
- Step-3. Update $\mathbf{c}$ as $\hat{\mathbf{c}}$ in Eq. (10).
- Step-4. Update $\mathbf{u}$ as $\hat{\mathbf{u}}$ in Eq. (16) for the case $q > 1$ or (18) for the case $q = 1$.
- Step-5. Check the convergence condition: if convergence has been obtained or the iteration number exceeds a predefined maximum number, terminate the iteration; otherwise, go to Step-2.

During the iteration procedure described above, each of the three variables is updated with the other two variables computed in the previous iteration. In Step-1 of the preceding iteration process, we only need to initialize two of the three variables, such as $\mathbf{u}$ and $\mathbf{c}$. In Step-5, the convergence criteria is $|\mathbf{c}^{(n)} - \mathbf{c}^{(n-1)}| < \varepsilon$, where $\mathbf{c}^{(n)}$ is the vector $\mathbf{c}$ updated in Step-3 at the n-th iteration, and ε is set to 0.001.

We used a synthetic image in Fig. 1a to show the robustness of our proposed technique to initialization, using three alternative initializations of the membership functions $u_1, \cdots, u_N$ and the constants $c_1, \cdots, c_N$. The initial membership function $\mathbf{u} = (u_1, \cdots, u_N)$ and the vector $\mathbf{c} = (c_1, \cdots, c_N)$ can be visualized as an image defined by $J_{\mathbf{u},\mathbf{c}}(x) = \sum_{i=1}^{N} c_i u_i(x)$. The images $J_{\mathbf{u},\mathbf{c}}$ for the three different initializations of $\mathbf{u}$ and $\mathbf{c}$ are shown in Fig. 1b, c, and d that show a wide range of patterns. The first initialization illustrated in Fig. 1b is achieved by randomly generating the membership functions $u_1(x), \cdots, u_N(x)$ and the constants $c_1, \cdots, c_N$. The bias field converges to the same function for these three alternative initializations of $\mathbf{u}$ and $\mathbf{c}$ up to a scalar multiple. The three estimated bias fields are the same, up to a minor difference, when the bias fields are normalized (e.g., dividing the bias field b by its maximum value $\max x\{b(x)\}$), as shown in Fig. 1e. Meanwhile, the membership function $\mathbf{u}$ converges to the same vector-valued function, with just a minor variation, providing the identical segmentation result as shown in Fig. 1f. The corrected bias field image is provided in Fig. 1g.

We display the energy minimization $F(\mathbf{u}, \mathbf{c}, \mathbf{w})$ of the variables $\mathbf{u}, \mathbf{c}$, and $\mathbf{w}$ computed at each iteration up to the 20 iterations in Fig. 1h. The energy $F(\mathbf{u}, \mathbf{c}, \mathbf{w})$ rapidly drops to the same value from three distinct initial values corresponding to three separate initializations. Figure 1h also presents the fast convergence of the iteration in MICO, as we can see that the energy is rapidly decreased and converges to the minimal value in less than 10 iterations. As a result, in our MICO applications, we often just perform 10 iterations.

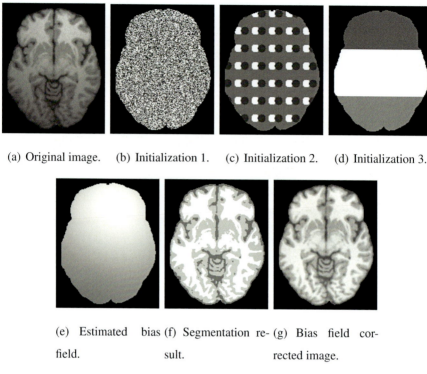

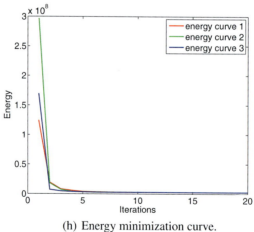

Fig. 1 Robustness of our proposed method to different initializations. (**a**) Original image, (**b**)–(**d**) three possible initializations of the membership functions are visualized, (**e**) estimated bias field, (**f**) segmentation result, (**g**) bias field correction result, (**h**) curves illustrating the energy F used in the iteration process for three different initializations (**b**), (**c**), and (**d**)

Some Extensions

Introduction of Spatial Regularization in MICO

The original MICO formulation described above can be easily extended by including a regularization term on the membership functions. Regularization of the membership functions can be accomplished using the MICO formulation by combining the total variations (TV) of the membership functions in the following energy:

$$\mathcal{F}(\mathbf{u}, \mathbf{c}, \mathbf{w}) = \lambda F(\mathbf{u}, \mathbf{c}, \mathbf{w}) + \sum_{i=1}^{N} TV(u_i), \quad (21)$$

where F is the energy defined in (8), $\lambda > 0$ is the weight of F, and TV is the total variations of u defined by the following:

$$TV(u) = \int_{\Omega} |\nabla u(x)| dx. \quad (22)$$

This energy should be minimized subject to the constraint that $0 \leq u_i(x) \leq 1$ and $\sum_{i=1}^{N} u_i(x) = 1$ for every point x. The variational formulation in (21) is referred to by *TVMICO* formulation. The definition of this energy (21) is simple; however, dealing with the aforementioned point-wise constraint is not straightforward in the context of energy minimization.

Many scholars have developed numerous numerical approaches (Goldstein and Osher 2009) to address variational problems in the context of image segmentation using a TV regularization term $TV(u)$ for a membership function u subject to the constraint $0 \leq u(x) \leq 1$ in recent years. These approaches can only segment images into two complementary regions, denoted by the membership functions u and $1-u$. In general, three or more membership functions $u_1, \cdots, u_N$ are employed to represent $N > 2$ regions for segmentation. Li et al. developed a numerical strategy to address the energy minimization problem with TV regularization on the membership functions in Li et al. (2010); they used the operator splitting method proposed by Lions and Mercier in (1979). The numerical technique provided in Li et al. (2010) can be used to minimize the energy $\mathcal{F}$ with respect to the membership functions $u_1, \cdots, u_N$ in Eq. (21). The energy minimizations with respect to the variables $\mathbf{c}$ and $\mathbf{w}$, which are independent of the TV regularization term of the membership functions, remain the same as described in section "Optimization of Energy Function and Algorithm".

The Proposed TV-Based MICO Model and Its Solver

Formulation of Proposed Model

Equation (21) can be modified with the help of the definition of total variation as follows:

$$\min_{\mathbf{u},\mathbf{c},\mathbf{w}} \lambda \int_{\Omega} \left| I(x) - \sum_{i=1}^{N} c_i \mathbf{w}^T G(x) \right|^2 u_i^q(x) dx + \sum_{i=1}^{N} \int_{\Omega} \|\nabla u_i^q(x)\| dx, \qquad (23)$$

where λ is a positive parameter which can balance the length of the boundary $\partial \Omega_i$ because Tv the second term in Eq. 23 equals to the length of the boundary Ω at ith position. We will discuss both cases for $q = 1$ and $q > 1$. When $q = 1$, u_i can only take values 0 and 1, and then the vector-valued function for bounded variation space can be defined as follows:

$$\mathcal{U}_0 \triangleq \left\{ \mathbf{u} = (u_1, \cdots, u_N)^T : u_i \in BV(\omega), u_i(x) \in \{0, 1\}, i = 1, \cdots, N, \right.$$

$$\left. \text{and} \sum_{i=1}^{N} u_i(x) = 1, \text{ for all } x \in \Omega \right\} \qquad (24)$$

At each point x, there is only one function with a value of 1, while all the other functions have a value of 0. As a result, set $\mathcal{U}_0$ is not continuous, which causes challenges and instability in numerical implementations. However, we may relax binary indicator function defined in Eq. 24 to fuzzy membership functions u_i that meet the nonnegativity and sum-to-one constraint, i.e., $(u_1, \ldots, u_N)$ belongs to the set described as $\mathcal{U}$ in Eq. (3). It is self-evident that $u_i(x) \in [0, 1]$ and is a simplex at any x. As a result, $u_i(x)$ may be thought of as the chance that pixel x belongs to the ith class.

The proposed model Eq. 23 is a convex with respect to $\mathbf{u}$, $\mathbf{c}$, and $\mathbf{w}$ independently, but not in together. The TV could be with L_2 (He et al. 2012) and L_1 (Li et al. 2016) fidelity terms. We can also use some nonlinear and nonconvex regularizations such as total generalized variation (Wali et al. 2019a) and Euler's elastica (Liu et al. 2019; Wali et al. 2019b) for further extensions; however, these models need more constrains to relax and require efficient algorithm such as ADMM. In this section, we only focus on L_1 fidelity term, and we called our proposed method as total variation-based multiplicative intrinsic component optimization (TVMICO).

ADMM and Its Numerical Analysis

In this subsection, ADMM is used to solve the proposed fuzzy-based MICO model (23). We introduce two additional variables $p = (p_1, \cdots, p_N)$ and $v = (v_1, \cdots, v_N)$ with constraints as $\nabla u_i = p_i$ and $u_i = v_i$. With these constraints the minimization problem Eq. (23) can be written as follows:

$$\min_{p,v,\mathbf{u},\mathbf{c},\mathbf{w}} \sum_{i=1}^{N} \left\{ \lambda \int_{\Omega} \left| I(x) - c_i \mathbf{w}^T G(x) \right|^2 v_i(x) dx + \int_{\Omega} \|p_i(x)\| dx \right\} + l_{\mathcal{U}}(\mathbf{v}),$$

$$\text{subject to } \nabla u_i = p_i, \ u_i = v_i, \ \forall i = 1, \cdots, N, \qquad (25)$$

where $l_{\mathcal{U}}$ is the indicator function, i.e.:

$$l_{\mathcal{U}}(\mathbf{v}) = \begin{cases} 0, & \mathbf{v} \in \mathcal{U}; \\ \infty, & \text{otherwise}. \end{cases}$$

The unconstrained augmented Lagrangian functional for Eq. (25) can be formulated as follows:

$$\mathcal{L}(\mathbf{p}, \mathbf{v}, \mathbf{u}, \mathbf{c}, \mathbf{w}; \mu_{\mathbf{p}}, \mu_{\mathbf{v}}) = \sum_{i=1}^{N} \left\{ \lambda \int_{\Omega} \left| I(x) - c_i \mathbf{w}^T G(x) \right|^2 v_i(x) dx + \int_{\Omega} \| p_i(x) \| dx \right\}$$

$$+ l_{\mathcal{U}}(\mathbf{v}), + \sum_{i=1}^{N} \left\{ \langle \mu_{p_i}, \nabla u_i - p_i \rangle + \frac{\gamma}{2} \int_{\Omega} \| \nabla u_i(x) - p_i(x) \|^2 dx \right\}$$

$$+ \sum_{i=1}^{N} \left\{ \langle \mu_{v_i}, u_i - v_i \rangle + \frac{\gamma}{2} \int_{\Omega} \left| u_i(x) - v_i(x) \right|^2 dx \right\}, \tag{26}$$

where $\mu_{p_i} = \mu_{p_1}, \cdots, \mu_{p_N}$ and $\mu_{v_i} = \mu_{v_1}, \cdots, \mu_{v_N}$ are Lagrange multipliers and γ is a positive constant. Here $\langle \mu_{p_i}, \nabla u_i - p_i \rangle = \int_{\Omega} \mu_{p_i}^T(x)(\nabla u_i(x) - p_i(x)) dx$ and $\langle \mu_{v_i}, u_i - v_i \rangle = \int_{\Omega} \mu_{v_i}^T(x)(u_i(x) - v_i(x)) dx$. The ADMM can update Lagrangian multipliers after solving primal variables in Gauss-Seidel manner. The ADMM for solving Eq. (26) can be described in the following Algorithm 1.

Algorithm 1 Proposed alternating direction method of multipliers for (26)

1. **Initialization**: primal and dual variables $\mathbf{p}^0, \mathbf{v}^0, \mathbf{u}^0, \mathbf{c}^0, \mathbf{w}^0$ and Lagrange multipliers $\mu_{\mathbf{p}}^0, \mu_{\mathbf{v}}^0$.
2. **Compute primal and dual variables**: for $k = 1, 2, \ldots$:

$$\mathbf{p}^{k+1} = \arg\min_{\mathbf{p}} \mathcal{L}(\mathbf{p}, \mathbf{v}^k, \mathbf{u}^k, \mathbf{c}^k, \mathbf{w}^k; \mu_{\mathbf{p}}^k, \mu_{\mathbf{v}}^k) \tag{27}$$

$$\mathbf{v}^{k+1} = \arg\min_{\mathbf{p}} \mathcal{L}(\mathbf{p}^{k+1}, \mathbf{v}, \mathbf{u}^k, \mathbf{c}^k, \mathbf{w}^k; \mu_{\mathbf{p}}^k, \mu_{\mathbf{v}}^k) \tag{28}$$

$$\mathbf{u}^{k+1} = \arg\min_{\mathbf{p}} \mathcal{L}(\mathbf{p}^{k+1}, \mathbf{v}^{k+1}, \mathbf{u}, \mathbf{c}^k, \mathbf{w}^k; \mu_{\mathbf{p}}^k, \mu_{\mathbf{v}}^k) \tag{29}$$

$$\mathbf{c}^{k+1} = \arg\min_{\mathbf{p}} \mathcal{L}(\mathbf{p}^{k+1}, \mathbf{v}^{k+1}, \mathbf{u}^{k+1}, \mathbf{c}, \mathbf{w}^k; \mu_{\mathbf{p}}^k, \mu_{\mathbf{v}}^k) \tag{30}$$

$$\mathbf{w}^{k+1} = \arg\min_{\mathbf{p}} \mathcal{L}(\mathbf{p}^{k+1}, \mathbf{v}^{k+1}, \mathbf{u}^{k+1}, \mathbf{c}^{k+1}, \mathbf{w}; \mu_{\mathbf{p}}^k, \mu_{\mathbf{v}}^k) \tag{31}$$

3. **Update the Lagrange multipliers**:

$$\mu_{p_i}^{k+1} = \mu_{p_i}^k + \gamma(\nabla u_i^{k+1} - p_i^{k+1})$$

$$\mu_{v_i}^{k+1} = \mu_{v_i}^k + \gamma(u_i^{k+1} - v_i^{k+1})$$

4. **Endfor** until some stopping criterion meets and get **output**.

In the following, we will present the solutions of subproblems individually.

p-Subproblem

We can write terms from (26) associated with primal variable **p** and fixed all other variables as follows:

$$\mathbf{p}^{k+1} = \arg\min_{\mathbf{p}} \sum_{i=1}^{N} \Big\{ \int_{\Omega} \|p_i(x)\| dx - \int_{\Omega} (\mu_{p_i}^k(x))^T p_i(x) dx$$
$$+ \frac{\gamma}{2} \int_{\Omega} \|\nabla u_i^k(x) - p_i(x)\|^2 dx \Big\}. \tag{32}$$

Equation (32) is equivalent to the following:

$$\mathbf{p}^{k+1} = \arg\min_{\mathbf{p}} \sum_{i=1}^{N} \Big\{ \int_{\Omega} \|p_i(x)\| dx + \frac{\gamma}{2} \int_{\Omega} \|p_i(x) - X^k\|^2 dx \Big\}, \tag{33}$$

where $X^k = \nabla u_i^k(x) + \frac{\mu_{p_i}^k(x)}{\gamma}$. Equation (33) has a close form solution, and it can be solved by shrinkage operator; we can compute $\mathbf{p}^{k+1}$ as follows:

$$\mathbf{p}^{k+1} = S\Big(X^k, \frac{1}{\gamma}\Big). \tag{34}$$

S denotes the shrinkage operator, which is defined as follows:

$$S(X, \gamma) = \frac{X}{\|X\|} * \max(\|X\| - \gamma, 0).$$

v-Subproblem

The subproblem for **v** is as follows:

$$\mathbf{v}^{k+1} = \arg\min_{\mathbf{v}} \sum_{i=1}^{N} \Big\{ \lambda \int_{\Omega} |I(x) - c_i^k(\mathbf{w}^k)^T G(x)|^2 v_i(x) dx - \int_{\Omega} (\mu_{v_i}^k(x))^T v_i(x) dx$$
$$+ \frac{\gamma}{2} \int_{\Omega} |u_i^k(x) - v_i(x)|^2 dx + l_{\mathscr{U}}(\mathbf{v}) \Big\}. \tag{35}$$

Equation (35) is equivalent to the following:

$$\mathbf{v}^{k+1} = \arg\min_{\mathbf{v}} \sum_{i=1}^{N} \Big\{ \frac{\gamma}{2} \int_{\Omega} |v_i(x) - Y^k|^2 \Big\} + l_{\mathscr{U}}(\mathbf{v}), \tag{36}$$

where $Y^K = u_i^k(x) + \frac{\mu_{v_i}^k(x)}{\gamma} - \frac{\lambda |I(x) - c_i^k(\mathbf{w}^k)^T G(x)|^2}{\gamma}$. Because $\mathscr{U}$ is a convex simplex at any x in domain Ω, the solution is given by the following:

$$\mathbf{v}^{k+1} = \Pi_{\mathcal{U}}\left(\left[Y^k\right]_{i=1}^N\right), \tag{37}$$

where Π denotes the projection onto the simplex $\mathcal{U}$; for more details, please see Chen and Ye (2011).

u-Subproblem

The subproblem for **u** is as follows:

$$\mathbf{u}^{k+1} = \arg\min_{\mathbf{u}} \sum_{i=1}^N \left\{ \int_\Omega (\nabla u_i^k(x))^T \mu_{p_i}(x) + \frac{\gamma}{2} \int_\Omega \left\| \nabla u_i^k(x) - p_i^{k+1}(x) \right\|^2 dx \right.$$
$$\left. + \int_\Omega (u_i^k(x))^T \mu_{v_i}^k(x) + \frac{\gamma}{2} \int_\Omega \left| u_i(x) - v_i^{k+1}(x) \right|^2 dx \right\}. \tag{38}$$

Its identical representation is as follows:

$$\mathbf{u}^{k+1} = \arg\min_{\mathbf{u}} \sum_{i=1}^N \frac{\gamma}{2} \left\{ \int_\Omega \left\| \nabla u_i(x) - Z_1^k \right\|^2 + \left| u_i(x) - Z_2^k \right|^2 \right\}, \tag{39}$$

where $Z_1^k = p_i^{k+1}(x) - \frac{\mu_{p_i}^k(x)}{\gamma}$ and $Z_2^k = v_i^{k+1}(x) - \frac{\mu_{v_i}^k(x)}{\gamma}$. By using first optimality condition for each $\mathbf{u}^{k+1}$, we have the following:

$$\nabla^T \left(\nabla u_i(x) - Z_1^k \right) + \left(u_i(x) - Z_2^k \right) = 0.$$

The closed-form solution of $\mathbf{u}^{k+1}$ can be produced from the following equation:

$$(\nabla^T \nabla + I) u_i^{k+1}(x) = \nabla^T p_i^{k+1}(x) + v_i^{k+1}(x) - \frac{\nabla^T \mu_{p_i}^k(x)}{\gamma} - \frac{\nabla^T \mu_{v_i}^k(x)}{\gamma}.$$

We follow Wang et al. (2008), where diagonalized technique is used to get the fast solution for $\mathbf{u}^{k+1}$.

Solutions for Subproblems c, w, and Bias Field Estimation b

The **c**-subproblem can be formulated as follows:

$$\mathbf{c}^{k+1} = \arg\min_{\mathbf{c}} \sum_{i=1}^N \left\{ \lambda \int_\Omega \left| I(x) - c_i \mathbf{w}^T G(x) \right|^2 v_i(x) dx \right\}. \tag{40}$$

To find $\mathbf{c}^{k+1}$, we compute the similar solution used in basic MICO described in section "Multiplicative Intrinsic Component Optimization" with regularization parameter λ.

$$c^{k+1} = \frac{\int_\Omega \lambda I(x)\mathbf{b}(x)u_i^{k+1}(x)dx}{\int_\Omega \lambda \mathbf{b}^2(x)u_i^{k+1}(x)dx}, \quad i = 1, \cdots, N. \qquad (41)$$

Here $\mathbf{b}(x) = (\mathbf{w}^k)^T G(x)$ is the bias field.

$$\mathbf{w}^{k+1} = \left(\int_\Omega \lambda G(x) G^T(x) \left(\sum_{i=1}^N (c_i^{k+1})^2 u_i^{k+1}(x) \right) dx \right)^{-1}$$

$$\times \int_\Omega \lambda G(x) I(x) \left(\sum_{i=1}^N c_i^{k+1} u_i^{k+1}(x) \right) dx. \qquad (42)$$

The estimated bias field is calculated by $\mathbf{b}^{k+1}$ using the optimum vector $\mathbf{w}^{k+1}$ given by Eq. (42).

$$\mathbf{b}^{k+1} = (\mathbf{w}^{k+1})^T G(x) \qquad (43)$$

Spatiotemporal Regularization for 4D Segmentation

The TVMICO formulation in (21) can be further extended to 4D MICO with spatiotemporal regularization of the tissue membership functions for segmentation of 4D data, which is a series of 3D scans of the same subject at different time points. While the basic MICO formulation described in section "Multiplicative Intrinsic Component Optimization" allows for multiple 4D extensions with different spatiotemporal regularization mechanisms, we only provide a simple and natural 4D extension of the basic MICO formulation as an example in the following.

We first outline a model of serial MR images collected from the same subject at different periods before presenting the 4D MICO formulation. By employing rigid registration with six degrees of freedom, we assumed that all images in a longitudinal series are registered to the first image in the series. As a result, all of the registered images in the series are in a same space, denoted by Ω, which can be represented by a 4D image $I(x, t)$ with spatial variable $x \in \Omega$ and temporal variable t in a time period $[0, L]$. Here $I(\cdot, t)$ can be modeled as a series of images.

$$I(x, t) = b(x, t) J(x, t) + n(x, t) \qquad (44)$$

where $J(\cdot, t)$ is the true image, $b(\cdot, t)$ is the bias field, and $n(\cdot, t)$ is additive noise.

We assume there are N types of tissues in the image domain Ω. The true image $J(x, t)$ can be approximated by $J(x, t) = \sum_{i=1}^N c_i(t) u_i(x, t)$, where N is the number of tissues in Ω, $u_i(\cdot, t)$ is the membership function of the i-th tissue, and the constant $c_i(t)$ is the value of the true image $J(x, t)$ in the i-th tissue. For convenience, we represent the constants $c_1(t), \cdots, c_N(t)$ with a column vector

$\mathbf{c}(t) = (c_1(t), \cdots, c_N(t))^T$. The membership functions $u_1(x,t), \cdots, u_N(x,t)$ are also represented by a vector-valued function $\mathbf{u}(x,t) = (u_1(x,t), \cdots, u_N(x,t))^T$.

The bias field $b(\cdot, t)$ at each time point t is estimated by a linear combination of a set of smooth basis functions $g_1(x), \cdots, g_M(x)$. Using the vector representation in section "Mathematical Description of Multiplicative Intrinsic Components", the bias field $b(\cdot, t)$ at the time point t can be expressed as follows:

$$b(x,t) = \mathbf{w}(t)^T G(x), \qquad (45)$$

with $\mathbf{w}(t) = (w_1(t), \cdots, w_M(t))^T$, where $w_1(t), \cdots, w_M(t)$ are the time-dependent coefficients of the basis function $g_j(x), j = 1, \cdots, M$.

The spatiotemporal regularization of the membership functions $u_i(x,t)$ can be naturally taken into account in the following variational formulation with a data term (image-based term) and a spatiotemporal regularization term as follows:

$$\mathcal{G}(\mathbf{u}, \mathbf{c}, \mathbf{w}) = \lambda \int_{[0,L]} F(\mathbf{u}(\cdot,t), \mathbf{c}(t), \mathbf{w}(t)) dt + \sum_{i=1}^{N} TV(u_i) \qquad (46)$$

where $\lambda > 0$ is a constant, $F(\mathbf{u}(\cdot,t), \mathbf{c}(t), \mathbf{w}(t))$ is the data term defined in (8) for the image $I(\cdot, t)$ at the time point t, namely:

$$F(\mathbf{u}(\cdot,t), \mathbf{c}(t), \mathbf{w}(t)) = \int_{\Omega} \sum_{i=1}^{N} |I(x,t) - \mathbf{w}(t)^T G(x) c_i(t)|^2 u_i^q(x,t) dx,$$

and $TV(u_i)$ is the spatiotemporal regularization term on the membership function $\mathbf{u}$, which can be expressed as follows:

$$TV(u_i) = \int |\nabla u_i(x,t)| dx dt, \qquad (47)$$

where the gradient operator ∇ is with respect to the spatial and temporal variables x and t. We call the above variational formulation a 4D MICO formulation.

The minimization of the energy $\mathcal{G}$ is subject to the constraints on the membership function. Therefore, we solve the following constrained energy minimization problem:

$$\text{Minimize } \mathcal{G}(\mathbf{u}, \mathbf{c}, \mathbf{w}) \qquad (48)$$

$$\text{subject to } 0 \leq u_i(x) \leq 1, i = 1, \cdots, N, \text{ and } \sum_{i=1}^{N} u_i(x) = 1$$

The minimization of the energy $\mathcal{G}$ with respect to $\mathbf{c}(t)$ and $\mathbf{w}(t)$ is independent of the spatiotemporal regularization term in (46). The optimal vectors $\mathbf{c}(t)$ and

$\mathbf{w}(t)$ can be computed for each time point t independently from the image $I(\cdot, t)$ as in the energy minimization for the basic MICO formulation described in section "Optimization of Energy Function and Algorithm". The numerical technique in Li et al. (2010) for variational formulations with TV regularization can be used to minimize $\mathcal{G}$ with respect to the 4D membership function $\mathbf{u}$ subject to the constraint in Eq. (48). In our future research work focusing on 4D segmentation based on the fundamental MICO formulation, we will provide a detailed explanation of the numerical approach for addressing the constrained energy minimization problem in Eq. (48) and its modified variants.

Modified MICO Formulation with Weighting Coefficients for Different Tissues

By inserting weighting coefficients $\lambda 1, \cdots, \lambda N$ for the N tissues in the specification of the energy function $F(\mathbf{u}, \mathbf{c}, \mathbf{w})$ in Eq. (8), the basic MICO formulation in section "Multiplicative Intrinsic Component Optimization" may be adjusted. The modified energy is defined as follows:

$$F(\mathbf{u}, \mathbf{c}, \mathbf{w}) = \int_{\Omega} \sum_{i=1}^{N} \lambda_i |I(x) - \mathbf{w}^T G(x) c_i|^2 u_i^q(x) dx, \qquad (49)$$

here λ_i is the coefficient for the i-th tissue. The parameters $\lambda_1, \cdots, \lambda_N$ provide users the option of improving the outcomes of the standard MICO formulation in 2. For instance, if the i-th tissue is over-segmented using the standard MICO formulation in section "Multiplicative Intrinsic Component Optimization", the above-modified formulation in Eq. (49) with a large $\lambda_i > 1$ can be used instead.

Results and Discussions

Our approach has been thoroughly validated on both synthetic and real MRI data, including 1.5T and 3T MRI data. In this part, we first provide experimental results for various synthetic and actual MR images, including those with significant intensity inhomogeneities. We also give quantitative evaluation findings and comparisons with other well-known methodologies.

In our MICO applications for 1.5T and 3T MR images, we employ 20 polynomials of the first three orders as the basis functions $g_1, \cdots, g_M$ with $M = 20$. For images obtained from 1.5T and 3T MRI scanners, our technique with these 20 basis functions works effectively. The intensity inhomogeneities in high-field (e.g., 7T) MRI scanners exhibit more complex profiles than 1.5T and 3T MR pictures. More basis functions are required in this circumstance so that a wider variety of bias fields may be well represented by linear combinations. Given an appropriately large number of basis functions, any function can be well approximated by a linear

Fig. 2 Our method's bias correction and tissue segmentation outcomes on 1.5T (upper row) and 3T (bottom row) MR scanner data. The original image, bias field corrected image, and segmentation result are displayed in the left, center, and right columns, respectively

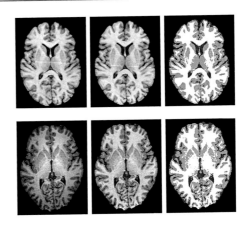

combination of a set of basis functions up to arbitrary precision (Powell 1981). The numerical stability of the computation of the inverse matrix A^{-1} in Eq. (14), with A being a $M \times M$ matrix, is a significant numerical challenge, especially when M is large. Thanks to the matrix analysis in section "Numerical Stability Using Matrix Analysis". We have demonstrated that the condition number of the matrix A is bounded by a constant as in Eq. (20), which is independent of the number of basis functions. This provides the numerical stability of the bias field computation, independent of the number of basis functions employed.

In our experiments, MICO has been used to 1.5T and 3T MRI data with promising results. In Fig. 2, we exhibit the bias field correction and segmentation outcomes of our technique for 1.5T and 3T MR images, accordingly. In the left, center, and right columns, the original images, bias field corrected images, and segmentation results are displayed, respectively. We tested MICO on the two images in the left column of Fig. 3 to show that our approach can deal with severe intensity inhomogeneities. The second, third, and fourth columns, respectively, show the estimated bias field, segmentation results, and bias field corrected images acquired by our approach. Despite the images' severe intensity inhomogeneities, our technique produces desirable bias field correction and tissue segmentation results, as demonstrated in Fig. 3.

The segmentation accuracy of our approach and the well-known software FSL, SPM, and FANTASM are quantitatively evaluated and compared in the following experiment. These three programs are available for free download at http://www.fmrib.ox.ac.uk/fsl/ (for FSL), http://www.fil.ion.ucl.ac.uk/spm/software/ (for SPM), and http://mipav.cit.nih.gov/ (for FANTASM). The data for our quantitative analysis was obtained from BrainWeb (http://www.bic.mni.mcgill.ca/brainweb/). BrainWeb also provides ground truth, which can be used to quantify segmentation accuracy.

It is worth noting that the intensity inhomogeneities created by BrainWeb are linear, which makes them reasonably straightforward to handle. To test segmentation algorithms in a more challenging scenario, we created simulated MR images with nonlinear intensity inhomogeneities as shown below. The range of values of the

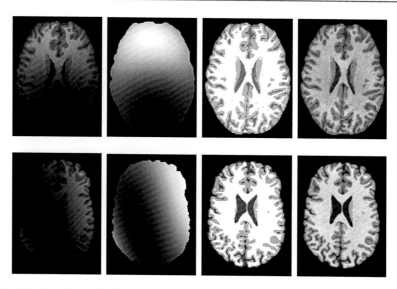

Fig. 3 The left column displays the results for images with extreme intensity inhomogeneity. Columns 2, 3, and 4 show the estimated bias fields, segmentation results, and bias field corrected images, respectively

bias field in the interval $[1-\alpha, 1+\alpha]$ with $\alpha > 0$ indicates the degree of intensity inhomogeneity. We created five image sets with $\alpha = 0.1, 0.2, 0.3, 0.4,$ and 0.5. We constructed six alternative bias fields with values in $[1-\alpha, 1+\alpha]$ for each α and multiplied them with the original image obtained from BrainWeb to obtain six images with varying intensity inhomogeneities. The images were then subjected to six different degrees of noise. Thus, the five sets of images have 30 images with varying degrees of intensity inhomogeneities and noise levels. We first show the segmentation results of the 4 tested methods for 2 of the 30 images in Fig. 4; we first show the segmentation results of the 4 tested methods for 2 of the 30 images, 1 with the lowest degree of intensity inhomogeneity (generated with $\alpha = 0.1$) and the other 1 the highest degree of intensity inhomogeneity (generated with $\alpha = 0.5$). By visual comparison, the segmentation results of the four approaches for an image with a low degree of intensity inhomogeneity seem similar, as shown in the upper row of Fig. 4. Our technique has a distinct benefit for images with a high degree of intensity inhomogeneity, as seen in the lower row of Fig. 4.

By evaluating the segmentation results using the Jaccard similarity (JS) index (Shattuck et al. 2001), a more objective and exact comparison of the segmentation accuracy of the four segmentation techniques can be done.

$$J(\mathcal{S}_1, \mathcal{S}_2) = \frac{|\mathcal{S}_1 \cap \mathcal{S}_2|}{|\mathcal{S}_1 \cup \mathcal{S}_2|}, \qquad (50)$$

here $|\cdot|$ indicates a region's area, $\mathcal{S}_1$ is the algorithm's segmented region, and $\mathcal{S}_2$ is the corresponding region generated from a reference segmentation result or the

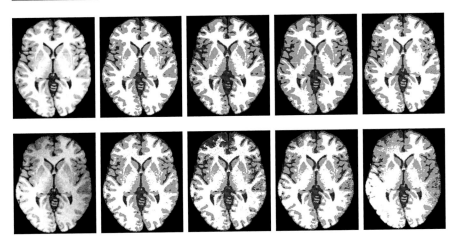

Fig. 4 Comparison of our method with SPM, FSL, and FANTASM on synthetic images with different degrees of intensity inhomogeneities. The input images are displayed in the first column from the left, containing one with a low degree of intensity inhomogeneity (in the top row) and one with a high degree of intensity inhomogeneity (in the lower row). The segmentation results of our technique, SPM, FSL, and FANTASM are displayed in the second, third, fourth, and fifth columns, respectively

ground truth. We have the ground truth of the segmentation of the WM, GM, and CSF for synthetic data from the BrainWeb, which can be directly utilized as S_2 in Eq. (50) to compute the JS index. The greater the JS value, the more similar the algorithm segmentation is to the reference segmentation.

The comparison of JS values of the 4 approaches on the 30 synthetic images with varying degrees of intensity inhomogeneities and different amounts of noise is shown in Fig. 5. The box plot of the JS values for the GM and WM generated by our approach (MICO and TVMICO), SPM, FSL, and FANTASM is shown in Fig. 5. In terms of segmentation accuracy and robustness, the box plot of the JS values in Fig. 5 clearly shows that MICO and TVMICO perform better than SPM, FSL, and FANTASM.

We see that the box in the box plot for the basic MICO is comparatively shorter, and there are no outliers in the JS values throughout all 30 test images. This demonstrates the basic MICO's intended robustness. The TVMICO is slightly more accurate than the regular MICO; however there are outliers in the TVMICO's JS values. The performance of TVMICO is determined by the parameter λ in Eq. (21), which must be modified in some circumstances. We set $\lambda = 0.01$ for all 30 test images in this experiment and observed that the results are generally favorable, except for one scenario, which results in outliers in the box plot in Fig. 5. In comparison, the basic MICO is more robust and has more steady performance than TVMICO, while the latter is somewhat more accurate in most circumstances. In reality, the difference in segmentation accuracy between MICO and TVMICO is not substantial for images with reasonable noise levels. When robustness is a priority and the image noise level is low, we recommend using

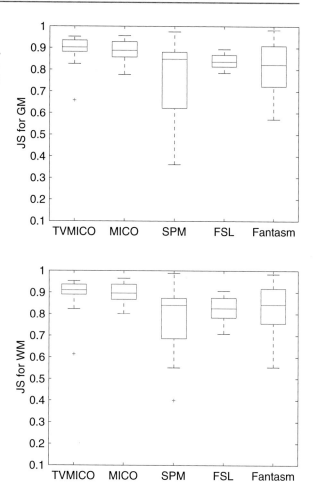

Fig. 5 Quantitative evaluation of TVMICO (with $\lambda = 0.01$), MICO, SPM, FSL, and FANTASM segmentation outcomes for 30 images using Jaccard similarity index with ground truth

the basic MICO. Otherwise, TVMICO has better segmentation and bias-corrected results; see Figs. 6, 7 and 8. When the noise level is high, the results obtained by our proposed TVMICO outperform the basic MICO. Figures 6, 7 and 8 show the progress in segmentation and bias field correction in zoomed images. We added various intensity inhomogeneities and noise to the images generated from the atrophy simulator to assess the performance of our technique in the presence of intensity inhomogeneities and noise. In this experimental result, we set $\lambda = 0.008$ in the TVMICO formulation in Eq. (23). We observed that the performance of the TVMICO formulation is affected by the parameter λ as well as certain extra parameters in the numerical method for energy minimization with respect to the membership functions. More information on the implementation and validations of the 4D MICO formulation in Eq. (46) and its modified variants will be published in a subsequent publication as an extension of this study. In the case of fully automatic

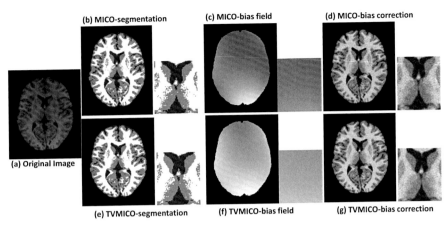

Fig. 6 On BrainWeb data, we obtained results for tissue segmentation and bias correction using our proposed MICO and TVMICO. Figure (**a**) shows the original image, (**b**) and (**e**) show the segmentation results, (**c**) and (**f**) show bias fields, and (**d**) and (**g**) provide bias field corrected images

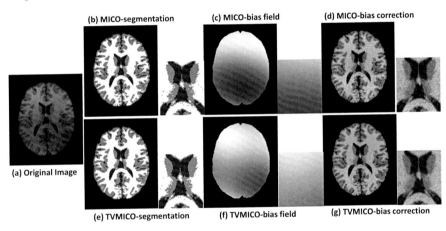

Fig. 7 On BrainWeb data, we obtained results for tissue segmentation and bias correction using our proposed MICO and TVMICO. Figure (**a**) shows the original image, (**b**) and (**e**) show the segmentation results, (**c**) and (**f**) show bias fields, and (**d**) and (**g**) provide bias field corrected images

segmentation of huge data sets, robustness and stability of performance are critical. The basic MICO is preferred to TVMICO because of its robustness and stability.

MICO's estimated bias field $\hat{b}$ can be used to compute the bias field corrected image $I/\hat{b}$. We examined the performance of MICO's bias field correction and compared it to two well-known bias field correction methods, namely, the N3 approach described in Sled et al. (1998) and the entropy minimization method proposed in Likar et al. (2001).

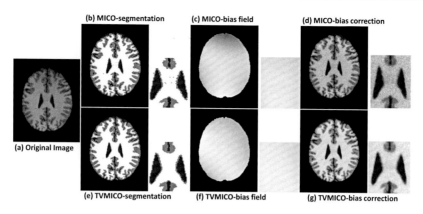

Fig. 8 On BrainWeb data, we obtained results for tissue segmentation and bias correction using our proposed MICO and TVMICO. Figure (**a**) shows the original image, (**b**) and (**e**) show the segmentation results, (**c**) and (**f**) show bias fields, and (**d**) and (**g**) provide bias field corrected images

The performance of bias field correction can be measured by calculating the coefficient of variations (CV) and coefficient of joint variation from the intensity inhomogeneities of the bias field corrected images (CJV).

The CV is defined for each tissue T (WM or GM) by the following:

$$CV(T) = \frac{\sigma(T)}{\mu(T)},$$

where $\sigma(T)$ and $\mu(T)$ denote the standard deviation and mean of the intensities in the tissue T, respectively. The CJV is defined as follows:

$$CJV = \frac{\sigma(WM) + \sigma(GM)}{|\mu(WM) - \mu(GM)|}.$$

The CV and CJV of the bias field corrected images are used to evaluate the performance of bias field correction, with lower CV and CJV values indicating better bias field correction outcomes.

We used our approach, as well as the N3 and entropy minimization algorithms included in the MIPAV software, to analyze 15 pictures from 3 Tesla MRI scanners. MIPAV software is freely accessible at http://mipav.cit.nih.gov/. The CV and CJV values of the 3 tested techniques for the 15 images are displayed in Fig. 9, demonstrating that our method outperforms the N3 and entropy minimization methods.

It is worth noting that the GM and WM are the ground truth in the conventional definition of CV and CJV in the literature on bias field correction (Vovk et al. 2007). We used an approximation of the ground truth of GM/WM by the intersection of the segmented GM/WM obtained by applying the K-means algorithm to the bias-

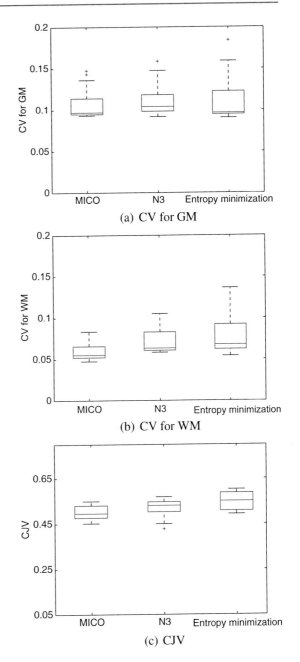

Fig. 9 In terms of CV and CJV, we compared the performance of our method MICO, the N3 algorithm, and the entropy reduction method on 15 images from 3T MR scanners (**a**) CV for GM. (**b**) CV for WM. (**c**) CJV corrected images by the three compared bias field correction methods: our method and the well-known N3 method (Sled et al. 1998) and the entropy minimization method (Likar et al. 2001).

As previously stated, we only employed 20 polynomials as basis functions in estimating the bias field. However, MICO's bias field rectification capabilities can be improved by adding additional and various types of basis functions, such as B-spline functions, to expand the spectrum of bias fields represented as linear combinations of the basis functions. It would allow MICO to be used to very high-field MRI (e.g., 7-Tesla) and other medical images with extreme intensity inhomogeneities.

Conclusion

First, in this chapter, we introduced an energy minimization-based technique named multiplicative intrinsic component optimization (MICO) for bias field estimation and segmentation of MR images. Second, we expanded the MICO formulation by using total variation (TV) as a convex regularization. Furthermore, we implemented the alternating direction method of multipliers (ADMM) for the TV-based MICO model, which can solve the model efficiently and guarantee its convergence. We computed the bias field using matrix and vector calculus, and we utilized matrix analysis to establish the numerical stability of the computation for bias field optimization. The evaluation and comparison of our technique with other methods on synthetic and actual MR data indicate its robustness, accuracy, and efficiency. Our approach has been applied effectively to 1.5T and 3T MR images with promising outcomes. In comparison to other popular software, the results of the experiments reveal that our technique provides essential improvements in terms of segmentation accuracy and robustness. We also demonstrated that the MICO formulation can be naturally extended to 3D/4D segmentation with spatial/spatiotemporal regularization, producing encouraging results.

References

Ahmed, M., Yamany, S., Mohamed, N., Farag, A., Moriarty, T.: A modified fuzzy c-means algorithm for bias field estimation and segmentation of MRI data. IEEE Trans. Med. Imaging **21**(3), 193–199 (2002)

Axel, L., Costantini, J., Listerud, J.: Intensity correction in surface-coil MR imaging. Am. J. Radiol. **148**(2), 418–420 (1987)

Barrow, H., Tenenbaum, J.: Recovering intrinsic scene characteristics from images. In: Hanson, A., Riseman, E. (eds.) Computer Vision Systems, pp. 3–26. Academic, Orlando (1978)

Bezdek, J.C., Ehrlich, R., Full, W., FCM: The fuzzy c-means clustering algorithm. Comput. Geosci. **10**(2–3), 191–203 (1984)

Chen, K.: Introduction to Variational Image-Processing Models and Applications (2013)

Chen, Y., Ye, X.: Projection onto a simplex, arXiv preprint arXiv:1101.6081 (2011)

Condon, B.R., Patterson, J., Wyper, D.: Image nonuniformity in magnetic resonance imaging: its magnitude and methods for its correction. Br. J. Radiol. **60**(1), 83–87 (1987)

Dawant, B., Zijdenbos, A., Margolin, R.: Correction of intensity variations in MR images for computer-aided tissues classification. IEEE Trans. Med. Imaging **12**(4), 770–781 (1993)

Gabay, D., Mercier, B.: A dual algorithm for the solution of nonlinear variational problems via finite element approximation. Comput. Math. Appl. **2**(1), 17–40 (1976)

Glowinski, R., Marroco, A.: Sur l'approximation, par éléments finis d'ordre un, et la résolution, par pénalisation-dualité d'une classe de problèmes de dirichlet non linéaires. ESAIM: Mathematical

Modelling and Numerical Analysis-Modélisation Mathématique et Analyse Numérique **9**(R2) 41–76 (1975)

Goldstein, T., Osher, S.: The split bregman method for L1 regularized problems, UCLA CAM Report 08-29 (2009)

Goldstein, T., Osher, S.: The split bregman method for l1-regularized problems. SIAM J. Imaging Sci. **2**(2), 323–343 (2009)

Golub, G., Loan, C.V.: Matrix Computations, 3rd edn. The Johns Hopkins University Press, Baltimore/London (1996)

Guillemaud, R., Brady, M.: Estimating the bias field of MR images. IEEE Trans. Med. Imaging **16**(3), 238–251 (1997)

He, Y., Hussaini, M.Y., Ma, J., Shafei, B., Steidl, G.: A new fuzzy c-means method with total variation regularization for segmentation of images with noisy and incomplete data. Pattern Recogn. **45**(9), 3463–3471 (2012)

Horn, R., Johnson, C.: Matrix Analysis. Cambridge University Press, Cambridge (1985)

Johnston, B., Atkins, M.S., Mackiewich, B., Anderson, M.: Segmentation of multiple sclerosis lesions in intensity corrected multispectral MRI. IEEE Trans. Med. Imaging **15**(2), 154–169 (1996)

Juntu, J., Sijbers, J., Van Dyck, D., Gielen, J.: Bias field correction for MRI images. In: Computer Recognition Systems, pp. 543–551. Springer, Berlin/Heidelberg (2005)

Kimmel, R., Elad, M., Shaked, D., Keshet, R., Sobel, I.: A variational framework for retinex. Int. J. Comput. Vis. **52**(1), 7–23 (2003)

Leemput, V., Maes, K., Vandermeulen, D., Suetens, P.: Automated model-based bias field correction of MR images of the brain. IEEE Trans. Med. Imag. **18**(10), 885–896 (1999)

Li, C., Kao, C., Gore, J.C., Ding, Z.: Minimization of region-scalable fitting energy for image segmentation. IEEE Trans. Imag. Proc. **17**(10), 1940–1949 (2008)

Li, C., Huang, R., Ding, Z., Gatenby, C., Metaxas, D., Gore, J.: A variational level set approach to segmentation and bias correction of medical images with intensity inhomogeneity. In: Proceedings of Medical Image Computing and Computer Aided Intervention (MICCAI). LNCS, vol. 5242, Part II, pp. 1083–1091 (2008)

Li, C., Xu, C., Anderson, A.W., Gore, J.C.: MRI tissue classification and bias field estimation based on coherent local intensity clustering: a unified energy minimization framework. In: International Conference on Information Processing in Medical Imaging, pp. 288–299. Springer (2009)

Li, F., Ng, M.K., Li, C.: Variational fuzzy Mumford–Shah model for image segmentation. SIAM J. Appl. Math. **70**(7), 2750–2770 (2010)

Li, C., Gore, J.C., Davatzikos, C.: Multiplicative intrinsic component optimization (mico) for MRI bias field estimation and tissue segmentation. Mag. Resonan. Imaging **32**(7), 913–923 (2014)

Li, F., Osher, S., Qin, J., Yan, M.: A multiphase image segmentation based on fuzzy membership functions and l1-norm fidelity. J. Sci. Comput. **69**(1), 82–106 (2016)

Likar, B., Viergever, M., Pernus, F.: Retrospective correction of MR intensity inhomogeneity by information minimization. IEEE Trans. Med. Imaging **20**(12), 1398–1410 (2001)

Lions, P., Mercier, B.: Splitting algorithms for the sum of two nonlinear operators. SIAM J. Numer. Anal. **16**(6), 964–979 (1979)

Liu, J., Huang, T.-Z., Selesnick, I.W., Lv, X.-G., Chen, P.-Y.: Image restoration using total variation with overlapping group sparsity. Inf. Sci. **295**, 232–246 (2015)

Liu, Z., Wali, S., Duan, Y., Chang, H., Wu, C., Tai, X.-C.: Proximal ADMM for Euler's elastica based image decomposition model. Numer. Math. Theory Methods Appl. **12**(2), 370–402 (2019)

McVeigh, E.R., Bronskil, M.J., Henkelman, R.M.: Phase and sensitivity of receiver coils in magnetic resonance imaging. Med. Phys. **13**(6), 806–814 (1986)

Narayana, P.A., Brey, W.W., Kulkarni, M.V., Sivenpiper, C.L.: Compensation for surface coil sensitivity variation in magnetic resonance imaging. Magn. Reson. Imaging **6**(3), 271–274 (1988)

Pham, D.: Spatial models for fuzzy clustering. Comput. Vis. Image Underst. **84**(2), 285–297 (2001)

Pham, D., Prince, J.: Adaptive fuzzy segmentation of magnetic resonance images. IEEE Trans. Med. Imaging **18**(9), 737–752 (1999)

Powell, M.J.D.: Approximation Theory and Methods. Cambridge University Press, Cambridge (1981)

Rudin, L.I., Osher, S., Fatemi, E.: Nonlinear total variation based noise removal algorithms. Phys. D: Nonlinear Phenom. **60**(1–4), 259–268 (1992)

Salvado, O., Hillenbrand, C., Wilson, D.: Correction of intensity inhomogeneity in MR images of vascular disease. In: EMBS'05, pp. 4302–4305. IEEE, Shanghai (2005)

Shattuck, D.W., Sandor-Leahy, S.R., Schaper, K.A., Rottenberg, D.A., Leahy, R.M.: Magnetic resonance image tissue classification using a partial volume model. Neuroimage **13**, 856–876 (2001)

Simmons, A., Tofts, P.S., Barker, G.J., Arrdige, S.R.: Sources of intensity nonuniformity in spin echo images at 1.5t. Magn. Reson. Med. **32**(1), 121–128 (1991)

Sled, J., Zijdenbos, A., Evans, A.: A nonparametric method for automatic correction of intensity nonuniformity in MRI data. IEEE Trans. Med. Imaging **17**(1), 87–97 (1998)

Stockman, G., Shapiro, L.G.: Computer Vision. Prentice Hall, Upper Saddle River (2001)

Styner, M., Brechbuhler, C., Szekely, G., Gerig, G.: Parametric estimate of intensity inhomogeneities applied to MRI. IEEE Trans. Med. Imaging **19**(3), 153–165 (2000)

Tappen, M., Freeman, W., Adelson, E.: Recovering intrinsic images from a single image. IEEE Trans. Pattern Anal. Mach. Intell. **27**(9), 1459–1472 (2005)

Tincher, M., Meyer, C.R., Gupta, R., Williams, D.M.: Polynomial modeling and reduction of RF body coil spatial inhomogeneity in MRI. IEEE Trans. Med. Imaging **12**(2), 361–365 (1993)

Tu, X., Gao, J., Zhu, C., Cheng, J.-Z., Ma, Z., Dai, X., Xie, M.: MR image segmentation and bias field estimation based on coherent local intensity clustering with total variation regularization, Med. Biol. Eng. Computi. **54**(12), 1807–1818 (2016)

Tustison, N.J., Avants, B.B., Cook, P.A., Zheng, Y., Egan, A., Yushkevich, P.A., Gee, J.C.: N4itk: improved n3 bias correction. IEEE Trans. Med. Imaging **29**(6), 1310–1320 (2010)

Tustison, N., Avants, B., Cook, P., Zheng, Y., Egan, A., Yushkevich, P., Gee, J.: N4itk: improved n3 bias correction. IEEE Trans. Med. Imaging **29**(6), 1310–1320 (2010)

Vovk, U., Pernus, F., Likar, B.: A review of methods for correction of intensity inhomogeneity in MRI. IEEE Trans. Med. Imaging **26**(3), 405–421 (2007)

Vovk, U., Pernus, F., Likar, B.: A review of methods for correction of intensity inhomogeneity in MRI. IEEE Trans. Med. Imaging **26**(3), 405–421 (2007)

Wali, S., Zhang, H., Chang, H., Wu, C.: A new adaptive boosting total generalized variation (TGV) technique for image denoising and inpainting. J. Vis. Commun. Image Represent. **59**, 39–51 (2019a)

Wali, S., Shakoor, A., Basit, A., Xie, L., Huang, C., Li, C.: An efficient method for Euler's elastica based image deconvolution. IEEE Access **7**, 61226–61239 (2019b)

Wang, Y., Yang, J., Yin, W., Zhang, Y.: A new alternating minimization algorithm for total variation image reconstruction. SIAM J. Imaging Sci. **1**(3), 248–272 (2008)

Weiss, Y.: Deriving intrinsic images from image sequences. In: Proceedings of 8th International Conference on Computer Vision (ICCV), vol. II, pp. 68–75 (2001)

Wells, W., Grimson, E., Kikinis, R., Jolesz, F.: Adaptive segmentation of MRI data. IEEE Trans. Med. Imaging **15**(4), 429–442 (1996)

Wicks, D.A.G., Barker, G.J., Tofts, P.S.: Correction of intensity nonuniformity in MR images of any orientation. Magn. Reson. Imag. **11**(2), 183–196 (1993)

Zheng, X., Lei, Q., Yao, R., Gong, Y., Yin, Q.: Image segmentation based on adaptive k-means algorithm. EURASIP J. Image Video Process. **2018**(1), 1–10 (2018)

Data-Informed Regularization for Inverse and Imaging Problems

35

Jonathan Wittmer and Tan Bui-Thanh

Contents

Introduction	1236
A Data-Informed Regularization (DI) Approach	1238
Data-Informed Regularization Derivation	1238
A Statistical Data-Informed (DI) Inverse Framework	1245
Properties of the DI Regularization Approach	1250
Applications to Imaging Problems	1256
Image Deblurring	1256
Image Denoising	1265
X-Ray Tomography	1265
Conclusions	1269
References	1271

Submitted to the editors DATE.

This work was partially funded by the National Science Foundation awards NSF-1808576 and NSF-CAREER-1845799; by the Defense Threat Reduction Agency award DTRA-M1802962; by the Department of Energy award DE-SC0018147; by KAUST; by 2018 ConTex award; and by 2018 UT-Portugal CoLab award. The authors are grateful to the supports.

J. Wittmer (✉)
Department of Aerospace Engineering and Engineering Mechanics, UT Austin, Austin, TX, USA
e-mail: jonathan.wittmer@utexas.edu

T. Bui-Thanh (✉)
Department of Aerospace Engineering and Engineering Mechanics, The Oden Institute for Computational Engineering and Sciences, UT Austin, Austin, TX, USA
e-mail: tanbui@ices.utexas.edu

© Springer Nature Switzerland AG 2023
K. Chen et al. (eds.), *Handbook of Mathematical Models and Algorithms in Computer Vision and Imaging*, https://doi.org/10.1007/978-3-030-98661-2_77

Abstract

This chapter presents a new regularization method for inverse and imaging problems, called data-informed (DI) regularization, that implicitly avoids regularizing the data-informed directions. Our approach is inspired by and has a rigorous root in disintegration theory. We shall, however, present an elementary and constructive path using the classical truncated SVD and Tikhonov regularization methods. Deterministic and statistical properties of the DI approach are rigorously discussed, and numerical results for image deblurring, image denoising, and X-ray tomography are presented to verify our findings.

Keywords

Inverse problems · Imaging · Tikhonov regularization · Truncated SVD · Data-informed regularization

Introduction

Regularization is often employed to facilitate the well-posedness of inverse (and imaging) problems. An inverse solution is thus a trade-off between the data misfit and the regularization. Due to noise and limited availability, available data typically informs limited directions in the parameter space where the inverse solution resides. A desired regularization, we argue, should minimally interfere with these data-informed directions. However, most regularization techniques regularize all parameter directions, including the data-informed ones, thus polluting the resulting inverse solution. Finding a "right" regularization remains an open problem in inverse and imaging communities.

Over the past decades, many different regularization approaches have been proposed including Tikhonov regularization (Tikhonov and Arsenin 1977), total variation regularization (Rudin et al. 1992; Beck and Teboulle 2009), and non-convex regularization strategies (Ramirez-Giraldo et al. 2011; Babacan et al. 2009; Nikolova 2005), to name a few. In the Bayesian statistical framework, these regularization strategies can be encoded as prior distributions for the inverse solutions. Perhaps the simplest and the most popular regularization strategy is the Tikhonov approach, which corresponds to a Gaussian prior in the Bayesian framework (Stuart 2010). One shortcoming of the Tikhonov prior is that it tends to be a smoothing prior (Mueller and Siltanen 2012), highly diffusing discontinuities. The total variation (TV) prior, which induces an anisotropic diffusion, seeks to minimally penalize discontinuities in the inverse solution (Rudin et al. 1992; Beck and Teboulle 2009; Mueller and Siltanen 2012). However, the TV prior is known to produce a staircasing effect due to non-differentiability of the TV functional

(Nikolova 2004). Because of the lack of differentiability, smooth approximations to the TV prior can be used, or more sophisticated optimization methods must be employed (Mueller and Siltanen 2012; Goldstein and Osher 2009). Similarly, inverse formulations using non-convex priors also require advanced optimization methods such as alternating direction method of multipliers (ADMM) (Chartr and Wohlberg 2013; Boley 2013; Boyd et al. 2010) or iteratively reweighted least-squares (IRLS) (Chartr and Yin 2008) to find the inverse solution.

This chapter presents a new regularization method for inverse and imaging problems, called data-informed (DI) regularization, that implicitly avoids regularizing the data-informed directions. *Our approach is inspired by and has a rigorous root in the disintegration theory.* We have, however, discovered a constructive path to understand our approach using the classical truncated SVD and Tikhonov regularization methods. The goal of this chapter is to share this constructive path to the DI approach and presents advantages/disadvantages of the DI approach on several existing applications in imaging. As will be shown theoretically and numerically, the DI approach avoids polluting the data-informed directions while regularizing the less data-informed ones.

Compared to existing approaches, our method has many distinct and advantageous features: (1) it automatically determines the directions equally informed by the data and any Tikhonov regularization while leaving the most informative directions untouched. In fact, we will show that, similar to the balanced truncation idea in control theory (see, e.g., Gugercin and Antoulas 2004; Antoulas 2005 and the references therein), this is done implicitly by seeking directions in parameter space that balance the information from regularization and data and removing the regularization on them. (2) We will show that our approach has an intuitive statistical interpretation, namely, it transforms both the data distribution (i.e., the likelihood) and prior distribution (induced by Tikhonov regularization) to the same Gaussian distribution whose covariance matrix is diagonal and the diagonal elements are exactly the singular values of a composition of the prior covariance matrix, the forward map, and the noise covariance matrix. (3) Though constructively derived and its insights obtained from the truncated singular value decomposition (SVD), the inverse solution resulting from our approach does not necessarily require the computation of an SVD, which may not be feasible for large-scale applications. We will present a nested matrix-free approach to obtain an approximate inverse solution. Our approach is thus more expensive than Tikhonov regularization when truncated SVD is not affordable. (4) By construction, features in DI solutions, dictated by the data-informed directions, are insensitive to the regularization parameter. For many inverse and imaging problems, these features dominate the solution, and thus the inverse solution resulting from our regularization technique is robust with respect to regularization parameter values. These findings will be demonstrated and supported by various numerical results from deblurring, denoising, and X-ray tomography problems.

A Data-Informed Regularization (DI) Approach

Data-Informed Regularization Derivation

In this section we review the key ideas behind regularization by truncation using the singular value decomposition (SVD). This provides the basic insights into the data-informed regularization technique. A statistical interpretation of the data-informed inverse framework will be discussed in section "A Statistical Data-Informed (DI) Inverse Framework". To begin, let us consider a linear inverse problem to determine $x \in \mathbb{R}^p$ given

$$y = Ax + e, \tag{1}$$

where $A \in \mathbb{R}^{d \times p}$, $e \sim \mathcal{N}\left(0, \lambda^2 I\right)$, $I \in \mathbb{R}^{d \times d}$, and $y \in \mathbb{R}^d$. In the following, the identity matrix I may have different size at different places and the actual size should be clear from the context. The simplest approach to attempt to solve this inverse problem is perhaps the least-squares approach:

$$\min_x \frac{1}{2} \|Ax - y\|^2, \tag{2}$$

where $\|\cdot\|$ denotes the standard Euclidean norm. Figure 1a plots the exact synthetic solution (black curve) against the least-squares solution (red curve) for a deconvolution problem with $d = p = 101$ and $\lambda = 0.05$. As can be seen, the least-squares solution blows up (or is unstable) due to the ill-conditioning of $A^T A$, which is not surprising since the inverse problem is (typically) ill-posed.

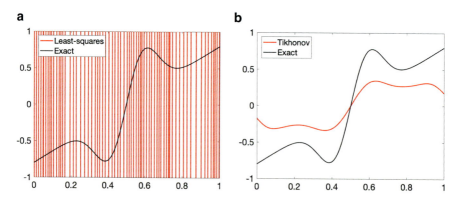

Fig. 1 Deconvolution using (**a**) the least-squares approach and (**b**) a Tikhonov regularization with regularization parameter $\alpha = 1$ and $x_0 = 0$

35 Data-Informed Regularization for Inverse and Imaging Problems

To overcome the ill-posedness, a classical Tikhonov regularization approach casts the above inverse problem into

$$\min_{x} \frac{1}{2} \|Ax - y\|^2 + \frac{\alpha}{2} \|x - x_0\|^2,$$

where x_0 is given. A Tikhonov solution is presented in Fig. 1b for $\alpha = 1$ and $x_0 = 0$. Though this approach stabilizes the solution, it also smooths out the solution everywhere.

Regularization by truncation does not require an explicit introduction of a regularization term as in Tikhonov regularization. For example, the truncated SVD starts with the SVD decomposition of A and then truncates all the singular vectors U_j and V_j corresponding to sufficiently small singular values, i.e.,

$$A = U\Sigma V^T$$

$$= \begin{pmatrix} U_1 & U_n U_{n+1} & U_d \\ \vdots & \cdots & \end{pmatrix} \begin{pmatrix} \sigma_1 & & & \\ & \ddots & & \\ & & \sigma_n & \\ & & & 0 \\ & & & & \ddots \\ & & & & & 0 \end{pmatrix} \begin{pmatrix} V_1^T \\ \vdots \\ V_n^T \\ V_{n+1}^T \\ \vdots \\ V_p^T \end{pmatrix}$$

$$= U^n \Sigma^n (V^n)^T,$$

where $U^n := [U_1, \ldots, U_n]$ (the first n columns of U corresponding to n nonzero singular values (This rank-n decomposition is often known as the reduced SVD.)), $\Sigma^n := \text{diag}[\sigma_1, \ldots, \sigma_n]$ ($\sigma_1 \geq \sigma_2 \geq \ldots \geq \sigma_n$), $n \leq \min\{d, p\}$, and $V^n := [V_1, \ldots, V_n]$ (the first n columns of V corresponding to n nonzero singular values). U^n forms an orthonormal basis for the column space of A, and V^n forms an orthonormal basis for the row space of A. The solution of (2) with this rank-n truncation using the pseudo-inverse $A^\dagger$ reads

$$x_{\text{SVD}}^n = A^\dagger y = V^n (\Sigma^n)^{-1} (U^n)^T y = \sum_{i=1}^{n} \frac{U_i^T y}{\sigma_i} V_i,$$

However, to avoid potentially dividing by very small singular values, a truncated SVD (TSVD) (Hansen 1990) with rank less than n is typically used. The rank-r TSVD solution using only the r largest singular values (with $r \leq n$) can be written as

$$x^r_{\text{TSVD}} := \sum_{i=1}^{r} \frac{U_i^T y}{\sigma_i} V_i, \quad (3)$$

Figure 2 applies the TSVD approach to the deconvolution problem and compares the results with the Tikhonov regularization. As can be seen, TSVD solutions are stable and do not seem to over-regularize the solution. However, as r increases, TSVD solutions tend to be more oscillatory (more unstable). How can this behavior of TSVD be explained?

The answer lies on the fact that the jth column of A is the observational vector when the parameter x is the jth canonical basis vector in $\mathbb{R}^p$. Thus the range space (column space) of A can be understood as the *observable subspace* in $\mathbb{R}^d$. Within this observable subspace, we say that the subspace spanned by U_j, i.e., $span\{U_j\}$, is more observable than the subspace spanned by U_i, i.e., $span\{U_i\}$, when $j < i$. Equivalently, $span\{U_i\}$ is less observable than $span\{U_j\}$. With this (relative) definition, $span\{U_1\}$ is most observable, while $span\{U_n\}$ is least observable. Clearly $j < i$ implies $1/\sigma_j \leq 1/\sigma_i$. Consequently, in the TSVD solution (3), less

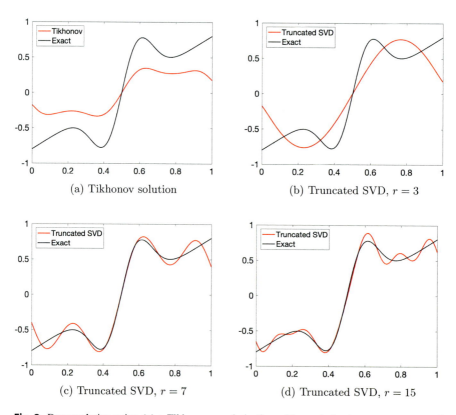

Fig. 2 Deconvolution using (**a**) a Tikhonov regularization with regularization parameter $\alpha = 1$ and $x_0 = 0$; (**b**) truncated SVD with $r = 3$; (**c**) truncated SVD with $r = 10$; and (**d**) truncated SVD with $r = 15$

observable modes (directions) U_i tend to promote oscillation and/or instability. In particular, U_n is the least stable (or most oscillatory) direction. As r increases, more oscillatory directions are amplified and added to the TSVD solution. This is exactly what Fig. 2 shows.

Given r (to be chosen based on the noise level ε), we define the **data-more-informed** parameter subspace as the row subspace spanned by $\{V_1, \ldots, V_r\}$ (corresponding to the **observable** subspace spanned by $\{U_1, \ldots, U_r\}$). Similarly, we define the **data-less-informed** parameter subspace as the row subspace spanned by $\{V_{r+1}, \ldots, V_n\}$ (corresponding to the less observable subspace spanned by $\{U_{r+1}, \ldots, U_n\}$). For brevity, we use **data-informed** and **data-uninformed** instead of **data-more-informed** and **data-less-informed**, though the latter is more precise as the definitions are relative. Additionally, we use *modes and directions* interchangeably to refer to the corresponding singular vectors V_j themselves, rather than the subspaces spanned by them.

The TSVD solution (3) clearly resides in the **data-informed** parameter subspace. The question is *where to truncate so that the solution is data-informed?* The result of Fig. 2 and its discussion suggest that r should be neither too large nor too small. That is, we seek to find r such that the solution captures information informed by the data while being least oscillatory. Clearly r is problem-dependent. For example, inspired by the Morozov's discrepancy principle (Morozov 1966), if the noise level ε is given (or can be estimated), r can be chosen such that $\sigma_j \geq \varepsilon$ for $j \leq r$.

A closer look at the TSVD solution (3) shows that the truncated SVD approach zeroes out the data-uninformed modes V_j for $j \geq r + 1$. We next show that this is equivalent to infinitely regularizing data-uninformed directions. To see this, let us now consider a regularization scheme where the data-uninformed modes are penalized infinitely, i.e., formally

$$\min \frac{1}{2} \|Ax - y\|^2 + \frac{1}{2} \|L(x - x_0)\|^2, \qquad (4)$$

where

$$L^T L := \infty \left[I - V^r (V^r)^T \right] = \infty (V^r)^\perp \left((V^r)^\perp \right)^T$$

$$= \left[V^r, (V^r)^\perp \right] \begin{bmatrix} 0 & 0 \\ 0 & \infty I \end{bmatrix} \left[V^r, (V^r)^\perp \right]^T,$$

and $\left[I - V^r (V^r)^T \right]$ is the orthogonal projection onto the data-uninformed subspace spanned by $\{V_j\}_{j=r+1}^d$. Here, multiplication by infinity is understood in the usual limit sense, e.g., $\infty I := \lim_{\alpha \to \infty} \alpha I$. Thus, regularization—an infinite amount in this case—is only added in data-uninformed directions. The solution of (4) is formally given by

$$x_{Inf} = \left\{A^T A + \infty \left(I - V^r (V^r)^T\right)\right\}^{-1} \left(A^T y + L^T L x_0\right)$$

$$= \left\{\left[V^r, (V^r)^\perp\right] \left(\begin{bmatrix} (\Sigma^r)^2 & 0 \\ 0 & D^2 \end{bmatrix} + \begin{bmatrix} 0 & 0 \\ 0 & \infty I \end{bmatrix}\right) \left[V^r, (V^r)^\perp\right]^T\right\}^{-1} A^T y$$

$$= V^r \left(\Sigma^r\right)^{-2} \left(V^r\right)^T A^T y = V^r \left(\Sigma^r\right)^{-1} \left(U^r\right)^T y =: x_{TSVD}^r,$$

where $(V^r)^\perp$ is the orthogonal complement of V^r in $\mathbb{R}^p$, $\Sigma^r := \text{diag}[\sigma_1, \ldots, \sigma_r]$, and $D := \text{diag}[\sigma_{r+1}, \ldots, \sigma_p]$. The second equality clearly shows that the regularization scheme adds infinity to all singular values that correspond to data-uninformed modes. The last equality proves that infinite regularization on data-uninformed parameter subspace is the same as the TSVD approach.

The beauty of the TSVD approach is that it avoids putting any regularization on data-informed parameter directions, and hence avoids polluting inverse solutions in these directions, while annihilating data-uninformed directions. However, it is often the case that there is no clear-cut between the data-informed and data-uninformed ones (i.e., $\sigma_k = 0$ for $k \geq r+1$) but gradual (sometimes exponential) decay of the singular values of A. In that case, completely removing less data-informed directions may not be ideal, as they may still contain valuable parameter information encoded in the data. Instead, we may want to impose finite regularization in the data-uninformed directions, i.e.,

$$\min \frac{1}{2} \|Ax - y\|^2 + \frac{1}{2} \|L(x - x_0)\|^2, \tag{5}$$

where

$$L^T L := \alpha \left(I - V^r (V^r)^T\right) = \alpha (V^r)^\perp \left((V^r)^\perp\right)^T$$

$$= \left[V^r, (V^r)^\perp\right] \begin{bmatrix} 0 & 0 \\ 0 & \alpha I \end{bmatrix} \left[V^r, (V^r)^\perp\right]^T.$$

Let us call this approach the **data-informed (DI) regularization method**. The inverse solution in this case reads

$$x_{DI} = \left\{A^T A + \alpha \left(I - V^r (V^r)^T\right)\right\}^{-1} \left(A^T y + L^T L x_0\right)$$

$$= \left\{\left[V^r, (V^r)^\perp\right] \left(\begin{bmatrix} (\Sigma^r)^2 & 0 \\ 0 & D^2 \end{bmatrix} + \begin{bmatrix} 0 & 0 \\ 0 & \alpha I \end{bmatrix}\right) \left[V^r, (V^r)^\perp\right]^T\right\}^{-1}$$

35 Data-Informed Regularization for Inverse and Imaging Problems

$$\times \left(A^T y + L^T L x_0\right)$$

$$= \left\{ \left[V^r, (V^r)^\perp\right] \left(\begin{bmatrix} (\Sigma^r)^2 & 0 \\ 0 & D^2 \end{bmatrix} + \alpha \begin{bmatrix} I & 0 \\ 0 & I \end{bmatrix} - \begin{bmatrix} \alpha I & 0 \\ 0 & 0 \end{bmatrix} \right) \left[V^r, (V^r)^\perp\right]^T \right\}^{-1}$$

$$\times \left(A^T y + L^T L x_0\right).$$

The last equality suggests that the DI approach can be considered as *first applying the same* (Note that α need not be the same for all directions.) *(finite) regularization for all parameter directions and then removing regularization in the data-informed directions.*

A few observations are in order: (1) When $r = 0$, DI becomes the standard Tikhonov regularization; (2) when $\alpha \to \infty$ DI approaches the truncated SVD; and (3) when $\alpha \ll \sigma_i$ for $i \leq r$ (i.e., regularization in the data-informed modes is negligible), the Tikhonov solution

$$x_{Tikhonov} = \left\{A^T A + \alpha I\right\}^{-1} \left(A^T y + L^T L x_0\right)$$

$$= \left\{ \left[V^r, (V^r)^\perp\right] \left(\begin{bmatrix} (\Sigma^r)^2 & 0 \\ 0 & D^2 \end{bmatrix} \right. \right.$$

$$\left. \left. + \alpha \begin{bmatrix} I & 0 \\ 0 & I \end{bmatrix} \right) \left[V^r, (V^r)^\perp\right]^T \right\}^{-1} \left(A^T y + L^T L x_0\right)$$

is close to the DI solution x_{DI} as the contribution of the regularization to data-informed modes is negligible. These observations are clearly demonstrated in Fig. 3 for a 1D deconvolution with $\lambda = 0.05$ with various combinations of regularization parameter α and the number of retained data-informed modes r. An important feature of the DI technique that can be seen from this result is that for each r the DI solution is robust with the regularization parameter, that is, the solution does not alter significantly, especially for moderate-to-large regularization, while Tikhonov solution is damped out as the regularization parameter increases. The last column of Fig. 3 shows that for $r = 20$ the DI solution retains high-frequency modes which are not regularized and is thus oscillatory.

In order to gain more insights into the behavior of DI regularization, we compute the relative error between the solutions using the DI approach and the truth for a wide range of regularization parameters and a few values of r. The results are shown in Fig. 4. As can be seen, when $r = 1$, DI is essentially Tikhonov, which is not surprising as all modes in the DI solution are regularized exactly the same as Tikhonov except for the first one (lowest frequency). For $r = \{5, 10\}$, the DI solution

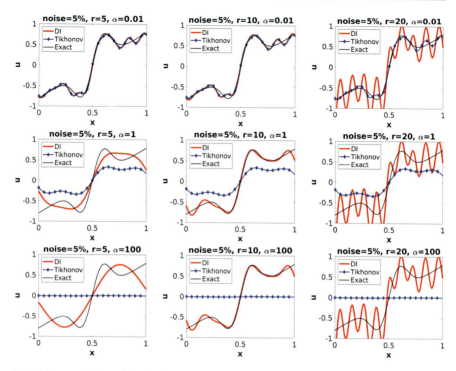

Fig. 3 Deconvolution with noise level $\lambda = 5\%$ using DI and Tikhonov regularization for various values of regularization parameter and r

behaves the same as Tikhonov for the under-regularization regime ($\alpha < 0.01$) as expected, and it outperforms Tikhonov for $\alpha > 0.01$ as the retained data-informed modes, which determine the quality of the deconvolution solution, are left untouched. For $r = 20$, the retained modes now also include high-frequency modes, and hence the DI approach is not as accurate as Tikhonov for $\alpha < 1$. For all cases with significant number of modes retained, i.e., $r > 5$, the DI solution quality is insensitive to a large range of the regularization parameter. Note that methods for choosing the regularization parameter α in practice include L-curve (Hansen and O'Leary 1993; Hansen 1992), the Morozov's discrepancy principle (Morozov 1966), and generalized cross-validation (Golub et al. 1979). These methods are inherently computationally costly, and this can be mitigated using the DI approach as it is robust with regularization parameter.

We have used a rank-r SVD approximation to derive and gain insights into the DI approach. For large-scale problems, this low rank-decomposition could be prohibitively expensive. To lead to an alternative computational approach (see Algorithm 2) and more importantly to provide a probabilistic view point of the DI approach, *let us take $r = n$ until the end of section "A Statistical Data-Informed (DI) Inverse Framework"*. In this case, since $V^n \left(V^n \right)^T$ is the orthogonal projection into

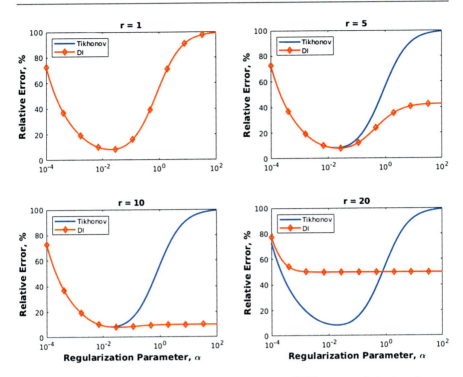

Fig. 4 Deconvolution with noise level $\lambda = 5\%$ using DI and Tikhonov regularizations for $\alpha = \left[10^{-4}, 10^2\right]$ and $r = \{1, 5, 10, 20\}$

the row space of A, i.e., $V^n \left(V^n\right)^T = A^T \left(AA^T\right)^\dagger A$, we can rewrite the inverse (optimization) problem (5) as

$$\min_x J := \frac{1}{2} \|Ax - y\|^2 + \frac{1}{2} \|L(x - x_0)\|^2, \qquad (6)$$

where

$$L^T L := \alpha \left(I - A^T \left(AA^T\right)^\dagger A\right).$$

In this form, the DI regularization approach (6) not only avoids using V^n explicitly but also brings us to a statistical data-informed inverse framework in the next section.

A Statistical Data-Informed (DI) Inverse Framework

The cost function in (6) can be rewritten as

$$\exp(-J) = \frac{\exp\left(-\tfrac{1}{2}\|Ax - y\|^2\right) \times \exp\left(-\tfrac{\alpha}{2}\|x - x_0\|^2\right)}{\exp\left(-\tfrac{\alpha}{2}(Ax - Ax_0)^T \left(AA^T\right)^\dagger (Ax - Ax_0)\right)}.$$

From a Bayesian inverse perspective (Kaipio and Somersalo 2005; Tarantola 2005; Franklin 1970; Lehtinen et al. 1989; Lasanen 2002; Stuart 2010; Piiroinen 2005), the numerator is the product of the likelihood

$$\pi_{\text{like}}(y|x) \propto \exp\left(-\frac{1}{2}\|Ax - y\|^2\right)$$

from the observational model (1) with the noise $e \sim \mathcal{N}(\mathbf{0}, \mathbf{I})$ and the Gaussian prior

$$\pi_{\text{prior}}(x) \propto \exp\left(-\frac{\alpha}{2}\|x - x_0\|^2\right) \quad (7)$$

with mean x_0 and I/α covariance matrix. In other words, the numerator is a Bayesian posterior with the aforementioned likelihood and Gaussian prior. *The key difference compared to the Bayesian approach is the denominator.*

We now show that the denominator is nothing more than the push-forward of the prior (7) via the forward map A. Indeed, let $\tilde{y} := Ax$ be a random variable induced by the forward map A. With $x \sim \mathcal{N}(x_0, I/\alpha)$, $\tilde{y}$ is also a Gaussian with mean $\tilde{y}_0$ and covariance matrix $\mathbb{C}$ where

$$\tilde{y}_0 := \mathbb{E}_x[Ax] = Ax_0$$

$$\mathbb{C} := \mathbb{E}_x\left[(\tilde{y} - \tilde{y}_0)(\tilde{y} - \tilde{y}_0)^T\right] = \mathbb{E}_x\left[A(x - x_0)(x - x_0)^T A^T\right] = \frac{1}{\alpha}AA^T.$$

Note that it is necessary to use the pseudo-inverse for the inverse of the covariance $\mathbb{C}$, i.e., $\mathbb{C}^{-1} := \alpha\left(AA^T\right)^\dagger$, since A may not have full row rank and thus the push-forward distribution can be a degenerate Gaussian.

Remark 1. The push-forward of the prior through the parameter-to-observable map A depends on x. It is through this push-forward term that the data-informed (DI) approach learns the data-informed parameter directions. Indeed, this new approach, through the push-forward term, changes the original prior

$$\exp\left(-\frac{\alpha}{2}(x - x_0)^T I (x - x_0)\right)$$

to the new one

35 Data-Informed Regularization for Inverse and Imaging Problems

$$\exp\left(-\frac{\alpha}{2}(x-x_0)^T\left[I-A^T\left(AA^T\right)^{\dagger}A\right](x-x_0)\right)$$

in such a way that the new prior leaves the data-informed directions, i.e., the row space of A, untouched, and hence only regularizes data-uninformed directions. The data-informed approach accomplishes this by the push-forward of the prior via the parameter-to-observable map A.

We can now define the DI posterior as

$$\pi_{DI}(x|y) = \frac{\pi_{like}(y|x) \times \pi_{prior}(x)}{A_{\#}\pi_{prior}(x)}, \tag{8}$$

where $A_{\#}\pi_{prior}(x)$ denotes the push-forward of $\pi_{prior}(x)$ via the parameter-to-observable map A.

We have constructively derived the DI approach by modifying the truncated SVD method and Gaussian prior with scaled-identity covariance matrix. In practice, the prior can be more informative about the correlations among components of x and in that case the covariance matrix is no longer an identity matrix. Let us denote by $\pi_{prior}(x) = \mathcal{N}(x_0, \Gamma/\alpha)$ the Gaussian prior with covariance matrix Γ/α. Let us also consider a more general data distribution where, for a given parameter x, the data is distributed by the Gaussian $\mathcal{N}(Ax, \Lambda)$. In order to use most of the above results, let us whiten both the parameter and observations. In particular, $\Lambda^{-\frac{1}{2}}y$ is the whitened observations (inducing $\Lambda^{-\frac{1}{2}}A$ as the new parameter-to-observable forward map), and $\Gamma^{-\frac{1}{2}}x$ is the prior-whitened parameter. (Here, the square roots for Γ and Λ are understood in the broader sense including: (1) if Γ and Λ are diagonal matrices, the square roots are simply diagonal matrices with square roots of the diagonal elements; (2) if Γ and Λ are not diagonal matrices, these square roots are understood in the spectral decomposition sense. For example: let $\Gamma = V\Sigma V^T$ be the spectral decomposition of Γ, then $\Gamma^{1/2} := V\Sigma^{1/2}V^T$. Note that this is meaningful as we assume the corresponding Gaussian distribution is non-degenerate and hence Σ is the diagonal matrix with positive diagonal elements; and (3) if Cholesky-type decomposition is available, i.e., $\Gamma = LL^T$ (L is not necessarily a Cholesky factorization), then $\Gamma^{1/2} = L$, and we simply add the "transpose" operator at appropriate places for one of the square roots.) The push-forward of the prior via $\Lambda^{-\frac{1}{2}}A$ now reads (Note that using the modified forward map $\Lambda^{-\frac{1}{2}}A$, though making the presentation clearer and constructive, is not necessary as using the original map A yields the same result.)

$$\Lambda^{-\frac{1}{2}}A_{\#}\pi_{prior}(x) = \mathcal{N}\left(\Lambda^{-\frac{1}{2}}Ax_0, \frac{1}{\alpha}\Lambda^{-\frac{1}{2}}A\Gamma A^T\Lambda^{-\frac{1}{2}}\right), \tag{9}$$

The DI posterior (8) with whitened parameter, whitened observations, and induced parameter-to-observable map now becomes

$$\pi_{DI}(x|y) = \frac{\pi_{like}(y|x) \times \pi_{prior}(x)}{\Lambda^{-\frac{1}{2}} A_{\#}\pi_{prior}(x)}$$

$$\propto \frac{\exp\left(-\frac{1}{2}\left\|\Lambda^{-\frac{1}{2}}Ax - \Lambda^{-\frac{1}{2}}y\right\|^2\right) \times \exp\left(-\frac{\alpha}{2}\left\|\Gamma^{-\frac{1}{2}}x - \Gamma^{-\frac{1}{2}}x_0\right\|^2\right)}{\exp\left(-\frac{\alpha}{2}\left\|\Lambda^{-\frac{1}{2}}Ax - \Lambda^{-\frac{1}{2}}Ax_0\right\|^2_{\left(\Lambda^{-\frac{1}{2}}A\Gamma A^T \Lambda^{-\frac{1}{2}}\right)^\dagger}}\right)}, \quad (10)$$

which, after writing the push-forward measure in terms of the whitened parameter, reads

$$\pi_{DI}(x|y) \propto \frac{\exp\left(-\frac{1}{2}\left\|\Lambda^{-\frac{1}{2}}Ax - \Lambda^{-\frac{1}{2}}y\right\|^2\right) \times \exp\left(-\frac{\alpha}{2}\left\|\Gamma^{-\frac{1}{2}}x - \Gamma^{-\frac{1}{2}}x_0\right\|^2\right)}{\exp\left(-\frac{\alpha}{2}\left\|\Gamma^{-\frac{1}{2}}x - \Gamma^{-\frac{1}{2}}x_0\right\|^2_{\Gamma^{\frac{1}{2}}A^T \Lambda^{-\frac{1}{2}}\left(\Lambda^{-\frac{1}{2}}A\Gamma A^T \Lambda^{-\frac{1}{2}}\right)^\dagger \Lambda^{-\frac{1}{2}}A\Gamma^{\frac{1}{2}}}\right)},$$

or equivalently

$$-\log\left(\pi_{DI}(x|y)\right) \propto \frac{1}{2}\left\|\Lambda^{-\frac{1}{2}}Ax - \Lambda^{-\frac{1}{2}}y\right\|^2 + \frac{1}{2}\left\|L\left(\Gamma^{-\frac{1}{2}}x - \Gamma^{-\frac{1}{2}}x_0\right)\right\|^2, \quad (11)$$

where

$$L^T L = \alpha \left(I - \Gamma^{\frac{1}{2}}A^T \Lambda^{-\frac{1}{2}}\left(\Lambda^{-\frac{1}{2}}A\Gamma A^T \Lambda^{-\frac{1}{2}}\right)^\dagger \Lambda^{-\frac{1}{2}}A\Gamma^{\frac{1}{2}}\right)$$

$$= \alpha \left(I - V^n V^{nT}\right) = \alpha \left(V^n\right)^\perp \left(\left(V^n\right)^\perp\right)^T$$

$$= \left[V^n, (V^n)^\perp\right] \begin{bmatrix} 0 & 0 \\ 0 & \alpha I \end{bmatrix} \left[V^n, (V^n)^\perp\right]^T, \quad (12)$$

where V^n contains the first n right singular vectors of the following SVD

$$\Lambda^{-\frac{1}{2}} A \Gamma^{\frac{1}{2}} := U \Sigma V^T. \quad (13)$$

As can be seen, the push-forward measure seeks to find the first n columns of V associated with the n nonzero singular values. The DI method then avoids regularizing these "data-informed directions" V^n. In other words, in the whitened parameter, the induced regularization by the prior is identity, and the DI approach removes regularization in the parameter subspace spanned by V^n. From (13) it is clear that V^n now depends on both the prior covariance Γ and the observational covariance Λ in addition to A. So how do we understand the parameter subspace spanned by V^n and hence the DI approach? To that end, let us define $\Sigma^{\frac{1}{2}}$ to be the same as Σ except on the main the diagonal where $\Sigma^{\frac{1}{2}}(i,i) = \sqrt{\Sigma(i,i)} = \sqrt{\sigma_i}$ (note that $\Sigma^{\frac{1}{2}}$ is nothing more than the square root of Σ when Σ is a square matrix). Let Ψ be the first n rows of $\Sigma^{\frac{1}{2}}$ and Φ be the first n columns of $\Sigma^{\frac{1}{2}}$. Clearly, by definition $\Psi(i,i) = \Phi(i,i) = \sqrt{\sigma_i}$ for $i \leq n$.

Let us define the following maps

$$z := Tx, \quad \text{where} \quad T := \Psi V^T \Gamma^{-\frac{1}{2}}, \tag{14}$$

$$w := Sy, \quad \text{where} \quad S := \Phi^T U^T \Lambda^{-\frac{1}{2}}. \tag{15}$$

where z are the first n coordinates of x in V, after whitening via $\Gamma^{-\frac{1}{2}}$ and then being scaled by Ψ. Similarly, w are the first n coordinates of y in U, after whitening via $\Lambda^{-\frac{1}{2}}$ and then being scaled by Φ. The map T pushes forward the prior in x to the prior in z as

$$\pi_{\text{prior}}(z) \sim \exp\left(-\frac{1}{2}\sum_{i=1}^{n}\sigma_i^{-1}(z_i - \bar{z}_i)^2\right), \tag{16}$$

where $\bar{z} = Tx_0$. Similarly, given x (and hence z), the induced likelihood in terms of w is given by

$$\pi_{\text{like}}(w|z) \sim \exp\left(-\frac{1}{2}\sum_{i=1}^{n}\sigma_i^{-1}(w_i - \sigma_i z_i)^2\right). \tag{17}$$

As can be seen from (16) and (17), the maps T and S transform the original parameter x and original data y to new parameter z and new data w. Two observations are in order: (1) though in general the original parameter and data dimensions are different, the new parameter and data have the same dimension; and (2) the new data w and new parameter z, up to the difference in the mean, have the same distribution. In particular, both z and w are $\mathbb{R}^n$-vectors of independent Gaussian distributions with diagonal covariance matrix $\Theta \in \mathbb{R}^{n \times n}$ with $\Theta_{ii} = \sigma_i$. Both z_i and w_i, up to the difference in the mean, are the same Gaussian distribution with variance σ_i. Since $\sigma_1 \geq \sigma_2 \geq \ldots \geq \sigma_n > 0$, the independent random variable z_i (and hence w_i) is ranked from the one with most variance to the one with least variance.

Let us call the ith column of U, namely U_i, the ith important direction in the data space and the ith column of V, namely V_i, the ith important direction in the parameter space. Let us also rank the degree of importance of U_i and V_i by the magnitude of σ_i. It follows that the transformations T and S map the original parameter x and data y into new parameter z and data w in which the corresponding parameter z_i and data w_i are equally important. This is similar to the concept of balanced transformation in control theory (see, e.g., Gugercin and Antoulas 2004; Antoulas 2005 and the references therein). The new parameter z is thus equally data-informed and prior-informed. In particular z_i is equally less data-informed and prior-informed relatively to z_j for $j < i$.

The DI method thus regularizes only the (equally) data-uninformed and prior-uninformed parameters/directions.

Properties of the DI Regularization Approach

Deterministic Properties

It is easy to see the optimality condition of the optimization problem $\max_x \log\left(\pi_{\mathrm{DI}}(x|y)\right)$ is given by

$$H x_{\mathrm{DI}} = b, \tag{18}$$

where

$$H := \left\{ A^T \Lambda^{-1} A + \alpha \left[\Gamma^{-1} - A^T \Lambda^{-\frac{1}{2}} \left(\Lambda^{-\frac{1}{2}} A \Gamma A^T \Lambda^{-\frac{1}{2}} \right)^\dagger \Lambda^{-\frac{1}{2}} A \right] \right\},$$

$$b := A^T \Lambda^{-1} y + \alpha \left[\Gamma^{-1} - A^T \Lambda^{-\frac{1}{2}} \left(\Lambda^{-\frac{1}{2}} A \Gamma A^T \Lambda^{-\frac{1}{2}} \right)^\dagger \Lambda^{-\frac{1}{2}} A \right] x_0$$

In order to solve the optimality condition (18) in practice, we can use the rank-r approximation

$$\Lambda^{-\frac{1}{2}} A \Gamma^{\frac{1}{2}} = U^n \Sigma^n (V^n)^T \approx U^r \Sigma^r (V^r)^T \tag{19}$$

for the push-forward matrix $A^T \Lambda^{-\frac{1}{2}} \left(\Lambda^{-\frac{1}{2}} A \Gamma A^T \Lambda^{-\frac{1}{2}} \right)^\dagger \Lambda^{-\frac{1}{2}} A$, where again n is the largest index for which $\sigma_n > 0$. Thus rank-r approximations (only for the regularization/prior) for H and y are given by

$$H^r := A^T \Lambda^{-1} A + \alpha \left(\Gamma^{-1} - \Gamma^{-\frac{1}{2}} V^r (V^r)^T \Gamma^{-\frac{1}{2}} \right),$$

$$b^r := \Gamma^{-\frac{1}{2}} V^n \Sigma^n (U^n)^T \Lambda^{-\frac{1}{2}} y + \alpha \left[\Gamma^{-1} - \Gamma^{-\frac{1}{2}} V^r (V^r)^T \Gamma^{-\frac{1}{2}} \right] x_0.$$

Note that we don't perform low-rank approximation for $A^T \Lambda^{-1} y$ in y as it requires only a matrix-vector product. We also leave the first term in H^r as is, since we invert

35 Data-Informed Regularization for Inverse and Imaging Problems

H^r using the conjugate gradient (CG) method which requires only matrix-vector products. In the numerical results section, we present a nested optimization method (see Algorithm 2) that avoids the low-rank approximation altogether. The analysis of such method is, however, more technical and thus left for future work. *The rank-r approximation to the solution of the optimality condition* (18) *is defined as*

$$H^r x_{DI}^r = b^r, \qquad (20)$$

for which the corresponding DI inverse formulation is given in (24) (by replacing r_ε with r), which reduces to (5) when $\Lambda = I$ and $\Gamma = I$. We can rewrite H^r in terms of n singular vectors corresponding to the n nonzero singular values as

$$H^r = \alpha \Gamma^{-\frac{1}{2}} \left[I + V^n D^n (V^n)^T \right] \Gamma^{-\frac{1}{2}},$$

where D^n is an $n \times n$ diagonal matrix with $D^n(i,i) = \left(\sigma_i^2 - \alpha\right)/\alpha$ for $i \leq r$ and $D^n(i,i) = \sigma_i^2/\alpha$ for $r < i \leq n$.

Lemma 1. *The DI solution with r data-informed modes reads*

$$x_{DI}^r := \Gamma^{\frac{1}{2}} V^n \Theta^n (U^n)^T \Lambda^{-\frac{1}{2}} y + \left[I - \Gamma^{\frac{1}{2}} V^n \overline{I}^n (V^n)^T \Gamma^{-\frac{1}{2}} \right] x_0, \qquad (21)$$

where Θ^n is an $n \times n$ diagonal matrix with $\Theta^n(i,i) = \sigma_i^{-1}$ for $i \leq r$ and $\Theta^n(i,i) = \sigma_i/\left(\sigma_i^2 + \alpha\right)$ for $r < i \leq n$. Here, $\overline{I}^n$ is an $n \times n$ diagonal matrix with $\overline{I}^n(i,i) = 1$ for $i \leq r$ and $\overline{I}^n(i,i) = \sigma_i^2/\left(\sigma_i^2 + \alpha\right)$ for $r < i \leq n$. Furthermore,

$$A x_{DI}^n = \Lambda^{\frac{1}{2}} U^n (U^n)^T \Lambda^{-\frac{1}{2}} y. \qquad (22)$$

Proof. Using a Woodbury formula, we have

$$\left(H^r\right)^{-1} = \frac{1}{\alpha} \Gamma^{\frac{1}{2}} \left[I - V^n \overline{d}_{DI}^{n,r} (V^n)^T \right] \Gamma^{\frac{1}{2}}, \qquad (23)$$

where $\overline{d}^{n,r}$ is an $n \times n$ diagonal matrix with $\overline{d}_{DI}^{n,r}(i,i) = \left(\sigma_i^2 - \alpha\right)/\sigma_i^2$ for $i \leq r$ and $\overline{d}_{DI}^{n,r}(i,i) = \sigma_i^2/\left(\sigma_i^2 + \alpha\right)$ for $r < i \leq n$. The computation of the product $\left(H^r\right)^{-1} y^r$ to arrive at the assertion is straightforward algebraic manipulation and hence omitted.

The result (22) shows that the image of the DI solution x_{DI} through the parameter-to-observable map is exactly the data if $U^n (U^n)^T = I$ or $\Lambda^{-\frac{1}{2}} y$ resides in the column space of U^n. This happens, for example, when A has full row rank and the number of data is not more than the dimension of the parameter, i.e., $d \leq p$. In this case, retaining all modes corresponding to nonzero singular values in the DI solution makes the data misfit vanish, that is, the DI solution in this case would match the noise, which is undesirable. As discussed in section "Data-Informed Regularization Derivation", r should be smaller than n for the solution to be meaningful. Let us define

$$r_\varepsilon := \max\{i : 1 \leq i \leq n \text{ and } \sigma_i \geq \varepsilon\},$$

for some $\varepsilon > 0$ (which, as discussed before, can be chosen using the Morozov's discrepancy principle), and the "reconstruction operator" (Colton and Kress 1998; Kirsch 2011)

$$\mathcal{R}_\varepsilon := \left(H^{r_\varepsilon}\right)^{-1} A^T \Lambda^{-\frac{1}{2}}.$$

Theorem 1. *For any $\varepsilon > 0$ and $\alpha > 0$, consider the inverse problem*

$$\min_x \mathcal{J} = \frac{1}{2} \left\| \Lambda^{-\frac{1}{2}} A x - \Lambda^{-\frac{1}{2}} y \right\|^2 + \frac{1}{2} \left\| L \Gamma^{-\frac{1}{2}} (x - x_0) \right\|^2, \tag{24}$$

using the DI approach with rank-r_ε approximation, where

$$L^T L = \alpha \left(I - \Gamma^{\frac{1}{2}} A^T \Lambda^{-\frac{1}{2}} \left(\Lambda^{-\frac{1}{2}} A \Gamma A^T \Lambda^{-\frac{1}{2}} \right)^\dagger \Lambda^{-\frac{1}{2}} A \Gamma^{\frac{1}{2}} \right).$$

The following hold:

(i) *The inverse problem with rank-r_ε DI approach, i.e., the optimization problem (24), is well-posed in the Hadamard sense.*
(ii) *Suppose that the nullspace of A is trivial, i.e., $\mathcal{N}(A) = \{0\}$, then the DI technique is a regularization strategy (Colton and Kress 1998; Kirsch 2011) in the following sense*

$$\lim_{\varepsilon \to 0} \mathcal{R}_\varepsilon \Lambda^{-\frac{1}{2}} A x = x.$$

(iii) *If $\alpha = \mathcal{O}(\varepsilon)$ and $\mathcal{N}(A) = \{0\}$, then the rank-r_ε DI technique is an admissible regularization method.*

Proof. From Lemma 1 we see that the DI solution $x_{\text{DI}}^{r_\varepsilon}$ is unique and furthermore

$$\left\| x_{\text{DI}}^{r_\varepsilon} \right\| \leq \beta(\varepsilon, \alpha) \left\| \Gamma^{\frac{1}{2}} \right\| \left\| \Lambda^{-\frac{1}{2}} \right\| \|y\| + \left(1 + \sqrt{\kappa(\Gamma)}\right) \|x_0\|,$$

where $\kappa(\Gamma)$ denotes the condition number of Γ, $\beta(\varepsilon, \alpha)$ is a constant defined as

35 Data-Informed Regularization for Inverse and Imaging Problems

$$\beta(\varepsilon, \alpha) := \frac{1}{\min_{r_\varepsilon < i \leq n} \{r_\varepsilon, \sigma_i + \alpha/\sigma_i\}},$$

which shows the DI solution is stable, which in turn proves i). To see assertion ii), we use the definition of $\mathcal{R}_\varepsilon$ and the SVD of $\Lambda^{-\frac{1}{2}} A \Gamma^{\frac{1}{2}}$ to arrive at

$$\mathcal{R}_\varepsilon \Lambda^{-\frac{1}{2}} A = \frac{1}{\alpha} \Gamma^{\frac{1}{2}} \left[I - V^n \overline{d}^{n,r_\varepsilon} (V^n)^T \right] V^n (\Sigma^n)^2 (V^n)^T \Gamma^{-\frac{1}{2}} =$$

$$\Gamma^{\frac{1}{2}} V^n \left[\begin{array}{c|c} I & 0 \\ \hline 0 & \mathrm{diag}\left(\frac{\sigma_i^2}{\sigma_i^2+\alpha}\right)_{r_\varepsilon < i \leq n} \end{array} \right] (V^n)^T \Gamma^{-\frac{1}{2}},$$

which implies

$$\lim_{\varepsilon \to 0} \mathcal{R}_\varepsilon \Lambda^{-\frac{1}{2}} A x = \Gamma^{\frac{1}{2}} V^n I (V^n)^T \Gamma^{-\frac{1}{2}} x = x,$$

where we have used the fact that $r_\varepsilon \to n$ as $\varepsilon \to 0$ and that $V^n (V^n)^T = I$ since $\mathcal{N}(A) = \{0\}$.

For assertion iii), it is sufficient to show that

$$\sup_y \left\{ \left\| \mathcal{R}_\varepsilon \Lambda^{-\frac{1}{2}} y - x \right\| : \left\| \Lambda^{-\frac{1}{2}} (Ax - y) \right\| \leq \varepsilon \right\} \to 0 \text{ as } \varepsilon \to 0,$$

for any x. We have

$$\left\| \mathcal{R}_\varepsilon \Lambda^{-\frac{1}{2}} y - x \right\| \leq \left\| \mathcal{R}_\varepsilon \Lambda^{-\frac{1}{2}} A x - x \right\| + \left\| \mathcal{R}_\varepsilon \Lambda^{-\frac{1}{2}} (Ax - y) \right\|$$

$$\leq \left\| \Gamma^{\frac{1}{2}} V^n \left[\begin{array}{c|c} 0 & 0 \\ \hline 0 & \mathrm{diag}\left(\frac{-\alpha}{\sigma_i^2+\alpha}\right)_{r_\varepsilon < i \leq n} \end{array} \right] (V^n)^T \Gamma^{-\frac{1}{2}} \right\| \|x\| + \|\mathcal{R}_\varepsilon\| \varepsilon$$

$$\leq \frac{\alpha}{\sigma_n^2 + \alpha} \sqrt{\kappa(\Gamma)} \|x\| + \varepsilon \left\| \Gamma^{\frac{1}{2}} \right\| \left\| \left[\begin{array}{c|c} \mathrm{diag}\left(\frac{1}{\sigma_i}\right)_{i \leq r_\varepsilon} & 0 \\ \hline 0 & \mathrm{diag}\left(\frac{\sigma_i}{\sigma_i^2+\alpha}\right)_{r_\varepsilon < i \leq n} \end{array} \right] \right\|$$

$$\leq \frac{\alpha}{\sigma_n^2 + \alpha} \sqrt{\kappa(\Gamma)} \|x\| + \varepsilon \sigma_n^{-1} \left\| \Gamma^{\frac{1}{2}} \right\|,$$

where we have used the result from (ii), definition of $\mathcal{R}_\varepsilon$, and the orthonormality of V and U. Using the assumption $\alpha = \mathcal{O}(\varepsilon)$ concludes the proof.

Remark 2. Note that most of the above arguments are still valid for infinite dimensional setting, i.e., $p = \infty$, assuming that Γ is a trace class. Indeed, $\Lambda^{-\frac{1}{2}} A \Gamma^{\frac{1}{2}}$ is then a compact operator, and we can invoke the infinite dimensional singular value decomposition (Colton and Kress 1983) for $\Lambda^{-\frac{1}{2}} A \Gamma^{\frac{1}{2}}$. Note that all the matrices are now interpreted as operators, transpose operator (superscript T) as adjoint operator, and $\Gamma^{-\frac{1}{2}}$ as pseudo-inverse if $\mathcal{N}(\Gamma) \neq \{0\}$. We leave out the details for the sake of brevity.

Statistical Properties

Now we discuss some probabilistic aspects of the DI prior and the DI posterior. Since the regularization parameter α plays no role in the following discussion, we absorb it into Γ. We define the DI prior as

$$\pi_{\text{DI-prior}}(x) \sim \exp\left\{-\frac{1}{2}\left\|L\Gamma^{-\frac{1}{2}}(x - x_0)\right\|^2\right\}. \tag{25}$$

From (12), the DI prior (pseudo-) inverse covariance is given by

$$(\mathbb{C}^n)^\dagger := \Gamma^{-\frac{1}{2}}\left[I - \Gamma^{\frac{1}{2}} A^T \Lambda^{-\frac{1}{2}} \left(\Lambda^{-\frac{1}{2}} A \Gamma A^T \Lambda^{-\frac{1}{2}}\right)^\dagger \Lambda^{-\frac{1}{2}} A \Gamma^{\frac{1}{2}}\right] \Gamma^{-\frac{1}{2}}$$

$$= \Gamma^{-\frac{1}{2}}\left[I - \Gamma^{\frac{1}{2}} A^T \left(A \Gamma A^T\right)^\dagger A \Gamma^{\frac{1}{2}}\right] \Gamma^{-\frac{1}{2}} = \Gamma^{-\frac{1}{2}} (V^n)^\perp \left((V^n)^\perp\right)^T \Gamma^{-\frac{1}{2}},$$

where we have used the fact that Λ is invertible in the second equality. Thus, Λ actually contributes to neither the DI prior nor its rank-r version

$$(\mathbb{C}^r)^\dagger := \Gamma^{-\frac{1}{2}} (V^r)^\perp \left((V^r)^\perp\right)^T \Gamma^{-\frac{1}{2}}.$$

The rank-r DI covariance thus reads

$$\mathbb{C}^r := \Gamma^{\frac{1}{2}} (V^r)^\perp \left[(V^r)^\perp\right]^T \Gamma^{\frac{1}{2}} = \Gamma^{\frac{1}{2}} \left(I - V^r (V^r)^T\right) \Gamma^{\frac{1}{2}} \tag{26}$$

which is clearly symmetric positive semidefinite in $\mathbb{R}^p$, though degenerate. (The nullspace of $\mathbb{C}^r$: $\mathcal{N}(\mathbb{C}^r) := \{x : \Gamma^{\frac{1}{2}} x \in \mathcal{R}(V^r)\}$, where $\mathcal{R}(\cdot)$ denotes the range space.) The DI prior (25) is not a well-defined density in $\mathbb{R}^p$, that is, it is not absolutely continuous with respect to the Lebesgue measure in $\mathbb{R}^p$. This is not surprising as we argue above that the DI prior is the prior on the less data-informed directions. Let us define

$$z^\perp := T^\perp x, \quad \text{where} \quad T^\perp := \left((V^r)^\perp\right)^T \Gamma^{-\frac{1}{2}}.$$

35 Data-Informed Regularization for Inverse and Imaging Problems

Theorem 2. *The following hold true:*

(i) z and $z^\perp$ are distributed by the push-forward density of the prior through T and $T^\perp$, respectively. In particular, $z \sim \mathcal{N}(Tx_0, I)$ and $z^\perp \sim \mathcal{N}(T^\perp x_0, I)$.
(ii) The DI prior density is the density of $z^\perp$ and hence is well-defined.
(iii) The DI prior density is the conditional density of x given z.

Proof. Assertion (i) is straightforward. To see the second assertion, we note that the density of $z^\perp$, ignoring the normalized constant, can be written as

$$\exp\left\{-\frac{1}{2}\left\|z^\perp - T^\perp x_0\right\|^2\right\} = \exp\left\{-\frac{1}{2}(x - x_0)\left(T^\perp\right)^T T^\perp (x - x_0)\right\}$$

$$= \exp\left\{-\frac{1}{2}(x - x_0)\Gamma^{-\frac{1}{2}}(V^r)^\perp\left((V^r)^\perp\right)^T \Gamma^{-\frac{1}{2}}(x - x_0)\right\},$$

which is exactly the DI prior (25). In other words, we have shown that the DI prior is a well-defined density on $z^\perp$. To see assertion (iii), we observe that

$$\pi_{\text{prior}}(x) = \pi_{\text{prior}}\left(V^r z + (V^r)^\perp z^\perp\right),$$

and owing to $z = Tx$, again ignoring the normalized constant, we have

$$\pi_{\text{prior}}(x|z) = \frac{\pi_{\text{prior}}(x)}{\pi(z)}$$

$$= \exp\left\{-\frac{1}{2}(x - x_0)\Gamma^{-\frac{1}{2}}(V^r)^\perp\left((V^r)^\perp\right)^T \Gamma^{-\frac{1}{2}}(x - x_0)\right\},$$

which is exactly the DI prior since $\pi(z) = \mathcal{N}(Tx_0, I)$ is exactly the push-forward density of $\pi_{\text{prior}}(x)$ via the map T.

Remark 3. Note that the above decomposition of x into z and $z^\perp$, through the maps T and $T^\perp$, is still valid for infinite dimensional settings. However, $z^\perp$ would be distributed by an infinite dimensional Gaussian measure with identity covariance operator, which is not a valid Gaussian measure. A more general understanding of the DI prior is through disintegration. Indeed, under mild conditions on the map T and its push-forward measure of the prior measure, the DI prior $\pi_{\text{prior}}(x|z)$ is nothing more than a disintegration of the prior measure via the map T, and this view is also valid for infinite dimensional settings.

To quantify the uncertainty in the DI inverse solution (21), we can use the covariance matrix of the DI posterior (10). For linear inverse problems with Gaussian prior and Gaussian noise—the problems considered in this chapter—the

Table 1 The difference between the DI and the Tikhonov covariance matrices

	$i \leq r$	$r < i \leq n$
DI posterior	$\overline{d}_{DI}^{n,r}(i,i) = \frac{\sigma_i^2 - \alpha}{\sigma_i^2}$	$\overline{d}_{DI}^{n,r}(i,i) = \frac{\sigma_i^2}{\sigma_i^2 + \alpha}$
Tikhonov posterior	$\overline{d}_{Tik}^{n,r}(i,i) = \frac{\sigma_i^2}{\sigma_i^2 + \alpha}$	$\overline{d}_{Tik}^{n,r}(i,i) = \frac{\sigma_i^2}{\sigma_i^2 + \alpha}$

covariance matrix is exactly the inverse of the Hessian. For rank-r DI approach, the DI posterior covariance matrix C_{DI}^{post} is given in (23), i.e.,

$$C_{DI}^{post} = \frac{1}{\alpha}\boldsymbol{\Gamma} - \frac{1}{\alpha}\boldsymbol{\Gamma}^{\frac{1}{2}} V^n \overline{d}_{DI}^{n,r} (V^n)^T \boldsymbol{\Gamma}^{\frac{1}{2}} \quad (27)$$

It is easy to see that the covariance matrix corresponding to the Tikhonov regularization is given by

$$C_{Tik}^{post} = \frac{1}{\alpha}\boldsymbol{\Gamma} - \frac{1}{\alpha}\boldsymbol{\Gamma}^{\frac{1}{2}} V^n \overline{d}_{Tik}^{n,r} (V^n)^T \boldsymbol{\Gamma}^{\frac{1}{2}}, \quad (28)$$

where both $\overline{d}_{DI}^{n,r}$ and $\overline{d}_{Tik}^{n,r}$ are diagonal matrices given in Table 1. Note that we have used α as the magnitude of the regularization to study the robustness and accuracy of all methods. If not needed, α can be straightforwardly absorbed into $\boldsymbol{\Gamma}$, and hence σ_i^2; in that case α is simply replaced by 1 everywhere it appears (including those in Table 1). As can be seen, $\overline{d}_{Tik}^{n,r}(i,i)$ is always non-negative for all i, while $\overline{d}_{DI}^{n,r}(i,i)$ is negative when $\sigma_i^2 < \alpha$ for $i \leq r$. That is, while the Tikhonov posterior uncertainty, C_{Tik}^{post} (Bayesian posterior with standard Gaussian prior), is always smaller than the prior uncertainty $\boldsymbol{\Gamma}$ no matter how much informed the data is, the DI posterior uncertainty could be higher than the prior counterpart if the data supports this. In other words, standard (or typical) Gaussian priors do not allow the data to increase the uncertainty and hence are prone to producing overconfident results (see section "Applications to Imaging Problems"). The DI prior, on the other hand, takes the parameter-to-observable map (the proxy to the data) into account, and thus along parameter directions that are more data-informed, i.e. $\sigma_i^2 \geq \alpha$, the posterior uncertainty is reduced relative to the prior uncertainty. Along parameter directions that are less data-informed, i.e., $\sigma_i^2 < \alpha$, the posterior uncertainty increases relative to the prior uncertainty.

Applications to Imaging Problems

Image Deblurring

One typical inverse problem in imaging is image deblurring. Given some blurry image, we want to recover the true, sharp image. To understand the deblurring process, we must first understand how an image becomes blurred in the first place.

A simple and effective mathematical model of the blurring process is convolution of a sharp image with a blurring kernel. This blurring kernel is often described mathematically as a point spread function (PSF). The PSF describes how energy from a point source (i.e., a single pixel) is *smeared* out among neighboring pixels, resulting in a blur.

Since convolution is a linear operation, it can be expressed mathematically as

$$\mathcal{A} X_{true} = B \qquad (29)$$

where $\mathcal{A}$ is the blurring (convolution) operator acting on the true image $X_{true} \in \mathbb{R}^{m_1 \times m_2}$ resulting in the blurred image $B \in \mathbb{R}^{m_1 \times m_2}$. By stacking (or *vectorizing*) the columns of X_{true}, we can write (29) as a linear algebraic equation. Let us denote by x_{true} the vectorized true image and by y the vectorized blurred image, i.e.,

$$x_{true} = \mathrm{vec}(X_{true}) \in \mathbb{R}^{m_1 m_2}, \quad y = \mathrm{vec}(B) \in \mathbb{R}^{m_1 m_2}$$

Also, since $\mathcal{A}$ is a linear operator acting on a vector, it has a matrix representation denoted by $A \in \mathbb{R}^{m_1 m_2 \times m_1 m_2}$. Finally, (29) becomes

$$A x_{true} = y \qquad (30)$$

Note that while this notation is convenient for manipulating mathematically, it is not efficient to construct the two-dimensional convolution matrix. A is a large sparse matrix, which, for large problems, cannot be stored in memory. Even on problems small enough to fit in memory, it is computationally expensive to explicitly construct this matrix. Fortunately, there are efficient methods for computing spectral decompositions of the matrices arising from convolution operators using the fast Fourier transform and discrete cosine transform. While interesting in their own right, these implementation details are not necessary for the following discussion. For a detailed treatment of image deblurring problems and algorithms, the interested reader is encouraged to consult (Hansen et al. 2006).

For all examples considered in this chapter

$$\Lambda = \lambda^2 I, \quad \text{and } \Gamma = I,$$

where λ is the noise level (the standard deviation).

Since truncated SVD (TSVD) and Tikhonov are spectral filtering methods, the regularized solution using these methods can be written using the following common form:

$$x_{filt} = \sum_{i=1}^{p} \phi_i \frac{U_i^T y}{\sigma_i} V_i, \qquad (31)$$

where ϕ_i is usually called the *filter factor* as it has the effect of filtering (damping) when ϕ_i is close to 0. It can be shown that the filter factor for rank-r TSVD is given by

$$\phi_i = \begin{cases} 1, & i \leq r \\ 0, & \text{otherwise.} \end{cases}$$

Likewise, the filter factor for Tikhonov regularization is given by

$$\phi_i = \frac{\sigma_i^2}{\sigma_i^2 + \alpha}$$

As discussed in section "A Data-Informed Regularization (DI) Approach", the DI method with rank-r approximation removes regularization on the first r directions V_i, $1 \leq i \leq r$, while being the same as Tikhonov on the other directions. For $\Gamma = I$ and $x_0 = 0$ the DI solution (see Lemma 1) can be written in the filtered form as

$$\phi_i = \begin{cases} 1, & i \leq r \\ \frac{\sigma_i^2}{\sigma_i^2 + \alpha}, & \text{otherwise.} \end{cases}$$

Remark 4. It should be emphasized that the DI method also shares the same spectral decomposition form in this case because $\Gamma = I$ and $x_0 = 0$. When $\Gamma \neq I$, singular vectors of $\Lambda^{-\frac{1}{2}} A$ do not necessarily diagonalize both A and Γ simultaneously. In other words, the filtered form (31) is not valid for the DI approach unless U and V are singular vectors of $\Lambda^{-\frac{1}{2}} A \Gamma^{\frac{1}{2}}$ and $x_0 = 0$. When $x_0 \neq 0$, there is an additional term contributed from x_0 as shown in the DI solution given in Lemma 1.

We can see here again that (1) when $r \to 0$, DI approaches Tikhonov; (2) when $\alpha \ll \sigma_i$ for $i \leq r$, Tikhonov is close to DI; and 3) when $\alpha \to \infty$, DI converges to TSVD. This can be clearly seen in Fig. 5a for a deblurring problem in which we plot the relative error between the deblurred images and the original ones for $m_1 = m_2 = 128$, $\lambda = 0.01$, $r = 400$, and a wide range of α. For the under-regularization regime, i.e., $\alpha < 1$, which should be avoided, the regularization is not sufficient to suppress the oscillations due to the high-frequency modes for both Tikhonov and DI methods, resulting in inaccurate reconstructions. For reasonable-to-over-regularization regimes, i.e., $\alpha > 1$, DI is the best compared to both Tikhonov and TSVD method as it combines the advantages from both sides. That is: (1) DI behaves similar to Tikhonov for reasonable (but small) regularization and outperforms Tikhonov in reasonable-to-over-regularization regimes; and (2) compared to TSVD, DI is more accurate for reasonable regularization parameters as it maintains the benefits of keeping useful information from all parameter directions while avoiding potential errors caused by over-regularization. Consequently, the DI

35 Data-Informed Regularization for Inverse and Imaging Problems

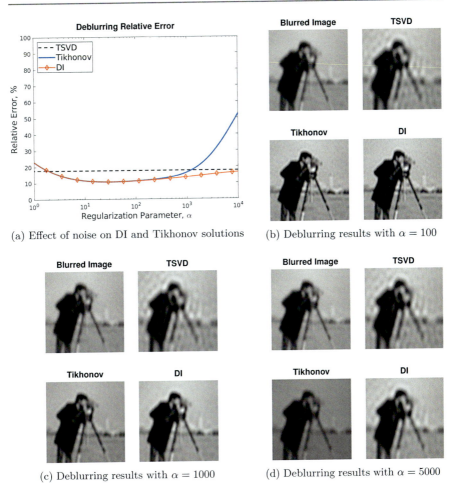

Fig. 5 Deblurring results for $m_1 = m_2 = 128$, $\lambda = 0.01$, $r = 400$. (**a**) relative error between deblurred images and the truth for a range of regularization parameter $\alpha \in [1, 10^4]$. (**b**) the DI deblurred image with $\alpha = 100$. (**c**) the DI deblurred image with $\alpha = 1000$. (**d**) the DI deblurred image with $\alpha = 5000$

error is the smallest of the three methods discussed for all $\alpha > 10^3$, and DI is robust with respect to the regularization parameter.

In Fig. 5b are the deblurred images for $\alpha = 100$ corresponding to the smallest deblurring error for both DI and Tikhonov. As can be seen, the Tikhonov result is similar to the DI one, while the truncated SVD result is blurry as it removes (putting infinite regularization on) useful information in directions V_i for $i > r$. Figure 5c, d show the deblurred images for $\alpha = 1000$ and $\alpha = 5000$, respectively, corresponding to cases where DI outperforms both Tikhonov and TSVD (see Fig. 5a). Indeed, the DI deblurred image has higher quality.

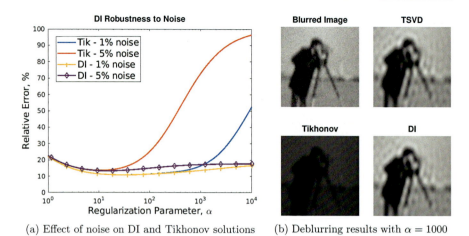

(a) Effect of noise on DI and Tikhonov solutions (b) Deblurring results with $\alpha = 1000$

Fig. 6 Deblurring results for $m_1 = m_2 = 128$, $\lambda = 0.05$, $r = 400$. (**a**) relative error of DI and Tikhonov solutions with respect to true solution for noise levels of 1% and 5% and $\alpha \in \left[1, 10^4\right]$. (**b**) the DI, Tikhonov, and TSVD deblurred images with $\alpha = 1000$

In order to see if the DI method is sensitive to noise, we now consider the case with $\lambda = 5\%$ noise. Deblurring accuracy for this case (purple) is shown in Fig. 6a together with the accuracy for the case of 1% noise (yellow). As can be seen, the solution quality of the DI method does not degrade significantly due to the presence of noise. Compare this to the difference seen in the Tikhonov method (red and blue curves) with the increase in noise level, we can see that the solution quality of the Tikhonov method degrades rapidly in the presence of noise. It can also be seen that Tikhonov regularization becomes more sensitive to the choice of α as the noise increases. Since the DI method regularizes only the data-uninformed directions, which also contain much of the noise, increasing the noise level has little effect on the solution quality.

For the rest of this section, we consider the more challenging cases with $\lambda = 5\%$ noise. To make the problem even more challenging, we consider images with missing pixels to simulate more interesting cases when images are damaged or incomplete. Figure 7 show the deblurring results using DI, TSVD, and Tikhonov (Tik) regularizations for damaged images with $m_1 = m_2 = 128$, $r = 400$. The first column contains four scenarios with 10% random data, 25% random data, 50% random data, and 100% data, all with noise. Note that we plot the damaged images by filling the missing data with 0. The second column contains the corresponding TSVD deblurring results. The last four columns contain the results from DI and Tikhonov with $\alpha = 10$ and 20. As can be observed, all methods are able to deblur and at the same time recover the true image quite well even with only 10% data. Both DI and Tikhonov yield clearer images compared to TSVD. The Tikhonov results are "darker," especially with $\alpha = 20$, indicating over-regularization, while the DI images are insensitive to regularization parameter as the data-informed modes are

35 Data-Informed Regularization for Inverse and Imaging Problems

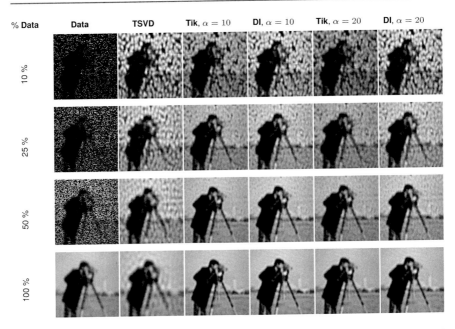

Fig. 7 Deblurring results using DI, TSVD, and Tikhonov (Tik) regularizations for damaged images with $m_1 = m_2 = 128$, $\lambda = 0.05$, $r = 400$. The first column consists of four scenarios with 10%, 25%, 50%, and 100% data. The second column is the corresponding TSVD deblurring results. The last four columns are the results from DI and Tikhonov with $\alpha = 10$ and 20

left untouched. Indeed, Fig. 8 clearly demonstrates these expected results for larger regularization parameters ($\alpha = 50$ and $\alpha = 100$).

Recall that the goal of sections "A Statistical Data-Informed (DI) Inverse Framework" and "Statistical Properties" is to gain insights into statistical properties of the DI prior. For linear parameter-to-observable maps—which are the cases for this chapter—with Gaussian observational noise, the posterior is also a Gaussian. As a result, the result at the end of section "Statistical Properties" also allows us to use the posterior covariances (27) and (28) to estimate the uncertainty in the corresponding inverse solutions. Since the posterior for either Tikhonov or DI prior is Gaussian, its diagonal contains the marginal pixel-wise variances, which can be used as a measure of uncertainty for each pixel. Clearly this does not take into account the correlation among pixels, but is straightforward to have a glimpse of uncertainty in high-dimensional (128^2-dimensional) spaces. We now study the uncertainty estimation in the solution of deblurring problems.

To begin, it is important to distinguish the following two cases:

- *Case I*: using only rank-r DI regularization in which rank-r approximation for the pseudo-inverse $\left(\Lambda^{-\frac{1}{2}} A \Gamma A^T \Lambda^{-\frac{1}{2}}\right)^{\dagger}$ is done as we have presented. The DI

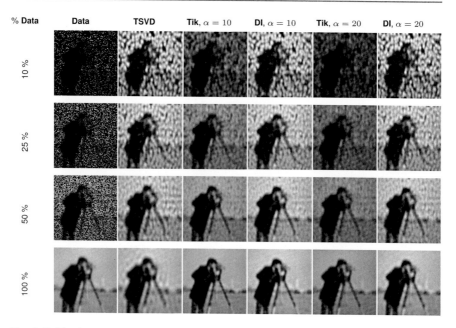

Fig. 8 Deblurring results using DI, TSVD, and Tikhonov (Tik) regularizations for damaged images with $m_1 = m_2 = 128$, $\lambda = 0.05$, $r = 400$. The first column consists of four scenarios with 10%, 25%, 50%, and 100% data. The second column is the corresponding TSVD deblurring results. The last four columns are the results from DI and Tikhonov with $\alpha = 50$ and 100

posterior covariance (27) thus involves the second and third columns in Table 1, and a rank-n SVD (13) is needed.
- *Case II*: performing rank-r low-rank approximation of the posterior covariance in addition to rank-r DI regularization. This amounts to using only the second column of Table 1 for the DI posterior covariance in (27). This case is typically more practical for large-scale problems as only a rank-r SVD (19) is needed.

In Fig. 9a are the minimum pixel-wise variances for four scenarios with 10% random data, 25% random data, 50% random data, and 100% random data for *Case II*. As can be seen, the uncertainty corresponding to the case of missing data is lower than the uncertainty for full data case! We expect the opposite, that is, more available (supposedly) informative data is expected to lead to lower uncertainty in the inverse solution. The observation is twofold: first, care needs to be taken for *Case II* results as rank-r approximation may not provide accurate uncertainty; second, for 10% data case, when $r > 500$ the uncertainty is larger compared to the full data case. This suggests that r needs to be sufficiently large for an accurate uncertainty estimation, and this will be confirmed in the discussion below for *Case I* in which we use the full rank (rank-n) decomposition (13). The criteria for estimating such a value of r are a subject for future research. (At the moment of writing this chapter, we have not yet found such a criteria.)

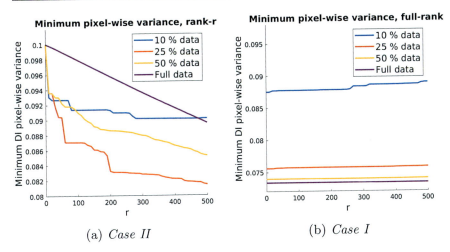

Fig. 9 Rank-r DI posterior pixel-wise uncertainty using rank-n SVD decomposition (*Case I* with both the second and third columns of Table 1) and using rank-r SVD decomposition (*Case II* with only the second column of Table 1)

We next discuss the results for *Case I*. Again, this requires a rank-n SVD (13), where n is the rank of A, to compute (27) using Table 1. Figure 9b shows that the minimum uncertainty for any missing data case is higher than the full data case regardless of any value of r in rank-r DI regularization. As also expected, the uncertainty scales inversely with the amount of available data, i.e., the more informative data we have, the smaller the uncertainty in the inverse solution. Note that the result and the conclusion for the largest pixel-wise variances are similar and hence omitted here.

We now compare the DI and Tikhonov posterior uncertainty estimations. Since *Case I*, though more expensive, provides more accurate uncertainty estimation, it is used for computing DI posterior pixel-wise variances. To be fair, we also use the full decomposition for Tikhonov regularization. In other words, the following comparison is based on (27) and (28) and Table 1. As discussed above in Figs. 6a and 7, $\alpha = 10$ corresponds to a case in the region where DI and Tikhonov give nearly the same reconstructions (in fact Tikhonov slightly over-regularizes), so let us start with this case first. Figure 10 shows that the DI posterior has higher pixel-wise variance than the Tikhonov posterior. This is consistent with the result and the discussion of Table 1 and Fig. 7, that is, the Tikhonov posterior is not only over-regularizing but also overconfident. For both methods, regions of higher uncertainty are visually discernible where data is missing. In the case of 100% data, the result is the same, namely, Tikhonov uncertainty estimation subjectively is less than the DI uncertainty estimation. In this case, the uncertainty estimate is not very interesting: both DI and Tikhonov have approximately uniform uncertainty everywhere as we have data everywhere. We next consider the case with $\alpha = 1000$ where Tikhonov significantly over-regularizes (see Fig. 6b). Figure 11, shows that while Tikhonov is

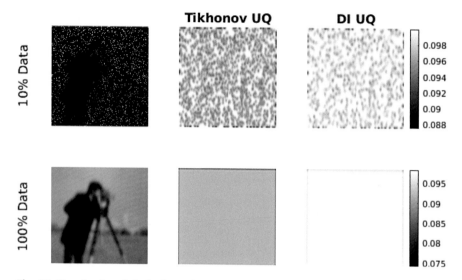

Fig. 10 Visualization of pixel-wise variance estimates for the deblurring problem with $\lambda = 0.05$, $r = 400$, and $\alpha = 10$. In the left column are the noisy images with 10% data and 100% data. In the second column are the Tikhonov uncertainty estimates for 10% data (top) and 100% data (bottom). Likewise, the third column contains the DI uncertainty estimates for 10% data (top) and 100% data (bottom)

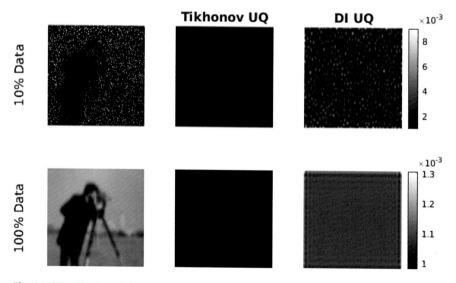

Fig. 11 Visualization of pixel-wise variances for the deblurring problem with $\lambda = 0.05$, $r = 400$, and $\alpha = 1000$. In the left column are the noisy images with 10% data and 100% data. In the second column are the Tikhonov uncertainty estimates for 10% data (top) and 100% data (bottom). The third column contains the DI uncertainty estimates for 10% data (top) and 100% data (bottom)

35 Data-Informed Regularization for Inverse and Imaging Problems

uniformly (very) overconfident, i.e., having small posterior uncertainty everywhere, DI gives informative UQ results. The latter can be clearly seen for the case with 10% data in which the uncertainty is higher for missing pixels. This implies that the DI priors could provide more useful UQ results than the Tikhonov (standard Gaussian) ones.

Image Denoising

We can extend the idea of data-informed (DI) regularization to the image denoising problem. Since noise typically resides in the high-frequency portion of the image, denoising can be performed by applying spectral filtering techniques directly to the noisy image. These noisy high-frequency modes are also the *less informative* modes in the DI setting. Taking the SVD of the noisy image, X_{noisy}, we have

$$X_{noisy} = U\Sigma V^T = \sum_i \sigma_i U_i V_i^T,$$

The denoised image can be obtained by "filtering" the noise as

$$X_{filt} = U\Sigma^{filt} V^T = \sum_i \phi_i \sigma_i U_i V_i^T,$$

where Σ^{filt} is the diagonal matrix with $\Sigma_{ii}^{filt} = \phi_i \sigma_i$. The filter factors ϕ_i are the same as those defined for the deblurring case. For a numerical demonstration, we pick a noisy image (Hansen et al. 2006) with 20% noise (see the top-left subfigure of Fig. 12a). Shown in Fig. 12a are denoised results using DI with $r = 20$ and $\alpha = 100$, TSVD with $r = 20$, and Tikhonov with $\alpha = 100$. Though the difference in the results is not clearly visible, the DI has smaller error compared the other two methods. This can be verified in Fig. 12b where the relative error between the denoised image and the true one for a wide range of "regularization parameter" $\alpha \in \left[10^{-2}, 10^4\right]$ is presented. Clearly, we would not choose $\alpha < 1$ as these correspond to under-regularization. For $\alpha > 1$, DI is the best compared to both Tikhonov and TSVD method as it combines the advantages from both methods. Indeed, the DI error is smallest for all $\alpha > 1$, and DI is robust with regularization parameter.

X-Ray Tomography

In the previous two examples, we have been able to implement spectral filtering methods directly by introducing filter factors which effectively modified the singular values to minimize the impact of noise on the inversion process. (Recall that the DI method also shares the same spectral decomposition form in this case because

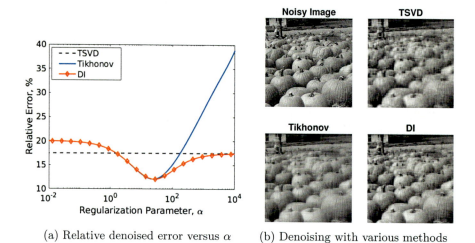

(a) Relative denoised error versus α (b) Denoising with various methods

Fig. 12 Denoising with DI, Tikhonov, and TSVD methods. (**a**) The relative error between the denoised image and the true one for a wide range of "regularization parameter". The DI error is smallest for all $\alpha > 1$ (corresponding reasonable to over-regularization regimes). (**b**) Denoised results using DI with $r = 20$ and $\alpha = 100$, TSVD with $r = 20$, and Tikhonov with $\alpha = 100$

$\boldsymbol{\Gamma} = \boldsymbol{I}$ and $\boldsymbol{x}_0 = 0$.) Each method relied on computing a full factorization of $\boldsymbol{\Lambda}^{-\frac{1}{2}}\boldsymbol{A}$ and then applying filters. While this is an effective and straightforward method to solve small-to-moderate inverse problems that helps provide insight into each approach, it can be cumbersome or even computationally infeasible to compute full factorizations for large-scale problems. It is not uncommon that inverse problems arising in imaging applications can lead to very large matrix operators. Indeed, we have seen even in the toy image deblurring problem in section "Image Deblurring" that matrix size of 16384×16384 is significantly large, and we have employed more sophisticated methods to compute the factorization of the convolution operator. For many problems, however, such efficient factorizations may not exist, or it is computationally prohibitive to compute a full factorization.

One way to overcome the challenge of factorizing large matrices is to solve the optimality condition (20) iteratively. Since $\boldsymbol{H}$ is symmetric positive definite, we choose the conjugate gradient (CG) method (see, e.g, Shewchuk 1994 and the references therein) which requires only matrix-vector products, which in turn avoids forming any matrices (including $\boldsymbol{A}$ or $\boldsymbol{H}$) completely. We consider two variants: (a) using CG to solve for (20), that is, we still require rank-r approximation of the DI regularization, and (b) using CG to solve for (18), that is, a rank-r approximation of the DI regularization is not required. In this case we use a least-squares optimization method to compute the pseudo-inverse $\left(\boldsymbol{\Lambda}^{-\frac{1}{2}}\boldsymbol{A}\boldsymbol{\Gamma}\boldsymbol{A}^T\boldsymbol{\Lambda}^{-\frac{1}{2}}\right)^\dagger$ acting on a vector for each CG iteration.

The detailed computational procedure for the a)-variant is given in Algorithm 1. Note that the viability of this method for large-scale problems relies on the availability of a randomized eigensolver to compute eigenvectors of $\Lambda^{-\frac{1}{2}}A\Gamma A^T\Lambda^{-\frac{1}{2}}$ (and thus right singular vectors of $\Lambda^{-\frac{1}{2}}A\Gamma^{\frac{1}{2}}$) which does not require explicit construction of a matrix, only access to matrix-vector products.

Algorithm 1 Data-informed inversion using randomized eigensolver and CG

Input: Data y, number of eigenvectors r, prior x_0, prior covariance matrix Γ, noise covariance matrix Λ, regularization parameter α

1: Define $F := \Lambda^{-\frac{1}{2}}A\Gamma^{\frac{1}{2}}$.
2: Create functions to compute matrix-vector products Fx and $F^T x$.
3: Compute the first r eigenvectors (V^r) of $F^T F$ using a randomized eigensolver.
4: Solve linear equation (20), i.e.,

$$\Gamma^{-\frac{1}{2}}\left[F^T F + \alpha(I - V_r V_r^T)\right]\Gamma^{-\frac{1}{2}}x = \Gamma^{-\frac{1}{2}}F^T y + \alpha\Gamma^{-\frac{1}{2}}(I - V_r V_r^T)\Gamma^{-\frac{1}{2}}x_0$$

using the conjugate gradient method.

To demonstrate the effectiveness of this approach for the DI method, we choose to solve the inverse problem of reconstructing an image from X-ray measurements. The forward model of generating X-ray measurements, A, is given by the Radon transform, and A^T is given by the inverse Radon transform. A more detailed description of the X-ray tomography inverse problem is given in Mueller and Siltanen (2012) (and the references therein). The problem setup in this section exactly follows the setup given in Mueller and Siltanen (2012). We use the MATLAB Image Processing Toolbox to compute the product of the Radon transform A and and its inverse A^T with a vector. Results using Algorithm 1 for a popular 256 × 256 phantom image with 256 measurement angles are shown in Fig. 13 for various values of the regularization parameter α and the rank r. Each row contains the results for each regularization parameter with different values of r. The corresponding values for α and r can also be found in the rows and columns of Table 2. Note that below each figure is the relative error of the corresponding reconstruction and the actual phantom image. These relative errors are collected in Table 2 for clarity. Note that for the last two images on the last row of Fig. 13, CG does not converge, and this issue is still under investigation. Other than that the observations are similar to the previous section. That is, compared to Tikhonov, DI is robust to the regularization parameter, and it is at least as good as Tikhonov regardless of the values of regularization parameter α and rank r.

Next we present the detailed computational procedure for the b)-variant in Algorithm 2. In order to compare variant b) with variant a), we compute the relative error of the reconstruction and the true image for various values of regularization parameter α. From the results in Fig. 13, we choose $r = 200$ to balance the accuracy and the cost of the eigensolver. The result is in Fig. 14, which shows that the b)-variant (red curve) is at least as good as the a)-variant (blue curve) while not requiring low-rank approximations. Indeed, to demonstrate this, we pick $\alpha = 100$

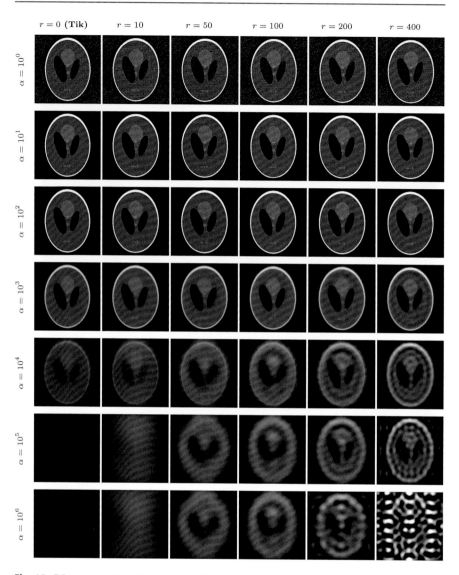

Fig. 13 DI reconstructions for various values of the regularization parameter α and the rank r. Each row contains the results for each regularization parameter with different values of r. The corresponding values for α and r can be found in the rows and columns of Table 2 along with the relative error between the reconstructed image and the true phantom

for which Fig. 14 shows that both variants give similar reconstruction quality, and the reconstruction from both variants is shown in Fig. 15. As can be seen, the result from the b)-variant looks much clearer, which is expected in this case, as $r = 200$ is not sufficient to capture all the data-informed modes for the a)-variant. *By using*

Table 2 Comparison of the relative errors of the DI solution estimate for various regularization parameters α and various values for r. The noise level here is $\lambda = 1\%$

α	Relative Error, %					
	$r = 0$ (Tik)	$r = 10$	$r = 50$	$r = 100$	$r = 200$	$r = 400$
1	33.52	33.52	33.52	33.52	33.52	33.52
10	31.73	31.73	31.73	31.73	31.73	31.73
100	24.44	24.45	24.45	24.45	24.45	24.45
1000	29.81	29.80	29.72	29.66	29.51	29.09
10^4	58.76	58.52	56.93	55.92	54.03	50.43
10^5	81.77	77.10	70.33	67.78	63.84	57.84
10^6	96.09	81.29	72.44	69.50	81.80	299.73

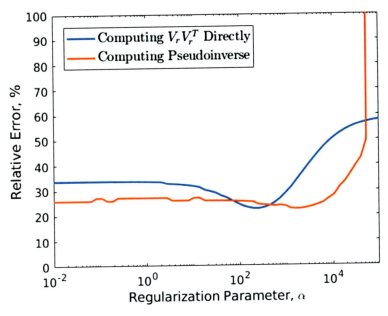

Fig. 14 A comparison between variant (**b**) (red curve) and variant (**a**) with $r = 200$ (blue curve). Here, we compute the relative error of the reconstruction and the truth image for various values of regularization parameter α

the pseudoinverse formulation, we can still get excellent results while avoiding the computation of a large factorization.

Conclusions

We have presented a new regularization technique called data-informed (DI) regularization that, though with disintegration origin, can be viewed as a combination of the classical truncated SVD and Tikhonov regularization. In particular, the DI

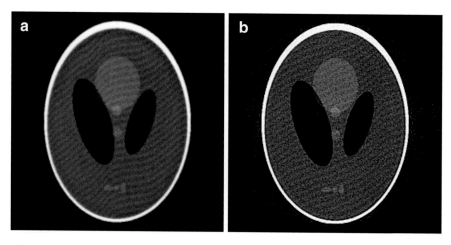

Fig. 15 X-Ray tomography reconstruction with 1% noise and $\alpha = 100$: (**a**) the result from the a)-variant with $r = 200$ and (**b**) the result from the b)-variant

---

Algorithm 2 Data-informed inversion using nested CG

Input: Data y, number of eigenvectors r, prior x_0, prior covariance matrix Γ, noise covariance matrix Λ, regularization parameter α

1: Define $F = \Lambda^{-\frac{1}{2}} A \Gamma^{\frac{1}{2}}$.
2: Create functions to compute matrix-vector products Fx and $F^T x$.
3: Solve linear equation (18), i.e.,

$$\Gamma^{-\frac{1}{2}}\left[F^T F + \alpha\left(I - F^T(FF^T)^\dagger F\right)\right]\Gamma^{-\frac{1}{2}} x = \Gamma^{-\frac{1}{2}} F^T y + \alpha \Gamma^{-\frac{1}{2}}\left(I - F^T(FF^T)^\dagger F\right)\Gamma^{-\frac{1}{2}} x_0$$

using the conjugate gradient method. For each CG iterations, compute the product of $F^T(FF^T)^\dagger F \Gamma^{-\frac{1}{2}}$ with any vector x using matrix-free Algorithm 3.

Algorithm 3 Compute the product of $F^T(FF^T)^\dagger F \Gamma^{-\frac{1}{2}}$ with any vector using optimization

---

Input: functions to compute Fx and $F^T x$, current estimate of x, prior covariance matrix Γ

1: Compute $b = F\Gamma^{-\frac{1}{2}} x$.
2: Using conjugate gradient method, solve linear equation

$$FF^T z = b.$$

3: Return $F^T z$.

---

approach does not pollute the data-informed modes and regularizes only less data-informed ones. As a direct consequence, the DI approach is at least as good as the Tikhonov method for any value of the regularization parameter, and it is more accurate than the TSVD (for reasonable regularization parameter). Due to the blending of these two classical methods, DI is expected to be robust with regularization parameter, and this is verified numerically. We have shown that DI is a regularization strategy. The DI approach has an interesting statistical interpretation, that is, it transforms both the data distribution (i.e., the likelihood) and prior distribution (induced by Tikhonov regularization) to the same Gaussian distribution whose covariance matrix is diagonal, and the diagonal elements are exactly the singular values of a composition of the prior covariance matrix, the forward map, and the noise covariance matrix. In other words, DI finds the modes that are most equally data-informed and prior-informed and leaves these modes untouched so that the inverse solution receives the best possible (balanced) information from both prior and the data. Furthermore, the DI approach takes the data uncertainty into account and hence can avoid overconfident uncertainty estimation. To demonstrate and to support our deterministic and statistical findings, we have presented various results for popular computer vision and imaging problems including deblurring, denoising, and X-ray tomography.

References

Antoulas, A.C.: Approximation of Large-Scale Systems. SIAM, Philadelphia (2005)

Babacan, S.D., Mancera, L., Molina, R., Katsaggelos, A.K.: Non-convex priors in bayesian compressed sensing. In: 2009 17th European Signal Processing Conference, pp. 110–114 (2009)

Beck, A., Teboulle, M.: Fast gradient-based algorithms for constrained total variation image denoising and deblurring problems. IEEE Trans. Image Process. **18**, 2419–2434 (2009)

Boley, D.: Local linear convergence of the alternating direction method of multipliers on quadratic or linear programs. SIAM J. Optim. **23**, 2183–2207 (2013)

Boyd, S., Parikh, N., Chu, E., Peleato, B., Eckstein, J.: Distributed optimization and statistical learning via the alternating direction method of multipliers. Found. Trends Mach. Learn. **3**, 1–122 (2010)

Chartrand, R., Wohlberg, B.: A nonconvex admm algorithm for group sparsity with sparse groups. In: 2013 IEEE International Conference on Acoustics, Speech and Signal Processing, pp. 6009–6013 (2013)

Chartrand, R., Yin, W.: Iteratively reweighted algorithms for compressive sensing. In: 2008 IEEE International Conference on Acoustics, Speech and Signal Processing, pp. 3869–3872 (2008)

Colton, D., Kress, R.: Integral Equation Methods in Scattering Theory. Wiley (1983)

Colton, D., Kress, R.: Inverse Acoustic and Electromagnetic Scattering, 2nd edn. Applied Mathematical Sciences, Vol. 93. Springer, Berlin/Heidelberg/New-York/Tokyo (1998)

Franklin, J.N.: Well-posed stochastic extensions of ill–posed linear problems. J. Math. Anal. Appl. **31**, 682–716 (1970)

Goldstein, T., Osher, S.: The slit Bregman method for L1-regularized problems. SIAM J. Imag. Sci. **2**, 323–343 (2009)

Golub, G., Heath, M., Wahba, G.: Generalized cross-validation as a method for choosing a good ridge parameter. **21**, 215–223 (1979)

Gugercin, S., Antoulas, A.C.: A survey of model reduction by balanced truncation and some new results. Int. J. Control. **77**, 748–766 (2004)

Hansen, P.C.: Truncated singular value decomposition solutions to discrete ill-posed problems with ill-determined numerical rank. SIAM J. Sci. Stat. Comput. **11**, 503–518 (1990)

Hansen, P.C.: Analysis of discrete ill-posed problems by means of the l-curve. SIAM Rev. **34**, 561–580 (1992)

Hansen, P.C., Nagy, J.G., O'Leary, D.P.: Deblurring Images: Matrices, Spectra, and Filtering. SIAM, Philadelphia (2006)

Hansen, P.C., O'Leary, D.P.: The use of the l-curve in the regularization of discrete ill-posed problems. SIAM J. Sci. Comput. **14**, 1487–1503 (1993)

Kaipio, J., Somersalo, E.: Statistical and Computational Inverse Problems, vol. 160 of Applied Mathematical Sciences. Springer, New York (2005)

Kirsch, A.: An Introduction to the Mathematical Theory of Inverse Problems, 2nd edn. Applied Mathematical Sciences, Vol. 120. Springer, New-York (2011)

Lasanen, S.: *Discretizations of generalized random variables with applications to inverse problems*, Ph.D. thesis, University of Oulu (2002)

Lehtinen, M.S., Päivärinta, L., Somersalo, E.: Linear inverse problems for generalized random variables. Inverse Prob. **5**, 599–612 (1989)

Morozov, V.A.: On the solution of functional equations by the method of regularization. Soviet Math. Dokl. (1966)

Mueller, J.L., Siltanen, S.: Linear and Nonlinear Inverse Problems with Practical Applications. SIAM, Philadelphia (2012)

Nikolova, M.: Weakly constrained minimization: Application to the estimation of images and signals involving constant regions. J. Math. Imaging Vision **21**, 155–175 (2004)

Nikolova, M.: Analysis of the recovery of edges in images and signals by minimizing nonconvex regularized least-squares. Multiscale Model. Simul. **4**, 960–991 (2005) (electronic)

Piiroinen, P.: *Statistical measurements, experiments, and applications*, Ph.D. thesis, Department of Mathematics and Statistics, University of Helsinki (2005)

Ramirez-Giraldo, J., Trzasko, J., Leng, S., Yu, L., Manduca, A., McCollough, C.H.: Nonconvex prior image constrained compressed sensing (ncpiccs): Theory and simulations on perfusion ct. Med. Phys. **38**, 2157–2167 (2011)

Rudin, L., Osher, S., Fatemi, E.: Nonlinear total variation based noise removal algorithms. Phys. D **60**, 259–268 (1992)

Shewchuk, J.R.: An introduction of the conjugate gradient method without the agonizing pain, Carnegie Mellon University (1994). https://www.cs.cmu.edu/~quake-papers/painless-conjugate-gradient.pdf

Stuart, A.M.: Inverse problems: A Bayesian perspective. Acta Numerica **19**, 451–559 (2010). https://doi.org/10.1017/S0962492910000061

Tarantola, A.: Inverse Problem Theory and Methods for Model Parameter Estimation. SIAM, Philadelphia (2005)

Tikhonov, A.N., Arsenin, V.A.: Solution of Ill-posed Problems. Winston & Sons, Washington, DC (1977)

Randomized Kaczmarz Method for Single Particle X-Ray Image Phase Retrieval

36

Yin Xian, Haiguang Liu, Xuecheng Tai, and Yang Wang

Contents

Introduction	1274
The Phase Retrieval Problem	1274
Challenges of X-Ray Data Processing	1275
Phase Retrieval with Noisy or Incomplete Measurements	1276
Outline	1276
Background: Phase Retrieval and Stochastic Optimization	1277
Phase Retrieval	1277
Stochastic Optimization and the Kaczmarz Method	1278
Variance-Reduced Randomized Kaczmarz (VR-RK) Method	1279
Application: Robust Phase Retrieval of the Single-Particle X-Ray Images	1281
Synthetic Single-Particle Data Recovery Experiment	1281
Recovery Efficiency Under Constraints	1282
Results of the PR772 Dataset	1283
Conclusion	1284
Appendix	1284
References	1286

Y. Xian (✉)
TCL Research Hong Kong, Hong Kong, SAR, China
e-mail: polinexian@tcl.com

H. Liu
Microsoft Research-Asian, Beijing, China
e-mail: haiguangliu@microsoft.com

X. Tai
Hong Kong Center for Cerebro-cardiovascular Health Engineering (COCHE), Shatin, Hong Kong, China
e-mail: xtai@hkcoche.org

Y. Wang
Hong Kong University of Science and Technology, Hong Kong, SAR, China
e-mail: yangwang@ust.hk

© Springer Nature Switzerland AG 2023
K. Chen et al. (eds.), *Handbook of Mathematical Models and Algorithms in Computer Vision and Imaging*, https://doi.org/10.1007/978-3-030-98661-2_112

Abstract

In this chapter, we investigate phase retrieval algorithm for the single-particle X-ray imaging data. We present a variance-reduced randomized Kaczmarz (VR-RK) algorithm for phase retrieval. The VR-RK algorithm is inspired by the randomized Kaczmarz method and the Stochastic Variance Reduce Gradient Descent (SVRG) algorithm. Numerical experiments show that the VR-RK algorithm has a faster convergence rate than randomized Kaczmarz algorithm and the iterative projection phase retrieval methods, such as the hybrid input output (HIO) and the relaxed averaged alternating reflections (RAAR) methods. The VR-RK algorithm can recover the phases with higher accuracy, and is robust at the presence of noise. Experimental results on the scattering data from individual particles show that the VR-RK algorithm can recover phases and improve the single-particle image identification.

Keywords

Stochastic optimization · Variance reduction · Phase retrieval · Randomized Kaczmarz algorithm

Introduction

The Phase Retrieval Problem

The mathematical formulation of phase retrieval is solving a set of quadratic equations. Methods to solve the phase retrieval problem can be classified into two categories: convex and non-convex approaches. Convex methods, like *PhaseLift* (Candès et al. 2013), convert the quadratic system equation to a linear system equation through a matrix-lifting technique. The *PhaseMax* method (Goldstein and Studer 2017; Bahmani and Romberg 2017) operates in the original signal space rather than lifting it to a higher dimensional space. It replaces the non-convex constraints with inequality constraints that define convex sets. The convex approaches have good recovery guarantees, but their computational complexities are usually high when the dimension of the signals is large.

On the other hand, the non-convex approaches turn the phase retrieval into an optimization problem. The most popular class of methods is based on alternate projection, such as the hybrid input output (HIO) method (Bauschke et al. 2003), the error reduction (ER) method (Fienup and Wackerman 1986), and the relaxed averaged alternating reflections (RAAR) method (Luke 2004). These methods are iterative projection methods, since they involve iterative projections onto the constraint sets. Unlike the convex approaches, convergence is not guaranteed for these algorithms, and stagnation may occur due to nonuniqueness of the solution (Fienup and Wackerman 1986). A unified evaluation of these iterative projection algorithms can be found in the paper of Marchesini (2007). Recently, a method called Wirtinger flow (Candès et al. 2015) is proposed. It works well with spectral method for initialization. The follow-up works include the truncated Wirtinger flow (Chen and Candès 2017), truncated amplitude flow (Wang et al. 2017), and reshaped Wirtinger flow (Zhang and Liang 2016). These methods have less computational complexities and have theoretical convergence guarantees.

The randomized Kaczmarz algorithm is introduced to solve the phase retrieval problem by Wei (2015). The randomized Kaczmarz method can be viewed as a special case of the stochastic gradient descent (SGD) (Needell et al. 2014). For the phase retrieval problem, the method is essentially SGD for the amplitude flow objective. It was shown numerically that the method outperforms the Wirtinger flow and the ER method (Wei 2015). The convergence rate of the randomized Kaczmarz method for the linear system is studied in the paper of Strohmer and Vershynin (2009). The theoretical justification of using randomized Kaczmarz method for phase retrieval has been presented in the paper of Tan and Vershynin (2019).

Challenges of X-Ray Data Processing

The structure of biological macromolecules is the key to understand the living cell function and behavior. The Protein Data Bank (PDB) (Bernstein et al. 1977) currently has more than 173,110 structures, but many structures of biological molecules and their complexes have not been determined. The cryo-electron microscopy (Cryo-EM) and the X-ray crystallography have been successfully applied in this field. The X-ray crystallography has solved about 90% of these structures. However, growing high-quality crystals of biomolecules is challenging, especially for biologically functional molecules. Therefore, determining structures from single molecules are appealing.

The use of the X-ray free electron lasers (XFEL) is a recent development in structure biology. The idea behind this method is to record the instantaneous elastic scattering from an ultrashort pulse. The pulse is so brief that it terminates before the onset of radiation damage ("diffract before destroy") (Liu and Spence 2016). With this application, the single-particle imaging becomes possible, even at room temperature. It allows one to understand the structures and dynamics of macromolecules.

The difference between the Cryo-EM and the X-ray crystallography is that the Cryo-EM data includes phase information of the structural factors, while the X-ray crystallographic diffraction data only provide amplitude information but lack phase information (Wang and Wang 2017; Scheres 2012). The illustrations and data processing examples are shown in the paper of Sorzano et al. (2004), Xian et al. (2018), and Gu et al. (2020). In order to solve the biological structures, the phase information is essential. It is normally obtained by experimental or computational means.

The challenges of XFEL single-particle imaging also include the following: (i) the signal-to-noise ratio (SNR) is low, and the information is influenced by noise; (ii) the orientation of each sample particle is unknown, leading to the difficulty in data merging and 3D reconstruction; (iii) conformational heterogeneity places a hurdle for single-particle identification and reconstruction (Wang and Wang 2017).

In this chapter, we investigate the phase retrieval algorithms of the XFEL data. The baseline for a good phase retrieval algorithm is its robustness against noises and the incompleteness of information (Shi et al. 2019).

Phase Retrieval with Noisy or Incomplete Measurements

The number of photons detected by the optical sensor is of Poisson distribution. For the phase retrieval problem contaminated by the Poisson noise, or has incomplete magnitude information, the prior information is crucial to process the data. Research for imposing prior information to image processing is shown in the literature (Le et al. 2007; Zhang et al. 2012; Hunt et al. 2018).

In order to better reconstruct the data, one can consider a variational model by introducing a total variation (TV) regularization, which is widely used in imaging processing community. TV regularization can enable recovery of signals from incomplete or limited measurements. The alternating direction of multipliers method (ADMM) (Glowinski and Le Tallec 1989; Wu and Tai 2010) and the split Bregman method (Goldstein and Osher 2009) is usually applied to solve the TV-regularization problem. They have been applied in the phase retrieval problem (Chang et al. 2016, 2018; Bostan et al. 2014; Li et al. 2016).

Besides TV regularization, Tikhonov regularization is another important smoothing techniques in variational image denoising. It is often applied in noise removal. The phase retrieval problem with a Tikhonov regularization has been solved by the Gauss-Newton method (Seifert et al. 2006; Sixou et al. 2013; Langemann and Tasche 2008; Ramos et al. 2019). Considering the sparsity constraints, the fixed point iterative approach (Fornasier and Rauhut 2008; Tropp 2006; Ma et al. 2018) has been applied for the problem with nonlinear joint sparsity regulation.

Outline

In this chapter, we further advance the convergence speed of the randomized Kaczmarz method for phase retrieval. The idea comes from the fact that the randomized Kaczmarz method is a weighted SGD, and the convergence rate of SGD is slower because of the random sampling variance. Therefore, reducing the sampling variance can improve the convergence rate of the randomized Kaczmarz method. Inspired by the stochastic variance reduce gradient (SVRG) method (Johnson and Zhang 2013), we present the variance-reduced randomized Kaczmarz method (VR-RK) for single-particle X-ray imaging phase retrieval. Considering the sparsity constraint and generality of the problem, we present the VR-RK method under both the L_1 and the L_2 constraints for computational analysis. Numerical results on the virus data show that the VR-RK method can recover information with higher accuracy at a faster convergence rate. It helps recover the lost information due to the beam stop for blocking the incidence X-ray beam.

The rest of the chapter is organized as follows. In the section "Background: Phase Retrieval and Stochastic Optimization," we give a general overview of phase retrieval and stochastic optimization. In the section "Variance-Reduced Randomized Kaczmarz (VR-RK) Method", the proposed variance-reduced randomized Kaczmarz method, and its variation under L_1 and L_2 constraints are presented. The evaluation of the algorithm is shown in the "Application: Robust Phase Retrieval of the Single-Particle X-Ray Images" section, and the single-particle X-ray image data are tested. The "Conclusion" section concludes the chapter.

Background: Phase Retrieval and Stochastic Optimization

Phase Retrieval

Formulation of the phase retrieval problem is as follows:

$$\min_x \sum_{k=1}^{m} (y_k - |\langle a_k, x \rangle|^2)^2. \tag{1}$$

where y is the measurement, x is the signal that need to be recovered, and a_k is the measurement operating vector. In the setting of forward X-ray scattering imaging at the far field, a_k is a Fourier vector, and y is a diffraction pattern of the target. The problem in phase retrieval is the limitation of optical sensors, which measures only the intensity.

The loss function of Eq. (1) is expressed as the squared difference between measurement intensities and the modelled intensities. It is a system of quadratic equation, and therefore, it is a non-convex problem.

To solve Eq. (1), the alternate projection methods are often used, such as HIO, ER, and RAAR methods as mentioned previously. These algorithms can be expressed in the form of fixed-point equation. They can be implemented jointly to better avoid local minima.

When the loss function is expressed as the squared loss of amplitudes, the formulation can be written as:

$$\min_x \sum_{k=1}^{m} (\sqrt{y_k} - |\langle a_k, x \rangle|)^2. \tag{2}$$

To solve Eq. (2), it is possible to apply the amplitude flow algorithm (Wang et al. 2017), which is essentially a gradient descent algorithm that can converge under good initialization.

Stochastic Optimization and the Kaczmarz Method

The phase retrieval problem can be solved by stochastic optimization approaches. For the problem:

$$\min_x \frac{1}{m} \sum_{k=1}^{m} f_k(x), \tag{3}$$

the gradient descent method updating rule is: $x_{k+1} = x_k - \frac{t_k}{m} \sum_{k=1}^{m} \nabla f_k(x_k)$, where t_k is the step size at each iteration and m is the number of samples, or the number of measurements in the phase retrieval setting. The gradient descent is expensive, and it requires evaluation of n derivatives at each iteration. To reduce the computational cost, the SGD is proposed:

$$x_{k+1} = x_k - t_k \nabla f_{i_k}(x_k) \tag{4}$$

where i_k is an index chosen uniformly in random from $\{1, \cdots, m\}$ at each iteration. The computational cost is $1/m$ of the standard gradient descent. The SVRG is proposed to reduce variance of SGD and has a faster convergence rate (Johnson and Zhang 2013). It is operated in epochs. In each epoch, the updating process is:

$$x_{k+1} = x_k - \eta \left(\nabla f_{i_k}(x_k) - \nabla f_{i_k}(\tilde{x}) + \frac{1}{m} \sum_{i=1}^{m} \nabla f_i(\tilde{x}) \right) \tag{5}$$

where η is the step size, and $\tilde{x}$ is a snapshot value in each epoch (Johnson and Zhang 2013).

The Kaczmarz method is a well-known iterative method for solving a system of linear equations $Ax = b$, where $A \in \mathbb{R}^{m \times n}$, $x \in \mathbb{R}^n$, and $b \in \mathbb{R}^m$. The classical Kaczmarz method sweeps through the rows in A in a cyclic manner and projects the current estimate onto a hyperplane associated with the row of A to get the new estimate. The randomized Kaczmarz method randomly chooses the row for projection in each iteration:

$$x_{k+1} = x_k + \frac{b_{i_k} - \langle a_{i_k}, x_k \rangle}{\|a_{i_k}\|_2^2} a_{i_k} \tag{6}$$

where a_{i_k} is the row of A. The randomized Kaczmarz can be viewed as a reweighted SGD with importance sampling for the least squares problem (Needell et al. 2014):

$$F(x) = \frac{1}{2} \|Ax - b\|_2^2 = \frac{1}{2} \sum_{i=1}^{m} (a_i^T x - b_i)^2. \tag{7}$$

The randomized Kaczmarz algorithm is essentially stochastic gradient descent for the amplitude flow problem in Eq. (2). This suggests that the acceleration schemes for SGD, such as the variance-reduced approach, can be applied to the algorithm and improve phase retrieval.

Variance-Reduced Randomized Kaczmarz (VR-RK) Method

Define $b_{i_k} = \sqrt{y_{i_k}}$; the formulation of Eq. (2) can be written as:

$$\min_{x} \sum_{k=1}^{m} (b_k - |\langle a_k, x \rangle|)^2. \tag{8}$$

The update scheme for randomized Kaczmarz for the phase retrieval objective of Eq. (8), according to the paper of Tan and Vershynin (2019), is:

$$x_{k+1} = x_k + \eta_k a_{i_k} \tag{9}$$

where

$$\eta_k = \frac{\text{sign}(\langle a_{i_k}, x_k \rangle) b_{i_k} - \langle a_{i_k}, x_k \rangle}{\|a_{i_k}\|_2^2}$$

i_k is drawn independently and identically distributed (i.i.d.) from the index set $\{1, 2, \cdots, m\}$ with the probability

$$g_k = \frac{\|a_{i_k}\|^2}{\|A\|_F^2}. \tag{10}$$

The VR-RK method is inspired by the randomized Kaczmarz method and the SVRG method. It is proposed originally to solve the linear system equation (Jiao et al. 2017). Let $f_i(x) = \frac{1}{2}(a_i^T x - b_i)^2$, and let

$$h_i(x) = \frac{f_i(x)}{g_i} = \frac{1}{2}(|a_i^T x| - b_i)^2 \frac{\|A\|_F^2}{\|a_i\|^2} \tag{11}$$

then,

$$\nabla h_i(x) = (a_i^T x - \text{sign}(a_i^T x) b_i) a_i \frac{\|A\|_F^2}{\|a_i\|^2} \tag{12}$$

Let $\mu_i(x) = \nabla h_i(x)$, and s be the size of the epoch. The variance-reduced randomized Kaczmarz algorithm for phase retrieval is shown in Algorithm 1.

Algorithm 1 Variance-reduced randomized Kaczmarz (VR-RK)

Initialize $\mu_i(\bar{x}) = 0$, and $\bar{\mu} = 0$, specify A, b, s.
At steps $k = 1, 2, \cdots$, if $k \mod s = 0$, then

$$\bar{x} = x_k \text{ and } \bar{\mu} = \mu(x_k)$$

Pick index i uniformly at random according to (10).
Update x_k by

$$x_{k+1} = x_k - \frac{m}{\|A\|_F^2}\left(\mu_{i_k}(x_k) - \mu_i(\bar{x}) + \bar{\mu}\right)$$

where $\bar{\mu} = \frac{1}{m}\sum_{i=1}^{m}\nabla h_i(\bar{x})$

Considering the generality of the problem, and L_2 constraint is imposed, the objective function is:

$$\min_{x} \frac{1}{2}\sum_{k=1}^{m}(b_k - |\langle a_k, x\rangle|)^2 + \gamma\|x\|_2. \tag{13}$$

Applying the randomized Kaczmarz method, according to Hefny et al. (2017), the updating process becomes:

$$x_{k+1} = x_k - \frac{(a_{i_k}^T x_k - \text{sign}(a_{i_k}^T x_k)b_{i_k})a_{i_k} + \gamma x_k}{\|a_{i_k}\|^2 + \gamma} \tag{14}$$

In the VR-RK setting, the updating process is:

$$\nabla c_{i_k}(x_k) = \frac{(a_{i_k}^T x_k - \text{sign}(a_{i_k}^T x_k)b_{i_k})a_{i_k} + \gamma x_k}{\|a_{i_k}\|^2 + \gamma} \tag{15}$$

$$x_{k+1} = x_k - \nabla c_{i_k}(x_k) + \nabla c_{i_k}(\bar{x}) - \frac{1}{m}\sum_{i=1}^{m}\nabla c_i(\bar{x}) \tag{16}$$

For the consideration of the sparsity, the L_1 instead of the L_2 constraint can be imposed; then the objective function becomes:

$$\min_{x} \frac{1}{2}\sum_{k=1}^{m}(b_k - |\langle a_k, x\rangle|)^2 + \lambda\|x\|_1. \tag{17}$$

To deal with this formula, the majorization-minimization (MM) technique and the C-PRIME method (Qiu and Palomar 2017) are employed. It is shown that the problem is equivalent to:

$$\min_{x}\left(C\|x-d\|_2^2+\lambda\|x\|_1\right) \qquad (18)$$

where C is a constant and $C \geq \rho_{\max}(A^H A)$, $\rho_{\max}$ is the largest eigenvalue of a matrix, and d is the constant vector that is defined as:

$$d := x_k - \frac{1}{C}A^H(Ax_k - b \odot e^{j\angle(Ax_k)}). \qquad (19)$$

Above, the notation $\odot$ is the element-wise Hadamard product of two vectors, and $\angle$ is the phase angle. The close form solution of x is:

$$x^* = e^{j\angle(d)} \odot \max\left\{|d| - \frac{\lambda}{2C}\mathbf{1}, \mathbf{0}\right\}.$$

Application: Robust Phase Retrieval of the Single-Particle X-Ray Images

In this section, we present numerical results of phase retrieval of the single-particle X-ray imaging data.

Synthetic Single-Particle Data Recovery Experiment

The first experiment is to test the reconstruction efficiency of the virus data, as shown in Fig. 1. The image size of Fig. 1a is 755 × 755 pixels, and the pixel values are normalized to [0,1]. The diffraction pattern (Fig. 1b) is created by taking the Fourier transform of Fig. 1a. In this experiment, X-ray scattering signals are mainly observed at low resolutions, corresponding to low frequencies in Fourier space. A gap is placed in the center of the diffraction pattern to allow the incident beam to pass through, to avoid damaging or saturating detector sensors. The gap results in an information loss at low-frequency regime, as shown in Fig. 1c. The low-frequency information corresponds to the overall shape of the object. Without which, it poses a challenge for reconstruction.

We reconstruct the sample virus image from the diffraction pattern with detector gap in Fig. 1c. The VR-RK, randomized Kaczmarz, HIO, and RAAR methods are tested in the MATLAB platform. In order to reconstruct the data, a reference signal is used as a priori for preprocessing as described in the paper of Barmherzig et al. (2019), and the numerical iteration is then performed. Comparison of convergence rates and the relative square errors is shown in Fig. 2 and Table 1. The relative square error is defined by: $\|x - \hat{x}\|^2 / \|x\|^2$, where x is the ground truth image and $\hat{x}$ is the reconstructed image. The experiment shows that the VR-RK algorithm has a faster convergence rate and a better reconstruction accuracy compared with the randomized Kaczmarz algorithm and the iterative projection algorithms.

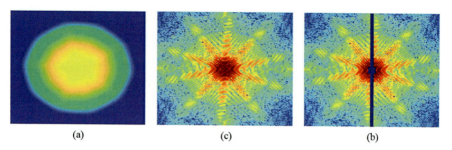

Fig. 1 Virus sample particle and its diffraction patterns (Li 2016). (**a**) Virus particle 2D projection imaging in real space. (**b**) Simulated X-ray data. (**c**) The simulated data with a gap. The size of pixels in the gap is 409

Fig. 2 Comparison of convergence rate

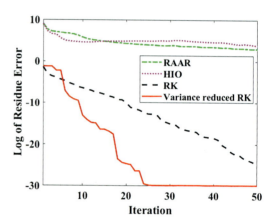

Table 1 Reconstruction error comparison

	VR-RK	RK	RAAR	HIO
Error	1.7540e-12	6.8635e-12	0.0307	0.1313

Recovery Efficiency Under Constraints

To further illustrate the convergence rate, we compare the VR-RK algorithm and the randomized Kaczmarz algorithm under L_1 and L_2 constraints on reconstructing the virus sample data. The cost function changes per iteration are shown in Fig. 3. From the figure, the loss function decays faster in VR-RK than randomized Kaczmarz method.

Considering that the single-particle X-ray imaging data are influenced by the Poisson noise, we examine the reconstruction accuracy at various noise levels, with ϵ from 0.005 to 0.1, and the measurement under the noise: $y = |Ax|^2(1+\epsilon)$.

Table 2 shows the relative square error of reconstruction using different phase retrieval algorithms in various noise levels. From Table 2, we can see that the VR-RK method outperforms other algorithms under noise.

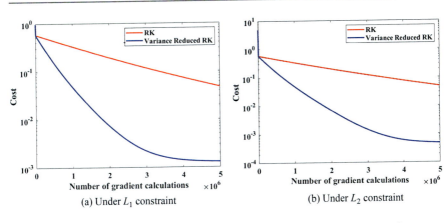

Fig. 3 Comparison of convergence rate. (**a**) Under L_1 constraint. (**b**) Under L_2 constraint

Table 2 Relative square error comparison

	VR-RK-L_2	VR-RK-L_1	RAAR	HIO
$\epsilon = 0.1$	0.3687	0.3685	1.2502	0.7315
$\epsilon = 0.05$	0.2438	0.2432	0.8775	0.5398
$\epsilon = 0.01$	0.1007	0.1013	0.3860	0.3150
$\epsilon = 0.005$	0.0707	0.0712	0.2703	0.2130

Results of the PR772 Dataset

We test the VR-RK algorithm on the PR772 particle dataset (Reddy et al. 2017). The image size is 256 × 256 pixels, and the pixel values are scaled to the range of [0, 255]. Illustration of the diffraction pattern of the single-particle data is shown in Fig. 4a and e.

For this dataset, the shrinkwrap method is applied to obtain a tight object support (Shi et al. 2019; Marchesini et al. 2003), and the square root of the diffraction intensities is used as a reference for the missing pixels during numerical iteration. A recovery example is shown in Fig. 4, and more recovery examples are presented in the supplementary materials.

We use the VR-RK algorithm and the RAAR and HIO methods to recover the data and classify the single-particle scattering pattern data and the non-single-particle scattering pattern data. We use the VR-RK for computation. There are 497 samples with labels in the validation set (Shi et al. 2019). Among them, 208 are single-particle samples, and 289 are non-single-particle samples. We use ISOMAP for data compression and clustering and KNN for classification. We use fourfold cross-validation. The VR-RK has the best result. The AUCs of the binary classification results are listed as follows (Table 3).

From the results, we can see that the VR-RK method can help recover the data and improve classification rate.

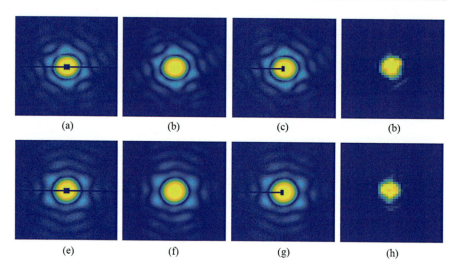

Fig. 4 PR772 single-particle scattering pattern phase retrieval. (**a**) and (**e**) are two single-particle diffraction patterns; (**b**) and (**f**) are the recovered diffraction patterns of (**a**) and (**e**), respectively; (**c**) and (**g**) show the comparison of the original and the recovered diffraction patterns, the left half is the original, and the right half is the recovered; (**d**) and (**h**) are the real-space images reconstructed using VR-RK algorithm from (**a**) and (**e**)

Table 3 AUC of binary classification

	VR-RK	RAAR	HIO
AUC	0.9501	0.9069	0.9231

Conclusion

In this chapter, we present the variance-reduced randomized Kaczmarz (VR-RK) method for XFEL single-particle phase retrieval. The VR-RK method is inspired by the randomized Kaczmarz method and the SVRG method. It is proposed in order to accelerate the convergence speed of the algorithm. Numerical results show that the VR-RK method has faster convergence rate and better accuracy under noises. Experiments on PR772 single-particle X-ray imaging data show that the VR-RK method can help recover and classify particles.

Appendix

For the PR772 dataset, further examples of phase retrieval recovery are shown here. Figure 5a and b are examples of 25 diffraction pattern sample reconstructions. Figure 5c shows the corresponding real-space recovered images.

Figures 6 and 7 are examples of 100 diffraction pattern samples reconstruction. Figure 8 shows the corresponding real space recovered images.

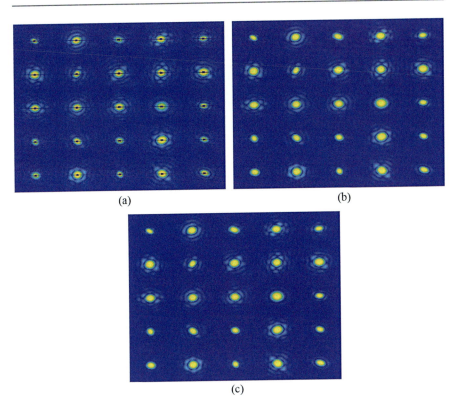

Fig. 5 Phase retrieval of the PR772 dataset. (**a**) Original data diffraction pattern illustrations. (**b**) Recovered image diffraction pattern illustrations. (**c**) Recovered real-space data illustrations

Fig. 6 Original data diffraction pattern illustrations

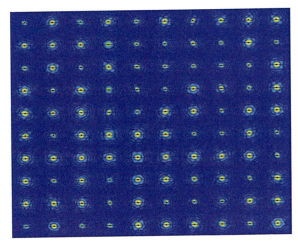

Acknowledgments Tai is supported by NSFC/RGC Joint Research Scheme (N_HKBU214/19), Initiation Grant for Faculty Niche Research Areas (RC-FNRA-IG/19-20/SCI/01) and CRF (C1013-21GF).

Fig. 7 Recovered image diffraction pattern illustrations

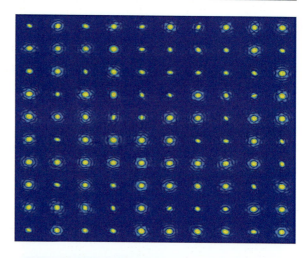

Fig. 8 Recovered real-space data illustrations

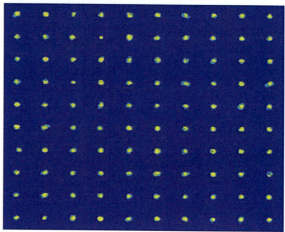

References

Bahmani, S., Romberg, J.: Phase retrieval meets statistical learning theory: a flexible convex relaxation. In: Artificial Intelligence and Statistics, pp. 252–260. PMLR (2017)

Barmherzig, D., Sun, J., Li, P., Lane, T.J., Candès, E.: Holographic phase retrieval and reference design. Inverse Probl. **35**(9), 094001 (2019)

Bauschke, H., Combettes, P., Luke, R.: Hybrid projection–reflection method for phase retrieval. JOSA A **20**(6), 1025–1034 (2003)

Bernstein, F., Koetzle, T., Williams, G., Meyer Jr, E., Brice, M., Rodgers, J., Kennard, O., Shimanouchi, T., Tasumi, M.: The protein data bank: a computer-based archival file for macromolecular structures. J. Mol. Biol. **112**(3), 535–542 (1977)

Bostan, E., Froustey, E., Rappaz, B., Shaffer, E., Sage, D., Unser, M.: Phase retrieval by using transport-of-intensity equation and differential interference contrast microscopy. In: 2014 IEEE International Conference on Image Processing (ICIP), pp. 3939–3943. IEEE (2014)

Candès, E., Strohmer, T., Voroninski, V.: Phaselift: exact and stable signal recovery from magnitude measurements via convex programming. Commun. Pure Appl. Math. **66**(8), 1241–1274 (2013)

Candès, E., Li, X., Soltanolkotabi, M.: Phase retrieval via Wirtinger flow: theory and algorithms. IEEE Trans. Inf. Theory **61**(4), 1985–2007 (2015)

Chang, H., Lou, Y., Ng, M., Zeng, T.: Phase retrieval from incomplete magnitude information via total variation regularization. SIAM J. Sci. Comput. **38**(6), A3672–A3695 (2016)

Chang, H., Lou, Y., Duan, Y., Marchesini, S.: Total variation–based phase retrieval for poisson noise removal. SIAM J. Imag. Sci. **11**(1), 24–55 (2018)

Chen, Y., Candès, E.: Solving random quadratic systems of equations is nearly as easy as solving linear systems. Commun. Pure Appl. Math. **70**(5), 822–883 (2017)

Fienup, J., Wackerman, C.: Phase-retrieval stagnation problems and solutions. JOSA A **3**(11), 1897–1907 (1986)

Fornasier, M., Rauhut, H.: Recovery algorithms for vector-valued data with joint sparsity constraints. SIAM J. Numer. Anal. **46**(2), 577–613 (2008)

Glowinski, R., Le Tallec, P.: Augmented Lagrangian and Operator-Splitting Methods in Nonlinear Mechanics. SIAM, Philadelphia (1989)

Goldstein, T., Osher, S.: The split bregman method for l1-regularized problems. SIAM J. Imag. Sci. **2**(2), 323–343 (2009)

Goldstein, T., Studer, C.: Convex phase retrieval without lifting via phasemax. In: International Conference on Machine Learning, pp. 1273–1281. PMLR (2017)

Gu, H., Xian, Y., Unarta, I., Yao, Y.: Generative adversarial networks for robust Cryo-EM image denoising. arXiv preprint arXiv:2008.07307 (2020)

Hefny, A., Needell, D., Ramdas, A.: Rows versus Columns: Randomized Kaczmarz or Gauss–Seidel for Ridge Regression. SIAM J. Sci. Comput. **39**(5), S528–S542 (2017)

Hunt, X., Reynaud-Bouret, P., Rivoirard, V., Sansonnet, L., Willett, R.: A data-dependent weighted LASSO under poisson noise. IEEE Trans. Inf. Theory **65**(3), 1589–1613 (2018)

Jiao, Y., Jin, B., Lu, X.: Preasymptotic convergence of randomized Kaczmarz method. Inverse Probl. **33**(12), 125012 (2017)

Johnson, R., Zhang, T.: Accelerating stochastic gradient descent using predictive variance reduction. Adv. Neural Inf. Process. Syst. **26**, 315–323 (2013)

Langemann, D., Tasche, M.: Phase reconstruction by a multilevel iteratively regularized gauss–newton method. Inverse Probl. **24**(3), 035006 (2008)

Le, T., Chartrand, R., Asaki, T.: A variational approach to reconstructing images corrupted by poisson noise. J. Math. Imag. Vis. **27**(3), 257–263 (2007)

Li, F., Abascal, J., Desco, M., Soleimani, M.: Total variation regularization with split Bregman-based method in magnetic induction tomography using experimental data. IEEE Sens. J. **17**(4), 976–985 (2016)

Li, P.: EE368 project: phase processing with a priori. http://github.com/leeneil/adm (2016)

Liu, H., Spence, J.: XFEL data analysis for structural biology. Quant. Biol. **4**(3), 159–176 (2016)

Luke, R.: Relaxed averaged alternating reflections for diffraction imaging. Inverse Probl. **21**(1), 37 (2004)

Ma, C., Wang, K., Chi, Y., Chen, Y.: Implicit regularization in nonconvex statistical estimation: Gradient descent converges linearly for phase retrieval and matrix completion. In: International Conference on Machine Learning, pp. 3345–3354. PMLR (2018)

Marchesini, S.: Invited article: a unified evaluation of iterative projection algorithms for phase retrieval. Rev. Sci. Instrum. **78**(1), 011301 (2007)

Marchesini, S., He, H., Chapman, H., Hau-Riege, S., Noy, A., Howells, M., Weierstall, U., Spence, J.: X-ray image reconstruction from a diffraction pattern alone. Phys. Rev. B **68**(14), 140101 (2003)

Needell, D., Ward, R., Srebro, N.: Stochastic gradient descent, weighted sampling, and the randomized Kaczmarz algorithm. Adv. Neural Inf. Process. Syst. **27**, 1017–1025 (2014)

Qiu, T., Palomar, D.: Undersampled sparse phase retrieval via majorization–minimization. IEEE Trans. Sig. Process. **65**(22), 5957–5969 (2017)

Ramos, T., Grønager, B., Andersen, M., Andreasen, J.: Direct three-dimensional tomographic reconstruction and phase retrieval of far-field coherent diffraction patterns. Phys. Rev. A **99**(2), 023801 (2019)

Reddy, H., Yoon, C., Aquila, A., Awel, S., Ayyer, K., Barty, A., Berntsen, P., Bielecki, J., Bobkov, S., Bucher, M.: Coherent soft X-ray diffraction imaging of coliphage PR772 at the Linac coherent light source. Sci. Data **4**(1), 1–9 (2017)

Scheres, S.: RELION: implementation of a bayesian approach to cryo-em structure determination. J. Struct. Biol. **180**(3), 519–530 (2012)

Seifert, B., Stolz, H., Donatelli, M., Langemann, D., Tasche, M.: Multilevel Gauss–Newton methods for phase retrieval problems. J. Phys. A: Math. General **39**(16), 4191 (2006)

Shi, Y., Yin, K., Tai, X., DeMirci, H., Hosseinizadeh, A., Hogue, B., Li, H., Ourmazd, A., Schwander, P., Vartanyants, I.: Evaluation of the performance of classification algorithms for XFEL single-particle imaging data. IUCrJ **6**(2), 331–340 (2019)

Sixou, B., Davidoiu, V., Langer, M., Peyrin, F.: Absorption and phase retrieval with Tikhonov and joint sparsity regularizations. Inverse Probl. Imag. **7**(1), 267 (2013)

Sorzano, C., Marabini, R., Velázquez-Muriel, J., Bilbao-Castro, J., Scheres, S., Carazo, J., Pascual-Montano, A.: XMIPP: a new generation of an open-source image processing package for electron microscopy. J. Struct. Biol. **148**(2), 194–204 (2004)

Strohmer, T., Vershynin, R.: A randomized Kaczmarz algorithm with exponential convergence. J. Fourier Anal. Appl. **15**(2), 262–278 (2009)

Tan, Y., Vershynin, R.: Phase retrieval via randomized Kaczmarz: theoretical guarantees. Inf. Infer. J. IMA **8**(1), 97–123 (2019)

Tropp, J.: Algorithms for simultaneous sparse approximation. Part II: Convex relaxation. Sig. Process. **86**(3), 589–602 (2006)

Wang, G., Giannakis, G., Eldar, Y.: Solving systems of random quadratic equations via truncated amplitude flow. IEEE Trans. Inf. Theory **64**(2), 773–794 (2017)

Wang, H., Wang, J.: How cryo-electron microscopy and X-ray crystallography complement each other. Protein Sci. **26**(1), 32–39 (2017)

Wei, K.: Solving systems of phaseless equations via kaczmarz methods: A proof of concept study. Inverse Probl. **31**(12), 125008 (2015)

Wu, C., Tai, X.: Augmented lagrangian method, dual methods, and split bregman iteration for ROF, vectorial TV, and high order models. SIAM J. Imag. Sci. **3**(3), 300–339 (2010)

Xian, Y., Gu, H., Wang, W., Huang, X., Yao, Y., Wang, Y., Cai, J.: Data-driven tight frame for Cryo-EM image denoising and conformational classification. In: 2018 IEEE Global Conference on Signal and Information Processing (GlobalSIP), pp. 544–548. IEEE (2018)

Zhang, H., Liang, Y.: Reshaped wirtinger flow for solving quadratic system of equations. Adv. Neural Inf. Process. Syst. **29**, 2622–2630 (2016)

Zhang, X., Lu, Y., Chan, T.: A novel sparsity reconstruction method from poisson data for 3d bioluminescence tomography. J. Sci. Comput. **50**(3), 519–535 (2012)

A Survey on Deep Learning-Based Diffeomorphic Mapping

37

Huilin Yang, Junyan Lyu, Roger Tam, and Xiaoying Tang

Contents

Introduction	1291
Background and Motivation	1291
Diffeomorphic Mapping	1291
Problem Statement and Framework Overview	1293
Deep Learning-Based Methods	1295
Related Deep Network Introduction	1296
Convolutional Neural Networks	1298
Fully Convolutional Network	1300
U-Net	1300
Autoencoders	1302

Huilin Yang and Junyan Lyu contributed equally with all other contributors.

H. Yang
Department of Electronic and Electrical Engineering, Southern University of Science and Technology, Shenzhen, Guangdong, China
School of Biomedical Engineering, The University of British Columbia, Vancouver, BC, Canada
e-mail: huiliny1@student.ubc.ca

J. Lyu
Department of Electronic and Electrical Engineering, Southern University of Science and Technology, Shenzhen, Guangdong, China
Queensland Brain Institute, The University of Queensland, St Lucia, QLD, Australia
e-mail: 12063003@mail.sustech.edu.cn

R. Tam
School of Biomedical Engineering, The University of British Columbia, Vancouver, BC, Canada
e-mail: roger.tam@ubc.ca

X. Tang (✉)
Department of Electronic and Electrical Engineering, Southern University of Science and Technology, Shenzhen, Guangdong, China
e-mail: tangxy@sustech.edu.cn

© Springer Nature Switzerland AG 2023
K. Chen et al. (eds.), *Handbook of Mathematical Models and Algorithms in Computer Vision and Imaging*, https://doi.org/10.1007/978-3-030-98661-2_108

Recurrent Neural Networks and Long Short-Term Memory Networks	1302
Unsupervised Methods	1303
Loss Function	1303
CNN-Based Methods	1305
VAE-Based Methods	1307
More Related Works	1309
Supervised Methods	1310
Loss Function	1310
CNN-Based Methods	1311
More Related Works	1314
Discussion and Future Direction	1314
Achievements and Applications	1314
Challenges	1315
Future Directions	1316
Conclusions	1316
References	1317

Abstract

Diffeomorphic mapping is a specific type of registration methods that can be used to align biomedical structures for subsequent analyses. Diffeomorphism not only provides a smooth transformation that is desirable between a pair of biomedical template and target structures but also offers a set of statistical metrics that can be used to quantify characteristics of the pair of structures of interest. However, traditional one-to-one numerical optimization is time-consuming, especially for 3D images of large volumes and 3D meshes of numerous vertices. To address this computationally expensive problem while still holding desirable properties, deep learning-based diffeomorphic mapping has been extensively explored, which learns a mapping function to perform registration in an end-to-end fashion with high computational efficiency on GPU. Learning-based approaches can be categorized into two types, namely, unsupervised and supervised. In this chapter, recent progresses on these two major categories will be covered. We will review the general frameworks of diffeomorphic mapping as well as the loss functions, regularizations, and network architectures of deep learning-based diffeomorphic mapping. Specifically, unsupervised ones can be further subdivided into convolutional neural network (CNN)-based methods and variational autoencoder-based methods, according to the network architectures, the corresponding loss functions, as well as the optimization strategies, while supervised ones mostly employ CNN. After summarizing recent achievements and challenges, we will also provide an outlook of future directions to fully exploit deep learning-based diffeomorphic mapping and its potential roles in biomedical applications such as segmentation, detection, and diagnosis.

Keywords

Diffeomorphic mapping · Deep learning · Unsupervised · Supervised

Introduction

Background and Motivation

In biomedical analysis, it is usually necessary and important to put biomedical manifolds of interest from different individuals into a common coordinate system for further analyses. Registration plays the role of putting two objects of interest into a common coordinate system, and it is usually a necessary pre-processing step before performing statistical analyses of anatomy. Diffeomorphic mapping provides one-to-one as well as smooth correspondences across different objects of interest, which serves as a powerful registration tool and has been successfully applied to a variety of biomedical applications (Louis et al. 2018; Tian et al. 2020; Debavelaere et al. 2020; Tang et al. 2019; Jiang et al. 2018; Yang et al. 2017a). However, most existing applications are based on a traditional numerical optimization scheme, which is time-consuming and could cost up to several hours for registering a single pair of 3D images or 2D meshes. In addition, registering only one template-and-target pair in a single optimization course could not learn any information from all available objects. In such context, utilizing deep learning to tackle this registration task has been extensively explored. Once the mapping function is obtained at the training phase, the network can perform registration within a few seconds given a pair of template and target, wherein only a forward pass is needed. This chapter aims at a comprehensive survey of recent progresses on deep learning-based diffeomorphic mapping methods addressing the two aforementioned issues: first, the low computational efficiency of traditional schemes that only optimize one template-and-target pair during a single optimization course and second, traditional schemes could not learn and utilize information of other available objects in the optimization process.

Diffeomorphic Mapping

Several specific properties are intuitively desirable for transformations across anatomical manifolds of interest from different individuals. First, the transformation is desired to be one-to-one, namely, an element in the template anatomy is supposed to have unique correspondence in the target anatomy. This property ensures the existence and uniqueness of the correspondence between the two anatomical manifolds of interest. In addition, it is critical that the deformed manifold obtained from the transforming process is close to the target anatomy, to ensure the accuracy of the transformation. Furthermore, since the folding of the deformation field over itself can destroy neighborhood structure which is essential for the study of anatomy, the transformation should be able to preserve the topology of template manifold before and after deformation. In this way, originally connected sets are still connected, and originally disconnected ones stay disconnected. As such, diffeomorphic mapping is of considerable interest in this regard. As shown in Fig. 1, in a diffeomorphic setting,

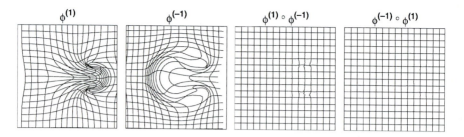

Fig. 1 Demonstration of inversion and composition in a diffeomorphic setting. Left-most: a forward deformation. Second: the corresponding inverse deformation. Both forward and inverse transformations are one-to-one. The last two: compositions of the forward and inverse transformations. (Taken from Ashburner 2007)

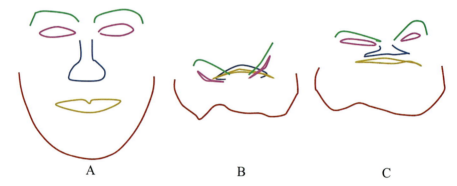

Fig. 2 Illustration of the topology-preserving property of diffeomorphic mapping. (**a**) is the original face, (**b**) is the deformed face obtained through a non-diffeomorphic mapping, and (**c**) is the deformed face obtained through a diffeomorphic mapping

both forward transformation and the corresponding inverse transformation are smooth and unfolding, and compositions of the forward and inverse transformations are very close to identity. Figure 2 illustrates the topology-preserving property of diffeomorphic mapping. The relative locations of different facial organs are well preserved after a diffeomorphic deformation, whereas those obtained from a non-diffeomorphic one are completely destroyed and unrecognizable.

Considering the aforementioned advantages, diffeomorphic mapping works well for mapping and analyzing anatomical information via various medical imaging media. Below, we will briefly introduce two widely used conventional diffeomorphic mapping methods: large deformation diffeomorphic metric mapping (LDDMM) (Beg et al. 2005; Vaillant et al. 2007; Glaunes et al. 2008) and stationary velocity field (SVF) (Arsigny et al. 2006; Modat et al. 2012). Most of the deep learning-based diffeomorphic mapping methods that we will cover in this chapter are based on the frameworks of the two methods.

Large Deformation Diffeomorphic Metric Mapping

LDDMM is a classical suite of algorithms within the academic discipline of computational anatomy, which provides not only a diffeomorphic mapping but also a geodesic metric induced on the group of diffeomorphisms. Under the LDDMM setting, manifolds to be registered could be volumes (Beg et al. 2005), curves (Glaunes et al. 2008), currents and surfaces (Vaillant and Glaunes 2005), landmarks (Joshi and Miller 2000), varifolds (Charon and Trouvé 2013), and tensors (Cao et al. 2006). The template manifold is mapped onto the target one by defining and solving a variational problem through a conditional ordinary differential equation (ODE) (Beg et al. 2005), wherein the diffeomorphism is obtained by minimizing a squared-error matching function between the deformed template and the target. To ensure diffeomorphism, the transforming flow should satisfy the Lagrangian and Eulerian specifications associated with the ODE. The diffeomorphism group is equipped with time-varying speed flows, with vector fields absolutely integrable in the Sobolev norm.

Stationary Vector Field

SVF-based diffeomorphic mapping (Arsigny et al. 2006) generalizes the principal logarithm to nonlinear geometric deformations. It is similar to the Log-Euclidean framework for tensors (Arsigny et al. 2005) in an infinite-dimensional way and aims at computing various statistics of general diffeomorphisms. SVF is defined by a stationary ODE in which the exponential of a vector field is the flow at time 1, which is solved based on nonlinear generalization of the "scaling and squaring" method. Different from LDDMM with time-varying speed flows, SVF provides flows of vector fields with stationary speed and a way to compute typical Euclidean statistics on diffeomorphisms via logarithms.

Problem Statement and Framework Overview

Given a moving object m and a fixed object f, it is preferable to find a diffeomorphic one-to-one correspondence between them so as to put them into a same reference system for further analyses. A general framework of diffeomorphic mapping for images or shapes can be, respectively, framed with the following objective functions:

$$J_{f,m}(v_t) = \min_{v_t:\dot{\phi}_t=v_t(\phi_t),\phi_0=id} \gamma R(\phi_1) + D(\phi_1^{-1} \circ m, f), \quad (1)$$

$$J_{f,m}(v_t) = \min_{v_t:\dot{\phi}_t=v_t(\phi_t),\phi_0=id} \gamma R(\phi_1) + D(\phi_1 \cdot m, f), \quad (2)$$

where $R(\phi_1)$ is a regularization term that ensures the mapping's smoothness and diffeomorphism property. The second term quantifies the overall discrepancy between the deformed moving object $\phi_1^{-1} \circ m$ (images) or $\phi_1 \cdot m$ (shapes) and the target/fixed object f. For simplicity, we denote as $\phi_1 \cdot m$ for both images and shapes in all subsequent contexts. After transformation, it is supposed that the deformed moving object $\phi_1 \cdot m$ should be very close to the fixed object f. v_t is the velocity of the transformation with respect to time t. We name it as a dynamic velocity field method when the velocity of the mapping varies across time and as a stationary velocity field method when the velocity of the mapping stays static during transformation. When $t = 0$, the registration field is identity such that $\phi_0 \cdot m = m$, and the optimal registration field ϕ_1 is obtained at $t = 1$. γ is a weight ranging from 0 to 1, serving as a trade-off coefficient between the regularization term and the overall discrepancy term. Increasing γ imposes more weight on the registration field enforcing a smoother transformation, whereas decreasing γ puts more attention on the discrepancy term making the deformed moving object closer to the fixed object.

It should be noted that, in traditional methods, we get only one optimal registration field at the time $t = 1$ after optimizing the objective function with respect to a pair of moving object and fixed object. At the top panel of Fig. 3, we present the flowchart of the typical optimization scheme in traditional methods. There are a variety of methods that can be categorized into this category, including large LDDMM (Beg et al. 2005; Vaillant et al. 2007; Glaunes et al. 2008) and SVF (Modat et al. 2012). During the past decade, they have been extensively and successfully applied to various biomedical applications (Tang et al. 2019; Jiang et al. 2018; Yang et al. 2015, 2017a; Bossa et al. 2010). Nevertheless, since these methods all make use of traditional optimization schemes and biomedical data usually have large size especially for 3D data such as MRI and CT, it usually takes such methods up to several hours to process one pair of objects of interest. In order

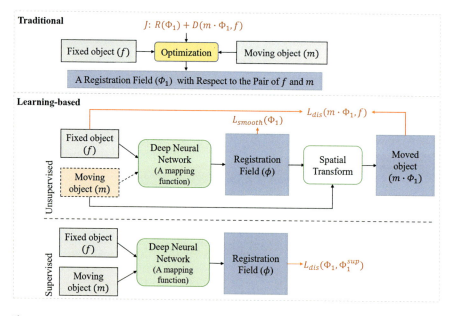

Fig. 3 An overview of traditional and deep learning-based registration methods. The top panel shows the flowchart of traditional methods, and the bottom panel shows the flowcharts of two types of deep learning-based methods (unsupervised ones and supervised ones)

37 A Survey on Deep Learning-Based Diffeomorphic Mapping

to address this problem, recent researches have started to focus on learning the registration field through deep learning. Endowed by powerful functional properties and computational efficiency on GPU of deep neural networks, the registration time has been largely reduced and is capable of predicting not only one but multiple registration fields. Once training has been finished, a pair of even large-size objects can be processed within several seconds. According to the learning style, deep learning-based diffeomorphic mapping can be categorized into two major classes, namely, unsupervised methods and supervised methods.

Deep Learning-Based Methods

Deep learning-based methods can be divided into unsupervised ones and supervised ones according to whether they require labels from traditional methods. A brief summary of key information for deep learning-based methods is illustrated in Fig. 4. Details will be described in the following subsections.

Unsupervised Methods

Unsupervised methods refer to such a kind of approach that trains a deep neural network without any information of registration fields obtained through traditional methods. It usually directly minimizes the discrepancy between the deformed moving object and the fixed object together with regularization on the registration field. In the upper part of the bottom panel in Fig. 3, we show the flowchart of unsupervised methods, which takes as inputs a fixed object and a moving object and

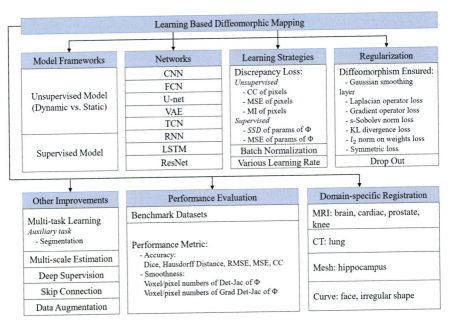

Fig. 4 A summary of deep learning-based diffeomorphic mapping

then feeds them into a deep neural network to predict the corresponding registration field. The subsequent spatial transform module takes the predicted registration field and the moving object as inputs to perform diffeomorphic deformation yielding the deformed moving object (also called moved object). Some methods only need to input fixed objects, and the moving object (in dashed box in Fig. 3) can be estimated during training. The whole training phase makes use of only two kinds of information, namely, the fixed objects and some methods the corresponding moving objects. Regularization is imposed on the registration field through the loss term $L_{\text{smooth}}(\phi_1)$, and measurement of similarity is conducted through minimizing the discrepancy loss $L_{\text{dis}}(m \cdot \phi_1, f)$. Unsupervised methods mimic the traditional optimization scheme except that they aim at predicting diffeomorphisms between a set of template-and-target pairs instead of between only one pair of template and target within one single training course.

Supervised Methods
Compared with unsupervised methods, supervised methods usually take three kinds of information as inputs including the fixed objects, the moving objects, and parameterizations of the corresponding registration fields such as the velocity or momentum acquired from performing traditional methods. This means that we first need to conduct traditional diffeomorphic registrations on all pairs of moving-and-fixed objects to obtain the corresponding registration fields, called ϕ_1^{sup}, as ground truth. We then use them together with all moving-and-fixed pairs of objects as training materials. The lower part of the bottom panel in Fig. 3 shows the training flow of supervised methods. The loss function $L_{\text{dis}}(\phi_1, \phi_1^{\text{sup}})$ minimizes the discrepancy between the predicted registration field ϕ_1 and the pre-obtained "ground truth" registration field ϕ_1^{sup}. Supervised methods assume the registration fields obtained through the traditional optimization scheme are optimal and try to make the learning-based predictions as close to them as possible.

The remainder of this chapter is organized as follows. Related deep network introduction and summary will be described in section "Related Deep Network Introduction". Unsupervised methods will be reviewed in section "Unsupervised Methods", followed by a survey of supervised methods in section "Supervised Methods". We will also cover current achievements and related applications in section "Discussion and Future Direction". After reviewing existing works, potential emerging topics and future directions will be elaborated. Finally, conclusion of this survey will be organized in section "Conclusions".

Related Deep Network Introduction

Leveraged by deep learning and neural networks, diffeomorphic mapping can be achieved in an efficient manner. Related neural network types that have been employed in learning-based diffeomorphic mapping approaches surveyed in this chapter are summarized in Fig. 4, and the specific approaches together with their corresponding adopted networks are, respectively, listed in Table 1 for unsupervised

Table 1 Summary of all reviewed unsupervised learning-based diffeomorphic mapping methods. (*ROI* region of interests, *CNN* convolutional neural network *FCN* fully convolutional network, *CVAE* conditional variational autoencoder, *VAE* variational autoencoder, *RNN* recurrent neural network, *TCN* temporal convolutional network, *ResNet* residual neural network, *SVF* static velocity field, *RDMM* region-specific diffeomorphic metric mapping, *LDDMM* large deformation diffeomorphic metric mapping, *NLCC* normalized local cross-correlation, *MSE* mean squared error, *SSD* sum of squared difference, *CD* the Chamfer distance, *EMD* the Earth mover's distance, d the displacement of the predicted registration field, v the velocity of the predicted registration field, *LOC* local orientation consistency, *KL divergence* Kullback-Leibler divergence, *OMT* optimal mass transport, *MRI* magnetic resonance imaging, *CT* computed tomography)

Reference	Network	Velocity	Similarity metric	Regularity term	Modality	ROI	Others
Balakrishnan et al. (2019)	U-Net	Static	NLCC; MSE	∇^2 on d	MRI	Brain	Auxiliary segmentation; instance-specific fine-tuning
Balakrishnan et al. (2018)	U-Net	Static	NLCC	∇^2 on d	MRI	Brain	–
Dalca et al. (2018)	U-Net	Static	MSE	∇^2 on inverse of covMatrix of v	MRI	Brain	Uncertainty analysis
Dalca et al. (2019a)	U-Net	Static	MSE	∇^2 on inverse of covMatrix of v	MRI	Brain	Surface information
Mok and Chung (2020a)	FCN	Static	NLCC	∇ on v; LOC (Mok and Chung 2020a)	MRI	Brain	Symmetric map & loss
Krebs et al. (2019)	CVAE	SVF	Symmetric NLCC	Gaussian smoothing layer on v	MRI	Cardiac	Symmetric loss; from healthy to pathological cases
Bône et al. (2019)	VAE	Dynamic	Norm on vector valued mesh metric	KL divergence	MRI; mesh	Brain; hippocampus	Current-splatting layer for meshes
Bône et al. (2019)	VAE	SVF	Likelihood probability	3-Sobolev norm (Zhang and Fletcher 2019) on v	MRI; mesh	Face; hippocampus	Current-splatting layer for meshes; private dataset

(continued)

Table 1 (continued)

Han et al. (2020)	CNN	SVF	NLCC	Differential operator on v	MRI	Brain	Brain with tumors
Shen et al. (2019a)	U-Net	SVF	NLCC	Differential operator on v	MRI	Knee	Symmetric loss
Louis et al. (2019)	RNN	Dynamic	SSD	Gaussian smoothing layer on v	MRI	Brain	Force small variance on latent space
Niethammer et al. (2019)	CNN	SVF	NLCC	OMT on multi-Gaussian kernel weights	MRI	Brain	OMT on local deformation
Krebs et al. (2021)	CVAE TCN	SVF	SSD	Gaussian smoothing layer on v	MRI	Cardiac	Regularity on both spatial and temporal domains
Mok and Chung (2020b)	CNN	SVF	NLCC	∇^2 on v	MRI	Brain	Coarse-to-fine; Laplacian pyramid framework
Shen et al. (2019b)	CNN	RD-MM	Multi-kernel NLCC	∇^2 on v; OMT regularity	CT	Lung; knee	Multi-kernel NLCC similarity metric
Hoffmann et al. (2020)	U-Net	SVF	Soft Dice	∇ on d	MRI	Brain	Train purely with synthetic data
Amor et al. (2021)	ResNet	LDD-MM	CD; EMD	LDDMM regularity on v	Mesh	Cortex; heart; liver; femur; hand	Train on one pair of data

methods and Table 2 for supervised methods. In this section, we will introduce in detail several main types of deep neural network (DNN) architectures that have been adopted in existing diffeomorphic mapping approaches.

Convolutional Neural Networks

Convolutional neural networks (CNNs) have made impressive progress in computer vision tasks including image recognition (Krizhevsky et al. 2012), object detection (Liu et al. 2020), and semantic segmentation (Lateef and Ruichek 2019). As shown in Fig. 5, a CNN typically consists of convolutional layers, pooling layers, activation layers, and fully connected layers. A convolutional layer contains a set of learnable

Table 2 Summary of all reviewed supervised learning-based diffeomorphic mapping methods. (*ROI* region of interests, *CNN* convolutional neural network, *FCN* fully convolutional network, *LSTM* long short-term memory module, *LDDMM* large deformation diffeomorphic metric mapping, *SSD* sum of squared difference, *MSE* mean squared error, v the velocity of the predicted registration field, *MRI* magnetic resonance imaging)

Reference	Network	Velocity	Similarity metric	Regularity term	Modality	ROI	Others
Yang et al. (2017c)	U-shape CNN	Static	SSD	LDDMM regularity on v	MRI	Brain	A correction network for momenta
Rohé et al. (2017)	U-Shape FCN	Static	SSD	–	MRI	Cardiac	Data augmentation; no regularity
Wang and Zhang (2020b)	CNN	LDDMM	SSD	Fourier domains of v; l_2 norm on network weights	MRI	Brain	Process in Fourier domain; two networks for real and imaginary parts separately
Ding et al. (2019)	CNN	LDDMM	SSD	∇^2 on v	MRI	Brain	Longitudinal registration study
Kwitt and Niethammer (2017)	CNN	LDDMM	SSD	∇^2 on v	MRI	Brain	–
Wang and Zhang (2020a)	CNN	LDDMM	MSE	l_2 norm on network weights	MRI	Brain	–
Pathan and Hong (2018)	LSTM; CNN	LDDMM	MSE	Momentum sequence	MRI	Brain	–
Krebs et al. (2017)	CNN	Dynamic	SSD	Fuzzy action control	MRI	Prostate	Reinforcement learning

kernels operating on local input regions to extract features. A pooling layer performs linear or nonlinear downsampling on feature maps to reduce spatial resolution and summarize local information. An activation layer can be the sigmoid function, the hyperbolic tangent function (Tanh), or the rectified linear unit (ReLU) (Sibi et al. 2013), introducing nonlinearity to CNNs. A fully connected layer is the same as multi-layer perceptron, providing final classification or regression predictions. By stacking these layers hierarchically, CNNs gain large receptive fields and thus can exploit and capture scale-invariant and translation-invariant features. Several novel CNN architectures including VGGNet (Simonyan and Zisserman 2014), Inception

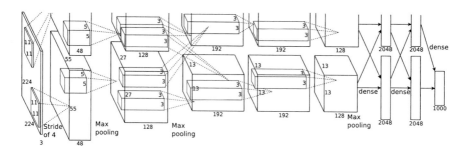

Fig. 5 A typical hierarchy of convolutional neural networks. (Taken from Krizhevsky et al. 2012)

(Szegedy et al. 2017), ResNet (He et al. 2016), and DenseNet (Iandola et al. 2014) have been proposed since 2012, after AlexNet achieved a big breakthrough on ImageNet classification; the error rate was reduced by half (Krizhevsky et al. 2012).

Fully Convolutional Network

Fully convolutional network (FCN) is proposed in 2015 and is originally designed for semantic segmentation. FCN first inputs images to an arbitrary backbone network to produce feature maps, the backbone network of which can be AlexNet or VGGNet. It then applies deconvolutional layers (Zeiler et al. 2010), which simply reverse the forward and backward pass of convolution to upsample feature maps. The upsampled dense-pixel feature maps are subsequently sent to a 1×1 convolution layer with desired channel dimensions to get pixel-level spatial predictions. As a result, FCN is able to take an input of arbitrary size and produce an output of the same size in an end-to-end manner, whereas previous CNNs cannot.

In order to refine spatial details, multi-scale feature maps from previous convolutional layers in the backbone network are deconvoluted and yield additional predictions. By aggregating those predictions, FCN can combine semantic (high-level) information from coarse-scale predictions and appearance (low-level) information from fine-scale predictions and thus further boost the final precision. As a result, FCN demonstrates state-of-the-art performance on PASCAL VOC, NYUDv2, and SIFT Flow. FCN and its variants have also been applied to medical image segmentation tasks such as lung segmentation (Kaul et al. 2019), whole brain segmentation (Roy et al. 2018), and retinal vessel segmentation (Lyu et al. 2019).

U-Net

Inspired by FCN, Ronneberger et al. (2015) propose a novel encoder-decoder network, named U-Net, for biomedical image segmentation. U-Net has a symmetrical

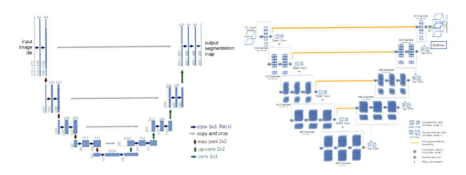

Fig. 6 Illustrations of U-Net (left) and V-Net (right). (Taken from Ronneberger et al. 2015 and Milletari et al. 2016)

U-shaped architecture with a contracting path and an expanding path, as illustrated in Fig. 6. The contracting path performs repetitions of two 3×3 convolutions followed by ReLU and 2×2 max pooling to encode contextual information, while the expanding path performs repetitions of two 3×3 convolutions with ReLU and 2×2 deconvolution to gradually decode feature maps and recover spatial resolution. Finally, a 1×1 convolution projects the feature maps to output space.

The feature maps from the contracting path are, respectively, skip connected to the corresponding feature maps from the expanding path with the same spatial resolutions. Compared with FCN that aggregates multi-level features in the output space, U-Net fuses these features via concatenation in the feature space. This enables U-Net to propagate more information from previous layers to subsequent layers, contributing to better gradient flows and faster convergence speed in the training course. Moreover, its light and compact architecture allows U-Net to better converge on biomedical image datasets which typically have fewer labeled training samples and higher spatial resolutions than natural images. U-Net is evaluated on the EM segmentation challenge (Arganda-Carreras et al. 2015) and the ISBI cell tracking challenge (Ulman et al. 2017), outperforming previous methods by large margins.

The success of U-Net promotes the development of its extensions. One of its most well-known variants is V-Net proposed by Milletari et al. in 2016. V-Net extends U-Net to work on volumetric medical image segmentation. There are several modifications other than simply changing 2D convolutions to 3D convolutions. V-Net learns additional residual functions at each stage: the input of each stage is element-wisely added to the corresponding output. This operation effectively simplifies the network and alleviates the vanishing gradient issue and thus solves the convergence problem on volumetric CNNs. In addition, $5 \times 5 \times 5$ convolutions are adopted to replace the original $3 \times 3 \times 3$ convolutions to gain larger receptive fields. Max pooling is replaced with strided convolution to perform parametric downsampling, introducing more nonlinearities. V-Net has been used to perform prostate segmentation on MRI volumes in a fast and accurate manner (Milletari et al. 2016).

Autoencoders

Autoencoders are first proposed for dimensionality reduction (Wang et al. 2014). An autoencoder learns an approximation of the identity mapping through an encoder-decoder structure: an encoder extracts the latent space representation of the input, and a decoder reconstructs the input with the extracted vector. The encoder and decoder can be multi-layer perceptrons, CNNs, or any feed-forward neural networks. Supervised by the identity loss (mean absolute error or mean square error) between the input and the output, the latent vector is able to provide a compact expression of the input in a lower-dimensional space. Autoencoders have been applied to principal component analysis (Kramer 1991), image denoising (Gondara 2016), and anomaly detection (Zhou and Paffenroth 2017).

Variational autoencoder (VAE) (Kingma and Welling 2013) is one of the most important variants of autoencoders. Unlike a traditional autoencoder, a VAE assumes that the latent space fits a certain probability distribution, such as a Gaussian distribution, and estimates the parameters of this probability distribution from the input data. Therefore, the approximated distribution of the latent space of a VAE matches the input space closer than that of a traditional autoencoder. In addition to minimizing the identity loss between the input and the output, a regularization term of Kullback-Leibler (KL) divergence between the desired distribution and the predicted distribution is used to train a VAE.

Recurrent Neural Networks and Long Short-Term Memory Networks

CNNs and the other aforementioned networks are unable to handle input sequences of various lengths and thus cannot model the temporal correlations within sequences. Recurrent neural networks (RNNs) (Karpathy et al. 2015) are proposed to solve this problem and have been widely used to process text, video, and time series. At each timestamp, a RNN collects the previous hidden state vector and the current input vector to update the current hidden state and produces output by sending the current hidden state vector to a feed-forward network.

However, RNNs suffer from vanishing or exploding gradients as the sequences grow longer (Karpathy et al. 2015), resulting in poor performance on capturing long-term dependencies. Long short-term memory (LSTM) (Karpathy et al. 2015) is explicitly designed to address such long-term dependency issue. LSTM introduces three gates to protect and control the cell state and the hidden state: the forget gate is used to determine how much information in the previous cell state should be kept; the input gate is used to collect useful information from the current input and the previous hidden state and add them to the filtered previous cell state so as to update the current cell state; and the output gate is used to output a filtered informative vector, namely, the current hidden state, from the updated cell state. All these three types of gates take the previous hidden state as well as the current input as inputs for calculations of their corresponding filter coefficients (Fig. 7).

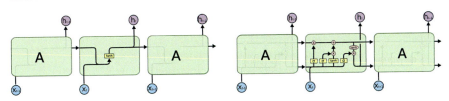

Fig. 7 Architectures of a RNN (left) model and a LSTM (right) model. (Taken from Karpathy et al. 2015)

Unsupervised Methods

In this section, we will survey the literature on methods of unsupervised deep learning-based diffeomorphic mapping. The goal of training an unsupervised network is to ensure it can predict the corresponding registration fields and perform diffeomorphic deformations when pairs of to-be-registered objects are given.

We will start the chapter with loss function and several representative similarity metrics. We then proceed to a variety of regularization approaches for diffeomorphic mapping, and finally several CNN-based (specifically U-Net-based and VAE-based) methods as well as more related works will be introduced.

Loss Function

In a deep learning-based unsupervised diffeomorphic mapping framework, the typical loss function is also composed of two parts, namely, the similarity term and the regularization term. However, the optimization procedure of deep learning-based methods is completely different from that of traditional methods. A typical loss function can be written as follows:

$$L^{\text{unsup}} = L_{\text{sim}}(m \cdot \phi_1, f) + \gamma L_{\text{reg}}(\phi_1), \qquad (3)$$

where L_{sim} is the similarity term measuring the difference between the deformed moving objects $m \cdot \phi_1$ and the fixed objects f and L_{reg} is the regularization term imposing certain constraint on the registration fields ϕ_1 to make them diffeomorphic. In the process of minimizing the loss function, the set of deformed moving objects is increasingly closer to the set of fixed objects, and the corresponding registration fields are becoming smoother. γ is a trade-off factor between the similarity term and the regularization term. A too large γ will result in inadequate registration fields that cause highly inaccurate registrations, whereas a too small γ will lead to overly flexible registration fields that might be irregular and not diffeomorphic anymore. In practice, γ is usually empirically chosen.

Similarity Metrics

For different data types, there are different metrics to quantify the similarity between the moved objects $m \cdot \phi_1$ and the fixed objects f. For image data, mean squared error

(MSE), normalized local cross-correlation (NLCC), and mutual information (MI) are often employed. MSE is computed by averaging squared pixel-wise (2D)/voxel-wise (3D) image intensity differences, which can be expressed as

$$MSE(m \cdot \phi_1, f) = \frac{1}{|\Omega|} \sum_{p \in \Omega} \left[[m \cdot \phi_1](p) - f(p) \right]^2. \quad (4)$$

In MSE, p indexes image pixels or voxels and Ω represents the whole image. Since loss function is minimized to train the deep network, a small MSE is desired to yield a good alignment result. The similarity loss term with MSE can be directly written as $L_{\text{sim}}(m \cdot \phi_1, f) = MSE((m \cdot \phi_1, f))$. Unlike MSE that measures a global difference, NLCC quantifies local cross-correlation and is commonly termed as CC, which is computed over the whole image. CC can be written as

$$CC(m \cdot \phi_1, f) = \sum_{p \in \Omega} NLCC$$

$$= \sum_{p \in \Omega} \frac{\left(\sum_{p_i} ([m \cdot \phi_1](p_i) - [m \cdot \phi_1](\bar{p}))(f(p_i) - f(\bar{p})) \right)^2}{\left(\sum_{p_i} (f(p_i) - f(\bar{p}))^2 \right) \left(\sum_{p_i} ([m \cdot \phi_1](p_i) - [m \cdot \phi_1](\bar{p}))^2 \right)} \quad (5)$$

where $f(\bar{p})$ and $[m \cdot \phi_1](\bar{p})$ denote images with local mean intensities, $f(\bar{p}) = \frac{1}{n^d} \sum_{p_i} f(p_i)$, with p_i iterates over a n^2 area (2D) or a n^3 volume (3D) around p. A higher CC indicates a better alignment, yielding the loss function: $L_{\text{sim}}(m \cdot \phi_1, f) = -CC(m \cdot \phi_1, f)$.

For shape data such as landmarks, curves, or meshes, l_2 norm or norm of differences on manifold-based vector-valued metrics (Vaillant et al. 2007; Glaunes et al. 2008) is usually taken as the similarity term.

Regularization for Diffeomorphic Mapping

In order to make the registration field diffeomorphic, imposing regularization on it is necessary to ensure the smoothness of the deformation. Typically, there are three types of vector fields parameterizing the registration field: displacement field, velocity field, and momentum field. Displacement directly gives the length and direction that the moving object should move. Integral of velocity over time gives the displacement. Momentum usually is a dual space of velocity, and thus velocity can be computed from momentum. These three types of vector fields characterize the registration field in different forms. Regularizations are usually conducted on displacement or velocity in a way of minimizing their differential spaces (such as the first-order or the second-order derivatives) with a vector norm. Let $\mathbf{u}$ denote displacement or velocity; then L_{reg} can be written as

$$L_{\text{reg}}(\phi) = \sum_{p \in \Omega} ||\nabla \mathbf{u}(p)||^2, \quad (6)$$

or

$$L_{\text{reg}}(\phi) = \sum_{p \in \Omega} ||\nabla^2 \mathbf{u}(p)||^2, \qquad (7)$$

where ∇ is the gradient operator. Other than formulating the regularization term as part of the loss function, employing a smoothing filter like Gaussian convolution right behind the layer for predicting displacement or velocity is also a useful way to smooth the registration field. Applying a Gaussian convolution layer is equivalent to imposing a diffusion-like regularization prior on the predicted velocity or displacement (Krebs et al. 2019).

CNN-Based Methods

Unsupervised learning-based diffeomorphic mapping has been extensively exploited since 2018. The most frequently used deep networks are CNN-based models including FCN and U-Net. We will detail several representative works in this subsection as well as briefly introduce other related works in section "More Related Works".

VoxelMorph VoxelMorph (Balakrishnan et al. 2019) takes one moving image and one fixed image as inputs and feeds them into a U-shape CNN to predict displacements $\mathbf{u}$ through a function $g_\theta(f, m)$ with network parameters θ. Then, a variant of spatial transform layer (Jaderberg et al. 2015) is applied to deform the moving image using the predicted displacement to register it onto the fixed image. Laplacian regularization is imposed on the predicted displacement. After finishing training VoxelMorph, a mapping that can predict a set of pairwise deformation fields across the population of interest is learned.

VoxelMorph has been extensively validated on 3731 T1-weighted brain MRI scans from eight publicly available datasets. All MRI data go through standard pre-processing steps (also applicable to other methods surveyed in this chapter), including affine spatial normalization and brain extraction using FreeSurfer (Fischl 2012). Considerable reduction on the registration time is achieved by VoxelMorph while holding comparable registration accuracy compared to the two classical methods (around 0.77 Dice for each of the three compared methods and 0.608 Dice for affine alignment). Both atlas-based registration and subject-to-subject registration are evaluated to validate the effectiveness of VoxelMorph. In addition, auxiliary tasks such as segmentation are also investigated in VoxelMorph. Extensive experiments show that incorporating auxiliary tasks in the training procedure is beneficial for the registration accuracy.

VoxelMorph-diff Unlike VoxelMorph which predicts deterministic displacement, VoxelMorph-diff (Dalca et al. 2018) assumes that each registration field between a pair of images fits a normal distribution and learns a generative model that is able to quantify the registration uncertainty. VoxelMorph-diff employs a similar framework as VoxelMorph does, but it predicts the mean and covariance matrix of the velocity field instead of the displacement. A posterior probability of multivariate normal distribution with a diagonal covariance matrix is imposed on the velocity field for a pair of given moving-and-fixed images, followed by seven squaring and scaling layers (Arsigny et al. 2006) to compute the deformation field ϕ_1. Then a subsequent spatial transform layer deforms the moving image using the obtained deformation field ϕ_1. The model is trained by optimizing the variational lower bound of KL divergence, making the predicted posterior probability distribution approximate the true posterior probability distribution. MSE is employed as the similarity metric in the loss function. A newly defined Laplacian operator on the inverse of the predicted covariance matrix of the velocity field with hyperparameter λ is employed for regularization. In the testing phase, given a pair of images, the predicted mean of the velocity field can be used as the optimal sample for the subsequent deformation process. Furthermore, numerous samples could be drawn from the learned distribution to evaluate the registration uncertainty.

For fair comparisons, the same datasets and settings as Balakrishnan et al. (2018) are used. VoxelMorph-diff outperforms VoxelMorph and ANTS SyN in terms of all metrics: 0.753, 0.75, and 0.75 on Dice; 0.7, 18096, and 6505 on the averaged number of voxels whose Jacobian determinants are less than or equal to 0; and 0.451s, 0.554s, and – on GPU (NVIDIA TitanX). Particularly, only VoxelMorph-diff can quantify the registration uncertainty. Experimental results indicate higher uncertainty appears at anatomical boundaries, while lower uncertainty appears at regions that are relatively far from anatomical boundaries.

SYMNet In order to ensure the preferable diffeomorphic mapping properties, SYMNet (Mok and Chung 2020a) is proposed. SYMNet presents a symmetric image registration method that maximizes the similarity between images with respect to the space of diffeomorphic mappings. In addition, it simultaneously estimates both forward and backward transformations with an additional local orientation consistency regularization term (Mok and Chung 2020a) that forces local deformations to be consistent and smooth. Specifically, SYMNet takes images X and Y as inputs; they are fed into a U-shape FCN to predict two symmetric velocity fields v_{XY} and v_{YX}. Meanwhile, it takes the negative of v_{XY} and v_{YX}, respectively, obtaining $-v_{XY}$ and $-v_{YX}$. After performing scaling and squaring (Arsigny et al. 2006) on each of these four velocity fields, four deformation fields $\phi_{XY}^{(0.5)}$, $\phi_{XY}^{(-0.5)}$, $\phi_{YX}^{(0.5)}$, and $\phi_{YX}^{(-0.5)}$ are yielded. Then, $\phi_{XY}^{(0.5)}$ and $\phi_{YX}^{(-0.5)}$ are composed and applied to X via a diffeomorphic spatial transformer derived from Jaderberg et al. (2015) to obtain $\phi_{XY}^{(1)}(X)$. $\phi_{YX}^{(0.5)}$ and $\phi_{XY}^{(-0.5)}$ are composed and applied to Y to obtain $\phi_{YX}^{(1)}(Y)$. $\phi_{XY}^{(0.5)}$ is applied to X yielding $\phi_{XY}^{(0.5)}(X)$, and $\phi_{YX}^{(0.5)}$ is applied to Y yielding $\phi_{YX}^{(0.5)}(Y)$. Thus, the similarity term consists of two parts: $L_{\text{sim}} = L_{\text{mean}} + L_{\text{pair}}$, in which $L_{\text{mean}} = -CC(\phi_{XY}^{(0.5)}(X), \phi_{YX}^{(0.5)}(Y))$ and $L_{\text{pair}} = -CC(\phi_{XY}^{(1)}(X), Y) - CC(\phi_{YX}^{(1)}(Y), X)$. For regularization purposes, SYMNet employs three terms: L_{Jdet}

measures the averaged number of voxels whose Jacobian determinants of the deformation field are less than 0, serving as a local orientation consistency regularity. L_{reg} measures the l_2 norm on the gradients of v_{XY} and v_{YX} across all voxels serving as a global smoothness regularity. L_{mag} measures the averaged discrepancy between the l_2 norm of v_{XY} and that of v_{YX}, explicitly guaranteeing the magnitudes of the two predicted symmetric velocity fields to be (approximately) the same. Both L_{mean} and L_{mag} enforce the mapping and the corresponding inverse mapping to be symmetric.

Comparing SYMNet with ANTs SyN (Avants et al. 2008), VoxelMorph (Balakrishnan et al. 2019), and VoxelMorph-diff (Dalca et al. 2018) are conducted via atlas-based registration using 425 T1-weighted brain MRI scans from OASIS (Fotenos et al. 2005). Different from the original experimental settings in VoxelMorph and VoxelMorph-diff, all learning-based methods involved in the comparisons are trained by pairwise registrations of all image pairs in the training set. The average Dice scores for SYMNet, VoxelMorph, VoxelMorph-diff, and ANTs SyN are, respectively, 0.738, 0.707, 0.693, and 0.680 (0.567 for affine only), and the corresponding numbers of voxels whose Jacobian determinants are less than or equal to 0 are, respectively, 0.471, 0.588, 346.712, and 0.047. The running time is 0.414s for SYMNet, 0.695 s for VoxelMorph, and 0.517 s for VoxelMorph-diff on a NVIDIA GTX 1080Ti GPU and 1039 s for ANTs SyN on an Intel Core i7-7700 CPU. SYMNet achieves the best performance on the evaluated dataset. Ablation studies successfully validate the effectiveness of the local orientation-consistency loss proposed by SYMNet.

VAE-Based Methods

Different from CNN-based methods which usually directly estimate the parameterizations (displacement, velocity, or momentum) of the registration field, VAE-based methods estimate a latent space that encodes the deformation space through an encoder and predict the velocity field through a decoder. Subsequent layers deform the moving image to reconstruct the input image (the fixed image) with the predicted velocity field, yielding the deformed moving image. Furthermore, the template, namely, the moving image, can be simultaneously estimated together with the latent space in the training phase. Two representative works will be described in detail, and more related works will be briefly covered in section "More Related Works".

ProbDR (Krebs et al. 2019) models registration in a probabilistic and generative framework by applying a conditional variational autoencoder (CVAE) with multi-scale deformations, denoted as ProbDR. ProbDR assumes that the transformations for a to-be-registered population could be represented using a compact low-dimensional latent space (follows a multivariate unit Gaussian distribution with spherical covariance) and assumes the velocity of the deformation could be decoded from this latent space. Given a pair of moving-and-fixed images, the corresponding low-dimensional representation in the latent space can be estimated and fed into the decoder for calculating the velocity field of the deformation, which is then smoothed by a Gaussian convolution layer to ensure the diffeomorphism property. After that, the moving image is fed into a dense warping layer implemented via STN together with the calculated velocity field to acquire the finally deformed moving image. ProbDR concurrently conducts registration in a multi-scale fashion to further boost the performance. The moving image at each scale is fed into the corresponding deconvolution layer of the same resolution in the decoder to learn more geometry-invariant representations in the latent space. A Boltzmann distribution likelihood with symmetric NLCC is employed as the posterior probability distribution of the input images given the moving image and the corresponding predicted latent space. Moreover, the trained decoder network can be used to sample and transport new deformations in the following way: sampling latent representations from the previously predicted mean and covariance and then applying the sampled representations to the new moving images through subsequent networks. To be noticed, although VoxelMorph-diff also learns a generative model, ProbDR learns a much more compact low-dimensional representation of the deformation field instead of predicting the velocity field which is of a much higher dimension.

Extensive evaluations are conducted on 3D intra-subject registration using 334 cardiac cine-MRIs. The training set are randomly shifted, rotated, scaled, and mirrored as data augmentation. It should be noted that the aforementioned methods are all trained without data augmentation. Comparisons are conducted with LCC-demons (Lorenzi et al. 2013), ANTs SyN (Avants et al. 2008), and VoxelMorph (Balakrishnan et al. 2019). ProbDR obtains the best results with respect to Dice, Hausdorff distance, and the averaged number of voxels whose Jacobian determinants are less than or equal to 0, respectively, being 0.812, 7.3 mm, and 1.4. VoxelMorph gets the best RMSE being 0.24, while ProbDR obtains a RMSE of 0.30. A five-disease classification accuracy of 83% is obtained by ProbDR when using the eight most discriminative components from canonical correlation analysis. ProbDR also demonstrates how to perform deformation transport from healthy to disease without inter-subject registration for pre-processing, which is needed by the other three methods.

LRShape Another method that learns low-dimensional representations of diffeomorphic mapping is proposed by Bone and published in 2019 (Bône et al. 2019), denoted as LRShape. Unlike all of the aforementioned methods that mainly focus on image data and prediction of static velocity field, LRShape focuses on shape data such as curves and surfaces and predicts a time-varying velocity field. A current-splatting layer (Durrleman 2010; Gori et al. 2017) that allows neural network architectures to process meshes is presented in this work. In contrast to ProbDR that takes both moving and fixed objects as the inputs, LRShape only takes the fixed shape as the input and estimates the template (namely, the moving shape) jointly with a low-dimensional representation. Specifically, the fixed shape is fed into the current-splatting layer to transform shape data into image type. Then the current-splatting expression is passed through an encoder to estimate the latent space that encodes deformations of the population of interest. The velocity field can be obtained as the output of the decoder and is applied to the estimated template to (approximately) reconstruct the input fixed shape. Noticeably, the template is the same for all fixed shapes. A s-Sobolev equivalent norm is adopted as the regularization term on the registration field to encourage smooth deformation.

Evaluations are conducted on the ADNI database (Jack et al. 2008). Compared with principal geodesic analysis (PGA) (Zhang and Fletcher 2014) on reconstruction (on training data) and generalization (on testing data unseen in the training phase), residuals of the hippocampus show that LRShape is better at reconstruction, while PGA is better at generalization. When using the learned 3D latent representations from PGA and LRShape as inputs for classifications of three classes, healthy control (HC: 54 cases), mild cognitive impairment (MCI: 53 cases), and Alzheimer's disease (AD: 53 cases), accuracies of 61.3% versus 58.8% for classifying CN/MCI/AD, 85.0% versus 84.1% for classifying CN/AD, 67.3% versus 67.3% for classifying CN/MCI, and 68.9% versus 71.7% for classifying MCI/AD are obtained from LRShape and PGA.

More Related Works

In addition to these detailedly described methods, there are a number of other related methods also built under unsupervised frameworks. Han et al. (2020) explores a CNN-based learning approach to register images with brain tumors to an atlas. It learns appearance mappings from images with tumors to the atlas and simultaneously predicts the corresponding transformations to the atlas space. Shen et al. (2019a) propose a method that jointly learns affine and diffeomorphic mappings through an end-to-end U-Net. In addition to the regular similarity and regularity terms, it is supervised by an additional symmetric loss. Riemannian manifold learning in association with a statistical task of longitudinal trajectory analyses is studied in Louis et al. (2019), which adopts a RNN to properly process the sequence of longitudinal data. Niethammer et al. (2019) jointly optimize over momenta and the parameters in a CNN of predicting regularizer, constructing a metric such that diffeomorphic transformations can be ensured in the continuum. A deep Laplacian pyramid image registration framework (Mok and Chung 2020b) is proposed in 2020, which is able to solve the optimization problem of image registration in a coarse-to-fine fashion within the space of diffeomorphism. Krebs et al. (2021) recently proposes learning a probabilistic motion model from image sequences for spatio-temporal registration. It encodes motion in a low-dimensional probabilistic space (a motion matrix), enabling various motion tasks such as simulation and interpolation of realistic motion patterns for faster data acquisition and data augmentation. This work is a variant of Krebs et al. (2019) by introducing a novel Gaussian process prior and employing a temporal convolutional network (Lea et al. 2016) for the temporal sequences. Another work Hinkle et al. (2018) aims to create atlas using a form of autoencoder, in which the encoder maps an image to a transformation and the decoder interpolates a deformable template to reconstruct the input. Shen et al. (2019b) describe a region-specific diffeomorphic mapping that allows for spatial-varying regularization advected via the estimated spatio-temporal velocity field, the framework of which is built based on CNN. Aiming to remove image-data dependency for learning-based methods, Hoffmann et al. (2020) exploit a new direction that leverages a generative model for diverse label maps and images, which exposes the networks to a wide range of variabilities during training. Besides, Detlefsen et al. (2018) employ continuous piecewise-affine-based (CPAB) (Freifeld et al. 2017) diffeomorphic mapping in the tasks of classifying digital numbers and face verification via CNN and show better results over methods without involving diffeomorphic mapping. Amor et al. (2021) proposes a method that uses deep residual networks (He et al. 2016) to implement LDDMM on surfaces and conducts evaluations on a variety of region of interests (ROIs) including the cortex, heart, liver, femur, and hand. Related information of all of the reviewed unsupervised learning works is organized in Table 1.

Supervised Methods

In this section, literature on supervised deep learning-based diffeomorphic mapping methods will be surveyed. The goal of training a supervised network is to obtain the parameterization of the registration field obtained through performing pairwise registration via traditional numerical optimization methods. This parameterization could be momentum, velocity, or displacement.

We will start reviewing the loss function with the most commonly used similarity metrics and several representative regularization ways to ensure diffeomorphism. After that, a variety of CNN-based methods as well as more related works will be covered.

Loss Function

For supervised learning-based diffeomorphic mapping, the loss function also consists of a similarity term and a regularization term. However, the similarity term is completely different from that in an unsupervised method. The typical loss function can be written as follows:

$$L^{\text{sup}} = L_{\text{sim}}(\mathbf{u}^{\text{sup}}, \mathbf{u}) + \gamma L_{\text{reg}}(\mathbf{u}), \tag{8}$$

where L_{sim} is the similarity term that measures the difference between the parameterization $\mathbf{u}$ of the predicted deformation and $\mathbf{u}^{\text{sup}}$ obtained from conducting one-to-one registration utilizing traditional methods. L_{reg} is the regularization term that imposes certain constraint on the parameterization of the deformation. When minimizing the loss function, the set of the estimated parameters of the deformation is increasingly closer to the set of the ground truth $\mathbf{u}_{\text{sup}}$, and the registration field is progressively smoother. γ is a trade-off factor between the similarity term and the regularization term, which behaves similarly as the unsupervised one.

Similarity Metrics

So far, supervised learning-based diffeomorphic mappings are focused on image data and usually employ sum of squared difference (SSD), also called MSE, as the similarity metric:

$$SSD(u^{\text{sup}}, u) = \frac{1}{|\Omega|} \sum_{p \in \Omega} ||u^{\text{sup}} - u||^2, \tag{9}$$

where p indexes image pixels or voxels and Ω represents the whole image. u and u^{sup}, respectively, represent the predicted parameterization and the one obtained from traditional methods. Since the loss function is minimized to train the framework, a small SSD is desired to yield a good alignment. The similarity loss term with SSD can be directly written as $L_{\text{sim}}(u^{\text{sup}}, u) = SSD(u^{\text{sup}}, u)$.

Regularization for Diffeomorphic Mapping

In a learning-based supervised framework, the LDDMM regularity term (Beg et al. 2005; Glaunes et al. 2008) is usually used when the prediction aims to obtain the registration field of LDDMM. l_2 norm on weights of the networks is also employed for smooth deformation purposes. In addition, a differential operator similar to Eq. 6 or Eq. 7 conducting regularization on momentum or velocity is also a common choice for ensuring diffeomorphism.

CNN-Based Methods

Quicksilver (Yang et al. 2016) proposes a fast predictive image registration method in 2016 which focuses only on atlas-based registration. A later version (Yang et al. 2017b) extends the former work to multi-modal image registration. Quicksilver (Yang et al. 2017c) is an enhanced version of the two previous works. It is a patch-based learning framework that mimics LDDMM by (approximately) predicting LDDMM's momentum through neural networks instead of employing traditional LDDMM. The predicted momentum is constrained by a LDDMM regularity term so as to ensure smooth mapping. Concretely, two patches of size $15 \times 15 \times 15$ of the same location, respectively, taken from the moving image and the fixed image are fed into the framework to learn feature maps, which encode spatial and contextual information of the inputs. The feature maps are subsequently passed through three independent decoding branches with identical network structure to predict the corresponding momentum at the three axes. SSD is employed as the similarity metric to train the network. An extra shooting procedure (Vialard et al. 2012) not included in the network is adopted to perform registration with the

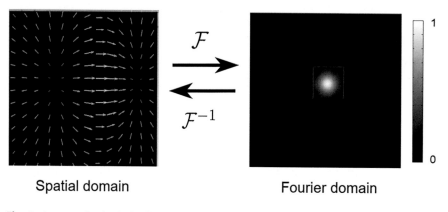

Fig. 8 An example of velocity field in spatial domain and Fourier domain. (Taken from Wang and Zhang 2020b)

predicted momentum. It is worth noting since the input patches are extracted from the whole MRI scans, a large stride 14 of the sliding window for the three axes is preferable considering the computational cost.

Besides, a probabilistic framework is presented to evaluate the registration uncertainty. It assumes the prior on the weights of each layer of the network is a diagonal matrix, each entry of which is drawn from a Bernoulli distribution (a way of drop out). A correction network is additionally proposed to further boost the registration accuracy. Specifically, the momentum predicted in the previous procedure is regarded as an initial prediction and is used to apply backward warp to the fixed patch. The moving patch and the backwardly warped fixed patch are subsequently fed into the correction network to estimate the residual momentum between the initial prediction and the true one (obtained from performing traditional LDDMM) with a residual connection. Results from LDDMM of traditional scheme implemented in PyCA (Singh et al. 2013) with GPU are employed to obtain the supervised labels. There are 3 types of evaluations, including atlas-to-image registration on 150 MRI scans from the OASIS longitudinal dataset (Fotenos et al. 2005), image-to-image registration on 373 MRI scans from the OASIS longitudinal dataset (Fotenos et al. 2005) for training and 2168 MRI scans from 4 datasets (LPBA40, IBSR18, MGH10, CUMC12) (Klein et al. 2009) for testing, and multi-modal registration (T1-weighted to T2-weighted) on 375 MRI scans from the IBIS 3D Autism Brain image dataset (Hazlett et al. 2017). Three metrics including target overlap (Yang et al. 2017c), number of voxels whose logarithm Jacobian determinant of the registration field are equal to or less than 0, and deformation errors (mm) are adopted for evaluation purposes. Comparisons are conducted with several related methods with respect to the three metrics.

SVF-Net Also in 2017, a U-shape FCN-based method is published. It takes as inputs pairs of moving and fixed images to predict SVF (Rohé et al. 2017) of the registration, with SSD as the similarity metric. Unlike Quicksilver which employs independent branches for predicting momentum of each axis, SVF-Net instead estimates the velocity using a 4D map, the last dimension of which, respectively, represents the velocity in x, y, z axes. No explicit regularization is presented in the loss function of SVF-Net. The true velocity labels are obtained by using an iterative log-approximation scheme with the scaling and squaring approach (Arsigny et al. 2006). It starts with the displacement field defined on the whole image grid and parameterizes a transformation that maps a set of selected landmarks from the moving image to the corresponding fixed image.

Inter-patient registration is conducted on 187 segmented 3D MRI cardiac scans acquired from multiple clinical centers. Small translations in x and y axes are performed as data augmentation for the training data. Results with respect to four evaluation metrics (Dice, HD, NLCC, relative variance of Log-Jacobian) for SVF-Net and LCC Log-Demons (Lorenzi et al. 2013) on four ROIs are shown. When performing on a NVIDIA TitanX GPU, SVF-Net takes less than 0.03 s for one pair of registration.

DeepFLASH The aforementioned VAE-based methods in section "VAE-Based Methods" follow a same mechanism: first, they learn a low-dimensional representation of the deformation field; subsequently, a decoder restores the corresponding registration field from the learned compact representation; and finally, registration is performed with the restored registration field. Distinctively, DeepFLASH (Wang and Zhang 2020b) distinguishes itself by predicting a low-dimensional Fourier representation of the velocity field, based on the fact that the velocity field does not develop high frequencies in the Fourier domain, as illustrated in Fig. 8. Thus, the training time and memory consumption can be drastically saved compared to other learning-based methods. To be concrete, DeepFLASH performs Fourier transform on the input moving-and-fixed images and then feeds the real parts of the Fourier representation of the two images into a R_{net} so as to estimate the real part of the Fourier representation for the to-be-predicted velocity field. Meanwhile, the imaginary parts of the Fourier representation of the two images are fed into another network that is parallel to R_{net}, called I_{net}, to estimate the corresponding imaginary part. Structures of R_{net} and I_{net} are identical to each other and are built based on CNN. Once the real and the imaginary parts are obtained, the velocity and the corresponding registration field can be recovered from the predicted low-dimensional representations in the Fourier domain. SSD is employed as the similarity metric which measures the difference between the predicted velocity field in Fourier domain and the ground truth obtained from conducting VM-LDDMM (Singh et al. 2013) as well as Fourier transformation. l_2 norm on weights of the network is adopted serving as the regularity term in the loss function.

Experiments are conducted on 3200 public T1-weighted 3D brain MRI scans from ADNI (Jack et al. 2008), OASIS (Fotenos et al. 2005), ABIDE (Di Martino et al. 2014), and LPBA40 (Shattuck et al. 2008) with 1000 subjects involved. Results are compared with three traditional methods, VM-LDDMM (Singh et al. 2013), ANTs SyN (Avants et al. 2008), and FLASH (Zhang and Fletcher 2019), as well as two learning-based methods: Quicksilver (Yang et al. 2017c) and VoxelMorph (Balakrishnan et al. 2019). When comparing Dice scores among these methods, 0.780 for DeepFLASH, 0.774 for VoxelMorph, 0.762 for Quicksilver, 0.788 for FLASH, 0.770 for ANTs SyN, and 0.760 for VM-LDDMM are obtained. Considering the training time, DeepFLASH takes 14.1 h, VoxelMorph takes 29.7 h, and Quicksilver takes 31.4 h under the same conditions. However, both DeepFLASH and Quicksilver need extra time for acquiring the registration labels through conducting conventional methods before the training procedure. The registration time on NVIDIA GTX 1080Ti GPUs is, respectively, 0.273 s for DeepFLASH, 0.571 s for VoxelMorph, 0.760 s for Quicksilver, 53.4 s for FLASH, and 262 s for VM-LDDMM.

More Related Works

Besides, Krebs et al. (2017) explores training a reinforcement learning model with a large number of synthetically deformed image pairs and a small number of real inter-subject pairs through agent-based action learning. Pathan and Hong (2018) combine LSTM and CNN to learn a predictive regression model based on LDDMM for longitudinal images with missing data. Ding proposes a framework similar to Quicksilver, called FPSGR (Ding et al. 2019; Kwitt and Niethammer 2017), to approximate a simplified geodesic regression model so as to capture longitudinal brain changes. To be specific, FPSGR predicts initial momenta supervised by the geodesic distance between images. The geodesic regression can be solved by approximately performing pairwise image registrations between the first image and all subsequent images of the longitudinal data. FPSGR-derived correlations with clinical indicators are also analyzed. A work on arXiv (Wang and Zhang 2020a) first estimates the regularity parameters of the image registrations for given image pairs using a CNN. Afterward, a new two-stream CNN-based network is trained to estimate the mapping from image pairs to their corresponding regularity parameters, under the supervision of the estimated regularity parameters from the previous step. Table 2 lists the related information of all reviewed supervised learning-based works.

Discussion and Future Direction

Achievements and Applications

Heretofore, deep learning-based diffeomorphic mapping, whether unsupervised or supervised, can achieve comparable or even better results than the state-of-the-art traditional methods (Lorenzi et al. 2013; Avants et al. 2008; Singh et al. 2013) when performing registrations within the same underlyingly assumed population as the training data. Besides, the time consumed by each pair of registration is considerably reduced, thanks to the efficient parallel computations of GPU and the ability of deep networks to learn and store registration mappings. In addition, atlases can be conveniently generated by VAE-based methods. Registration uncertainty and sampling new deformation as well as conducting deformation transport can also be achieved by training a probabilistic generative model. Furthermore, incorporating time sequence data and temporal convolutional networks can jointly predict registration fields for sequential data and perform progression analyses of diseases such as Alzheimer's disease. The aforementioned methods mainly focus on using deep neural networks to perform registration tasks in a diffeomorphic way. The works that focus on specific applications making use of these deep learning-based registration methods have also emerged recently.

Dalca et al. (2019b) propose a strategy that combines a conventional probabilistic atlas-based segmentation method with a deep learning-based registration method, being able to train a model for segmenting new testing MRI scans without any manually segmented images involved in the training phase. An efficient method for yielding either universal or conditional templates and jointly performing registration between images and templates is presented in Dalca et al. (2019a). In Evan et al. (2020), a model which learns to compute an attribute-specific spatial deformation is proposed. This model can deform a brain template in certain ways that take a wide range of ages, presence of diseases, and different genders into consideration. Cheng et al. (2020) focus on cortical surface registration utilizing unsupervised learning. Olut et al. (2020) use deformations obtained from deep registration models to conduct data augmentation. Specifically, it builds statistical deformation models based on unlabeled data using principal component analysis and subsequently uses the acquired statistical deformation space to augment training samples with labels.

Challenges

Nevertheless, there are still a variety of challenges presented to researchers. Learning-based techniques can only accurately register objects that come from the same population as in training. To be concrete, these methods can merely register images whose image contrast and geometric content are similar to those of the training data. This limitation comes from the inherent property of deep learning; it can only capture and store characteristics of data involved during training. For instance, when we use a deep registration network trained on T1-weighted MRI scans to register T2-weighted MRI scans or other modalities such as CT scans, the performance is usually inferior and much lower than that of performing registration between T1-weighted pairs. Besides, medical imaging scans of the same ROI obtained from different machines or different sites could be of various distributions even within the same modality. Thus, challenges still need to be solved to acquire the desirable property of conventional methods, namely, being able to register any type of data rather than only those involved in training. This limitation is also applicable to shape data. Furthermore, when comes to shape data, existing deep registration frameworks usually first transform shapes into image representations and then feed the transformed image representations into deep neural networks for subsequent procedures. However, this kind of image representation may deteriorate the resolution of the original shape. This is due to the fact that only values on the grid are considered and no strong constraint of the original shape is involved in the network structures. Thus, how to design a more suitable deep registration framework for shape data remains a challenging topic to explore.

Future Directions

As a newly emerging topic, deep learning-based diffeomorphic mapping demonstrates promising potentials to improve or exploit in several directions. For cross-modality or cross-ROI registrations (train on one modality or ROI but perform registration on another modality or ROI), the recent approaches of domain adaptation (Sun et al. 2015; Wilson and Cook 2020) and domain generalization (Li et al. 2018; Zhou et al. 2021) might serve as potential solutions. To be specific, if training data from a new domain are available, domain adaption methods can be used to fine-tune the deep registration networks so as to make the tuned networks applicable for the new data. On the contrary, domain generalization methods can serve as a technique to handle data from an unseen domain if training data are unavailable. Additionally, a generative adversarial network (Yi et al. 2019) might be used to improve the performance of the registration. A generator produces the deformation fields, while the discriminator evaluates whether they are good or not. The zero-sum game can significantly contribute to improving the quality and authenticity of the deformation fields. As for applications, deep learning-based registration frameworks, especially for surfaces and curves, can explicitly incorporate geometrical information into neural networks. This is potentially beneficial for other tasks such as more regular and smooth organ segmentations or more accurate landmark detections. As far as deep learning-based shape registration is concerned, an elaborately designed network suitable for handling meshes could further improve the registration performance.

Conclusions

In this chapter, we firstly describe the conventional diffeomorphic registration problem and its general objectives. Afterward, several deep neural networks used in learning-based diffeomorphic mapping are briefly introduced. Subsequently, the general loss functions, similarity metrics, regularity terms, and recent works of deep registration frameworks, both unsupervised and supervised, are examined in detail. Several data types such as MRI, CT, surface, and curve are covered in these works. In addition, we summarize current achievements, applications, and challenges in this field. Finally, we provide several potential future directions to explore at the end of this chapter.

Acknowledgments This study was supported by the National Natural Science Foundation of China (62071210); the Shenzhen Basic Research Program (JCYJ20200925153847004, JCYJ20190809120205578); and the High-Level University Fund (G02236002). The authors would like to thank Yuanyuan Wei from the University of British Columbia for his help on this chapter.

References

Amor, B.B., Arguillère, S., Shao, L.: Resnet-LDDMM: advancing the LDDMM framework using deep residual networks (2021). arXiv preprint arXiv:210207951

Arganda-Carreras, I., Turaga, S.C., Berger, D.R., Cireşan, D., Giusti, A., Gambardella, L.M., Schmidhuber, J., Laptev, D., Dwivedi, S., Buhmann, J.M., et al.: Crowdsourcing the creation of image segmentation algorithms for connectomics. Front. Neuroanat. **9**, 142 (2015)

Arsigny, V., Fillard, P., Pennec, X., Ayache, N.: Fast and simple calculus on tensors in the log-Euclidean framework. In: International Conference on Medical Image Computing and Computer-Assisted Intervention, pp. 115–122. Springer (2005)

Arsigny, V., Commowick, O., Pennec, X., Ayache, N.: A log-Euclidean framework for statistics on diffeomorphisms. In: International Conference on Medical Image Computing and Computer-Assisted Intervention, pp. 924–931. Springer (2006)

Ashburner, J.: A fast diffeomorphic image registration algorithm. Neuroimage **38**(1), 95–113 (2007)

Avants, B.B., Epstein, C.L., Grossman, M., Gee, J.C.: Symmetric diffeomorphic image registration with cross-correlation: evaluating automated labeling of elderly and neurodegenerative brain. Med. Image Anal. **12**(1), 26–41 (2008)

Balakrishnan, G., Zhao, A., Sabuncu, M.R., Guttag, J., Dalca, A.V.: An unsupervised learning model for deformable medical image registration. In: Proceedings of the IEEE Conference on Computer Vision and Pattern Recognition, pp. 9252–9260 (2018)

Balakrishnan, G., Zhao, A., Sabuncu, M.R., Guttag, J., Dalca, A.V.: Voxelmorph: a learning framework for deformable medical image registration. IEEE Trans. Med. Imaging **38**(8), 1788–1800 (2019)

Beg, M.F., Miller, M.I., Trouvé, A., Younes, L.: Computing large deformation metric mappings via geodesic flows of diffeomorphisms. Int. J. Comput. Vis. **61**(2), 139–157 (2005)

Bône, A., Louis, M., Colliot, O., Durrleman, S., Initiative, A.D.N., et al.: Learning low-dimensional representations of shape data sets with diffeomorphic autoencoders. In: International Conference on Information Processing in Medical Imaging, pp. 195–207. Springer (2019)

Bossa, M., Zacur, E., Olmos, S., Initiative, A.D.N., et al.: Tensor-based morphometry with stationary velocity field diffeomorphic registration: application to ADNI. Neuroimage **51**(3), 956–969 (2010)

Cao, Y., Miller, M.I., Mori, S., Winslow, R.L., Younes, L.: Diffeomorphic matching of diffusion tensor images. In: 2006 Conference on Computer Vision and Pattern Recognition Workshop (CVPRW'06), pp. 67–67. IEEE (2006)

Charon, N., Trouvé, A.: The varifold representation of nonoriented shapes for diffeomorphic registration. SIAM J. Imaging Sci. **6**(4), 2547–2580 (2013)

Cheng, J., Dalca, A.V., Fischl, B., Zöllei, L., Initiative, A.D.N., et al.: Cortical surface registration using unsupervised learning. NeuroImage **221**, 117161 (2020)

Dalca, A.V., Balakrishnan, G., Guttag, J., Sabuncu, M.R.: Unsupervised learning for fast probabilistic diffeomorphic registration. In: International Conference on Medical Image Computing and Computer-Assisted Intervention, pp. 729–738. Springer (2018)

Dalca, A.V., Rakic, M., Guttag, J., Sabuncu, M.R.: Learning conditional deformable templates with convolutional networks (2019a). arXiv preprint arXiv:190802738

Dalca, A.V., Yu, E., Golland, P., Fischl, B., Sabuncu, M.R., Iglesias, J.E.: Unsupervised deep learning for Bayesian brain MRI segmentation. In: International Conference on Medical Image Computing and Computer-Assisted Intervention, pp. 356–365. Springer (2019b)

Debavelaere, V., Durrleman, S., Allassonnière, S., Initiative, A.D.N.: Learning the clustering of longitudinal shape data sets into a mixture of independent or branching trajectories. Int. J. Comput. Vis. **128**, 2794–2809 (2020)

Detlefsen, N.S., Freifeld, O., Hauberg, S.: Deep diffeomorphic transformer networks. In: Proceedings of the IEEE Conference on Computer Vision and Pattern Recognition, pp. 4403–4412 (2018)

Di Martino, A., Yan, C.G., Li, Q., Denio, E., Castellanos, F.X., Alaerts, K., Anderson, J.S., Assaf, M., Bookheimer, S.Y., Dapretto, M., et al.: The autism brain imaging data exchange: towards a large-scale evaluation of the intrinsic brain architecture in autism. Mol. Psychiatry **19**(6), 659–667 (2014)

Ding, Z., Fleishman, G., Yang, X., Thompson, P., Kwitt, R., Niethammer, M., Initiative, A.D.N., et al.: Fast predictive simple geodesic regression. Med. Image Anal. **56**, 193–209 (2019)

Durrleman, S.: Statistical models of currents for measuring the variability of anatomical curves, surfaces and their evolution. PhD thesis, Université Nice Sophia Antipolis (2010)

Evan, M.Y., Dalca, A.V., Sabuncu, M.R.: Learning conditional deformable shape templates for brain anatomy. In: International Workshop on Machine Learning in Medical Imaging, pp. 353–362. Springer (2020)

Fischl, B.: Freesurfer. Neuroimage **62**(2), 774–781 (2012)

Fotenos, A.F., Snyder, A., Girton, L., Morris, J., Buckner, R.: Normative estimates of cross-sectional and longitudinal brain volume decline in aging and ad. Neurology **64**(6), 1032–1039 (2005)

Freifeld, O., Hauberg, S., Batmanghelich, K., Fisher, J.W.: Transformations based on continuous piecewise-affine velocity fields. IEEE Trans. Pattern Anal. Mach. Intell. **39**(12), 2496–2509 (2017)

Glaunes, J., Qiu, A., Miller, M.I., Younes, L.: Large deformation diffeomorphic metric curve mapping. Int. J. Comput. Vis. **80**(3), 317–336 (2008)

Gondara, L.: Medical image denoising using convolutional denoising autoencoders. In: 2016 IEEE 16th International Conference on Data Mining Workshops (ICDMW), pp. 241–246. IEEE (2016)

Gori, P., Colliot, O., Marrakchi-Kacem, L., Worbe, Y., Poupon, C., Hartmann, A., Ayache, N., Durrleman, S.: A Bayesian framework for joint morphometry of surface and curve meshes in multi-object complexes. Med. Image Anal. **35**, 458–474 (2017)

Han, X., Shen, Z., Xu, Z., Bakas, S., Akbari, H., Bilello, M., Davatzikos, C., Niethammer, M.: A deep network for joint registration and reconstruction of images with pathologies. In: International Workshop on Machine Learning in Medical Imaging, pp. 342–352. Springer (2020)

Hazlett, H.C., Gu, H., Munsell, B.C., Kim, S.H., Styner, M., Wolff, J.J., Elison, J.T., Swanson, M.R., Zhu, H., Botteron, K.N., et al.: Early brain development in infants at high risk for autism spectrum disorder. Nature **542**(7641), 348–351 (2017)

He, K., Zhang, X., Ren, S., Sun, J.: Deep residual learning for image recognition. In: Proceedings of the IEEE Conference on Computer Vision and Pattern Recognition, pp. 770–778 (2016)

Hinkle, J., Womble, D., Yoon, H.J.: Diffeomorphic autoencoders for LDDMM atlas building (2018)

Hoffmann, M., Billot, B., Eugenio Iglesias, J., Fischl, B., Dalca, A.V.: Learning image registration without images (2020). arXiv e-prints arXiv–2004

Iandola, F., Moskewicz, M., Karayev, S., Girshick, R., Darrell, T., Keutzer, K.: Densenet: implementing efficient convnet descriptor pyramids (2014). arXiv preprint arXiv:14041869

Jack, C.R. Jr, Bernstein, M.A., Fox, N.C., Thompson, P., Alexander, G., Harvey, D., Borowski, B., Britson, P.J., Whitwell, J.L., Ward, C., et al.: The Alzheimer's disease neuroimaging initiative (ADNI): MRI methods. J. Magn. Reson. Imaging: Off. J. Int. Soc. Magn. Res. Med. **27**(4), 685–691 (2008)

Jaderberg, M., Simonyan, K., Zisserman, A., Kavukcuoglu, K.: Spatial transformer networks (2015). arXiv preprint arXiv:150602025

Jiang, Z., Yang, H., Tang, X.: Deformation-based statistical shape analysis of the corpus callosum in mild cognitive impairment and Alzheimer's disease. Curr. Alzheimer Res. **15**(12), 1151–1160 (2018)

Joshi, S.C., Miller, M.I.: Landmark matching via large deformation diffeomorphisms. IEEE Trans. Image Process. **9**(8), 1357–1370 (2000)

Karpathy, A., Johnson, J., Fei-Fei, L.: Visualizing and understanding recurrent networks (2015). arXiv preprint arXiv:150602078

Kaul, C., Manandhar, S., Pears, N.: Focusnet: An attention-based fully convolutional network for medical image segmentation. In: 2019 IEEE 16th International Symposium on Biomedical Imaging (ISBI 2019), pp. 455–458. IEEE (2019)

Kingma, D.P., Welling, M.: Auto-encoding variational Bayes (2013). arXiv preprint arXiv:13126114

Klein, A., Andersson, J., Ardekani, B.A., Ashburner, J., Avants, B., Chiang, M.C., Christensen, G.E., Collins, D.L., Gee, J., Hellier, P., et al.: Evaluation of 14 nonlinear deformation algorithms applied to human brain mri registration. Neuroimage **46**(3), 786–802 (2009)

Kramer, M.A.: Nonlinear principal component analysis using autoassociative neural networks. AIChE J. **37**(2), 233–243 (1991)

Krebs, J., Mansi, T., Delingette, H., Zhang, L., Ghesu, F.C., Miao, S., Maier, A.K., Ayache, N., Liao, R., Kamen, A.: Robust non-rigid registration through agent-based action learning. In: International Conference on Medical Image Computing and Computer-Assisted Intervention, pp. 344–352. Springer (2017)

Krebs, J., Delingette, H., Mailhé, B., Ayache, N., Mansi, T.: Learning a probabilistic model for diffeomorphic registration. IEEE Trans. Med. Imaging **38**(9), 2165–2176 (2019)

Krebs, J., Delingette, H., Ayache, N., Mansi, T.: Learning a generative motion model from image sequences based on a latent motion matrix. IEEE Trans. Med. Imaging **40**(5), 1405–1416 (2021)

Krizhevsky, A., Sutskever, I., Hinton, G.E.: Imagenet classification with deep convolutional neural networks. Adv. Neural Inf. Process. Syst. **25**, 1097–1105 (2012)

Kwitt, R., Niethammer, M.: Fast predictive simple geodesic regression. In: Third International Workshop DLMIA, p. 267 (2017)

Lateef, F., Ruichek, Y.: Survey on semantic segmentation using deep learning techniques. Neurocomputing **338**, 321–348 (2019)

Lea, C., Vidal, R., Reiter, A., Hager, G.D.: Temporal convolutional networks: a unified approach to action segmentation. In: European Conference on Computer Vision, pp. 47–54. Springer (2016)

Li, D., Yang, Y., Song, Y.Z., Hospedales, T.: Learning to generalize: meta-learning for domain generalization. In: Proceedings of the AAAI Conference on Artificial Intelligence, vol. 32 (2018)

Liu, L., Ouyang, W., Wang, X., Fieguth, P., Chen, J., Liu, X., Pietikäinen, M.: Deep learning for generic object detection: a survey. Int. J. Comput. Vis. **128**(2), 261–318 (2020)

Lorenzi, M., Ayache, N., Frisoni, G.B., Pennec, X., ADNI, et al.: LCC-demons: a robust and accurate symmetric diffeomorphic registration algorithm. NeuroImage **81**, 470–483 (2013)

Louis, M., Charlier, B., Durrleman, S.: Geodesic discriminant analysis for manifold-valued data. In: Proceedings of the IEEE Conference on Computer Vision and Pattern Recognition Workshops, pp. 332–340 (2018)

Louis, M., Couronné, R., Koval, I., Charlier, B., Durrleman, S.: Riemannian geometry learning for disease progression modelling. In: International Conference on Information Processing in Medical Imaging, pp. 542–553. Springer (2019)

Lyu, J., Cheng, P., Tang, X.: Fundus image based retinal vessel segmentation utilizing a fast and accurate fully convolutional network. In: International Workshop on Ophthalmic Medical Image Analysis, pp. 112–120. Springer (2019)

Milletari, F., Navab, N., Ahmadi, S.A.: V-net: Fully convolutional neural networks for volumetric medical image segmentation. In: 2016 Fourth International Conference on 3D Vision (3DV), pp. 565–571. IEEE (2016)

Modat, M., Daga, P., Cardoso, M.J., Ourselin, S., Ridgway, G.R., Ashburner, J.: Parametric non-rigid registration using a stationary velocity field. In: 2012 IEEE Workshop on Mathematical Methods in Biomedical Image Analysis, pp. 145–150. IEEE (2012)

Mok, T.C., Chung, A.: Fast symmetric diffeomorphic image registration with convolutional neural networks. In: Proceedings of the IEEE/CVF Conference on Computer Vision and Pattern Recognition, pp. 4644–4653 (2020a)

Mok, T.C., Chung, A.C.: Large deformation diffeomorphic image registration with laplacian pyramid networks. In: International Conference on Medical Image Computing and Computer-Assisted Intervention, pp. 211–221. Springer (2020b)

Niethammer, M., Kwitt, R., Vialard, F.X.: Metric learning for image registration. In: Proceedings of the IEEE/CVF Conference on Computer Vision and Pattern Recognition, pp. 8463–8472 (2019)

Olut, S., Shen, Z., Xu, Z., Gerber, S., Niethammer, M.: Adversarial data augmentation via deformation statistics. In: European Conference on Computer Vision, pp. 643–659. Springer (2020)

Pathan, S., Hong, Y.: Predictive image regression for longitudinal studies with missing data (2018). arXiv preprint arXiv:180807553

Rohé, M.M., Datar, M., Heimann, T., Sermesant, M., Pennec, X.: SVF-Net: learning deformable image registration using shape matching. In: International Conference on Medical Image Computing and Computer-Assisted Intervention, pp. 266–274. Springer (2017)

Ronneberger, O., Fischer, P., Brox, T.: U-Net: convolutional networks for biomedical image segmentation. In: International Conference on Medical Image Computing and Computer-Assisted Intervention, pp. 234–241. Springer (2015)

Roy, A.G., Navab, N., Wachinger, C.: Concurrent spatial and channel 'squeeze & excitation' in fully convolutional networks. In: International Conference on Medical Image Computing and Computer-Assisted Intervention, pp. 421–429. Springer (2018)

Shattuck, D.W., Mirza, M., Adisetiyo, V., Hojatkashani, C., Salamon, G., Narr, K.L., Poldrack, R.A., Bilder, R.M., Toga, A.W.: Construction of a 3d Probabilistic Atlas of Human Cortical Structures. Neuroimage **39**(3), 1064–1080 (2008)

Shen, Z., Han, X., Xu, Z., Niethammer, M.: Networks for joint affine and non-parametric image registration. In: Proceedings of the IEEE/CVF Conference on Computer Vision and Pattern Recognition, pp. 4224–4233 (2019a)

Shen, Z., Vialard, F.X., Niethammer, M.: Region-specific diffeomorphic metric mapping (2019b). arXiv preprint arXiv:190600139

Sibi, P., Jones, S.A., Siddarth, P.: Analysis of different activation functions using back propagation neural networks. J. Theor. Appl. Inf. Technol. **47**(3), 1264–1268 (2013)

Simonyan, K., Zisserman, A.: Very deep convolutional networks for large-scale image recognition (2014). arXiv preprint arXiv:14091556

Singh, N., Hinkle, J., Joshi, S., Fletcher, P.T.: A vector momenta formulation of diffeomorphisms for improved geodesic regression and atlas construction. In: 2013 IEEE 10th International Symposium on Biomedical Imaging, pp. 1219–1222. IEEE (2013)

Sun, S., Shi, H., Wu, Y.: A survey of multi-source domain adaptation. Inf. Fusion **24**, 84–92 (2015)

Szegedy, C., Ioffe, S., Vanhoucke, V., Alemi, A.: Inception-v4, inception-resnet and the impact of residual connections on learning. In: Proceedings of the AAAI Conference on Artificial Intelligence, vol. 31 (2017)

Tang, X., Ross, C.A., Johnson, H., Paulsen, J.S., Younes, L., Albin, R.L., Ratnanather, J.T., Miller, M.I.: Regional subcortical shape analysis in premanifest Huntington's disease. Hum. Brain Map. **40**(5), 1419–1433 (2019)

Tian, L., Puett, C., Liu, P., Shen, Z., Aylward, S.R., Lee, Y.Z., Niethammer, M.: Fluid registration between lung CT and stationary chest tomosynthesis images. In: International Conference on Medical Image Computing and Computer-Assisted Intervention, pp. 307–317. Springer (2020)

Ulman, V., Maška, M., Magnusson, K.E., Ronneberger, O., Haubold, C., Harder, N., Matula, P., Matula, P., Svoboda, D., Radojevic, M., et al.: An objective comparison of cell-tracking algorithms. Nat. Methods **14**(12), 1141–1152 (2017)

Vaillant, M., Glaunes, J.: Surface matching via currents. In: Biennial International Conference on Information Processing in Medical Imaging, pp. 381–392. Springer (2005)

Vaillant, M., Qiu, A., Glaunès, J., Miller, M.I.: Diffeomorphic metric surface mapping in subregion of the superior temporal gyrus. NeuroImage **34**(3), 1149–1159 (2007)

Vialard, F.X., Risser, L., Rueckert, D., Cotter, C.J.: Diffeomorphic 3D image registration via geodesic shooting using an efficient adjoint calculation. Int. J. Comput. Vis. **97**(2), 229–241 (2012)

Wang, J., Zhang, M.: Deep learning for regularization prediction in diffeomorphic image registration (2020a). arXiv preprint arXiv:201114229

Wang, J., Zhang, M.: Deepflash: an efficient network for learning-based medical image registration. In: Proceedings of the IEEE/CVF Conference on Computer Vision and Pattern Recognition, pp. 4444–4452 (2020b)

Wang, W., Huang, Y., Wang, Y., Wang, L.: Generalized autoencoder: a neural network framework for dimensionality reduction. In: Proceedings of the IEEE Conference on Computer Vision and Pattern Recognition Workshops, pp. 490–497 (2014)

Wilson, G., Cook, D.J.: A survey of unsupervised deep domain adaptation. ACM Trans. Intell. Syst. Technol. (TIST) **11**(5), 1–46 (2020)

Yang, X., Li, Y., Reutens, D., Jiang, T.: Diffeomorphic metric landmark mapping using stationary velocity field parameterization. Int. J. Comput. Vis. **115**(2), 69–86 (2015)

Yang, X., Kwitt, R., Niethammer, M.: Fast predictive image registration. In: Deep Learning and Data Labeling for Medical Applications, pp. 48–57. Springer, Cham (2016)

Yang, H., Wang, J., Tang, H., Ba, Q., Yang, G., Tang, X.: Analysis of mitochondrial shape dynamics using large deformation diffeomorphic metric curve matching. In: 2017 39th Annual International Conference of the IEEE Engineering in Medicine and Biology Society (EMBC), pp. 4062–4065. IEEE (2017a)

Yang, X., Kwitt, R., Styner, M., Niethammer, M.: Fast predictive multimodal image registration. In: 2017 IEEE 14th International Symposium on Biomedical Imaging (ISBI 2017), pp. 858–862. IEEE (2017b)

Yang, X., Kwitt, R., Styner, M., Niethammer, M.: Quicksilver: fast predictive image registration–a deep learning approach. NeuroImage **158**, 378–396 (2017c)

Yi, X., Walia, E., Babyn, P.: Generative adversarial network in medical imaging: a review. Med. Image Anal. **58**, 101552 (2019)

Zeiler, M.D., Krishnan, D., Taylor, G.W., Fergus, R.: Deconvolutional networks. In: 2010 IEEE Computer Society Conference on Computer Vision and Pattern Recognition, pp. 2528–2535. IEEE (2010)

Zhang, M., Fletcher, P.T.: Bayesian principal geodesic analysis in diffeomorphic image registration. In: International Conference on Medical Image Computing and Computer-Assisted Intervention, pp. 121–128. Springer (2014)

Zhang, M., Fletcher, P.T.: Fast diffeomorphic image registration via fourier-approximated lie algebras. Int. J. Comput. Vis. **127**(1), 61–73 (2019)

Zhou, C., Paffenroth, R.C.: Anomaly detection with robust deep autoencoders. In: Proceedings of the 23rd ACM SIGKDD International Conference on Knowledge Discovery and Data Mining, pp. 665–674 (2017)

Zhou, K., Liu, Z., Qiao, Y., Xiang, T., Loy, C.C.: Domain generalization: a survey (2021). arXiv preprint arXiv:210302503